Cover image: From left to right and top to bottom:
- *Meloidogyne incognita* egg with developed second-stage juvenile
- Tomato root galls severely infected by *Meloidogyne incognita*
- Infective second-stage juvenile of *Meloidogyne oleae*
- Mature female of *Meloidgyne javanica*
- Olive galled root infected by *Meloidogyne javanica*, egg mass stained in red
- Mature female lip region of *Meloidogyne africana*
- Perineal pattern of *Meloidogyne aegracyperi*

SYSTEMATICS OF ROOT-KNOT NEMATODES
(NEMATODA: MELOIDOGYNIDAE)

SYSTEMATICS OF ROOT-KNOT NEMATODES (NEMATODA: MELOIDOGYNIDAE)

Cover image: From left to right and top to bottom:

- *Meloidogyne incognita* egg with developed second-stage juvenile
- Tomato root galls severely infected by *Meloidogyne incognita*
- Infective second-stage juvenile of *Meloidogyne oleae*
- Mature female of *Meloidgyne javanica*
- Olive galled root infected by *Meloidogyne javanica*, egg mass stained in red
- Mature female lip region of *Meloidogyne africana*
- Perineal pattern of *Meloidogyne aegracyperi*

SYSTEMATICS OF ROOT-KNOT NEMATODES (NEMATODA: MELOIDOGYNIDAE)

Sergei A. Subbotin
Juan E. Palomares-Rius
Pablo Castillo

David J. Hunt and Roland N. Perry (Series Editors)

NEMATOLOGY MONOGRAPHS AND PERSPECTIVES
VOLUME 14

TUTA SUB AEGIDE PALLAS · 1683 ·

BRILL
LEIDEN-BOSTON
2021

This book is printed on acid-free paper.

Library of Congress Cataloging-in-Publication Data

Names: Subbotin, Sergei A., author. | Palomares Rius, Juan Emilio, author. | Castillo, Pablo, author.
Title: Systematics of root-knot nematodes (nematoda: Meloidogynidae) / Sergei A. Subbotin, Juan Emilio Palomares Rius, Pablo Castillo; David J. Hunt and Roland N. Perry (series editors).
Description: Leiden ; Boston : Brill, 2021. | Series: Nematology monographs and perspectives, 1573-5869 ; volume 14 | Includes bibliographical references and index.
Identifiers: LCCN 2021030915 (print) | LCCN 2021030916 (ebook) | ISBN 9789004366343 (hardback) | ISBN 9789004387584 (ebook)
Subjects: LCSH: Root-knot nematodes. | Nematode diseases of plants.
Classification: LCC SB998.M45 S83 2021 (print) | LCC SB998.M45 (ebook) | DDC 632/.65182–dc23
LC record available at https://lccn.loc.gov/2021030915
LC ebook record available at https://lccn.loc.gov/2021030916

ISBN: 978 90 04 36634 3
E-ISBN: 978 90 04 38758 4

Contents

Preface

Root-knot nematodes of the genus *Meloidogyne* represent one of the most damaging and agriculturally important groups of plant-parasitic nematodes. These nematodes are obligate sedentary endoparasites infecting most species of higher plants and have a cosmopolitan distribution. Annual worldwide economic losses due to nematode infection of crops have been estimated at several hundred billion US dollars. Although 98 valid *Meloidogyne* species are known at present, the majority of research articles and books has focused on four species that are commonly regarded as the major root-knot nematodes species: *M. arenaria*, *M. hapla*, *M. incognita* and *M. javanica*. These four species are extremely widespread, infecting and damaging a wide range of major crops. The far-reaching presence of these species has led to frequent misidentifications and limited attention given to less important species, some of which are currently listed as quarantine pests in various countries. In writing this book we have attempted to fill a need for a reference book on the systematics of the root-knot nematodes. We tried to summarise facts from many publications for all species. The book includes introductions to the morphology, biology, biogeography, genomics, phylogeny and host-parasite relationships of root-knot nematodes. Methods of biochemical and molecular diagnostics presently used for identification of root-knot nematodes are also briefly described. The main part of this book is devoted to species descriptions, with photos and drawings of all developing stages. Information on distribution, plant hosts, pathogenicity, polytomous identification codes and available biochemical and molecular characterisation are also given for each valid species. We believe that this book will be useful for nematologists and plant pathologists and hope that it will stimulate further research on the systematics of this important nematode group.

Sergei A. SUBBOTIN, Juan E. PALOMARES-RIUS
and Pablo CASTILLO
Sacramento, USA, and Cordoba, Spain
October, 2020

Acknowledgements

The authors greatly appreciate the support from Drs W. Bert (Belgium), J.A. Brito (USA), V.N. Chizhov (Russia), J.D. Eisenback (USA), Z.A. Handoo (USA), R.N. Inserra (USA) and A. Yu. Ryss (Russia) for helpful advice and comments on the book draft and valuable information and materials. We also gratefully acknowledge the journals and organisations for allowing reproduction of illustrations. Deepest appreciation is expressed to the series editors, David Hunt and Roland Perry, for assistance, guidance and consistent help.

From left to right: Juan E. Palomares-Rius, Pablo Castillo and Sergei A. Subbotin.

Nematology Monographs & Perspectives, 2021, Vol. 14, 1-3

Chapter 1

Taxonomic History

Although knotting of crop roots has probably been observed since the earliest days of cultivation, one of the recorded reports was an observation of small worms occurring within galled roots of glasshouse-grown cucumber in England by the mycologist Berkeley (1855), who described and figured the plant roots. Cornu (1878) was the first to describe a population of these worms infesting roots of sainfoin in France as *Anguillula marioni*, for which he adopted the common name of root-knot nematode (RKN). Müller (1884) illustrated a perineal pattern while describing RKN causing galls on the roots of *Dodartia orientalis* L., which he erroneously referred to as *Heterodera radicicola* (Greeff, 1872), confusing nematodes previously described as *Anguillula radicicola* by Greeff (1872). Treub (1885) briefly described a RKN on the roots of diseased sugarcane from Java. He named it *Heterodera javanica* and distinguished it from *H. radicicola* Müller, 1884 by a few measurements. The genus *Meloidogyne* was proposed by Göldi (1887) for the description of *Meloidogyne exigua*, a nematode causing root galls on coffee in Rio de Janeiro state, in Brazil.

Early studies on life cycle, development pathology and control methods were published by Atkinson (1889), Neal (1889) and Stone & Smith (1898). Bessey (1911) published a monograph entitled '*Root-knot and its control*' dealing with life history, control and hosts (500 plant species) for RKN. Experiments made by Christie & Albin (1944) and Christie (1946) demonstrated the existence of host races in RKN. Chitwood (1949) showed that there were morphological differences between RKN and he was able to differentiate morphologically and redescribe four species and propose a new species. Chitwood (1949) also re-erected the genus *Meloidogyne* and emphasised the taxonomic significance of the pattern striations in the perineal area of females. Christie (1959) published a chapter on classification, biology, parasitic habits, plant hosts and control of RKN in his book entitled '*Plant nematodes, their bionomics and control*'.

Sledge & Golden (1964) proposed the genus *Hypsoperine* Sledge & Golden, 1964 for RKN species where the mature female was characterised by a thicker cuticle and an elevated, cone-like posterior region. It was considered as an intermediate genus between *Heterodera* and *Meloidogyne*. Whitehead (1968) did not recognise this genus and noticed that the aforementioned distinguishing morphological characters of *Hypsoperine* were also observed independently in several species of *Meloidogyne*. This opinion was also shared by Jepson (1987), Luc *et al.* (1988) and Eisenback & Triantaphyllou (1991). Siddiqi (1986) recognised as valid the genus *Hypsoperine* with *H. acronea* (Coetzee, 1956) Sledge & Golden, 1964, *H. graminis* Sledge & Golden, 1964, *H. megriensis* Poghossian, 1971, *H. mersa* Siddiqi & Booth, 1991, *H. ottersoni* Thorne, 1969 and *H. propora* Spaull, 1977, and included *H. spartinae* Rau & Fassuliotis, 1965 in a new genus named *Spartonema* Siddiqi, 1986. Subsequently, Siddiqi (2000) synonymised *Hypsoperine* with *Meloidogyne* but still maintained the genus *Spartonema* with two species: *S. spartinae* (Rau & Fassuliotis, 1965) Siddiqi, 1986, and *S. kikuyense* (De Grisse, 1961) Siddiqi, 2000. Based on the analysis of 18S rRNA gene sequences, Plantard *et al.* (2007) refuted the generic status of both *Hypsoperine* and *Spartonema* as previously proposed by Jepson (1987).

Whitehead (1968) published a compendium with detailed descriptions of 23 species with identification keys and contributed invaluable data to many species in the genus, including four new species. In a diagnostic compendium of the genus, Esser *et al.* (1976) presented tabular morphometric and morphological data supported by illustrations to facilitate identification of 35 species. Franklin (1957, 1971, 1972, 1976, 1979) provided a series of reviews of *Meloidogyne*. In a book entitled '*Root-knot nematodes (*Meloidogyne *species). Systematics, biology and control*', edited by Lamberti & Taylor (1979), Franklin (1979) listed 36 RKN species.

Further significant progress in research on systematics, biology and management of RKN was attained by the 'International *Meloidogyne* project' (IMP) funded by the United States Agency for International Development in 1975 and headquartered at North Carolina State University, USA. The goal of the IMP was to assist developing nations in increasing the yields of their economic food crops through research into RKN. This project involved over 100 nematologists from more than 70 developing countries (Sasser *et al.*, 1983) and resulted in the publication of several articles and books. The book edited by Taylor & Sasser (1978) and entitled '*Biology, identification and control of root-knot nematodes (*Meloidogyne *species)*' was designed primarily as an aid to the project and gave essential information on all aspects of *Meloidogyne* based on the literature and the wide experience of the authors. Eisenback *et al.* (1981) presented another book: '*A guide to the four most common species of root-knot nematodes (*Meloidogyne *species) with a pictorial key*'. Several years later, the two-volume book entitled '*An advanced treatise on* Meloidogyne' was also published; volume I (Sasser & Carter) was devoted to taxonomy, morphology, biology and control, and volume II (Barker, Carter & Sasser) dealt with the methodology of sampling, extraction, identification and design of experiments. In this book Hirschmann (1985) listed 56 species of RKN.

A book entitled '*Identification of root-knot nematodes (*Meloidogyne *species)*' by Jepson (1987) published by CABI became a fundamental and essential practical guide for identification of 54 species and subspecies. The book contained detailed species descriptions with information on hosts, geographical distribution, a new diagnosis for the genus and a comprehensive species list with synonyms and descriptions of two new species. Other important sources of information on systematics of RKN include a chapter written by Eisenback & Triantaphyllou (1991) for the '*Manual of agricultural nematology*' book and also the '*Root-knot nematode taxonomic database*' compiled by Eisenback (1997) and containing reprints of original *Meloidogyne* species descriptions. The book by Karssen (2002) entitled '*The plant-parasitic nematode genus* Meloidogyne *Göldi (Tylenchida) in Europe*' included a historical review of the genus, followed by a revision and short description of the European RKN species, and finished with a study of some morphological characteristics. '*Root-knot nematodes*' edited by Perry *et al.* (2009), a chapter on these nematodes written by Karssen *et al.* (2013) for the book '*Plant nematology*' edited by Perry & Moens (2013), and the paper by Ghaderi & Karssen (2020) are the latest most comprehensive compendia for systematic information on RKN. The taxonomic history of *Meloidogyne* is given in detail in several publications: Whitehead (1968), Franklin (1979), Hewlett & Tarjan (1983), Hirschmann (1985), Jepson (1987), Karssen (2002) and Hunt & Handoo (2009).

Nematology Monographs & Perspectives, 2021, Vol. 14, 5-12

Chapter 2

Diagnosis of the Genus *Meloidogyne*

Genus *Meloidogyne* Göldi, 1887
= *Caconema* Cobb, 1924
= *Hypsoperine* Sledge & Golden, 1964
= *Hypsoperine* (*Hypsoperine*) Sledge & Golden, 1964 (Siddiqi, 1986)
= *Hypsoperine* (*Spartonema*) (Siddiqi, 1986)
= *Spartonema* Siddiqi, 1986

DIAGNOSIS (after Siddiqi, 2000 with modifications). Meloidogyninae. Root-gall inciting (Fig. 1). **Mature female:** Round to pear-shaped with short projecting neck, white, sedentary up to 3 mm long. No cyst stage. Vulva and anus close together, terminal, perineum with a fingerprint-like cuticular pattern, elevated in some species. Phasmidial apertures dot-like, slightly anterior to and on either side of anus. Cuticle thick, striated. Stylet slender, generally 8-25 μm long, with small knobs. Excretory pore anterior to median bulb, often closely posterior to base of stylet. Ovaries paired prodelphic, convoluted. Rectal gland six large, secrete gelatinous material in which eggs are deposited; eggs not retained in body. **Male:** Vermiform, up to 2.5 mm long, tail end twisted, develops by metamorphosis within a swollen juvenile. Cuticle strongly annulated, lateral field usually with four incisures, outer bands often areolated. Labial region round, not sharply offset, with distinct labial disc and few annuli, lateral sectors wider than submedians. Stylet robust (10.5-29.0 μm), with large knobs. Pharyngeal glands mostly ventral to intestine. Spicules slender, generally 19-44 μm long, gubernaculum 7-11 μm long. Testis single, or rarely paired. Tail short, rounded. Phasmids dot-like, near cloacal aperture, which is subterminal. Bursa absent. **Juveniles:** First stage with a blunt tail tip, moulting within egg, second and third moults occurring within cuticle of second stage. Second stage vermiform, 250-600 μm long, migratory, infective, straight to arcuate habitus upon death. Labial region with coarse annuli (1-4), a distinct labial disc, framework lightly sclerotised, lateral sectors wider than submedian sectors, stylet slender

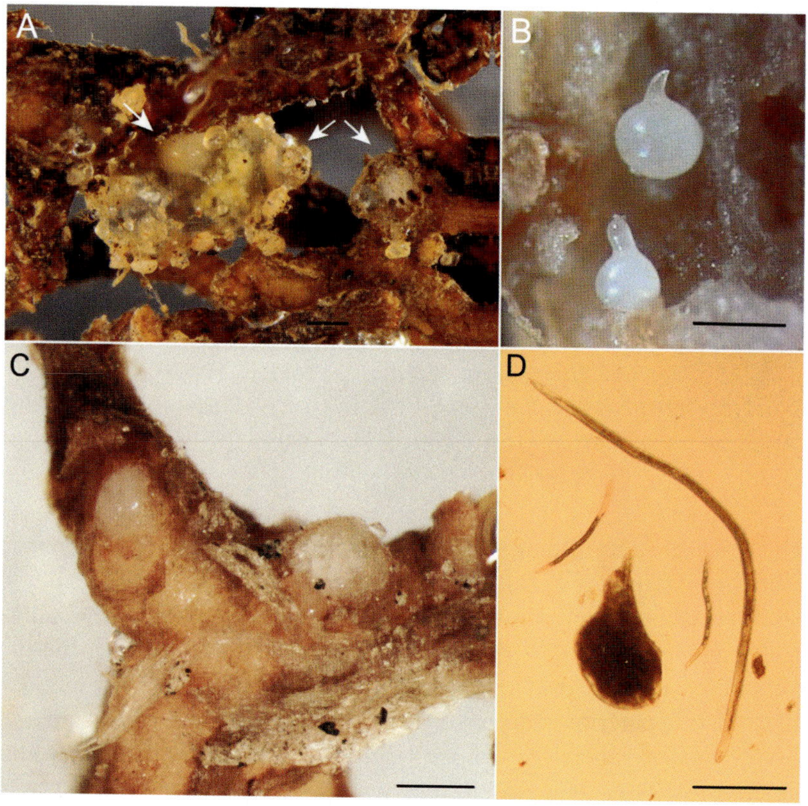

Fig. 1. *Root-knot nematodes. A: Females of* Meloidogyne nataliei *on grape roots (after Álvarez-Ortega* et al., *2019); B: White females (Courtesy of University of Florida, USA); C: Females of* M. floridensis *on a peach rootstock root; D: Female, male and second-stage juvenile and egg of* M. floridensis. *(Scale bars: A = 500 μm; B, D = 300 μm; C = 400 μm.) (After Reighard* et al., *2019.)*

(7.5-22.5 μm long), excretory pore posterior to hemizonid. Median bulb with large oval refractive thickenings. Three pharyngeal glands ventrally overlapping intestine. Lateral field with four incisures. Tail with conspicuous hyaline region, tip narrow, irregular in outline. Third and fourth stage lacking a functional stylet, sedentary and swollen.

TYPE SPECIES

1. *Meloidogyne exigua* Göldi, 1887
   = *Heterodera exigua* (Göldi, 1887) Marcinowski, 1909

OTHER SPECIES

2. *M. aberrans* Tao, Xu, Yuan, Wang, Lin, Zhuo & Liao, 2017
3. *M. acronea* Coetzee, 1956
   = *Hypsoperine acronea* (Coetzee, 1956) Sledge & Golden, 1964
   = *Hypsoperine* (*Hypsoperine*) *acronea* (Coetzee, 1956) Sledge & Golden, 1964 (Siddiqi, 1986)
4. *M. aegracyperi* Eisenback, Holland, Schroeder, Thomas, Beacham, Hanson, Paes-Takahashi & Vieira, 2019
5. *M. africana* Whitehead, 1960
   = *M. decalineata* Whitehead, 1968
   = *M. oteifae* Elmiligy, 1968
6. *M. aquatilis* Ebsary & Eveleigh, 1983
7. *M. arabicida* López & Salazar, 1989
8. *M. ardenensis* Santos, 1968
   = *M. deconincki* Elmiligy, 1968
   = *M. litoralis* Elmiligy, 1968
9. *M. arenaria* (Neal, 1889) Chitwood, 1949
   = *Anguillula arenaria* Neal, 1889
   = *Tylenchus arenarius* (Neal, 1889) Cobb, 1890
   = *Heterodera arenaria* (Neal, 1889) Marcinowski, 1909
   = *M. arenaria arenaria* (Neal, 1889) Chitwood, 1949
   = *M. arenaria thamesi* Chitwood, 1952 *in* Chitwood, Specht & Havis, 1952
   = *M. thamesi* Chitwood, 1952 *in* Chitwood, Specht & Havis, 1952
   = *M. thamesi gyulai* Amin, 1993
   = *M. gyulai* Amin, 1993
10. *M. artiellia* Franklin, 1961
11. *M. baetica* Castillo, Vovlas, Subbotin & Troccoli, 2003
12. *M. brevicauda* Loos, 1953
13. *M. californiensis* Abdel-Rahman & Maggenti, 1987
14. *M. camelliae* Golden, 1979
15. *M. caraganae* Shagalina, Ivanova & Krall, 1985
16. *M. carolinensis* Eisenback, 1982
17. *M. chitwoodi* Golden, O'Bannon, Santo & Finley, 1980
18. *M. chosenia* Eroshenko & Lebedeva, 1992
19. *M. christiei* Golden & Kaplan, 1986
20. *M. citri* Zhang, Gao & Weng, 1990
21. *M. coffeicola* Lordello & Zamith, 1960

= *Meloidodera coffeicola* (Lordello & Zamith, 1960) Kirjanova, 1963
22. *M. cruciani* García-Martinez, Taylor & Smart, 1982
23. *M. cynariensis* Pham, 1990
24. *M. daklakensis* Trinh, Le, Nguyen, Nguyen, Liebanas & Nguyen, 2018
25. *M. donghaiensis* Zheng, Lin & Zheng, 1990
26. *M. dunensis* Palomares-Rius, Vovlas, Troccoli, Liebanas, Landa & Castillo, 2007
27. *M. duytsi* Karssen, Aelst & van der Putten, 1998
28. *M. enterolobii* Yang & Eisenback, 1983
= *M. mayaguensis* Rammah & Hirschmann, 1988
29. *M. ethiopica* Whitehead, 1968
= *M. brasilensis* Charchar & Eisenback, 2002[1]
30. *M. fallax* Karssen, 1996
31. *M. fanzhiensis* Chen, Peng & Zheng, 1990
32. *M. floridensis* Handoo, Nyczepir, Esmenjaud, van der Beek, Castagnone-Sereno, Carta, Skantar & Higgins, 2004
33. *M. fujianensis* Pan, 1985
34. *M. graminicola* Golden & Birchfield, 1965
= *M. hainanensis* Liao & Feng, 1995 **syn. n.**
= *M. lini* Yang, Hu & Xu, 1988 **syn. n.**
35. *M. graminis* (Sledge & Golden, 1964) Whitehead, 1968
= *Hypsoperine graminis* Sledge & Golden, 1964
= *Hypsoperine* (*Hypsoperine*) *graminis* Sledge & Golden, 1964 (Siddiqi, 1986)
36. *M. hapla* Chitwood, 1949
37. *M. haplanaria* Eisenback, Bernard, Starr, Lee & Tomaszewski, 2003
38. *M. hispanica* Hirschmann, 1986
39. *M. ichinohei* Araki, 1992
40. *M. incognita* (Kofoid & White, 1919) Chitwood, 1949
= *Oxyuris incognita* Kofoid & White, 1919
= *M. acrita* Chitwood, 1949 (Esser, Perry & Taylor, 1976)
= *M. incognita acrita* Chitwood, 1949

[1] Note: The specific epithet is consistently and erroneously spelled as '*brasiliensis*' by Monteiro *et al.* (2017) when they synonymised the taxon. This spelling error should be regarded as a *lapsus calami* and does not enter into synonymy (D.J. Hunt).

= *M. kirjanovae* Terenteva, 1965
= *M. elegans* da Ponte, 1977
= *M. grahami* Golden & Slana, 1978
= *M. wartellei* Golden & Birchfield, 1978
= *M. polycephannulata* Charchar, Eisenback, Vieira, Fonseca-Boiteux & Boiteux, 2009

41. *M. indica* Whitehead, 1968
42. *M. inornata* Lordello, 1956
    = *M. incognita inornata* Lordello, 1956
43. *M. izalcoensis* Carneiro, Almeida, Gomes & Hernandez, 2005a
44. *M. javanica* (Treub, 1885) Chitwood, 1949
    = *Heterodera javanica* Treub, 1885
    = *Tylenchus* (*Heterodera*) *javanicus* (Treub, 1885) Cobb, 1890
    = *M. javanica javanica* (Treub, 1885) Chitwood, 1949
    = *M. javanica bauruensis* Lordello, 1956
    = *M. bauruensis* Lordello, 1956 (Esser, Perry & Taylor, 1976)
    = *M. dimocarpus* Liu & Zhang, 2001 **syn. n.**
    = *M. lordelloi* da Ponte, 1969
    = *M. lucknowica* Singh, 1969
45. *M. jianyangensis* Yang, Hu, Chen & Zhu, 1990
46. *M. jinanensis* Zhang & Su, 1986
47. *M. kikuyensis* de Grisse, 1961
    = *Spartonema kikuyense* (De Grisse, 1961) Siddiqi, 2000
48. *M. konaensis* Eisenback, Bernard & Schmitt, 1995
49. *M. kongi* Yang, Wang & Feng, 1988
50. *M. kralli* Jepson, 1984
51. *M. lopezi* Humphreys-Pereira, Flores-Chaves, Gomez, Salazar, Gomez-Alpizar & Elling, 2014
52. *M. luci* Carneiro, Correa, Almeida, Gomes, Mohammad Deimi, Castagnone-Sereno & Karssen, 2014
53. *M. lusitanica* Abrantes & Santos, 1991
54. *M. mali* Itoh, Ohshima & Ichinohe, 1969
    = *M. ulmi* Marinari-Palmisano & Ambrogioni, 2000
55. *M. maritima* Jepson, 1987
56. *M. marylandi* Jepson & Golden, 1987 *in* Jepson, 1987
57. *M. megadora* Whitehead, 1968
58. *M. megatyla* Baldwin & Sasser, 1979
59. *M. mersa* Siddiqi & Booth, 1991
    = *Meloidogyne* (*Hypsoperine*) *mersa* Siddiqi & Booth, 1991

60. *M. microcephala* Cliff & Hirschmann, 1984
61. *M. microtyla* Mulvey, Townshend & Potter, 1975
62. *M. mingnanica* Zhang, 1993
63. *M. minor* Karssen, Bolk, van Aelst, van den Beld, Kox, Korthals, Molendijk, Zijlstra, van Hoof & Cook, 2004
64. *M. moensi* Le, Nguyen, Nguyen, Liebanas, Nguyen & Trinh, 2019
65. *M. morocciensis* Rammah & Hirschmann, 1990
66. *M. naasi* Franklin, 1965
67. *M. nataliei* Golden, Rose & Bird, 1981
68. *M. oleae* Archidona-Yuste, Cantalapiedra-Navarrete, Liebanas, Rapoport, Castillo & Palomares-Rius, 2018
69. *M. oryzae* Maas, Sanders & Dede, 1978
70. *M. ottersoni* (Thorne, 1969) Franklin, 1971
    = *Hypsoperine ottersoni* Thorne, 1969
    = *Hypsoperine* (*Hypsoperine*) *ottersoni* Thorne, 1969 (Siddiqi, 1986)
71. *M. ovalis* Riffle, 1963
72. *M. panyuensis* Liao, Yang, Feng & Karssen, 2005
73. *M. paranaensis* Carneiro, Carneiro, Abrantes, Santos & Almeida, 1996a
74. *M. partityla* Kleynhans, 1986
75. *M. petuniae* Charchar, Eisenback & Hirschmann, 1999
76. *M. phaseoli* Charchar, Eisenback, Charchar & Boiteux, 2008b
77. *M. pini* Eisenback, Yang & Hartman, 1985
78. *M. piperi* Sahoo, Ganguly & Eapen, 2000
79. *M. pisi* Charchar, Eisenback, Charchar & Boiteux, 2008a
80. *M. platani* Hirschmann, 1982
81. *M. propora* Spaull, 1977
    = *Hypsoperine propora* (Spaull, 1977) Siddiqi, 1986
    = *Hypsoperine* (*Hypsoperine*) *propora* (Spaull, 1977) Siddiqi, 1986
82. *M. querciana* Golden, 1979
83. *M. salasi* Lopez, 1984
84. *M. sasseri* Handoo, Huettel & Golden, 1994
85. *M. sewelli* Mulvey & Anderson, 1980
86. *M. silvestris* Castillo, Vovlas, Troccoli, Liébanas, Palomares-Rius & Landa, 2009
87. *M. sinensis* Zhang, 1983

88. *M. spartelensis* Ali, Tavoillot, Mateille, Chapuis, Besnard, El Bakkali, Cantalapiedra-Navarrete, Liebanas, Castillo & Palomares-Rius, 2015
89. *M. spartinae* (Rau & Fassuliotis, 1965) Whitehead, 1968
    = *Hypsoperine spartinae* Rau & Fassuliotis, 1965
    = *Hypsoperine* (*Spartonema*) *spartinae* Rau & Fassuliotis, 1965 (Siddiqi, 1986)
    = *Spartonema spartinae* (Rau & Fassuliotis, 1965) Siddiqi, 1986
90. *M. subarctica* Bernard, 1981
91. *M. suginamiensis* Toida & Yaegashi, 1984
92. *M. tadshikistanica* Kirjanova & Ivanova, 1965
93. *M. thailandica* Handoo, Skantar, Carta & Erbe, 2005
94. *M. trifoliophila* Bernard & Eisenback, 1997
95. *M. triticoryzae* Gaur, Saha & Khan, 1993
96. *M. turkestanica* Shagalina, Ivanova & Krall, 1985
97. *M. vandervegtei* Kleynhans, 1988
98. *M. vitis* Yang, Hu, Liu, Chen, Peng, Wang & Zhang, 2021

Species Inquirendae

1. *M. actinidiae* Li & Yu, 1991
2. *M. cirricauda* Zhang & Weng, 1991
3. *M. marioni* (Cornu, 1879) Chitwood & Oteifa, 1952
    = *Anguillula marioni* Cornu, 1879
    = *Heterodera marioni* (Cornu, 1879) Marcinowski, 1909
4. *M. megriensis* (Poghossian, 1971) Esser, Perry & Taylor, 1976
    = *Hypsoperine megriensis* Poghossian, 1971
    = *Hypsoperine* (*Hypsoperine*) *megriensis* Poghossian, 1971 (Siddiqi, 1986)
5. *M. pakistanica* Shahina, Nasira, Salma, Mehreen & Bhatti, 2015
6. *M. poghossianae* Kirjanova, 1963
    = *M. acronea apud* Poghossian, 1961 *nec M. acronea* Coetzee, 1956
7. *M. vialae* (Lavergne, 1901) Chitwood & Oteifa, 1952
    = *Anguillula vialae* Lavergne, 1901
    = *Heterodera vialae* (Lavergne, 1901) Marcinowski, 1909

Nomina Nuda

1. *M. californiensis* Abdel-Rahman, 1981
2. *M. carolinensis* Fox, 1967

3. *M. goeldii* Santos, 1997
4. *M. panyuensis* Liao, 2001
5. *M. shunchangensis* Wu, 2011
6. *M. zhanjiangensis* Liao, 2001

Nematology Monographs & Perspectives, 2021, Vol. 14, 13-17

Chapter 3

Morphology

The general morphology of RKN was described in detail by Chitwood (1949), Esser *et al.* (1976), Taylor & Sasser (1978), Eisenback (1985), Hirschmann (1985), Jepson (1987), Eisenback & Triantaphyllou (1991), Eisenback & Hunt (2009), Karssen *et al.* (2013), Ghaderi & Karssen (2020) and other authors, and is briefly given below.

Female

Body swollen, saccate and very variable in size. Neck protruding anteriorly. Vulva and anus located terminally in perineal region (Fig. 2). In most species bodies are symmetrical and neck and perineal region are in a straight line. However, neck can project from longitudinal axis at an angle of up to 90° to one side. Most species have mean stylet lengths of 8-25 μm, cone in most species slightly curved dorsally, shaft straight with three knobs. Shape of knobs species-specific in many cases. Dorsal pharyngeal gland orifice (DGO) located posteriorly to stylet knobs. DGO distance has a broad range among species (2-12 μm). Pharynx with a large muscular median bulb with conspicuous valve plates and three ventrally overlapping pharyngeal glands. Two subventral gland orifices opening into the pharyngeal lumen immediately posterior to valve plates. Excretory pore located anterior to median bulb. Females didelphic with two very long and greatly convoluted gonads occupying a major part of total body content. Approximately 60% of female gonad is made up of the proper ovaries. Cuticle in perineal region forming a fingerprint-like pattern. Perineal region including tail terminus, phasmids, lateral lines, anus and vulva surrounded by cuticular folds or striae. Vulva located transversely on posterior end of body wall and surrounded by two, slightly elevated, vulval lips. Perineal pattern rather species-specific and does not change significantly during female maturation. Six large, unicellular rectal glands in posterior body region, connected to rectum and producing a very large amount of gelatinous matrix material. Eggs

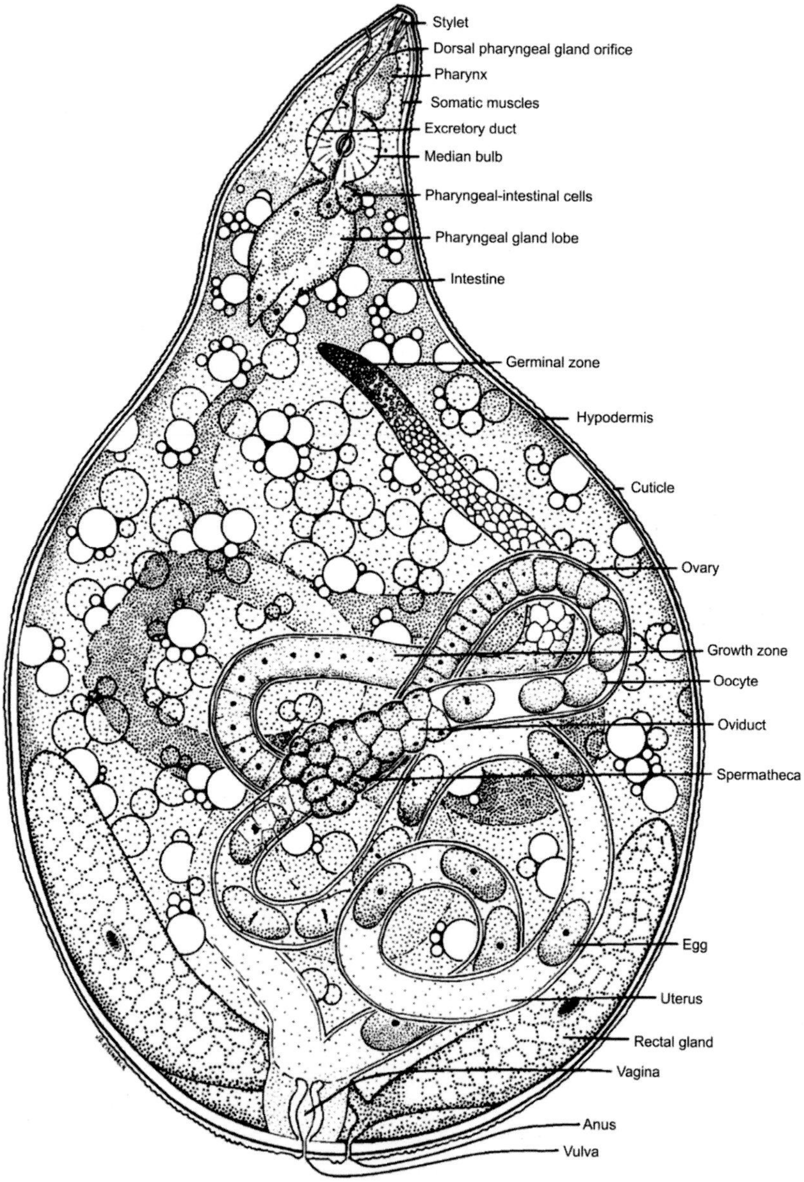

Fig. 2. *Morphology of female. (After Eisenback, 1985.)*

deposited in this protective egg sac. Rectal glands not developed and no egg mass formed in *M. spartinae* and *M. kikuyensis*.

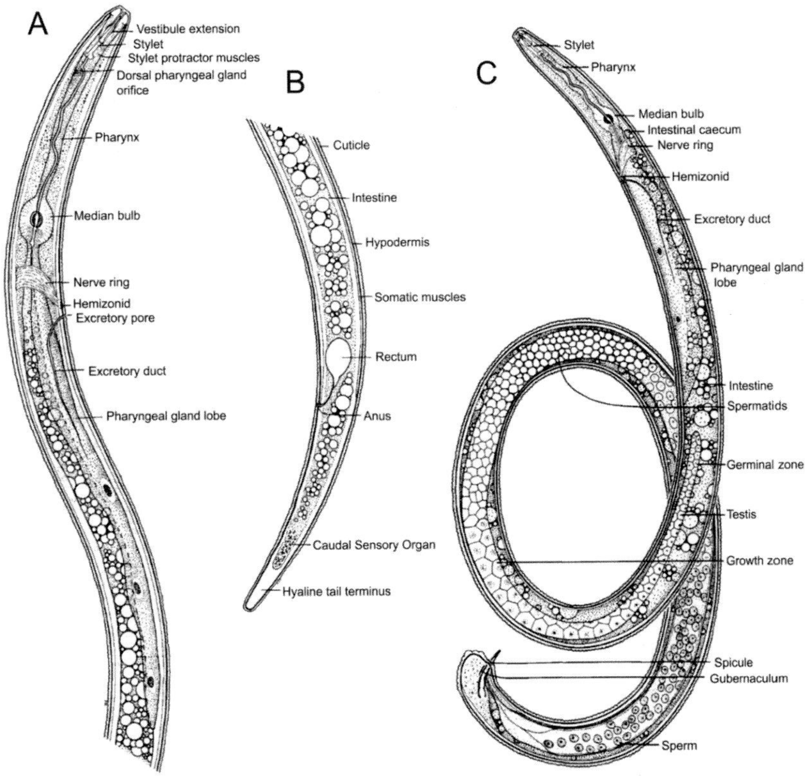

Fig. 3. *Morphology. A: Anterior region of second-stage juvenile (J2); B: Posterior region of J2; C: Male. (After Eisenback, 1985.)*

Male

Vermiform males vary greatly in body size from 700 to 2890 μm. Male anterior end composed of labial cap and labial region, providing many good diagnostic features (Fig. 3). Labial cap including a labial disc surrounded by lateral and median labials. Four sensory organs terminating on median labials (labial sensilla), and six others surrounding stoma area (labial sensilla). Labial region may or may not be offset from body. Stylet length has a broad range within genus (10.5-29 μm), although most species have an average stylet length of 18-24 μm. DGO located 1.5-13 μm posterior to stylet knob base. Pharynx with a slender procorpus and oval-shaped median bulb with distinct valve plates. Isthmus is short; ventrally overlapping gland lobe in most species has two instead of usual three pharyngeal nuclei. Obscure pharyngo-intestinal junction situated at level of nerve ring. Excretory pore position exhibiting large intraspecific variation and of limited value as a differential character. Hemizonid usually located anterior to excretory pore. One gonad present in normal males, but when environment affects sex expression, some individuals may develop two testes. Testis usually outstretched with germinal and growth zones. Most of gonad consists of a long *vas deferens* packed with developing sperm. *Vas deferens* ending posteriorly in a glandular region and forming a cloaca with intestine in region of spicules. Lateral field in most species with four incisures, but up to 15 in certain species, outer bands often areolated. Spicule length ranging from 19 to 47 μm among species, gubernaculum crescentic. Tail bluntly rounded and short, without caudal alae.

Juvenile

First-stage juvenile (J1) with a blunt tail tip, moulting within egg, second and third moults occurring within cuticle of second stage. Second-stage juvenile (J2) vermiform, migratory, infective, straight to arcuate habitus upon death with body length 250-600 μm (Fig. 3). Labial region with coarse annuli (1-4), a distinct labial disc, framework lightly sclerotised, lateral sectors wider than submedian sectors, stylet slender, 7-23 μm long. Excretory pore posterior to hemizonid. Median bulb with large oval refractive thickenings. Lateral field with four incisures. J2 tail 13-115 μm long, tapering towards conspicuous hyaline region, tip narrow, irregular in outline. Third and fourth stages are sedentary and swollen, lacking a functional stylet.

Egg

Oval, varying in size and shape but in most species averaging 77-139 μm long $\times$ 31-68 μm wide, exceptionally smaller in *M. ichinohei* (19-29 μm long $\times$ 10-12 μm wide). Eggshell consisting of an outer vitelline layer *ca* 30 μm thick, a middle chitinous layer *ca* 400 μm thick and an inner glycolabialid layer of varying thickness. Glycolabialid layer making egg very resistant to harsh chemicals, consequently, this stage not sensitive to toxins such as common nematicides. Egg development beginning within a few hours of deposition, cell division occurring until a fully formed J1 with a visible stylet lying coiled in egg membrane. First moult takes place in egg to develop infective J2.

Measurements and Indices for Morphological Identification

RKN are characterised by a combination of measurements and ratios derived from the various body parts. Such morphometric characters are usually abbreviated, the most common being listed below. Measurements of, for example, the body, pharynx and tail are taken along the mid-line of the relevant structure. Measurements of the spicule, a curved structure, are usually taken along the median line. Body diam. should be measured at a right angle to the longitudinal body axis.

Most commonly used abbreviations:

L = Total body length (labial region to tail tip).

W = Total body diam.

a = Total body length divided by max. body diam.

b = Total body length divided by pharyngeal length (the pharynx is defined as anterior end to the pharyngo-intestinal junction, *i.e.*, not to the posterior tip of the overlapping gland lobes).

b′ = Total body length divided by distance from anterior end of body to posterior end of pharyngeal glands.

c = Total body length divided by tail length.

c′ = Tail length divided by body diam. at the anal/cloacal aperture.

V = Position of vulva from anterior end expressed as percentage of body length.

V′ = Position of vulva from anterior end expressed as percentage of distance from labial region to anal aperture.

T = Distance between cloacal aperture and anteriormost part of testis expressed as percentage of body length.

m = Length of conical part of stylet as percentage of total stylet length.

o = Distance of dorsal pharyngeal gland opening posterior to stylet knobs expressed as a percentage of stylet length.

DGO = Dorsal pharyngeal gland orifice and distance of orifice posterior to basal knobs, depending on context.

EP/ST = Distance from anterior end to excretory pore divided by female stylet length.

Nematology Monographs & Perspectives, 2021, Vol. 14, 19-24

Chapter 4

Biology

Development and Life Cycle

The life cycle of RKN is completed in 3-6 weeks, depending on the species and environmental conditions, such as temperature, host species, *etc.* Ontogeny comprises eggs, four juvenile stages in addition to the adult stages, *i.e.*, male and egg-laying female (Fig. 4). During parasitism, RKN establish and maintain an intimate relationship with their host. The J2 then undergo three further moults to develop into adults. Sexual dimorphism with round females and vermiform males is associated with the sedentary lifestyle. Females remain sedentary, whereas males become vermiform and motile and finally leave the plant root (Rohini *et al.*, 1986; Castagnone-Sereno *et al.*, 2013).

Reproduction and Cytogenetics

One remarkable feature of RKN is their outstanding diversity in terms of mode of reproduction and various degrees of polyploidy and aneuploidy. Species can reproduce sexually or *via* different modes of parthenogenesis (meiotic or mitotic), and some species are even able to reproduce both sexually (amphimixis) and *via* meiotic parthenogenesis (Triantaphyllou, 1985a, b; Castagnone-Sereno *et al.*, 2013). Most RKN species of major agricultural importance (*e.g.*, *M. arenaria*, *M. incognita* and *M. javanica*) reproduce exclusively by mitotic parthenogenesis, which involves a single mitotic division with no reduction in chromosome number (Triantaphyllou, 1981, 1985a, b; Castagnone-Sereno *et al.*, 2013). In such mitotic species, no gametes with half the number of chromosomes are produced in the absence of meiosis. However, males can be sporadically observed in these species and sex is determined by environmental conditions, the frequency of males increasing in conditions of crowding or poor nutrition (Triantaphyllou, 1985a). They are able to produce sperm and mate with females, but although sperm

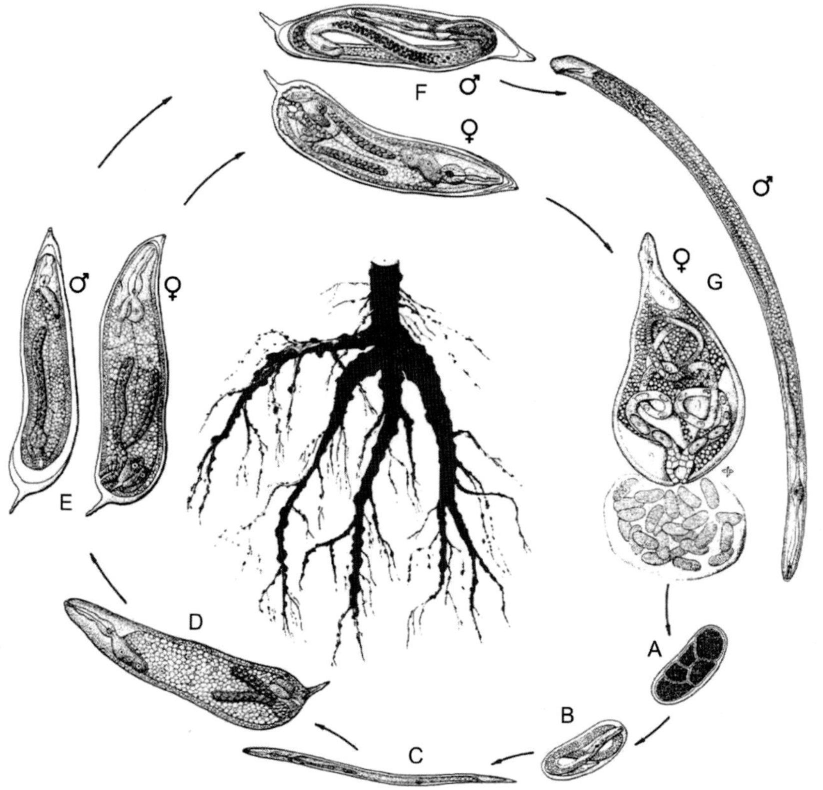

Fig. 4. *Life history of the root-knot nematode. A, B: Egg; C: Second-stage juvenile; D: Third-stage juvenile; E, F: Fourth-stage juvenile; G: Adults. (After C. Papp, CDFA, USA.)*

can be occasionally observed in the female spermatheca and the sperm nucleus can reach the egg, it has been reported that the sperm nucleus disintegrates in the egg cytoplasm during or just following mitotic division and thus apparently never fuses with the egg nucleus (Triantaphyllou, 1981; Castagnone-Sereno *et al.*, 2013). Meiotic parthenogenesis in species such as *M. hapla*, *M. chitwoodi* and *M. fallax* appears to follow a similar scheme. In these species, in the absence of fertilisation by a male, the egg nucleus undergoes classical meiosis with two successive divisions (the first is reductional, and the second is equational), and the somatic number of chromosomes is restored by the fusion of the second polar body with the egg pronucleus (Triantaphyllou, 1966, 1985a, b; van der Beek *et al.*, 1998; Castagnone-Sereno *et al.*, 2013). One interesting case is *M. floridensis*, which exhibits a cytological mechanism of meiotic parthenogenesis distinct from the one described above. *Meloidogyne floridensis* is clearly distinct from *M. incognita*, the latter being a mitotic parthenogenetically reproducing species. The oogenesis of *M. floridensis*, however, deviates from the meiotic parthenogenesis described by Triantaphyllou (1966) and van der Beek *et al.* (1998) by the absence of a second maturation division in the female body. The occurrence of meiotic parthenogenesis and suppression of the second maturation division could point toward an intermediate type of parthenogenesis between the meiotic form with two maturation divisions and mitotic parthenogenesis (Handoo *et al.*, 2004). A second interesting case is *M. hapla*, which represents two cytological races: race A reproduces by facultative meiotic parthenogenesis, whereas race B reproduces exclusively by mitotic parthenogenesis (Table 1).

The chromosomal complement of *Meloidogyne* spp. reflects the complexity of their reproduction (Fig. 5). The chromosome number is quite variable (Table 1) and it may differ greatly in populations of the same parthenogenetic species. Most of the amphimictic and automictic species are diploid with a haploid chromosome number of 18. The smallest haploid chromosome number, n = 4, was found in *M. nataliei* (Triantaphyllou, 1985b), followed by n = 7 in *M. spartinae* and *M. kikuyensis* (Triantaphyllou, 1987a, 1990), all of these species being amphimictic. Triantaphyllou (1987a, b) believed that the smaller chromosome num-

Table 1. *Chromosome number and mode of reproduction of some* Meloidogyne *species.*

Species	Chromosome number	Mode of reproduction	Reference
M. arenaria	2n = 30-38, 40-48, 3n = 51-56	Obligatory mitotic parthenogenesis	Triantaphyllou (1985a)
M. africana	2n = 21	Obligatory mitotic parthenogenesis	Janssen *et al.* (2017)
M. ardenensis	2n = 51-54	Obligatory mitotic parthenogenesis	Janssen *et al.* (2017)
M. carolinensis	n = 18	Amphimixis	Triantaphyllou (1985a)
M. chitwoodi	n = 18	Facultative meiotic parthenogenesis	van der Beek & Karssen (1997)

Table 1. *(Continued.)*

Species	Chromosome number	Mode of reproduction	Reference
M. cruciani	2n = 42-48	Obligatory mitotic parthenogenesis	Esbenshade & Triantaphyllou (1985a, b)
M. enterolobii	2n = 44-46	Obligatory mitotic parthenogenesis	Yang & Eisenback (1983); Rammah & Hirschmann (1988)
M. ethiopica	2n = 36-44	Obligatory mitotic parthenogenesis	Carneiro *et al.* (2004a)
M. exigua	n = 18	Facultative meiotic parthenogenesis	Triantaphyllou (1985b); Muniz *et al.* (2009)
M. fallax	n = 18	Facultative meiotic parthenogenesis	van der Beek & Karssen (1997)
M. floridensis	n = 18, 19, 20	Facultative meiotic parthenogenesis	Handoo *et al.* (2004)
M. graminicola	n = 18	Facultative meiotic parthenogenesis	Triantaphyllou (1985b)
M. graminis	n = 18	Facultative meiotic parthenogenesis	Triantaphyllou (1985b)
M. hapla cytological race A	n = 13-17, 28-34	Facultative meiotic parthenogenesis	Triantaphyllou (1966); Triantaphyllou (1985b)
M. hapla cytological race B	2n = 30-32, 43-48	Obligatory mitotic parthenogenesis	Triantaphyllou (1966); Triantaphyllou (1985b)
M. hispanica	2n = 33-36	Obligatory mitotic parthenogenesis	Triantaphyllou (1985b)
M. incognita	2n = 32-38, 3n = 41-46	Obligatory mitotic parthenogenesis	Triantaphyllou (1985b), Eisenback & Triantaphyllou (1991)
M. inornata	3n = 54-58	Obligatory mitotic parthenogenesis	Carneiro *et al.* (2008)
M. izalcoensis	2n = 44-48	Obligatory mitotic parthenogenesis	Carneiro *et al.* (2005a)
M. javanica	2n = 42-48	Obligatory mitotic parthenogenesis	Triantaphyllou (1985b)
M. kikuyensis	n = 7	Amphimixis	Triantaphyllou (1990)

Table 1. *(Continued.)*

Species	Chromosome number	Mode of reproduction	Reference
M. konaensis	2n = 44	Obligatory mitotic parthenogenesis	Eisenback *et al.* (1994)
M. luci	2n = 42-46	Obligatory mitotic parthenogenesis	Carneiro *et al.* (2014)
M. mali	2n = 22	Amphimixis	Janssen *et al.* (2017)
M. megatyla	n = 18	Amphimixis	Goldstein & Triantaphyllou (1982)
M. microcephala	2n = 37, 74	Obligatory mitotic parthenogenesis	Triantaphyllou & Hirschmann (1997)
M. microtyla	n = 18-19	Amphimixis	Triantaphyllou (1985a)
M. minor	n = 17	Facultative meiotic parthenogenesis	Karssen *et al.* (2004)
M. morocciensis	2n = 42-49	Obligatory mitotic parthenogenesis	Rammah & Hirschmann (1990b); Carneiro *et al.* (2008)
M. naasi	n = 18	Facultative meiotic parthenogenesis	Triantaphyllou (1985a)
M. nataliei	n = 4	Amphimixis	Triantaphyllou (1985b)
M. oryzae	3n = 50-56	Obligatory mitotic parthenogenesis	Eisenback & Triantaphyllou (1991); Mattos *et al.* (2018)
M. ottersoni	n = 18	Facultative meiotic parthenogenesis	Triantaphyllou (1985a)
M. paranaensis	3n = 50-52	Obligatory mitotic parthenogenesis	Carneiro *et al.* (1996a)
M. partityla	2n = 40-42	Obligatory mitotic parthenogenesis	Marais and Kruger (1991)
M. petuniae	2n = 41, 47	Obligatory mitotic parthenogenesis	Charchar *et al.* (1999); Chitwood & Perry (2009)
M. pini	n = 18	Amphimixis	Triantaphyllou (1979)
M. platani	2n = 42-44	Obligatory mitotic parthenogenesis	Hirschmann (1982)
M. querciana	2n = 30-32	Obligatory mitotic parthenogenesis	Triantaphyllou (1985a)
M. spartinae	n = 7	Amphimixis	Triantaphyllou (1987a)
M. subarctica	n = 18	Amphimixis	Bernard (1981)

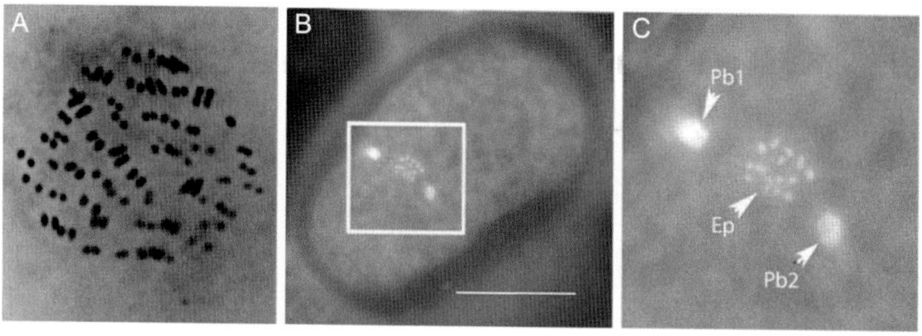

Fig. 5. *A: Chromosomes during single maturation division of oocytes of* Meloidogyne javanica *(after Eisenback & Triantaphyllou, 1991); B: Oocytes of* M. hapla *at anaphase of meiosis II; C: Enlarged view of boxed area with 16 chromosomes counted in egg pronucleus (Ep) with polar bodies (Pb1. Pb2). (Scale bar = 4 μm.) (After Liu & Williamson, 2006.)*

ber may more accurately represent the ancestral chromosomal form from which the rest of the RKN evolved. The majority of the apomictic species have a high chromosome number and represent states of polyploidy: either triploids or hypotriploids.

Nematology Monographs & Perspectives, 2021, Vol. 14, 25-30

Chapter 5

Distribution of Root-knot Nematode Species

Knowledge of RKN diversity and prevalence, as well as the major environmental and agronomical factors for understanding their distribution in specific areas, is of vital importance for designing control measures to reduce significant damage. *Meloidogyne* species are among nature's most successful plant parasites and have a significant economic impact on host plants due to their wide host range and distribution throughout temperate and tropical environments (Perry *et al.*, 2009).

RKN have been found in all agricultural regions of the world and any crop is likely to suffer damage from these parasites. Distribution maps and host range data of some species are available and updated regularly as a useful source for determining nematode damage potential (http://www.cabi.org/dmpd). On a global scale, the distribution of RKN species varies greatly. Some are cosmopolitan, such as the four most common *Meloidogyne* spp. (*M. arenaria*, *M. hapla*, *M. incognita*, *M. javanica*), whilst others are particularly restricted geographically. Despite the global distribution of the genus, 49 of the described *Meloidogyne* species have so far only been recorded from their type locality: These are:

a) Africa: *M. spartelensis*, *M. propora* and *M. vandervegtei*.
b) Asia: *M. aberrans*, *M. caraganae*, *M. chosenia*, *M. citri*, *M. cynariensis*, *M. daklakensis*, *M. donghaiensis*, *M. fanzhiensis*, *M. fujianensis*, *M. ichinohei*, *M. jianyangensis*, *M. jinanensis*, *M. kongi*, *M. mersa*, *M. microcephala*, *M. mingnanica*, *M. moensi*, *M. panyuensis*, *M. piperi*, *M. sinensis*, *M. tadshikistanica*, *M. thailandica*, *M. triticoryzae* and *M. vitis*.
c) Europe: *M. baetica*, *M. dunensis* and *M. silvestris*.
d) North America: *M. aegracyperi*, *M. aquatilis*, *M. californiensis*, *M. christiei*, *M. megatyla*, *M. microtyla*, *M. nataliei*, *M. ovalis*, *M. pini*, *M. platani*, *M. querciana*, *M. sasseri*, *M. sewelli* and *M. subarctica*.

e) South and Central America: *M. arabicida*, *M. lopezi*, *M. petuniae*, *M. phaseoli* and *M. pisi*.

The highest global biodiversity of the genus *Meloidogyne* occurs in Asia where 44 species have been reported, North America with 35, Central and South America 31, followed by Europe with 24, Africa with 23 and Oceania with 11 (Table 2). The most widely distributed and commonest species are *M. arenaria*, *M. hapla*, *M. incognita* and *M. javanica*, which have been reported on every continent with the exception of Antarctica. Other species with a wide geographical distribution include: *M. enterolobii*, *M. exigua*, *M. fallax*, *M. hispanica*, *M. graminicola*, and *M. naasi*, whilst other species are restricted to one continent, *e.g.*, *M. acronea*, *M. ardenensis*, *M. brevicauda*, *M. camelliae*, *M. coffeicola*, *M. cruciani*, *M. duytsi*, *M. haplanaria*, *M. kikuyensis*, *M. kralli*, *M. maritima*, *M. oryzae*, *M. paranaensis*, *M. salasi* and *M. spartinae*.

The geographic distribution of *Meloidogyne* species is mostly dependent on the prevalence of host plants supporting reproduction, abiotic factors (mainly temperature) and their introduction to new areas by means of infected plant material or infested soil. Temperature is the factor most easily recognised as important in determining their distribution, especially in relation to the ability to survive the effects of heat and cold extremes. Based on the ability to survive labile-phase transitions that occur at 10°C within the genus *Meloidogyne*, two groups can be distinguished: *i*) thermophils, which do not survive in soils at temperatures below 10°C, *i.e.*, *M. javanica*, *M. arenaria*, and *M. exigua*; and *ii*) cryophils, which can survive in soil at temperatures down to and below 0°C, *i.e.*, *M. chitwoodi*, *M. hapla* and probably *M. naasi* (Van

Table 2. *Occurrence of* Meloidogyne *species in world regions arranged by alphabetical order.*

Region (species number)	Species
Africa (23)	*M. acronea*, *M. africana*, *M. arenaria*, *M. artiellia*, *M. chitwoodi*, *M. enterolobii*, *M. ethiopica*, *M. exigua*, *M. fallax*, *M. graminicola*, *M. hapla*, *M. hispanica*, *M. incognita*, *M. izalcoensis*, *M. javanica*, *M. kikuyensis*, *M. megadora*, *M. morocciensis*, *M. naasi*, *M. partityla*, *M. propora*, *M. spartelensis*, *M. vandervegtei*

Table 2. *(Continued.)*

Region (species number)	Species
Asia (44)	*M. aberrans, M. africana, M. arenaria, M. artiellia, M. brevicauda, M. camelliae, M. caraganae, M. chosenia, M. citri, M. cynariensis, M. daklakensis, M. donghaiensis, M. enterolobii, M. exigua, M. fanzhiensis, M. fujianensis, M. graminicola, M. graminis, M. hapla, M. hispanica, M. ichinohei, M. incognita, M. indica, M. izalcoensis, M. javanica, M. jianyangensis, M. jinanensis, M. kongi, M. mali, M. marylandi, M. mersa, M. microcephala, M. mingnanica, M. moensi, M. naasi, M. panyuensis, M. piperi, M. sinensis, M. suginamiensis, M. tadshikistanica, M. thailandica, M. triticoryzae, M. turkestanica, M. vitis*
Europe (24)	*M. ardenensis, M. arenaria, M. artiellia, M. baetica, M. chitwoodi, M. dunensis, M. duytsi, M. enterolobii, M. exigua, M. fallax, M. graminicola, M. hapla, M. hispanica, M. incognita, M. javanica, M. kralli, M. luci, M. lusitanica, M. mali, M. maritima, M. minor, M. naasi, M. oleae, M. silvestris*
North America (35)	*M. aegracyperi, M. aquatilis, M. arenaria, M. californiensis, M. carolinensis, M. chitwoodi, M. christiei, M. cruciani, M. enterolobii, M. fallax, M. floridensis, M. graminicola, M. graminis, M. hapla, M. haplanaria, M. hispanica, M. incognita, M. javanica, M. mali, M. marylandi, M. megatyla, M. microtyla, M. naasi, M. nataliei, M. ottersoni, M. ovalis, M. partityla, M. pini, M. platani, M. querciana, M. sasseri, M. sewelli, M. spartinae, M. subarctica, M. trifoliophila*
South and Central America (31)	*M. africana, M. arabicida, M. arenaria, M. chitwoodi, M. coffeicola, M. cruciani, M. enterolobii, M. ethiopica, M. exigua, M. graminicola, M. graminis, M. hapla, M. hispanica, M. incognita, M. inornata, M. izalcoensis, M. javanica, M. konaensis, M. lopezi, M. luci, M. marylandi, M. minor, M. morocciensis, M. naasi, M. oryzae, M. ottersoni, M. paranaensis, M. petuniae, M. phaseoli, M. pisi, M. salasi*
Oceania (11)	*M. arenaria, M. fallax, M. hapla, M. hispanica, M. incognita, M. javanica, M. konaensis, M. minor, M. naasi, M. paranaensis, M. trifoliophila*

Gundy, 1985). Similarly, soil texture, moisture, aeration, oxygen availability, toxins and osmotic potential are interacting factors and it is difficult to determine the effect of each one separately on the distribution of *Meloidogyne* species. RKN are active in soils with moisture levels at 40-60% of field capacity. The gelatinous matrix of the egg sac appears to maintain a high moisture level and provides a barrier to water loss from eggs (Karssen & Moens, 2006).

The spatial pattern of RKN populations in agricultural or natural ecosystems is influenced by factors such as the length of time the nematode population has been present in the ecosystem. RKN depend on host plants for their feeding (obligate parasites), their dispersal over short distances being mainly due to plant-to-plant infestation through root networks in the soil (Perry *et al.*, 2009). RKN dispersal over long distances is necessarily passive, *via* propagation material (seedlings, tubers, *etc.*), machinery (soil adhering to machinery will not only cause secondary foci in the same field but also initiate primary foci in previously non-infested fields tilled by the contaminated machinery), soil, water (rain transports soil and RKN to ditches; using the water from these ditches for irrigation will further spread these nematodes) (Vovlas *et al.*, 2005; Been & Schomaker, 2006). Medium scale distribution may result from active and mechanical redistribution of soil by machinery, mixed and displaced, either in the direction of cultivation (cultivation, ploughing and harvesting) or at right angles to it (ploughing, winter ploughing). This redistribution may adopt distinct shapes such as the development of the so-called infestation focus or hotspot, which in RKN nematodes cause large infestations in a short time (Been & Schomaker, 2006).

The literature available on the vertical distribution of RKN is limited. The root system of the host is the most important factor influencing the vertical distribution of RKN. Wesemael & Moens (2008) studied the vertical distribution of *M. chitwoodi* in several crops (summer barley, bean, beet, carrot, marigold and black fallow) to a depth of 70 cm, and a logistic model was fitted to the mean cumulative percentages of nematodes at increasing soil depth without any significant effect on the host. Similarly, Rodríguez-Kabana & Robertson (1987) studied the vertical distribution of *M. arenaria* in a groundnut field in Alabama, USA, and found that the distribution, as adjusted to a quasi-hyperbolic function, had the highest nematode populations in the top 30-40 cm of the soil profile (where the soil texture was lightest and the majority of

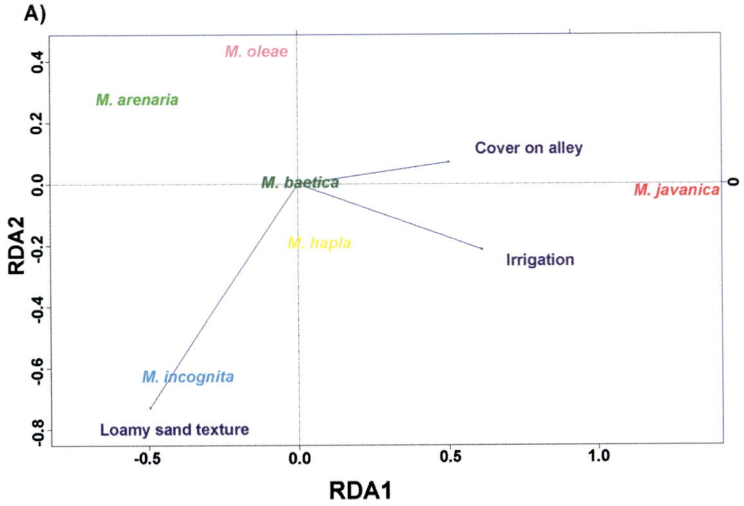

Fig. 6. *Canonical ordination analyses of Hellinger-transformed prevalence of* Meloidogyne *spp. data on cultivated olives with respect to environmental variables. The response variables are the fitted explanatory variables from the four data sets* (i.e., *climate, soil, topography and agronomic management) by forward selection procedure. Ordination plots of the redundancy analyses were used to investigate the relationships between prevalence of* Meloidogyne *spp. and explanatory variables. Ecological predictors are represented on the plots as agronomic practices related to irrigation regimen (Irrigation) and cover vegetation between rows (Cover on alley), and soil loamy sand texture (Loamy sand texture) on olive orchards. The first canonical axis (RDA axis 1) explains 85.9% of species-variables relationships, and the second axis (RDA axis 2) explains 11.2%.*

groundnut roots were concentrated). All of these findings indicate that management practices must take into consideration the fact that RKN are distributed deeply in soil and successful management must be able to reduce populations found deep in the soil profile (Rodríguez-Kabana & Robertson, 1987).

Recently, a study was done in Spain to obtain data and a better understanding of the distribution and abundance of RKN associated with wild and cultivated olives in Andalusia, southern Spain, including the environmental variables associated with their distribution. Three major parameters (from a large number of variables) drive the distribution of *Meloidogyne* spp. in cultivated olives in southern Spain: covered vegetation on alley, irrigation and soil texture, but different species responded differently to them (Archidona-Yuste *et al.*, 2018). In particular, the presence of *M. incognita* was highly correlated with sandy loamy soils, the presence of *M. javanica* with agronomic management practices such as irrigated soils and cover vegetation, while the presence of *M. arenaria* was correlated with the absence of cover vegetation between rows and absence of irrigation (Fig. 6). These parameters likely influence the selection of each particular *Meloidogyne* species from a major dispersal source, such as the rooted plantlets used to establish orchards (Aït Hamza *et al.*, 2017; Archidona-Yuste *et al.*, 2018).

Nematology Monographs & Perspectives, 2021, Vol. 14, 31-39

Chapter 6

Host-Parasite Relationships

Symptoms of Infection

Meloidogyne species are obligate plant parasites able to reproduce on nearly every species of higher plants inducing thickenings, small to large galls or root-knots. Figure 7 gives some examples of visible symptoms of galls; some of them are very conspicuous but others are hardly visible.

Above-ground symptoms observed on RKN-infected plants are non-specific and similar to those produced on any plant having a damaged root system. These symptoms include suppression of shoot growth and complementary decreased shoot-root ratio, nutritional deficiencies in the foliage, particularly chlorosis, temporary wilting during periods of mild water stress or during midday, even when adequate soil moisture is available, and suppressed plant yield (Lu *et al.*, 2014). Nematode-damaged plants are usually located in patches or along the planting row, reflecting an aggregate pattern (Nyczepir & Becker, 1998). Disease symptoms are more pronounced and yield reduction greater when abnormally high temperatures occur early in the growing season, soil fertility is low or imbalanced, and/or root-rot organisms attack the plants. Usually, the amount of these symptoms is also related to the inoculum density in soil and the number of J2 penetrating and becoming established within the root tissue of plants.

Giant Cell Formation

RKN initiate a subtle interaction with their hosts through intercellular migration after sensing chemical gradients of root diffusates (Teillet *et al.*, 2013). The J2 enter the elongation zone of the root and, using cell wall hydrolytic enzymes such as endoglucanases, endoxylanases, pectatelyases, *etc.*, from their subventral glands secreted into the apoplast (Perry & Moens, 2011), they reach the vascular cylinder by entering through the root meristem area (Fig. 8). Once established,

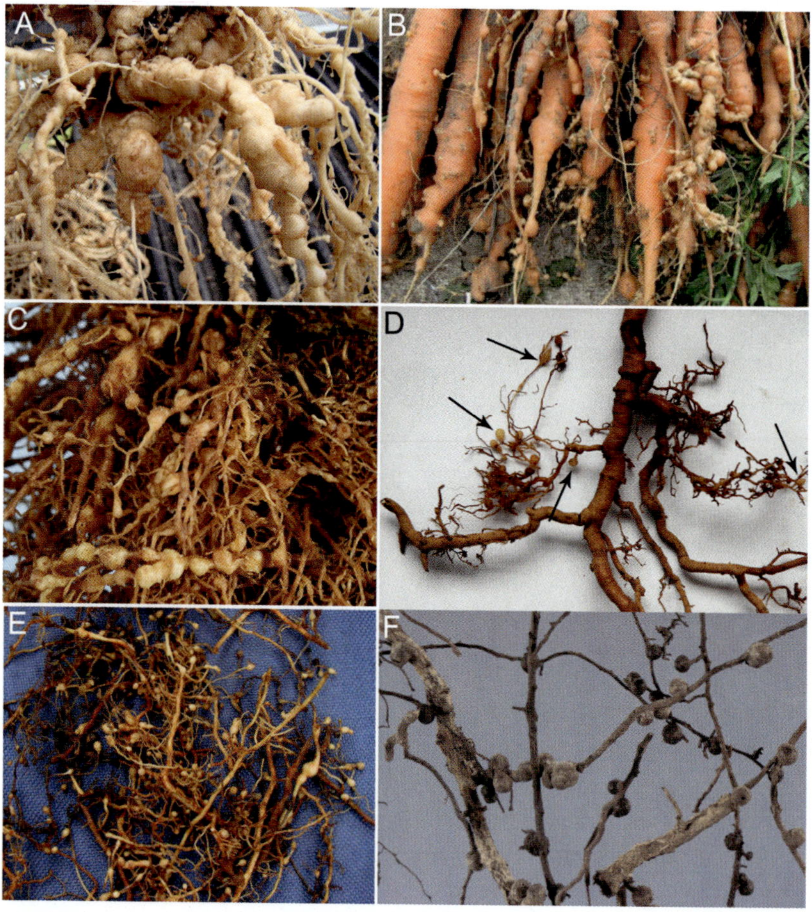

Fig. 7. *Severe galling symptoms induced by root-knot nematodes. A: Galls on tomato caused by* Meloidogyne *sp. (after University of Florida, USA); B: Galls on carrot caused by* Meloidogyne *sp. (after Lindsey du Toit, Washington State University, USA); C: Galls on tomato caused by* M. mali *(after Ahmed* et al., *2013); D: Galls (arrows) on wild kiwifruits caused by* M. aberrans *(after Tap* et al., *2017); E: Galls on* Coffea arabica *caused by* M. africana *(after Janssen* et al., *2017); F: Galls on Turkey oak caused by* M. christiei *(after Brito* et al., *2015).*

a group of 5-8 cells in the vascular cylinder develop into feeding cells, called giant cells (GCs) or coenocytes (Escobar *et al.*, 2015). Additionally, cortex cells surrounding the GCs divide and become hypertrophied and the pericycle cells proliferate (Berg *et al.*, 2008). The

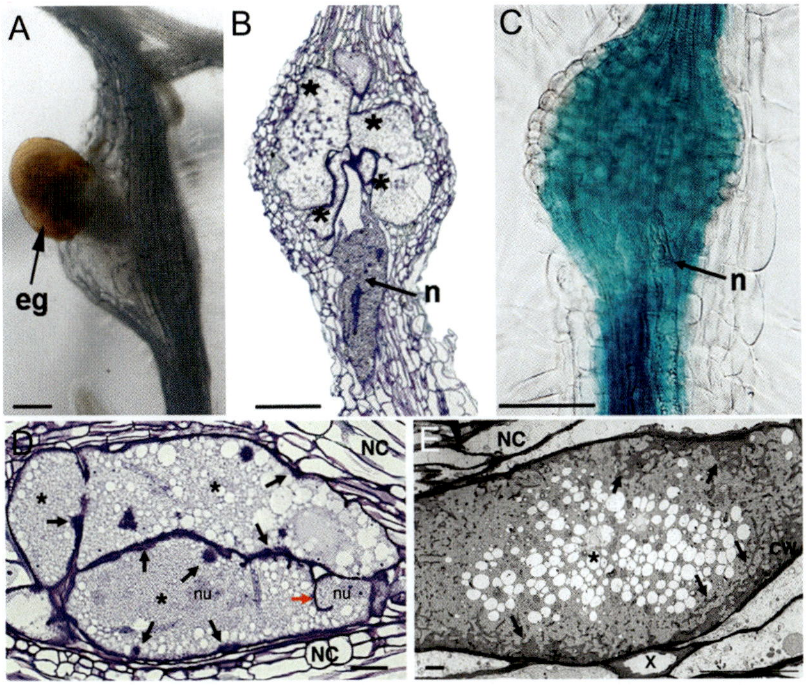

Fig. 8. *Galls induced by* Meloidogyne *spp. and ultrastructure of induced giant cells. A: Egg mass (eg) protruding from a gall in a* Cucumis sativus *root infected by* M. javanica*; B: Longitudinal section of a gall from* Arabidopsis thaliana *showing multinucleate giant cells and anterior region of the nematode; C: Vascular tissue of a gall (after Palomares-Rius* et al.*, 2017b); D: Light microscopy image of sectioned giant cells embedded in a gall and stained with Toluidine blue cells in* A. thaliana *roots. Cell wall thickenings (black arrows), and a cell wall stub (red arrow) indicating arrest of cytokinesis; E: Ultrastructure of giant cell sections showing cell wall ingrowths (black arrows) along regions predominantly flanking the vascular tissue (after Rodiuc* et al.*, 2014). Abbreviations: ⋆ = multinucleate giant cell; NC = neighbouring cells; eg = egg-mass; n = nematode; x = xylem; CW = cell wall; nu = nucleus. (Scale bars: A-C = 100 μm; D = 25 μm; E = 5 μm.)*

xylem in the vicinity is grossly disrupted and GCs are encaged by a newly developed intricate xylem network (Christie, 1936; Bartlem *et al.*, 2014). Profuse and mostly asymmetric division of vascular cells, partially resembling the divisions occurring during lateral root formation (Cabrera *et al.*, 2014a), as well as hyperplasia of the surrounding tissues

(Escobar *et al.*, 2015), increase root girth resulting in the formation of galls (Palomares-Rius *et al.*, 2017b).

Inside the galls, GCs undergo repeated mitosis with partial cytokinesis and endoreduplication or equivalent processes such as defective mitosis or nuclear fusion, leading to DNA amplification, which is thought to be necessary for GC expansion (Escobar *et al.*, 2015). GCs expand, increasing their volume by 100-fold from 3-40 days post-infection on average (Cabrera *et al.*, 2015a). Interestingly, the volume of individual GCs does not always correlate with the stage of gall development as GCs probably grow asynchronously. The GC becomes a typical highly metabolically active cell with dense cytoplasm containing abundant organelles (endoplasmic reticulum, ribosomes, mitochondria and Golgi bodies) (Christie, 1936; Berg *et al.*, 2008). The large central vacuole is also fragmented into smaller ones and chloroplast-like structures with starch accumulation are observed (Escobar *et al.*, 2015). The average volume occupied by all GCs as a pool within a gall shows a strong correlation to that of the infection stage (Cabrera *et al.*, 2015a). GCs also become transfer cells with cell wall ingrowths and irregular thickenings that increase the effective solute exchange area (Jones & Gunning, 1976; Berg *et al.*, 2008), demonstrating the molecular signatures of this cell type (Cabrera *et al.*, 2014b; Palomares-Rius *et al.*, 2017b).

Host Races

The ability of populations within a given nematode species to parasitise specific hosts or cultivars of crop plants has often been used to classify variants of species into races, pathotypes or biotypes (Triantaphyllou, 1987b). Host suitability to *Meloidogyne* species can be assessed by measuring their reproduction on plants after artificial inoculations (Lewis, 1987). The reproduction factor (*Rf*) has been widely used in nematological studies to define resistance and susceptibility of plants to plant-parasitic nematodes including *Meloidogyne* spp. (Greco & Di Vito, 2009).

Sasser (1954) was the first to propose a simple and practical test for the identification of the four major RKN species based on their response to several differential hosts and the type of galling induced. This host range studies revealed that there were one or more crop plants that were not infected by some RKN species, and that the non-hosts varied

Table 3. *The North Carolina differential host race test for four root-knot nematode species (Hartman & Sasser, 1985; Rammah & Hirschmann, 1990a; Carneiro* et al.*, 2003a; Carneiro & Cofcewicz, 2008; Robertson* et al.*, 2009).*

Meloidogyne species and race	Cotton, 'Deltapine 16'	Tobacco, 'NC 95'	Pepper, 'California Wonder'	Peanut, 'Florunner'	Tomato, 'Rutgers'
M. incognita					
Race 1	–	–	+	–	+
Race 2	–	+	+	–	+
Race 3	+	–	+	–	+
Race 4	+	+	+	–	+
Race 5	–	–	–	–	+
Race 6	–	+	–	–	+
M. arenaria					
Race 1	–	+	+	+	+
Race 2	–	+	–	–	+
Race 3	–	+	+	–	+
M. javanica					
Race 1	–	+	–	–	+
Race 2	–	+	+	–	+
Race 3	–	+	–	+	+
Race 4	–	+	+	+	+
Race 5	–	–	–	–	+
M. exigua					
Race 1	–	–	+	–	–
Race 2	?	–	–	–	+
Race 3	–	–	–	–	–

with the nematode species. Hartman & Sasser (1985) proposed the formal recognition of 'host races' within the major species and proposed a standardised set of differential hosts for distinguishing these host races and as an aid to species identification (Table 3). This differential host test, in conjunction with perineal pattern morphology, had several objectives: *i*) to give a preliminary identification of the four common species based on the host response; *ii*) to detect mixed populations; *iii*) to detect new and undescribed species; and *iv*) to detect host races and other pathogenic variation within common species. This method identified four races of *M. incognita*, two races of *M. javanica*, two races of *M. arenaria*, and three races of *M. hapla*. Further studies reported on new races of some species in different parts of world. Two additional host races of *M. javanica* – *M. javanica* race 2, which reproduces on pepper, and race 3, which reproduces on peanut but not pepper – were described using the North Carolina differential host test (Rammah & Hirschmann, 1990). Later, Carneiro *et al.* (2003a) described *M. javanica* race 4, which reproduced on both pepper and peanut. In a wide study on 140 RKN populations from representative horticultural regions of Spain, Robertson *et al.* (2009) identified two new races of *M. incognita* races 5 and 6, *M. javanica* race 5, and *M. arenaria* race 3 (Table 3). Atypical host reactions have been also found for main species (Taylor *et al.*, 1982; Stanton & O'Donnell, 1998). Host races were also proposed for *M. exigua*, *M. hapla*, *M. naasi* and *M. graminicola*.

Because the term 'race', when used in the context of plant disease, generally refers to populations of a pathogen that differ in virulence on host species that carry specific genes for resistance to other populations of that pathogenic species, Roberts (1995) developed an improved scheme for characterising the large variability in host range of *Meloidogyne*. It relied on the reaction of a number of differentials, each with a single resistance gene, to the nematode population. Nevertheless, given the wide variability within the genus, it may not be possible to develop a single scheme to characterise the reactions of all nematode populations to a wide range of crops throughout the world (Stanton & O'Donnell, 1998). No correlation could be found between cytogenetic, isoenzymatic, or molecular infraspecific polymorphism and races (*i.e.*, host ranges) in the four major species (Triantaphyllou, 1985b; Cenis, 1993). This observation raises questions about the genetic basis of such races, which probably do not represent monophyletic groups but rather a result from convergent evolution (Castagnone-Sereno *et al.*, 2013).

Moens *et al.* (2009) suggested that the formal recognition of the host races should be discontinued since the host race concept has been controversial, although the recognition of variation in host ranges is important. However, in this book we accept the race concept but note that race delimiting can be problematic for some species. Recognition of race could be a step for further study and erection of a new species. Nowadays, it will be more useful when dealing with a RKN population that has a variant host range to confirm that the population has been identified adequately by integrative taxonomy, and then simply to acknowledge the variant host preference of the population.

Plant Host Range

Many RKN are important pests of agricultural crops (Table 4). Some RKN species have a limited host range, such as those parasitising cereals and legumes, grasses, rice, citrus, *etc.*, while others are polyphagous (Sasser & Carter, 1982; Jepson, 1987; Di Vito & Greco, 1988c; Sikora, 1988; Eisenback & Triantaphyllou, 1991; Castillo *et al.*, 2003a; Bridge *et al.*, 2005). Although many *Meloidogyne* spp. are identified as polyphagous plant-parasitic nematodes with a wide host range, some good examples have shown that the host range of *Meloidogyne* spp. can be a useful tool for the management of these species. Knowl-

Table 4. *Some major agricultural crops as host for* Meloidogyne *spp.*

Crop	Species
Alfalfa	*M. arenaria, M. artiellia, M. chitwoodi, M. incognita, M. javanica, M. hapla, M. microtyla, M. minor, M. naasi*
Apple	*M. arenaria, M. incognita, M. hapla, M. javanica, M. mali*
Banana	*M. arenaria, M. cruciani, M. enterolobii, M. graminicola, M. hispanica, M. incognita, M. izalcoensis, M. javanica, M. megadora, M. oryzae*
Beans	*M. acronea, M. arenaria, M. artiellia, M. enterolobii, M. exigua, M. floridensis, M. graminicola, M. hapla, M. haplanaria, M. hispanica, M. incognita, M. inornata, M. javanica, M. konaensis, M. luci, M. petunia*
Citrus	*M. arenaria, M. citri, M. donghaiensis, M. fujianensis, M. hapla, M. incognita, M. indica, M. jianyangensis, M. javanica, M. kongi, M. mingnanica*
Coffee	*M. africana, M. arabicida, M. arenaria, M. coffeicola, M. daklakensis, M. enterolobii, M. exigua, M. hapla, M. incognita, M. inornata, M. izalcoensis, M. javanica, M. kikuyensis, M. lopezi, M. megadora, M. moensi, M. paranaensis*
Corn	*M. arenaria, M. chitwoodi, M. cruciani, M. incognita, M. javanica, M. floridensis, M. hispanica, M. konaensis, M. luci*
Cotton	*M. arenaria, M. acronea, M. incognita, M. enterolobii, M. indica*
Grape	*M. arenaria, M. chitwoodi, M. incognita, M. hapla, M. javanica, M. enterolobii, M. ethiopica, M. floridensis, M. hispanica, M. luci, M. mali, M. morocciensis, M. nataliei, M. vitis*
Olive	*M. arenaria, M. baetica, M. hapla, M. incognita, M. javanica, M. lusitanica, M. oleae*

Table 4. *(Continued.)*

Crop	Species
Peanut	*M. arenaria, M. hapla, M. haplanaria, M. javanica, M. panyuensis, M. pisi*
Potato	*M. africana, M. arenaria, M. chitwoodi, M. fallax, M. fanzhiensis, M. enterolobii, M. hapla, M. incognita, M. javanica, M. luci, M. minor, M. oryzae, M. petunia, M. sinensis*
Rice	*M. graminicola, M. graminis, M. incognita, M. javanica, M. oryzae, M. ottersoni, M. salasi, M. triticoryzae*
Sorghum	*M. acronea, M. artiellia, M. graminis, M. naasi, M. paranaensis, M. triticoryzae, M. incognita, M. enterolobii, M. graminicola*
Soybean	*M. arenaria, M. enterolobii, M. ethiopica, M. graminicola, M. hapla, M. incognita, M. inornata, M. javanica, M. luci, M. mali, M. morocciensis, M. naasi, M. triticoryzae*
Strawberry	*M. arenaria, M. hapla, M. incognita, M. javanica, M. fallax*
Sugarbeet	*M. arenaria, M. chitwoodi, M. floridensis, M. hapla, M. incognita, M. javanica, M. luci, M. naasi*
Sugarcane	*M. arenaria, M. incognita, M. javanica, M. enterolobii, M. ethiopica, M. hispanica, M. kikuyensis*
Sweet potato	*M. arenaria, M. cruciani, M. enterolobii, M. hapla, M. incognita, M. javanica, M. petunia*
Tobacco	*M. arenaria, M. cruciani, M. enterolobii, M. ethiopica, M. exigua, M. hapla, M. hispanica, M. incognita, M. inornata, M. javanica, M. konaensis, M. luci, M. microcephala, M. paranaensis, M. petunia, M. phaseoli, M. platani*
Tomato	*M. acronea, M. africana, M. ardenensis, M. arenaria, M. camelliae, M. carolinensis, M. cruciani, M. dunensis, M. enterolobii, M. ethiopica, M. exigua, M. fallax, M. floridensis, M. graminicola, M. hapla, M. haplanaria, M. hispanica, M. incognita, M. inornata, M. izalcoensis, M. javanica, M. konaensis, M. luci, M. mali, M. microtyla, M. minor, M. morocciensis, M. oryzae, M. ovalis, M. petunia, M. phaseoli, M. platani, M. spartelensis, M. suginamiensis, M. trifoliophila*
Turfgrasses	*M. arenaria, M. chitwoodi, M. fallax, M. graminicola, M. graminis, M. incognita, M. marylandi, M. microtyla, M. minor, M. naasi*
Wheat	*M. aegracyperi, M. arenaria, M. artiellia, M. chitwoodi, M. fallax, M. graminicola, M. hapla, M. hispanica, M. incognita, M. javanica, M. konaensis, M. marylandi, M. microtyla, M. minor, M. naasi, M. oryzae, M. sasseri, M. triticoryzae*

edge of the host range is essential when developing management systems based on resistant cultivars and non-host rotation crops. Thus, appropriate identification of *Meloidogyne* species and estimation of their population density in soil are essential for designing effective control measures in the context of sustainability and integrated pest management. For example, *M. artiellia* reproduces and causes severe damage only to leguminous plants, cereals and crucifers (Di Vito *et al.*, 1985b). Therefore, rotating legumes with plant crops belonging to other botanical families may result in satisfactory control of this nematode. Similarly, *M. arenaria* reproduces on, and causes damage to, peanut, soybean, vegetables and other crops, and can be managed by rotating with cotton, a non-host (Rodríguez-Kabana *et al.*, 1987). Forage crops may be useful to use as rotation alternatives for RKN susceptible to vegetables. A 3-year rotation to bermudagrass (*Cynodon dactylon*) reduced population densities of *M. incognita* to extremely low levels in a vegetable production system (Johnson *et al.*, 1995).

RKN infections on *Citrus* spp. are frequently found in Asia, but are rare and of limited economic importance in other citrus-growing areas, such as North, Central and South America or the Mediterranean Basin (Vovlas & Inserra, 1996). Twelve *Meloidogyne* species have been reported to infect citrus roots worldwide (Table 4); however, reproduction of some species has been observed only in certain regions: *M. citri*, *M. donghaiensis*, *M. fujianensis*, *M. jiangyangensis*, *M. kongi*, and *M. mingnanica* in China, *M. indica* in India, *M. exigua* in Surinam and Guadeloupe, and *M. incognita* in Queensland, Australia (Inserra *et al.*, 1978; Vovlas & Inserra, 1996; Inserra *et al.*, 2003a-2003g). *Citrus* spp. are also probably invaded by some RKN species that usually reproduce on other hosts: for example, *M. javanica* has been reported on citrus more frequently than the other species of *Meloidogyne* but failed to complete its life cycle, which was attributed to a lack of giant cell formation following nematode penetration, essential for juvenile maturation and egg production (Orion & Cohn, 1975; Inserra *et al.*, 1978). When a compatible host-parasite interaction has been established by a nematode, infection of the plant can be followed by a sequence of disruptions in the physiological processes that lead to pathogenesis (Melakeberhan & Webster, 1993). Reductions of normal growth plant parameters may be used as indicators of plant physiology impairment related to pathogenesis.

Nematology Monographs & Perspectives, 2021, Vol. 14, 41-52

Chapter 7

Pathogenicity and Threshold Damage

The RKN are devastating parasites of crop plants in the agricultural production of major food crops, vegetables, fruit and ornamental plants growing in tropical, subtropical, and temperate climates, and certainly contribute significantly to a net reduction in yield, although assessing the true amount of the problem is challenging (Fig. 9). One difficulty with assessing nematode impact is that damage resulting from nematode infection is often less obvious than that caused by many other pests or diseases and can even be confused with other physiological stresses like, for example, excess of water or lack of nitrogen fertilisation. Damage can differ depending on factors such as climate, host and specific nematode species.

Pathogenicity is used in plant pathology to express the capacity of an organism to induce disease, or the amount of physiological damage caused to the host plant by the presence of a pathogen (Shaner *et al.*, 1992). Although controversy persists as to whether plant-parasitic nematodes should be considered as parasites or pathogens, in plant nematology the terminology used in plant pathology for fungi and bacteria is applied (Triantaphyllou, 1987b). For that reason, the term 'pathogenicity' in plant nematology has an ambivalence that expresses, on the one hand, a qualitative concept (characterisation of the capacity of a given nematode species to establish a compatible host-parasite relationship) and, on the other, a quantitative concept (definition of differences in the expression of the damage caused in the plant). Reproductive fitness together with virulence are major components of pathogenicity in *Meloidogyne* species (Shaner *et al.*, 1992) and thus important for the assessment and understanding of disease reactions of plants to pathogens.

The effects of parasitism by RKN on plant growth and yield result from disruption of the normal process of plant root growth and function, the root system system failing adequately to explore the soil due to nematode attack (Karssen & Moens, 2006). *Meloidogyne* infection af-

Fig. 9. *Root systems of lettuce plants healthy (left) and severely damaged by* Meloidogyne arenaria.

fects water and nutrient uptake by the root system. Several studies have demonstrated that *Meloidogyne* infection increased N, P and Ca concentrations in below-ground plant parts, but decreased K, Zn and Mn (Melakeberhan *et al.*, 1987; Karssen & Moens, 2006). Bergeson (1966) put forward the hypothesis that RKN may have a direct parasitic effect by withdrawing and accumulating N at the expense of the above-ground part and thus limiting its development. Accumulation of N in RKN-infected roots has been observed in pepper (Shaffiee & Jenkins, 1963) and tomato (Hunter, 1958; Maung & Jenkins, 1959; Bergeson, 1966). Results from pathogenicity studies of *M. graminicola* on rice suggest that a withdrawal and immobilisation of nutrients by the nematode can be one of the causes of growth and yield reductions (Prot *et al.*, 1994).

The few studies on the water relationships of *Meloidogyne*-infected plants suggest that water consumption is not affected by nematode infection if soil moisture is not a limiting factor but under periodic stress consumption is reduced. Plant development is suppressed by infections that inhibit root growth during moisture stress and prevent roots from extending into moist soil (Davis *et al.*, 2014). *Meloidogyne* infection of roots also decreases the rate of chlorophyll and photosynthesis rates in leaves (Loveys & Bird, 1973; Haseeb *et al.*, 1990; Lu *et al.*, 2014). Leaf chlorophyll content provides a measure of photosynthetic capacity and is related to the nitrogen concentration in the plant (Evans, 1989), which *Meloidogyne* spp. can influence by interfering with water and nutrient transport (Melakeberhan *et al.*, 1987; Kirkpatrick *et al.*, 1991; Carneiro *et al.*, 2002). Therefore, because chlorophyll content is affected by nitrogen concentration, it can be an indicator of the damage caused to the plant by *Meloidogyne* spp. These nematodes interfere with the production of root-derived factors regulating photosynthesis. For example, both cytokinins and gibberellins in tomato root tissue and xylem exudates can be decreased in plants infected with *M. incognita* compared with non-infected plants (Brueske & Bergeson, 1972). The high concentrations of auxins and cytokinins in galls have been described in detail by the use of reporter genes driven by specific promoters as 'sensors' of both phytohormones, such as DR5, ARR5 or TCS (Cabrera *et al.*, 2015b). Plant hormones are central to the nematode responsiveness of genes and involved in two early steps in the compatible interaction, *viz.*, the modulation of plant immunity and the initiation of a feeding cell (Goverse & Bird, 2011). The increased metabolic activity of giant cells stimulates mobilisation of photosynthetic products from shoots to roots and, in particular, to the giant cells where they are removed and utilised by the feeding nematode.

Several environmental and edaphic factors are known to influence the pathogenicity of plant-parasitic nematodes to host plants, *e.g.*, soil temperature, texture, or moisture, all of which may affect population densities of nematodes (Wallace, 1983). Soil type and texture influences plant development and provides a habitat that can affect crop productivity and RKN (Noe & Barker, 1985). Soil texture largely determines soil moisture holding capacity and aeration and influences the ability of RKN to hatch, move through soil, locate and penetrate a host (Noe & Barker, 1985). A major global challenge in the coming years will be to ensure food security for the human population. Climate change will favour the spread of plant pests and pathogens due to 'tropicalisation' of temperate agricultural areas in the near future (Nicol *et al.*, 2011). Increasing temperatures are expected to enhance plant growth rates and yields, but conversely they will also provide better conditions for the development of pests/diseases (Colagiero & Ciancio, 2011).

The damage a RKN species may cause, especially the yield loss, is basic to implementation of appropriate control strategies. For example, to include a nematode in the list of quarantine organisms, the pathogenic potential of the nematode in a country must be demonstrated. Also, national policy makers will consider the impact a nematode may have on given crops and areas for their decisions. At the farm level, information on the potential of the nematode populations to cause yield loss is a prerequisite to deciding if and when to apply a treatment, and choosing the most appropriate control strategy. RKN cause indirect damage because of the quarantine status of some species of *Meloidogyne* in several countries or regions. For example, *M. chitwoodi* A2/227, *M. enterolobii* A2/361, *M. fallax* A2/295, and *M. mali* A2/409 are included in the A2 EPPO list. Listing species as quarantine organisms can reduce the risks of spread through international trade. In general, RKN are not regulated as a group because the major economically important species are already widely distributed (Hockland *et al.*, 2006). *Meloidogyne chitwoodi* is a severe parasite of economically important crops such as potatoes and carrots, and is on the lists of prohibited organisms of many countries (Canada, the EU, Mexico, and other countries in Latin America and the Far East). *Meloidogyne enterolobii* is polyphagous and has many host plants including cultivated crops and weeds, attacking herbaceous as well as woody plants. Currently, in the USA several states have imposed external quarantine for this nematode. Although no specific quarantine requirements for *M. enterolobii* are currently in force in Europe, measures such as preventing the introduction of the nematode, establishing if the species is present in the country and, if so, determining its distribution, preventing its spread, and eradicating incursions, need to be applied in the EPPO region (Anon., 2016c).

Damage by RKN may also influence other plant stress factors and can induce predisposition of their hosts to attack by other pathogens, such as fungi or bacteria, or modifying the host response to the latter (Back *et al.*, 2002). RKN interactions with other plant pathogens are important biological phenomena of great importance in agriculture. When a plant is infected with one RKN its response to other pathogens may be altered (Back *et al.*, 2002). These alterations exert significant influences upon disease development, aetiology of pathogens involved and, ultimately, on disease control. Interactions between RKN and fungi have been studied and documented in several host crops including chickpea, cotton, cowpea, potato, tomato, *etc.* (Back *et al.*, 2002; Castillo *et al.*, 2003b). Synergistic interactions can be summarised as being positive where an association between RKN and pathogen results in plant damage exceeding the sum of individual damage ($1 + 1 > 2$). Conversely, where an association between RKN and pathogen results in plant damage less than that expected from the sum of the individual organisms, the interaction may be described as antagonistic ($1 + 1 < 2$). Where RKN and pathogen are known to interact and are shown to cause plant damage that equates to the sum of individual damage, the association may be described as neutral ($1 + 1 = 2$) (Back *et al.*, 2002). Since many plant diseases can be managed by plant host resistance, the most important interactions between RKN and other pathogens is the breakdown resistance (France & Abawi, 1994; Castillo *et al.*, 2003b). RKN and bacterial interactions have usually been regarded as providing wounds in the roots through which bacteria may enter the plant. This was documented in the association between *Ralstonia solanacearum* (Smith, 1896) Yabuuchi *et al.*, 1995 and *M. incognita*, or *Agrobacterium tumefaciens* (Smith & Townsend, 1907) Conn, 1942 with *M. hapla* (Taylor, 1990). However, the interactions between *Meloidogyne* spp. and Fusarium wilt fungi seem more biological or physiological rather than physical (Mai & Abawi, 1987).

In addition, these interactions may also involve above-ground pests such as whitefly. Guo & Ge (2016) found that tomato roots infected with *M. incognita* reduce whitefly fitness. Nematode infection activated two different types of phytohormone-dependent systemic defence signalling: the salicylic acid-dependent systemic defence (SAR), which was enhanced in leaves, and the jasmonic acid-dependent systemic defence (ISR), which was activated after whitefly infestation. Rhizosphere microbiomes modulated by pre-crops could also assist plants in defence against plant-parasitic nematodes (Elhady *et al.*, 2018).

Since many studies on RKN have focused on the pathogenicity on numerous host plants, the present chapter will concentrate on giving a summary of the host range and damage thresholds of the up-to-date studied host plant/*Meloidogyne* spp. combinations (Table 5). Severity of damage and threshold levels caused by RKN can be species-specific and vary by host, crop rotation, environmental conditions and soil type (Greco & Di Vito, 2009).

Plant growth impairment caused by *Meloidogyne* spp. to crops is influenced by nematode species, as well as the initial nematode population density in the soil at sowing (Sasanelli, 1994). Appropriate

Table 5. *Damage threshold densities of different host plant*-Meloidogyne *combinations.*

Meloidogyne sp.-host plant	Damage threshold (eggs + J2 cm^{-3} or g^{-1} of soil)	Reference
M. arenaria-carrot	$0.25\text{-}0.75^T$	Medina-Canales *et al.* (2012)
M. arenaria-grafted melon onto squash hybrid Shintozoa	0.14	Kim & Ferris (2002)
M. arenaria-peanut	0.01^T	McSorley *et al.* (1992)
M. arenaria-sweet basil	0.15^T	Vovlas *et al.* (2008a)
M. arenaria-tomato hybrids 'ARTH-3', 'ARTH-4', 'Avinash' and variety 'Selection-7'	0.1	Kumar *et al.* (2016)
M. arenaria-white mulberry	1.10^T	Castillo *et al.* (2001)
M. artiellia-chickpea (spring)	0.02^T	Di Vito & Greco (1988a)
M. artiellia-chickpea (winter)	0.14^T	Di Vito & Greco (1988a)
M. artiellia-wheat	$0.23\text{-}0.43^T$	Di Vito & Greco (1988b)
M. chitwoodi-carrot	0.012^T	Heve *et al.* (2015)
M. chitwoodi-sugarbeet	2.8^T	Griffin *et al.* (1982)
M. exigua-coffee	1.20^T	Di Vito *et al.* (2000)
M. graminicola-rice	$<1.0^T$	Plowright & Bridge (1990); Sharma Poudyal *et al.* (2005)
M. graminicola-rice	0.01	Prot & Matias (1995)
M. graminicola-rice	0.01	Kumar *et al.* (2017)
M. hapla-alfalfa	1.60^T	Inserra *et al.* (1983)
M. hapla-carrot (microplots, glasshouse)	$1.18\text{-}4.3^T$	Vrain (1982)
M. hapla-carrot (organic, mineral soil)	$0.70\text{-}0.60^T$	Gugino *et al.* (2006)
M. hapla-lettuce (microplots, glasshouse)	$1.00\text{-}7.00^T$	Viaene & Abawi (1996)

Table 5. *(Continued.)*

Meloidogyne sp.-host plant	Damage threshold (eggs + J2 cm^{-3} or g^{-1} of soil)	Reference
M. hapla-onion	2.0-10.7	Olthof & Potter (1972); MacGuidwin *et al.* (1987)
M. hapla-rose rootstock	0.011^T	Meressa *et al.* (2016)
M. hapla-sugarbeet	0.6	Griffin *et al.* (1982)
M. hapla-tomato	2.6	Potter & Olthof (1977)
M. incognita-artichoke	1.10^T	Di Vito & Zaccheo (1981)
M. incognita-bean	$0.02\text{-}0.25^T$	Crozzoli *et al.* (1997); Di Vito *et al.* (2004)
M. incognita-cabbage	0.50^T	Sasanelli *et al.* (1982a)
M. incognita-cantaloupe	0.19^T	Di Vito *et al.* (1983)
M. incognita-cassava	1.0^T	Crozzoli & Parra (1999)
M. incognita-catalpa	0.78^T	Sasanelli *et al.* (1996)
M. incognita-celery	0.15^T	Vovlas *et al.* (2008b)
M. incognita-coffee	2.09^T	Vovlas & Di Vito (1991)
M. incognita-corn	10.0^T	Di Vito *et al.* (1980)
M. incognita-cotton	0.0002^T	Zhou & Starr (2003)
M. incognita-cowpea	$0.03\text{-}0.74^T$	Crozzoli *et al.* (1997, 1998)
M. incognita-cucumber	0.004^T	Giné *et al.* (2017)
M. incognita-eggplant	0.054^T	Di Vito *et al.* (1986)
M. incognita-grapevine	$0.5\text{-}0.78^T$	Anwar & Van Gundy (1989); Sasanelli *et al.* (2006)
M. incognita-grapevine	0.025	Quader *et al.* (2002)
M. incognita-guava	0.05	Casassa *et al.* (1998)
M. incognita-kenaf	0.13^T	Di Vito *et al.* (1997)
M. incognita-kiwi	0.43^T	Di Vito *et al.* (1988)
M. incognita-lentil	0.5	Hisamuddin & Azam (2010)
M. incognita-lettuce	0.13^T	Asuaje *et al.* (2004)
M. incognita-melon	0.025^T	Ploeg & Phillips (2001)
M. incognita-okra	0.5	Ganaie *et al.* (2011)
M. incognita-papaya	0.16^T	Bustillo *et al.* (2000)
M. incognita-parsley	0.03^T	Aguirre *et al.* (2003)
M. incognita-pepper (glasshouse, microplots)	$0.165\text{-}0.30^T$	Di Vito *et al.* (1985a, 1992); Di Vito (1986);
M. incognita-radish	20.0^T	Duncan & Ferris (1983)
M. incognita-soybean	0.31	Niblack *et al.* (1986)

Table 5. *(Continued.)*

Meloidogyne sp.-host plant	Damage threshold (eggs + J2 cm^{-3} or g^{-1} of soil)	Reference
M. incognita-spinach	0.156-0.25^{T}	Dammini Premachandra & Gowen (2015); Di Vito *et al.* (2004b)
M. incognita-sugarbeet	1.19^{T}	Di Vito *et al.* (1981)
M. incognita-sunflower	1.85^{T}	Sasanelli & Di Vito (1992)
M. incognita-tobacco	1.25^{T}	Di Vito *et al.* (1983); Vovlas *et al.* (2004a)
M. incognita-tomato	0.55^{T}	Di Vito *et al.* (1981); Di Vito & Ekanayake (1983); Di Vito *et al.* (1991)
M. incognita-tomato hybrids 'ARTH-3', 'ARTH-4', 'Avinash' and variety' Selection-7'	0.1	Kumar *et al.* (2016)
M. incognita-watermelon	0.41^{T}	Xing & Westphal (2012)
M. incognita-yucca	0.45^{T}	Crozzoli & Parra (1999)
M. incognita-zucchini	8.1-1.5^{T}	Vela *et al.* (2014)
M. inornata-bean	9.9^{T}	Dadazio *et al.* (2016)
M. javanica-banana	0.15^{T}	Vovlas *et al.* (1983)
M. javanica-bean	0.6-4.0^{T}	Di Vito *et al.* (2007); Sharma (1981); Machado *et al.* (2004)
M. javanica-chickpea	0.10^{T}	Sharma *et al.* (1995)
M. javanica-coffee	1.34^{T}	Vovlas & Di Vito (1991)
M. javanica-cucumber	0.1^{T}	Giné *et al.* (2014)
M. javanica-groundnut	1.8^{T}	Di Vito *et al.* (1999)
M. javanica-olive	0.10-0.49^{T}	Jahanshahi Afshar *et al.* (2014); Sasanelli *et al.* (2002)
M. javanica-peach rootstock GF 677	0.57^{T}	Di Vito *et al.* (2005)
M. javanica-peanut	0.67	El-Sherif *et al.* (2009)
M. javanica-pepper	0.28^{T}	Mekete *et al.* (2003); Moosavi (2015)
M. javanica-potato	0.50^{T}	Vovlas *et al.* (2005)

Table 5. *(Continued.)*

Meloidogyne sp.-host plant	Damage threshold (eggs + J2 cm^{-3} or g^{-1} of soil)	Reference
M. javanica-rice (Asian, African)	0.261.0-8.0[T]	Di Vito *et al.* (1996a)
M. javanica-common sesban (loamy, sandy, clay soil)	0.01-1.00[T]	Desaeger & Rao (2000)
M. incognita-radish	10.0[T]	Duncan & Ferris (1983)
M. javanica-sunflower (glasshouse, microplots)	3.03-0.45[T]	Sasanelli *et al.* (1982b); Di Vito *et al.* (1996b)
M. javanica-taro	0.008	Sipes & Arakaki (1997)
M. javanica-tomato	0.36[T]	Mekete *et al.* (2003)
M. javanica-tomato hybrids 'ARTH-3', 'ARTH-4', 'Avinash' and variety 'Selection-7'	0.1	Kumar *et al.* (2016)
M. javanica-watermelon	0.2[T]	López-Gómez *et al.* (2014)
M. javanica-zucchini	0.2[T]	López-Gómez *et al.* (2015)

[T] Tolerance limit estimated with the Seinhorst model for plant growth or yield. Tolerance limit = *Pi* (initial nematode density cm^{-3} of soil) below which no effect could be measured.

identification of *Meloidogyne* species and estimation of their population density in soil are essential for designing effective control measures in the context of sustainability and integrated pest management. This is especially important for RKN, since host plant resistance, which could reduce the initial nematode population density to tolerance threshold levels, is scarce among vegetable crops (Sasser & Carter, 1985). In this sense, a minimum population density (T) is required before measurable yield loss occurs (tolerance limit) (Seinhorst, 1965, 1979). Pathogenicity studies on *Meloidogyne* species indicated that they are very well adapted to parasitism, as extremely high populations in soils usually do not kill their host plants. Nevertheless, damage thresholds are quite different among *Meloidogyne* host plant combinations and range from 0.0002 nematodes cm^{-3} of soil in the combination *M. javanica*-cucumber (Giné *et al.*, 2014) to 35 nematodes cm^{-3} of soil in the combination *M. hapla*-carrot (Vrain, 1982) (Table 5).

As *Meloidogyne* population densities are related to the number of generations the nematode completes per growing cycle of the crop, the length of the growing cycle is important. Short-cycle crops will allow the nematode to complete fewer generations, perhaps only one, compared with long-cycle crops, and thus limit nematode populations. In addition, temperature is an important factor that influences whether there is more than one generation per crop. Several crop plants can be either sown or transplanted. In pots, transplanted tomatoes showed more than doubled reproduction rates and final soil population densities of *M. incognita* compared with sown tomatoes of the same cultivar and growing in infested soils at the same soil density for the same time period (Ekanayake & Di Vito, 1984). Similarly, perennial crops such as trees allow RKN to reproduce continuously for several consecutive years, so ceilings of the nematode population will be reached only after several years. Thereafter, nematode population declines and increases will alternate due to rotations of seasons, and severe root damage and flushes of new root growth (Souza *et al.*, 2008).

The damage threshold for a nematode species is defined as that population density at which a detectable yield loss occurs, whereas the economic threshold is the population density at which the cost of the yield loss equals the cost of control (Chiang, 1979). The importance of damage threshold determination is due to its role in the decision making process and in development of management strategies. Damage thresholds may differ for the same nematode species from one environment to another, among nematode species, among populations of a species, and among host cultivars (McSorley & Phillips, 1993).

Damage thresholds are effectively used for economic decisions on the control of nematodes (Ferris, 1981). The most widely accepted damage threshold is the tolerance limit (T) estimated by Seinhorst (1965) in a biologically descriptive mathematical model relating nematode population density and plant growth or yield, $y = m + (1 - m)Z^{P-T}$, where y is the ratio between plant weight at nematode density (P) at the time of sowing and that in the absence of nematodes, $m =$ the minimum value of y (y at a very large initial nematode population density), T is the density below which no reduction of plant growth or yield occurs and Z is a constant < 1 reflecting nematode damage, with $z^{-T} = 1.05$ (Seinhorst, 1965, 1979, 1998; Viaene *et al.*, 1997).

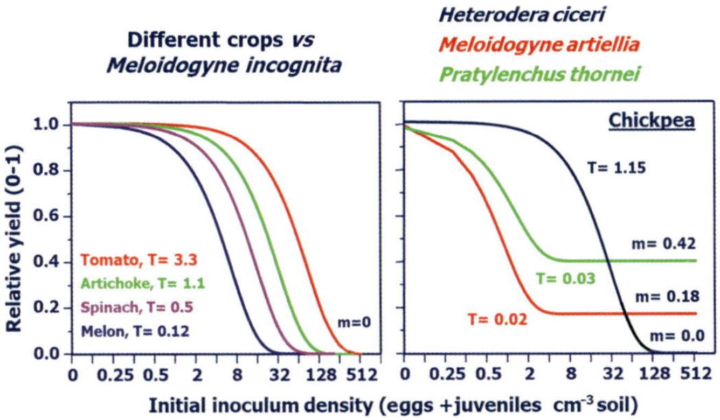

Fig. 10. *Relationships between initial population densities of* Meloidogyne incognita *and relative yield on tomato, artichoke, spinach and melon, showing different tolerance limits; and relationships between initial population densities of* Heterodera ciceri, M. artiellia, *and* Pratylenchus thornei *and relative yield of chickpea, showing different tolerance limits.*

Seinhorst's equation has been accepted as a useful basis for describing the relationship between the initial population density and productivity or yield in many host plant-*Meloidogyne* associations (Table 5). Results from these studies indicated that *Meloidogyne* species are very well adapted to parasitism because minimum value of *Y* was hardly ever zero, indicating that even with extremely high nematode populations in soils *Meloidogyne* spp. do not kill their host plants. Nevertheless, these results also indicated that tolerance limits may vary depending on host plant and nematode species (Table 5; Fig. 10). One of the advantages of yield loss estimation by the Seinhorst curve is that with an initial inoculum density in soil it is possible to estimate the potential damage in a particular host plant-nematode combination (Sasanelli, 1994). The model proposed by Seinhorst fits well to data for annual plants and especially with nematodes that develop only one generation per growing season. With nematodes developing several generations per crop cycle, such as the majority of *Meloidogyne* spp., the effects of the succeeding generations may be small at low initial nematode densities and low reproduction rates, whilst at larger nematode densities and with large nematode reproduction rates the effect of a second and later generations can be substantial (Seinhorst, 1995). With perennials, the same model is appropriate only during the first growing season; in subsequent seasons it will be dependent on the nematode populations that occur at the beginning of each growing season. In the following years, the plant may even suffer less damage. Also, if the effect on yield is considered, and yield is represented by fruits, it must be kept in mind that root damage suffered by a plant in a given year will usually affect the yield of the succeeding year(s). All this makes it difficult to fit a model to the damage suffered by perennials (Greco & Di Vito, 2009).

Nematology Monographs & Perspectives, 2021, Vol. 14, 53-61

Chapter 8

Genomes

Cells of nematodes have both nuclear and mitochondrial genomes. Of all plant-parasitic nematodes, RKN currently have the most genomes sequenced, with the major crop species sequenced and compared, and some species even sequenced among different populations (Blanc-Mathieu *et al.*, 2017; Szitenberg *et al.*, 2017). The new technologies and the reduction in costs for sequencing with the new computational capabilities have helped to increase knowledge about genomes. The understanding of the genomic basis could help us to understand their variable success on different crops and the knowledge of their evolution, which could benefit agriculture in the future. Additionally, the use of genomes has important implications in species identification, effector knowledge, and population variability, among others. Several genomes have been sequenced (Table 6). However, other species infecting woody plants, with more restricted host ranges and/or more ancestral in the *Meloidogyne* phylogenetic tree, are still missing from this knowledge.

Nuclear Genome

RKN biology exhibits an important variation in mode of reproduction (from amphimixis to mitotic parthenogenesis), cytogenetics and host range (Moens *et al.*, 2009). The differences in reproduction are linked to the structure and function of their genomes. Szitenberg *et al.* (2017) compared 19 genomes from five nominal species (*M. incognita*, *M. arenaria*, *M. javanica*, *M. enterolobii* and *M. floridensis*) and found several main characteristics in the genomes: *i*) the genome of the apomictic Incognita group (MIG), (*M. incognita*, *M. arenaria*, *M. javanica*, and *M. floridensis*) have spans and gene counts much greater than observed in the diploid *M. hapla*; *ii*) MIG genomes harbour a substantial number of pairs of divergent gene copies with congruent evolutionary histories, probably arising from a duplication event before the speciation of the four MIG taxa analysed (this number of pairs

Table 6. *Nuclear genomes sequenced for root-knot nematode (*Meloidogyne*) species and their main characteristics (adapted from Susič* et al.*, 2020a).*

Species	Strain/ isolate designation	Accession (DDBJ/ ENA/GenBank)	Assembly size (Mb)	Genome coverage	Number of contigs/ scaffolds	N50	GC content (%)	Number of predicted genes	CEGMA		Reference
									C	A	
M. arenaria	HarA	GCA_003693565.1	163.75	100	46 436	10 504	30.3	30 308	91	2.7	Szitenberg *et al.* (2017)
M. arenaria	-	GCA_900003985.1	258.07	100	26 196	16 462	29.8	103 001	95	n/a	Blanc-Mathieu *et al.* (2017)
M. arenaria	A2-O	GCA_003133805.1	284.05	60	2224	204 551	30.0	n/a	94.8	n/a	Sato *et al.* (2018)
M. chitwoodi	Mc1, Mc2, Mc1Roza	JACZZP000000000, JACZZP000000000, JACZZP000000000	47.47, 46.98, 47.78	-	30, 39, 38	-	25.0, 24.9, 25.0	10 441, 10 424, 10 660	-	-	Bali *et al.* (2021)
M. enterolobii	L30	GCA_003693675.1	162.97	200	42 008	10 552	30.2	31 051	81	2.6	Szitenberg *et al.* (2017)
M. enterolobii	Swiss	GCA_903994135.1	240	60	4437	143 330	30.0	63 841	94.8	3.3	Koutsovoulos *et al.* (2019)
M. floridensis	-	GCA_000751915.1	96.67	200	58 696	3698	30.0	15 327	58.1	n/a	Lunt *et al.* (2014)
M. floridensis	SJF1	GCA_003693605.1	74.85	100	8887	13 261	30.2	14 144	84	1.6	Szitenberg *et al.* (2017)
M. graminicola	IARI	GCA_002778205.1	38.19	180	4304	20 482	23.1	10 196	84.3	n/a	Somvanshi *et al.* (2018)

Table 6. *(Continued.)*

Species	Strain/ isolate designation	Accession (DDBJ/ ENA/GenBank)	Assembly size (Mb)	Genome coverage	Number of contigs/ scaffolds	N50	GC content (%)	Number of predicted genes	CEGMA		Reference
									C	A	
M. graminicola	Mg-VN18	GCA_014773135.1	41.5	288	283	294 907	23	10 331	96	n/a	Phan *et al.* (2020)
M. hapla	VW9	GCA_000172435.1	53.01	10	3450	37 608	27.4	14 420	94.8	n/a	Opperman *et al.* (2008)
M. incognita	Morelos	GCA_000180415.1	82.10	5	9538	12 786	31.4	19 212	77	n/a	Abad *et al.* (2008)
M. incognita	W1	GCA_003693645.1	121.96	100	33 351	16 520	30.6	24 714	83	2.4	Szitenberg *et al.* (2017)
M. incognita	V3	GCA_900182535.1	183.53	100	12 091	38 588	29.8	45 351	97	n/a	Blanc-Mathieu *et al.* (2017)
M. javanica	VW4	GCA_003693625.1	150.35	300	34 316	14 128	30.2	26 917	90	2.5	Szitenberg *et al.* (2017)
M. javanica	-	GCA_900003945.1	235.80	100	31 341	10 388	29.9	98 578	96	n/a	Blanc-Mathieu *et al.* (2017)
M. luci	SI-Smartno V13	ERS3574357	209.16	200	327	1 711 905	30.2	n/a	95.2	2.88	Susič *et al.* (2020a)

CEGMA = Core Eukaryotic Genes Mapping Approach (Parra *et al.*, 2007); C = % complete CEGMA orthologues; A = average copy number per CEGMA orthologue. 'n/a': not assessed.

of divergent copies not found in *M. hapla*); *iii*) the genome showed a partial triploidy of the MIG species, with two copies of one genome (very small differences or identical among them) and a single copy of the other (*ca* 3% different in coding regions and much more in non-coding regions) (alpha and beta homologues, respectively); and *iv*) not all genes were present as divergent duplicate copies; homozygous genes were frequently shared between MIG species, suggesting that there has been stochastic loss of heterozygosity in some divergent gene copies, implying that these events have been ongoing through speciation. Exploring them, Szitenberg *et al.* (2017) found evidence for frequent recombination events, most likely non-crossover events, where discordant similarities mapped along assembly scaffolds showed replacement of one copy by its sister. These genome characteristics are supposed to be produced by hybridisation between two diverse RKN strains followed by loss of meiosis (Szitenberg *et al.*, 2017).

The genome sizes and gene numbers of the MIG species do not fully correlate with exact ploidy, and the prediction of these species as mentioned before showed hypotriploidy (Szitenberg *et al.*, 2017); other assemblies showed peaks of protein-coding sequences (CDS) mapping of three, 3-4 and four loci in the genomes of *M. incognita*, *M. javanica* and *M. arenaria* (Blanc-Mathieu *et al.*, 2017). Several mechanisms are suggested to shape the genome of MIG species: *i*) gene conversion can be associated with crossovers or occur without crossing over, depending on how the break is repaired, and this process of one allele overwriting another enables gene conversion to alter allele frequencies (Korunes & Noor, 2017) and it could influence genome structure and diversity (Chen *et al.*, 2007; Pessia *et al.*, 2012; Korunes & Noor, 2017); *ii*) homogenisation of sequence copies as concerted evolution in highly similar sequences, than between more divergent sequences in the same genome (Chen *et al.*, 2007); *iii*) deletion of one of the divergent copies; and *iv*) transposable elements (TE); these elements cover about a 1.7 times higher proportion of the genomes of MIG compared to the sexual relative and might also participate in their plasticity (Blanc-Mathieu *et al.*, 2017; Szitenberg *et al.*, 2017). Transposable elements span 50.0, 50.8 and 50.8% of the genome assemblies of the asexual *M. incognita*, *M. javanica* and *M. arenaria*, respectively (Blanc-Mathieu *et al.*, 2017), while in other species (*M. hapla* and *M. enterolobii*) the number of transposable elements is lower in comparison to MIG species (Szitenberg *et al.*, 2017). However, the proportions of loci present as divergent pairs differ among the MIG species suggesting independent changes with some of the above mechanisms in each lineage. Blanc-Mathieu *et al.* (2017) showed that >60% of homologous gene pairs display diverged expression patterns across developmental life stages in *M. incognita*, suggesting a substantial functional impact of the genome structure on the biology of these species. There were signs of positive selection between these genes, and usually the gene pairs detected as under positive selection show a significantly higher proportion of diverged expression profiles. This could lead to a functional divergence at the biochemical level, having these new mutations adaptive advantages.

A diversity of Pfam domains and associated gene ontology (many related to the enzymatic and other catalytic functions) has been predicted by positively selected genes. Different domains and functions were specifically enriched in positively selected genes in each species, suggesting that the functional consequences of the hybrid genome structure were different in each species. Interestingly, the automictic MIG species, *M. floridensis*, differs from the apomictic species in that it has become homozygous throughout much of its genome. *Meloidogyne enterolobii* is an apomictic nematode and also contains divergent genome copies, but these arose from a different progenitor from the one of the MIG (Szitenberg *et al.*, 2017). *Meloidogyne enterolobii* thus likely represents an independent origin of apomixes.

The different available genomes have important differences in the sizes and gene numbers (Table 6), probably due to differences among the population studied and the different methods of study. Additionally, Szitenberg *et al.* (2017) found more genetic diversity within *M. arenaria* than between the other MIG species examined. In this sense, other studies have also indicated that *M. arenaria* contains considerable diversity (Blok *et al.*, 1997; Adam *et al.*, 2005; Carneiro *et al.*, 2008).

Mitochondrial Genome

The basic circular mitochondrial genome of *Meloidogyne* spp. varies from 18 000 to 20 000 bp and consists of 12 protein-coding genes, the large and small rRNA genes (*rrnS* and *rrnL*), and 22 tRNA (transfer RNA) genes (Fig. 11A, B). The *Meloidogyne* spp. mitochondrial genome lacks the *atp8* gene, as found in other nematodes (Hu & Gasser, 2006;

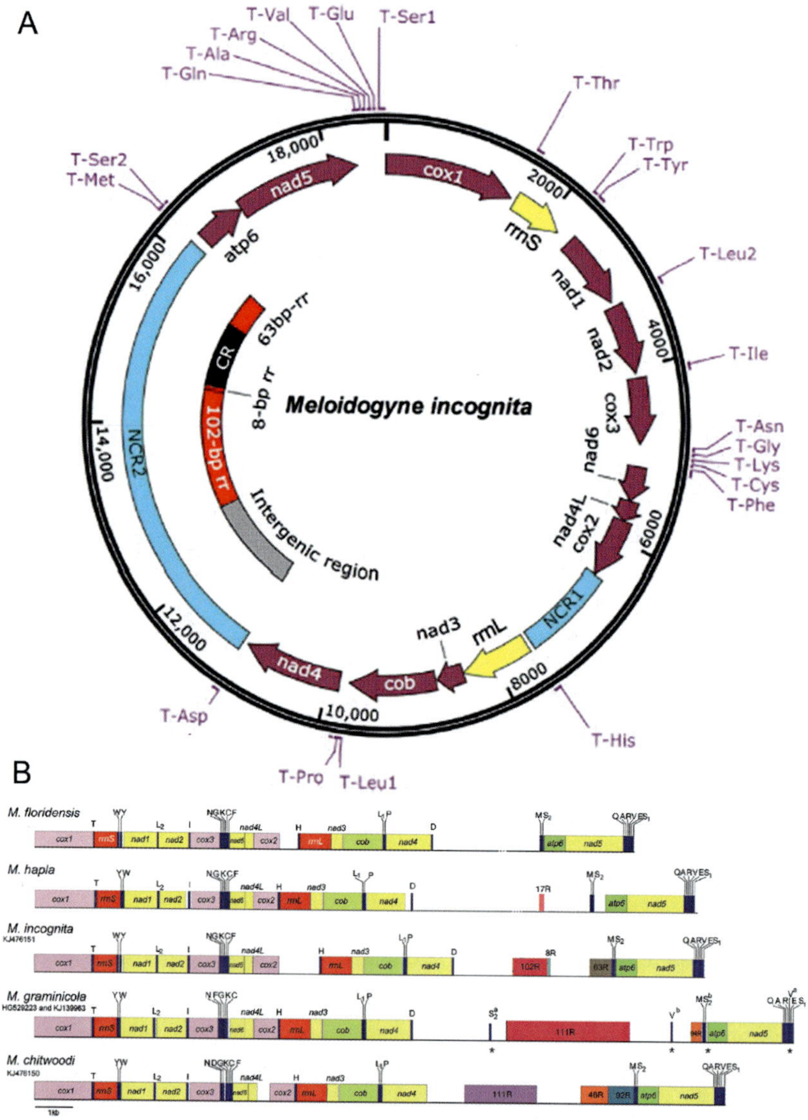

Fig. 11. *A: Circular mitochondrial genome of* Meloidogyne incognita *(after Humphreys-Pereira* et al., *2014b); B: Linear maps of the mitochondrial genomes of five* Meloidogyne *species. Dots interrupting the main genomic line indicate unknown sequences. tRNA genes are designated by single letter abbreviation. Tandem repeats are represented by boxes labelled by the repeat length in kb followed by 'R'. (After García & Sánchez-Puerta, 2015.)*

Table 7. *Root-knot nematode (*Meloidogyne*) species' mitochondrial genomes sequenced and their main characteristics (adapted from García & Sánchez-Puerta, 2015).*

Species	Authors	Strain or isolate code	Genome size[a] (bp)	A-T content (%)	Protein-coding sequence length (%)	Number of non-coding regions	Repeats		
M. floridensis	García & Sánchez-Puerta (2015)	isolate 5	>15 811	83.5	nd[c]	2	no repeats found		
M. hapla	García & Sánchez-Puerta (2015)	VW9	>17 355	81.4	nd[c]	1	17 bp (7.6)[b]		
M. incognita	García & Sánchez-Puerta (2015)	Morelos	17 985-18 332	83.3	53-54%	2	102 bp (12.8)	8 bp (9.4)	63 bp (7.9-13.4)
M. incognita	García & Sánchez-Puerta (2015)	NCMI4	17 662-19 100	83.0	51-54%	2	102 bp (19-24)	8 bp (10.4)	63 bp (9)
M. chitwoodi	García & Sánchez-Puerta (2015)	CAMC2	18 201	85.0	49.4%	2	111 bp (16.9)	48 bp (15.6)	92 bp (8.1)
M. graminicola	Besnard *et al.* (2014)	Batangas	20 030	84.3	49.0%	2	111 bp (29)	94 bp (3)	

Table 7. *(Continued.)*

Species	Authors	Strain or isolate code	Genome size[a] (bp)	A-T content (%)	Protein-coding sequence length (%)	Number of non-coding regions	Repeats		
M. graminicola	Sun *et al.* (2014)	Hainan	19 589	83.5	50.6%	2	111 bp (25)	94 bp (3)	
M. graminicola	Somvanshi *et al.* (2018)	IARI	19 020	82.9	43.9%	5	-[d]	-[d]	-[d]
M. arenaria	Humphreys-Pereira & Elling (2015)	Ma	17 580	83.0	52%	2	102 bp (27)	6 bp (2)	63 bp (6)
M. enterolobii	Humphreys-Pereira & Elling (2015)	Men	18 900	83.0	52%	3	102 bp (27)	53 bp (7.6)	63 bp (2.3)
M. javanica	Humphreys-Pereira & Elling (2015)	Mj	19 600	83.0	50%	2	102 bp (27)	8 bp (4)	63 bp (10)

[a] The genome size depends on the number of copies of the R63 repeat, which varies from 7 to 13 in the genomic assembly.
[b] Numbers in parentheses indicate the copy number of each repeat.
[c] Not determined because the total genome length is unknown.
[d] Not determined.

Palomares-Rius *et al.*, 2017a). All genes are encoded on the same strand of the mtDNA with very few and often short intergenic sequences. Position of coding genes are well conserved among all the species sequenced (Table 7), while the position of tRNAs could differ among species (Tryptophan (W), Tyrosine (Y), Phenylalanine (F), Aspartic acid (D), and Glycine (G)), and in some cases more than one tRNA are found for the same amino acid (*i.e.*, Leucine (L) and Serine (S)) (García & Sánchez-Puerta, 2015). The mitochondrial genomes show the presence of non-coding regions (up to three) and in the majority of the species studied with the presence of repeated sequences. These non-coding regions could be distributed among different regions of the genome (Fig. 11B). Several sets of different-sized repeats have been found (Table 7) and are clustered apart from the protein-coding genes. Some non-coding sequences could be found differentially among the different species sequenced, as well as the repeat regions with different repetition sizes. Genome composition is high in A + T content, ranging from 81.4% (*M. hapla*) to 85.0% (*M. chitwoodi*) (Table 7).

Comparisons of mitochondrial coding regions have revealed closely related haplotypes globally distributed, in some cases indicating a recent speciation and evidence for reticulate evolution within *M. arenaria* (Janssen *et al.*, 2016). Janssen *et al.* (2016) indicate that the barcode region *nad5* can reliably identify the major lineages of tropical RKN. Data using phylogenomics showed the congruent phylogenetic history between nuclear genomic and mitochondrial genomic data for a restricted set of species (Szitenberg *et al.*, 2017).

Nematology Monographs & Perspectives, 2021, Vol. 14, 63-83

Chapter 9

Methods of Biochemical and Molecular Diagnostics

Biochemical and molecular diagnostics are terms used specifically for the characterisation of an organism based on protein isoforms and DNA or RNA sequences. These techniques provide more rapid, reliable and cheaper identification of RKN than morphological approaches. Diagnostics of RKN using different biochemical and molecular methods were given in detail by Williamson (1991), Abrantes *et al.* (2004) and Blok & Powers (2009).

Protein-based Diagnostics

Protein electrophoresis was one of the first of the molecular techniques to be applied in diagnostics of nematodes. Dickson *et al.* (1971), Bergé & Dalmasso (1975, 1976), Dalmasso & Bergé (1975, 1977, 1978, 1979, 1983), Janati *et al.* (1982), Fargette (1984, 1987a, b), Fargette & Braaksma (1990), and Esbenshade & Triantaphyllou (1985a, 1987b, 1990) developed a biochemical diagnostic technique reliant on isozyme profiles for RKN. Variations in esterase (EST) and malate dehydrogenase (MDH) isozyme profiles provided patterns differentiating the majority of *Meloidogyne* species (Fig. 12). Patterns of superoxide dismutase (SOD) and glutamate-oxaloacetate transaminase (GOT) are also given for some species (Carneiro *et al.*, 2000; Molinari *et al.*, 2005). Presently, isozyme profiles are known for the majority of species (Table 8). This method is still used for *Meloidogyne* species identification despite some limitations. Progress in electrophoretic procedures has made possible nematode identification from the protein extract of a single female in less than 2 h from the time the females are collected from infected plant roots (Esbenshade & Triantaphyllou, 1990). Females are prepared by grinding in an ice bath with a homogeniser. Adults dis-

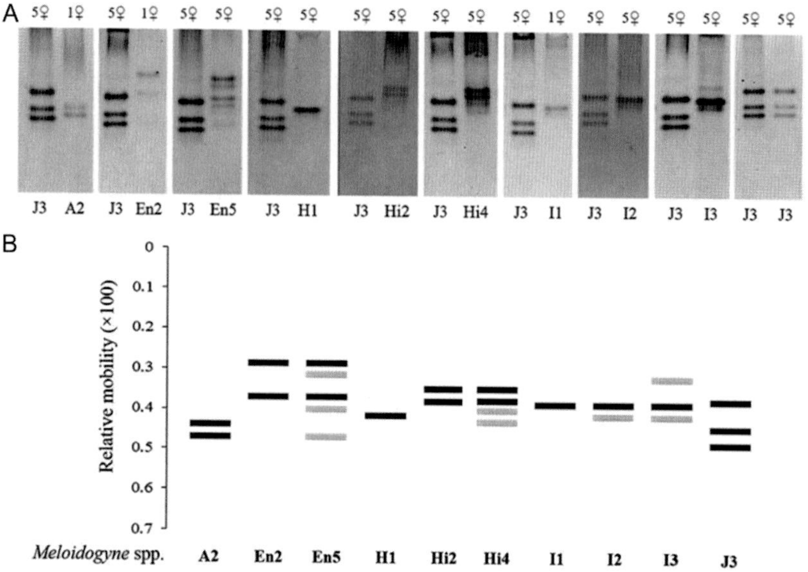

Fig. 12. *Esterase phenotypes (A) and schematic representation (B) of* Meloidogyne *spp. isolates identified. A2* = M. arenaria*; En2, En5* = M. enterolobii*; H1* = M. hapla*; Hi2, Hi4* = M. hispanica*; I1, I2, I3* = M. incognita*; J3* = M. javanica. *Protein extract from* M. javanica *females were used as reference phenotype. Number of females used in each protein homogenate is indicated above in (A). Black lines correspond to major bands and grey lines indicate minor/faint bands in (B). (After Santos* et al.*, 2019.)*

sected from roots can be stored in saline solution at room temperature for up to 3 weeks and crushed females can be stored at −15°C. Comparison of samples with standard is done after separation by polyacrylamide gel electrophoresis. Isozymes can be localised using the appropriate staining procedures. Patterns are usually compared with a known standard, frequently isozymes from *M. javanica*. Isozymes are used primarily with the female egg-laying stage using single individuals (Esbenshade & Triantaphyllou, 1990). Isozyme phenotypes are named by different letters with numbers. One of the main disadvantages of this method is its sensitivity to nematode age. Only young adult females are used for such diagnostics. It has been also noticed that intraspecific diversity or differences in the patterns obtained from different laboratories may also contribute to variations in phenotypes (Blok & Powers, 2009).

Table 8. *Esterase and malate dehydrogenase isozyme phenotypes for* Meloidogyne *species.*

Species	Esterase phenotype	Malate dehydrogenase phenotypes	Reference
M. aberrans	S2	N1	Tao *et al.* (2017)
M. africana	AF2	H1	Janssen *et al.* (2017)
M. arabicida	AR2	N1	Carneiro *et al.* (2004a); Hernandez *et al.* (2004); Villain *et al.* (2013)
M. ardenensis	-	N1a	Karssen & van Hoenselaar (1998)
M. arenaria	A1, A2	N1, N3	Esbenshade & Triantaphyllou (1985a, b); Brito *et al.* (2008); Carneiro *et al.* (2000, 2008); Santos *et al.* (2019)
M. artiellia	M2-VF1	N1b	Karssen & van Hoenselaar (1998); Castillo *et al.* (2003a)
M. baetica	B1	-	Castillo *et al.* (2003a)
M. carolinensis	VS1-S1a	H1	Esbenshade & Triantaphyllou (1985a)
M. chitwoodi	S1	N1a	Esbenshade & Triantaphyllou (1985a, 1990); Karssen & van Hoenselaar (1998); Humphreys-Pereira & Elling (2013)
M. christiei	-	N1a	Brito *et al.* (2015)
M. coffeicola	C2	C1	Carneiro *et al.* (2000); Castro *et al.* (2004)
M. cruciani	M3a (Cr3)	N1	Esbenshade & Triantaphyllou (1985a, b); Cofcewicz *et al.* (2005)
M. dunensis	VS1	N1c	Palomares-Rius *et al.* (2007)
M. duytsi	VS1	N2	Esbenshade & Triantaphyllou (1985a)
M. enterolobii	En2, M2 (VS1-S1), Hi2, Hi4	N1a	Esbenshade & Triantaphyllou (1985a); Carneiro *et al.* (2001); Brito *et al.* (2004, 2008); Santos *et al.* (2019)
M. ethiopica	E3	N1	Carneiro *et al.* (2004a)

Table 8. *(Continued.)*

Species	Esterase phenotype	Malate dehydrogenase phenotypes	Reference
M. exigua	E1 (VF1), E2, E2a, E3	N1	Esbenshade & Triantaphyllou (1990); Carneiro *et al.* (1996b, 2000, 2005b); Oliveira *et al.* (2005); Muniz *et al.* (2008, 2009)
M. fallax	F3	N1b	van der Beek & Karssen (1997)
M. floridensis	Mf3 (P3)	N1	Carneiro *et al.* (2000); Brito *et al.* (2008, 2010)
M. graminicola	G1 (VS1), G2 (R2), G3 (R3)	N1a	Esbenshade & Triantaphyllou (1985a); Carneiro *et al.* (2000); Brito *et al.* (2008); Soares *et al.* (2020)
M. graminis	VS1	N4	Esbenshade & Triantaphyllou (1987)
M. hapla	H1	H1, H2	Esbenshade & Triantaphyllou (1985a, b); Carneiro *et al.* (2000); Ali *et al.* (2016)
M. haplanaria	HA1	HA1	Eisenback *et al.* (2003)
M. hispanica	Hi2, Hi4 (S2-M1)	N1	Esbenshade & Triantaphyllou (1985a); Santos *et al.* (2019)
M. ichinohei	IC2	N1	Janssen *et al.* (2017)
M. incognita	I1, I2, S2, I3	N1	Esbenshade & Triantaphyllou (1985a, b); Carneiro *et al.* (2000); Santos *et al.* (2012, 2019)
M. indica	-	-	Khan *et al.* (2018)
M. inornata	In3	N1	Carneiro *et al.* (2000, 2008)
M. izalcoensis	Iz4	N1	Carneiro *et al.* (2005a); Santos *et al.* (2018a); Stefanelo *et al.* (2018)
M. javanica	J2a, J2b, J3	N1	Esbenshade & Triantaphyllou (1985a, b); Carneiro *et al.* (2000); Brito *et al.* (2008); Ali *et al.* (2016); Santos *et al.* (2019)
M. jianyangensis	JI3	-	Baojun *et al.* (1990)

Table 8. *(Continued.)*

Species	Esterase phenotype	Malate dehydrogenase phenotypes	Reference
M. konaensis	K3	N1	Monteiro *et al.* (2016)
M. kralli	-	Nlc	Karssen & van Hoenselaar (1998)
M. lopezi	L2	N3	Humphreys-Pereira *et al.* (2014a)
M. luci	L3	N1	Carneiro *et al.* (2000)
M. lusitanica	P1 (A1)	P3 (N1c)	Pais & Abrantes (1989); Karssen & van Hoenselaar (1998)
M. mali	VS1	H1, H3	Esbenshade & Triantaphyllou (1985a); Ahmed *et al.* (2013)
M. maritima	VS1-S1	N1a	Karssen *et al.* (1998a)
M. marylandi	VS1	N1c	Oka *et al.* (2003)
M. megadora	Me3	Me1	Maleita *et al.* (2012c, 2016)
M. microcephala	A1	N1	Esbenshade & Triantaphyllou (1985a)
M. microtyla	M1	H1	Esbenshade & Triantaphyllou (1985a)
M. minor	VS1	N1a	Karssen *et al.* (2004)
M. morocciensis	A3	N1	Rammah & Hirschmann (1990b); Carneiro *et al.* (2008); Silva *et al.* (2020)
M. naasi	VF1	N1a	Esbenshade & Triantaphyllou (1985a)
M. nataliei	S1	-	Álvarez-Ortega *et al.* (2019)
M. oleae	A1	N1c	Archidona-Yuste *et al.* (2018)
M. oryzae	O1	O3	Carneiro *et al.* (2000); Mattos *et al.* (2018)
M. ottersoni	Ot0	N1a	Leite *et al.* (2020)
M. panyuensis	S1-F1	N1b	Liao *et al.* (2005)
M. paranaensis	P1, P2, P2a	N1	Carneiro *et al.* (1996b, 2000, 2004b); Santos *et al.* (2018b)
M. partityla	Mp3	N1	Brito *et al.* (2008)
M. petuniae	VS1-S1	N1	Charchar *et al.* (1999)
M. phaseoli	E3	-	Charchar *et al.* (2008b)
M. pisi	E5	-	Charchar *et al.* (2008a)

Table 8. *(Continued.)*

Species	Esterase phenotype	Malate dehydrogenase phenotypes	Reference
M. platani	S1	N1a	Esbenshade & Triantaphyllou (1985a)
M. querciana	F1	N3a	Esbenshade & Triantaphyllou (1985a)
M. salasi	VS1-2	S3	Mattos *et al.* (2019)
M. silvestris	A1	N1c	Castillo *et al.* (2009a)
M. spartelensis	S1	N1b	Ali *et al.* (2015, 2016)
M. trifoliophila	T1	T1, T2	Mercer *et al.* (1997)
M. vitis	VF1	N3d	Yang *et al.* (2021)

DNA-based Diagnostics

Molecular techniques rely on the occurrence of nucleotide polymorphisms in DNA sequences among species. The first report of a DNA technique used for taxonomic purposes was probably in 1985 when Curran *et al.* (1985) analysed genomic DNA with restriction enzymes to differentiate *M. arenaria* from *M. javanica* and other non-plant-parasitic nematodes. However, this method was time-consuming, required a substantial amount of DNA and it was not sensitive. The polymerase chain reaction (PCR) has become one of the most widely used techniques for studying the genetic diversity of nematodes and their identification. PCR is a rapid, inexpensive and simple means of producing relatively large numbers of copies of DNA molecules *via* an enzyme catalyst. The PCR method requires a DNA template (starting material) containing the region to be amplified, two oligonucleotide primers flanking this target region (Table 9), DNA polymerase and four deoxynucleotide triphosphates mixed in a buffer containing magnesium ions. PCR is performed in a tube in a thermocycler with programmed heating and cooling. The resulting amplified products are electrophoretically separated according to their size on agarose or polyacrylamide gels and visualised using ethidium bromide or other DNA fluorophors, which interact with double-stranded DNA and cause it to fluoresce under UV radiation. Once identified, nematode target DNA generated by PCR amplification can be

Table 9. *Primer combinations used for molecular diagnostics and phylogenetic studies of root-knot nematodes.*

Primer combination and code (direction)*	Primer sequence (5′-3′)	Amplified region	Reference
TW81 (f)	GTT TCC GTA GGT GAA CCT GC	ITS rRNA	Curran *et al.* (1994)
AB28 (r)	ATA TGC TTA AGT TCA GCG GGT		
F194 (f)	CGT AAC AAG GTA GCT GTA G	ITS rRNA	Ferris *et al.* (1993)
F195 (r)	TCC TCC GCT AAA TGA TAT G		
18S (f)	TTG ATT ACG TCC CTG CCC TTT	ITS rRNA	Vrain *et al.* (1992)
26S (r)	TTT CAC TCG CCG TTA CTA AGG		
MelF5 (f)	TAC GGA CTG AGA TAA TGG T	18S rRNA	Tigano *et al.* (2005)
MelR5 (r)	GGT TCA AGC CAC TGC GA		
1096F (f)	GGT AAT TCT GGA GCT AAT AC	18S rRNA	Holterman *et al.* (2006)
2646R (r)	GCT ACC TTG TTA CGA CTT TT		
988F (f)	CTC AAA GAT TAA GCC ATG C	18S rRNA	Holterman *et al.* (2006)
1912R (r)	TTT ACG GTC AGA ACT AGG G		
1813F (f)	CTG CGT GAG AGG TGA AAT	18S rRNA	Holterman *et al.* (2006)
2646R (r)	GCT ACC TTG TTA CGA CTT TT		
194 (f)	TTA ACT TGC CAG ATC GGA CG	IGS rRNA	Blok *et al.* (1997)
195 (r)	TCT AAT GAG CCG TAC GC		
D2A (f)	ACA AGT ACC GTG AGG GAA AGT TG	28S rRNA	Nunn (1992)
D3B (r)	TCG GAA GGA ACC AGC TAC TA		
U831B (f)	AAY AAR ACM AAG CCN ATY TGG AC	*hsp90*	Skantar & Carta (2000)
L1110B (r)	TCG CAR TTC TCC ATR ATC AA		

Table 9. *(Continued.)*

Primer combination and code (direction)*	Primer sequence (5′-3′)	Amplified region	Reference
RKN-d1F (f)	TCG AAC ATG TCA AAA GGA GC	*hsp90*	Skantar & Carta (2004)
RKN-5R (r)	GCY GAT CTT GTY AAC AAC CYT GGA AC		
m1 (f)m2 (r)	CGT GTA ACA GAG ATG CCA GAGTG GAG GAA CAG TAA GTG AG	*map-1*	Semblat *et al.* (2001)
COI-F5-Mel (f)	TGA TTG ATT TAG GTT CTG GAA CTK SWT GAA C	*COI*	Powers *et al.* (2018)
COI-R9-Mel (r)	CAT AAT GAA AAT GGG CAA CAA CAT AAT AAG TAT C		
JB3 (f)	TTT TTT GGG CAT CCT GAG GTT TAT	*COI*	Derycke *et al.* (2005)
JB4 (r)	TAA AGA AAG AAC ATA ATG AAA ATG		
NAD5F2 (f)	TAT TTT TTG TTT GAG ATA TAT TAG	*nad5*	Janssen *et al.* (2015)
NAD5R1 (r)	CGT GAA TCT TGA TTT TCC ATT TTT		
C2F3 (f)	GGT CAA TGT TCA GAA ATT TGT GG	*COII*-intergenic spacer-16S rRNA	Powers & Harris (1993)
1108 (r)	TAC CTT TGA CCA ATC AC GCT		
TRNAH (f)	TGA ATT TTT TAT TGT GAT TAA	16S rRNA	Stanton *et al.* (1997)
MRH106 (r)	AAT TTC TAA AGA CTT TTC TTA GT		
MORF (f)	ATC GGG GTT TAA TAA TGG G	Intergenic spacer mtDNA	Stanton *et al.* (1997)
MTHIS (r)	AAA TTC AAT TGA AAT TAA TAG C		
COIIF1 (f)	TAR ATT KNT TTC ATR RTT TTA ATT GT	*COII*	Kiewnick *et al.* (2014)
COIIR1 (r)	CAC AAA TTT CTG AAC ATT GMC C		

* f-forward, r-reverse.

further characterised by various analyses, including restriction fragment length polymorphism (RFLP) or sequencing.

Genes Used for Molecular Diagnostics

The eukaryotic cell contains two different genomes: that of the nucleus and that of mitochondria. Molecular diagnostics of RKN use data from both genomes.

NUCLEAR GENES

Historically, the only nuclear genes with a high enough copy number for easy study were ribosomal genes. These genes code ribosomal RNAs (rRNA), which are nearly two-thirds of the mass of the ribosome. The genes encoding rRNA are arranged in tandem, in several hundred copies, and are organised in a cluster that includes a small subunit (SSU or 18S) and a large subunit (LSU or 28S) gene, which are separated by a small 5.8S gene. Another ribosomal gene, a ubiquitous component of large ribosomal subunits in eukaryotic cells, is 5S rRNA linked to the intergenic spacer (IGS) region. The IGS region contains many repeats. In addition to the coding sequences, the rDNA array also contains spacer: an external transcribed spacer (ETS) and two internal transcribed spacers, ITS1 and ITS2. Diagnostics of RKN species is presently based on the ITS1-5.8S-ITS2 rRNA gene, the D2-D3 expansion fragment of 28S rRNA gene, the IGS rRNA gene and 18S rRNA gene sequences. Several protein coding genes, including the RNA polymerase II gene (*rpb1*) (Lunt *et al.*, 2008; Rybarczyk-Mydlowska *et al.*, 2013), *hsp90* (Skantar & Carta, 2000) and *map-1* (gene family encodes expansin-like proteins that are secreted into plant tissues during parasitism) (Tomalova *et al.*, 2012), are also successfully used for diagnostics of some RKN, but not those belonging to the Incognita group.

MITOCHONDRIAL GENES

Nematode mtDNA sequences accumulate substitution changes much more quickly than the rRNA sequences and tends to be extremely A + T rich. The relatively rapid rate of evolution and rearrangements that occur in mtDNA has limited the design of universal primers for several genes for RKN. The region between the *COII* and the large subunit RNA

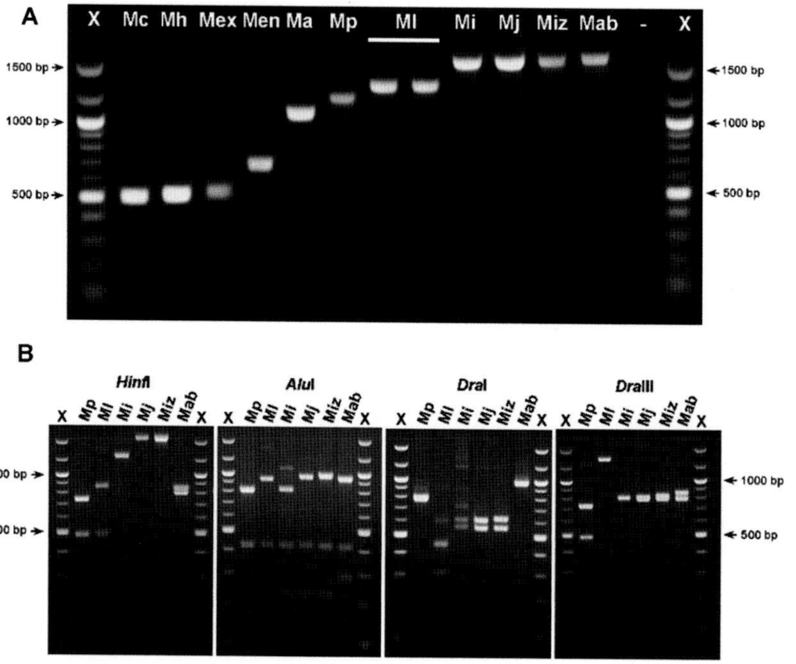

Fig. 13. COII-16S rRNA *PCR amplification products and PCR-RFLP patterns. A: Amplification products of DNA from* Meloidogyne *spp. using primers C2F3 and 1108 primers; B: Restriction patterns generated with* Hinf*I,* Alu*I,* Dra*I and* Dra*III. Mc* = M. chitwoodi*; Mh* = M. hapla*; Mex* = M. exigua*; Men* = M. enterolobii*; Ma* = M. arenaria*; Mp* = M. paranaensis*; Ml* = M. lopezi*; Mi* = M. incognita*; Mj* = M. javanica*; Miz* = M. izalcoensis*; Mab* = M. arabicida*; – = negative control; X = 100 bp DNA ladder. (After Humphreys-Pereira* et al.*, 2014a.)*

gene containing an intergenic region with unique size and nucleotide polymorphism has been utilised for distinguishing different species and host races of *Meloidogyne* (Powers & Harris, 1993; Orui, 1998; McClure *et al.*, 2012; Humphreys-Pereira *et al.*, 2014a; Pagan *et al.*, 2015; Smith *et al.*, 2015) (Fig. 13). The C2F3 and 1108 primers amplified product of different lengths for species, for example: *M. christiei*, *M. graminis*, *M. graminicola*, *M. hapla*, *M. chitwoodi*, *M. partityla* – 530-540 bp, *M. enterolobii* ~ 700 bp, *M. floridensis*, *M. arenaria* ~ 1.2 kb, *M. paranaensis* ~ 1250 bp, *M. lopezi* ~ 1370 bp, *M. incognita*, *M. javanica*, *M. izalcoensis*, *M. arabicida* – 1.6-1.7 kb. Recently, sequencing of two other genes, *nad*5 (Janssen *et al.*, 2016; Kolombia *et al.*, 2017) (Fig. 14)

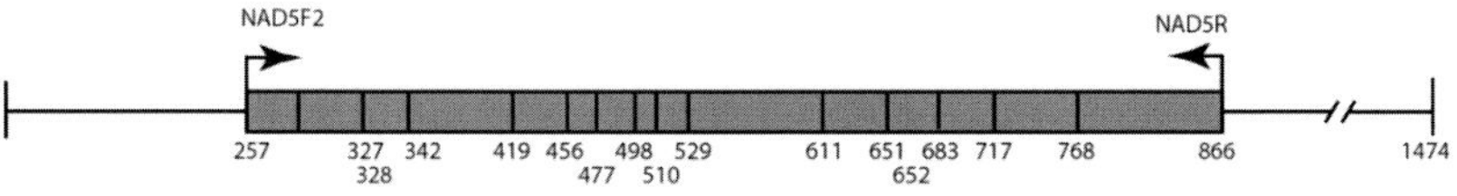

Position within gene	327	328	342	419	456	477	498	510	529	611	651	652	683	717	768
Position within amplified region	71	72	86	163	200	221	242	254	273	355	395	396	427	461	512
M. enterolobii	A	C	A	G	G	T	A	T	G	A	A	A	T	A	G
M. javanica H1	A	T	G	G	A	A	G	T	G	G	G	T	T	A	A
M. javanica H2	A	T	G	G	A	A	G	T	G	G	G	T	T	A	G
M. incognita H1	A	T	G	G	A	A	A	T	G	A	A	T	T	A	A
M. incognita H2	A	T	G	G	A	A	A	T	G	A	G	T	T	A	A
M. incognita H3	A	T	G	G	A	A	G	T	G	A	R	T	T	A	A
M. incognita H4	A	T	A	G	A	A	A	T	G	A	A	T	T	A	A
M. arenaria H1	A	T	G	G	A	A	A	C	A	G	G	T	T	A	A
M. arenaria H2	A	T	G	G	A	A	A	C	G	G	G	C	A	A	A
M. arenaria H3	A	T	G	G	A	A	A	C	G	G	G	T	T	A	A
M. arenaria H4	A	T	G	G	A	A	A	C	G	G	G	C	T	A	A
M. floridensis	A	T	G	A	T	T	A	C	G	A	G	T	T	T	A
M. luci H1	A	T	G	G	A	A	A	C	G	G	G	T	T	A	A
M. luci H2	A	A	G	G	A	A	A	C	G	G	G	T	T	G	A
M. inornata	A	T	G	G	A	A	A	C	G	G	G	T	T	A	A
M. ethiopica	A	T	G	G	A	A	A	C	G	G	G	T	T	A	A
Meloidogyne sp.1 H1	T	T	G	G	A	A	A	C	G	G	G	T	T	A	A
Meloidogyne sp.1 H2	A	T	G	G	A	A	A	C	G	G	G	T	T	A	A
Meloidogyne sp.2	A	T	G	G	A	A	A	C	G	G	A	T	T	A	A

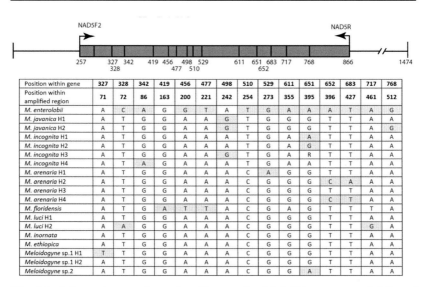

Fig. 14. *Schematic overview of* nad5 *gene for* Meloidogyne enterolobii *and* Meloidogyne *species from the Incognita and Ethiopica groups showing the positions of primers and polymorphic nucleotide sites. (After Janssen* et al.*, 2016; Kolombia* et al.*, 2017.)*

and *COI* (Kiewnick *et al.*, 2014; Powers *et al.*, 2018), were also proposed as a reliable method for RKN diagnostics (Figs 13, 14). *COI* is a useful marker for non-MIG RKN and *M. enterolobii*, whereas *nad5* can be applied for identification of several species from the Incognita group.

PCR with Specific Primers

PCR with species-specific primer is a popular technology for species identification due to its high accuracy, sensitivity and convenience. This technique enables the detection of one or several species in a nematode mixture by a single PCR test, thus decreasing diagnostic time and costs. Species-specific sequence fragment of mtDNA, rRNA gene or specific sequence for certain species called sequence characterised amplified regions (SCAR) derived from Random Amplified Polymorphic DNA (RAPD) fragment can be used as molecular markers for species identification. Species-specific primer for the PCR is designed to be bound to the target specific region sequence. Detection of a specific size amplicon in a gel indicates the presence of a certain species within a sample (Fig. 15). Diagnostics using PCR with species-specific primers

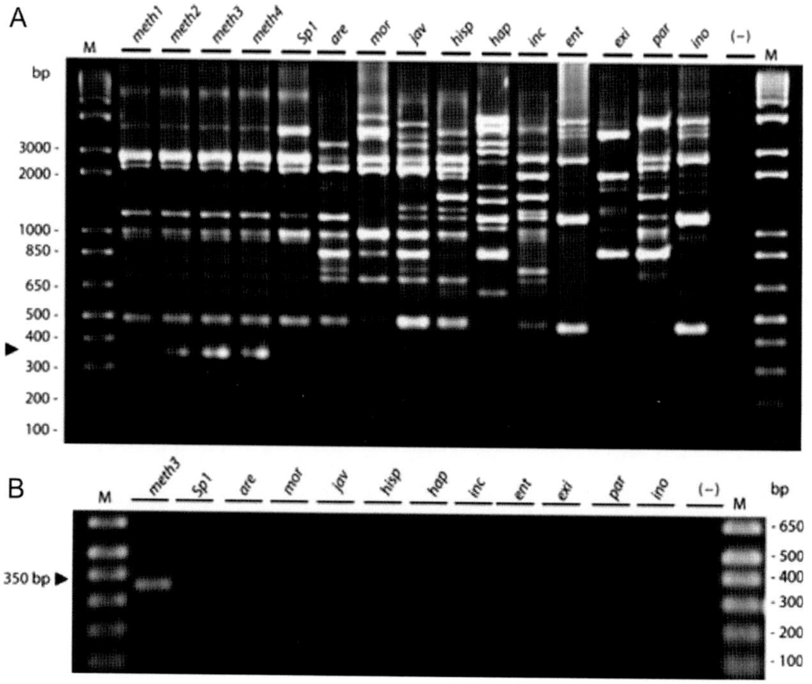

Fig. 15. *A: RAPD patterns for* Meloidogyne ethiopica *and other* Meloidogyne *spp. obtained with primer OPE-07. The* M. ethiopica-*specific 350 bp fragment is shown with an arrowhead; B: SCAR-PCR amplification pattern for* M. ethiopica *and other root-knot nematode species generated with primer pair methF/R. Arrowhead indicates the specific fragment detected only in* M. ethiopica *(meth3). M = 1 kb DNA ladder; meth =* M. ethiopica*; Sp1 =* Meloidogyne *sp.; are =* M. arenaria*; mor =* M. morocciensis*; jav =* M. javanica*; hisp =* M. hispanica*; hap =* M. hapla*; inc =* M. incognita*; ent =* M. enterolobii*; exi =* M. exigua*; par =* M. paranaensis*; ino =* M. inornata. *(After Correa* et al.*, 2014.)*

have been developed for a wide range of RKN (Table 10); however, these specific primers should be used wisely considering the fact that most of them were not tested with a wide range of samples and false positive and false negative results might be expected.

Real-time PCR

Real-time PCR, also called quantitative PCR or qPCR, provides a simple and elegant method for quantification of nematodes in samples.

Table 10. *Species-specific primers for conventional PCR used for diagnostics of* Meloidogyne *spp.*

Species	Primer code and sequence (5′-3′)	Gene fragment	Amplified size	Reference
M. arabicida	ar-A12F-TCG GCG ATA GTA CGT ATT TAG CG	SCAR	~ 300 bp	Correa *et al.* (2013)
	ar-A12R-TAG TGA TTT CGG CGA TAG GC			
M. arenaria	Far-TCG GCG ATA GAG GTA AAT GAC	SCAR	~ 420 bp	Zijlstra *et al.* (2000)
	Rar-TCG GCG ATA GAC ACT ACA AACT			
M. arenaria	MaF-TCG AGG GCA TCT AAT AAA GG	SCAR	~ 950 bp	Dong *et al.* (2001)
	MaR-GGG CTG AAT AAT CAA AGG AA			
M. chitwoodi	Fc-TGG AGA GCA GCA GGA GAA AGA	SCAR	~ 800 bp	Zijlstra (2000)
	Rc-GGT CTG AGT GAG GAC AAG AGT A			
M. chitwoodi	MC3F-CCA ATG ATA GAG ATA GGA AC	SCAR	~ 400 bp	Williamson *et al.* (1997)
	MC1R-CTG GCT TCC TCT TGT CCA AA			
M. chitwoodi	C64-GAT CTA TGG CAG ATG GTA TGG A	IGS rRNA	~ 900 bp	Petersen *et al.* (1997)
	1839-AGC CAA AAC AGC GAC CGT CTA C			
M. ethiopica	Meth-F-ATG CAG CCG CAG GGA ACG TAG TTG	SCAR	~ 350 bp	Correa *et al.* (2014)
	Meth-R-TGT TGT TTC ATG TGC TTC GGC ATC			
M. graminicola	SCAR-MgFW-GGG GAA GAC ATT TAA TTG ATG ATC AAC	SCAR	~ 640 bp	Bellafiore *et al.* (2015)
	SCAR- MgRev-GGT ACC GAA ACT TAG GGA AAG			
M. graminicola	Mg-F3-TTA TCG CAT CAT TTT ATT TG	ITS rRNA	~ 369 bp	Htay *et al.* (2016)
	Mg-R2-CGC TTT GTT AGA AAA TGA CCC T			
M. graminicola	GRAJ17-1F-TTC GAC TCT GTA CGA AAG CC	SCAR	~ 230 bp	Mattos *et al.* (2019)
	GRAJ17-1R-CAA AAG TAA CCG GAC ACT CTT TT			
M. oryzae	ORYA12 F-CCA GCA TCC GCT GTT GTAT	SCAR	~ 120 bp	Mattos *et al.* (2019)
	ORYA12 R-AAC AGG CTC CAG GTG AAA AG			

Table 10. *(Continued.)*

Species	Primer code and sequence (5′-3′)	Gene fragment	Amplified size	Reference
M. salasi	SAL R12-1F-CAA AGA ACG GGG TTT ATT CG SAL R12-1R-GGT TAT CCG AAC TCC CCA AT	SCAR	~ 160 bp	Mattos *et al.* (2019)
M. izalcoensis	iz-AB2F-GGA AAC CCC TAA TTA GGA TAC ACT iz-AB2R-CG CTT GAT TTG AGC AGT AGG	SCAR	~ 670 bp	Correa *et al.* (2013)
M. enterolobii	63VNL-GAA ATT GCT TTA TTG TTA CTA AG 63VTH-TAG CCA CAG CAA AAT AGT TTT C	mtDNA	~ 322 bp	Blok *et al.* (2002)
M. enterolobii	Me-F-AAC TTT TGT GAA AGT GCC GCT G Me-R-TCA GTT CAG GCA GGA TCA ACC	IGS rRNA	~ 200 bp	Long *et al.* (2006)
M. enterolobii	MK7F-GAT CAG AGG CGG GCG CAT TGC GA MK7R-CGA ACT CGC TCG AAC TCG AC	SCAR	520 bp	Tigano *et al.* (2010)
M. exigua	Ex-D15-F-CAT CCG TGC TGT AGC TGC GAG Ex-D15-R-CTC CGT GGG AAG AAA GAC TG	SCAR	562 bp	Randig *et al.* (2002)
M. fallax	Ff-CCA AAC TAT CGT AAT GCA TTA TT Rf-GGA CAC AGT AAT TCA TGA GCT AG	SCAR	~ 515 bp	Zijlstra *et al.* (2000)
M. fallax	F64-TGG GTA GTG GTC CCA CTC TG 1839-AGC CAA AAC AGC GAC CGT CTA C	IGS rRNA	1100 bp	Petersen *et al.* (1997)
M. hapla	Fh-TGA CGG CGG TGA GTG CGA Rh-TGA CGG CGG TAC CTC ATA G	SCAR	~ 610 bp	Zijlstra (2000)
M. hapla	MH0F-CAG GCC CTT CCA GCT AAA GA MH1R-CTT CGT TGG GGA ACT GAA GA	SCAR	~ 960 bp	Wiliamson *et al.* (1997)
M. hapla	MhF-GGC TGA GCA TAG TAG ATG ATG TT MhR-ACC CAT TAA AGA GGA GTT TTG C	SCAR	~ 1500 bp	Dong *et al.* (2001)
M. hapla	JMV1-GGA TGG CGT GCT TTC AAC JMV hapla-AAA AAT CCC CTC GAA AAA TCC ACC	IGS rRNA	~ 440 bp	Wishart *et al.* (2002)

Table 10. *(Continued.)*

Species	Primer code and sequence (5′-3′)	Gene fragment	Amplified size	Reference
M. hapla	Mha17f-TGA ATA GTT GGT GGC CTC TG	16D10 effector	110 bp	Gorny *et al.* (2019)
	Mha17f-TGT GCT ATT TCC AAG GGT AAA G			
M. incognita	Finc-CTC TGC CCA ATG AGC TGT CC	SCAR	~ 1200 bp	Zijlstra *et al.* (2000)
	Rinc-CTC TGC CCT CAC ATT AGG			
M. incognita	MiF-TAG GCA GTA GGT TGT CGG G	SCAR	~ 1350 bp	Dong *et al.* (2001)
	MiR-CAG ATA TCT CTG CAT TGG TGC			
M. incognita	Inc-K14-F-GGG ATG TGT AAA TGC TCC TG	SCAR	~ 399 bp	Randig *et al.* (2002)
	Inc-K14-R-CCC GCT ACA CCC TCA ACT TC			
M. incognita	MIE-for-TCC GTG CTG TAG CTT GCC C	SCAR	~ 900 bp	El-Ghore *et al.* (2004)
	MIE-rev-CAC CAT CCG TTA TAA GCT CTG			
M. incognita	MI-F-GTG AGG ATT CAG CTC CCC AG	SCAR	~ 955 bp	Meng *et al.* (2004)
	MI-R-ACG AGG AAC ATA CTT CTC CGT CC			
M. incognita	Mi2F4-ATG AAG CTA AGA CTT TGG GCT	SCAR	~ 300 bp	Kiewnick *et al.* (2013)
	Mi1R1-TCC CGC TAC ACC CTC AAC TTC			
M. javanica	Fjav-GGT GCG CGA TTG AAC TGA GC	SCAR	~ 620 bp	Zijlstra *et al.* (2000)
	Rjav-CAG GCC CTT CAG TGG AAC TAT AC			
M. javanica	MjF-CCT TAA TGT CAA CAC TAG AGC C	SCAR	~ 1650 bp	Dong *et al.* (2001)
	MjR-GGC CTT AAC CGA CAA TTA GA			
M. javanica	ACG CTA GAA TTC GAC CCT GG	SCAR	~ 517 bp	Meng *et al.* (2004)
	GGT ACC AGA AGC AGC CAT GC			
M. javanica	MJE-for-GTC CGT TAT CTG AGC TTA T	SCAR	~ 1000 bp	El-Ghore *et al.* (2004)
	MJE-rev-AGT CAC TCC ATC ACC TTC A			

Table 10. *(Continued.)*

Species	Primer code and sequence (5′-3′)	Gene fragment	Amplified size	Reference
M. naasi	N-ITS-CTC TTT ATG GAG AAT AAT CGT R195-CCT CCG CTT ACT GAT ATG	ITS rRNA	~ 433 bp	Zijlstra *et al.* (2004)
M. naasi	Mn28SFs-GTC TGA TGT GCG ACC TTT CAC TAT RK28SUR-CCC TAT ACC CAA GTC AGA CGA T	D2-D3 of 28S rRNA	~ 272 bp	Ye *et al.* (2015)
M. paranaensis	par-C09F-GCC CGA CTC CAT TTG ACG GA par-C09R-CCG TCC AGA TCC ATC GAA GTC	SCAR	~ 208 bp	Randig *et al.* (2002)
M. graminis	Mg28SFs-GAT GTG CGA TAT TTT CCG TCA AGG RK28SUR-CCC TAT ACC CAA GTC AGA CGA T	D2-D3 of 28S rRNA	~ 198 bp	Ye *et al.* (2015)
M. graminis	MgmITSF-GAT CGT AAG ACT TAA TGA GCC MgITSRs-TGC ATA AGG CAA CAT AAT GT	ITS rRNA	~ 612 bp	Ye *et al.* (2015)
M. marylandi	Mm28SFs-GAT GTG CGA TAT TTT TTT TTC GAA RK28SUR-CCC TAT ACC CAA GTC AGA CGA T	D2-D3 of 28S rRNA	~ 198 bp	Ye *et al.* (2015)
M. marylandi	MgmITSF-GAT CGT AAG ACT TAA TGA GCC MmITSRs-CT GAT CTG ATT TAC ATT ACA CGG	ITS rRNA	~ 323 bp	Ye *et al.* (2015)
M. partityla	ITS-1 F-CGC AGT GGC TTG AAC CGG MpSpec-TGA ACT TTT ATT GGT GAA AG	ITS rRNA	~ 630 bp	Brito *et al.* (2016)
M. ethiopica group	Me309F-CTA ATT TGG GTG AAT TT Me549R-AAT CAA AAT CTT CTC CT	*COII*-16S rRNA region	~ 241 bp	Gerič Stare *et al.* (2019)
M. incognita group	Mt575R-AGA ACT TAA ACT CTA AAT AAC C2F3-GGT CAA TGT TCA GAA ATT TGT GG	*COII*-16S rRNA region	~ 621 bp	Gerič Stare *et al.* (2019)
M. incognita group	MIGF-ACA CAG GGG AAA GTT TGC CA MIGR-GAG TAA GGC GAA GCA TAT CC	SCAR	~ 500 bp	Qiu *et al.* (2006)
M. vitis	Mv-F-CTG GTT CAG GGT CAT TTA TAA AC Mv-R-TAT ACG CTT GTG TGG ATG AC	ITS rRNA	~ 170 bp	Yang *et al.* (2021)

Real-time PCR requires an instrumentation platform that consists of a thermal cycler, optics for fluorescence excitation and emission collection, and computerised data acquisition and analysis software. The PCR quantification technique measures the number of nematodes indirectly by assuming that the number of target DNA copies in the sample is proportional to the number of targeted nematodes. Quantitative information in a PCR comes from those few cycles where the amount of DNA grows logarithmically from just above background to the plateau. The real-time technique allows continuous monitoring of the sample during PCR using hybridisation probes (TaqMan) enabling simultaneous quantification of several nematode species in one reaction, or double-stranded dyes such as SYBR Green. Real-time PCR is easy to perform and results are readily available in 45 min to 1.5 h. Compared with traditional PCR methods, real-time PCR has advantages. It allows for faster, simultaneous detection and quantification of target DNA. The automated system overcomes the laborious process of estimating the quantity of the PCR product after gel electrophoresis. The system can easily be adapted for high-throughput analyses of many samples at a time (96er or 384er format) (Braun-Kiewnick & Kiewnick, 2018). Real-Time has been used for detection and quantification of several main RKN species (Table 11).

Loop-mediated Isothermal Amplification (LAMP)

The invention of LAMP by Notomi *et al.* (2000) provided a new approach to molecular diagnosis since it does not need a thermal cycler. This method is commercially available and has been widely used for the detection of a variety of pathogens, including bacteria, viruses, fungi and nematodes. LAMP uses a set of 4-6 primers that form loops to generate new priming sites and a highly processive DNA polymerase with strand displacement activity (*Bst* polymerase) to separate the DNA strands and amplify DNA with high specificity under isothermal conditions at 60-65°C within 0.5-1.5 h. DNA products can be visualised either by gel electrophoresis or by addition of a fluorescent dye to a positive LAMP reaction, which produces a colour change and allows detection with the naked eye or UV light. In addition, it has been improved by the use of a lateral flow dipstick (LFD) method to confirm visually the presence of amplicons; this was termed the LAMP-LFD technique (Kiatpathomchai

Table 11. *Identification of the* Meloidogyne *spp. by real-time PCR, loop-mediated isothermal amplification (LAMP) and recombinase polymerase amplification (RPA).*

Method	Species	Reference
Real-time PCR	*M. chitwoodi*	Zijlstra & Van Hoof (2006); De Haan *et al.* (2014)
	M. fallax	Zijlstra & Van Hoof (2006); De Haan *et al.* (2014)
	M. arenaria	Agudelo *et al.* (2011)
	M. minor	De Weerdt *et al.* (2011)
	M. enterolobii	Kiewnick *et al.* (2015)
	M. hapla	Watanabe *et al.* (2013); Sapkota *et al.* (2016); Gorny *et al.* (2019)
	M. javanica	Berry *et al.* (2007)
	M. incognita	Toyota *et al.* (2008); Zhao *et al.* (2010); Watanabe *et al.* (2013)
	M. graminicola	Katsuta *et al.* (2016); He *et al.* (2021)
LAMP	*M. incognita*	Niu *et al.* (2011)
	M. enterolobii	Niu *et al.* (2012); He *et al.* (2013)
	M. mali	Wei *et al.* (2016); Zhou *et al.* (2017)
	M. hapla	Peng *et al.* (2017)
	M. chitwoodi and *M. fallax*	Zhang & Gleason (2019)
	M. graminicola	He *et al.* (2021)
	M. camelliae	Cai *et al.* (2016)
	M. partityla	Waliullah *et al.* (2020)
RPA	*M. enterolobii*	Subbotin (2019); Ju *et al.* (2019)
	M. javanica	Ju *et al.* (2019); Chi *et al.* (2020)
	M. incognita	Ju *et al.* (2019)
	M. arenaria	Ju *et al.* (2019)
	M. hapla	Song *et al.* (2021); Subbotin & Burbridge (2021)

et al., 2008; Ding *et al.*, 2010). This method clearly holds potential for testing in the field or in under-equipped laboratories (Fig. 16). The sensitivity of some LAMP assays is 100 or more times higher than the conventional PCR (Niu *et al.*, 2011; Zhang & Gleason, 2019). LAMP diagnostic technique has been developed for several *Meloidogyne* species (Table 11).

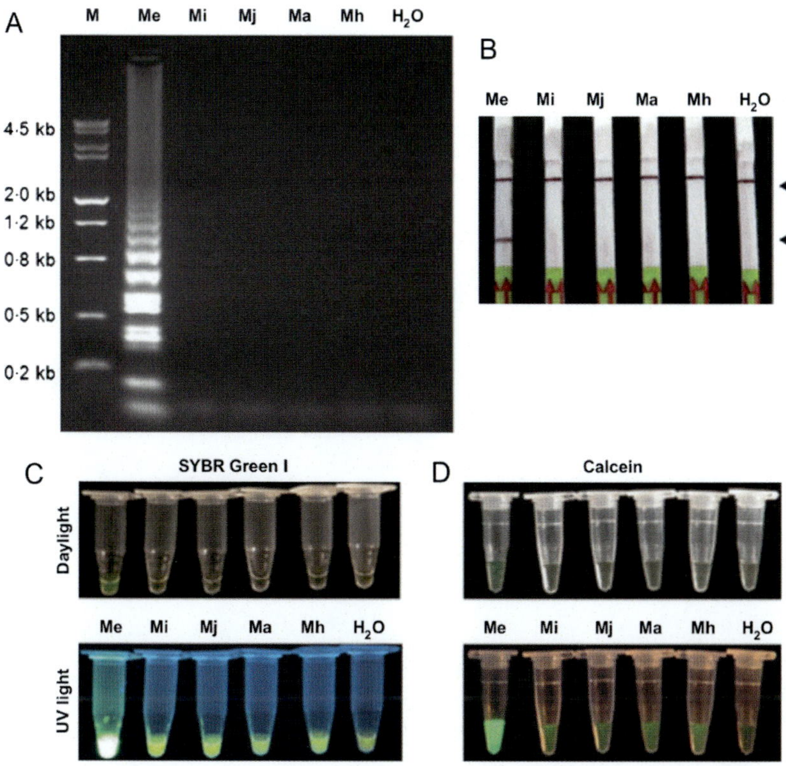

Fig. 16. *Specificity of* Meloidogyne enterolobii *(Me) LAMP detection and product confirmation. A: Specificity of Me-LAMP assay on a gel; B: Specificity of LAMP-LFD for* Meloidogyne *spp. detection. Control bands and specific bands for* M. enterolobii *are shown by arrows; C, D: Specificity of LAMP assay products visualised by adding SYBR Green I (C) and calcein (D). Top row: direct visualisation by naked eye. Bottom row: observation under UV transillumination. Samples Me, Mi, Mj, Ma and Mh represent* M. enterolobii, M. incognita, M. javanica, M. arenaria, *and* M. hapla, *respectively. H_2O represents a no-template negative control. (After Niu* et al., *2012.)*

Recombinase Polymerase Amplification (RPA)

RPA is a relatively new isothermal methodology for amplifying DNA and represents a hugely versatile alternative to PCR (Piepenburg *et al.*, 2006; James & Macdonald, 2015; Daher *et al.*, 2016). RPA uses a highly efficient displacement polymerase that amplifies a few copies of target nucleic acid in 20 min at a constant temperature (37-42°C). It does so

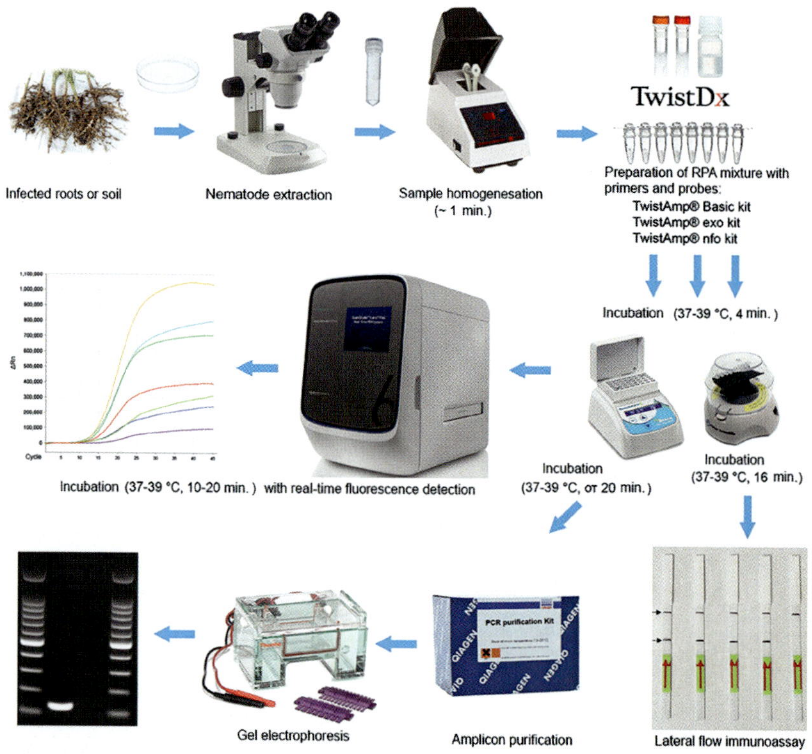

Fig. 17. *Workflow of RPA assays for root-knot nematode diagnostics with TwistAmp® kits (TwistDx, UK). (After Subbotin* et al., *2019 with modifications.)*

by utilising three core enzymes: recombinase, single-stranded binding protein (SSB), and strand-displacing polymerase. The recombinase enzyme forms a complex with a primer to facilitate their binding to the targeted DNA template. Then, the SSB binds to the displaced strands of DNA and prevents the displacement of recombinase-primer complex by branch migration. The strand-displacing polymerase then recognises the bound recombinase-primer complex and initiates DNA synthesis. Like PCR, RPA produces an amplicon constrained in size to the binding sites of the primers. The advantages of RPA include highly efficient and rapid amplification and a low constant operating temperature. RPA can be carried out by using fluorescent probes in real time or detected by agarose gel electrophoresis or lateral flow assay (Fig. 17). Several tests using RPA have demonstrated high sensitivity and specificity for detecting small amounts of nematode DNA (Ju *et al.*, 2019; Subbotin *et al.*, 2019; Chi *et al.*, 2020). RPA has good flexibility to be adapted to various detection systems. RPA has some important advantages over PCR methods, the first being that it uses crude nematode extract for the analysis instead of DNA extracts, which are required for PCR and other assays. The second advantage is that results are available in 15-20 min for RPA *vs* 1.5-3.0 h for PCR assays. RPA is the easiest to perform and the reagents are very stable.

Nematology Monographs & Perspectives, 2021, Vol. 14, 85-97

Chapter 10

Phylogenetic Relationships

Phylogeny and Molecular Grouping of Root-knot Nematodes

Tandingan De Ley *et al.* (2002) were the first to use 18S rRNA gene sequences for a rigorous reconstruction of *Meloidogyne* phylogeny. This analysis, which included only 12 species of *Meloidogyne* and four outgroup taxa, revealed three clades (I, II and III) within the genus. Later, Tigano *et al.* (2005), Plantard *et al.* (2007), Holterman *et al.* (2009), Kiewnick *et al.* (2014) and Janssen *et al.* (2017), using the same gene but with more species, confirmed the presence of these three major clades within *Meloidogyne*. Several other genes were also successfully used for reconstruction of *Meloidogyne* single gene phylogeny: the ITS rRNA (De Ley *et al.*, 1999; Castillo *et al.*, 2003a, 2009a; Landa *et al.*, 2008; McClure *et al.*, 2012; Trisciuzzi *et al.*, 2014; Ali *et al.*, 2015; Tao *et al.*, 2017; Archidona-Yuste *et al.*, 2018; Trinh *et al.*, 2018); the D2-D3 of 28S rRNA (Castillo *et al.*, 2003a, 2009; Tenente *et al.*, 2004; Landa *et al.*, 2008; McClure *et al.*, 2012; Kiewnick *et al.*, 2014; Trisciuzzi *et al.*, 2014; Ali *et al.*, 2015; Tao *et al.*, 2017; Archidona-Yuste *et al.*, 2018; Trinh *et al.*, 2018); the RNA polymerase II gene (*rpb1*) (Lunt, 2008; Rybarczyk-Mydlowska *et al.*, 2014); the region between the *COII* and 16S rRNA genes of mtDNA (McClure *et al.*, 2012; Humphreys-Pereira *et al.*, 2014; Ali *et al.*, 2015; Tao *et al.*, 2017; Archidona-Yuste *et al.*, 2018; Trinh *et al.*, 2018); *hsp90* (Nischwitz *et al.*, 2013); *COI* (Kiewnick *et al.*, 2014; García & Sánchez-Puerta, 2015; Janssen *et al.*, 2017; Archidona-Yuste *et al.*, 2018; Powers *et al.*, 2018; Trinh *et al.*, 2018); IGS rRNA (Onkendi & Moleleki, 2013), and *COII* (Kiewnick *et al.*, 2014). Several authors also reconstructed the multigene *Meloidogyne* phylogeny using supertree (Adams *et al.*, 2009) or supermatrix (Brito *et al.*, 2015; Blanc-Mathieu *et al.*, 2017; Janssen *et al.*, 2017) approaches for better resolution for some groups of species relationships.

Álvarez-Ortega *et al.* (2019) reconstructed a multigene tree from the dataset containing four gene sequences of 56 *Meloidogyne* species, which represents more than half the valid species of RKN. The phylogenetic analysis and several published phylogenies of *Meloidogyne* clearly revealed that not all species can be distributed in the three major clades. In this study Álvarez-Ortega *et al.* (2019) proposed a grouping and a new clade and group numbering as an attempt to allocate all studied species. The RKN species can be distributed among 11 mainly highly or moderately supported clades or groups; five groups compose the superclade with 75% of the studied species (Fig. 18). The superclade contained the clade I and the clade III *sensu* Tandingan De Ley *et al.* (2002). Species from the former clade II were designated into new clades II, IV, VI and VII as proposed by Álvarez-Ortega *et al.* (2019).

The evolutionary trends of the RKN that compose the superclade have been characterised by Tandingan De Ley *et al.* (2002), Tigano *et al.* (2005) and Holterman *et al.* (2009). After adding more taxa Álvarez-Ortega *et al.* (2019) were able to characterise these observed patterns more robustly.

Molecular group I, or clade I, includes *Meloidogyne* species distributed in warmer climates and contains *M. incognita*, *M. javanica*, *M. arenaria* and 17 other species that are commonly referred to as the tropical RKN complex (Incognita and Ethiopica groups). Three species of the Incognita group, *M. arenaria*, *M. incognita* and *M. javanica*, are globally distributed, polyphagous pests of many agricultural crops. *Meloidogyne phaseoli* and *M. morocciensis* are found in Brazil and North Africa, whereas *M. floridensis* is known only from USA. *Meloidogyne arabicida*, *M. izalcoensis*, *M. lopezi*, *M. paranaensis* and *M. konaensis* parasitise coffee tree and other dicotyledons in North, Central and South America. *Meloidogyne thailandica*, and the representatives of the Ethiopica group (Gerič Stare *et al.*, 2019) (*M. luci*, *M. ethiopica*, *M. inornata* and *M. hispanica*), infect different dicotyledons and have been reported from Asia, Africa, South America and Europe.

It has been shown by nuclear genome sequencing that *M. incognita*, *M. javanica* and *M. arenaria* contain divergent copies of many loci. The different evolutionary histories of these copies, likely arising *via* historical hybridisation and genome duplications, complicate both phylogenetic analyses and species identifications that rely on nuclear gene sequences (Hugall *et al.*, 1999; Lunt, 2008; Szitenberg *et al.*, 2017). Álvarez-Ortega *et al.* (2019) presented a statistical parsimony network showing the phylogenetic relationships between the ITS rRNA gene haplotypes of the Incognita, Ethiopica groups and *M. enterolobii*.

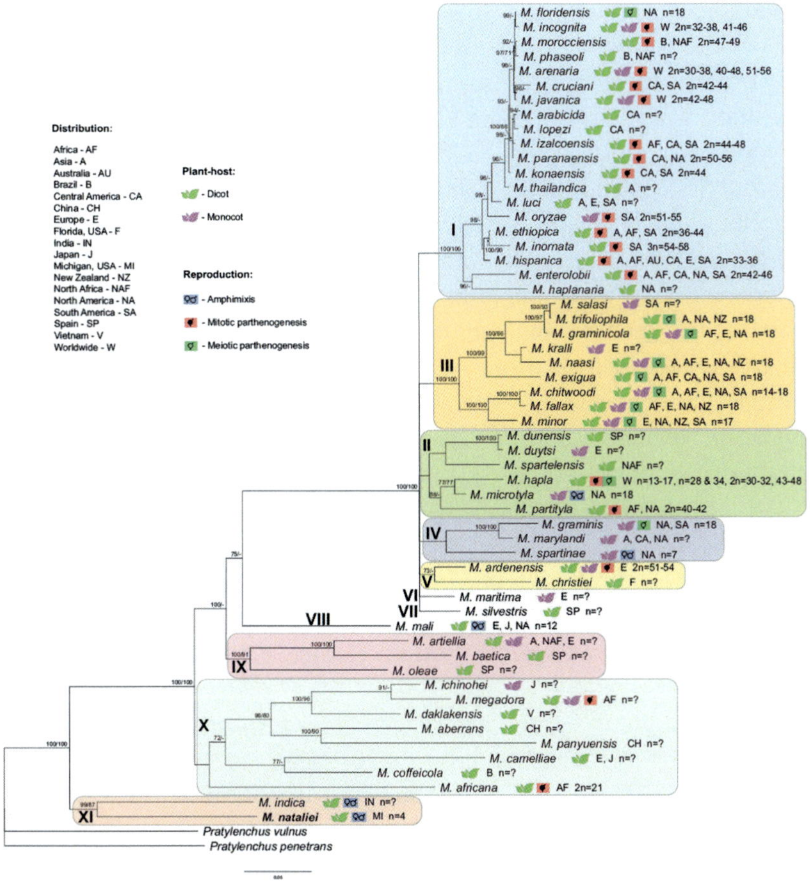

Fig. 18. *Bayesian 50% majority rule consensus tree for* Meloidogyne *spp. as inferred from 18S rRNA, ITS1 rRNA, D2-D3 expansion segments of 28S rRNA,* COI *gene and* COII*-16S rRNA sequence alignment under the GTR + I + G model. Branch support of over 70% is given for appropriate clades and it is indicated as: posterior probabilities value in Bayesian inference analysis/bootstrap value from maximum-likelihood analysis. Clade number, or molecular group number is also given. (n = ?: chromosome number information unknown). (After Álvarez-Ortega* et al., *2019.)*

This ITS rRNA gene fragment does not allow for the discrimination of the species within the Incognita, Ethiopica groups, but only the groups themselves. This analysis also confirmed a reticular character of rRNA gene evolution. Diagnostics of the majority of the species from clade I are mainly based on mtDNA genes, namely, *nad5*, the region between *COII* and 16S of mtDNA or isozymes. All species of this clade, for which the reproduction mode is known, are polyploids and exclusively comprise mitotic parthenogenetic species, except for the meiotic parthenogenetic *M. floridensis*.

Molecular group II, or clade II, contains six species that parasitise moncotyledons and dicotyledons. It includes *M. microtyla* from North America and *M. hapla*, which is a major pest of many crops worldwide. The first species has different modes of reproduction: mitotic parthenogenesis and amphimixis, whereas *M. hapla* has two reproduction modes: meiotic and mitotic parthenogenesis depending on race. *Meloidogyne partityla*, from Africa and North America, reproduces by mitotic parthenogenesis, whereas *M. spartelensis*, *M. duytsi*, and *M. dunensis* have unknown reproductive strategies and are presently found in Europe or North Africa.

Molecular group III, or clade III, contains nine species parasitising monocotyledons (Graminicola group: *M. graminicola*, *M. ottersoni*, *M. salasi*, *M. aegracyperi*, and *M. trifoliophila*) and dicotyledons and is primarily distributed in several continents, except for *M. salasi* found in South America and *M. kralli* from Europe. Species from this group are exclusively meiotic parthenogenetic.

Molecular group IV, or clade IV, contains *M. spartinae*, *M. marylandi* and *M. graminis* and its host range seems to be limited to the Poaceae only. These species are likely have a North American origin, although *M. graminis* was reported from Europe and South America and *M. marylandi* is also found in Asia. *Meloidogyne spartinae* is an amphimictic species, whereas *M. graminis* is exclusively meiotic parthenogenetic.

Molecular group V, or clade V, includes *M. christiei*, which has been found only parasitising Turkey oak roots, and *M. ardenensis*. *Meloidogyne christiei* forms small galls involving the tissues immediately surrounding the nematode. Although the mode of reproduction for this species is unknown, the presence of many males in populations may indicate amphimixis.

Both molecular groups VI and VII, or clades VI and VII, are monotypic; VI contains *M. maritima* that infects *Ammophila arenaria* (L.) Link on dunes at Perranporth, Cornwall, UK, while VII is represented by *M. silvestris*, which parasitises the roots of European holly, *Ilex aquifolium* L., in Soria province, Spain.

Molecular group VIII, or clade VIII, only includes the amphimictic species *M. mali* which parasitises trees and woody plants of the genera *Ulmus*, *Euonymus*, *Acer* and others in Japan. This species was likely introduced to North America and Europe from Asia.

Molecular group IX, or clade IX, includes *M. artiellia*, *M. baetica* and *M. oleae*. The first species was found in Europe, Asia and North Africa, parasitising moncotyledons and dicotyledons, whereas the other two species are reported in Spain from olive trees and other plants. Mode of reproduction and chromosome number for this group are still not described.

Molecular group X, or clade X, includes eight species: *M. aberrans*, *M. panyuensis*, *M. camelliae*, *M. ichinohei*, *M. daklakensis*, from East Asia, *M. megadora* and *M. africana* from Africa, and *M. coffeicola*, from Brazil, which parasitises coffee trees. These species parasitise moncotyledons and dicotyledons. The reproductive mode is only known for two representatives of this clade: *M. megadora* and *M. africana* both exhibit mitotic parthenogenesis. *Meloidogyne africana* has the lowest number of chromosomes (2n = 21) of RKN known to reproduce by mitotic parthenogenesis (Janssen *et al.*, 2017).

Finally, molecular group XI, or clade XI, represents the earliest branching lineage of RKN and includes two species, *M. indica* and *M. nataliei*.

Based on analysis of many morphological characters, Jepson (1987) proposed 12 morphological groups within the J2, six groups of perineal patterns within females, and seven groups based on labial region morphology in males. She placed all studied species in several *Meloidogyne* groups named according to the oldest described species within them, namely, '*graminis*', '*acronea*', '*exigua*', '*nataliei*' and others (Table 12). Although none of these groupings fits exactly with the molecular groups proposed in the present study, there are some patterns of overlap (Álvarez-Ortega *et al.*, 2019).

Evolutionary Relationships of *Meloidogyne nataliei* and *M. indica* with Other Root-knot Nematodes

The phylogenetic trees obtained by Álvarez-Ortega *et al.* (2019) from the D2-D3 of 28S and ITS rRNA gene datasets and the multigene tree showed that *M. nataliei* clustered with *M. indica* in a highly-supported clade. Both species are morphologically similar. Jepson

Table 12. *Grouping of some* Meloidogyne *species.*

Species	After Jepson (1987)				Molecular clade, group
	J2 group	Female group	Male group	Morphology and host group	
M. aberrans	-	-	-	-	X
M. acronea	2	3	7	*acronea*	-
M. aegracyperi	-	-	-	-	III, Graminicola group
M. africana	3	3	7	*acronea*	X
M. aquatilis	8	3	7	*graminis*	-
M. arabicida	-	-	-	-	I, Incognita group
M. ardenensis	4	6	7	*exigua*, parasitising the Oleaceae	V
M. arenaria	7	4	7	*exigua*, parasitising herbaceous host, group 2	I, Incognita group
M. artiellia	2	1	1	*exigua*, parasitising herbaceous host, group 1	IX
M. baetica	-	-	-	-	IX
M. brevicauda	1	1	1/5	*exigua*, parasitising *Camellia* spp.	-
M. camelliae	5	1	1	*exigua*, parasitising *Camellia* spp.	X
M. carolinensis	6	6	7	-	-
M. chitwoodi	6	6	5	*exigua*, parasitising herbaceous host, group 1	III
M. chosenia	-	-	-	-	-
M. christiei	-	-	-	-	V

Table 12. *(Continued.)*

Species	After Jepson (1987)				Molecular clade, group
	J2 group	Female group	Male group	Morphology and host group	
M. coffeicola	5	2	7	*exigua*, parasitising *Coffea* spp.	X
M. cruciani	7	5	7	*exigua*, parasitising herbaceous host, group 2	I, Incognita group
M. daklakensis	-	-	-	-	X
M. dunensis	-	-	-	-	II
M. duytsi	-	-	-	-	II
M. enterolobii	10	4	1	*exigua*, parasitising tree hosts	I
M. ethiopica	5	3	7	*exigua*, parasitising herbaceous host, group 2	I, Ethiopica group
M. exigua	7	1	7	*exigua*, parasitising *Coffea* spp.	III
M. fallax	-	-	-	-	III
M. floridensis	-	-	-	-	I, Incognita group
M. graminicola	11	3	5	*graminis*	III, Graminicola group
M. graminis	8	5	7	*graminis*	IV
M. hapla	8	1	7	*exigua*, parasitising herbaceous host, group 2	II
M. haplanaria	-	-	-	-	I
M. hispanica	-	-	-	-	I, Ethiopica group
M. ichinohei	-	-	-	-	X

Table 12. *(Continued.)*

Species	After Jepson (1987)				Molecular clade, group
	J2 group	Female group	Male group	Morphology and host group	
M. incognita	6	6	6	*exigua*, parasitising herbaceous host, group 2	I, Incognita group
M. indica	1	2	1	*exigua*, parasitising tree hosts	XI
M. inornata	-	-	-	-	I, Ethiopica group
M. izalcoensis	-	-	-	-	I, Incognita group
M. javanica	7	5	3	*exigua*, parasitising herbaceous host, group 2	I, Incognita group
M. kikuyensis	2	5	1/5/7	*exigua*, parasitising herbaceous host, group 1	-
M. konaensis	-	-	-	-	I, Incognita group
M. kralli	10	3	5	*graminis*	III
M. lopezi	-	-	-	-	I, Incognita group
M. luci	-	-	-	-	I, Ethiopica group
M. mali	2	4	7	*exigua*, parasitising tree hosts	VIII
M. maritima	8	5	7	*graminis*	VI
M. marylandi	8	3	-	*graminis*	IV
M. megadora	9	3	7	*acronea*	X
M. megatyla	6	6	7	*exigua*, parasitising tree hosts	-

Table 12. *(Continued.)*

Species	After Jepson (1987)				Molecular clade, group
	J2 group	Female group	Male group	Morphology and host group	
M. microcephala	9	-	7	*exigua*, parasitising herbaceous host, group 2	-
M. microtyla	3	5	7	*exigua*, parasitising herbaceous host, group 1	II
M. minor	-	-	-	-	III
M. morocciensis	-	-	-	-	I, Incognita group
M. naasi	11	3	3	*graminis*	III
M. nataliei	1	2	2	*nataliei*	XI
M. oleae	-	-	-	-	IX
M. oryzae	11	3	5	*graminis*	I, Incognita group
M. ottersoni	11	3	4	*graminis*	III, Graminicola group
M. ovalis	-	3	7	*exigua*, parasitising tree hosts	-
M. panyuensis	-	-	-	-	X
M. paranaensis	-	-	-	-	I, Incognita group
M. partityla	-	-	-	-	II
M. phaseoli	-	-	-	-	I, Incognita group
M. platani	7	4	7	*exigua*, parasitising tree hosts	-
M. propora	1	2	4	*exigua*, parasitising herbaceous host, group 1	-

Table 12. *(Continued.)*

Species	After Jepson (1987)				Molecular clade, group
	J2 group	Female group	Male group	Morphology and host group	
M. querciana	4	6	7	*exigua*, parasitising tree hosts	-
M. salasi	-	-	-	-	III, Graminicola group
M. sewelli	10	1	4	*graminis*	-
M. silvestris	-	-	-	-	VII
M. spartelensis	-	-	-	-	II
M. spartinae	12	3	7	*graminis*	IV
M. subarctica	4	3	1	*exigua*, parasitising herbaceous host, group 1	-
M. tadshikistanica	7	4	6	*exigua*, parasitising herbaceous host, group 2	-
M. thailandica	-	-	-	-	I, Incognita group
M. trifoliophila	-	-	-	-	III, Graminicola group
M. vitis	-	-	-	-	VIII

(1987) distinguished 12 morphological groups within J2 of *Meloidogyne* and placed *M. nataliei*, *M. indica*, *M. brevicauda* and *M. propora* in group 1 (Table 13). This group is characterised by a tapering J2 tail terminus with a very broad and rounded tip. Jepson (1987) noticed that there were small qualitative differences between species within the group, but they were not easily defined in practice. *Meloidogyne indica* and *M. brevicauda* were reported from the Indian subcontinent, and *M. propora* was found in the Outer Islands of the Seychelles, an archipelago in the Indian Ocean off East Africa. If these three species are from the tropics, *M. nataliei*, by contrast, is from Michigan, with a humid continental climate, and one of the coldest regions in the USA. Álvarez-Ortega *et al.* (2019) hypothesise that *M. nataliei* is an invasive species for Michigan, and was likely introduced with its plant host, *Vitis labrusca* L., from south-eastern USA, which is a centre of grapevine diversity. *Vitis labrusca* is one of 17 native species that originated from this region (Wan *et al.*, 2013). Perhaps future intensive nematological surveys in south-eastern USA regions may reveal new localities for *M. nataliei.*

Álvarez-Ortega *et al.* (2019) suggested that *M. nataliei* together with *M. indica* may represent primitive species among other RKN. Molecular results rejected a hypothesis by Goldstein & Triantaphyllou (1986), who suggested that the Michigan grape RKN might not belong to the same phyletic group as *Meloidogyne*. Phani *et al.* (2018) provided a molecular characterisation of *M. indica* and noted that this RKN species should be considered the most ancestral taxon of the genus, based on molecular data.

Meloidogyne nataliei is a diploid amphimictic species that has a haploid complement of only four chromosomes (Triantaphyllou, 1985b). The four chromosomes of *M. nataliei* are relatively larger, being quite different from those other *Meloidogyne* species. Moreover, *M. nataliei* reproduces exclusively by cross-fertilisation. Thus, our results confirmed the hypothesis of Triantaphyllou (1985b, 1987b) that amphimictic *Meloidogyne* species with a small chromosome number are closer to the ancestry of the genus, from which the mitotic parthenogenetic RKN species have evolved.

Ryss (1988) considered pratylenchids as the most closely related to *Meloidogyne* by details of the labial region and pharyngeal structure and also believed these morphological similarities to be indicative of common ancestry between Meloidogynidae and Pratylenchidae. A close relationship of *Meloidogyne* with Pratylenchidae was shown by Subbotin *et al.* (2006), who presented a molecular phylogeny reconstructed based on the D2-D3 of 28S rRNA gene sequences. Using 18S rRNA gene for reconstruction of the tylenchid phylogeny, Holterman *et al.* (2009) also suggested that that RKN have evolved from a *Pratylenchus*-like ancestor. It has been noticed that the morphology of the labial region of *M. nataliei* J2 and males, being heavily sclerotised, resembles the *Pratylenchus* labial region pattern. The strong stylet and tail with a widely rounded terminus in *M. nataliei* J2 are also similar to those of some *Pratylenchus*. Therefore, these features may represent a plesiomorphic condition in *Meloidogyne*.

In the phylogenetic trees *Meloinema* clustered with *Meloidogyne*, in a basal position and more closely with *Meloidogyne indica* and *M. nataliei*. The maximum likelihood testing does not exclude that these two genera are sister taxa. Molecular analysis also suggested that a sedentary parasitism has independently appeared twice in the Meloidogynidae + Pratylenchidae clade: *i*) the *Meloidogyne* + *Meloinema* lineage; and *ii*) the *Nacobbus* lineage (Subbotin & Kim, 2021).

Evolutionary Trends within the Root-knot Nematodes

Triantaphyllou (1985b) summarised the cytogenetic information about RKN and suggested that the obligate amphimictic species with n = 18 or 19 should be considered as closely related to the ancestral predecessors of *Meloidogyne* spp. He also believed that the low chromosomal numbers in most other nematodes offered support for a polyploidy origin of most *Meloidogyne* species. Janssen *et al.* (2017) concluded that the basic haploid chromosome number of the genus *Meloidogyne* could possibly be even n = 7. The placing of amphimictic *M. nataliei* with n = 4 at a basal position to all *Meloidogyne* species confirms the hypothesis on small chromosome number in ancestral species (Álvarez-Ortega *et al.*, 2019).

Castagnone-Sereno *et al.* (2013) noticed that the extensive species diversification within RKN demonstrated in terms of chromosome complement is the reflection of a complex evolution within the genus involving genome duplication, polyploidisation, introgression and hybridisation. High rate of species diversification of *Meloidogyne* observed in the superclade showed that these evolutionary processes allowed the nematodes to become extremely polyphagous and significantly increased their dispersal potential. The occurrence of parthenogenesis has been correlated in RKN with their increasing importance as crop parasites (Holterman *et al.*, 2009).

It has been suggested that the current apomictic species derived from diploid sexual ancestors and obligatory parthenogenetic mitotic species evolved from facultatively parthenogenetic meiotic species, following suppression of meiosis during oocyte maturation (Castagnone-Sereno, 2006). However, the mapping of reproduction mode on a multigene phylogenetic tree made by Janssen *et al.* (2017) and Álvarez-Ortega *et al.* (2019) did not confirm this hypothesis, but showed that the transition to mitotic parthenogenesis may have occurred earlier in RKN evolution than meiotic parthenogenesis.

Mapping of geographical reports of species on the phylogenetic tree also allows some speculation about RKN origin. The distribution of three basal groups of these nematodes in South and Eastern Asia-Africa and Africa-South America may suggest that the RKN originated from these regions (Álvarez-Ortega *et al.*, 2019).

Nematology Monographs & Perspectives, 2021, Vol. 14, 99-111

Chapter 11

Polytomous Key to the Species of *Meloidogyne*

Morphological identification of *Meloidogyne* species has been based on study of the female perineal pattern and many characters of the different life stages and was discussed in detail by Whitehead (1968), Franklin (1972, 1978), Esser *et al.* (1976), Taylor & Sasser (1978), Jepson (1987), Eisenback & Triantaphyllou (1991) and Eisenback & Hunt (2009). Given the large number of *Meloidogyne* species and their high variability in morphological and morphometric characteristics, the identification of nematodes from this group is often a very difficult process. Published literature on RKN contains some regional identification keys or keys for certain RKN species groups, but lacks an identification key to all RKN species. One of the problems in constructing such a comprehensive identification key for *Meloidogyne* spp. is the large number of valid species and lack of detailed descriptions for some species. To overcome this problem, in this book we have developed a polytomous or matrix key by utilising the major and most important differential characters for the identification of *Meloidogyne* species, allowing the separation of the species into different groups and facilitating the identification process and comparison among related species. The species have been arranged in alphabetical order in the polytomous key (Table 13). The *Meloidogyne* groupings based on morphological and plant host criteria proposed by Jepson (1987) and based on molecular datasets are given in Table 12.

Although morphology and morphometrics continue to be important for *Meloidogyne* spp. identification, biochemical and molecular methods provide more reliable, rapid and convenient solutions for their diagnostics. The development of new approaches in identification involve the integration of molecular with morphological-morphometric and host range data.

Stages and Characters Used for the Polytomous Key

FEMALE

Body length varies significantly from as short as 419 μm in *M. californiensis* to as large as 1850 μm in *M. mersa*. This character is not very useful for the majority of species having a similar body length. Within species, body length and shape may be modified by biotic and abiotic factors (Perry *et al.*, 2009). **Stylet length** is available for all species and varies from a mean value of *ca* 14.5 μm in *M. fallax* to as short as 8.0 μm in *M. salasi* or as large as 25.0 μm in *M. brevicauda*. In all species, the stylet cone length is over half the total stylet length (Jepson, 1987).The morphology of the stylet in females is a species-specific character (Eisenback *et al.*, 1980; Eisenback, 1985). However, the usefulness of stylet morphology of females for routine identifications is limited only by the difficulty of specimen preparation for examination using light microscopy (LM) (Eisenback, 1985). The shape of stylet knobs and that of their junction with the shaft have important diagnostic significance as they range from offset, small and round (*M. hapla*), offset, transversely elongate (*M. javanica*), not offset, posteriorly sloping, merging with shaft (*M. arenaria*) (Eisenback, 1985; Jepson, 1987), but there is no information for some species. The **excretory pore position in the female in relation to the stylet length (EP/ST)** in *Meloidogyne* varies from a mean value of *ca* 2.0 in *M. luci*, to as low as 0.4 in *M. chosenia* or as high as 5.0 in *M. tadshikistanica*. This character is very useful as an identification aid as many species can be separated in groups based upon this character (Taylor, 1987). The most characteristic morphological feature of *Meloidogyne* is the **perineal pattern**, located at the posterior body region of adult females (Figs 19, 20). The perineal pattern is a unique configuration on the surface of the cuticle in the posterior body area comprising vulva-anus area, tail terminus, phasmids, lateral lines and surrounded by cuticular striae (Franklin, 1965b; Hirschmann, 1985; Karssen & van

Fig. 19. *Basic morphology of the perineal pattern of* Meloidogyne *females. A: After Esser* et al. *(1976); B: After Eisenback* et al. *(1981); C: After Jepson (1987). Zone 1 = area in pattern centre containing perineum and adjacent area usually free of continuous striae; 2. Zone 2 = striated area just below (anterior to) vulva; Zone 3 = striated area lateral to but bounded by perineum; Zone 4 = striated area above (posterior to) anus and tail area. P = perineum, T = tail area.*

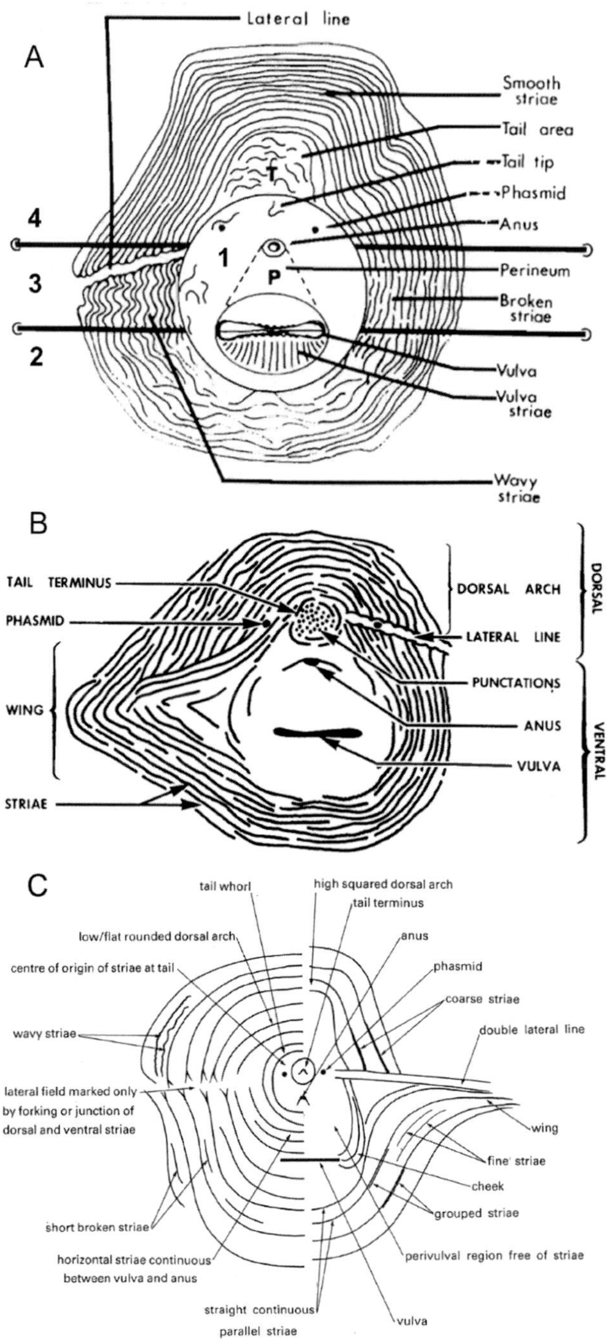
A
Lateral line
Smooth striae
Tail area
Tail tip
Phasmid
Anus
Perineum
Broken striae
Vulva
Vulva striae
Wavy striae
T
P
1
2
3
4
B
TAIL TERMINUS
PHASMID
WING
STRIAE
DORSAL ARCH
LATERAL LINE
PUNCTATIONS
ANUS
VULVA
DORSAL
VENTRAL
C
tail whorl
high squared dorsal arch
tail terminus
low/flat rounded dorsal arch
anus
centre of origin of striae at tail
phasmid
coarse striae
wavy striae
double lateral line
lateral field marked only by forking or junction of dorsal and ventral striae
wing
fine striae
cheek
grouped striae
short broken striae
horizontal striae continuous between vulva and anus
perivulval region free of striae
straight continuous parallel striae
vulva

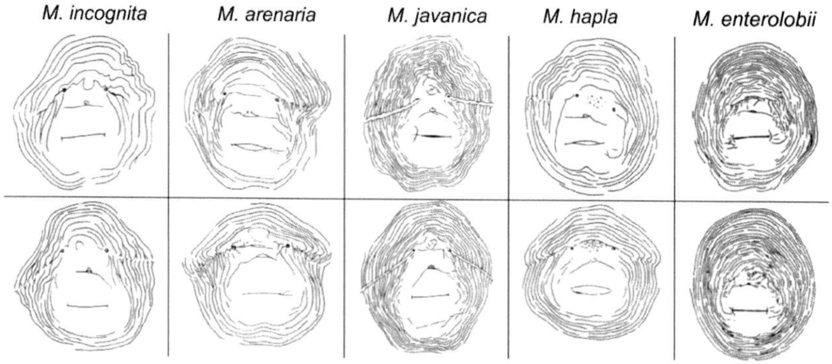

Fig. 20. *Perineal pattern morphology of five main* Meloidogyne *species (Anon., 2016c).*

Aelst, 2001). Müller (1884) observed and illustrated the perineal pattern for the first time. Chitwood (1949) introduced it as an identification character for a few RKN. Illustrations of perineal patterns have been added to every species description since then (Karssen & van Aelst, 2001). Several authors (Taylor *et al.*, 1955; Whitehead, 1968; Esser *et al.*, 1976; Jepson, 1987) compared in detail morphological differences between perineal patterns of many RKN. Esser *et al.* (1976) proposed that, for the more effective utilisation of the features of the posterior cuticular pattern, it should be divided into four definitive zones. Zone 1 includes the perineum, vulva, anus and phasmids. It represents a roughly circular area in the centre of the pattern usually free of continuous striae. Striae of Zone 1 are usually few, broken, and scattered. Zone 2 is the area under the vulva, and specifically refers to the mass or band of striae directly below the perineum. Zone 3 encompasses the group of striae lateral to the perineum including the lateral lines. Zone 4 is marked by a mass of arch-like striae anterior to the anus and contains a striated circular area with the tail and its tip. Analysis of patterns revealed that certain characteristics of each zone could be of value in differentiating patterns (Esser *et al.*, 1976). Based on general perineal pattern morphology, Jepson (1987) proposed to divide *Meloidogyne* species into six distinct groups and this categorisation is used in the polytomous key (Fig. 21).

Male

Body length varies from a mean value of *ca* 1475 μm in *M. nataliei*, to as short as 820 μm in *M. aquatilis* or as large as 2400 μm in *M.*

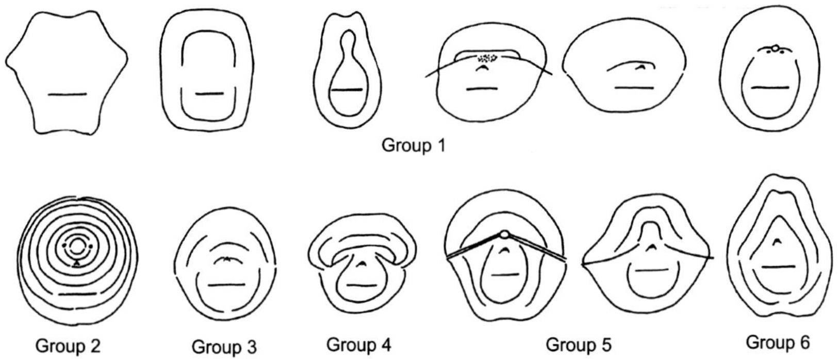

Fig. 21. *Perineal pattern grouping of root-knot nematodes according to Jepson (1987).*

mersa. This character is not very useful for the majority of species having a rather similar body length. As can be seen in Table 13, it is difficult to separate groups of species based upon this character, but it may help together with other characters. **Stylet length** varies from a mean value of *ca* 20.5 μm in *M. megadora*, to as short as 12.2 μm in *M. fanzhiensis* or as large as 28.9 μm in *M. nataliei*. This character shows a broad range in the genus (Jepson, 1983b, 1987). Overall shape of the knobs may be an important character for species identification. Measurements may help identification in some instances, particularly in those species with much larger knobs than most *Meloidogyne* (*e.g.*, *M. megadora*, *M. megatyla* and *M. acronea*) (Jepson, 1983b), but for many species these data are not available. Similarly, the DGO distance exhibits much variation within species and ranges from 2.0 to 13.0 μm; in general, this character cannot distinguish species and is not used for species differentiation in this polytomous key. **Spicule length** varies from a mean value of *ca* 30.5 μm in *M. enterolobii* to as short as 21.2 μm in *M. ottersoni* or as large as 43.0 μm in *M. nataliei*. Although this character shows a continuous range, it is easy to measure and might be useful when combined with other male characters. The lateral fields extend from the anterior end to the tail and usually have four equidistant straight lines forming three bands or ridges of the same width, although up to 15 lines may occur. The **number of lines in the lateral field** and presence or absence of **areolation** in the outer bands may also help in identification of *Meloidogyne*. The **labial region** may also be useful for identification based upon the type of junction of the labial region and the body, shape of labial annulus, number of post-labial annuli, and general shape of anterior end in relation to body (Jepson, 1983c, 1987). In the majority of species it is offset from the body contour, but in some it is continuous with the body contour. However, its use is limited since it is not clearly defined for many species. Initially, it was used to differentiate *M. arenaria*, *M. incognita*, *M. javanica* and *M. hapla* (Eisenback & Hirschmann, 1981; Jepson, 1987), but on the basis of the available information we only recognise two groups based on the junction with body contour.

SECOND-STAGE JUVENILE

Body length varies from a mean value of *ca* 437 μm in *M. enterolobii*, to as short as 320 μm in *M. kikuyensis* or as large as 773 μm in *M. spartinae*. Body length shows a continuous gradation among species. **Stylet length** ranges from 7.5 μm in *M. minor* to 23.0 μm in *M. nataliei*. This character shows low variability among species, although it may be helpful in identifying certain species. Jepson (1983b) believed that in many specimens the labial skeleton obscures the end of the stylet and stylet length may therefore be underestimated. She proposed using the method of Hooper (1977) for stylet observation. Jepson (1983a, 1987) distinguished three most important J2 tail characters for species identification: tail length, hyaline region and tail shape, and proposed separating all species into 12 groups. Using a similar approach, Ghaderi & Karssen (2020) recently distinguished eight groups. However, the group boundaries are rather complicated due to the high variability of these characters and compounded by the increasing number of species descriptions. For this reason, only a few characters: **tail length**, **hyaline region length**, **c** and **c′ ratio** for species identification were selected for the polytomous key. The tail length varies from a mean value of *ca* 50 μm in *M. incognita*, to as short as 11 μm in *M. mingnanica* or as long as 100 μm in *M. spartinae*. The hyaline region varies from a mean value of *ca* 12.5 μm in *M. coffeicola*, to as short as 3.5 μm in *M. acronea* or as long as 28.0 μm in *M. spartelensis*. The c and c′ ratios vary from a mean value of *ca* 9.4 and 4.8 in *M. paranaensis* and *M. christiei*, respectively, to as low as 5.2 and 1.5 in *M. subarctica* and *M. propora*, respectively, or as high as 24.9 and 9.0 in *M. indica* and *M. spartinae*, respectively.

The polytomous key has been completed for 98 valid species using the available original descriptions and others. Some values not given by the author have been obtained from published drawing or photos of specimens.

Morphometric and Morphological Characters Used to Distinguish *Meloidogyne* Spp. in the Polytomous Key

FEMALE

A) Body length
- **Group 1:** >850 μm
- **Group 2:** 600-849 μm
- **Group 3:** <600 μm

B) Stylet length
- **Group 1:** >17.5 μm
- **Group 2:** 14.5-17.5 μm
- **Group 3:** 12.0-14.4 μm
- **Group 4:** <12.0 μm

C) EP/ST ratio
- **Group 1:** 3.5-5.0
- **Group 2:** 2.0-3.4
- **Group 3:** 1.0-1.9
- **Group 4:** <1.0

D) Perineal pattern morphology (Fig. 21)

Group 1: Species with very distinct patterns and unlike any other species (*M. artiellia*, *M. baetica*, *M. brevicauda*, *M. camelliae*, *M. lusitanica*, *M. partityla* and others).

Group 2: Species with patterns having the centre of origin of striae at the tail and with more or less continuous striae between the vulva and anus; the perivulval region is free of striae (*M. coffeicola*, *M. indica*, *M. inornata*, *M. kongi*, *M. nataliei* and others).

Group 3: Species with round and oval patterns without marked lateral lines (*M. africana*, *M. ethiopica*, *M. graminicola*, *M. minor*, *M. naasi*, *M. trifoliophila* and others).

Group 4: Species with round and oval patterns with a flat dorsal arch, striae very close together, either with a 'shoulder' where the dorsal and ventral striae meet at the lateral field, or having an inner lateral line region with either irregular closely looped striae or appearing as a raised ridge (*M. arenaria*, *M. enterolobii*, *M. mali* and others).

Group 5: Species with patterns with double lateral lines or very prominent single lines (*M. javanica*, *M. graminis*, *M. hispanica*, *M. microtyla* and others).

Group 6: Species with oval patterns with a high squared dorsal arch (*M. incognita*, *M. chitwoodi*, *M. megatyla*, *M. paranaensis* and others).

MALE

A) Body length
Group 1: >1800 μm
Group 2: 1550-1800 μm
Group 3: 1125-1549 μm
Group 4: <1125 μm

B) Stylet length
Group 1: >24.0 μm
Group 2: 20.5-24.0 μm
Group 3: 17.5-20.4 μm
Group 4: <17.5 μm

C) Spicule length
Group 1: >34.5 μm
Group 2: 26.5-34.5 μm
Group 3: <26.5 μm

D) Number of lines in lateral field
Group 1: 4
Group 2: >4

E) Lateral field areolation
Group 1: areolated
Group 2: not areolated

F) Labial region shape
Group 1: continuous with body
Group 2: offset

SECOND-STAGE JUVENILES

A) Body length
Group 1: >500 μm
Group 2: 400-500 μm
Group 3: <400 μm

B) Stylet length
Group 1: >14.0 μm
Group 2: 11.5-14.0 μm
Group 3: <11.5 μm

C) Tail length
Group 1: >69.5 μm
Group 2: 60.0-69.5 μm
Group 3: 35.0-59.9 μm
Group 4: <35.0 μm

D) Hyaline region length
Group 1: >15.5 μm
Group 2: 12.1-15.5 μm
Group 3: 8.5-12.0 μm
Group 4: <8.5 μm

E) c ratio
Group 1: >12.5
Group 2: 8.5-12.5
Group 3: <8.5

F) c′ ratio
Group 1: >6.2
Group 2: 5.1-6.2
Group 3: 3.5-5.0
Group 4: <3.5

Table 13. *Polytomous key for the identification of* Meloidogyne *spp. based on females, males and second-stage juveniles arranged alphabetically.*

	Female				Male						Second-stage juvenile					
Species	A	B	C	D	A	B	C	D	E	F	A	B	C	D	E	F
1. *M. aberrans*	*12	23	2	3	12	3	213	2	1	2	2	1	3	4	23	23
2. *M. acronea*	213	34	1	3	324	34	213	1	1	1	23	32	3	4	23	3
3. *M. aegracyperi*	3	3	3	3	43	4	3	1	-	1	23	32	12	1	3	1
4. *M. africana*	23	32	2	3	324	43	23	1	2	1	213	23	3	324	213	3
5. *M. aquatilis*	21	43	3	3	4	4	3	1	1	2	2	32	23	3	32	2
6. *M. arabicida*	213	34	2	6	3124	324	312	1	1	1	23	312	32	213	23	32
7. *M. ardenensis*	23	21	4	6	324	234	21	1	1	1	23	23	34	23	2	34
8. *M. arenaria*	213	2	2	4	213	23	2	1	1	2	2	3	213	324	3	2
9. *M. artiellia*	2	32	3	1	43	3124	23	1	2	1	3	12	4	4	1	4
10. *M. baetica*	12	21	4	1	123	43	213	1	1	2	23	2	3	32	3	2
11. *M. brevicauda*	12	12	3	1	34	2	21	1	1	2	1	1	4	4	1	4
12. *M. californiensis*	3	312	3	3	3124	213	213	1	1	2	12	23	1	1	3	1
13. *M. camelliae*	12	21	2	1	12	2	12	1	1	2	12	23	3	4	2	3
14. *M. caraganae*	213	12	4	3	3	32	2	1	1	2	2	12	43	3	12	4
15. *M. carolinensis*	32	2	3	6	312	324	21	1	1	2	21	23	3	2	21	3
16. *M. chitwoodi*	32	43	3	6	423	3	23	1	1	2	32	3	3	32	23	4
17. *M. chosenia*	32	21	4	3	4	32	32	1	1	1	3	23	3	3	32	43
18. *M. christiei*	23	32	3	6	34	34	3	1	2	2	23	32	3	32	2	3
19. *M. citri*	2	2	2	3	1	1	1	1	1	1	2	2	3	1	2	3
20. *M. coffeicola*	1	21	3	2	32	12	32	1	1	2	32	3	4	2	21	4

Table 13. *(Continued.)*

	Female				Male						Second-stage juvenile					
Species	A	B	C	D	A	B	C	D	E	F	A	B	C	D	E	F
21. *M. cruciani*	213	324	2	5	32	213	21	1	2	2	2	32	3	3	2	3
22. *M. cynariensis*	2	2	3	5	-	-	-	-	-	-	32	2	3	1	3	1
23. *M. daklakensis*	23	23	3	3	34	34	32	1	2	2	3	213	34	312	32	32
24. *M. donghaiensis*	21	324	213	3	123	21	12	1	1	1	2	21	3	4	23	34
25. *M. dunensis*	32	32	3	3	312	324	12	1	2	2	2	23	213	213	3	12
26. *M. duytsi*	123	3	2	5	324	3	32	1	1	2	2	32	12	32	3	21
27. *M. enterolobii*	213	213	1	4	213	213	2	1	1	2	2	23	32	1	32	3
28. *M. ethiopica*	32	324	2	3	34	2134	21	1	1	1	23	3	3	213	23	2
29. *M. exigua*	32	324	3	1	4	3	3	1	2	1	3	3	3	213	3	2
30. *M. fallax*	32	23	3	6	34	32	23	1	1	2	23	3	3	21	32	2
31. *M. fanzhiensis*	312	43	2	6	3124	4	32	1		2	23	3	4	3	12	4
32. *M. floridensis*	213	23	2	6	324	324	213	1	1	1	3	3	3	34	23	43
33. *M. fujianensis*	213	34	3	3	213	21	12	1	2	1	23	2	3	-	2	3
34. *M. graminicola*	32	4	2	3	34	43	2	2	2	2	2	32	12	12	3	3
35. *M. graminis*	23	34	4	5	32	3	2	1	1	2	21	2	12	12	3	1
36. *M. hapla*	23	43	213	1	34	324	32	1	1	1	3	3	34	324	32	3
37. *M. haplanaria*	12	32	1	3	123	324	1	1	2	2	23	32	21	213	32	3
38. *M. hispanica*	213	32	2134	5	213	2	2	1	2	2	32	32	3	2	23	3
39. *M. ichinohei*	21	34	12	1	32	43	2	2	1	2	21	32	312	4123	23	213
40. *M. incognita*	23	32	3	6	2134	12	12	1	2	1	32	32	3	32	32	23
41. *M. indica*	213	32	3	2	34	43	32	1	2	1	23	23	4	4	1	4

Table 13. *(Continued.)*

	Female				Male						Second-stage juvenile					
Species	A	B	C	D	A	B	C	D	E	F	A	B	C	D	E	F
42. *M. inornata*	23	2	2	2	2134	213	213	1	2	2	23	23	3	23	213	3
43. *M. izalcoensis*	21	2	2	6	213	12	213	1	2	2	2	2	3	23	23	3
44. *M. javanica*	23	213	3	5	34	23	23	1	2	1	23	2	3	43	23	21
45. *M. jianyangensis*	32	32	3	1	324	213	23	1	2	2	23	12	32	43	2	1
46. *M. jinanensis*	32	324	1	5	324	32	213	1		2	213	213	3	2	32	32
47. *M. kikuyensis*	231	23	2	5	324	34	21	1	2	2	3	21	4	4	2	4
48. *M. konaensis*	123	213	2	3	312	213	23	1	2	1	12	21	312	12	32	213
49. *M. kongi*	2	23	2	2	213	2	12	1	2	2	23	21	3	324	23	3
50. *M. kralli*	3	32	4	3	43	3	32	1	2	1	2	32	21	12	3	12
51. *M. lopezi*	21	12	213	6	12	21	21	1	2	2	12	32	32	32	23	2
52. *M. luci*	23	2	2	2	2134	2	213	1	2	1	32	2	3	32	23	3
53. *M. lusitanica*	12	21	213	1	234	12	12	1	1	2	213	12	3	32	2	43
54. *M. mali*	21	23	2	4	32	32	21	1	2	2	23	21	43	423	12	34
55. *M. maritima*	23	32	3	5	34	23	213	1	2	1	23	2	23	213	3	21
56. *M. marylandi*	213	32	4	3	-	-	-	-	-	-	23	32	23	32	3	32
57. *M. megadora*	23	23	2	3	1234	32	213	1	2	1	21	23	312	123	3	1
58. *M. megatyla*	32	213	3	6	34	21	21	1	2	2	23	12	34	3	21	43
59. *M. mersa*	1	32	12	4	1	23	1	2	1	2	1	12	12	324	23	12
60. *M. microcephala*	23	32	1	5	34	23	23	1	2	1	2	3	3	1	21	34
61. *M. microtyla*	32	32	3	5	34	3	2	1	1	2	32	23	213	3	23	3
62. *M. mingnanica*	1	3	2	1	1	2	1	1	2	1	2	3	4	4	1	4

Table 13. *(Continued.)*

Species	Female A	Female B	Female C	Female D	Male A	Male B	Male C	Male D	Male E	Male F	Second-stage juvenile A	Second-stage juvenile B	Second-stage juvenile C	Second-stage juvenile D	Second-stage juvenile E	Second-stage juvenile F
63. *M. minor*	32	32	3	3	43	34	32	1	2	1	32	3	32	123	3	213
64. *M. moensi*	32	213	2	3	34	4	32	1	1	1	2	32	34	1	21	32
65. *M. morocciensis*	213	21	1	6	213	12	12	1	1	2	23	23	3	3	32	32
66. *M. naasi*	32	324	3	3	34	34	23	1	1	2	2	21	123	1	3	12
67. *M. nataliei*	12	1	4	2	32	1	1	1	2	2	1	1	4	43	1	4
68. *M. oleae*	32	3	34	3	324	43	23	1	1	2	3	23	4	43	1	4
69. *M. oryzae*	23	213	2	3	213	3	23	2	2	2	12	12	1	12	32	1
70. *M. ottersoni*	3	43	3	3	4	4	3	1	2	1	2	2	2	1	3	1
71. *M. ovalis*	23	12	3	3	32	213	2	1	2	1	3	32	34	2	32	34
72. *M. panyuensis*	23	32	2	3	12	21	213	1	1	2	23	12	32	34	3	3
73. *M. paranaensis*	23	2	1	6	1234	123	312	1	1	1	213	2	3	3	23	23
74. *M. partityla*	21	21	3	1	4123	324	2	1	1	1	23	32	32	213	2	3
75. *M. petuniae*	21	32	2	3	1234	21	21	1	1	2	32	3	3	23	3	3
76. *M. phaseoli*	23	213	1	6	2134	123	21	1	1	2	21	32	32	213	23	3
77. *M. pini*	213	213	3	2	324	23	213	1	1	2	23	213	3	2	23	3
78. *M. piperi*	213	32	2	3	-	-	-	-	-	-	23	3	32	213	23	3
79. *M. pisi*	21	23	2	5	1234	213	21	1	1	2	32	23	3	4123	32	2
80. *M. platani*	213	2	3	4	3124	213	23	1	1	2	23	2	32	23	32	3
81. *M. propora*	12	213	3	2	34	23	21	2	1	2	32	1	4	4	1	4
82. *M. querciana*	123	21	2	6	213	3	2	1	1	2	21	32	3	324	213	423
83. *M. salasi*	32	43	3	3	2134	324	32	1	1	2	21	32	213	123	3	213

Table 13. *(Continued.)*

	Female				Male						Second-stage juvenile					
Species	A	B	C	D	A	B	C	D	E	F	A	B	C	D	E	F
84. *M. sasseri*	12	32	324	5	213	32	21	1	1	2	12	21	1	1	3	1
85. *M. sewelli*	3	23	3	1	32	3	2	1	2	2	12	23	12	32	3	1
86. *M. silvestris*	3	1	43	3	12	12	21	1	1	2	12	21	3	21	21	34
87. *M. sinensis*	21	12	2	3	2134	12	21	1	2	1	12	1	213	2	32	213
88. *M. spartelensis*	23	2	32	3	32	3	23	1	1	2	23	21	12	1	3	1
89. *M. spartinae*	12	324	3	3	12	324	213	1	1	1	1	12	1	1	32	1
90. *M. subarctica*	2	32	23	3	23	34	12	1	1	2	213	12	3	4	3	2
91. *M. suginamiensis*	12	32	2	3	324	324	213	1	2	2	23	21	4	4	1	4
92. *M. tadshikistanica*	23	2	1	4	34	21	21	1	2	1	32	21	3	3	2	4
93. *M. thailandica*	213	32	213	6	34	34	213	1	1	1	21	3	23	12	32	2
94. *M. trifoliophila*	32	32	32	3	43	34	2	2	2	2	32	2	12	12	3	12
95. *M. triticoryzae*	3	3	3	3	324	34	23	1	2	1	32	2	23	1	3	21
96. *M. turkestanica*	213	32	3	3	43	34	23	1	2	2	23	12	43	4	12	4
97. *M. vandervegtei*	12	12	2	3	123	12	1	1	1	1	3	23	43	3	2	3
98. *M. vitis*	12	1234	2	4	324	324	21	1	1	2	32	21	32	213	32	324

* The first number in each code indicate the value character showing the mean, and the following numbers cover the range within each species.

- male not found.

Nematology Monographs & Perspectives, 2021, Vol. 14, 113-757

Chapter 12

Descriptions and Diagnoses of *Meloidogyne* Species

1. *Meloidogyne exigua* Göldi, 1887
(Figs 22-25)

COMMON NAME: Brazilian pyroid coffee root-knot nematode.

Meloidogyne exigua is emerging as a significant problem for tropical agriculture. This species is an important plant pest on coffee, *Coffea arabica* L., in Brazil and other countries (Campos & Villain, 2005). The nematode was first identified by Jobert (1878) as a cause of a malady of coffee roots in Brazil and was described by Göldi (1887). The description by Göldi is erroneous in many respects but is adequate to place it in the genus (Chitwood, 1949). Chitwood (1949) re-erected the genus *Meloidogyne* and briefly redescribed *M. exigua* from preserved specimens. This species is widely distributed in all coffee-growing countries of the world and causes large losses in productivity (Sasser, 1977). Hernandez *et al.* (2004) showed that it is likely to be the dominant species in Costa Rican coffee plantations. An atypical population of *M. exigua* has been recorded from the rubber tree (*Hevea brasiliensis* Muell. Arg.) in Matto Grosso State, Brazil (Santos *et al.*, 1992; Bernardo *et al.*, 2003; Fonseca *et al.*, 2003). According to Santos (1997) this species was also found on *H. brasiliensis* in Pará and Amazonas States, Brazil (Muniz *et al.*, 2009).

Descriptions and measurements of some *M. exigua* populations are given by several authors (Chitwood, 1949; Lordello & Zamith, 1958; Whitehead, 1968; Lima & Ferraz, 1985; López, 1985; Flores & López, 1989a, b, c; Eisenback & Triantaphyllou, 1991; Muniz *et al.*, 2009).

MEASUREMENTS

See Table 14.

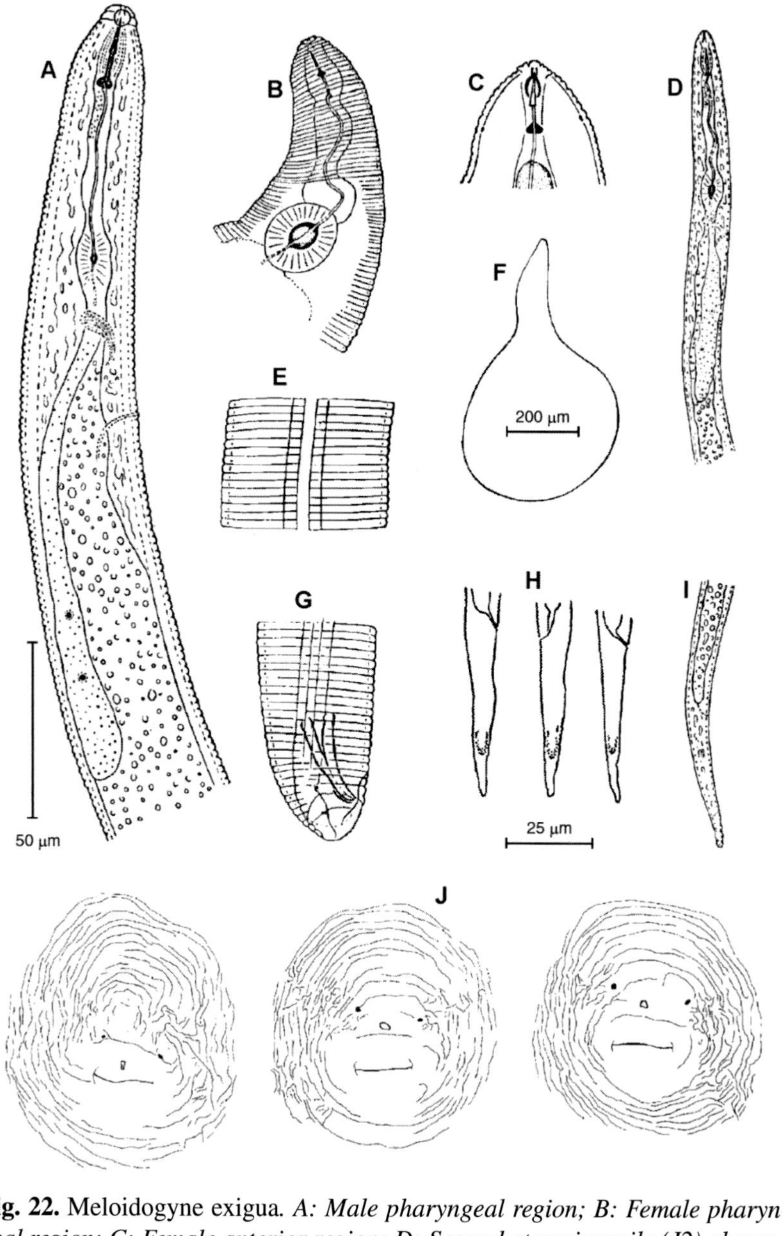

Fig. 22. Meloidogyne exigua. *A: Male pharyngeal region; B: Female pharyngeal region; C: Female anterior region; D: Second-stage juvenile (J2) pharyngeal region; E: Male lateral field; F: Entire female; G: Male tail region; H, I: J2 tail regions; J: Perineal patterns. (Scale bars: A-C, E, G = 50 μm; F = 200 μm; D, H-J = 25 μm.) (After Lordello & Zamith, 1958; Whitehead, 1968.)*

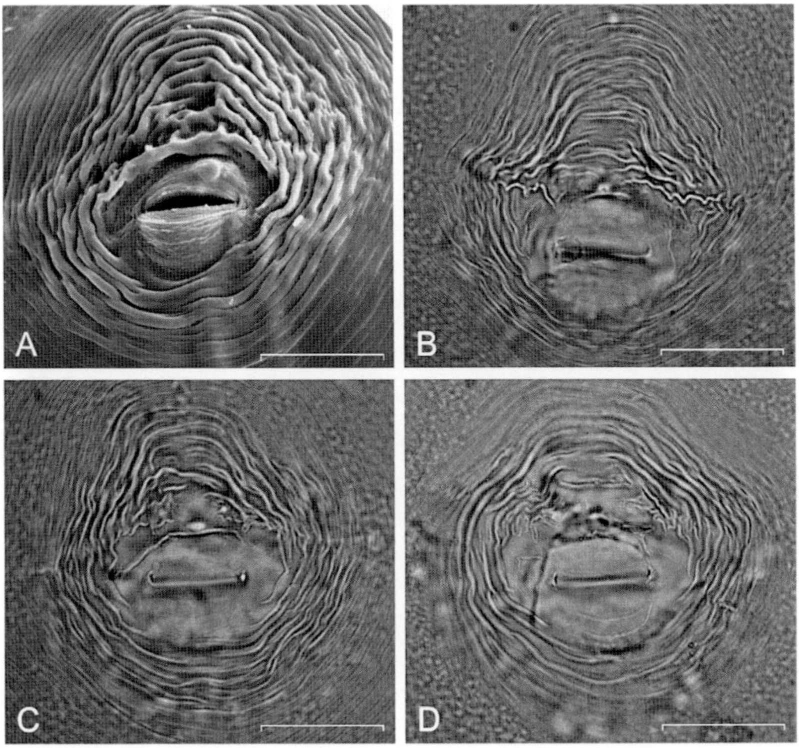

Fig. 23. *Perineal patterns of* Meloidogyne exigua *race 3 from rubber tree. A: SEM; B-D: LM. (Scale bars: A-D = 25 μm.) (After Muniz* et al.*, 2009.)*

DESCRIPTION (*AFTER WHITEHEAD, 1968; JEPSON, 1987; MUNIZ* ET AL., *2009*)

Female

Body white, variable in size, typically rounded with a defined neck varying from short to elongated. Labial disc rounded, slightly raised, fused with median and lateral labials with almost rectangular amphidial apertures located between labial disc and lateral labials. Stylet moderately slender, cone with slight dorsal curvature. Knobs transversely ovoid and somewhat offset from shaft. Posterior cuticular pattern roughly circular in shape, striae wide apart, short disordered striae at sides of pattern, dorsal arch low, rounded but also compressed dorsolaterally, lateral line not prominent, marked by wavy, broken striae, phasmids wide apart. Perivulval region free from striae and transversely ovoid. A variant of the posterior cuticular pattern has a higher dorsal arch.

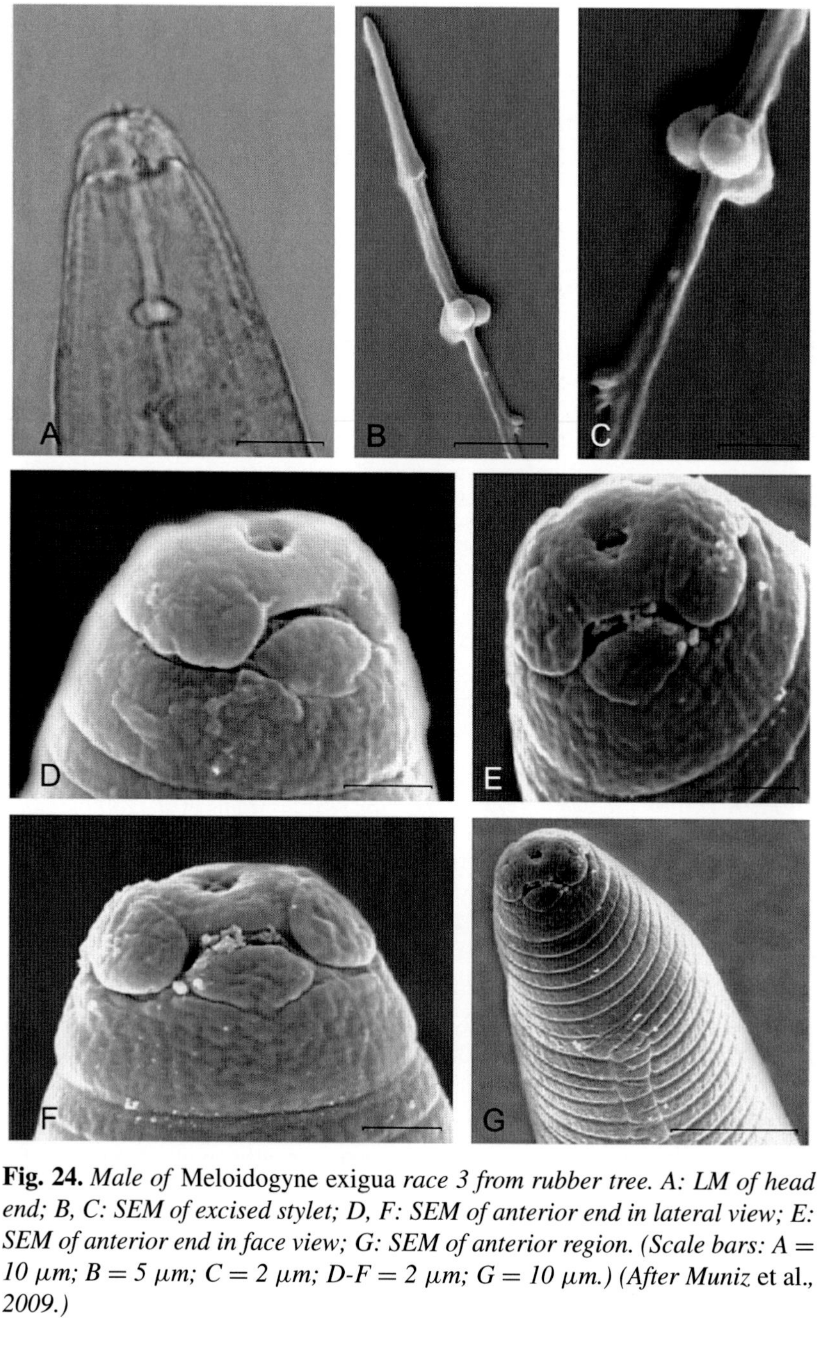

Fig. 24. *Male of* Meloidogyne exigua *race 3 from rubber tree. A: LM of head end; B, C: SEM of excised stylet; D, F: SEM of anterior end in lateral view; E: SEM of anterior end in face view; G: SEM of anterior region. (Scale bars: A = 10 μm; B = 5 μm; C = 2 μm; D-F = 2 μm; G = 10 μm.) (After Muniz* et al.*, 2009.)*

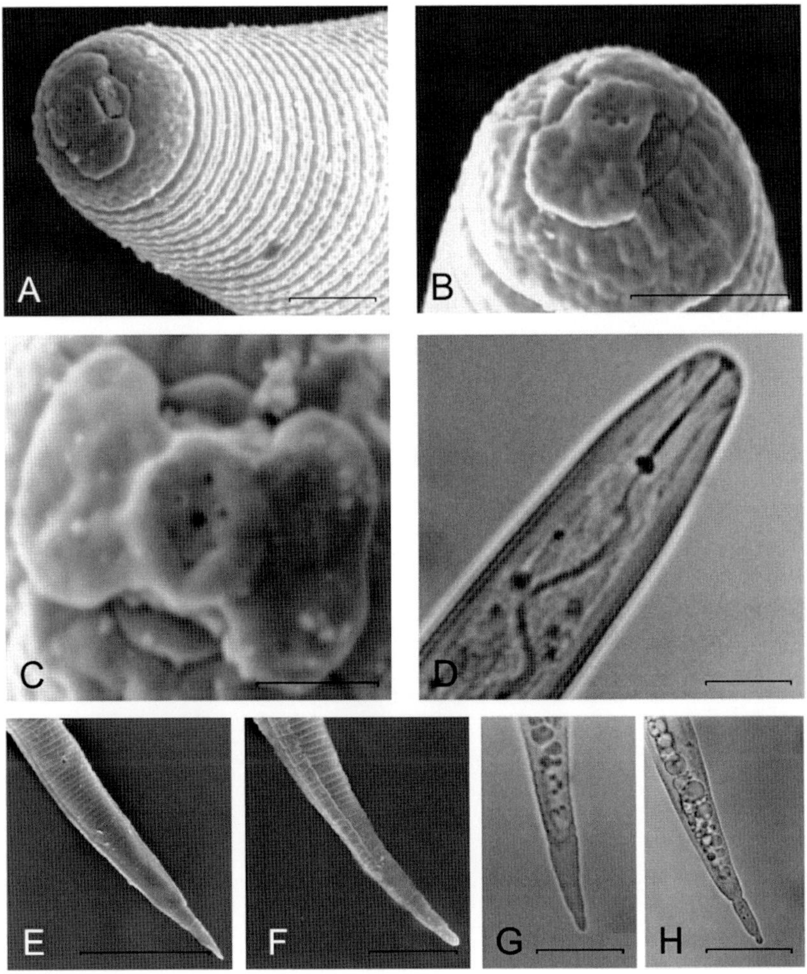

Fig. 25. *Second-stage juvenile of* Meloidogyne exigua *race 3 from rubber tree. A: SEM of anterior region; B: SEM of anterior end in lateral view; C: SEM of anterior end in face view; D: LM of head; E, F: SEM of tails; G, H: LM of tails. (Scale bars: A, B = 2 μm; C = 1 μm; D, G, H = 14 μm; E = 20 μm; F = 10 μm.) (After Muniz* et al.*, 2009.)*

Male

Body vermiform, not twisted, bluntly rounded posteriorly, labial region sclerotised. Labial disc and median labials fused, forming a smooth, continuous cap. Amphidial apertures elongate, located posterior to lat-

Table 14. *Morphometrics of female, male and second-stage juvenile (J2) of* Meloidogyne exigua. *All measurements are in μm and in the form: mean ± s.d. (range).*

Character	Coffee Chitwood (1949)	Coffee Whitehead (1968)	Coffee Lordello & Zamith (1958)	Coffee Lima & Ferraz (1985)	Coffee Flores & López (1989a)	Rubber tree Race 3 Muniz *et al.* (2009)
Female (n)	-	-	-	160	30	16
L	-	-	387-496	499 (345-620)	417.4	541 ± 83.2 (356-664)
W	-	-	279-372	337 (220-455)	242.7	310 ± 20.3 (280-344)
Stylet	14	-	10.7	12.9 (8.2-14.8)	12.2	13.6 ± 0.7 (12.5-15.0)
Stylet knob width	-	-	-	3.5 (2.4-4.5)	2.7	2.8 ± 0.2 (2.5-3.0)
Stylet knob height	-	-	-	2.2 (1.5-3.0)	2.0	1.8 ± 0.2 (1.5-2.3)
DGO	3	-	4.6-7.7	4.6 (3.8-6.5)	4.7	5.4 ± 0.8 (4.5-7.5)
Median bulb length	-	-	30.6-33.6	35.2 (30.0-38.5)	35.4	45.6 ± 6.9 (36-60)
Median bulb width	-	-	24.5-26.0	27.5 (25.2-30)	28.5	45.5 ± 6.3 (31-55)
Median bulb valve width	-	-	-	-	9.7	9.3 ± 0.6 (8.5-10.0)
Median bulb valve length	-	-	-	-	10.9	12.0 ± 0.6 (11-13)
Ant. end to median bulb valve	-	-	-	-	69	72 ± 10.4 (47-88)
Ant. end to excretory pore	-	-	-	19 (10-59)	28	38 ± 9.4 (20-56)
a	-	-	-	1.5 (1.1-2.0)	1.8	1.7 ± 0.3 (1.2-2.2)
Vulval slit	-	-	-	20.5 (15.0-26.5)	18.9	19.5 ± 1.6 (16.3-22.5)
Vulval slit to anus	-	-	-	15.3 (10.0-23.2)	15.2	13.1 ± 1.5 (10-15)
Interphasmid distance	-	-	-	22.6 (11.5-31.0)	21	17.3 ± 3.7 (12.5-25.0)
EP/ST				1.5		
Male (n)	-	-	-	100	30	10
L	1100	-	832-1092	1128 (660-1450)	1015	1068 ± 77.9 (990-1240)

Table 14. *(Continued.)*

Character	Coffee Chitwood (1949)	Coffee Whitehead (1968)	Coffee Lordello & Zamith (1958)	Coffee Lima & Ferraz (1985)	Coffee Flores & López (1989a)	Rubber tree Race 3 Muniz *et al.* (2009)
W	-	-	26-46	34 (20-48)	38	36 ± 3.8 (32-42)
Stylet	17.5-18	-	18.4-19.9	18.3 (13.9-22.5)	18.3	18.7 ± 0.9 (17-20)
Stylet knob width	4.5-5.0	-	4.0-6.1	4.6 (3.5-6.0)	4.0	3.1 ± 0.6 (2.5-4.0)
Stylet knob height	2.5-3.0	-	3.0	3.1 (2.2-4.5)	2.4	2.9 ± 0.2 (2.5-3.0)
DGO	3.0-3.5	-	-	4.6 (3.0-6.5)	4.3	5.2 ± 0.3 (5.0-5.5)
Ant. end to median bulb valve	-	-	-	-	72	73 ± 5.2 (67-84)
Ant. end to excretory pore	-	-	-	129 (76-155)	114	115 ± 30.2 (55-146)
Spicules	27	-	20-26	28.2 (22.5-37.0)	24.8	28.7 ± 2.2 (25-32)
Gubernaculum	-	-	7.7	-	7.0	8.0 ± 1.3 (6.5-10.0)
Tail	-	-	6.1-10.0	9.2 (5.2-14.0)	10.5	10.5 ± 1.7 (9.0-11.0)
a	27	-	23.8-32.0	32.9 (21.9-48.8)	26.9	29.9 ± 3.8 (24.8-35.0)
c	-	-	95.8-110	127 (58.7-248.1)	98.1	103.8 ± 15.8 (93.1-124.0)
J2 (n)	-	25	-	160	30	16
L	281-337	329 ± 20 (289-370)	333-358	376 (330-450)	366.9	375 ± 26.5 (330-435)
W	-	-	13.7-15.3	16.6 (13-22)	14.8	14.9 ± 1.0 (13.5-17.5)
Stylet	-	9.9 ± 0.6 (8.6-11.4)	9.2	11.7 (9.5-14.0)	9.8	11.6 ± 0.4 (11.0-12.5)
DGO	-	3.2 ± 0.4 (2.5-4.1)	-	3.8 (2.6-5.0)	3.5	3.9 ± 0.5 (3.0-5.0)
Ant. end to median bulb valve	-	-	-	57 (45-69)	51	49 ± 2.2 (44-53)
Ant. end to excretory pore	-	-	-	75 (62-92)	75	72 ± 4.8 (66-79)
Tail	-	44 ± 3.0 (39-50)	44.4-46.0	49.5 (38.0-62.5)	46.6	46.9 ± 1.6 (45.0-49.5)

Table 14. *(Continued.)*

Character	Coffee Chitwood (1949)	Coffee Whitehead (1968)	Coffee Lordello & Zamith (1958)	Coffee Lima & Ferraz (1985)	Coffee Flores & López (1989a)	Rubber tree Race 3 Muniz *et al.* (2009)
a	31-33	5.6 ± 1.8 (22.2-28.9)	22.2-26.0	22.1 (17.6-30.6)	25	25.2 ± 2.1 (21.4-29.3)
c	8.6	7.5 ± 0.36 (6.4-8.0)	7.3-7.8	8.1 (5.8-10.7)	7.9	8.0 ± 0.7 (6.8-9.1)
Anal body diam.	-	-	7.7-9.2	11.8 (8.8-16.0)	9.3	9.7 ± 0.5 (9.0-10.5)
Hyaline region	-	-	-	-	13.9	12.9 ± 1.9 (10-18)

eral edges of labial disc, amphids often producing exudates. Labial region continuous with body contour, lacking annulation. Lateral fields beginning as two incisures eight annuli posterior to labial region, four areolated incisures usually present in other areas of body. Stylet cone bluntly pointed, gradually increasing in diam. posteriorly, knobs rounded, offset from shaft, shaft straight and cylindrical, never narrowing at junction with knobs.

J2

Body vermiform, slender with weakly sclerotised labial framework, tail narrow, elongate, ending in a bluntly rounded tip, labial disc and median labials fused and dumbbell-shaped. Labial disc rounded, slightly raised above median labials, each median labial larger than labial disc, lateral labials small, lower than median labials, labial region smooth, continuous with body contour. Lateral fields with four incisures. Tail narrow, tapering to a finely rounded tip.

TYPE PLANT HOST: Coffee, *Coffea* sp.

OTHER PLANTS: Nematode infects rubber tree *Hevea brasiliensis* (Carneiro *et al.*, 2000; Muniz *et al.*, 2009), *Grevilea robusta* A. Cunn. (Santos, 1988), watermelon (*Citrullus vulgaris* Schrad) and onion (*Allium cepa* L.) (Moraes *et al.*, 1972), pepper (Lordello, 1964; Oliveira *et al.*, 2005), tomato, common bean (*Phaseolus vulgaris* L.), cacao (*Theobroma cacao* L.) and soybean (*Glycine max* (L.) Merr.) (Oliveira *et al.*, 2005). Lima *et al.* (1985) found that weed hosts of *M. exigua* included *Amaranthus deflexus* L., *Poinsettia heterophylla* (L.) Klotzsch & Garcke, *Ipomoea acuminata* (Vahl) Roemer & Schult., *Leonurus sibiricus* L., *Stachys arvensis* L., and *Taraxacum officinale* (L.) Weber *ex* F.H.Wigg., Prim. Fl. Holsat. Other weed hosts of *M. exigua* include *Commelina diffusa* Burm. f., *Cyperus rotundus* L., *Solanum nigrum* L., and *Spananthe paniculata* Jacq. (Moraes *et al.*, 1972; Aragon *et al.*, 1978; López & Vilchez, 1991; Rich *et al.*, 2008).

TYPE LOCALITY: Province of Rio de Janeiro, Brazil.

DISTRIBUTION: *Asia*: Turkey, India, Thailand; *Central America and Caribbean*: Costa Rica, Dominican Republic, El Salvador, Guadeloupe, Guatemala, Honduras, Martinique, Nicaragua, Panama, Puerto Rico, Trinidad & Tobago; *South America*: Bolivia, Brazil, Colombia, French Guiana, Peru, Suriname, Venezuela; *Africa*: Mozambique; *Europe*: Greece, Italy (Cain, 1974; Lehman & Lordello, 1982; Eisenback & Triantaphyllou, 1991; Campos & Villain, 2005; Coyne *et al.*, 2006; Herrera *et al.*, 2011; Wesemael *et al.*, 2011; Anon., 2015; Artavia-Carmona & Peraza-Padilla, 2020). Reports from Europe were accidental and local. Findings of this species in the USA (Florida; New York) require confirmation.

SYMPTOMS: According to Campos & Villain (2005), *M. exigua* causes typical rounded galls, mostly on newly formed roots. The galls are initially white to yellowish brown and turn dark brown as the root becomes older. Egg masses are produced either in the cortex beneath the root epidermis or protruding outside the cortex.

PATHOGENICITY: Yield losses in full sun-exposed plantations with appropriate agronomical management have been estimated as 10-15% in Costa Rica (Bertrand *et al.*, 1997) and 45% in Brazil (Barbosa *et al.*, 2004).

CHROMOSOME NUMBER: Populations of *M. exigua* reproduce by facultative meiotic parthenogenesis and have a haploid chromosome number of $n = 18$ (Triantaphyllou, 1985b; Muniz *et al.*, 2009).

POLYTOMOUS KEY CODES: *Female*: A32, B324, C3, D1; *Male*: A34, B32, C23, D1, E1, F2; *J2*: A3, B3, C3, D213, E3, F2.

RACES: Carneiro *et al.* (1996b) reported that some *M. exigua* populations collected on rubber were not able to develop on tomato plants, while those collected on coffee could. These observations indicated the presence of host races within *M. exigua*. Carneiro & Almeida (2000) and Muniz *et al.* (2008, 2009) distinguished three races: race 1 reproduced on coffee and pepper plants but not on tomato; race 2 reproduced on coffee, pepper and tomato; and race 3 reproduced only on rubber tree.

BIOCHEMICAL AND MOLECULAR CHARACTERISATION: This species can be distinguished by its four esterase phenotypes (E1, E2, E2a, E3) and three malate dehydrogenase phenotypes (N1, N1a, N2) (Carneiro *et al.*, 1996b, 2000, 2005b; Oliveira *et al.*, 2005; Muniz *et al.*, 2008, 2009). Enzymatic variability was reported by Oliveira *et al.* (2005) within 57 populations on coffee. Thirteen populations (22.8%) showed the typical single band (E1) esterase phenotype, whereas the remaining (77.2%) presented an additional band phenotype (E2). Similar results were observed by Muniz *et al.* (2008) when a new phenotype, E3, was detected for *M. exigua* from coffee. PCR-SCAR with specific markers was developed by Randig *et al.* (2002, 2004). Sequences of the 18S rRNA, the D2-D3 of 28S rRNA genes, and the *COII*-16S rRNA fragment were provided by Tenente *et al.* (2004), Tigano *et al.* (2005) and Humphreys-Pereira *et al.* (2014).

RELATIONSHIPS (DIAGNOSIS): The species belongs to Molecular group III and is clearly molecularly differentiated from all other *Meloidogyne* species. *Meloidogyne exigua* has small females with a unique perineal pattern.

2. *Meloidogyne aberrans* Tao, Xu, Yuan, Wang, Lin, Zhuo & Liao, 2017

(Figs 26-29)

COMMON NAME: Kiwifruit root-knot nematode.

High infection rates of roots of wild kiwifruit, *Actinidia chinensis* Planch. and soil infestation by a RKN were found in Anshun, GuiZhou

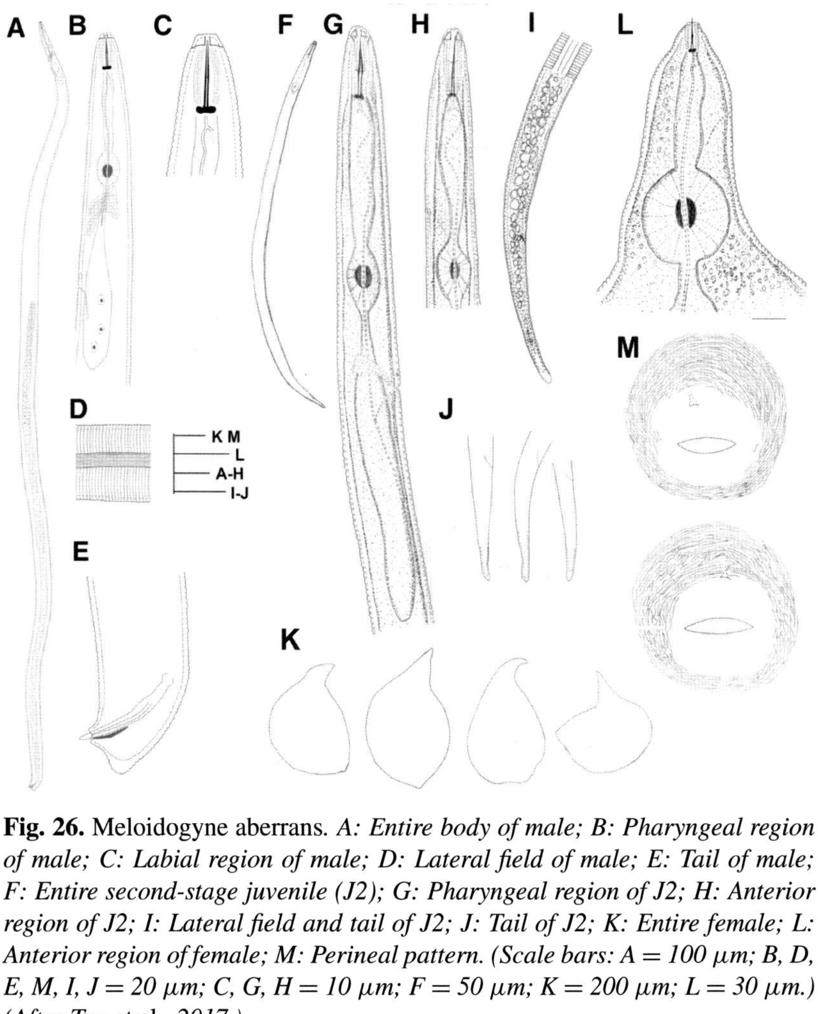

Fig. 26. Meloidogyne aberrans. *A: Entire body of male; B: Pharyngeal region of male; C: Labial region of male; D: Lateral field of male; E: Tail of male; F: Entire second-stage juvenile (J2); G: Pharyngeal region of J2; H: Anterior region of J2; I: Lateral field and tail of J2; J: Tail of J2; K: Entire female; L: Anterior region of female; M: Perineal pattern. (Scale bars: A = 100 μm; B, D, E, M, I, J = 20 μm; C, G, H = 10 μm; F = 50 μm; K = 200 μm; L = 30 μm.) (After Tao* et al.*, 2017.)*

Province, China. Morphology, esterase phenotype and molecular analyses revealed that this nematode was different from other RKN and it was described as a new species by Tao *et al.* (2017).

MEASUREMENTS *(AFTER TAO* ET AL., *2017)*

- *Holotype female*: L = 884.5 μm; W = 544.7 μm; neck length = 306.4 μm; stylet = 14.4 μm; DGO = 4.8 μm; median bulb length = 41.8 μm; median bulb width = 40.6 μm; stylet knob height =

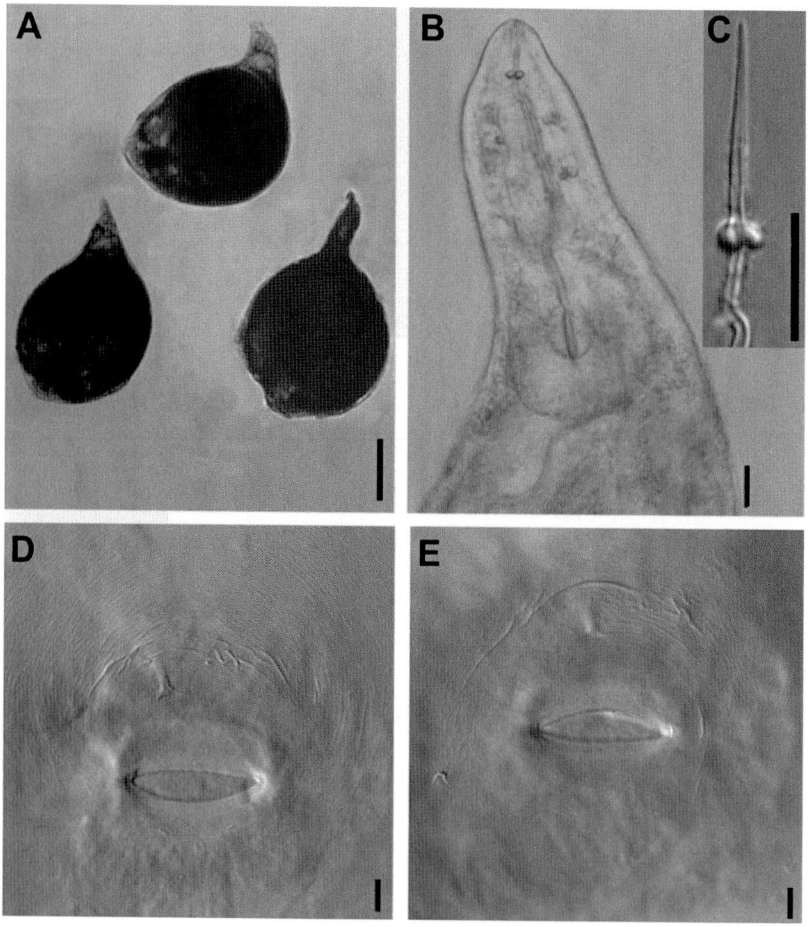

Fig. 27. Meloidogyne aberrans. *LM. A: Entire body of female; B: Anterior region of female; C: Stylet of female; D, E: Perineal pattern. (Scale bars: A = 200 μm; B-E = 10 μm.) (After Tao* et al.*, 2017.)*

2.04 μm; stylet knob width = 3.93 μm; anterior end to excretory pore = 50.4 μm; anterior end to median bulb valve = 84.5 μm; vulva length = 31.8 μm; vulva-anus distance = 23.3 μm; a = 1.6.

- *Paratype females* (n = 25): L = 938.1 ± 91.7 (806.2-1119.1) μm; W = 581.6 ± 78.5 (441.3-712.6) μm; neck length = 282.8 ± 57.8 (184.4-378) μm; stylet = 14.5 ± 0.6 (13.6-15.5) μm; DGO = 4.5 ± 0.5 (3.7-5.8) μm; median bulb length = 39.3 ± 3.7 (32.1-45.1) μm; median bulb width = 38.4 ± 4.2 (31.5-46.6) μm; stylet knob height =

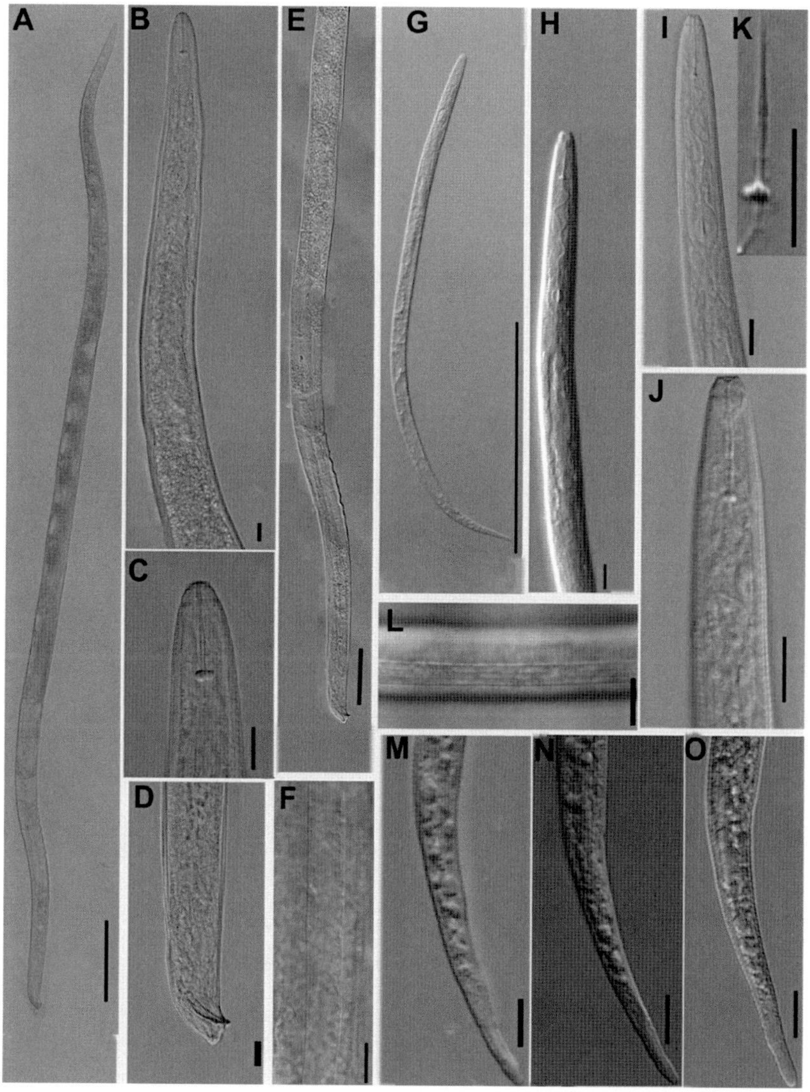

Fig. 28. Meloidogyne aberrans. *A: Entire body of male; B: Pharyngeal region of male; C: Anterior region of male; D: Tail of male; E: Posterior region and testis of male; F: Lateral field of male; G: Entire second-stage juvenile (J2). H, I: Pharyngeal region of J2; J: Anterior region of J2; K: Stylet of J2; L: Lateral field of J2; M-O: Tail of J2. (Scale bars: A = 200 μm; B-E = 10 μm; F, L = 200 μm; G-I, K, M-O = 10 μm; J = 100 μm.) (After Tao* et al.*, 2017.)*

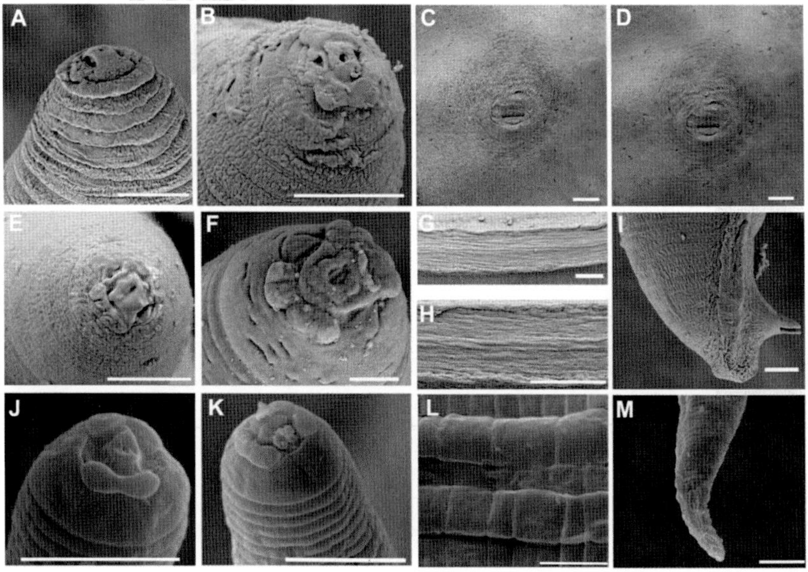

Fig. 29. Meloidogyne aberrans. *SEM. A, B: Female* en face *view; C, D: Perineal pattern; E, F: Male* en face *view; G, H: Male lateral field; I: Male tail; J, K: Second-stage juvenile (J2)* en face *view; L: J2 lateral field; M: J2 tail. (Scale bars: A, B, E-M = 5 μm; C, D = 20 μm.) (After Tao* et al., *2017.)*

2.5 ± 0.3 (2.0-2.8) μm; stylet knob width = 4.3 ± 0.3 (3.9-4.8) μm; anterior end to excretory pore = 48.1 ± 7.9 (32.0-57.8) μm; anterior end to median bulb valve = 85.6 ± 13.7 (67.4-106.8) μm; vulva length = 33.6 ± 4.4 (23.7-41.1) μm; vulva-anus distance = 23.1 ± 2.5 (17.8-27.1) μm; a = 1.6 ± 0.2 (1.3-2.0); EP/ST = 2.8.

- *Paratype males* (n = 10): L = 1882.2 ± 162.7 (1701.5-2162.6) μm; W = 54.5 ± 3.7 (49.2-59.5) μm; labial region height = 4.4 ± 1.1 (2.7-5.7) μm; labial region diam. = 9.2 ± 0.3 (8.8-9.5) μm; stylet = 18.9 ± 0.6 (18.2-19.6) μm; DGO = 4.6 ± 0.6 (3.8-5.3) μm; median bulb length = 18.2 ± 1.2 (16.7-19.9) μm; median bulb width = 13.7 ± 1.8 (11.8-17.0) μm; stylet knob height = 2.5 ± 0.3 (2.0-2.9) μm; stylet knob width = 4.7 ± 0.4 (4.1-5.0) μm; anterior end to excretory pore = 137.9 ± 5.1 (131.3-143.4) μm, anterior end to median bulb valve = 78.9 ± 4.7 (73.4-86.3) μm; anterior end to pharyngeal gland base = 224.2 ± 18.5 (210.6-260.6) μm; anal body diam. = 22.1 ± 2.6 (19.8-27.2) μm; tail = 9.4 ± 0.6 (8.8-10.2) μm; spicules = 31.5 ± 5 (22.7-36.8) μm; gubernaculum = 8.5 ± 0.8 (7.0-9.4) μm; a = 34.6 ± 2.7

(29.8-37.0); b′ = 8.4 ± 1 (7.4-10.1); c = 202.2 ± 27.4 (167.1-240.3); c′ = 0.4 ± 0.1 (0.3-0.5).

- *Paratype J2* (n = 27): L = 451.7 ± 17.4 (419.2-473.8) μm; W = 14.7 ± 0.2 (14.4-15.2) μm; labial region height = 2.4 ± 0.4 (1.6-2.8) μm; labial region diam. = 5 ± 0.4 (4.3-5.4) μm; stylet = 16.3 ± 0.3 (15.9-16.8) μm; DGO = 3.3 ± 0.3 (3.0-3.9 μm); median bulb length = 14.9 ± 1.2 (12.6-16.4) μm; median bulb width = 8.2 ± 0.6 (7.4-9.3) μm; stylet knob height = 1.8 ± 0.2 (1.5-2.1) μm; stylet knob width = 2.1 ± 0.1 (1.8-2.2) μm; anterior end to excretory pore = 86.7 ± 3.2 (80.5-91.6) μm; anterior end to median bulb valve = 58 ± 2.6 (53.4-61.6) μm; anterior end to end of cardia = 106.1 ± 5.4 (94.3-113.5) μm; anterior end to pharyngeal gland base = 191.8 ± 11.6 (175.2-210.1) μm; anal body diam. = 9.4 ± 0.6 (8.2-10.4) μm; tail = 53 ± 2.6 (48.5-57.0) μm; hyaline region = 3.6 ± 1.1 (2.2-5.5) μm; a = 30.7 ± 1.3 (28.9-32.7); b = 4.3 ± 0.3 (3.8-5.0); b′ = 2.4 ± 0.1 (2.2-2.5); c = 8.5 ± 0.4 (7.9-9.3); c′ = 5.7 ± 0.4 (4.9-6.2).

DESCRIPTION

Female

Body completely embedded in galled tissue and pearly white, pear-shaped to ovoid with neck projecting at different angles. Posterior end of body with distinct, elevated perineum. Labial region slightly offset. Labial cap distinct, labial disc elevated. Under SEM, labial disc appearing round-squared, slightly elevated, fused with median labials, dumbbell-shaped. Six inner labial sensilla surrounding ovoid prestoma, stoma slit-like. Lateral labials large, triangular, separated from labial disc. Amphidial apertures elongated. Stylet moderately long, with round knobs, conus slightly curved and shaft straight. Excretory pore distinct, typically located 2.0-3.5 stylet lengths posterior to stylet knobs. Median bulb developed, rounded, with heavily sclerotised valve. Pharyngeal gland with a large dorsal lobe and two subventral gland lobes. Perineal pattern oval, striae extremely faint, broken. Vulval slit wider than vulva-anus distance. Anal fold visible in several specimens. Phasmid not visible.

Male

Body vermiform, tapering anteriorly. Labial region slightly offset from body, with an obvious labial cap. Labial framework sclerotised.

Under SEM, labial disc appearing round-squared, elevated. Large stoma-like slit located in oval prestoma and surrounded by six inner labial sensilla. Median labials large, separated from labial disc, forming a deep slit. Lateral labials large, triangular, separated from labial disc. Amphidial apertures elongated, located between labial disc and lateral labials. Stylet straight, cone narrow, sharply pointed, shaft widened slightly. Stylet knobs distinct, rounded and slightly concave anteriorly. Lateral fields narrow, occupying about one-fifth of the body diam., with 11-15 lateral lines at mid-body, outer bands areolated in some specimens under SEM. Excretory pore distinct, located posterior to nerve ring. Hemizonid conspicuous, located about 3-4 annuli anterior to excretory pore. Median bulb oval. Single testis extending anteriorly. Spicules of variable length, arcuate, slender, two pores clearly visible at tip under SEM. Gubernaculum simple, almost straight. Tail short, hemispherical, with a humped end and twisted posterior body portion.

J2

Body vermiform, tapering at both ends, ventrally curved after killing with heat. Labial region smooth, continuous with body, depressed in outline at oral aperture in lateral view. Under SEM, labial disc appearing round-squared, oral aperture located in middle of labial disc surrounded by six inner labial sensilla. Median labials distinctly protruded, extending further than lateral labials and labial disc, resulting in an oral depression. Amphidial apertures appearing as a wide slit between labial disc and lateral labials. Stylet long, straight or conus slightly curved, cone narrow, sharply pointed, shaft widened slightly posteriorly, knobs distinct, sloping posteriorly. Body annuli distinct, fine. Lateral fields with four lines, areolated completely under SEM. Excretory pore distinct, located posterior to nerve ring. Hemizonid conspicuous, located 1-2 annuli anterior to excretory pore or immediately anterior to excretory pore. Median bulb oval, with heavily sclerotised valve. Pharyngeal gland lobe long, ventrally overlapping intestine. Tail tapering gradually toward a bluntly round terminus. Hyaline tail short, sometimes not clearly defined. Phasmids indistinct.

TYPE PLANT HOST: Kiwifruit, *Actinidia chinensis* Planch.

OTHER PLANTS: No other hosts reported.

TYPE LOCALITY: Anshun City, Guizhou Province, China (26°13′N, 106°13′E).

DISTRIBUTION: *Asia*: China.

SYMPTOMS: Most galls induced by *M. aberrans* on kiwifruit roots were on root tips and were oval or rounded and relatively large. Typically, a simple gall contained 1-10 females that each deposited an egg mass within the root tissue. Above-ground symptoms were nutritional deficiency with dwarf plants and small-sized fruits (Tao *et al.*, 2017).

CHROMOSOME NUMBER: No available data.

POLYTOMOUS KEY CODES: *Female*: A12, B23, C2, D3; *Male*: A12, B3, C213, D2, E1, F2; *J2*: A2, B1, C3, D4, E23, F23.

BIOCHEMICAL AND MOLECULAR CHARACTERISATION: The isozyme electrophoretic analysis of young, egg-laying females of *M. aberrans* showed a rare Est phenotype, S2, *i.e.*, two Est bands at Rm = 40.5 and 44.5. The band of Mdh phenotype of *M. aberrans* was similar in size to that of *M. javanica* N1 Mdh phenotype (Tao *et al.*, 2017). Tao *et al.* (2017) provided sequences of 18S rRNA, D2-D3 of 28S rRNA, ITS rRNA genes and *coxII*-16S rRNA gene region.

RELATIONSHIPS (DIAGNOSIS): The species belongs to Molecular group X and is clearly molecularly differentiated from all other *Meloidogyne* species. Females of *M. aberrans* have a round and faint perineal pattern, a medium-length stylet (13.6-15.5 μm) and a prominent posterior protuberance, which is rare in the genus. Males have more than four lateral incisures, which is not common in the genus, and the J2 has a long stylet.

3. *Meloidogyne acronea* Coetzee, 1956
(Fig. 30)

COMMON NAME: African cotton root-knot nematode.

The first short description was made by Coetzee (1956) based on material collected from roots of *Sorghum bicolor* (L.) Moench. var.

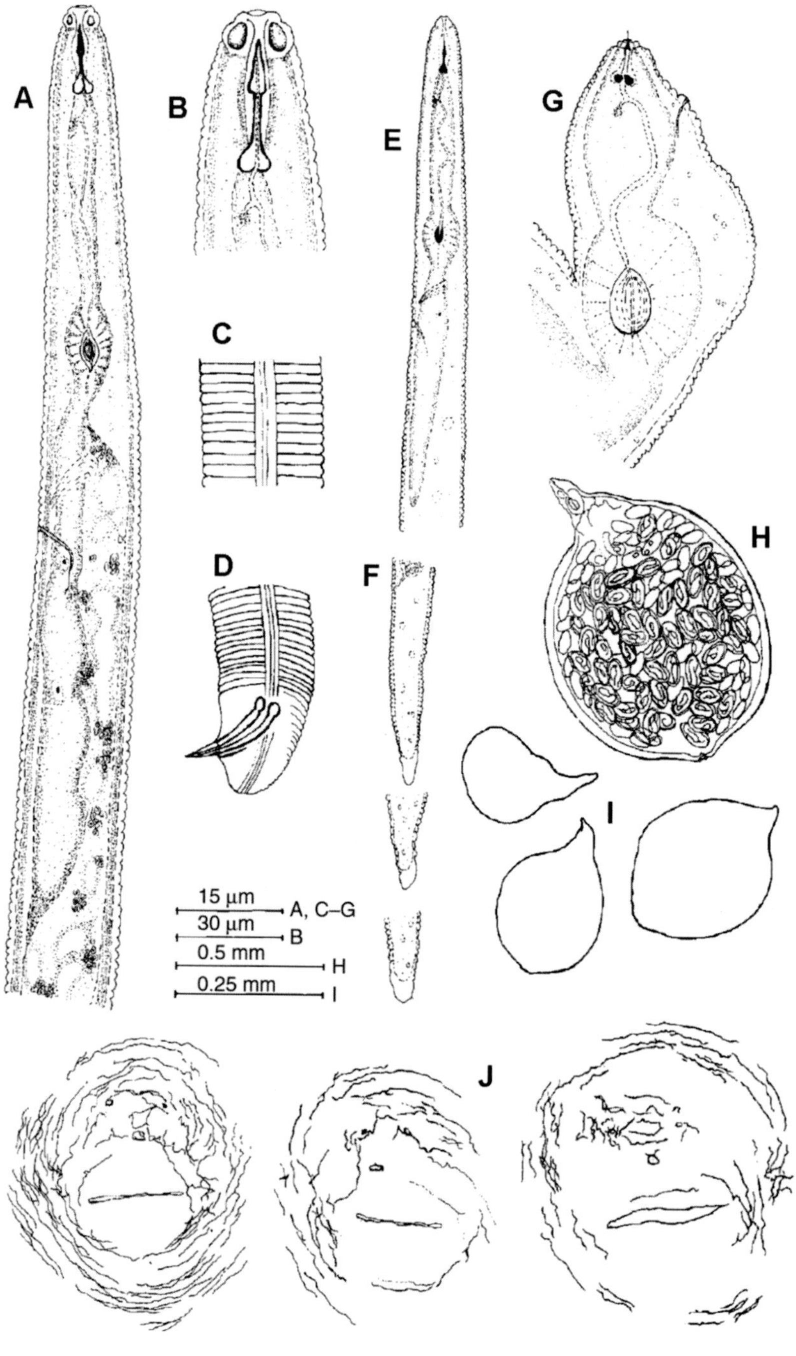
A
B
E
G
C
D
F
H
I
15 µm
A, C–G
30 µm
B
0.5 mm
H
0.25 mm
I
J

Fig. 30. Meloidogyne acronea. *A: Male pharyngeal region; B: Male anterior region; C: Male lateral field; D: Male tail region; E: Second-stage juvenile (J2) pharyngeal region; F: J2 tail regions; G: Female pharyngeal region; H: Mature female; I: Entire females; J: Perineal patterns. (After Page, 1985.)*

Radar on a farm in North West Province, South Africa. Coetzee & Botha (1965) later used the original material for a re-description and illustration of this species. This root-knot nematode was also reported from the lower Shire valley of Malawi (Bridge *et al.*, 1976). The species is restricted to alluvial soils with high water-holding capacity (Page, 1983).

MEASUREMENTS

See Table 15; holotype and allotype after Coetzee & Botha (1966).

- *Holotype female*: L = 1330 μm; W = 630 μm; stylet = 13 μm.
- *Allotype male*: L = 1720 μm; a = 49; b = 14; c = 138; stylet = 18 μm.

DESCRIPTION *(AFTER WHITEHEAD, 1968)*

Female

Body of very variable size, shape ovate, saccate or rounded. Body cuticle thickening posteriad and very thick posterior to neck, posterior end of body on a protuberance. Labial region with one, two or three annuli posterior to labial cap. Stylet knobs rounded, with anterior margins rounded, flattened or posteriorly sloped. Females often full of embryonated eggs. Anus and vulva situated on a flattened, posterior protuberance. Posterior cuticular pattern generally circular shape, dorsal arch very low, rounded or slightly flattened dorsally, tending to be concave laterally, often with short striae bordering lateral line region, no clear lateral lines observed, cuticle on tail between phasmids occasionally with small striae, striae rarely directed towards angles of vulva, often a curved, strongly marked fold or stria passing anteriad and outwards from vicinity of anal fold, phasmids usually fairly close together.

Male

Labial region not offset, labial disc not elevated, lateral labials usually present. Stylet knobs strong, equal, rounded with markedly posteriorly-

Table 15. *Morphometrics of female, male and second-stage juvenile (J2) of* Meloidogyne acronea. *All measurements are in* μm *and in the form: mean* ± *s.d. (range).*

Character	South Africa, Coetzee (1956)	South Africa, Coetzee & Botha (1965)	South Africa, Whitehead (1968)	Malawi, Page (1985)
Female (n)	-	4	20	-
L	840-950	980-1040	719 ± 142.2 (527-980)	787 ± 34 (597-1243)
W	400-600	530-750	486 ± 128 (351-764)	-
Stylet	11-13	12	12 (10-14)	11.5 ± 0.2 (9.9-12.6)
Stylet knob width	4	-	2 (2-4)	3.0 ± 0.1 (2.2-4.4)
Stylet knob height	2	-	-	2.1 ± 0.1 (1.6-3.0)
DGO	4	3-4	3 (2-3)	3.9 ± 0.2 (2.7-5.5)
Median bulb length	-	-	32 (28-35)	-
Median bulb width	-	-	27 (23-30)	-
Median bulb valve width	-	-	12 (11-14)	-
Median bulb valve length	-	-	9 (8-10)	12.7 ± 0.2 (10.6-14.4)
Labial end to excretory pore	-	-	-	28.5 ± 1.5 (14.9-41.4)
EP/ST*	3.8			
a			1.51 ± 0.2 (1.20-1.98)	-
Male (n)	-	4	26	-
L	>1660	1450-1870	1352 ± 204 (953-1703)	1241 ± 32.3 (946-1419)
W	40	27-41	-	-
Stylet	16-18	17-18	18 ± 1.08 (16.5-20.1)	17.7 ± 0.2 (15.9-18.6)

Table 15. *(Continued.)*

Character	South Africa, Coetzee (1956)	South Africa, Coetzee & Botha (1965)	South Africa, Whitehead (1968)	Malawi, Page (1985)
Stylet knob width	4-5	-	4.4 ± 0.8 (3.2-6.1)	4.4 ± 0.1 (3.2-5.4)
Stylet knob height	2	-	-	3.3 ± 0.1 (2.7-4.3)
DGO		3.5	4.1 ± 1.31 (2.2-7.2)	3.9 ± 0.3 (2.1-6.3)
Median bulb length	-	-	16.8 ± 2.27 (13.7-21.6)	-
Median bulb width	-	-	9.7 ± 1.59 (7.2-14.7)	-
Median bulb valve length	-	-	6.7 ± 0.95 (5.0-9.4)	6.7 ± 0.4 (4.4-10.6)
Ant. end to excretory pore	-	-	-	112.7 ± 5.7 (100-132.2)
Spicules	32-34	33-35	30.3 ± 3.78 (23.7-36.0)	-
Gubernaculum	-	8	8.7 ± 1.09 (7.2-10.1)	-
Tail	-	10-14	-	-
a	38	39-55	47.7 ± 4.37 (40.1-55.0)	-
c	16	138-150	125 ± 32.5 (86-237)	-
J2 (n)	-	10	25	-
L	440-460	490 (420-490)	401 ± 17 (354-427)	354 ± 2.0 (341-370)
W	14.6	14.7 (14-16)	-	-
Stylet	10	11.1 (9.7-12.0)	10.4 ± 0.66 (9.6-11.9)	12.4 ± 0.3 (10.0-14.6)
Stylet knob width	2	-	-	2.2 ± 0.1 (1.7-3.1)
Stylet knob height	1	-	-	1.0 ± 0.1 (0.6-1.7)

Table 15. *(Continued.)*

Character	South Africa, Coetzee (1956)	South Africa, Coetzee & Botha (1965)	South Africa, Whitehead (1968)	Malawi, Page (1985)
DGO	3	-	-	2.7 ± 0.1 (2.2-4.5)
Median bulb length	-	-	12.5 ± 1.09 (10.1-14.4)	-
Median bulb valve width	-	-	6.8 ± 0.65 (5.4-7.9)	-
Median bulb valve length	-	-	4.4 ± 0.88 (3.2-6.1)	3.8 ± 0.1 (2.8-5.0)
Ant. end to excretory pore	-	-	-	73.2 ± 1.6 (61.8-85.5)
Tail	-	49 (44-56)	43 ± 3 (35-47)	39.2 ± 0.8 (33.2-48.6)
Hyaline region	-	3.5 (2.0-4.0)	-	5.4 ± 0.2 (4.4-6.7)
a	-	32 (30-35)	26.4 ± 2.04 (22.0-31.3)	-
b	-	5.4 (5.1-5.7)	2.12 ± 0.96 (1.93-2.25)	-
c	-	9.4 (8.4-10.3)	9.4 ± 0.56 (8.4-10.7)	-

* Estimated from drawing.

sloped anterior margins. Posterior pharyngeal region overlapping intestine ventrally, isthmus about one median bulb length long. Hemizonid 72 (64-87) annuli posterior to labial region. Lateral fields with four main incisures for greater part of its length, occasionally a number of subsidiary incisures present at mid-body, outer bands of lateral field usually plain, sometimes areolated mid-body and often towards posterior end, inner band plain except on tail end. Tail rather bluntly rounded, smooth, terminus unstriated. Phasmids lateral, subterminal. Spicules thin-walled with offset, expanded labial region, open proximally and dorsoventrally expanded shaft, blade thicker-walled, tapering to subacute slightly swollen hyaline, bifid terminus, ventral-lateral flange running from shaft to near stylet terminus. Gubernaculum angular crescent shape.

J2

Juvenile labial region in lateral view truncated cone-shaped, not offset, labial cap usually not marked off clearly from first labial annulus, which is clearly separated from usually two faintly separated basal annuli. Length of stylet blade more or less equal to length of shaft with rounded knobs. Hemizonid about length of median bulb posterior to median bulb, excretory pore observed in one specimen, immediately posterior to hemizonid. Posterior pharyngeal region overlapping intestine ventrally, lateral field with four incisures for greater part of its length, not marked by clear cross-striation except on tail. Tail finely annulated from anus to vicinity of bluntly rounded unstriated terminus.

TYPE PLANT HOST: Sorghum, *Sorghum bicolor* (L.) Moench.

OTHER PLANTS: Pearl millet, *Pennisetum glaucum* (L.) R. Br. Nematodes successfully reproduced on beans and tomato, but these plants could be considered as unsuitable hosts (Coetzee & Botha, 1966). In Malawi, Bridge *et al.* (1976) found this species on cotton (*Gossypium hirsutum* L.). Host range trials revealed that this nematode also infected the roots of pigeon pea (*Cajanus cajan* (L.) Millsp.), sunflower (*Helianthus annuus* L.), senna (*Cassia* spp.), cluster bean (*Cyamopsis tetragonoloba* (L.) Taub.), groundnut (*Arachis hypogaea* L.) and roselle (*Hibiscus sabdariffa* L.).

TYPE LOCALITY: Farm Lindeshof, Dr Ruth Segomotsi Mompati district, North West Province, South Africa.

DISTRIBUTION: *Africa*: South Africa, Malawi (Anon., 2010a). Unconfirmed reports from Tanzania, Angola and Kenya (Whitehead & Kariuki, 1960; Whitehead, 1969).

SYMPTOMS: Nematodes induce rather inconspicuous galls on the roots. Symptoms include proliferation of the lateral roots and distortion of the taproot leading to stunting and delayed flowering. These symptoms were exacerbated when the cotton was grown under water-stressed conditions (Page & Bridge, 1994). Page & Bridge (1994) believed that the distribution of this nematode was restricted to soils with a high water-holding capacity, which allows the eggs to survive over the 6-7 month dry season. Therefore, it is unlikely that *M. acronea* poses a serious threat to small holder cotton while it continues to be grown under non-intensive and rainfed conditions.

PATHOGENICITY: *Meloidogyne acronea* can cause significant damage to cotton (Bridge *el al.*, 1976). Yield losses in *M. acronea*-infested cotton fields have been reported to be up to 50% at Ngabu, Malawi (Page & Bridge, 1994).

CHROMOSOME NUMBER: No available data.

POLYTOMOUS KEY CODES: *Female*: A213, B34, C1, D3; *Male*: A324, B34, C213, D1, E1, F1; *J2*: A23, B32, C3, D4, E23, F3.

BIOCHEMICAL AND MOLECULAR CHARACTERISATION: No available data.

RELATIONSHIPS (DIAGNOSIS): It differs from other related species by females having a fine, fairly obscure perineal pattern, which is difficult to define. Jepson (1987) considered this species, together with *M. africana* and *M. megadora*, as belonging to the *acronea* group of species, which is endemic to Africa. It can be separated from *M. africana* and *M. megadora* by the longer EP/ST ratio in females (3.8 *vs* 3.1 and 2.2, respectively).

4. *Meloidogyne aegracyperi* Eisenback, Holland, Schroeder, Thomas, Beacham, Hanson, Paes-Takahashi & Vieira, 2019

(Figs 31-33)

COMMON NAME: Nutsedge root-knot nematode.

This species was described from roots of purple nutsedge in southern New Mexico, USA.

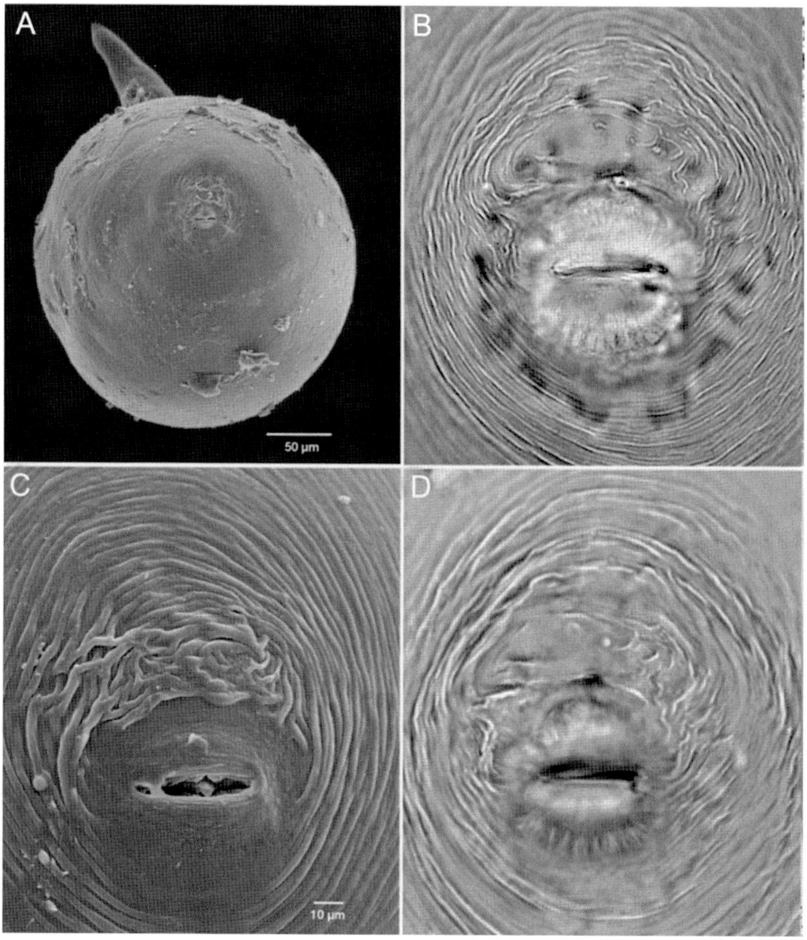

Fig. 31. Meloidogyne aegracyperi. *LM and SEM. A: Entire female; B-D: Perineal patterns (scale bar in C). (After Eisenback* et al.*, 2019.)*

MEASUREMENTS *(AFTER EISENBACK* ET AL.*, 2019)*

- *Holotype female*: L = 360 μm; W = 292 μm; a = 1.3; neck length = 148 μm; stylet = 12 μm; stylet knob height = 1.5 μm; stylet knob width = 2.4 μm; DGO = 4.9 μm; anterior end to median bulb valve = 60.1 μm.
- *Paratype females* (n = 30): L = 373 ± 44 (310-460) μm; W = 306 ± 46 (210-420) μm; a = 1.2 ± 0.2 (0.8-1.7); neck length = 153 ± 27 (100-210) μm; stylet = 12 μm; stylet knob height = 1.5 ± 0.2 (1.2-

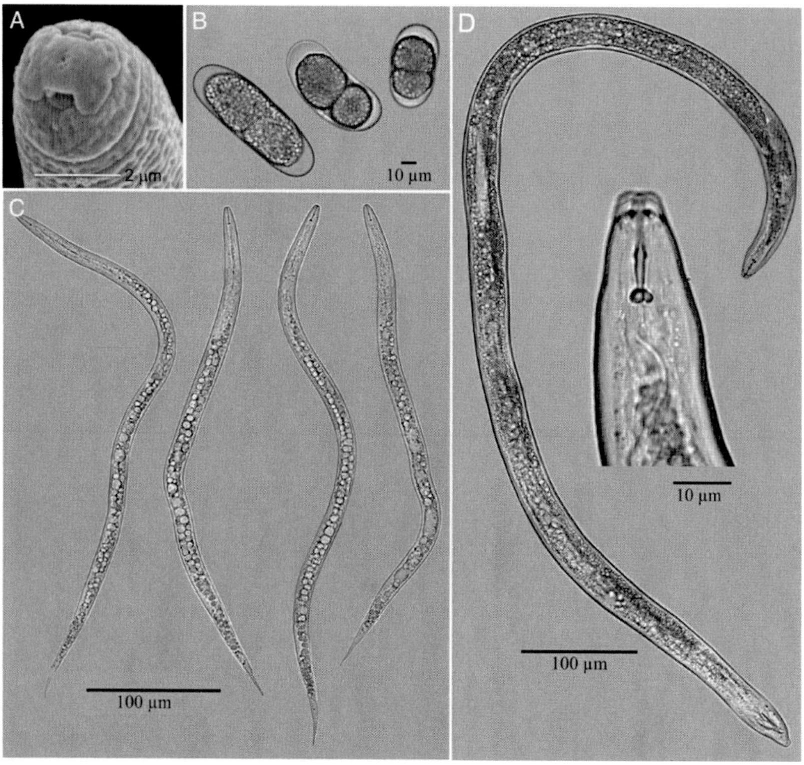

Fig. 32. Meloidogyne aegracyperi. *LM and SEM. A: Second-stage juvenile (J2) labial region; B: Eggs; C: Entire J2; D: Entire male and anterior region of male. (After Eisenback* et al., *2019.)*

1.9) μm; stylet knob width = 2.6 ± 0.3 (2.0-3.0) μm; DGO = 4.8 ± 0.6 (4.0-6.1) μm; anterior end to median bulb valve = 63.8 ± 2.2 (59-67) μm; interphasmid distance = 18 ± 2.7 (14.2-24.4) μm; vulva length = 21 ± 2.9 (13.5-26.0) μm; vulva-anus distance = 15.7 ± 2.3 (10.9-21.0) μm.

- *Paratype males* (n = 2): L = 1113, 1134 μm; W = 29.5, 32.3 μm; stylet = 14.6, 16.5 μm; stylet knob height = 2.0, 2.5 μm; stylet knob width = 4.0, 4.2 μm; DGO = 3.0, 3.3 μm; anterior end to excretory pore = 93.1, 96.3 μm; tail = 10.8, 14.5 μm; spicules = 21.8, 25.4 μm.
- *Paratype J2* (n = 30): L = 426 ± 24.7 (388-484) μm; W = 15 ± 0.9 (13.6-17.4) μm; a = 28.4 ± 2.8 (23.6-34.6); c = 5.8 ± 0.4 (5.1-6.5); stylet = 10.9 ± 0.4 (10.1-11.8) μm; stylet knob height = 1.4 ± 0.2 (1.1-1.9) μm; stylet knob width = 2.1 ± 0.2 (1.7-2.5) μm; DGO =

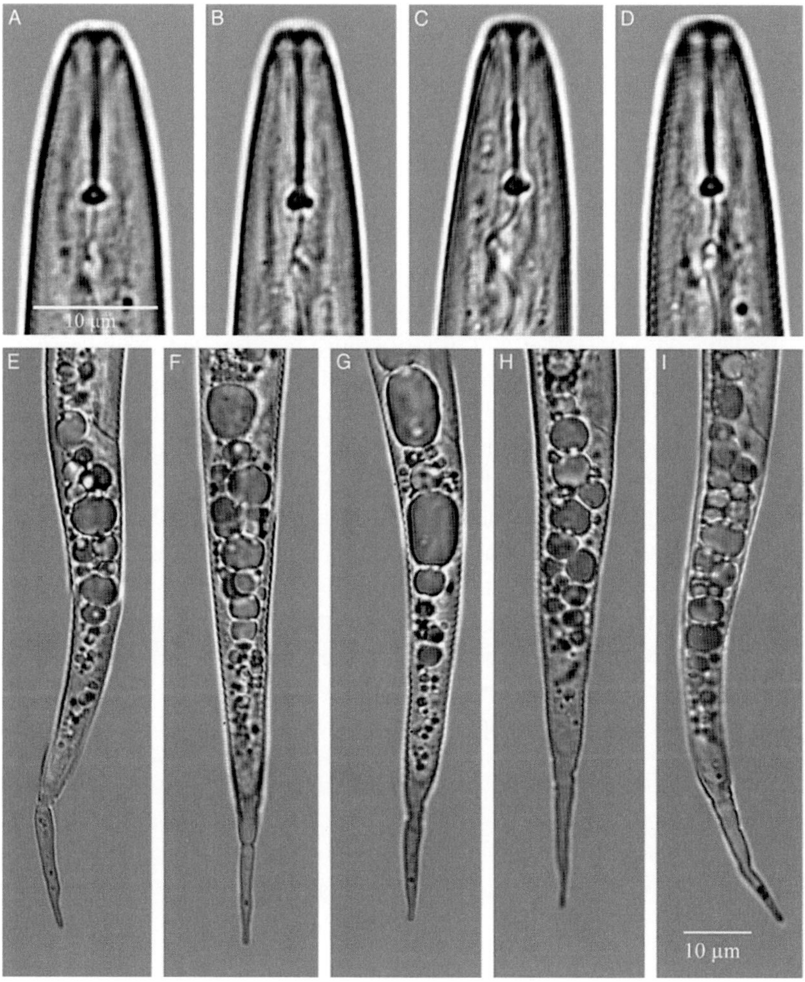

Fig. 33. Meloidogyne aegracyperi. *LM. A-D: Second-stage juvenile (J2) anterior region (scale bar in A); E-I: J2 posterior region (scale bar in I). (After Eisenback* et al.*, 2019.)*

3.7 ± 0.5 (2.7-4.8) μm; anterior end to excretory pore = 68.5 ± 7.9 (52.0-80.4) μm; tail = 73.1 ± 4.9 (63.6-88.7) μm; width at anus level = 11 ± 0.6 (10.2-11.9) μm; hyaline region = 22 ± 2.0 (18.5-26.6) μm.

- *Eggs*: L = 91.6 ± 2.3 (85.2-99.8) μm; W = 39.7 ± 0.1 (37.1-48.1) μm; L/W = 2.3 ± 0.1 (2.1-2.6).

Description

Female

Mature females very small and pearly white with egg masses usually contained completely inside galled root tissues. Body shape differing from many other species as neck often at a 90-130° angle to protruding posterior end containing perineal pattern. Labial region low, labial framework weakly developed, with one labial annulus. Cone of stylet slightly curved dorsally, posterior edges of knobs angular, tapering onto shaft. Excretory pore level with base of stylet. Lining of median bulb triradiate, with posterior and anterior portions rounded. Numerous (3-10) small vesicles present in anterior median bulb. Two, small, rounded pharyngo-intestinal cells at base of median bulb, followed by a large nucleated dorsal pharyngeal gland lobe with two smaller nucleated subventral gland cells. Six, large rectal gland cells connecting to rectum and producing gelatinous matrix forming the egg mass. Perineal pattern raised on a protuberance at posterior end of body, consisting of a rounded dorsal arch with a tail terminal area that is usually smooth, but may be marked with thick lines and many horizontal, rope-like striae. Phasmids typical for genus. Vulval lips usually flattened, but may be rounded and slightly protruding. Smooth, regular striae surrounding vulva and tail terminal area and giving appearance of a dorsoventrally elongated oval pattern.

Male

Two males were found. Anterior end tapering, labial disc slightly concave around stoma, one distinct labial annulus, labial framework slight, stylet shaft tapering posteriorly. Body twisting 90° throughout its length. Stylet knobs rounded, offset from shaft. Pharyngeal glands overlapping intestine ventrally. Four lines in lateral field. Paired spicules with gubernaculum typical for genus. Tail tip slightly offset from remainder of body.

J2

Body with a very long tail and tail terminus. Labial framework weak, stylet small, with a constriction near junction of shaft and knobs. In SEM, labial region has a slit-like oral opening placed on rounded labial disc and surrounded by six small pit-like openings of inner labial sensilla. Small depressions in cuticle on dorsal and ventral labial pairs mark outer labial sensilla. Rounded knobs tapering onto shaft. DGO well posterior to base of stylet. Pharyngeal glands overlapping intestine ventrally. Four lines in lateral field. Tail very long, hyaline region very narrow, making tail finely pointed. Phasmids located mid-way between anus and tail tip.

TYPE PLANT HOST: Purple nutsedge, *Cyperus rotundus* L.

OTHER PLANTS: Yellow nutsedge (*Cyperus esculentus* L.), perennial ryegrass (*Lolium perenne* L.), wheat (*Triticum aestivum* L.), bentgrass (*Agrostis canina* L.) and barley (*Hordeum vulgare* L.).

TYPE LOCALITY: Lower Rio Grande Valley, Dona Ana County, New Mexico, USA, (Rincon/Hatch Hwy 185, onion field, N32 39.431 W107 07.801).

DISTRIBUTION: *North America*: USA (New Mexico).

CHROMOSOME NUMBER: No available data.

POLYTOMOUS KEY CODES: *Female*: A3, B3, C3, D3; *Male*: A43, B4, C3, D1, E-, F1; *J2*: A23, B32, C12, D1, E3, F1.

BIOCHEMICAL AND MOLECULAR CHARACTERISATION: The 18S rRNA, D2-D3 of 28S rRNA, ITS rRNA, intergenic *COII*-16S region and *hsp90* genes were sequenced and analysed by Eisenback *et al.* (2019). This species formed a clade with *M. graminicola*.

RELATIONSHIPS (DIAGNOSIS): The species belongs to Molecular group III, Graminicola group (*M. aegracyperi*, *M. graminicola*, *M. salasi* and *M. trifoliophila*). *Meloidogyne aegracyperi* is characterised by the small female with a perineal pattern that occurs on a posterior protuberance that is at a 90-130° angle with the neck. Males are very rare in *M. aegracyperi*, which may be a useful diagnostic character since males are common in *M. naasi* and *M. graminicola*.

5. *Meloidogyne africana* Whitehead, 1960
(Figs 34-37)

COMMON NAME: African coffee root-knot nematode.

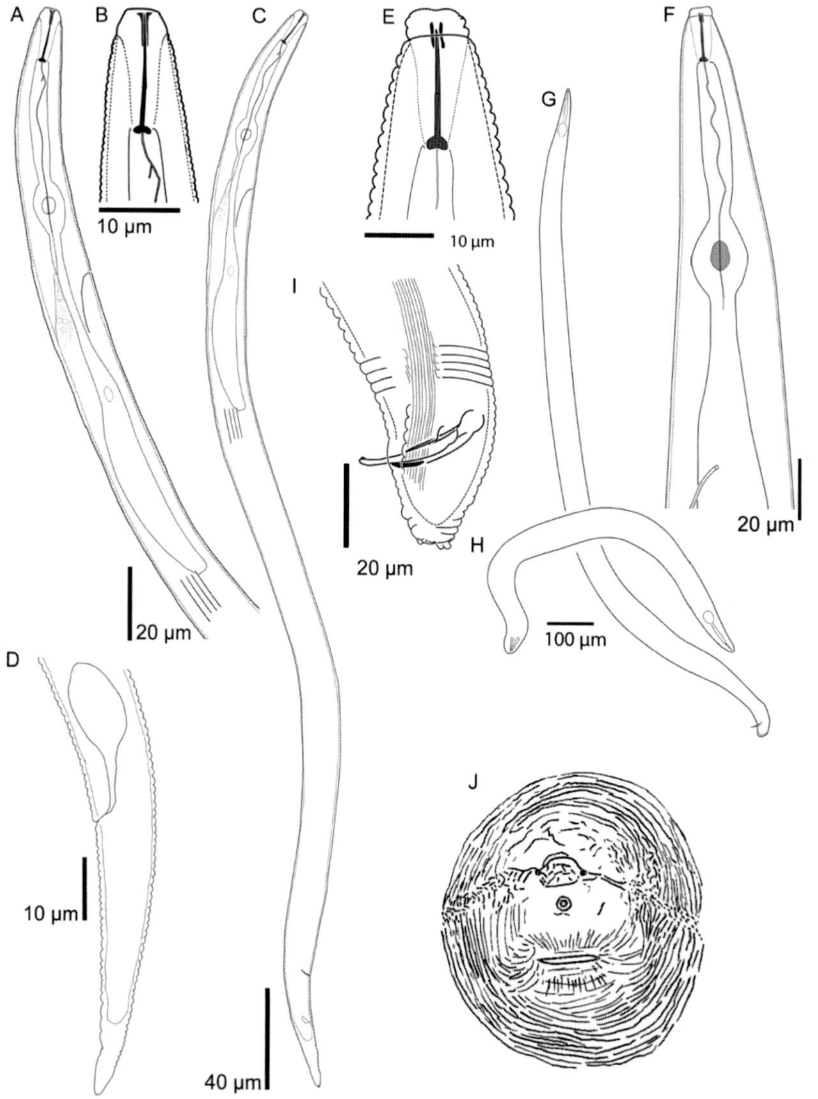

Fig. 34. Meloidogyne africana. *A: Second-stage juvenile (J2) anterior body; B: J2 juvenile head; C: J2 habitus; D: J2 tail; E: Male head; F: Male anterior body; G, H: Variable male habitus during development as sex-reversed females; I: Male tail; J: Perineal pattern. (A-I after Janssen* et al., *2017; J after Whitehead, 1968.)*

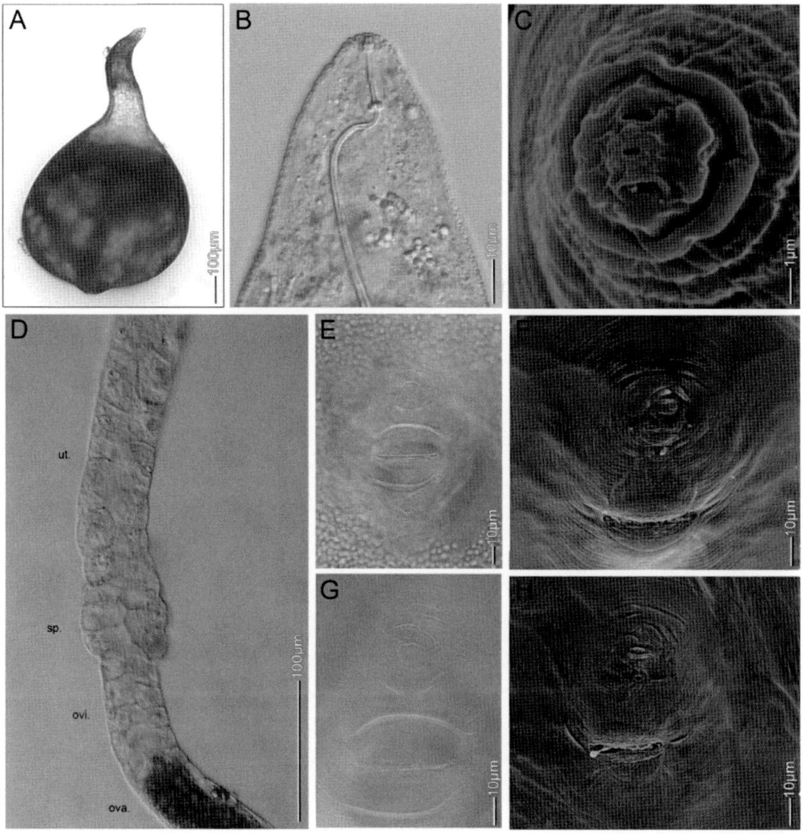

Fig. 35. Meloidogyne africana *female. LM and SEM. A: General habitus with characteristic protuberance; B: Head, lateral view; C: Labial region (*en face *view); D: Gonad morphology of uterus (ut.), spermatheca (sp.), oviduct (ovi.) and ovarium (ova.); E, G: Photomicrographs of perineal pattern; F, H: SEM of perineal pattern. (After Janssen* et al.*, 2017.)*

This nematode was found parasitising coffee nurseries in Kenya and Tanzania and later reported from other African countries and also from India and Argentina. Descriptions of this nematode were provided by Whitehead (1960, 1968). Janssen *et al.* (2017) compared the original type material of *M. africana*, *M. oteifae*, *M. megadora* and *M. decalineata* and concluded that *M. oteifae* and *M. decalineata* should be considered as synonyms of *M. africana*. Janssen *et al.* (2017) also provided detailed morphological re-description, molecular and biochemical characterisation and karyology of *M. africana*.

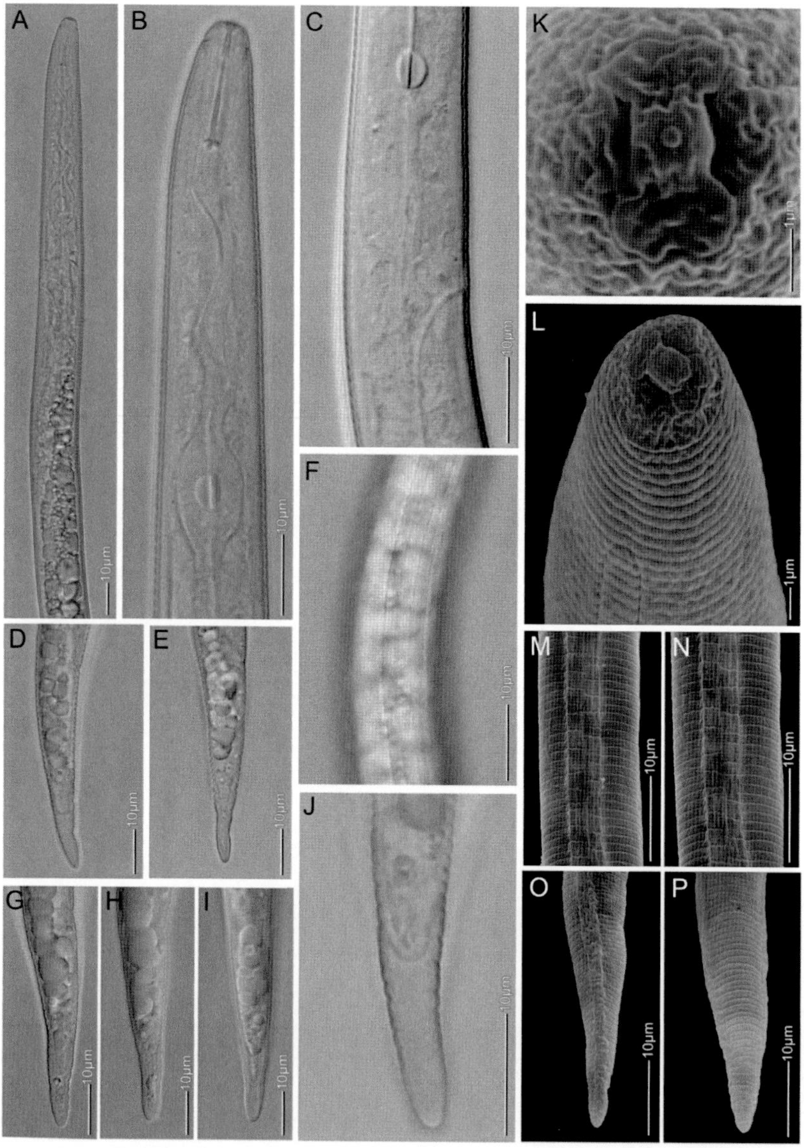

Fig. 36. Meloidogyne africana *second-stage juvenile. LM and SEM. A, B: Anterior body; C: Meta- and post-corpus region; D, E, G-I: Tail variation; F: Mid-body lateral field; J: Hyaline tail region; K: Labial region,* en face *view; L: Labial region, lateral view; M, N: SEM of mid-body lateral field; O: Tail, lateral view; P: Tail, ventral view. (After Janssen* et al., *2017.)*

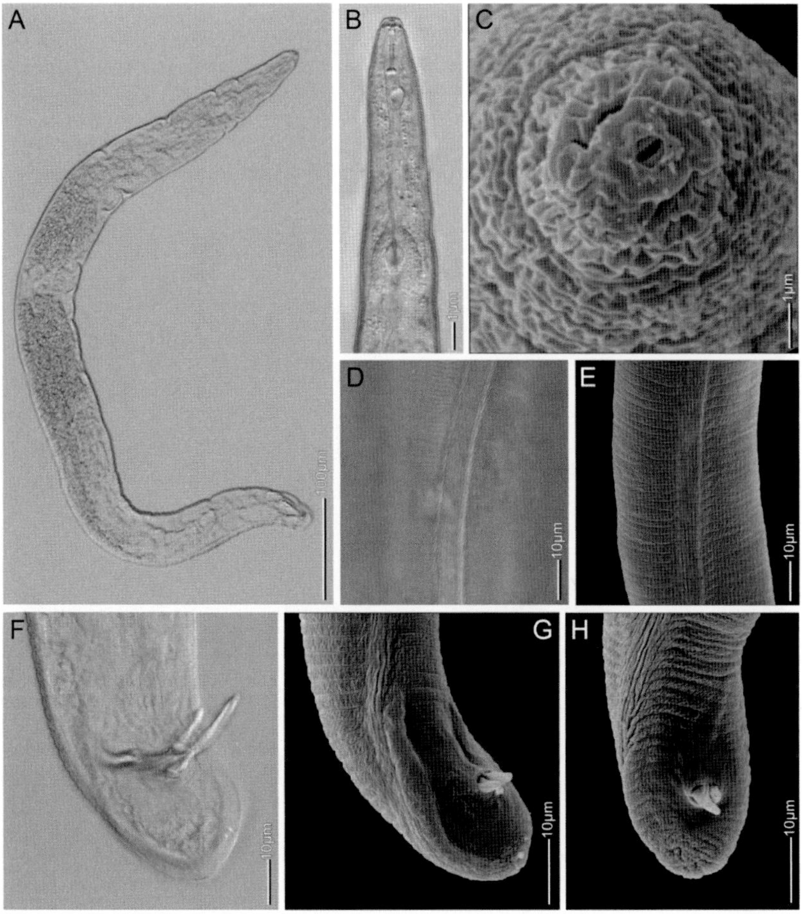

Fig. 37. Meloidogyne africana. *Males. A: Habitus of dwarf sex-reversed female; B: Anterior body in lateral view; C: Labial region,* en face *view; D, E: Mid-body lateral field; F-H: Tail. (After Janssen* et al., *2017.)*

MEASUREMENTS

See Table 16.

Holotype male: L = 1320 μm; a = 39; b = 5.1; c = 88; stylet = 21 μm.

Table 16. *Morphometrics of female, male and second-stage juvenile (J2) of* Meloidogyne africana. *All measurements are in μm and in the form: mean ± s.d. (range).*

Character	Kenya, Whitehead (1968)	Congo, Elmiligy (1968)	Tanzania, Whitehead (1968)	Tanzania, Janssen *et al.* (2017)
Female (n)	17	10	20	22
L	760 ± 73 (660-910)	600 (520-680)	819 ± 133 (649-1041)	615 ± 95 (400-770)
W	-	475 (400-550)	-	375 ± 59 (300-540)
Stylet	15	13.5 (13-14)	14 (12-17)	14.3 ± 0.8 (13.0-15.5)
Stylet knob width	-	-	3 (3-4)	3.1 ± 0.3 (3.0-4.0)
DGO	4-9	3.5 (3-4)	4 (4-6)	5.7 ± 0.8 (4.5-7.0)
Median bulb length	-	39 (35-45)	45 (36-54)	-
Median bulb width	-	32 (30-35)	43 (34-49)	-
Median bulb valve length	-	15 (13-17)	15 (14-17)	-
Median bulb valve width	-	10 (9-13)	12 (10-14)	-
EP/ST**	3.2	-	-	-
a	1.6 ± 0.2 (1.4-1.9)	-	1.62 ± 0.2 (1.2-2.1)	1.6 ± 0.2 (1.3-2.4)
Male (n)	18*	10	2	21
L	1470 ± 197 (1200-1850)	1160 (980-1270)	1630, 1700	1285 ± 245 (816-1750)
a	38.9 ± 4.7 (31-50)	26 (25-28)	29.6, 42.5	26.0 ± 4.0 (19.2-34.3)
b	5.1 ± 1.0 (3.8-6.7)	-	-	-
c	86 ± 9.7 (69-103)	128 (111-159)	-	-
Body diam.	-	-	-	50.0 ± 7.6 (36-66)

Table 16. *(Continued.)*

Character	Kenya, Whitehead (1968)	Congo, Elmiligy (1968)	Tanzania, Whitehead (1968)	Tanzania, Janssen *et al.* (2017)
Labial region height	-	6	5, 6.5	3.3 ± 0.5 (2.5-4.0)
Labial region diam.	-	12	10.5	8.3 ± 0.4 (7.5-9.0)
Stylet	20.7 ± 1.1 (19-22)	22 (19-23)	20, 19	15.7 ± 1.1 (14-18)
Stylet knob width	-	-	-	3.5 ± 0.4 (3.0-4.0)
DGO	4-6	3.5 (3.0-4.5)	4	5 ± 0.4 (4.0-6.0)
Median bulb length	-	-	20, 25	-
Median bulb width	-	-	14	-
Median bulb valve length	-	-	12	-
Ant. end to median bulb	-	-	-	64.0 ± 8.2 (54-78)
Ant. end to excretory pore	-	-	-	150 ± 24.9 (110-197)
Spicules	26-35	33 (29-37)	33-37	26.5 ± 2.3 (24-31)
Gubernaculum	7-9	11 (10-12)	7	7.6 ± 1.8 (6.0-10.0)
J2 (n)	25	30	25	30
L	420 ± 23 (380-470)	370 (320-400)	543 ± 24 (471-573)	422 ± 39 (352-536)
a	24.4 ± 1.7 (22-28)	26.5 (22-29)	36.3 ± 1.9 (32.8-40.0)	25.5 ± 3.1 (19.5-31.1)
b	-	-	2.55 ± 0.3 (1.95-2.97)	-
c	10.8 ± 1.8 (7.3-14.3)	8 (7.5-9.2)	11.2 ± 0.5 (10.3-12.2)	10.1 ± 1.0 (7.8-12.7)
c′	-	4.0 (3.0-4.8)	-	4.1 ± 0.3 (3.5-4.7)

Table 16. *(Continued.)*

Character	Kenya, Whitehead (1968)	Congo, Elmiligy (1968)	Tanzania, Whitehead (1968)	Tanzania, Janssen *et al.* (2017)
Body diam.	-	-	-	16.8 ± 2.3 (14-22)
Labial region height	-	-	-	2.7 ± 0.5 (2.0-4.0)
Labial region diam.	-	-	-	5.5 ± 0.3 (5.0-6.0)
Stylet	14.8 ± 1.5 (12-18)	12 (11-13)	12.4 ± 0.7 (10.7-13.7)	11.5 ± 0.5 (10.5-12.5)
Stylet knob width	-	-	-	1.9 ± 0.2 (1.5-2.0)
DGO	-	3	-	4.0 ± 0.6 (3.0-5.5)
Median bulb length	-	-	13.8 ± 0.9 (11.5-15.5)	-
Median bulb valve length	-	-	5.9 ± 0.4 (5.0-6.8)	-
Ant. end to median bulb	-	-	-	46.0 ± 5.6 (38-61)
Ant. end to excretory pore	-	-	-	75.5 ± 6.9 (55-84)
Tail	-	-	48 ± 2 (44-52)	42.1 ± 1.9 (39-46)
Hyaline region	-	-	-	10.5 ± 1.3 (8.0-13.0)

[*] Possibly mixed with males of *M. hapla* according to Janssen *et al.* (2017).
[**] Estimated from drawing.

DESCRIPTION (*AFTER WHITEHEAD, 1968; JANSSEN* ET AL.*, 2017*)

Female

Body pyroid, posterior end with protuberance bearing vulva and anus. Labial region truncated cone, one or two annuli posterior to labial cap. Stylet knobs rounded or tending to flatten anteriorly. Excretory pore 16-30 annuli posterior to labial region. Hemizonid observed in one specimen immediately anterior to excretory pore. Posterior cuticular pattern composed of smooth striae forming low dorsal arch. Tail terminus wide, generally covered with short, coarse, disordered striae and surrounded by concentric circles of striae forming distinct raised tail pattern. Phasmids adjacent to tail terminus. Lateral fields wide, without incisures but very short, faint striae in each field. Perineal pattern on a raised perineum as a consequence of a clear protuberance, and vulva surrounded by circles of striae, which are sometimes crossed by other striae radiating from vulva.

Male

Labial region not offset, in lateral view fairly low, truncated cone shape with small labial cap followed by very short first annulus and longer second annulus (which may be very faintly subdivided). In ventral view labial cap small, separated by amphid openings from lateral labial sectors and bearing one or two faintly separated annuli. Labial framework thin. Anterior cephalid on second body annulus, posterior cephalid three or four annuli posterior to base of relaxed stylet. Stylet knobs laterally rounded with posteriorly sloping anterior margins or drawn out somewhat laterad, pharyngo-intestinal junction *ca* 0.5 median bulb length posterior to bulb. Posterior pharyngeal region a long, narrow glandular lobe running mostly ventral to intestine. Hemizonid *ca* two body annuli long 73 or 81 annuli posterior to labial region and two or three annuli anterior to excretory pore. Lateral field arising about level of base of relaxed stylet, width *ca* one-seventh body diam. mid-body, marked by ten evenly spaced incisures, outermost crenate, number of incisures reducing to one or two in vicinity of tail and extending virtually to tail terminus. Deirids and phasmids not seen. Annuli extending to tail terminus, tail somewhat rostrate. One testis. Spicules with large, thin-walled, offset heads, shaft thicker-walled, blades thick-walled, tapering to subacute termini, gubernaculum crescentic.

J2

Labial region with two annuli posterior to labial cap in lateral view, one annulus in dorsal or ventral view. Stylet prominent with rounded basal knobs. DGO *ca* 3 μm posterior to stylet base. Genital primordium ventral at 53-62% of body length from anterior end of body. Lateral field prominent with four lines, which may be areolated in mid-body region. SEM observations indicating that more than four lines can be present in mid-body region. Tail with rounded terminus and long hyaline region.

TYPE PLANT HOST: Coffee, *Coffea arabica* L.

OTHER PLANTS: Host tests revealed that *M. africana* parasitised *Solanum lycopersicum* L. ('Moneymaker') and *Sansevieria* sp. (Janssen *et al.*, 2017). Other reported hosts included *Zea mays* L. in India (Chitwood & Toung, 1960), *Capsicum annuum* L. in Sudan (Yassin & Zeidan, 1982), *Pueraria phaseoloides* (Roxb.) Benth., *Coffea canephora* Pierre *ex* A. Froehner in Congo (Elmiligy, 1968), and *Chrysanthemum cinerariaefolium* L., *Vigna unguiculata* (L.) Walp., *Syzygium aromaticum* (L.) Merrill & Perry, and *Solanum tuberosum* L. in East Africa (Whitehead, 1969) and *Phaseolus vulgaris* and *Echinochloa crus-galli* (L.) P. Beauv. in Argentina (Doucet & Pinochet, 1992).

TYPE LOCALITY: Kamaara coffee nursery, Meru district, Kenya.

DISTRIBUTION: *Africa*: Kenya (Whitehead, 1960; Campos & Villain, 2005), Sudan (Yassin & Zeidan, 1982), Tanzania (Whitehead, 1968; Bridge, 1984; Janssen *et al.*, 2017), Congo (Elmiligy, 1968; Campos & Villain, 2005), São Tomé and Príncipe (Lordello & Fazuoli, 1980; Campos & Villain, 2005), *Asia*: India (Chitwood & Toung, 1960); *South America*: Argentina (Doucet & Pinochet, 1992).

SYMPTOMS: The above-ground symptoms include stunting, chlorosis and necrosis of leaves. In coffee roots, galls are usually positioned on the apical tip of the root, resulting in impeded root extension. Young galls are rounded, while older galls tend to be oval, 1-3 mm in size, and contain more than one female with egg sacs always embedded within the gall (Janssen *et al.*, 2017). On tomato roots, the symptoms caused by *M. africana* vary slightly; galls are much smaller and are more evenly distributed throughout the root system, with root tip galls less frequent (Janssen *et al.*, 2017).

CHROMOSOME NUMBER: From 20 favourable late-prophase or early-metaphase chromosomal planes, the chromosome number of *M. africana* was determined to be $2n = 21$, the lowest known number of chromosomes in RKN reproducing by mitotic parthenogenesis (Janssen *et al.*, 2017).

POLYTOMOUS KEY CODES: *Female*: A23, B32, C2, D3; *Male*: A324, B43, C23, D1, E2, F1; *J2*: A213, B23, C3, D324, E213, F3.

BIOCHEMICAL AND MOLECULAR CHARACTERISATION: Electrophoretic isozyme analysis of single young egg-laying females of *M. africana* revealed a unique esterase phenotype, consisting of two fast migrating esterase bands, designated as AF2. The malate dehydrogenase isozyme analysis revealed a single broad band in a similar position as the H1 phenotype of *M. hapla* (Janssen *et al.*, 2017). Janssen *et al.* (2017) also provided sequences of 18S rRNA, the D2-D3 of 28S rRNA and *COI* genes.

RELATIONSHIPS (DIAGNOSIS): The species belongs to Molecular group X and is clearly molecularly differentiated from all other *Meloidogyne* species. Jepson (1987) considered *M. africana* in the *acronea* group. *Meloidogyne africana* is clearly differentiated from all other species by the unique esterase phenotype and DNA sequences. *Meloidogyne africana* can be differentiated from *M. megadora* by a shorter J2 tail (39-52 *vs* 47-76 μm), and female stylet knobs (rounded *vs* posteriorly sloped).

6. *Meloidogyne aquatilis* Ebsary & Eveleigh, 1983
(Figs 38-39)

COMMON NAME: Aquatic root-knot nematode.

This nematode was found in roots and soil samples from prairie cordgrass collected in 1981 at Deschênes, Quebec, Canada.

MEASUREMENTS *(AFTER EBSARY & EVELEIGH, 1983 AND JEPSON, 1987)*

- *Holotype female*: L (excluding neck) = 623 μm; W = 289 μm; L/W = 2.2; neck = 114 μm; stylet = 11 μm; DGO = 6 μm.
- *Paratype females* (n = 20): L (excluding neck) = 780 (775-874) μm; W = 295 (247-334) μm; L/W = 2.3 (2.1-2.4); neck = 110 (99-129) μm; stylet = 11.5 (10.5-12.5) μm; stylet knob width = 2.5-3.0 μm; DGO = 4.0-6.6 μm; vulval slit length = 23 (21-25) μm;

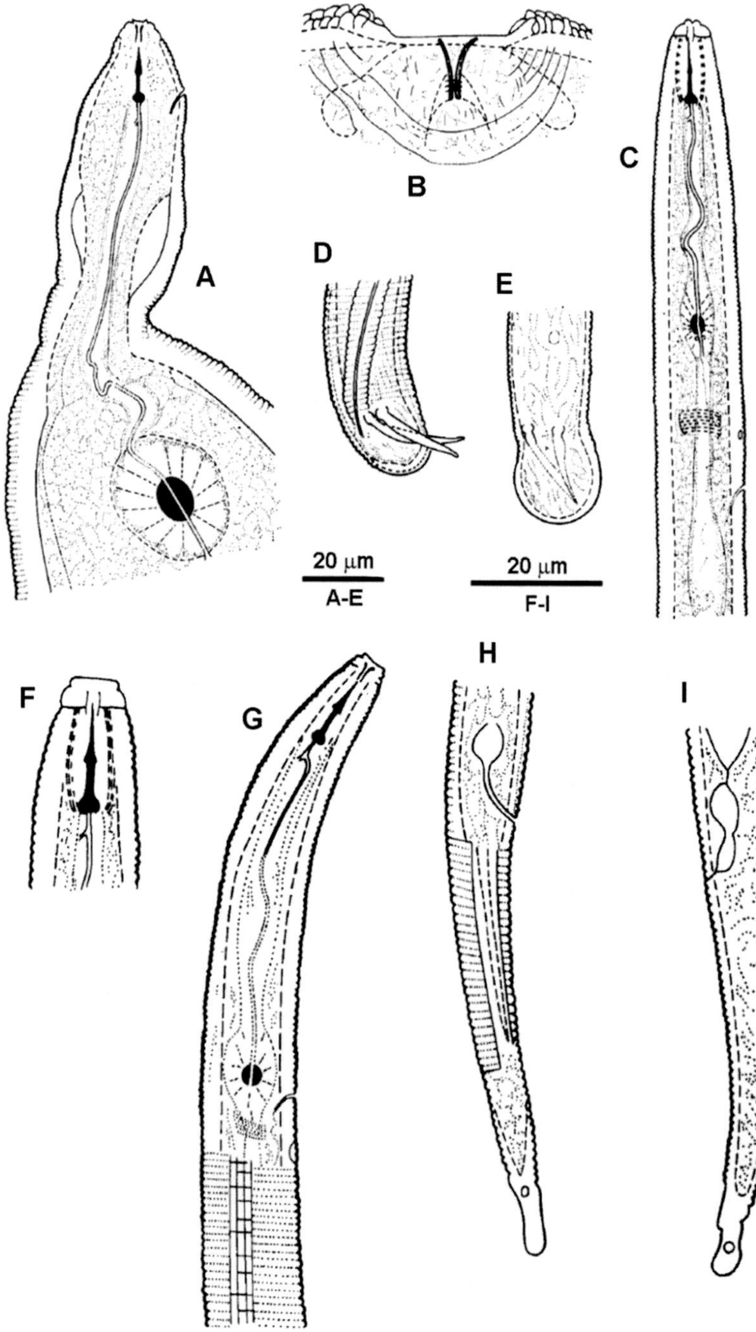
A
B
C
D
E
20 μm
A-E
20 μm
F-I
F
G
H
I

Fig. 38. Meloidogyne aquatilis. *A: Female anterior region; B: Perineal region in lateral view; C: Male anterior region; D, E: Male tail; F: Labial region of second-stage juvenile (J2); G: Anterior region of J2; H, I: Tail of J2. (After Ebsary & Eveleigh, 1983.)*

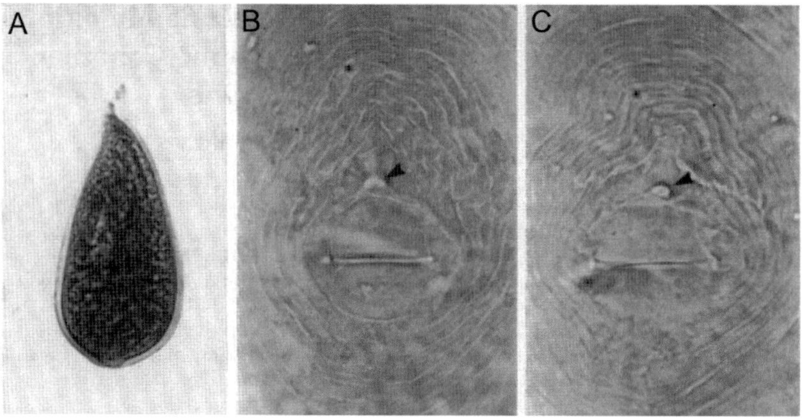

Fig. 39. Meloidogyne aquatilis. *A: Female; B, C: Perineal patterns (anus arrowed). (After Ebsary & Eveleigh, 1983.)*

vulva-anus distance = 14 (12-16) μm; interphasmid distance = 14 (12.5-17.0) μm; distance mid-way between phasmids to anus = 6.5 (6.0-7.5) μm; cuticle thickness at mid-body = 29 μm; annulus wide = 4-5 μm; EP/ST = 1.6 (estimated from drawings).

- *Allotype male*: L = 975 μm; a = 42; b = 7; c = 78; stylet = 16 μm, spicules = 24 μm.
- *Paratype males* (n = 3): L = 820 (775-874) μm; a = 40 (36-45); b = 4.5 (4-5); c = 103 (96-108); stylet = 14 (13-16) μm; stylet knob width = 3 μm; DGO = 3-4 μm; anterior end to excretory pore = 96 (78-111) μm; annulus width = 4 μm; spicules = 26 (24-26) μm; gubernaculum = 6 μm.
- *Paratype J2* (n = 27): L = 459 (418-490) μm; a = 33 (30-36); b = 2.0 (1.9-2.1); c = 7.5 (7-10); stylet = 11 (10-12) μm; stylet knob width = 2 μm; DGO = 3 (2-4) μm; anterior end to excretory pore = 70 (62-76) μm; tail = 61 (48-68) μm; hyaline region = 12 (10-12) μm; distance phasmid-terminus = 11 (9-17) μm.
- *Eggs*: L = 90 (80-106) μm; W = 43 (42-44) μm.

DESCRIPTION

Female

Young and mature females light brown in colour, elongate ovoid to pear-shaped, with a slight posterior protuberance. Annuli particularly conspicuous on pharyngeal and caudal regions, less conspicuous over remainder of body. Labial region slightly offset with two annuli and a prominent labial cap. Labial framework weak. Stylet slender, knobs small, rounded. Excretory pore distinct near base of stylet knobs or slightly posterior. Vulval area recessed into posterior protuberance. Anus covered by a fold of cuticle in vulval depression. Perineal pattern coarse, transversely ovoid-rounded with a high truncate dorsal arch. Striae discontinuous and evenly but widely spaced. Curved dorsoventral striae forming a slight tail whorl. Perivulval region free of striae, but irregularly bound by a fold of cuticle. Lateral lines marked only by forking of dorsal and ventral striae. Phasmids small, indistinct and closer together than vulval width. Egg sac external to root, 3-4 times larger than female.

Male

Body slender, vermiform, tapering anteriorly and posteriorly. Labial region offset by constriction from body, labial cap prominent with one or two smooth post-labial annuli. Annulation prominent. Stylet slender; knobs rounded, sloping posteriorly. Vesicles present posterior to median bulb valve. Hemizonid 7-9 annuli anterior to excretory pore, occupying *ca* two annuli. Lateral field with four incisures, outer incisures crenate. Outer sectors of field completely areolated, inner sector with infrequent transverse striae. Spicules arcuate, head cephalated. Gubernaculum spindle-shaped. Tail rounded, clavate in ventral view.

J2

Labial region with two or three fine annuli, oral disc small. Low labial region slightly offset from body. Labial framework weak. Stylet fine, slender, knobs rounded, posteriorly sloping. Hemizonid 7-9 annuli posterior to excretory pore and occupying *ca* two annuli. Lateral incisures incompletely areolated. Rectum inflated. Phasmids minute, offset to ventral margin of lateral field. Seventy percent of sample with a disc-like structure near slightly clavate terminus.

TYPE PLANT HOST: Prairie cordgrass, *Spartina pectinata* Bosc *ex* Link.

OTHER PLANTS: No other hosts were reported.

TYPE LOCALITY: Ottawa River at Deschênes, Quebec, Canada.

DISTRIBUTION: *North America*: Canada (Quebec).

SYMPTOMS: Females embedded completely in the roots without gall formation.

CHROMOSOME NUMBER: No available data.

POLYTOMOUS KEY CODES: *Female*: A21, B43, C3, D3; *Male*: A4, B4, C3, D1, E1, F2; *J2*: A2, B32, C23, D3, E32, F2.

BIOCHEMICAL AND MOLECULAR CHARACTERISATION: No available data.

RELATIONSHIPS (DIAGNOSIS): Jepson (1987) considered *M. aquatilis* to be in the *graminis* group. *Meloidogyne aquatilis* is most similar in perineal pattern to *M. graminis* and *M. maritima*. It differs by a perineal pattern without a lateral field *vs* having a deeply marked lateral field. The J2 are readily distinguished by having the hemizonid 7-9 annuli posterior to the excretory pore *vs* four annuli posterior in *M. graminis*, a shorter tail (48-68 μm) with a refractive disc-like structure near the tail terminus *vs* a longer tail (52-88 μm) without the disc-like structure in *M. graminis*. The male body length of *M. aquatilis* is shorter than in *M. maritima* (775-874 *vs* 1037-1462 μm).

7. *Meloidogyne arabicida* López & Salazar, 1989
(Figs 40, 41)

COMMON NAME: Costa Rican coffee root-knot nematode.

This species was originally described from coffee in the Turrialba Valley, and later in the San Isidro valley of Costa Rica and presently has

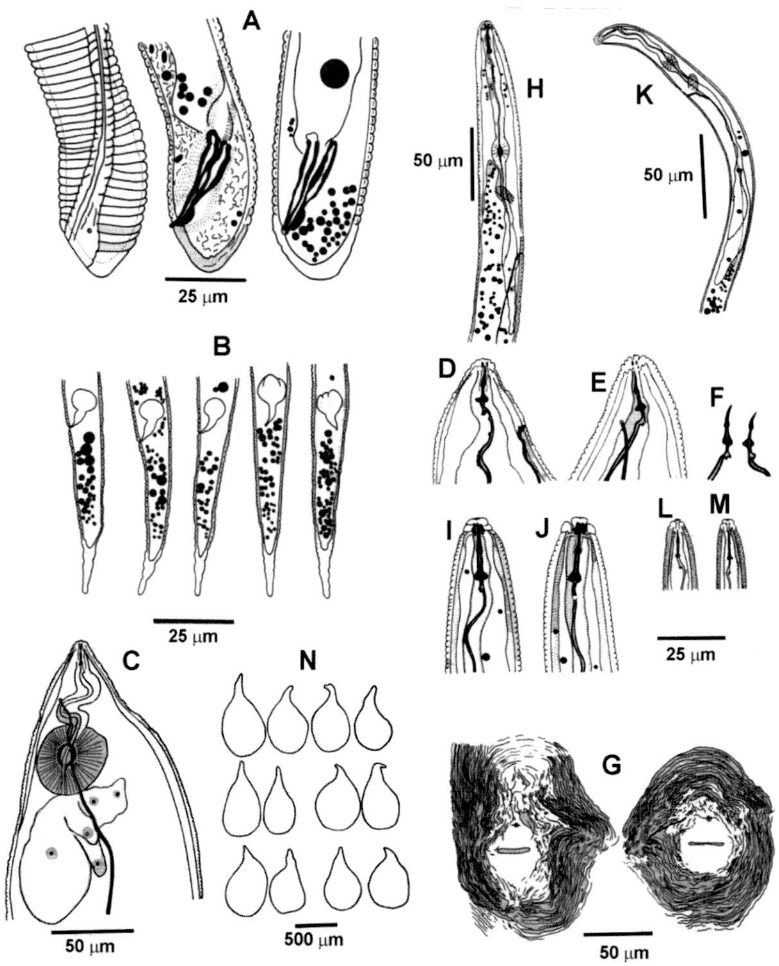

Fig. 40. Meloidogyne arabicida. *A: Male tail region; B: Second-stage juvenile (J2) tail; C-E: Female anterior region; F: Female stylet; G: Perineal pattern; H: Male anterior region; I, J: Male labial region; K: J2 Anterior region; L, M: J2 labial region; N: Entire females. (After López & Salazar, 1989.)*

a limited distribution (Villain *et al.*, 2014). Descriptions of this species were provided by López & Salazar (1989) and Chaves (1994).

MEASUREMENTS *(AFTER LÓPEZ & SALAZAR, 1989)*

- *Holotype female*: L = 665 μm; W = 385 μm; a = 1.72; stylet = 12 μm; DGO = 3 μm; anterior end to excretory pore = 31 μm, to

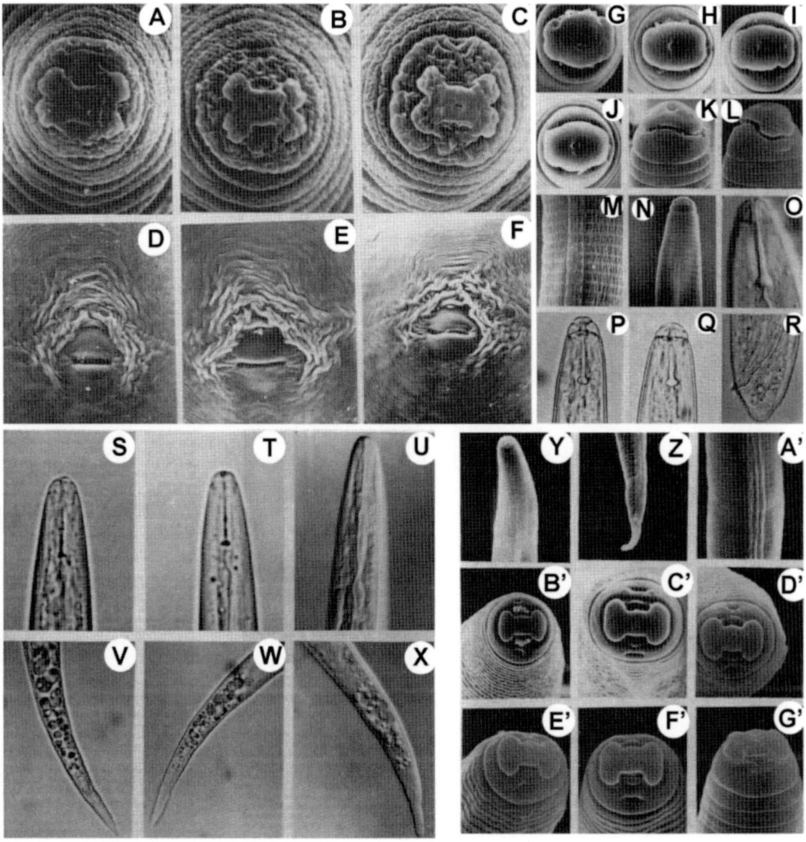

Fig. 41. Meloidogyne arabicida. *LM and SEM. A-C: Female* en face *view; D-F: Perineal pattern; G-L: Male* en face *view; M: Male lateral field; N-Q: Male labial region; R: Male tail; S-U, Y: Second-stage juvenile (J2) labial region; V-X, Z: J2 tail; A′: J2 lateral field; B′-G′: J2* en face *view. (After López & Salazar, 1989.)*

median bulb = 88 μm; median bulb width = 28 μm; median bulb length = 31 μm; median bulb valve length = 15 μm; median bulb valve width = 11 μm.

- *Paratype females* (n = 30): L = 773 ± 131.9 (543-1206) μm; W = 445 ± 62.1 (310-586) μm; stylet = 12 ± 1.0 (10-13) μm; stylet knob height = 2.1 ± 0.4 (1.4-2.8) μm; stylet knob width = 3.6 ± 0.4 (2.8-4.5) μm; anterior end to median bulb = 84 ± 8.9 (70-109) μm; median bulb width = 36 ± 6.1 (22-46) μm; median bulb length = 37 ± 4.9

(24-44) μm; median bulb valve width = 11 ± 1.0 (9-13) μm; median bulb valve length = 13 ± 1.3 (10-15) μm; anterior end to excretory pore = 30 ± 7.8 (18-59) μm; vulva = 24 ± 2.2 (20-27) μm; anus-vulva distance = 19 ± 2.5 (14-26) μm; interphasmid distance = 19 ± 2.6 (13-24) μm; a = 1.7 ± 0.2 (1.4-2.6); stylet knob width/height = 1.7 ± 0.3 (1.3-2.5); median bulb width/height = 1.0 ± 0.1 (0.7-1.6); median bulb valve length/width = 1.2 ± 0.1 (0.8-1.5); EP/ST = 2.8.

- *Allotype male*: L = 1768 μm; body diam. = 23.6 μm; labial region height = 4.1 μm; labial region diam. = 10.9 μm; stylet = 18.4 μm; stylet knob height = 2.7 μm; stylet knob width = 4.1 μm; DGO = 4.1 μm; median bulb width = 10.2 μm; median bulb valve width = 3.9 μm; median bulb valve length = 6.1 μm; anterior end to median bulb = 82 μm, to excretory pore = 147 μm; testis = 553 μm; spicules = 31.4 μm; gubernaculum = 9.6 μm; tail = 13.7 μm; distance cloacal aperture to phasmids = 2.3 μm; distance phasmids to terminus = 6.8 μm; a = 75; c = 129.
- *Paratype males* (n = 30): L = 1414 ± 0.25 (905-1881) μm; body diam. = 26 ± 4.6 (16-35) μm; labial region height = 5 ± 0.7 (4-7) μm; labial region diam. = 11 ± 1.3 (9-12) μm; stylet = 19 ± 1.6 (16-22) μm; stylet knob height = 3 ± 0.5 (2-4) μm; stylet knob width = 5 ± 0.6 (3-6) μm; DGO = 4 ± 0.7 (3-5) μm; median bulb width = 11 ± 2.4 (7-16) μm; median bulb valve width = 4 ± 0.9 (3-6) μm; median bulb valve length = 6 ± 1.3 (5-10) μm; anterior end to median bulb = 86 ± 8.7 (57-106) μm, to excretory pore = 140 ± 19.8 (87-178) μm; testis = 624 ± 123.8 (403-846) μm; spicules = 27 ± 4.4 (19-36) μm; gubernaculum = 8 ± 0.9 (7-10) μm; tail = 15 ± 3.0 (9-20) μm; distance cloacal aperture to phasmids = 5 ± 2.7 (2-7) μm; distance phasmids to terminus = 8 ± 2.7 (3-14) μm; a = 55.2 ± 9.2 (41-75); c = 100.1 ± 22.6 (57-143); L/anterior end to median bulb ratio = 17.2 ± 4.3 (10.4-33.1); stylet knob width/height = 1.5 ± 0.3 (0.8-2.3).
- *Paratype J2* (n = 30): L = 443 ± 24.8 (372-480) μm; body diam. = 14 ± 1.4 (12-18) μm; labial region diam. = 5.4 ± 0.6 (4.2-6.7) μm; labial region height = 2.5 ± 0.4 (2.0-3.9) μm; stylet = 11 ± 1.2 (9-15) μm; stylet knob width = 1.8 ± 0.2 (1.4-2.5) μm; stylet knob height = 1.2 ± 0.2 (0.8-2.0) μm; DGO = 3.1 ± 0.6 (2.0-4.7) μm; median bulb valve length = 4.6 ± 0.5 (3.9-6.1) μm; median bulb valve width = 3.9 ± 0.4 (3.1-4.5) μm; pharynx = 113 ± 15 (94-149) μm; anterior end to excretory pore = 83 ± 6.7 (75-102) μm; tail = 52 ±

4.3 (40-62) μm; hyaline region = 14 ± 2.4 (9-21) μm; body diam. at anus level = 10 ± 0.9 (9-12) μm; a = 32.4 ± 3.1 (26.1-40.0); b = 4.0 ± 0.5 (3.0-5.4); c = 8.6 ± 0.7 (7.6-11.6); c′ = 4.9 ± 0.5 (3.8-6.0); tail/hyaline region = 3.6 ± 0.6 (2.5-5.9).
- *Eggs*: L = 90 ± 4.8 (80-101) μm; W = 49 ± 1.8 (45-53) μm; L/W = 1.8 ± 0.1 (1.6-2.2).

DESCRIPTION

Female

Pearl white colour, oval, with pointed anterior part and slightly flat or rounded posterior part. Annulated cuticle, frequently with incomplete annulation in anterior region of body. Labial region slightly separated from rest of body. Under SEM, labial region showing a relatively rectangular labial disc, slightly raised over median labials. Prestoma oval and oriented in dorsoventral plane, surrounded by inner labial sensilla which are difficult to see and open externally through small rounded pores. Stoma with a slit-like opening oriented in dorsoventral plane. Median labials divided in centre by a deep indentation. Amphidial apertures with rectangular slit-like shape and located immediately beneath lateral sides of labial disc. Lateral labials inconspicuous. Labial region formed by only one ring. Labial region appearing sclerotised under light microscope. Stoma vestibule and its extension easily observed. Stylet relatively short and delicate, cone dorsally curved with triangular base as it joins shaft. Shaft relatively short, with same diam. along its length. Three basal knobs, oval and posteriorly projecting. Stylet lumen gradually decreasing when it reaches cone. DGO duct branched. Lumen of pharynx in procorpus relatively wide, frequently showing rounded swellings. Excretory pore located at level of anterior part of procorpus. Excretory duct curved, sometimes with a triangular swelling near excretory pore, duct not very evident when it reaches intestine. Median bulb large, round, muscular with oval and sclerotised valvular apparatus in its centre. Basal glandular portion of pharynx appearing as a massive structure composed of at least three lobules with five large nuclei distinguishable. Perineal patterns with relatively angular contours, with thick striae in centre and thinner in periphery, dorsal arch relatively high and rectangular. Most patterns with lateral projections of striae forming wings, which can be present in both or only one lateral portions. Phasmids small. Vulva elongated, smooth, slit-like, without prominent striae originating from it.

Male

Labial region slightly narrower than rest of body, smooth, with only one ring and labial cap easily distinguished. Under SEM, males showing a labial cap composed of a large and rounded labial disc, amalgamated with median labials, slightly raised over these, with rounded lateral borders. Oval prestoma located in centre of labial disc, surrounded by six inner labial sensilla with tiny pore-like apertures. Stoma aperture slit-like, oriented in dorsoventral plane. Median labials slightly narrower than labial disc. Amphidial apertures narrow slit-like, located immediately beneath lateral borders of labial disc. Cuticle strongly annulated. Lateral fields beginning 10-12 annuli posterior to labial region as two longitudinal incisures with arcuate borders. Transverse annulations crossing lateral fields to form areolation over entire length of field, areolation in central band frequently not agreeing totally with transverse annulation. Labial framework strongly sclerotised. Stylet strong, cone pointed, longer than shaft, with aperture in its anterior portion and a triangular widening where it joins with shaft. Three rounded basal knobs. Stylet lumen almost as wide as procorpus. Procorpus three times longer than oval and muscular median bulb. Median bulb with strong valve apparatus in its central portion. Nerve ring surrounding isthmus. Excretory duct arcuate, indistinguishable when it reaches median level of intestine. Basal lobule of pharynx ventrally overlapping intestine and with three visible nuclei. Hemizonid located 4-5 annuli anteriorly to excretory pore, two annuli long. Intestine dorsally extending in anterior part of body up to lower level of median bulb. Spermatozoids globular, granular. Gubernaculum simple, with an arcuate lower border.

J2

Labial region almost continuous with body, lateral sections slightly narrower than rest of body and labial cap slightly raised. Under LM, labial region appearing smooth, with only one annulus. Under SEM, anterior region of body with an elongated labial cap formed by labial disc that is slightly risen over the median labials. Prestoma oval, located in centre of labial disc, surrounded by six inner labial sensilla which open as minute rounded pores. Amphidial apertures appearing as rectangular slit-like openings located immediately beneath lateral borders of labial disc. Lateral labials narrow, with slightly arcuate borders generally not exceeding line of lateral borders of median labials and at a lower level than labial cap. Labial framework weakly developed.

Body annulated from anterior region up to terminal portion of tail. Lateral fields beginning at level of procorpus as two incisures, increasing to three and then to four. Both outer incisures slightly arcuate. All four incisures reaching anus level where one of two central incisures terminating and other extending slightly further, both outer incisures continuing almost up to terminal portion of tail. Areolation of lateral fields incomplete, especially in central band. Stylet weakly developed with three rounded and posteriorly sloping basal knobs. Procorpus 2.0-2.5 longer than oval and muscular median bulb, which has a sclerotised valvular apparatus in its centre. Nerve ring surrounding narrow isthmus. Hemizonid located 1-2 annuli anterior to excretory pore and *ca* one annulus long. Excretory pore located at level of nerve ring, excretory duct curved and indistinguishable when it reaches intestine. Basal part of pharynx with three nuclei and two distinguishable lobes ventrally overlapping intestine. Anus aperture a small pore, located in central portion of body, rectum dilated. Tail relatively long, ending in a digitate shape and terminus rounded.

TYPE PLANT HOST: Coffee, *Coffea arabica* L.

OTHER PLANTS: No other hosts were reported.

TYPE LOCALITY: Juan Viñas, province of Cartago, Costa Rica.

DISTRIBUTION: *Central America*: Costa Rica.

SYMPTOMS: *Meloidogyne arabicida* causes severe die-back symptoms on coffee and is frequently associated with the coffee disease complex known as corky root (López & Salazar, 1989). Coffee corky-root disease, also called corchosis, was first detected in 1974 in a small area of Costa Rica where the RKN *M. arabicida* is the dominant species. Diseased plants showed a progressive decline, starting with leaf chlorosis, followed by the fall of flowers and fruits and leading to plant death in a few years. The root systems of the diseased plants show reduced growth and many galls, leading to an extensive development of corky tissue on both main and secondary roots. Further testing and observation revealed that typical corky-root symptoms were observed only in plants inoculated with a combination of *Fusarium oxysporum*

Schltdl. and *M. arabicida*. *Meloidogyne arabicida* on its own caused galls and a reduction in shoot height, but no corky-root symptoms (Bertrand *et al.*, 2000).

PATHOGENICITY: It is recommended that local authorities pay careful attention to the coffee planting material being transported from this region because of the high degree of pathogenicity of this nematode on *C. arabica* and linked with its association with *F. oxysporum* (Villain *et al.*, 2013).

CHROMOSOME NUMBER: No available data.

POLYTOMOUS KEY CODES: *Female*: A213, B34, C2, D6; *Male*: A3124, B324, C312, D1, E1, F1; *J2*: A23, B312, C32, D213, E23, F32.

BIOCHEMICAL AND MOLECULAR CHARACTERISATION: *Meloidogyne arabicida* has a unique esterase phenotype (EST AR2, Rm: 1.20, 1.40) (Carneiro *et al.*, 2004c; Hernandez *et al.*, 2004; Villain *et al.*, 2013). Correa *et al.* (2013) developed SCAR specific primers for diagnostics of *M. arabicida*. Sequences of the 18S rRNA, the D2-D3 of 28S rRNA, *COII*-16S rRNA genes and PCR-*COII*-16S rRNA-RFLP profile for *M. arabicida* (Fig. 13) were provided in several publications (Tigano *et al.*, 2005; Tomalova *et al.*, 2012; Humphreys-Pereira *et al.*, 2014).

RELATIONSHIPS (DIAGNOSIS): The species belongs to Molecular group I, the Incognita group. This species was distinguished from other known species by esterase phenotype. *COII*-16S rRNA gene sequences enable differentiation of this nematode from related *M. lopezi* and others. *Meloidogyne arabicida* can be distinguished from *M. coffeicola* by the shape and size of the body and characteristics of the perineal pattern, smaller max. body diam. and shorter stylet, higher values for the a and c indexes of the male and longer body length and longer J2 tail (López & Salazar, 1989).

8. *Meloidogyne ardenensis* Santos, 1968
(Figs 42, 43)

COMMON NAME: Arden root-knot nematode.

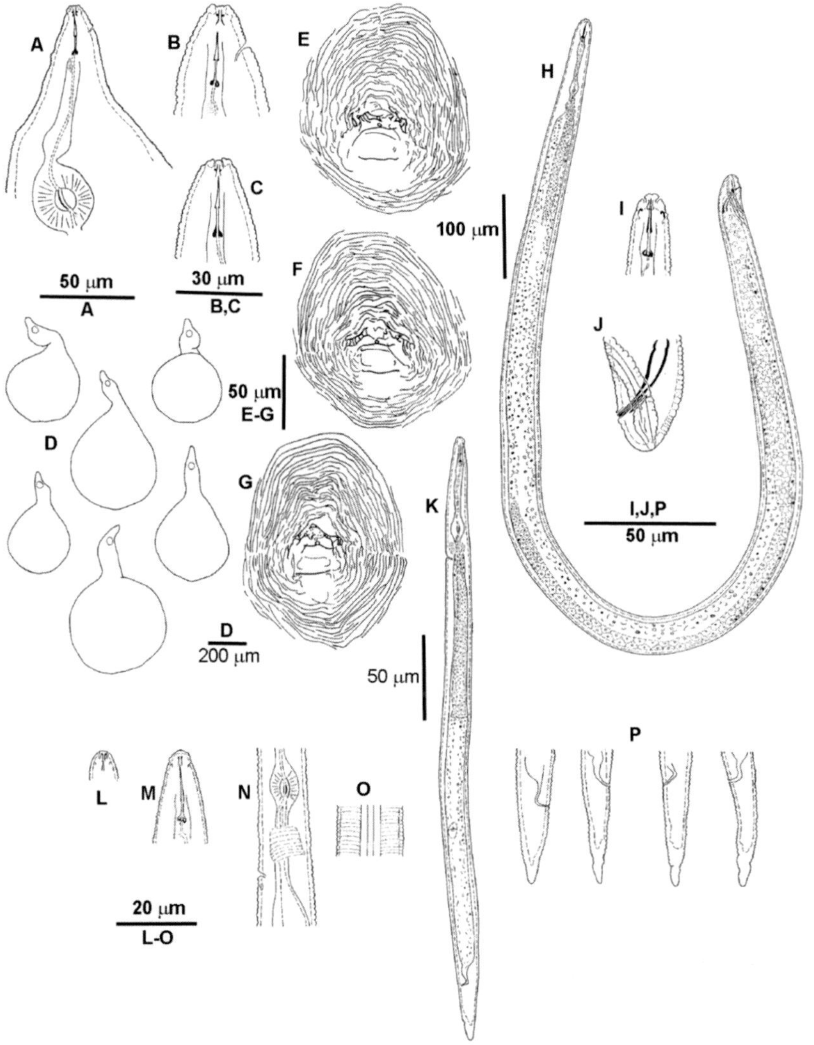

Fig. 42. Meloidogyne ardenensis. *A-C: Female anterior region; D: Entire females; E-G: Perineal pattern; H: Entire male; I: Male anterior region; J: Male tail; K: Entire second-stage juvenile (J2); L, M: J2 labial region; N: Pharyngeal region of J2; O: Lateral field at mid-body of J2; P: J2 tail. (After Santos, 1965.)*

This species was first described from *Vinca minor* L. and woodland plants in the UK and was later found in other plants across Europe. After analysis, Karssen & van Hoenselaar (1998) found that *M. deconincki*

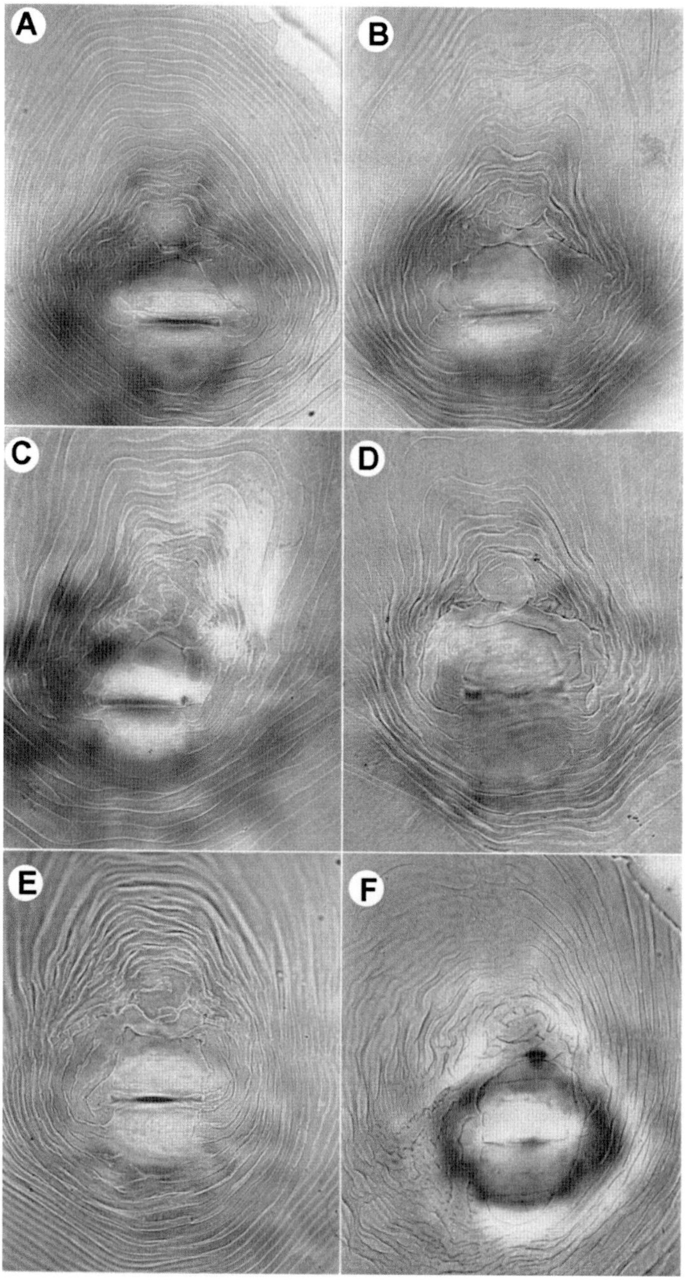

Fig. 43. Meloidogyne ardenensis. *LM. A-F: Perineal patterns. (After Santos, 1965.)*

Elmiligy, 1968 and *M. litoralis* Elmiligy, 1968 did not differ from *M. ardenensis* in morphology, host reactions and tested isozymes and concluded that these descriptions were based on a mixture of *M. ardenensis* and *M. hapla*. These authors considered *M. deconincki* and *M. litoralis* as synonyms of *M. ardenensis* and we accept this synonymy.

MEASUREMENTS

See Table 17.
After Santos, 1968:

- *Holotype female*: L (including neck) = 595 μm; neck length = 200 μm; W = 410 μm; median bulb length = 35 μm; median bulb width = 27 μm; stylet = 18 μm; DGO = 4 μm.
- *Allotype male*: L = 1420 μm; W = 43 μm; stylet = 22 μm; DGO = 3 μm; spicules = 37 μm.
- *Eggs* (n = 25): L = 113 (97-125) μm; W = 37 (33-46) μm; L/W ratio = 2.9 (2.6-3.4).

DESCRIPTION

Female

Body pyroid with long, well-defined neck and slight posterior prominence bearing vulva and anus. Cuticle in mature female 9 (7-12) μm thick on spherical part of body. Labial region low, trapezoidal, with two annuli posterior to labial cap. In dorsoventral view amphid openings visible between cap and first annulus. Stylet curved dorsally, anterior part a little longer than posterior, stylet knobs fairly prominent, rounded, posteriorly sloping. Excretory pore on 9th (8-11) annulus posterior to labial region. Posterior cuticular pattern oval in shape, made up of wavy striae well separated and forming a medium to high trapezoidal arch. Lateral fields marked by change in direction or breaks in the striae, sometimes by short lateral lines, but some striae are continuous from dorsal to ventral region. There are fringes of striae within inner lateral line region on at least one side. Wide anal fold appearing as a curved line between anus and vulva. A stippled area sometimes present between anus and tail tip.

Male

Labial region truncated cone-shaped, apparently four labial annuli in dorsal or ventral views and three in lateral view. Anterior part of

Table 17. *Morphometrics of female, male and second-stage juvenile (J2) of* Meloidogyne ardenensis. *All measurements are in μm and in the form: mean ± s.d. (range).*

Character	England, UK, (Santos, 1968)	Scotland, UK, (Thomas & Brown, 1981)	(Karssen & van Hoenselaar, 1998)	Moscow region, Russia, (Chizhov & Turkina, 1986)
Female (n)	25	10	9	4
L	634 (472-795)	-	-	430-500
W	405 (309-520)	-	-	390-450
Neck length	209 (153-300)	-	-	-
Stylet	17 (15-19)	13 ± 2.2 (11-17)	18.2 ± 0.7 (17.1-19.0)	-
Stylet knob width	-	-	-	-
DGO	4 (3-6)	-	5.6 ± 0.5 (5.1-6.3)	-
Ant. end to median bulb	-	-	76 ± 12.2 (63-95)	-
Median bulb length	35 (30-41)	-	38.5 ± 3.0 (34.8-45.5)	-
Median bulb width	30 (26-35)	-	30.2 ± 1.4 (28.4-32.9)	-
Median bulb valve length	12 (10-13)	-	-	-
Median bulb valve width	9 (7-11)	-	-	-
Vulval width	26 (23-30)	-	-	25-29
Interphasmid distance	29 (24-34)	-	-	-
Distance from vulva centre to anus	20 (17-25)	-	-	20-25
EP/ST*	0.8			
Male (n)	7	3	10	1
L	1479 (965-1755)	1522 (765-2243)	1609 ± 272 (1062-1939)	1650

Table 17. *(Continued.)*

Character	England, UK, (Santos, 1968)	Scotland, UK, (Thomas & Brown, 1981)	(Karssen & van Hoenselaar, 1998)	Moscow region, Russia, (Chizhov & Turkina, 1986)
a	-	33.0 (17.8-46.7)	40.7 ± 4.7 (31.6-46.5)	41.3
b	-		-	8.0
c	-	95.2 (69.5-124.6)	-	100.3
Body diam.	39 (25-48)	45 (43-48)	39.5 ± 3.8 (32.9-44.2)	-
Labial region height	6 (5-7)	-	-	5
Labial region diam.	12 (10-13)	-	-	13
Stylet	22 (17-24)	24 (23-24)	22.5 ± 0.6 (21.5-23.5)	20
Stylet knob width	4 (4-5)	-	-	4
DGO	4 (3-4)	-	5.4 ± 0.3 (5.1-5.7)	-
Median bulb length	-	-	-	-
Median bulb width	10 (10-11)	-	-	-
Median bulb valve length	-	-	-	-
Ant. end to median bulb	-	95 (83-103)	74 ± 5.0 (64-82)	90
Ant. end to excretory pore	-	279 (271-288)	-	-
Tail	-	15 (11-18)	-	13
Spicules	34 (28-38)	38 (37-38)	36.5 ± 1.0 (35.4-37.9)	32
Gubernaculum	10 (7-11)		9.8 ± 0.4 (9.5-10.7)	6.5
J2 (n)	25	15	11	20
L	417 (372-453)	412 ± 11.3 (391-426)	407 ± 27.5 (365-451)	410 (370-430)

Table 17. *(Continued.)*

Character	England, UK, (Santos, 1968)	Scotland, UK, (Thomas & Brown, 1981)	(Karssen & van Hoenselaar, 1998)	Moscow region, Russia, (Chizhov & Turkina, 1986)
a	25.7 (21.8-31.5)	22.9 ± 0.8 (21.8-24.5)	18.5 ± 2.9 (15.3-24.3)	24.4 (20.7-29.1)
b	-	-	-	2.3 (2.0-2.7)
c	10.6 (8.6-12.3)	10.6 ± 0.7 (9.7-12.2)	10.3 ± 0.7 (9.6-11.2)	9.4 (8.7-10.9)
c′	-	3.5 ± 0.5 (2.9-4.8)	3.4 ± 0.3 (2.9-3.9)	-
Body diam.	16 (15-18)	18 ± 0.8 (17-19)	22.5 ± 3.4 (18.3-27.2)	17 (14-19)
Labial region height	-	-	-	2.4-3.2
Labial region diam.	-	-	-	5-6
Stylet	12 (9-14)	12 ± 1.4 (10-14)	12.4 ± 0.5 (12.0-13.2)	-
DGO	2 (2-3)	-	3.5 ± 0.3 (3.2-3.8)	2-4
Median bulb length	-	-	-	9-11
Median bulb valve length	-	-	-	12-15
Ant. end to median bulb	-	-	-	50-70
Ant. end to excretory pore	-	-	-	55-85
Tail	39 (32-45)	39 ± 2.3 (35-41)	39.7 ± 1.8 (36.7-41.7)	35-50
Hyaline region	-	-	-	12-15

* Estimated from drawing.

stylet a little longer than posterior and stylet knobs rounded. Lateral field marked for greater length of the body by four incisures, outer bands partially areolated. Excretory pore one or two annuli posterior to hemizonid. Phasmids opposite cloacal aperture. Spicules curved ventrally and gubernaculum is crescentic. Stylet curved dorsally, anterior part a little longer than posterior, stylet knobs fairly prominent, rounded, posteriorly-sloping. Excretory pore on 9th (8-11) annulus posterior to labial region. Tail bluntly rounded with terminus not striated, length *ca* 60% of anal body diam.

J2

Body clearly annulated. Labial region truncated cone-shaped, not offset, in lateral view three annuli posterior to labial cap but none seen in dorsal or ventral views, distal sclerotisation of vestibule lining knob-like. Stylet slender, anterior part somewhat longer than posterior, which has posteriorly-sloping knobs. Pharynx gland lobe lying latero-ventrally over intestine. Hemizonid two or three annuli long, usually 3-5 annuli posterior to excretory pore. Lateral field marked by three bands (four incisures), starting anteriorly as a narrow band posterior to stylet and narrowing gradually posterior to anus, terminating about halfway along tail. Tail conoid with rounded, unstriated terminus. Hyaline tail part with cuticular constriction. Rectum not swollen.

TYPE PLANT HOST: Lesser periwinkle, *Vinca minor* L.

OTHER PLANTS: Parasitises trees, shrubs and dicotyledon weeds. The host list includes: *Astilbe* sp., *Anemone angulosa* Lam., *Avena sativa* L., *Betula pendula* Roth., *Betula* sp., *Carpinus betulus* L., *Crataegus monogyna* L., *Cucumis sativus* L., *Daucus carota* L., *Fragaria* × *ananassa* Duchesne, *Fraxinus excelsior* L., *Hordeum vulgare*, *Ligustrum vulgare* L., *Lolium multiflorum* Lam., *Lonicera nitida* Wils., *Solanum lycopersicum*, *Pinus spinosa* L., *Rubus fruticosis* L., *Rosa canina* L., *Sambucus nigra* L., *Samburus* sp., *Solanum nigrum*, *Triticum aestivum*, *Quercus robur* L., and other plants (Santos, 1968; Franklin, 1976; Sturhan, 1976; Thomas & Brown, 1981; Stephan & Trudgill, 1982; Chizhov & Turkina, 1986; Jepson, 1987; Southey, 1993; Karssen & van Hoenselaar, 1998).

TYPE LOCALITY: Ranby, 5.6 km west of Retford, Nottinghamshire, on east side of the main road from Nottingham to Doncaster, UK.

DISTRIBUTION: *Europe*: Belgium, France, Germany, Norway, Poland, The Netherlands, Russia, Slovakia, UK (Wesemael *et al.*, 2011).

BIOLOGY: In Scotland *M. ardenensis* has only one generation per year. In pots on tomato, *M. ardenensis* completed its life cycle in 60 days at 18°C and development from female to female was continuous (Stephan & Trudgill, 1982). In Russia it may develop 2-3 generations (Chizhov & Turkina, 1986).

SYMPTOMS: This nematode induces elongate and small (1-3 mm) galls with development of few secondary roots.

PATHOGENICITY: Sturhan (1976) considered that the nematode could be a harmful parasite in tree culture. The experiments showed that *M. ardenensis* can invade tomato roots and cause some damage (Stephan, 1983).

CHROMOSOME NUMBER: 2n = 51-54, a triploid mitotic parthenogenetic species. Despite being parthenogenetic, males of *M. ardenensis* appear to be sexually active, as spermatheca in all specimens studied were filled (Janssen *et al.*, 2017).

POLYTOMOUS KEY CODES: *Female*: A23, B21, C4, D6; *Male*: A324, B234, C21, D1, E1, F1; *J2*: A23, B23, C34, D23, E2, F34.

BIOCHEMICAL AND MOLECULAR CHARACTERISATION: Multiple banding esterase pattern and malate dehydrogenase pattern (N1a) are given by Karssen & van Hoenselaar (1998). Sequence information is given for 18S rRNA (Holterman *et al.*, 2009), *COI* (Janssen *et al.*, 2017), RNA polymerase II (Rybarczyk-Mydłowska *et al.*, 2013) genes.

RELATIONSHIPS (DIAGNOSIS): The species belongs to Molecular group V and is clearly differentiated molecularly from all other *Meloidogyne* species. Jepson (1987) considered this species to be in the *exigua* group. It differs from other species by small EP/ST and perineal pattern.

9. *Meloidogyne arenaria* (Neal, 1889) Chitwood, 1949
(Figs 44, 45)

COMMON NAME: Peanut root-knot nematode.

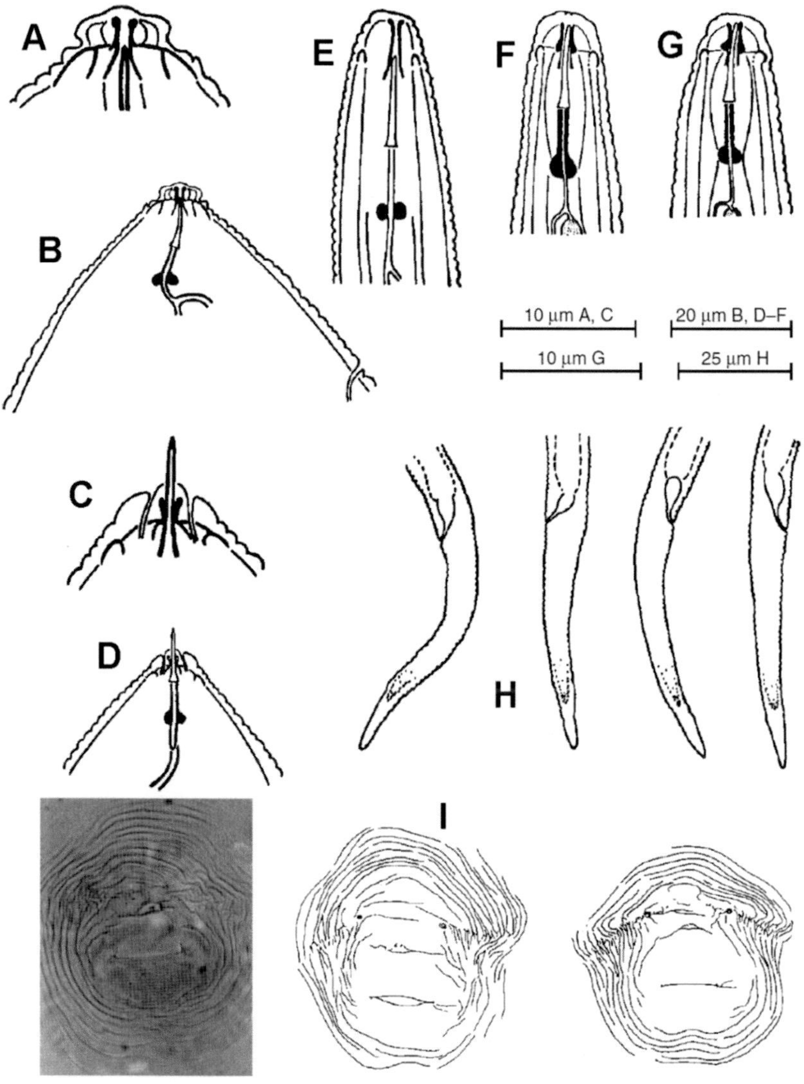

Fig. 44. Meloidogyne arenaria. *A-D: Female anterior region; E, F: Male labial region; G: Labial region of second-stage juvenile (J2); H: Tail of J2; I: Perineal pattern. (After Orton Williams, 1975.)*

Meloidogyne arenaria is considered to be one of the major economically important RKN because of its wide distribution and damage. It occurs in warmer regions and is seldom found in areas where the aver-

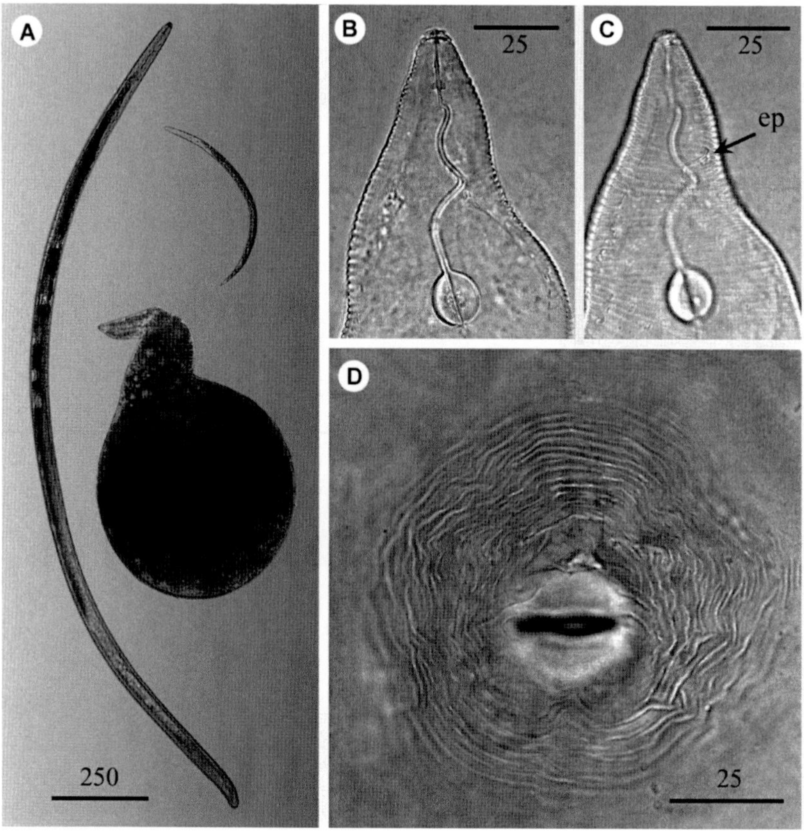

Fig. 45. Meloidogyne arenaria. *LM. A: Entire female, male and second-stage juvenile; B, C: Female anterior region; D: Perineal pattern. Abbreviation: ep = excretory pore. (Scale bars in μm.) (Courtesy of N. Vovlas.)*

age monthly temperatures approach freezing. It is the most morphologically, isoenzymatically, cytologically and molecularly variable of the *Meloidogyne* species (Eisenback & Triantaphyllou, 1991; Rammah & Hirschmann, 1993; Carneiro *et al.*, 2008). Descriptions of *M. arenaria* are given by many authors (Chitwood, 1949; Whitehead, 1968; Eisenback & Hirschmann, 1979, 1980, 1981; Eisenback *et al.*, 1981; Eisenback, 1982; Cliff & Hirschmann, 1984; Osman *et al.*, 1985; Jepson, 1987; Hirschmann & Rammah, 1993; Carneiro *et al.*, 2008; Skantar *et al.*, 2008; García & Sánchez-Puerta, 2012 and others).

Meloidogyne thamesi was first identified as *M. arenaria* from *Boehmeria nivea* (L.) Gaud. from Florida and illustrated by Chitwood (1949), and then designated as a subspecies of *M. arenaria* by Chitwood *et al.* (1952). Peanut was also recognised as not a good host (Chitwood *et al.*, 1952). This nematode was described as a new species by Goodey (1963) and included in the work of Whitehead (1968) with measurements and description. Siddiqi (2000) considered this species as valid; however, it is a synonym of *M. arenaria* according to Hunt & Handoo (2009) and we follow this synonymy.

MEASUREMENTS *(AFTER EISENBACK* ET AL., *1981; CLIFF & HIRSCHMANN, 1985; CARNEIRO* ET AL., *2008)*

See Table 18.

DESCRIPTION

Female

Body creamy-white, pyriform with prominent neck, without tail protuberance. Body annuli fine. Labial region slightly offset, very small relative to total body. Labial disc rounded, raised above median labials in lateral view. Stylet very robust; both cone and shaft broad. Shaft increasing in width posteriorly and gradually merging with stylet knobs. Stylet knobs large, rounded to tear drop-shaped, gradually merging with shaft, in some populations knobs more offset from shaft. Perineal patterns highly variable and not very useful for identification. Dorsal arch low and rounded to high and squarish. Striae in arch indented at lateral lines forming a shoulder on arch. Dorsal and ventral striae often meeting at an angle at lateral lines. Some striae forking and short and irregular near lateral lines. Striae smooth to wavy and some may bend toward vulva. Patterns may also have striae that extend laterally to form one or two wings.

Male

Body of variable length, usually slender, tapering to bluntly rounded ends, posterior region twisted through 90 degrees. Lateral field with four incisures, areolated throughout. Size of labial region variable, usually short. Shape of labial region is a useful diagnostic character for isolates A1N1, A2N1, A2N3 race 1 and race 2. In face view, labial disc large and rounded and fused with crescentic median labials. Lateral labials usually absent and only occasionally marked by short remnant lines. In lateral view, large, smooth, labial cap variable in height, but usually low to moderately raised and extending a short distance onto

Table 18. *Morphometrics of female, male and second-stage juvenile (J2) of* Meloidogyne arenaria. *All measurements are in μm and in the form: mean ± s.d. (range).*

Character	Chitwood (1949)	Osman *et al.* (1985), Florida, USA, Race 1	Osman *et al.* (1985), Florida, USA, Race 2	Hirschmann & Rammah (1993), Hypotriploid population	Hirschmann & Rammah (1993), Triploid populations
Female (n)	-	-	-	25	150
L	510-1000	828 ± 81.5 (666-981)	739 ± 100 (565-962)	-	-
a	-	1.6 ± 0.3 (1.0-2.4)	1.4 ± 0.2 (1.1-2.1)	-	-
W	400-600	545 ± 114.7 (381-791)	524 ± 85 (378-736)	-	-
Stylet	14-16	15.2 ± 0.9 (13.8-17.2)	15.5 ± 1.0 (13.4-17.2)	16.3 ± 0.1* (14.8-17.8)	15.1 ± 0.1* (13.4-16.7)
Stylet knob width	4-5	4.6 ± 0.5 (3.4-5.3)	4.3 ± 0.5 (3.1-5.0)	4.5 ± 0.1 (4.1-5.0)	4.7 ± 0.0 (3.8-5.5)
Stylet knob height	2	3.0 ± 0.4 (1.9-3.8)	2.8 ± 0.3 (2.2-3.4)	2.3 ± 0.0 (1.9-2.6)	2.8 ± 0.0 (2.1-3.8)
DGO	4-6	4.8 ± 0.7 (3.4-6.6)	4.1 ± 0.9 (2.5-6.3)	4.8 ± 0.2 (3.1-8.2)	4.8 ± 0.1 (3.1-6.6)
Median bulb length	-	49.3 ± 5.2 (40.6-66.6)	44.1 ± 3.4 (38.4-50.9)	-	-
Median bulb width	-	43.2 ± 3.9 (35-50)	42.5 ± 4.2 (35-52.5)	-	-
Median bulb valve width	-	11.5 ± 1.3 (9.1-13.8)	11.1 ± 1.0 (9.4-13.1)	-	-
Median bulb valve length	-	17.4 ± 1.8 (13.4-20.9)	14.4 ± 1.7 (10.0-17.8)	-	-
Ant. end to excretory pore	-	47 ± 12.5 (24.7-75.9)	33.5 ± 7.5 (21.9-56.3)	-	-
Vulval slit	-	27.1 ± 2.8 (20.9-32.8)	24.0 ± 2.5 (18.1-31.3)	-	-
Vulval slit to anus distance	-	19.8 ± 2.5 (15.0-25.9)	19.6 ± 2.3 (15.6-25.3)	-	-
Interphasmid distance	28-31	-	-		-
EP/ST*	2.4				
Male (n)	-	-	-	30	150
L	1270-2000	1544 ± 233 (987-2018)	1996 ± 205 (1556-2446)	1598 ± 28.1 (1215-1903)	1720 ± 23.4 (979-2278)

Table 18. *(Continued.)*

Character	Chitwood (1949)	Osman *et al.* (1985), Florida, USA, Race 1	Osman *et al.* (1985), Florida, USA, Race 2	Hirschmann & Rammah (1993), Hypotriploid population	Hirschmann & Rammah (1993), Triploid populations
a	44-65	48.6 ± 4.9 (37.6-57.8)	51.5 ± 6.2 (24.7-60.5)	47.2 ± 0.9 (37.3-56.5)	48.1 ± 0.8 (30-63.7)
c	-	104 ± 15.9 (76-145)	155 ± 25.7 (111-212)	124.9 ± 3.5 (94.4-167.8)	126.6 ± 2.4 (87.9-190.4)
Body diam.	-	31.7 ± 2.8 (25.6-36.3)	38.1 ± 2.8 (30.9-45.0)	34.1 ± 0.7 (25.4-45.7)	36.4 ± 0.4 (27.4-47.5)
Stylet	20-24	22.6 ± 1.8 (17.5-25.9)	24.3 ± 1.7 (20.0-26.9)	22.2 ± 0.1 (20.6-23.5)	22.7 ± 0.1 (19.8-28.4)
Stylet knob width	4-5	4.5 ± 0.5 (3.4-5.9)	5.5 ± 0.4 (4.4-6.3)	5.1 ± 0.1 (4.4-5.7)	4.9 ± 0.0 (2.8-4.6)
Stylet knob height	3	3.3 ± 0.4 (2.5-3.8)	3.6 ± 0.5 (2.8-4.7)	2.7 ± 0.0 (2.2-3.1)	3.5 ± 0.0 (2.8-4.6)
DGO	4-7	6.0 ± 0.7 (4.7-7.2)	3.8 ± 0.8 (2.5-6.3)	5.5 ± 0.2 (3.6-7.4)	5.8 ± 0.1 (3.7-7.9)
Median bulb width	-	14.0 ± 2.2 (9.4-18.8)	14.2 ± 1.6 (10.0-17.2)	-	-
Median bulb valve length	-	8.9 ± 1.7 (6.9-14.7)	7.3 ± 1.0 (5.9-10.0)	-	-
Ant. end to median bulb valve	-	100 ± 12.7 (64-116)	97 ± 11.0 (65-119)	97.6 ± 1.2 (74.0-114.7)	105.3 ± 0.9 (82.8-121.3)
Ant. end to excretory pore	-	166 ± 24.1 (118-213)	174 ± 19.1 (130-206)	161 ± 1.7 (144.3-185)	172.8 ± 1.6 (119.3-213.2)
Spicules	31-34	26.8 ± 3.1 (21.3-33.8)	31.1 ± 3.2 (25.9-38.1)	33.6 ± 0.3 (29.6-37.0)	31.7 ± 0.2 (26.7-39.4)
Gubernaculum	-	8.2 ± 0.4 (7.2-8.8)	9.0 ± 2.0 (6.6-13.1)	8.7 ± 0.2 (7.4-10.5)	9.0 ± 0.2 (7.3-10.4)
Tail	-	14.9 ± 2.3 (9.7-19.4)	13.1 ± 1.5 (10.3-15.6)	13.0 ± 0.3 (10.2-18.9)	13.5 ± 0.1 (10.7-16.7)
J2 (n)	-	-	-	30	150
L	450-490	494 ± 3.4 (413-556)	430 ± 18.1 (400-484)	452 ± 4.1 (400-490)	504 ± 4.3 (392-605)
a	26-32	31.2 ± 2.2 (24.9-36.7)	28.3 ± 1.2 (25.5-31.8)	30.9 ± 0.3 (27.6-34.1)	33.1 ± 0.3 (22.4-40.5)
b	7.2-7.8	-	-		-
c	6.0-7.5	7.5 ± 0.3 (6.9-8.4)	7.8 ± 0.3 (7.3-8.5)	8.4 ± 0.1 (7.6-9.4)	9.0 ± 0.1 (7.5-10.9)
Body diam.	-	15.8 ± 0.8 (14.4-17.8)	15.2 ± 0.7 (14.1-16.9)	14.6 ± 0.1 (13.8-15.9)	15.3 ± 0.1 (12.8-17.8)

Table 18. *(Continued.)*

Character	Chitwood (1949)	Osman *et al.* (1985), Florida, USA, Race 1	Osman *et al.* (1985), Florida, USA, Race 2	Hirschmann & Rammah (1993), Hypotriploid population	Hirschmann & Rammah (1993), Triploid populations
Stylet	10	12.8 ± 0.9 (10.9-14.4)	13.5 ± 0.9 (10.9-15.6)	11.4 ± 0.1 (10.8-11.8)	11.1 ± 0.0 (10.1-11.9)
Stylet knob width	2	2.0 ± 0.3 (1.3-2.5)	2.1 ± 0.3 (1.3-2.8)	-	-
Stylet knob height	1	1.5 ± 0.3 (0.9-2.5)	1.3 ± 0.2 (0.9-1.9)	-	-
DGO	3	4.0 ± 0.7 (2.5-5.0)	3.1 ± 0.4 (2.2-4.1)	3.4 ± 0.1 (2.8-4.2)	3.7 ± 0.0 (2.8-4.7)
Median bulb valve width	-	4.0 ± 0.3 (3.4-4.7)	4.0 ± 0.3 (3.1-4.7)	-	-
Median bulb valve length	-	4.8 ± 0.4 (4.1-6.0)	4.7 ± 0.4 (3.8-5.3)	-	-
Ant. end to median bulb valve	-	61.6 ± 4.3 (49.4-69.7)	58 ± 2.8 (52.5-63.8)	60.3 ± 0.3 (56.5-64.6)	60.9 ± 0.4 (49.4-71.2)
Ant. end to excretory pore	-	90 ± 6.7 (71-102)	84 ± 4.4 (72-96)	86.7 ± 0.6 (79.2-91.0)	89.8 ± 0.6 (75.0-105.2)
Tail	-	66.1 ± 4.6 (53.1-75.0)	55 ± 3.3 (48.4-63.4)	54.1 ± 0.5 (48.1-60.3)	56 ± 0.5 (43.6-69.4)
Hyaline region	-	10.8 ± 1.6 (8.1-14.4)	13.4 ± 1.2 (10.9-16.3)	-	-
Anal body diam.	-	11.5 ± 0.7 (10.0-13.1)	10.7 ± 0.8 (9.4-13.4)	-	-

* Estimated from drawing.

labial region. Labial region usually smooth and often marked by two or three incomplete annulations (A2N1) or rarely marked by one or two incomplete annulations (A2N3, race 1 and race 2) (Carneiro *et al.*, 2008). Hemizonid 1-6 body annuli anterior to excretory pore. One testis or two testes, outstretched. Sperm large, globular, granular. Spicules slender, arcuate, size of spicule head varying among populations. Gubernaculum small, slender, crescentic. Tail short, bluntly rounded. Phasmids prominent, located at cloacal aperture level.

J2

Body generally long, slender, tapering to bluntly rounded anterior and sharply pointed posterior ends. Body annulations fine becoming irregular and larger in tail region. Lateral field with four incisures. In SEM, labial disc rounded to elongate and slightly rectangular, sometimes distinctly raised above median labials. Median labials large, ovoid to triangular. In face view fused labial disc and median labials forming a dumbbell-shaped labial cap. Labial region usually smooth, occasionally with one or two incomplete annulations (A2N3 race 1 and race 2) (Carneiro *et al.*, 2008). Hemizonid 1-3 annuli anterior to excretory pore. Phasmids small, 8-12 annuli posterior to anal opening. Tail length and shape variable among populations. Long slender tail similar in all populations, terminus narrow and tapering and may bear several distinct annulations. Tail tip finely rounded to pointed. Hyaline region not well defined.

TYPE PLANT HOST: Groundnut, *Arachis hypogaea* L.

OTHER PLANTS: Recorded hosts include numerous vegetables and food crops (*e.g.*, aubergine, Brussels sprouts, cabbage, carrot, celery, endive, lettuce, onion, pea, peppers, potato, spinach, sweet potato and tomato), cucurbits, cereals (*e.g.*, barley, oats, corn, rye and wheat), grasses and pasture legumes (Orton Williams, 1975; Jepson, 1987; Rich *et al.*, 2008; López-Pérez *et al.*, 2011).

TYPE LOCALITY: Archer, Florida, or Lake City, Florida, USA.

DISTRIBUTION: This species is cosmopolitan in distribution, being found in most of the warmer regions of the world ranging from parts of Canada, the USA, Central and South America, through the countries bordering on the Mediterranean, Central and South Africa, the Middle East, India, Malaysia to Japan and Australia, In cooler climates it is frequently encountered in glasshouses. Although widely spread, it does not seem as common as *M. incognita* and *M. javanica*.

CHROMOSOME NUMBER: Reproduction is by mitotic parthenogenesis. Species includes populations with chromosome numbers varying from $2n = 30$ to $3n = 56$. A triploid form with somatic chromosome numbers higher than 50 is the most common and, consequently, is typical of the species. Diploid populations with $2n = 30$-38 and hypotriploid ones with $2n = 40$-48 chromosomes are found in many parts of the world (Triantaphyllou, 1985a; Carneiro *et al.*, 2008).

POLYTOMOUS KEY CODES: *Female*: A213, B2, C2, D4; *Male*: A213, B23, C2, D1, E1, F2; *J2*: A2, B3, C213, D324, E3, F2.

HOST RACES: Three host races are recognised: race 1, which infects peanut, and races 2 and 3, which do not (Table 3) (Taylor & Sasser, 1978; Hartman & Sasser, 1985).

BIOCHEMICAL AND MOLECULAR CHARACTERISATION: *Meloidogyne arenaria* isolates showed three enzymatic phenotypes (esterase and malate dehydrogenase: A2N1, A1N1, A2N3) (Fig. 12). Isolates with enzymatic phenotype A2N3 race 1 belonged to the *M. arenaria* described in 1949 by Chitwood. Isolates with phenotypes A2N1 and A1N1 were considered typical of *M. arenaria* race 2 and they differed morphologically from isolate A2N3 race 1. Isolates of A2N3 race 2 were identified as an atypical *M. arenaria* (Carneiro *et al.*, 2000, 2008). The Est phenotype A3 of *M. arenaria* was identified as *M. morocciensis* (Carneiro *et al.*, 2008; Silva *et al.*, 2013; Mattos *et al.*, 2016).

PCR with species-specific SCAR primers for identification of *M. arenaria* was developed by Zijlstra *et al.* (2000) and others. Carneiro *et al.* (2008) noticed that the reliability of these was not 100%, considering that three isolates with the typical enzymatic phenotypes A2N1 were not identified by this marker. *Meloidogyne arenaria* can be differentiated from many other species by sequences of *COII*-16S rRNA (Pagan *et al.*, 2015) and *nad*5 (Janssen *et al.*, 2016) genes.

RELATIONSHIPS (DIAGNOSIS): The species belongs to Molecular group I, the Incognita group. Considerable variation in perineal pattern was observed both within and among populations of *M. arenaria*. Perineal patterns can be used as a complement to identification of isozyme phenotypes, which provide reliable characters for making diagnostic conclusions of *Meloidogyne* species (Carneiro *et al.*, 1996b, 2000, 2004a, 2008).

10. *Meloidogyne artiellia* Franklin, 1961
(Figs 46-48)

COMMON NAME: British root-knot nematode.

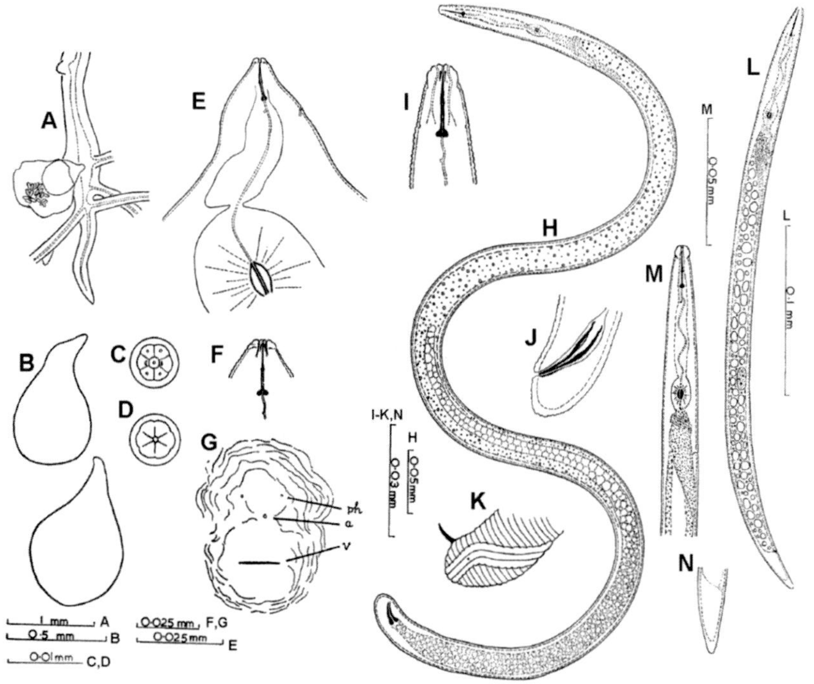

Fig. 46. Meloidogyne artiellia. *A: Cabbage root with mature female; B: Entire females; C:* En face *view of labial region; D: Labial skeleton; E, F: Female labial region; G: Detail of perineal pattern (a = anus; ph = phasmid; V = vulval slit); H: Entire male; I: Male labial region; J, K: Male tail; L: Entire second-stage juvenile (J2); M: Anterior region of J2; N: Tail of J2. (After Franklin, 1961.)*

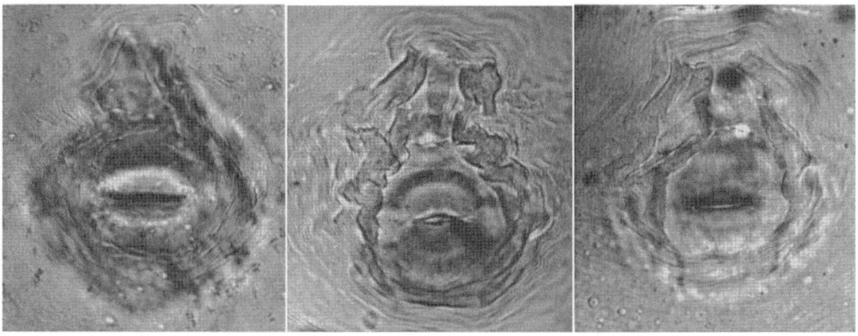

Fig. 47. Meloidogyne artiellia *LM. Perineal patterns. (After Franklin, 1961.)*

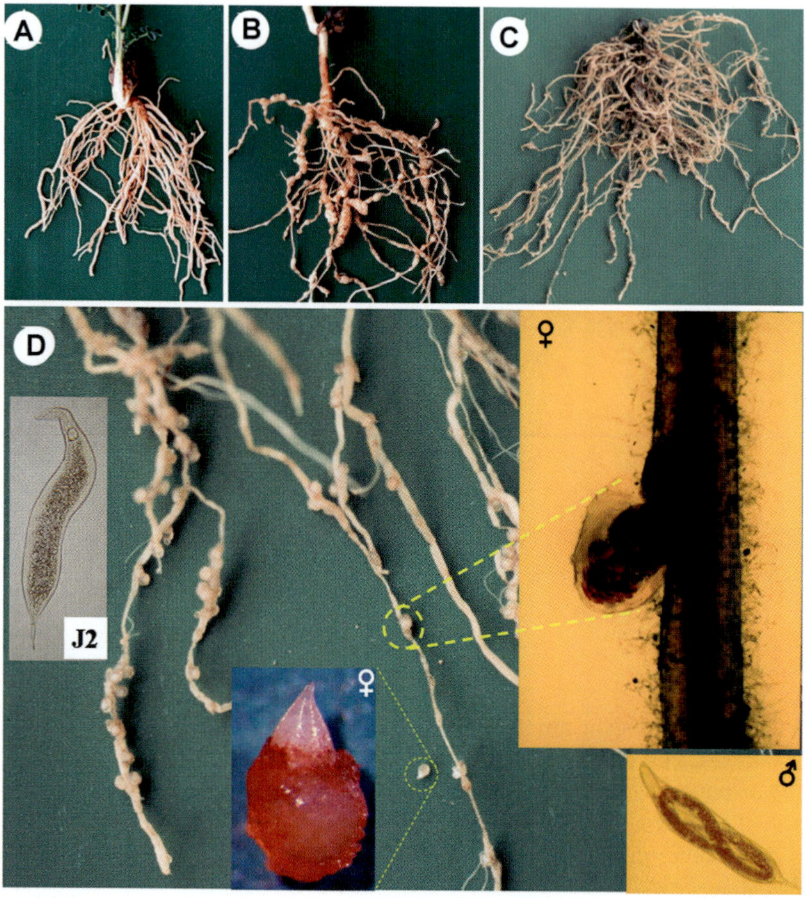

Fig. 48. *Root-knot nematodes (RKN) infecting chickpea roots. A: Healthy; B-D: Chickpea root systems heavily infected by RKN; B:* Meloidogyne incognita*; C, D:* Meloidogyne artiellia*; galled roots and white mature females of* M. artiellia *and higher magnification views of second-stage juvenile (J2), mature female (♀) and immature male (♂). (After Castillo* et al., *2010.)*

Meloidogyne artiellia was recorded for the first time on oats in England (Franklin, 1961) and later reported from several countries in Europe, the Middle East and northern Africa. This nematode can cause severe damage on cereals and vegetables (Greco, 1984; Greco *et al.*, 1984; Di Vito & Greco, 1988a, b; Lombardo *et al.*, 2011; Buisson *et al.*, 2014).

MEASUREMENTS

See Table 19 (after Franklin, 1961; Whitehead, 1968; Jepson, 1987; Karssen & van Hoenselaar, 1998).

After Franklin (1961):

- *Holotype female*: L = 700 μm; W = 400 μm; stylet = 14 μm.
- *Allotype male*: L = 952 μm; W = 41 μm; stylet = 17 μm; spicules = 28 μm; a = 32; b = 13; c = 92; T = 65%.
- *Eggs* (n = 20): L = 95 (75-111) μm; W = 37 (34-43) μm.

DESCRIPTION

Female

Body pear-shaped with broad, short neck and without posterior protuberance. Labial region with two annuli posterior to labial cap. Stylet relatively short, cone straight, knobs relatively small, ovoid and posteriorly sloping. Excretory pore located between stylet knobs and median bulb level. Perineal pattern distinct. General outline of pattern with an upper smaller area enclosing usually quite distinct phasmids. Anus situated at centre and vulva occupying diam. of lower, large part of pattern. Striae fine, continuous and very close together with tail and perivulval area free of striae, except for those that curve to each end of vulva. Dorsal arch angular. No clear lateral lines.

Male

Labial region offset from body. Labial cap distinct, labial disc not elevated, lateral labials present. Labial region without transverse incisures. Short stylet, knobs posteriorly sloping and ovoid-shaped. DGO to stylet knobs relatively long. Lateral field with four incisures for greater length of body, in mid-body bands may be marked by subsidiary incisures (5-7 total incisures), lateral field areolated towards anterior and posterior ends of body, inner band rarely and irregularly striated mid-body, outer bands occasionally areolated mid-body. Spicules not very thick-walled, heads thin-walled, slightly offset, shaft and blade thicker-walled with ventral lateral flange, shaft slightly swollen ventrad, blade of spicule tapering to subacute bifid terminus. Gubernaculum thin, crescentic. Tail rather rostrate.

Table 19. *Morphometrics of female, male and second-stage juvenile (J2) of* Meloidogyne artiellia. *All measurements are in μm and in the form: mean ± s.d. (range).*

Character	UK, (Franklin, 1961)	UK, (Whitehead, 1968)	Turkey, (Imren *et al.*, 2014)	(Karssen & van Hoenselaar, 1998)
Female (n)	8	6	10	9
L	700 (650-760)	-	355.9 ± 6.0 (344-362.6)	-
W	400 (340-460)	-	385.9 ± 7.2 (344-392.3)	-
Stylet	12-16	14 (12-15)	12.9 ± 0.3 (11.5-14.6)	13.9 ± 0.3 (13.3-14.5)
Stylet knob width	-	4 (3-5)	-	-
DGO	4-7	4 (4-5)	-	4.0 ± 0.3 (3.8-4.4)
Ant. end to median bulb	-	-	-	54 ± 3.2 (51-60)
Median bulb length	-	52 (48-55)	-	47.3 ± 3.2 (44.2-53.7)
Median bulb width	-	50 (47-57)	-	45 ± 3.5 (37.9-48.0)
Median bulb valve length	-	13 (13-14)	-	-
Median bulb valve width	-	9 (9-10)	-	-
Vulval width	17.5 (15-22)	-	16.1 ± 0.2 (14.4-18.1)	-
EP/ST*	1.8			
Male (n)	15	-	-	6
L	1070 (820-1370)	-	-	1008 ± 45 (937-1040)
a	-	-	-	37.9 ± 2.9 (34.5-41.7)
c	83 (60-100)	-	-	-
Body diam.	29 (23-36)	-	-	26.7 ± 1.1 (25.3-27.8)
Labial region height	-	-	-	-
Stylet	19 (17-27)	-	-	16.3 ± 0.7 (15.2-16.4)

Table 19. *(Continued.)*

Character	UK, (Franklin, 1961)	UK, (Whitehead, 1968)	Turkey, (Imren *et al.*, 2014)	(Karssen & van Hoenselaar, 1998)
DGO	5-7	-	-	4.9 ± 0.4 (4.4-5.1)
Ant. end to median bulb	-	-	-	56 ± 4.6 (52-60)
Spicules	27 (25-30)	-	-	26.7 ± 0.9 (25.3-27.8)
Gubernaculum	(6-9)	-	-	8.2 ± 0.7 (7.6-8.8)
J2 (n)	20	25	10	10
L	354 (334-370)	336 ± 15 (301-361)	355.9 ± 8 (344-369.6)	345 ± 11.6 (326-367)
a	23.5 (22.0-25.5)	22.8 ± 1.5 (19.7-24.8)	23.7 ± 1.5 (20.5-25.7)	23.2 ± 1.2 (21.5-25.3)
b	-	2.3 ± 0.1 (2.1-2.7)	-	-
c	14.3 (13-16)	-	14.3 ± 0.9 (12.3-15.4)	15.6 ± 1.2 (13.3-17.2)
c′	-	-	-	2.6 ± 0.3 (2.1-3.1)
Body diam.	15.4 (10-16)	-	15.0 ± 0.8 (14.4-16.8)	14.9 ± 0.4 (14.5-15.2)
Stylet	14.7 (14-16)	11.5 ± 0.7 (10.4-12.9)	14.1 ± 0.7 (14.0-15.6)	12.6 ± 0.3 (12.1-13.3)
DGO	3.7 (2.5-4.5)	3.4 ± 0.7 (2.0-4.6)	2.2 ± 0.1 (2.2-2.8)	3.6 ± 0.3 (3.2-3.8)
Median bulb length	-	12.1 ± 1.2 (10.1-15.1)	-	-
Median bulb valve length	-	3.9 ± 0.4 (3.2-5.0)	-	-
Tail	24.5	22.0 ± 2.0 (16.0-25.0)	27.4 ± 1.5 (24.0-28.8)	22.2 ± 1.8 (19.0-25.3)
Hyaline region	-	-	6.7 ± 1.0 (4.8-8.0)	6.5 ± 1.4 (4.4-8.2)

* Estimated from drawing.

J2

Body relatively short. Labial region slightly offset, hemispherical shape in lateral view. Labial cap very short. *En face* view distinctive. Stylet knobs small with posteriorly sloping anterior margins. Lateral field with four incisures for greater part of its length, outer bands mid-body usually cross-striated every other body stria, occasionally every body stria, inner band occasionally cross-striated. Hemizonid anterior and adjacent to secretory-excretory pore. Tail relatively short and conical. Rectum rarely or not inflated. Hyaline tail part very short, ending in rounded tip.

TYPE PLANT HOST: Cabbage, *Brassica oleracea* L. var. *capitata*.

OTHER PLANTS: This nematode may damage all kind of cereals and also leguminous and cruciferous crops. *Meloidogyne artiellia* infects barley (*Hordeum vulgare*), sorghum (*Sorghum bicolor*), wheat (*Triticum durum* L. and *T. aestivum*), cabbage, cauliflower, radish (*Raphanus sativus* L.), rashad (*Nasturtium fontanum* Asch.), turnip (*Brassica rapa* L.), lentil (*Lens culinaris* Medik.), alfalfa (*Medicago sativa* L.), broad bean (*Vicia faba* L.), chickpea (*Cicer arietinum* L.), clovers (*Trifolium incarnatum* L., *T. pratense* L., and *T. repens* L.), dogtooth pea (*Lathyrus sativus* L.) and vetch (*Vicia sativa* L.) (Di Vito *et al.*, 1985b; Greco *et al.*, 1992).

TYPE LOCALITY: Wells, Norfolk, England, UK.

DISTRIBUTION: *Europe*: Belgium, France, Greece, Italy, Russia, Spain, UK; *Asia*: China, Israel, Syria, Turkey; *Africa*: Algeria, Morocco, Tunisia (Anon., 2013a).

BIOLOGY: Nematodes complete one generation per year (Di Vito & Greco, 1988c). However, in France the second generation may develop from June, but it is unclear whether it will have sufficient time to complete the entire cycle before harvest (Buisson *et al.*, 2014). Newly formed eggs require a period of chilling for the hatching of the J2. This RKN is well adapted to survive both cold winters, as eggs or active J2, and dry summers, as eggs or anhydrobiotic J2 (Di Vito & Greco, 1988c). Populations fell to 48% in a growing season of wheat and in the absence of hosts (Di Vito & Greco, 1988b, c).

SYMPTOMS: Root galls are very small and often are covered by large egg masses. Infected plants show chlorotic leaves and poor growth (Greco *et al.*, 1992).

PATHOGENICITY: In Sicily, Italy, Syria and other Middle Eastern countries, especially in sandy soils, *M. artiellia* damages various crops such as durum wheat, barley, chickpea, broad bean, lentil and vetch (Di Vito & Greco, 1988a, b; Greco *et al.*, 1992; Rivoal & Cook, 1993; Lombardo *et al.*, 2011). In microplot experiments, grain losses of 90% occurred at initial population densities of 32 eggs and J2 (ml soil)$^{-1}$ with durum wheat and 8 and 1 eggs and J2 (ml soil)$^{-1}$ with winter and spring chick pea, respectively (Di Vito & Greco, 1988a, b).

CHROMOSOME NUMBER: No available data.

POLYTOMOUS KEY CODES: *Female*: A2, B32, C3, D1; *Male*: A43, B3124, C23, D1, E2, F1; *J2*: A3, B12, C4, D4, E1, F4.

BIOCHEMICAL AND MOLECULAR CHARACTERISATION: *Meloidogyne artiellia* populations from France, Italy and the UK, all show a malate dehydrogenase N1b type and an unusual esterase M2-VF type (two close bands and a weaker one). Castillo *et al.* (2003a) revealed that the esterase phenotype of *M. artiellia* had two bands at Rm 0.52 and 0.54. The malate dehydrogenase type is the same type as was detected for *M. fallax* (Karssen & van Hoenselaar, 1998).

Sequence information is given for IGS and ITS rRNA, 18S rRNA and 28S rRNA genes (De Giorgi *et al.*, 2002; Castillo *et al.*, 2003a; Damme *et al.*, 2013; Rybarczyk-Mydłowska *et al.*, 2013; Gamel *et al.*, 2014; Janssen *et al.*, 2017), *COI* (Janssen *et al.*, 2017), RNA polymerase II (Rybarczyk-Mydłowska *et al.*, 2013) and *hsp90* (De Luca *et al.*, 2009) genes. The diagnostic ITS-RFLP profile generated by two enzymes was given by Gamel *et al.* (2014).

RELATIONSHIPS (DIAGNOSIS): The species belongs to Molecular group IX and is clearly differentiated molecularly from all other *Meloidogyne* species. *Meloidogyne artiellia* is clearly separated from all other *Meloidogyne* species by perineal pattern and the short and relatively broad tail of the J2.

11. *Meloidogyne baetica* Castillo, Vovlas, Subbotin & Troccoli, 2003
(Figs 49-52)

COMMON NAME: Mediterranean olive root-knot nematode.

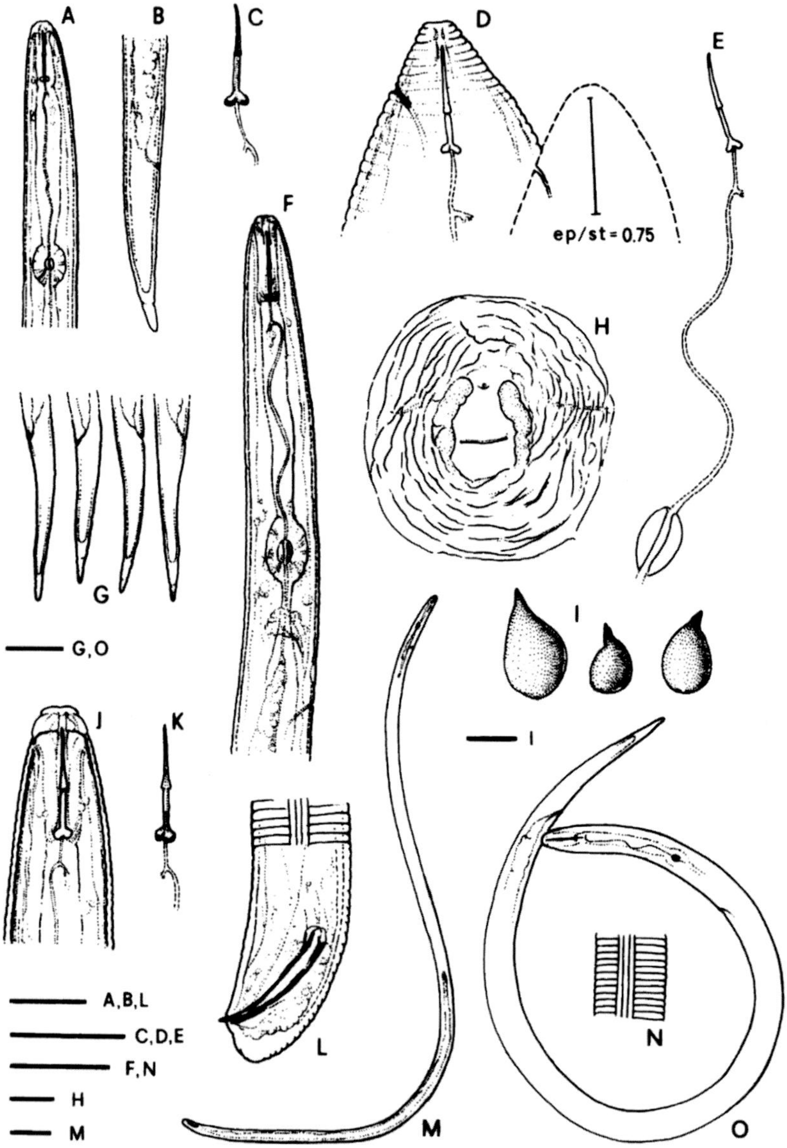

Fig. 49. Meloidogyne baetica. *A, F: Anterior body region of second-stage juvenile (J2); B, G: Tail region of J2; C: J2 stylet; O: Entire J2; D: Female labial region; E: Female stylet; H: Perineal pattern; I: Entire female; Male: J: Male labial region; K: Male stylet; L: Male tail; M: Entire male; N: Lateral field of J2 at mid-body. (Scale bars: A-O = 20 μm; M = 100 μm; I = 500 μm.) (After Castillo* et al.*, 2003a.)*

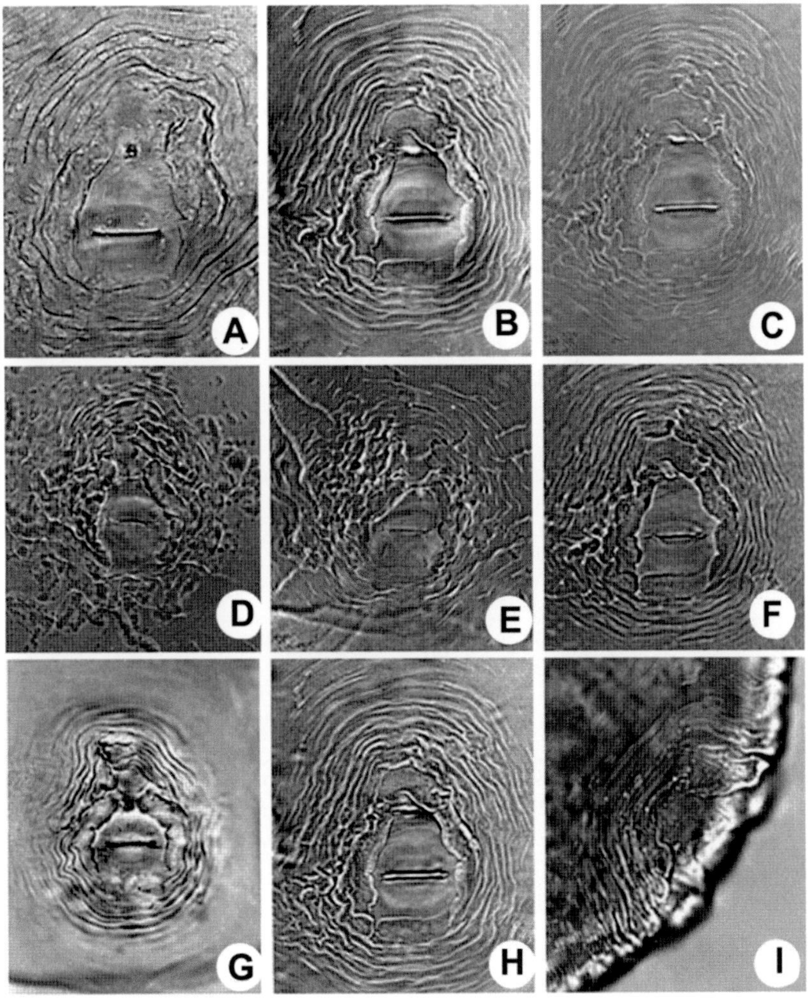

Fig. 50. Meloidogyne baetica. *LM. A-H: Typical variation of perineal pattern; I: Terminal cone. (After Castillo* et al., *2003a.)*

This species was described infecting wild olive at southern Spain. It has a narrow host-range and is considered as a parasite of woody plants.

MEASUREMENTS *(AFTER CASTILLO* ET AL., *2003A)*

- *Holotype female*: L = 845 μm; W = 516 μm; a = 1.64; stylet = 18 μm; DGO = 4.0 μm; anterior end to excretory pore = 16 μm;

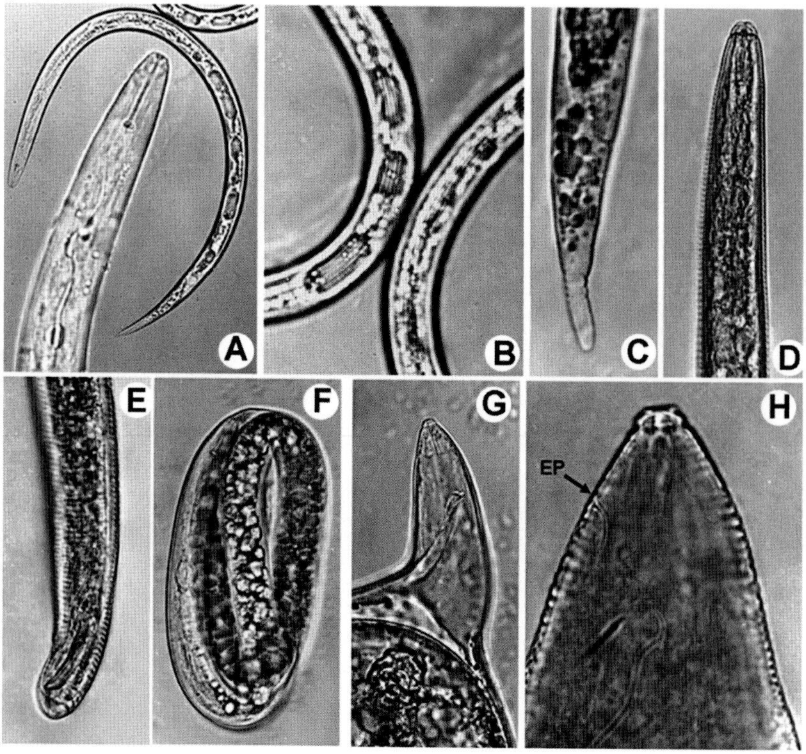

Fig. 51. Meloidogyne baetica. *LM. A: Entire second-stage juvenile (J2) and anterior region; B: Lateral field of J2; C: J2: tail; D: Male labial region; E: Male tail; F: Embryonated egg; Female: G: Female anterior region; H: Female labial region; (EP = excretory pore). (After Castillo* et al., *2003a.)*

excretory pore distance from anterior end/length of stylet; vulva length = 24 μm; vulva to anus = 22 μm; EP/ST ratio = 0.9.

- *Paratype females* (n = 20): L = 911 ± 163 (775-1263) μm; W = 523 ± 88 (469-559) μm; a = 1.6 ± 0.1 (1.5-1.7); stylet = 17.5 ± 0.8 (17-19) μm; EP/ST ratio = 0.7 ± 0.1 (0.5-0.8); vulva length = 20 ± 2.7 (17-24) μm; vulva to anus = 21 ± 1.6 (19-25) μm.
- *Eggs* (n = 14): L = 119 ± 9.1 (109-139) μm; W = 42 ± 2.4 (38-46) μm; L/W = 2.6 ± 0.5 (2.5-3.2).
- *Allotype male*: L = 1810 μm; a = 58; b = 13.2; c = 129; stylet = 18 μm; DGO = 4.0 μm; labial region height = 6.0 μm; labial region diam. = 8.5 μm; body annulation = 2.7 μm; lateral field width = 10 μm anterior end to centre of median bulb = 103 μm; anterior

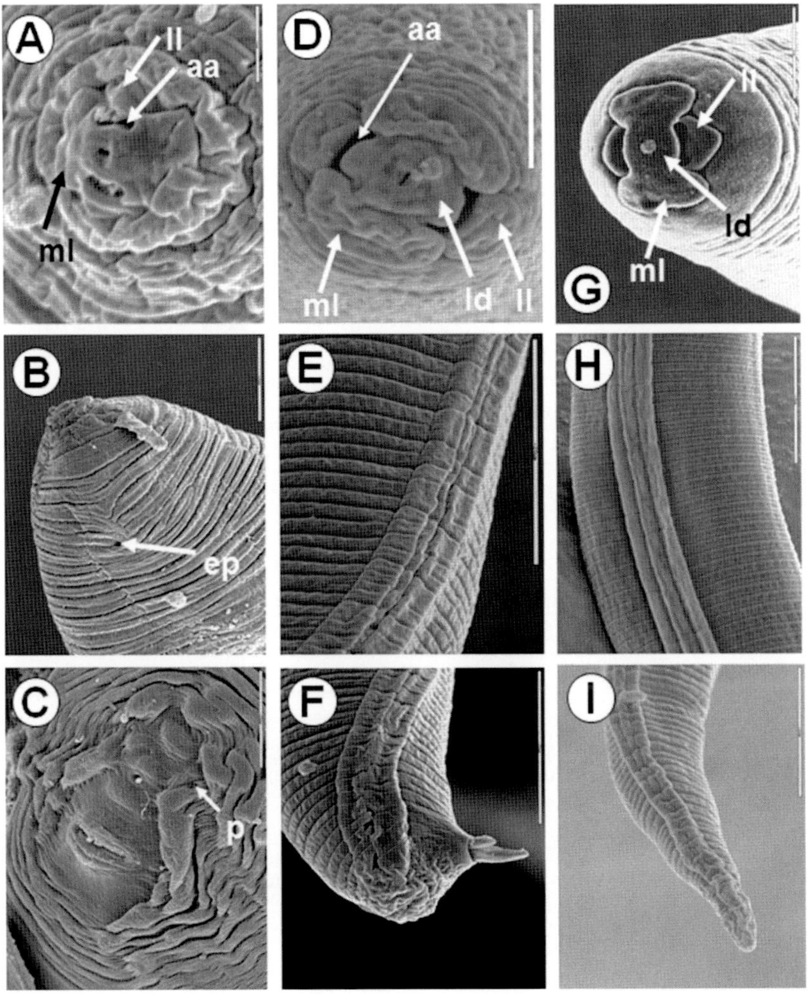

Fig. 52. Meloidogyne baetica. *SEM. A: Female* en face *view; B: Female anterior region showing excretory pore; C: Perineal pattern; D: Male* en face *view; E: Male lateral field at mid-body; F: Male tail; G: Second-stage juvenile (J2)* en face *view; H: J2 lateral field at mid-body; I: J2 tail. Abbreviations: aa = amphidial aperture; ep = excretory pore; ld = labial disc; ll = lateral labial; ml = medial labials; p = phasmid. (Scale bars: A, G = 2 μm; D = 5 μm; B, H, I = 10 μm; C, E, F = 20 μm.) (After Vovlas* et al.*, 2004b.)*

end to excretory pore = 141 μm; tail = 14 μm; spicules = 28 μm; gubernaculum = 9 μm.

- *Paratype males* (n = 22): L = 1811 ± 406 (1545-2205) μm; a = 62 ± 7.0 (58-75); b = 12.4 ± 2.2 (9.8-15.7); b′ = 8.2 ± 1.4 (6.5-10.1); c = 171 ± 37 (131-242); stylet = 17.0 ± 1.6 (16-19) μm; stylet knob width = 2.5 ± 0.1 μm; labial region height = 6.0 ± 0.2 (5.5-6.0) μm; labial region diam. = 9.0 ± 0.6 (8.5-11.0) μm; DGO = 4.0 ± 0.5 (3.5-5.5) μm; anterior end to centre of median bulb = 109 ± 17 (95-132) μm; anterior end to excretory pore = 157 ± 18 (137-179) μm; T = 36 ± 4.8 (31-48)%; annulation = 2.5 ± 0.1 μm wide at mid-body; lateral field = 9 ± 1.0 (8-11) μm; spicules = 27 ± 4.0 (24-36) μm; gubernaculum = 12 ± 1.5 (10-14) μm; tail = 11 ± 1.7 (9-14) μm.
- *Paratype J2* (n = 26): L = 403 ± 136 (394-422) μm; a = 29 ± 1.2 (28-32); c = 7.9 ± 0.4 (7.3-8.4); stylet = 13.5 ± 0.5 (13-14) μm; W = 13.5 ± 0.5 (13-14) μm; stylet knob width = 1.5 ± 0.2 (1.5-2.0) μm; DGO = 4.0 ± 0.7 (3.0-5.0) μm; anterior end to excretory pore = 93 ± 5.3 (85-100) μm; tail = 50 ± 2.3 (47-54) μm; hyaline region = 12 ± 1.2 (10-13) μm.

Description (*after Castillo* et al., *2003a; Vovlas* et al., *2004b*)

Female

Body completely enclosed by galled tissue; body pearly white, pear-shaped and anterior body portion commonly off-centre from a median plane and with almost terminal vulva. Labial disc and median labials fused to form a narrow labial structure in face view. Amphidial openings oval-shaped, located between labial disc and lateral labials. Neck usually short, not bent. Pharyngeal gland with one large dorsal lobe with one nucleus, two small nucleated subventral gland lobes, variable in shape, position, and size, usually located posterior to dorsal gland lobe. Two large pharyngo-intestinal cells near junction of median bulb and intestine. Labial framework distinct, labial region offset and bearing one annulus. Excretory pore distinct, constantly anterior to level of stylet knobs. Stylet strong and small, slightly curved dorsally. Perineal patterns typically formed of striae and ridges in cuticle, latter being more pronounced nearer vulva and anus. Dorsal arch enclosing usually quite distinct phasmids, and anus, situated at centre. Vulval slit centrally located in unstriated area, nearly as wide as vulva-anus distance. Commonly, large egg sac occurs outside small root gall, containing up to 435 eggs.

Male

Body vermiform, tapering anteriorly; tail rounded, with twisting posterior body portion. Labial region slightly offset, with large labial annulus and a prominent post-labial annulus. In SEM, labial disc oval and fused with elongate-oval median labials forming a rectangular structure. Lateral labials rounded-oval. Amphidial apertures elongate slits located between labial disc and lateral labials. Body annulation distinct. Lateral field, composed of four incisures, forming three bands, wide, about one-third of body diam., faintly and irregularly areolated at mid-body and tail region. Stylet knobs rounded. Testis one, occupying 31-48% of body cavity. Spicules and gubernaculum rather short. Tail bluntly rounded and annulated.

J2

Body slender, tapering to an elongated tail. Labial framework weak, vestibule and vestibule extension distinct but not strongly sclerotised, except at anterior terminus. In face view, labial disc and median labials fused into one structure. Labial disc small and rounded, slightly elevated above median labials. Median labials elongate-oval, not indented medianly with rounded margin. Amphidial apertures appearing as elongate slits between labial disc and lateral labials. Lateral labials triangular with rounded margins. Labial region smooth. Stylet slender, cone weakly expanding at junction with shaft, knobs rounded. Lateral fields with four incisures, irregularly areolated along entire body. Tail elongate-conoid, tapering to a slender, terminal, digitiform process.

TYPE PLANT HOST: Wild olive, *Olea europaea* spp. *sylvestris* (Miller) Hegi.

OTHER PLANTS: Two natural woody host plants, lentisc (*Pistacia lentiscus*) and pipe vine (*Aristolochia baetica*) (Vovlas *et al.*, 2004b).

TYPE LOCALITY: Vejer de la Frontera (Cádiz), southern Spain.

DISTRIBUTION: *Europe*: Spain.

CHROMOSOME NUMBER: No available data.

POLYTOMOUS KEY CODES: *Female*: A12, B21, C4, D1; *Male*: A123, B43, C213, D1, E1, F2; *J2*: A23, B2, C3, D32, E3, F2.

BIOCHEMICAL AND MOLECULAR CHARACTERISATION: The isozyme electrophoretic analysis of single- and five-specimen groups of young egg-laying females of *M. baetica* revealed a single esterase band at Rm 0.31 that did not occur in the esterase phenotypes of *M. artiellia* or *M. javanica*. The ITS, 18S and D2-D3 28S rRNA gene sequences are provided by Castillo *et al.* (2003a) and Ali *et al.* (2015).

RELATIONSHIPS (DIAGNOSIS): The species belongs to Molecular group IX and is clearly molecularly differentiated from all other *Meloidogyne* species. *Meloidogyne baetica* can be distinguished from other *Meloidogyne* spp. by the perineal pattern, which is almost similar to that of *M. artiellia*, characterised by distinct inner striae forming two distinct longitudinal bands, extending throughout the perineum to just posterior to the vulva and by the female excretory pore located anterior to stylet knob level, EP/ST ratio extremely small (0.5-0.8) and short tail in J2.

12. *Meloidogyne brevicauda* Loos, 1953
(Figs 53, 54)

COMMON NAMES: Tea or Indian root-knot nematode.

Meloidogyne brevicauda was described as a species parasitising tea plants in Sri Lanka (Loos, 1953). It was the first description of a new species of RKN after Chitwood (1949) described or delineated five species and placed them in the genus *Meloidogyne*. This species has been reported only in several Asian countries.

MEASUREMENTS

After Loos (1953):

- *Females*: L = 680-1860 μm; W = 310-1060 μm; stylet = 22.1 μm; stylet knob width = 4-5 μm.
- *Males*: L = 970-1440 μm; W = 28-52 μm; a = 26-44; total length/length body from anterior end to posterior end of median bulb = 12.6-17.0; stylet = 19.5-20.7 μm; stylet knob width = 5.2 μm;

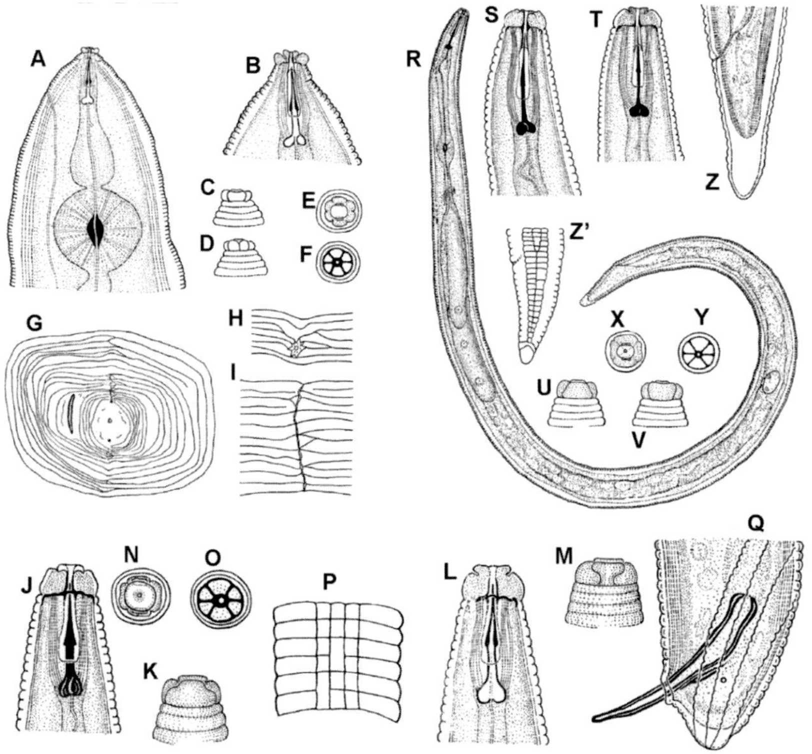

Fig. 53. Meloidogyne brevicauda. *A: Female anterior region; B-D: Female labial region; E:* En face *view; F: Labial skeleton; G: Perineal pattern; H: Excretory pore; I: Lateral line and transverse striae; J-M: Male labial region; N:* En face *view; O: Labial skeleton; P: Male lateral field; Q: Male tail; R: Entire second-stage juvenile (J2); S-V: J2 labial region; W:* En face *view; X:* En face *view; Y: Labial skeleton; Z, Z′: J2 tail. (After Loos, 1953.)*

median bulb length = 14.0-17.5 μm; median bulb width = 8-10 μm; spicules = 34.0-42.5 μm; gubernaculum = 10.0-10.5 μm.
- *J2*: L = 460-590 μm; W = 14.7-20.6 μm; tail = 17.5-28.0 μm; a = 23-33; total length/length body from anterior end to posterior end of median bulb = 6.2-7.3; c = 21-29.
- *Eggs*: L = 108.5-133 μm; W = 41.7-59.5 μm.

After Whitehead (1968):

- *Females* (n = 20): L = 2802 (1673-4388) μm; W = 1007 $\pm$ 142 (714-1265) μm; stylet (n = 8) = 21 (17-25) μm; stylet knob width = 5

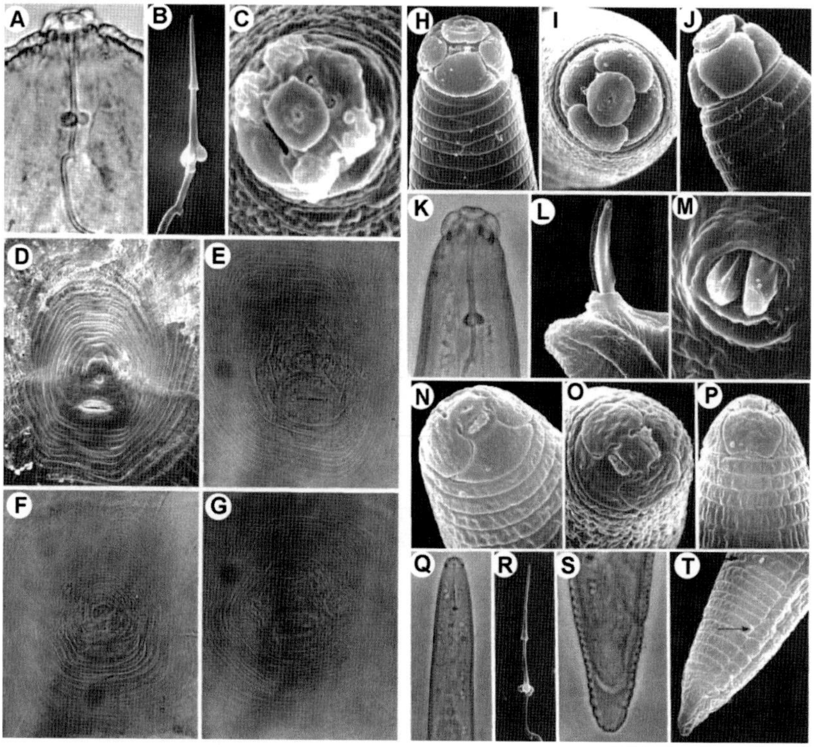

Fig. 54. Meloidogyne brevicauda. *LM and SEM. A: Female labial region; B: Female excised stylet; C: Female* en face *view; D-G: Perineal patterns; H-K: Male labial region; L, M: Detail of spicules; N-Q: Second-stage juvenile (J2) labial region; R: J2 excised stylet; S, T: J2 tail (arrow points to phasmid). (After Eisenback & Gnanapragrasam, 1992.)*

(3-8) μm; DGO (n = 8) = 6 (4-8) μm; median bulb length (n = 6) = 50 (40-62) μm; median bulb width (n = 6) = 36 (26-49) μm; median bulb valve length (n = 6) = 14 (13-16) μm; median bulb valve width (n = 6) = 10 (9-10) μm; EP/ST = 1.4 (estimated from drawings).

- *Males* (n = 4): L = 1216 $\pm$ 129 (1061-1365) μm; a = 40.5 $\pm$ 5.3 (35.1-47.5); labial length = 6.7 $\pm$ 0.36 (6.5-7.2) μm; stylet = 23.6 $\pm$ 0.36 (23.0-23.7) μm; stylet knob width = 4.6 $\pm$ 0.51 (4.3-5.4) μm; DGO = 6.2 $\pm$ 1.07 (5.0-7.6) μm; b′ = 16.1 $\pm$ 2.31 (13.7-18.8); c = 125 $\pm$ 7.7 (117-135); median bulb length = 17.1 $\pm$ 0.91 (15.8-18.0) μm; median bulb width = 9.4 $\pm$ 2.10 (6.5-11.5) μm; median bulb valve length = 6.1 $\pm$ 0.26 (5.0-7.2) μm; spicules (n = 8) = 32.4 $\pm$ 2.75 (29.5-36.0) μm.

- *J2* (n = 13): L = 522 ± 31 (459-583) μm; a = 28.5 ± 2.37 (24.1-32.0); b′ = 8.0 ± 0.4 (7.2-8.7); c (n = 4) = 23.6 ± 1.43 (21.7-24.9); c′ (n = 4) = 1.9 ± 0.1 (1.7-2.0); length body to middle of genital primordium = 362 ± 22 (312-397) μm; stylet = 14.9 ± 0.6 (13.7-16.0) μm; median bulb length = 13.9 ± 1.5 (10.1-16.5) μm; median bulb width = 8.7 μm; median bulb valve length = 4.0 ± 0.4 (3.2-4.7) μm; tail (n = 4) = 23 ± 1.0 (22-24) μm.

DESCRIPTION *(AFTER LOOS, 1953; WHITEHEAD, 1968; EISENBACK & GNANAPRAGASAM, 1992)*

Female

Body shape globular to pear-shaped with short necks or broadly cylindrical with long narrow necks, well-filled egg sacs rare. Anterior end of female containing pharynx which includes a narrow procorpus, a large median bulb, a narrow isthmus, and a large glandular region that has one dorsal and two subventral pharyngeal gland lobes, each with a large nucleolus and nucleus. Two large nucleated pharyngo-intestinal cells lying at dorsal base of median bulb at junction with intestine. Labial disc and labials are prominent and protruding from usual body contour. Slit-like stoma located in oval-shaped prestoma, which is situated centrally on labial disc and surrounded by six small pore-like openings of inner labial sensilla. Labial disc raised above surrounding labials. Subventral and subdorsal labial pairs separate, not fusing to form one dorsal and one ventral labial. These labials are often termed median labials because specimen orientation is difficult to discern in SEM. Lateral labials are large and fused with lateral edges of median labials. Stylet slender, but knobs are large and rounded, often sloping posteriorly. Cone gradually increasing in width posteriorly and overlapping shaft, which also widens gradually posteriorly. Excretory pore 20-28 annuli posterior to labial region, usually in vicinity of spear and always anterior to median bulb. Perineal pattern quite distinctive. Striae are smooth, coarse, and continuous, completely encircling perineum and extending anteriorly for some distance. Dorsal arch high and squarish to rounded. Lateral fields usually indistinct, but may be marked by irregularities, or small, wing-like projections. Striae may completely encircle the large, rounded tail terminus.

Male

Large and round labial disc separated from labials by a deep indentation. Large subdorsal and subventral labial pairs fused, forming one dorsal and one ventral labial, extending posteriorly and lying next to first body annulus. Lateral labials also large and adjacent to first body annulus. First body annulus also narrower than remaining body annuli. Hemizonid three or four annuli anterior to excretory pore. Excretory pore at nearly two-thirds of pharynx length. Lateral field with four incisures, usually fully areolated. Terminus of spicule bluntly rounded, gubernaculum tapering anteriorly. Tail marked by a crescentic fold in cuticle surrounding posterior portion of cloacal aperture.

J2

Labial disc squarish to rectangular, connected to median labials by a narrow piece of cuticle. Subdorsal and subventral labials fused and forming one dorsal and ventral labial, extending posteriorly, may be indented medianly. Large median labials apparent in both LM and SEM. Lateral labials also large and adjacent to first body annulus. Amphidial openings appearing as small elongate oval openings between labial disc and median labials. Labial annulations absent. Stylet knobs rounded with posteriorly directed anterior margins. Hemizonid prominent, about five annuli anterior to excretory pore. Hemizonion observed about 14 annuli posterior to hemizonid. Lateral field with four incisures (sometimes also a central fifth) for greater part of its length, irregularly cross-striated mid-body and on tail. Rectum not inflated. Tail bluntly rounded, striae coarsening towards terminus, which is mostly smooth. Tail with clear hyaline region nearly one-third of its length. Phasmids about half length of tail posterior to anus.

TYPE HOST: Tea plant, *Camellia sinensis* (L.) Kuntze.

OTHER PLANT: Saffron, *Crocus sativus* L.

TYPE LOCALITY: Nuwara Eliya, Ceylon. Elevation approximately 1980 m a.s.l.

DISTRIBUTION: *Asia*: Azerbaijan, China (Fujian), India (Tamil Nadu and West Bengal) and Sri Lanka (Rao, 1970; Sivapalan, 1972, 1978; Lamberti *et al.*, 1987; Eisenback & Gnanapragrasam, 1992; Anon., 2012).

SYMPTOMS: Tea bushes infected by *M. brevicauda* are stunted with pale, dull leaves. Severe infection may result in defoliation and bushes may fail to recover (Loos, 1953; Sivapalan, 1972).

CHROMOSOME NUMBER: No available data.

POLYTOMOUS KEY CODES: *Female*: A12, B12, C3, D1; *Male*: A34, B2, C21, D1, E1, F2; *J2*: A1, B1, C4, D4, E1, F4.

BIOCHEMICAL AND MOLECULAR CHARACTERISATION: No available data.

RELATIONSHIPS (DIAGNOSIS): The J2 tail of *M. brevicauda* is bluntly rounded and similar to those of *M. indica*, *M. nataliei* and *M. propora*. The perineal pattern of *M. brevicauda* is quite distinctive from these species.

13. *Meloidogyne californiensis* Abdel-Rahman & Maggenti, 1987
(Figs 55, 56)

COMMON NAME: Californian root-knot nematode.

Bulrush plants with large white root galls were collected in 1979 from the edge of a freshwater stream of Pomponio Beach, Half Moon Bay, California, USA. This RKN was described as a new species.

MEASUREMENTS *(AFTER ABDEL-RAHMAN & MAGGENTI, 1987)*

- *Holotype female*: L (with neck) = 628 μm; W = 364 μm; L/W ratio = 1.7; neck length = 28 μm; neck greatest diam. = 76 μm; stylet = 13 μm; stylet knob width = 2 μm; stylet knob height = 1 μm; DGO = 4 μm; median bulb length = 38 μm; median bulb width = 28 μm; valve length = 14 μm; valve width = 10 μm; anterior end to excretory pore = 26 μm.
- *Females* (n = 18-36): L (without neck) = 325 $\pm$ 59.7 (264-492) μm; neck length = 94 $\pm$ 20.6 (72-140) μm; W = 226 $\pm$ 60.7 (168-403) μm; neck diam. = 43 $\pm$ 7 (33-56) μm; cuticle thickness in mid-

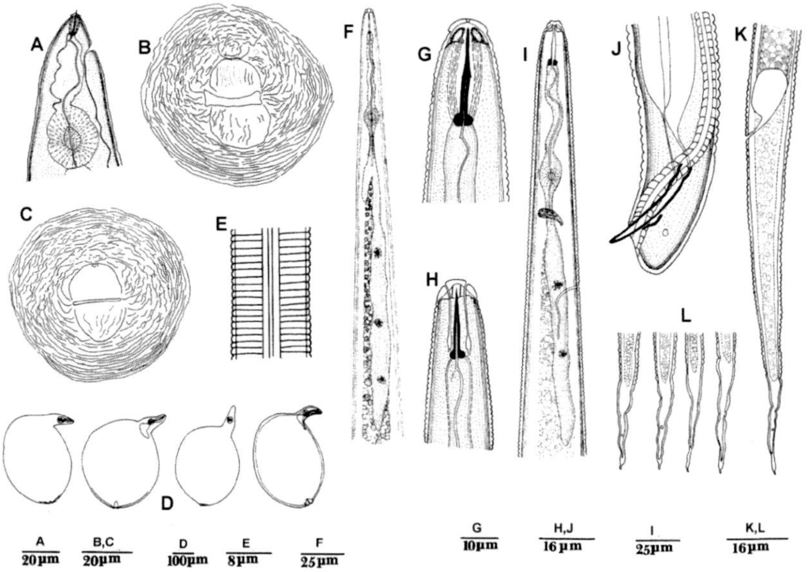

Fig. 55. Meloidogyne californiensis. *A: Female anterior region; B, C: Perineal pattern; D: Entire female; E: Second-stage juvenile (J2) lateral field; F: J2 anterior region; G, H: Male labial region; I: Male anterior region; J: Male tail; K: J2 tail; L: J2 tail tip. (After Abdel-Rahman & Maggenti, 1987.)*

body = 3 ± 0.6 (2-6) μm; stylet = 14 ± 2.1 (12-18) μm; stylet knob width = 2.3 ± 0.3 (2-3) μm; stylet knob height = 1.5 ± 0.3 (1.0-2.0) μm; DGO = 4.3 ± 0.7 (3.6-6.0) μm; anterior end to stylet base = 15 ± 2.2 (12-20) μm; anterior end to extremity to excretory pore = 24 ± 4.7 (19-33) μm; anterior end to end median bulb valve = 72 ± 8.6 (65-94) μm; median bulb length = 29 ± 4.3 (23-38) μm; median bulb width = 24 ± 2.5 (22-29) μm; vulval slit width = 23 ± 2.9 (19-27) μm; interphasmid width = 18 ± 3.0 (15-25) μm; distance from anus to vulva (centre) = 17 ± 2.9 (12-23) μm; distance from anus to centre of imaginary line between phasmids = 5.5 ± 1.2 (4-8) μm; EP/ST = 1.9.

- *Allotype male*: L = 1095 μm; W = 35; stylet = 19 μm; stylet knob height = 3 μm; stylet knob width = 4 μm; a = 31.1; b = 13.7; c = 97.7; c′ = 63.6; DGO = 4 μm; anterior enter to median bulb valve = 80 μm; anterior end to excretory pore 126 μm; anterior end to hemizonid = 114 μm, anterior end to pharyngeal gland = 237 μm; hemizonid length = 5 μm; hemizonid to excretory pore = 8 μm; body

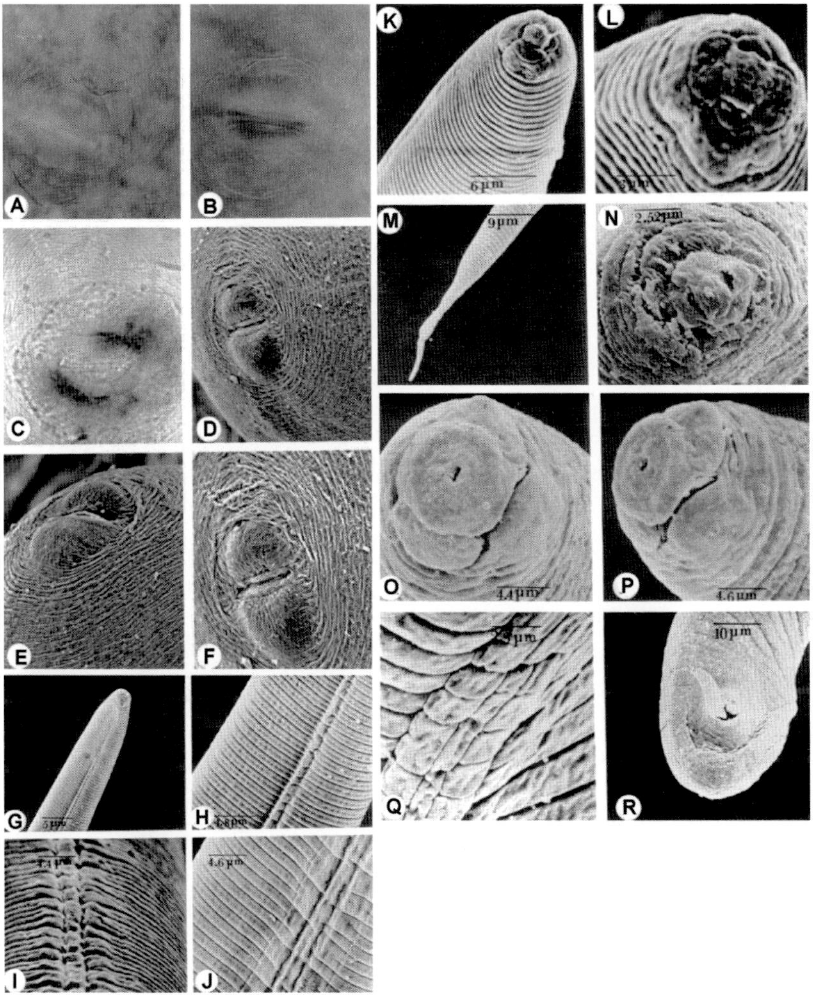

Fig. 56. Meloidogyne californiensis. *LM and SEM. A-F: Perineal patterns; G: Second-stage juvenile (J2) anterior region; H-J: J2 lateral field; K, L: J2* en face *view; M: J2 tail; N: Female* en face *view; O, P: Male* en face *view; Q: Male lateral field; R: Male tail. (Scale bars: A-F = 20 μm; G = 5 μm; H-J = 4.8, 4.4, 4.6 μm, respectively; K-R = 6, 3, 9, 2.52, 4.4, 4.6, 2.3, 10 μm, respectively). (After Abdel-Rahman & Maggenti, 1987.)*

annulation = 2 μm wide; testes length = 654 μm; spicules = 20 μm; gubernaculum = 5 μm; anal body diam. = 17 μm; tail = 11 μm; phasmids to tail tip = 11 μm (at cloacal aperture level); T = 59.8%.

- *Males* (n = 33): L = 1362 ± 358 (712-1952) μm; W = 35 ± 4.8 (24-42) μm; stylet = 22 ± 2.7 (18-28) μm; a = 35 ± 5.7 (28-45); b = 11 ± 0.4 (10-12); c = 100 ± 32 (65-140); DGO = 4 ± 1.1 (2-6) μm; anterior end to pharyngo-intestinal junction = 111 ± 9.4 (102-128) μm, to excretory pore = 127 ± 20 (110-171) μm, to pharyngeal gland end = 213 ± 24.4 (188-265) μm; tail = 13 ± 3.7 (7-23) μm; spicules = 28 ± 5.6 (20-40) μm; gubernaculum = 8 ± 1.7 (5-9) μm; testis length = 670 ± 233 (290-990) μm; phasmids to tail tip = 13 ± 2.8 (10-17) μm; T = 51 ± 8 (36-65)%.
- *J2* (n = 35): L = 559 ± 29.8 (448-628) μm; W = 17 ± 0.8 (16-18) μm; stylet = 12.1 ± 0.6 (11-13) μm; a = 32.9 ± 0.9 (32-34); b′ = 9 ± 0.8 (8-10); c = 6.2 ± 0.2 (6-7); DGO = 2-3 μm; anterior end to excretory pore = 86 ± 4.7 (77-91) μm; anterior end to median bulb valve = 61 ± 4.8 (55-70) μm; anterior end to pharyngeal gland base = 222 ± 29.1 (173-264) μm, anterior end to hemizonid = 79 ± 17.6 (74-88) μm; median bulb length = 12.7 ± 0.8 (12-14) μm; median bulb width = 8 ± 0.8 (6-9) μm; tail = 88.6 ± 4.2 (82-98) μm; hyaline region = 24 ± 4.4 (16-30) μm.
- *Eggs* (n = 29): L = 112 ± 8.0 (95-124) μm; W = 48 ± 5.8 (34-56) μm.

Description

Female

Body pearly white, globular, pear-shaped, sometimes saccate, with prominent posterior protuberances. Neck distinct, tapering anteriorly, directed oblique to longitudinal axis of body. Stylet delicate, basal knobs rounded, sloping posteriorly. Labial annuli, two or three, not offset from neck. Labial framework weakly developed. Excretory pore located posterior to stylet, about two stylet lengths, or 16-26 annuli from anterior end. Cuticle thin, 2-6 μm at mid-body. Neck striation distinct, body striation fine except around vulva. Perineal pattern with two posterior cuticular protuberances on each side of vulva. Lateral lines indistinct, many with wavy broken striae marking lateral areas of pattern, arch low, rounded, with spaced broken striae. Ventral region marked with smooth discontinuous striae. Anus covered dorsally by prominent fold, phasmids large, conspicuous. Tail area with a prominent whorl.

Male

Body vermiform, slender, tapering, rounded at both extremities. Labial region slightly offset, with two or three annuli, spear heavy with massive rounded knobs sloping posteriorly. Lateral field marked with four areolated incisures. Incisures beginning 12 annuli from labial region, with LM outer bands only, with or without areolation. Lateral field bands three, equal in width, forming 20% of body diam. Body cuticle coarsely annulated, annuli about 2 μm wide. Hemizonid anterior to excretory pore, about three body annuli long. Phasmids at level of cloacal aperture. Spicules arcuate, gubernaculum as illustrated. Body length varying, with long slender males more than twice as long as short ones. Long males possess two testes, short males one.

J2

Body vermiform, tapering slightly anteriorly and pronounced posteriorly. Labial region slightly offset, with two or three annuli, labial framework weak. Stylet delicate, knobs small, sloping posteriorly. Lateral field beginning 29-30 annuli posterior to labial region, anteriorly starting with two areolated incisures, as seen with SEM, then four areolated incisures along most of body, no areolation seen with LM, occupying 0.25-0.32 of body at mid-body. Outer two bands wider than middle band. Hemizonid two body annuli anterior to excretory pore, about three body annuli long. Pharyngo-intestinal valve inconspicuous, posterior to level of excretory pore. Rectum inflated, phasmids small, conspicuous, located mid-way between hyaline area and tail tip. Tail shape consistent, terminal part of tail deformed, terminus pointed, tip with mucron-like projection.

TYPE PLANT HOST: Bulrush, *Bolboschoenus robustus* (Pursh) Sojak.

OTHER PLANTS: Several infection tests on tomato, rice, barley, sugarbeet and other crops by this nematode gave negative results, no galls being observed or nematodes recovered.

TYPE LOCALITY: Pomponio Beach, Half Moon Bay, California, USA. At the edge of a fresh water creek, as well as from completely submerged plants in the water.

DISTRIBUTION: *North America*: USA (California).

SYMPTOMS: Galls produced on bulrush by this species of *Meloidogyne* have a distinct shape. Terminal galls are peanut-shaped, clubbed, or spindle-shaped. Non-terminal galls are sometimes curved or spiral-shaped like those produced by *M. naasi*, but they are not accompanied by root branching. *Meloidogyne naasi* was also found in this area.

CHROMOSOME NUMBER: No available data.

POLYTOMOUS KEY CODES: *Female*: A3, B312, C3, D3; *Male*: A3124, B213, C213, D1, E1, F2; *J2*: A12, B23, C1, D1, E3, F1.

BIOCHEMICAL AND MOLECULAR CHARACTERISATION: No available data.

RELATIONSHIPS (DIAGNOSIS): *Meloidogyne californiensis* differs from other described species by its distinct perineal pattern with two prominent protuberances bordering the vulva and it could be placed in the *graminis* group (Jepson, 1987) on body and tail length. *Meloidogyne californiensis* is morphologically similar to *M. ottersoni* and *M. spartinae*. The J2 of *M. californiensis* have a larger ratio a and smaller stylet than *M. ottersoni*; juvenile tail shapes also differ. The female of *M. californiensis* differs from *M. ottersoni* in the position of the excretory pore, which is far anterior and opposite to the stylet base in *M. ottersoni*, and the perineal pattern shape. In *M. californiensis* the excretory pore is located on the 16-25th annulus posterior to the labial region. The J2 of *M. californiensis* are shorter, 448-628 μm, than juveniles of *M. spartinae*, 612-912 μm. *Meloidogyne californiensis* has a smaller stylet, 11-13 *vs* 14-17 μm in *M. spartinae* (Abdel-Rahman & Maggenti, 1987).

14. *Meloidogyne camelliae* Golden, 1979
(Figs 57, 58)

COMMON NAME: Camellia root-knot nematode.

This RKN was first described on common camellia plants from Japan intercepted by officials of the Unites States Department of Agriculture (Golden, 1979), and later detected on camellia bonsai plants by the Italian Phytosanitary Control Service (Trisciuzzi *et al.*, 2014). The

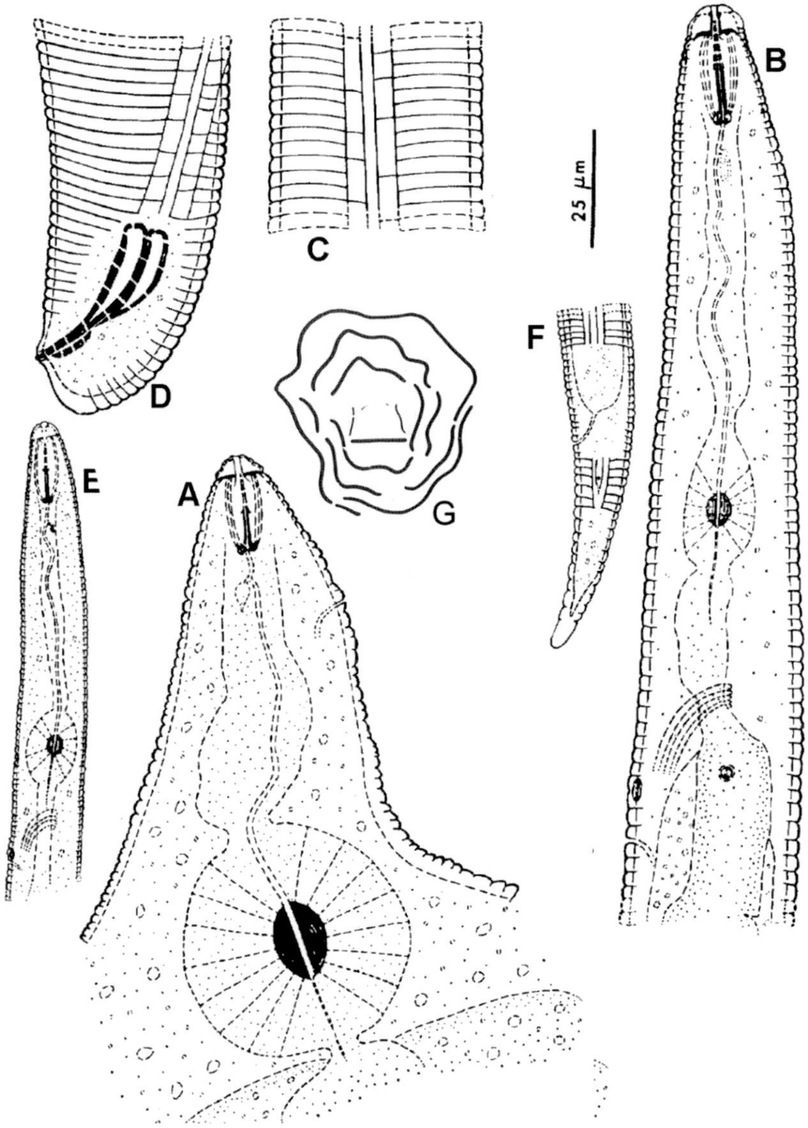

Fig. 57. Meloidogyne camelliae. *A: Female anterior region; B: Male anterior region; C: Male lateral field; D: Male tail; E: Anterior region of second-stage juvenile (J2); F: J2 tail; G: Perineal pattern. (A-F after Golden, 1979; G after Jepson, 1987.)*

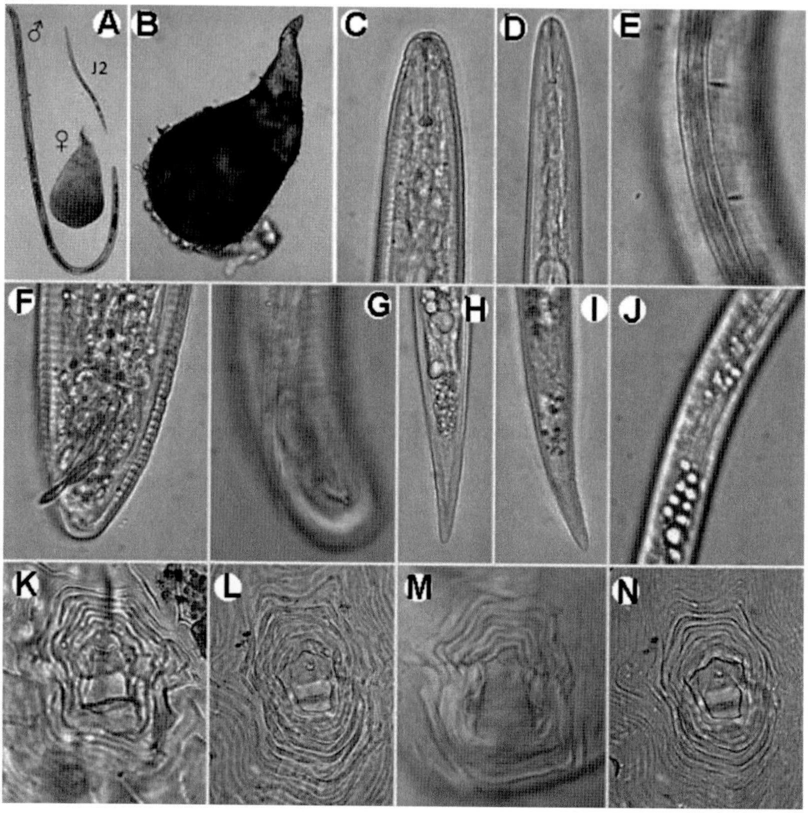

Fig. 58. Meloidogyne camelliae. *LM. A: Entire female, male and second-stage juvenile (J2); B: Entire female; C: Male labial region; D: J2 labial region; E: Male lateral field; F, G: Male tail; H, I: J2 tail; J: J2 lateral field; K-N: Perineal pattern. (Courtesy of N. Vovlas.)*

camellia RKN was identified in *Camelliae* plants from Thailand (Long *et al.*, unpubl.). *Meloidogyne camelliae* was found on cultivars of common and Japanese camellias in Sagamihara, Kanagawa, Japan (Aihara *et al.*, 1981). This species was morphologically characterised by Golden (1979), Aihara *et al.* (1981), Okamoto & Yaegashi (1981a, b) and Yaegashi & Okamoto (1981).

MEASUREMENTS

After Golden (1979):

- *Holotype female*: L = 989 μm; W = 530 μm; a = 1.8; b = 5.2; stylet = 18 μm; DGO = 4.3 μm; stylet knob width = 4.3 μm; vulval slit length = 33 μm; distance from vulval slit to anus = 22.4 μm.
- *Females* (n = 30): L = 905 ± 95 (700-1081) μm; W = 545 ± 75 (357-674) μm; a = 1.7 ± 0.2 (1.4-2.2); b = 5.2 ± 0.6 (4.1-6.2); stylet = 17.5 ± 0.3 (17.2-18.1) μm; stylet knob width = 4.9 ± 0.4 (4.0-5.6) μm; DGO = 4.4 ± 0.4 (3.4-5.6) μm; anterior end to median bulb valve = 88 ± 9.5 (72-110) μm, anterior end to excretory pore = 31 μm; vulval slit length = 30 ± 3 (26-36) μm; distance from vulval slit to anus = 20 ± 2.9 (17-27) μm; EP/ST = 2.2.
- *Allotype male*: L = 2034 μm; a = 43; b = 8.7; c = 170; stylet = 22.9 μm; DGO = 5.2 μm; spicules = 35 μm; gubernaculum = 8.6 μm; tail = 12 μm.
- *Males* (n = 55): L = 1945 ± 151 (1587-2180) μm; W = 44 ± 3.5 (34-55) μm; a = 44 ± 3.8 (36-54): b = 7.6 ± 0.8 (5.6-9.8); c = 182 ± 28 (127-321); stylet = 22.4 ± 0.7 (20.7-23.7) μm; DGO = 5.3 ± 0.8 (4-7) μm; anterior end to median bulb valve = 102 ± 6.6 (86-112) μm; spicules = 35 ± 1.2 (33-39) μm; gubernaculum = 9.6 ± 0.6 (8-12) μm; tail = 11 ± 1.4 (6-13) μm.
- *J2* (n = 70): L = 501 ± 21 (443-576) μm; a = 26 ± 1.8 (21-30); b = 3.1 ± 0.4 (2.3-3.8); c = 10.7 ± 0.6 (9.5-12.0); stylet = 11.6 ± 0.2 (11.2-12.0) μm; DGO = 3.7 ± 0.4 (3.0-4.5) μm; anterior end to median bulb valve = 65 ± 2.7 (58-73) μm; labial width = 6.2 ± 0.2 (5.2-6.9) μm; labial height = 3.0 ± 0.1 (2.6-3.4) μm; lateral field width = 19.5 ± 1.2 (17-24) μm; tail = 47 ± 3.1 (40-56) μm; hyaline region = 6.3 ± 1.4 (4.0-8.9) μm; a = 1.6 ± 0.3 (1.0-2.7); b = 1.8 ± 0.5 (1.0-2.9); distance from phasmids to tail tip = 35 ± 2.3 (32-38) μm.
- *Eggs* (n = 25): L = 110 ± 5 (100-120) μm; W = 48.6 ± 2.3 (46-53) μm; L/W = 2.3 ± 0.2 (2.0-2.6).

After Aihara *et al.* (1981):

- *J2*: L = 455-504 μm; c = 11.0-11.8; c′ = 3.5-4.0; tail = 40.8-44.9 μm; position of excretory pore = 18.6-19.4%.

Description

Female

Body globular to pear-shaped, without a posterior protuberance and with prominent neck situated anteriorly on a median plane with terminal vulva. Females pearly white, often becoming cream-coloured in older specimens. Labial region offset from neck, bearing a labial cap and two labial annuli. Labial framework distinct but weak, stylet fairly strong, with posteriorly-sloping knobs. Excretory pore clearly visible, variable in exact position. Massive egg sac extruded posteriorly, often two or more times size of female. Perineal pattern highly distinctive, with heavy rope-like striae forming a squarish to rectangular outline having shoulders or projections, sometimes appearing almost star-like. Vulva and anus sunken in a squarish area devoid of striae.

Male

Body long, slender, vermiform, tapering slightly at both extremities. Cuticular annuli prominent. Labial region only slightly offset, with large labial annulus (labial cap), and without postlabial annuli. Lateral field forming about 25% body diam. at mid-body, commonly with four lines; two outer bands occasionally areolated but not completely so, centre band slightly smaller, sometimes showing a fifth line, especially in older specimens. Stylet, knobs, cephalids, hemizonid, excretory pore, and anterior portion commonly as illustrated. Testis one. Spicules arcuate, tips rounded. Distinct phasmids at level of or posterior to cloacal aperture. Tail short, rounded.

J2

Body vermiform, tapering at both extremities but much more so posteriorly. Labial region continuous with body, with weak labial framework, but without postlabial annuli. Body annuli distinct, becoming coarser in posterior portion. Lateral field with four incisures, not areolated, and forming 25% of body diam. Cephalids indistinct and not shown. Phasmids small but distinct, located about one anal body diam. posterior to anus. Rectum not dilated. Tail with rather coarse annuli extending almost to end, with fine rounded tip.

TYPE PLANT HOST: Roots of *Camellia japonica* L.

OTHER PLANTS: *Camellia sasanqua* Thunb., *Cleyera japonica* Thunb., *Eurya japonica* Thunb., *E. emarginata* (Thunb.) Makino, *Camellia sinensis* (L.) Kuntze, *Oxalis* sp., *Lagerstroemia indica* (L.) Pers. (Golden, 1979; Aihara *et al.*, 1981, 1983; Trisciuzzi *et al.*, 2014; Cai *et al.*, 2016). In the host range test, no infection of citrus, corn, rose, soybean, and wheat by *M. camelliae* was detected. Tomato had a few scattered tiny galls in which a single small adult female was occasionally found and a small number of eggs was seen rarely. Some of the galls contained only an immature female. On camellia, however, during the test period large numbers of females developed on the roots. Also, a volunteer oxalis plant (*Oxalis* sp.) found growing with one of the camellia replicates had a moderate infection (Golden, 1979).

TYPE LOCALITY: Unknown location in Japan.

DISTRIBUTION: *Asia*: Japan, Thailand.

SYMPTOMS: On the roots of its type host and on oxalis, *M. camelliae* does not appear like a usual root-knot species. There is little or no swelling of the root at the infection site and half or more of the female body protrudes from the root (Golden, 1979).

CHROMOSOME NUMBER: No available data.

POLYTOMOUS KEY CODES: *Female*: A12, B21, C2, D1; *Male*: A12, B2, C12, D1, E1, F2; *J2*: A12, B23, C3, D4, E2, F3.

BIOCHEMICAL AND MOLECULAR CHARACTERISATION: No isozyme phenotype is available for this species. PCR-*COII*-16S rRNA-RFLP profile for *M. camelliae* was given by Orui (1998). LAMP method was recently developed by Cai *et al.* (2016). Sequence information is given for 18S, ITS, partial 28S, *COII* and *COI* genes (Wang *et al.*, 2013; Trisciuzzi *et al.*, 2014; Gu, unpubl.).

RELATIONSHIPS (DIAGNOSIS): The species belongs to Molecular group X and is clearly molecularly differentiated from all other *Meloidogyne* species. The perineal pattern with coarse, rope-like striae readily distinguishes *M. camelliae* from other species of the genus.

15. *Meloidogyne caraganae* Shagalina, Ivanova & Krall, 1985
(Fig. 59)

COMMON NAME: Peashrub root-knot nematode.

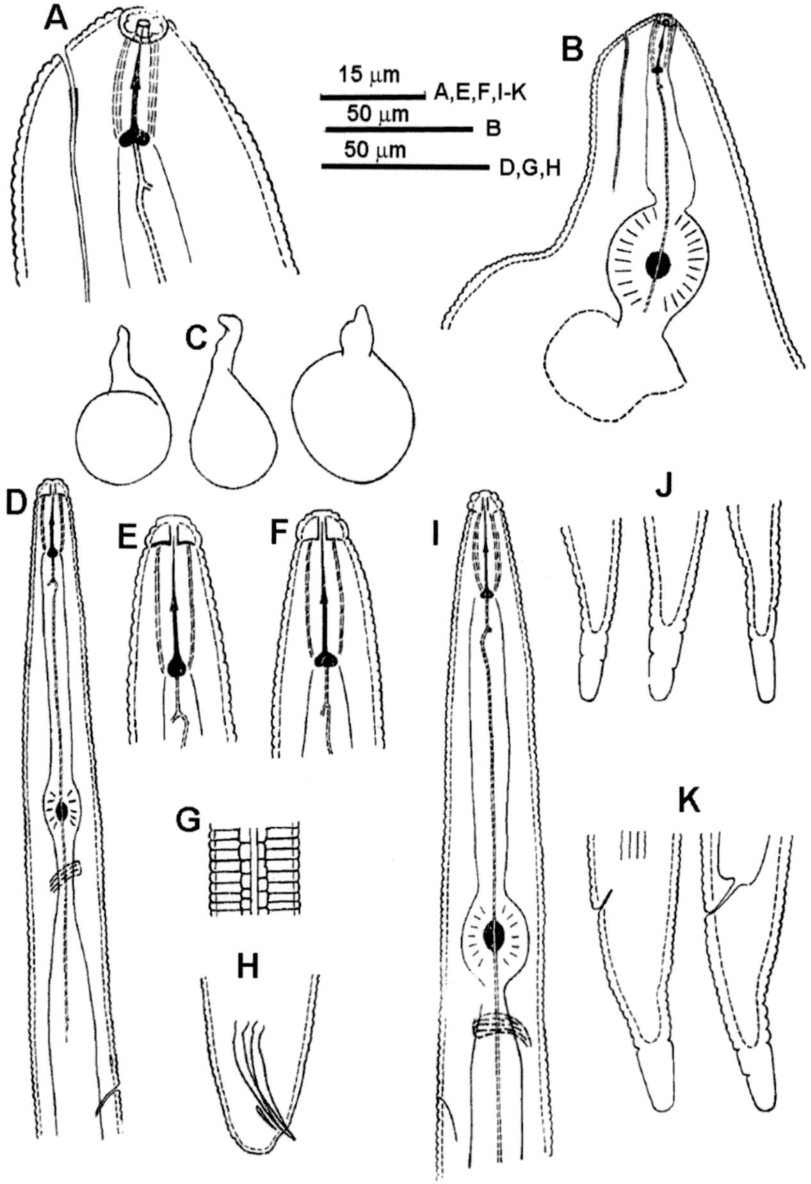

Fig. 59. Meloidogyne caraganae. *A, B: Female anterior region; C: Entire female; D: Male anterior region; E, F: Male labial region; G: Male lateral field; H: Male tail; I: Second-stage juvenile (J2) anterior region; J, K: J2 tail. (After Shagalina* et al., *1985.)*

This species was described from Turkish peashrub in Tajikistan.

MEASUREMENTS (*AFTER SHAGALINA* ET AL., *1985*)

- *Holotype female*: L (with neck) = 630 μm; W = 398.8 μm; neck length = 246.5 μm; stylet = 16.6 μm.
- *Paratypes females* (n = 16): L (with neck) = 739 (588-967) μm; W = 413.9 (325-529.3) μm; neck length = 262.1 (155.7-362.5) μm; stylet = 17.6 (15.9-18.6) μm; stylet knob height = 2.2 μm; stylet knob width = 4.0 μm; DGO = 5.6 (5.2-6.2) μm; anterior end to excretory pore = 7.5-9.7 μm; median bulb length = 35.1 (33.5-37.5) μm; median bulb width = 28.5 (24.2-31.1) μm; vulval slit = 24.2 (20.7-26.2) μm; distance vulva-anus = 21.1 (16.4-25.5) μm; distance between phasmids = 19.3 (15.6-25.4) μm; EP/ST = 0.8.
- *Allotype male*: L = 1399.3 μm; W = 28.7 μm; a = 48.7; b′ = 14.0; c = 103.5; stylet = 20.1 μm; spicules = 28.7 μm; gubernaculum = 6.2 μm; T = 46.9%.
- *Paratype males* (n = 3): L = 1338 (1140-1495) μm; W = 33.6 (27-43.1) μm; a = 41 (34.4-51.6); b′ = 15 (13.2-16.4); c = 115.5 (102.1-138.7); stylet = 19.8 (19.3-20.7) μm; DGO = 5.2-6.0 μm; anterior end to the base of median bulb = 88.9 (86.2-90.9) μm, anterior end to excretory pore = 144.2 (124.2-156.5) μm; spicules = 30.9 (28.7-33.1) μm; gubernaculum = 6.6 (6.2-6.9) μm; T = 49.5 (45.3-55.3)%.
- *J2* (n = 15): L = 450 (417-482) μm; W = 17.7 (15.2-19.2) μm; a = 25.4 (22.4-28.6); b′ = 7.3 (6.2-8.0); c = 15.1 (12.3-18.3); c′ = 2.6 (2.0-3.4); stylet = 14.9 (13.5-15.6) μm; labial region height = 3.2 (2.5-3.7) μm; labial region diam. = 6.3 (6.0-7.2) μm; DGO = 4.1 (3.6-4.8) μm; anterior end to median bulb valve = 61.9 (55.4-72.0) μm; anterior end to excretory pore = 85.7 (77.0-92.4) μm; width at anal level = 11.3 (9.1-12.7) μm; tail = 30.2 (26.4-36.0) μm.
- *Eggs* (n = 14): L = 115.2 (105-125) μm; W = 49.7 (48-54) μm.

DESCRIPTION

Female

Adults with pear-shaped, seldom spherical, body shape. Neck varying from short to long. Labial region not separated from rest of body with six labials and well-developed labial disc. Cuticle thick. Coarse cuticle rings clearly visible on whole body surface. Stylet slightly bent with rounded, posteriorly-sloping basal knobs. Metenchium longer than telenchium. Median bulb rounded with developed valve. Cuticle pattern in anal-vulvar area often oval-shaped. Lines thin, slightly wave-like, on dorsal side forming rectangular arc, on ventral side sometimes forming wings. Lateral fields smooth, of varying size and lacking cuticular folds, located from rudiment of tail to vulval opening on both sides. Phasmids clearly visible. Tail rudiment slightly projecting. Lateral fields not present in all specimens.

Male

Lateral field forming 20-25% of body diam., consisting of four incisures, outermost of which are slightly jagged. All incisures reaching end of tail and intermediate part of body may have five incisures. Stylet thin with rounded and posteriorly sloping basal knobs. Median bulb oval shape with developed valve. Hemizonid located 6-7 annuli anterior to excretory pore. Nerve ring located right posterior to base of median bulb. Tail very short with broadly-rounded smooth terminus. Phasmids located at cloacal aperture level. Spicules curved, gubernaculum small and simple.

J2

Body slightly curved dorsoventrally. Labial region slightly separated from body with two cuticle annuli. Lateral field 25% of body diam., consisting of four incisures, outermost of which are slightly jagged. Stylet thin with rounded, posteriorly directed, basal knobs. Hemizonid not visible. Nerve ring located in anterior part of isthmus. Tail very short and obtusely conical. Tail narrowing sharply in intermediate part and often directed to dorsal side. hyaline region forming 25-33% of its length. There are 16-20 cuticle annuli on ventral side of tail. Phasmids located on middle of tail.

TYPE PLANT HOST: Turkish peashrub, *Caragana turkestanica* Kom.

OTHER PLANTS: No other hosts were reported.

TYPE LOCALITY: Safedra village, Vossei district, Kulob, south-east slope of Vaksh Mountain range (1600 m a.s.l.), Tajikistan.

DISTRIBUTION: *Asia*: Tajikistan.

CHROMOSOME NUMBER: No available data.

POLYTOMOUS KEY CODES: *Female*: A213, B12, C4, D3; *Male*: A3, B32, C2, D1, E1, F2; *J2*: A2, B12, C43, D3, E12, F4.

BIOCHEMICAL AND MOLECULAR CHARACTERISATION: No available data.

RELATIONSHIPS (DIAGNOSIS): The species is similar to *M. artiellia* and *M. turkestanica* and differs by the hyaline region of the tail (33 *vs* 20 and 25% of tail length, respectively) and longer female stylet (15.9-18.6 *vs* 12-15 and 13.8-15.4 μm, respectively). The low EP/ST ratio could help in separation from similar species. This species is also very similar to *M. chosenia* and *M. marylandi*.

16. *Meloidogyne carolinensis* Eisenback, 1982

(Figs 60, 61)

COMMON NAME: Blueberry root-knot nematode.

A biological study of the blueberry RKN parasitising cultivated and wild blueberries in North Carolina, USA, was conducted by Fox (1967). Observations on the morphology, cytology, mode of reproduction, and host range revealed many unusual features peculiar to this species and this nematode was only formally described much later by Eisenback (1982). The nematode was found in 24 of 63 North Carolina commercial blueberry plantings sampled in 1967 and is probably indigenous to that state.

MEASUREMENTS *(AFTER EISENBACK, 1982)*

- *Holotype female*: L = 566.7 μm; W = 426 μm; neck length = 124.7 μm; neck width = 110 μm; stylet = 16.4 μm; stylet knob height = 2.4 μm; DGO = 4.9 μm; anterior end to excretory pore = 10 μm.
- *Females* (n = 30): L = 508.6 ± 75.5 (373.5-690.3) μm; L (without neck) = 377.9 ± 58.7 (297-507.6) μm; neck length = 130.6 ± 37.3

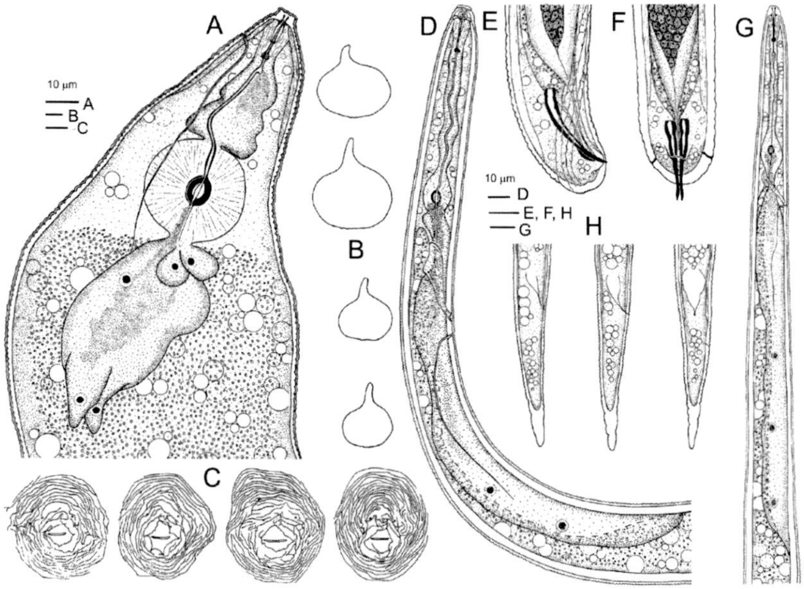

Fig. 60. Meloidogyne carolinensis. *A: Female anterior region; B: Entire female; C: Perineal pattern; D: Male anterior region; E, F: Male tail; G: Second-stage juvenile (J2) anterior region; H: J2 tail. (After Eisenback, 1982.)*

(71.0-236.7) μm; neck width = 96.2 ± 21.9 (68.4-174.6) μm; W = 444.4 ± 74 (315-593.1); L/W = 4.0 ± 0.8 (1.9-5.6); L (without neck)/W = 1.2 ± 0.1 (0.9-1.5); stylet = 15.9 ± 0.1 (14.9-16.9) μm; stylet knob height = 2.3 ± 0.2 (1.9-2.8) μm; stylet knob width = 4.0 ± 0.2 (3.3-4.4) μm; DGO = 4.6 ± 0.8 (2.4-5.9) μm; anterior end to excretory pore = 13.9 ± 3.6 (8.6-28.2) μm; vulval length = 20.2 ± 1.8 (14.9-23.7) μm; anus to vulva (centre) distance = 16.4 ± 3.4 (5.6-24.2) μm; interphasmid distance = 27.8 ± 4.4 (14.0-34.7) μm; EP/ST = 1.2.

- *Allotype male*: L = 1324.2 μm; W = 34.7 μm; stylet = 20.9 μm; stylet knob height = 2.5 μm; stylet knob width = 4.2 μm; DGO = 4.2 μm; anterior end to median bulb valve = 86.4 μm, anterior end to excretory pore = 150.7 μm; body diam. at stylet base = 16.3 μm; body diam. at excretory pore = 29.2 μm; tail = 6.8 μm; spicules = 28.1 μm.
- *Males* (n = 30): L = 1487.5 ± 145.5 (1253.7-1980) μm; W = 36.6 ± 2.7 (32.1-44.5) μm; stylet = 20 ± 1.1 (17.4-21.9) μm; stylet knob

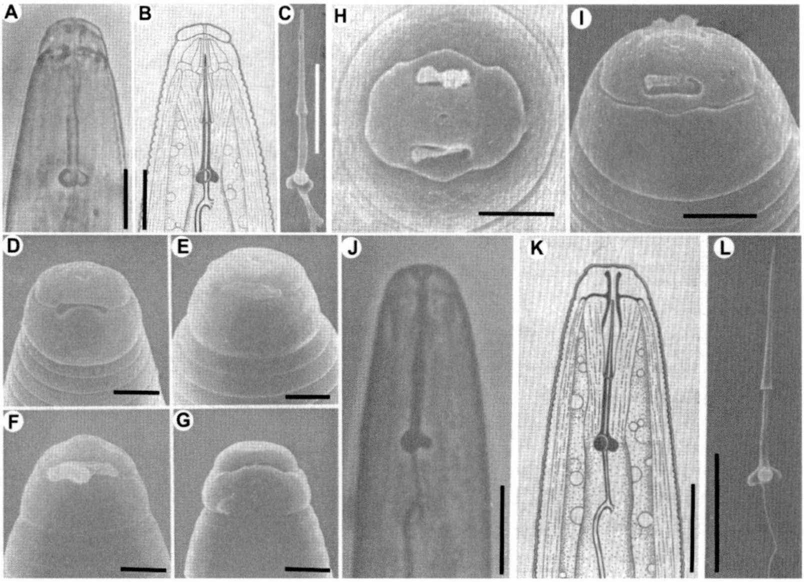

Fig. 61. Meloidogyne carolinensis. *LM and SEM. A, B, D-G: Male labial region; C: Male stylet; H-K: Second-stage juvenile (J2) labial region; L: J2 stylet. (Scale bars: C-I, L = 5 μm; A, B, J, K = 10 μm.) (After Eisenback, 1982.)*

height = 2.7 ± 0.2 (2.3-3.0) μm; stylet knob width = 4.4 ± 0.3 (3.6-5.2) μm; DGO = 4.0 ± 0.7 (1.9-5.2) μm; anterior end to median bulb valve = 86.8 ± 6.2 (75.5-98.0) μm, anterior end to excretory pore = 151.8 ± 13.9 (126.3-180.9) μm; body diam. at stylet base = 17.5 ± 0.9 (15.7-19.1) μm; body diam. at excretory pore = 28.7 ± 1.7 (25.6-31.3) μm; tail = 7.4 ± 1.4 (4.9-10.4) μm; spicules = 31.7 ± 2.4 (27.0-35.3) μm; a = 40.7 ± 3.5 (32.3-49.9).

- *J2* (n = 30): L = 463.7 ± 26.8 (416.7-515.7) μm; W = 15.1 ± 0.9 (13.4-17.6) μm; stylet = 11.9 ± 0.6 (10.9-13.1) μm; DGO = 3.9 ± 0.5 (3.0-4.7) μm; anterior end to median bulb valve = 60 ± 4.5 (53.1-78.2) μm, anterior end to excretory pore = 80.4 ± 3.6 (73.4-86.1) μm; tail = 42.5 ± 3.9 (34.7-49.0) μm; body diam. at anus = 10.8 ± 0.5 (9.7-11.8) μm; a = 30.7 ± 2.2 (27.2-36.3); b = 7.8 ± 0.0 (5.7-8.6); c = 10.9 ± 0.0 (9.7-13.3).
- *Eggs* (n = 30): L = 90.7 ± 1.1 (79.6-104.2) μm; W = 37.1 ± 0.4 (34.0-41.6) μm; L/W = 2.5 ± 0.1 (1.9-3.0).

Description

Female

Body translucent white, variable in size, pear-shaped to ovoid, width often greater than length without neck. Neck prominent, cuticular annulations on body finer posteriorly. Body posteriorly flattened, with slight protuberance. In SEM, stoma slit-like, located in ovoid prestomatal cavity, surrounded by pit-like openings of six inner labial sensilla. Raised labial disc, separated from labials by a deep groove. Labial disc often rectangular, may be indented medianly on one or both sides, often marked by two or four bumps. Lateral and median labials often fused, forming one structure. Labial region offset, with 2-3 annuli. In LM, labial framework weak, hexaradiate, lateral sectors slightly enlarged, vestibule and extension prominent. Excretory pore anterior to stylet base. Stylet delicate; cone slightly curved dorsally; shaft enlarged posteriorly, knobs tapering onto shaft, separate, rounded posteriorly. Distance of dorsal pharyngeal gland orifice to stylet base *ca* one shaft length, branched into three ducts, ampulla large. Subventral gland orifices branched, located immediately posterior to enlarged lumen of median bulb. Pharyngeal gland with one large dorsal lobe with one nucleus, two small nucleated subventral gland lobes, variable in shape, position, and size, usually posterior to dorsal gland lobe, two small rounded, pharyngo-intestinal cells with nuclei attached to dorsal lobe between median bulb and intestine. Perineal pattern rounded to hexagonal, striae coarse, sometimes continuous, smooth to wavy. One large, nearly continuous cuticular ridge surrounding perivulval region. Perivulval region free of striae. Phasmids small, directly on either side of anus, surface structure not apparent in SEM.

Male

Body vermiform, tapering anteriorly, bluntly rounded posteriorly, tail twisting through 90°. Labial region shape extremely variable. Typically, labial cap low, rounded, tapering posteriorly, labial region narrower than first body annulus. In SEM, stoma slit-like, located in ovoid to hexagonal prestomatal cavity surrounded by pit-like openings of six inner labial sensilla. Labial disc rounded, slightly raised near prestoma. Rounded median labials fused with labial disc forming elongate labial cap. Four labial sensilla marked by cuticular depressions on median labials. Amphidial apertures appearing as elongate slits between labial disc and lateral sectors of labial region. Lateral labials absent. Labial region not annulated. Body annuli distinct. Lateral field with four incisures, areolated, beginning near level of stylet base. In LM, labial framework moderately developed, hexaradiate, lateral sectors slightly enlarged. Vestibule and extension distinct. Stylet morphology extremely variable. Typical stylet with opening marked by slight protuberance several microns from stylet tip, cone pointed, gradually increasing in diam. posteriorly, junction of cone and shaft uneven. Shaft cylindrical, often slightly wider near middle. Knobs broadly elongate, offset from shaft, indented slightly anteriorly, rounded posteriorly. Distance of dorsal pharyngeal gland orifice to stylet base moderately long, orifice branching into three ducts, ampulla indistinct. Procorpus distinct, median bulb ovoid, triradiate lining of enlarged lumen of median bulb thinner than in female. Subventral gland orifices posterior to lining of median bulb, branched. Pharyngo-intestinal junction at level of nerve ring, indistinct. Two nuclei in gland lobe, lobe variable in length. Intestinal caecum extending anteriorly near level of medium bulb. Excretory pore distinct. Hemizonid 3-4 annuli anterior to excretory pore. Usually one testis, sometimes two, outstretched or anteriorly reflexed. Spicules arcuate, gubernaculum distinct. Tail short, phasmids at level of cloacal aperture. Variants: more than 50% of all males exhibit distinct morphological variation in labial region morphology. Differences exist in height and shape of labial cap and diam. of labial region relative to width of first body annulus. Variant 1, similar to typical male except labial cap much higher, more rounded. Variant 2, one or both median labials lower than labial disc. Variant 3, labial disc distinctly elevated near prestomatal cavity. Variant 4, labial cap narrower than in typical male; labial region wider than first body annulus. Stylet morphology also variable. Variant 1, stylet much longer, knobs larger, more rounded, tapering anteriorly onto shaft. Variant 2, knobs anteriorly indented, each knob often appearing as two, knobs tapering onto shaft.

J2

Body vermiform, tapering more posteriorly than anteriorly. In SEM, stoma slit-like located in ovoid prestomatal cavity, surrounded by pit-like openings of six inner labial sensilla. Labial disc, median labials, and lateral labials fused into one structure. Labial disc elevated above stoma. Median labials with rounded margins, labial sensilla distinct. Amphidial apertures located between labial disc and lateral labials. Labial region smooth. Body annuli distinct. Lateral field with four incisures, areolated. In LM, labial framework weak, hexaradiate. Vestibule and vestibule extension more distinct than rest of framework. Stylet cone gradually increasing in width, shaft cylindrical, knobs rounded and offset from shaft. Distance of dorsal pharyngeal gland orifice to stylet base long, orifice branching into ducts, ampulla indistinct. Median bulb ovoid, triradiate lining strongly sclerotised, subventral gland orifices branched, located immediately posterior to enlarged lumen of median bulb. Pharyngo-intestinal junction indistinct, at level of nerve ring. Gland lobe of variable length with three nuclei. Excretory pore distinct, hemizonid 1-2 annuli anterior to excretory pore. Tail annuli larger and more irregular posteriorly. Hyaline region distinct. Phasmids small, always posterior to anus.

TYPE PLANT HOST: Roots of cultivated highbush blueberry 'Wolcott' (cultivar derived from hybrids of *Vaccinium corymbosum* L. and *V. lamarekii* Camp).

OTHER PLANTS: Highbush, lowbush (*V. angustifolium* Ait.), and creeping blueberry (*V. crassifolium* Andr.), as well as several cultivars of cultivated blueberry. *Rhododendron* sp. was also a good host. A few mature females developed with some reproduction possible on beet (*Beta vulgaris* L.), radish (*Raphanus sativus*), cabbage (*Brassica oleracea* var. *capitata*), carrot (*Daucus carota* var. *sativa* D.C.), and tomato (*Solanum lycopersicum*).

TYPE LOCALITY: Frank Blanchard farm on state Highway 11 near Charity, Rose Hill, Duplin County, North Carolina, USA.

DISTRIBUTION: *North America*: USA (North Carolina).

CHROMOSOME NUMBER: Haploid chromosome n = 18 (Triantaphyllou, 1985b).

POLYTOMOUS KEY CODES: *Female*: A32, B2, C3, D6; *Male*: A312, B324, C21, D1, E1, F2; *J2*: A21, B23, C3, D2, E21, F3.

BIOCHEMICAL AND MOLECULAR CHARACTERISATION: No available data.

RELATIONSHIPS (DIAGNOSIS): This species differs from others in the perineal pattern. This species may be polymorphic with respect to male morphology. However, the morphological characters of this species make it difficult to distinguish from similar species.

17. *Meloidogyne chitwoodi* Golden, O'Bannon, Santo & Finley, 1980 (Figs 62-64)

COMMON NAME: Columbia root-knot nematode.

The Columbia RKN was described from potatoes originating from Washington state, USA. The common name derived from the Columbia River between Oregon and Washington states. This species is considered to be the major nematode pest of potatoes in the Pacific Northwest states of the USA and the annual predicted loss there would be approximately 40 million $US, if control measures were not applied (Santo, 1994). Presently, *M. chitwoodi* is an EPPO A2 quarantine pest and is regulated in Canada and some states of the USA. It was first detected in the EPPO region in the 1980s in The Netherlands, but a review of old illustrations and old specimens of *Meloidogyne* suggests that it may have occurred earlier (in the 1930s) and may have been present throughout the intervening period (Anon., 2006, 2013b, 2016a). It infects a broad range of plants, including potatoes, vegetables, wheat, corn, alfalfa, and numerous weeds. The ability of *M. chitwoodi* to reproduce on both monocotyledons and dicotyledons limits feasible crop rotation strategies as a control measure. This species is a major problem for potato growers and requires control measures in virtually every field where it is found (Lehman *et al.*, 1983; Elling, 2013).

MEASUREMENTS

See Table 20.

DESCRIPTION *(AFTER GOLDEN* ET AL., *1980)*

Female

Body pearly white, globular to pear-shaped, with slight posterior protuberance visible occasionally, and with distinct neck situated anteriorly

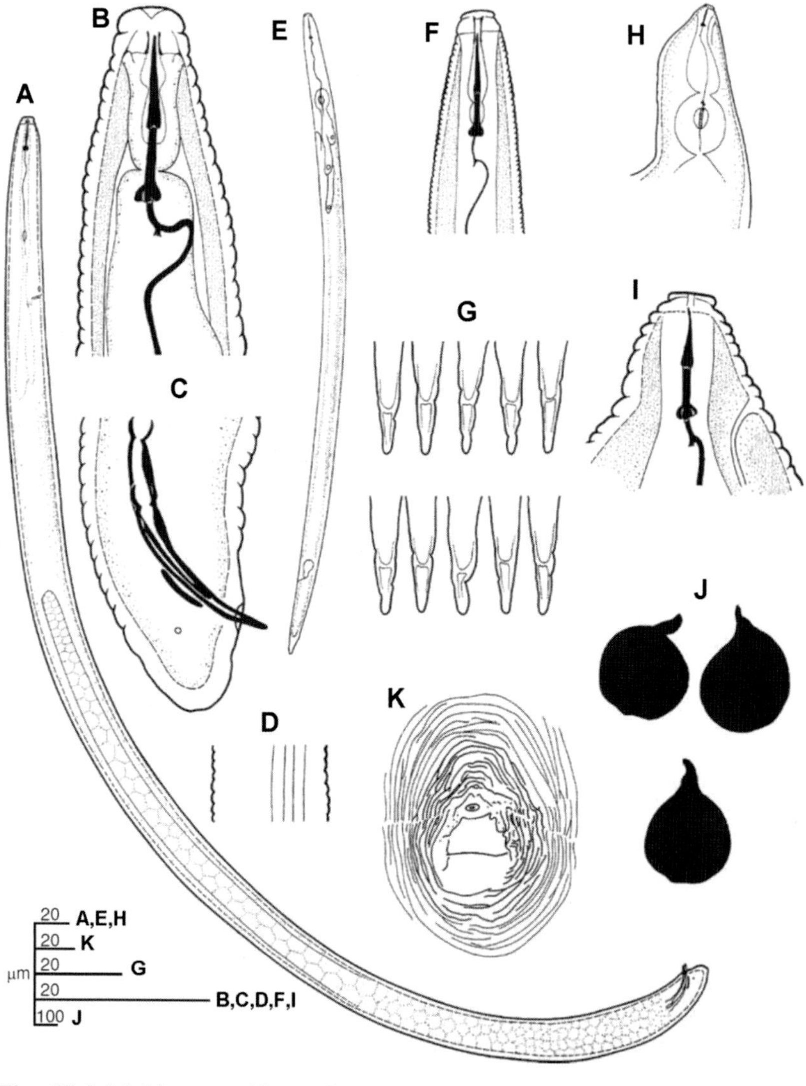

Fig. 62. Meloidogyne chitwoodi. *A: Entire male; B: Male labial region; C: Male tail; D: Male lateral field; E: Entire second-stage juvenile (J2); F: J2 labial region; G: J2 tail; H, I: Female anterior region; J: Entire female; K: Perineal pattern. (After Jepson, 1985.)*

on a median plane with terminal vulva. Pharyngeal and anterior region often appearing as illustrated. Several small vesicles or vesicle-like structures usually present within median bulb and clustered around lumen an-

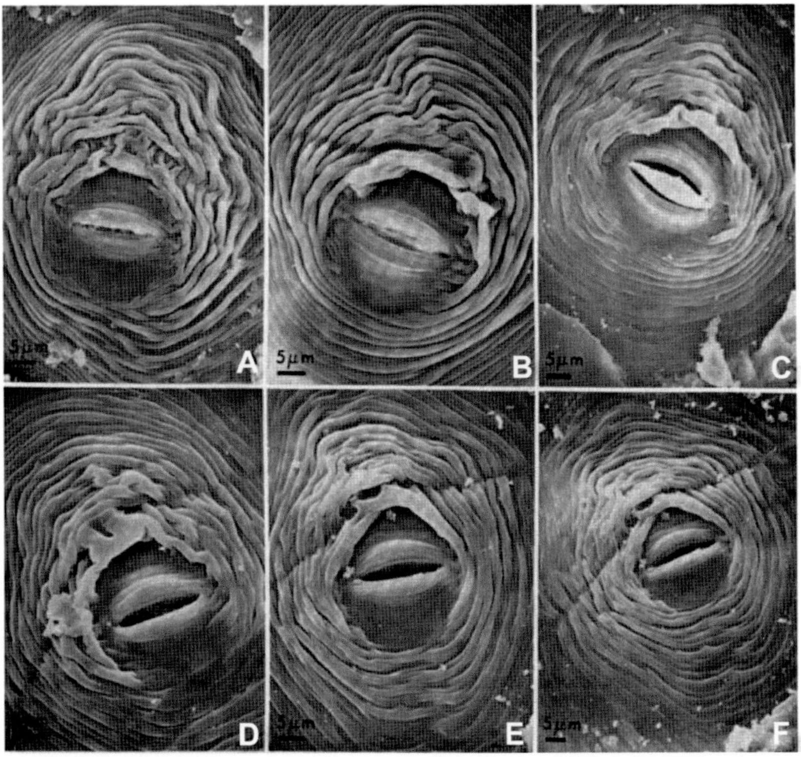

Fig. 63. Meloidogyne chitwoodi. *SEM. A-F: Perineal patterns. (After Golden* et al., *1980.)*

terior to valve plates of median bulb. Labial region with distinct but weak labial framework, offset from neck but variable in exact shape, bearing a labial cap and usually one labial annulus. Excretory pore clearly visible and commonly located at a distance equal to about 1.5 stylet lengths from anterior end. Stylet small but strong, having a dorsal curvature and rounded knobs sloping posteriorly. Perineal pattern quite distinctive, with striae around and above anal area being broken, curved, twisted, or curled as illustrated, overall shape of pattern appearing round to oval. Dorsal arch variable, ranging from low and round to high and square. Punctations sometimes visible in a small area between anus and first inner striae (tail terminus) and in this area striae were absent (Humphreys-Pereira & Elling, 2014a). Vulva sunken in an area variable in shape and devoid of striae.

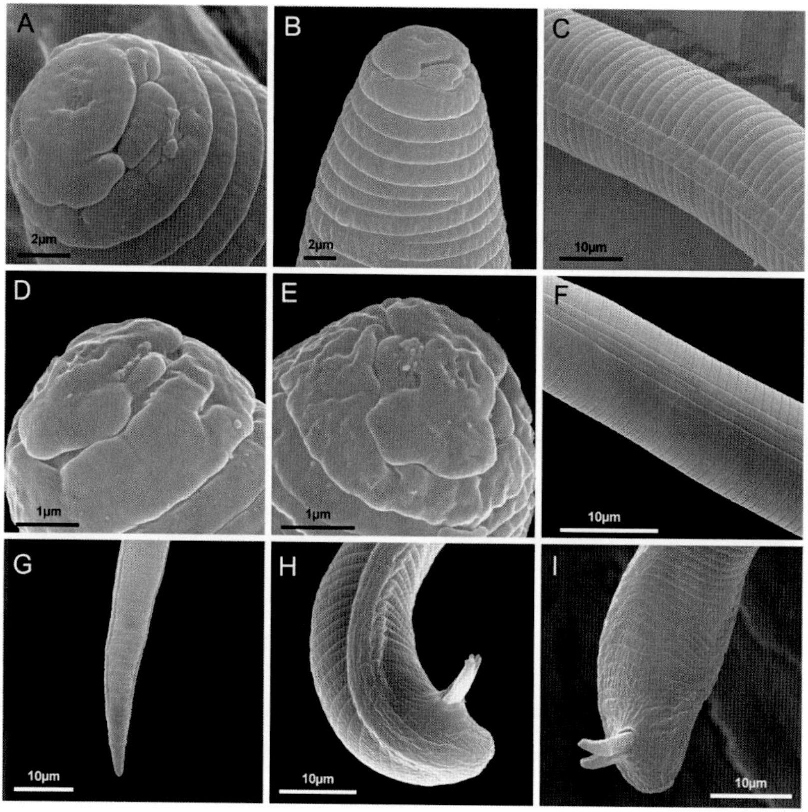

Fig. 64. Meloidogyne chitwoodi. *SEM of males and second-stage juvenile (J2). A, B: Male lip region; C: Male lateral field; D, E: J2 lip region; F: J2 lateral field; G: J2 tail region; H, I: Male tail region. (After Humphreys-Pereira & Elling, 2014a.)*

Male

Body slender, vermiform, tapering slightly at both extremities. Labial disc large with median labials forming a labial cap, similar to juveniles. Labial disc slightly higher than median labials and post-labial annulus lacking annulations. In labial region, two large, not well-delimited, lateral labials about two-thirds of size of median labials. Cuticular annuli distinct, becoming more prominent a short distance from either end. Lateral field with four lines, centre band smaller than outer two, some areolation evident with SEM but difficult to resolve with optical microscopy. Testes one or two. Spicules arcuate, under SEM can be seen

Table 20. *Morphometrics of females, males and second-stage juveniles (J2) of* Meloidogyne chitwoodi. *All measurements are in μm and in the form: mean ± s.d. (range).*

Character	Quincy, Washington, USA, Holotype, allotype, Golden *et al.* (1980)	Quincy, Washington, USA, Paratypes, Golden *et al.* (1980)	Quincy, Washington, USA, Humphreys-Pereira & Elling (2013, 2014a)	Prosser, Washington, USA, Humphreys-Pereira & Elling (2013, 2014a)	Washington, Humphreys-Pereira & Elling (2013, 2014a)	Tulelake, California, USA, Humphreys-Pereira & Elling (2013, 2014a)
Female (n)	-	60	30	30	30	30
L	598	591 ± 60 (430-740)	-	-	-	-
a	1.3	1.4 ± 0.2 (1.1-1.8)	-	-	-	-
W	474	422 ± 42 (344-518)	-	-	-	-
Stylet	12	11.9 ± 0.3 (11.2-12.5)	13.4 ± 0.8 (11.9-14.5)	13.2 ± 0.6 (11.6-14.6)	13.0 ± 0.8 (11.4-14.6)	13.51 ± 0.7 (11.8-14.5)
Stylet knob width	-	3.8 ± 0.3 (3.4-4.3)	3.9 ± 0.3 (3.2-4.4)	3.7 ± 0.3 (3.3-4.5)	3.9 ± 0.3 (3.4-4.4)	3.7 ± 0.3 (3.2-4.4)
DGO	4.3	4.2 ± 0.6 (3.4-5.5)	3.7 ± 0.5 (2.8-5.1)	3.7 ± 0.5 (2.9-4.7)	3.6 ± 0.5 (2.8-4.5)	3.8 ± 0.5 (3.1-5.0)
Ant. end to median bulb valve	-	63 ± 7 (52-80)	-	-	-	-
Ant. end to excretory pore	-	18 ± 5 (10-27)	-	-	-	-
Vulval slit	29.2	27 ± 3.0 (19-32)	20.9 ± 1.2 (18.5-23.6)	20.4 ± 1.5 (18.0-23.7)	22.8 ± 1.5 (19.4-25.9)	20.8 ± 1.1 (18.7-23.4)
Vulva-anus	18.1	18.0 ± 2.0 (13-22)	15.5 ± 1.5 (13.0-18.4)	15.0 ± 1.1 (13.1-17.8)	16.4 ± 1.4 (13.2-19.1)	14.8 ± 1.5 (12.9-18.1)
EP/ST		1.5	-	-	-	-
Male (n)	1	30	20	20	20	20
L	1046	1068 ± 100 (887-1268)	1211 ± 170 (860-1457)	1257 ± 123 (965-1470)	1206 ± 152 (892-1477)	1304 ± 114 (1004-1488)
a	39	36 ± 4.0 (28-46)	34.8 ± 7.7 (21.6-48.8)	36.8 ± 6.2 (22-45)	30.2 ± 2.7 (25.2-36.4)	34.6 ± 5.2 (20.2-43.2)
b	5.6	7.2 ± 1.0 (6-9)	-	-	-	-
c	174	162 ± 20 (140-226)	134.3 ± 24.8 (104.9-220.7)	144.8 ± 21.6 (102.2-182.5)	136.5 ± 31.2 (91.8-203.5)	158.6 ± 27 (111.8-201.5)
W	-	30 ± 3.9 (22-37)	35 ± 4 (29-40)	35 ± 4 (29-44)	40 ± 3 (31-45)	38 ± 5 (32-50)
Labial region diam.	-	18.3 ± 0.2 (18.1-18.5)	-	-	-	-
Stylet	18.1	-	18.2 ± 1.1 (16.2-19.4)	18.2 ± 0.7 (16.6-19.4)	18 ± 0.9 (16.3-19.1)	17.3 ± 1.0 (16.0-19.3)

Table 20. *(Continued.)*

Character	Quincy, Washington, USA, Holotype, allotype, Golden *et al.* (1980)	Quincy, Washington, USA, Paratypes, Golden *et al.* (1980)	Quincy, Washington, USA, Humphreys-Pereira & Elling (2013, 2014a)	Prosser, Washington, USA, Humphreys-Pereira & Elling (2013, 2014a)	Washington, Humphreys-Pereira & Elling (2013, 2014a)	Tulelake, California, USA, Humphreys-Pereira & Elling (2013, 2014a)
Stylet knob width	-	-	4.2 ± 0.3 (3.4-4.5)	4.1 ± 0.3 (3.4-4.5)	4.1 ± 0.3 (3.5-4.5)	3.9 ± 0.2 (3.3-4.4)
DGO	2.6	3.0 ± 0.4 (2.2-3.4)	3.2 ± 0.4 (2.6-4.2)	3.1 ± 0.5 (2.3-4.1)	3.1 ± 0.4 (2.3-4.0)	3.1 ± 0.4 (2.4-4.1)
Ant. end to median bulb	-	71 ± 5 (61-77)	74 ± 6 (63-81)	72 ± 5 (64-82)	77 ± 5 (66-85)	77 ± 6 (65-85)
Ant. end to excretory pore	-	-	127 ± 15 (97-153)	126 ± 11 (105-145)	140 ± 13 (104-161)	137 ± 13 (113-169)
Spicules	26	27 ± 1.2 (26-29)	28 ± 2.3 (23-31)	28 ± 1.8 (25-31)	28 ± 1.6 (25-31)	28 ± 1.6 (26-32)
Gubernaculum	7.3	7.7 ± 0.6 (6.5-8.2)	7.6 ± 0.6 (6.3-8.6)	7.5 ± 1.0 (6.1-9.6)	7.7 ± 0.7 (6.2-8.9)	8.1 ± 0.7 (6.3-9.3)
Tail	6	6.8 ± 0.9 (4.7-9.0)	9.1 ± 1.3 (6.6-14.8)	8.8 ± 1.0 (7.1-11.4)	9.0 ± 1.1 (6.1-10.5)	8.4 ± 1.3 (6.5-10.5)
J2 (n)	-	60	30	30	30	30
L	-	390 ± 16 (336-417)	370 ± 15 (339-397)	366 ± 20 (335-417)	377 ± 18 (341-417)	379 ± 20 (340-422)
a	-	27.5 ± 1.2 (24.5-29.8)	26.1 ± 1.3 (23.4-29.3)	26.2 ± 3.4 (21.4-32.7)	26.7 ± 2.5 (20.8-31.6)	26.7 ± 1.8 (22.3-30.2)
b	-	3.6 ± 0.2 (3.3-3.8)	-	-	-	-
c	-	8.9 ± 0.4 (7.9-9.6)	8.1 ± 0.6 (7.2-9.2)	8.3 ± 0.4 (7.5-9.3)	8.5 ± 0.6 (7.2-9.5)	8.1 ± 0.5 (7.3-9.6)
W	-	14.2 ± 0.6 (12.5-15.5)	14.2 ± 0.4 (13.4-14.9)	14.1 ± 1.6 (11.5-17.1)	14.2 ± 0.9 (12.7-16.8)	14.3 ± 0.8 (12.4-16.0)
Labial region height	-	2.3 ± 0.2 (1.7-2.6)	-	-	-	-
Labial region diam.	-	5 ± 0.2 (4.7-5.2)	-	-	-	-
Stylet	-	9.9 ± 0.3 (9.0-10.3)	9.9 ± 0.5 (9.0-10.6)	9.9 ± 0.6 (9.0-10.7)	10 ± 0.4 (9.1-10.6)	9.8 ± 0.3 (9.2-10.4)
Stylet knob width	-	-	2.1 ± 0.2 (1.7-2.5)	2.0 ± 0.2 (1.8-2.5)	2.0 ± 0.2 (1.8-2.5)	2.0 ± 0.2 (1.7-2.4)
DGO	-	3.2 ± 0.2 (2.6-3.9)	3.0 ± 0.4 (2.4-3.6)	2.8 ± 0.3 (2.1-3.4)	2.9 ± 0.4 (2.2-3.5)	3.2 ± 0.5 (2.4-4.0)
Ant. end to median bulb	-	51 ± 3 (43-56)	47 ± 3 (42-53)	46 ± 3 (42-57)	48 ± 3 (43-54)	50 ± 3 (44-54)
Ant. end to excretory pore	-	-	72 ± 6 (63-84)	70 ± 6 (61-82)	73 ± 5 (63-83)	72 ± 5 (65-83)

Table 20. *(Continued.)*

Character	Quincy, Washington, USA, Holotype, allotype, Golden *et al.* (1980)	Quincy, Washington, USA, Paratypes, Golden *et al.* (1980)	Quincy, Washington, USA, Humphreys-Pereira & Elling (2013, 2014a)	Prosser, Washington, USA, Humphreys-Pereira & Elling (2013, 2014a)	Washington, Humphreys-Pereira & Elling (2013, 2014a)	Tulelake, California, USA, Humphreys-Pereira & Elling (2013, 2014a)
Tail	-	43 ± 1.8 (39-47)	46 ± 2.6 (41-51)	44 ± 2.7 (40-50)	45 ± 2.6 (40-49)	47 ± 2.4 (40-51)
Hyaline region	-	11 ± 1 (8.6-13.8)	11.8 ± 1.2 (9.3-13.9)	11.6 ± 1.4 (9.6-13.6)	12.4 ± 0.9 (10.2-13.9)	12.2 ± 0.9 (10.3-13.9)

to have dentate tips ventrally. Phasmids located at or anterior to cloacal aperture. Tail short, rounded.

J2

Body small, vermiform, tapering at both extremities but more so posteriorly. Labial region not offset, with weak framework, bearing a labial disc and a large post-labial annulus lacking striations. Stylet knobs round to irregular. Cuticular annulation on most of body very fine. Lateral field with four lines, areolated. Cephalids indistinct or not seen. Phasmids small, difficult to see, located in anterior one-third of tail. Rectum not inflated. Tail commonly appearing as illustrated, having a short, blunt hyaline region remaining almost same diam. for its length and with little or no taper. Tail terminus rounded, with slightly clavate form or with oval appendix-like structure (Humphreys-Pereira & Elling, 2014a).

TYPE PLANT HOST: Roots and tubers of potato, *Solanum tuberosum* L.

OTHER PLANTS: According to the CABI/EPPO quarantine pest descriptions, hosts of *M. chitwoodi* are found in several plant families including crop plants and common weeds (Ferris *et al.*, 1993; Smith *et al.*, 1997). Potatoes and tomatoes are good hosts, whereas barley, corn, oats, sugarbeet, wheat, and various Poaceae (grasses and weeds) are moderate hosts (den Nijs *et al.*, 2004). Moderate to poor hosts occur in the Brassicaceae, Cucurbitaceae, Fabaceae, Lamiaceae, Liliaceae, Umbelliferae and Vitaceae. In an extensive host range study, O'Bannon *et al.* (1982) found that 53 of 68 weed and cultivated plant species were hosts to *M. chitwoodi*. Among the weed species listed as hosts were *Cirsium vul-*

gare (Savi) Ten., *Dactylis glomerata* L., *Panicum capillare* L., *Setaria viridis* (L.) P. Beauv. and *Sonchus asper* (L.) Hill. Similarly, Griffin *et al.* (1984) reported several range grasses as hosts to *M. chitwoodi*, and these included *Agropyron desertorum* (Fisch. *ex* Link) Schult., *Bromus inermis* Leysser. and *Pascopyrum smithii* (Rydb.) Barkworth & D.R. Dewey. Kutywayo & Been (2006) listed *Capsella bursa-pastoris* L., *Senecio vulgaris* L. and *Solanum nigrum* as hosts of *M. chitwoodi*, and *S. sarrachoides* Sendtn. was also reported as a host of this nematode (Boydston *et al.*, 2008; Rich *et al.*, 2008).

HOST RACES: At least two races (race 1 and race 2) are distinguished in Pacific Northwest states (Santo & Pinkerton, 1985). Race 2 seems to be absent in The Netherlands (van der Beek *et al.*, 1999). In Mexico, race 2 was found to be the predominant isolate in Tlaxcala State (Cuevas, 1995). Both of these races can reproduce on a wide range of crops commonly grown in the Columbia Basin, including potatoes, corn and wheat. Differential host tests showed that these races have different abilities to reproduce on carrot and alfalfa (Mojtahedi *et al.*, 1988). Of these, race 1 was identified first and is more prevalent in the Columbia Basin, while race 2 is typically found when potatoes are grown in rotation with alfalfa (Mojtahedi *et al.*, 1994).

TYPE LOCALITY: A field near Quincy, Washington, USA.

DISTRIBUTION: *Europe*: Belgium, France, Germany, Italy, The Netherlands, Sweden, Switzerland, Portugal; *Asia*: Turkey; *Africa*: South Africa, Mozambique; *North America*: Mexico, USA (California, Colorado, Idaho, Nevada, New Mexico, Oregon, Texas, Utah, Washington) (reported from Virginia, but later survey did not reveal the presence of this species); *South America*: Argentina (Wesemael *et al.*, 2011; Anon., 2012a, 2016a; Onkendi *et al.*, 2014).

SYMPTOMS: Above-ground symptoms of heavily infected plants include stunting and yellowing. Symptoms caused on potato roots are highly variable. High infections frequently cause irregular swellings or galls on roots, stolons and tubers. Some potato cultivars, although heavily infested, may be free from visible external symptoms, while the internal potato tissue is necrotic and brownish, just below the skin. *Meloidogyne chitwoodi* reduces the market value of potatoes as a result of internal necrosis and external galling. Necrotic spots in the flesh of tubers of as little as 5% of a crop make it commercially unacceptable.

PATHOGENICITY: Santo & O'Bannon (1981) showed that this species also suppressed yield of Fielder spring wheat in the microplot study by 41% and 73% in plots inoculated with 0.75 and 9.0 eggs (cm soil)$^{-3}$, respectively (Nyczepir *et al.*, 1984). In the experiments with initial *M. chitwoodi* densities of 25 J2 (100 g soil)$^{-1}$, the percentage of damaged carrot taproots increased from 10% when harvested 100 days after sowing to 70% when harvested 140 days after sowing. In a field trial, 11.5% of the carrots were damaged after a field period of 139 days and the initial *M. chitwoodi* population increased from 3 to 111 J2 (100 g soil)$^{-1}$ (Wesemael & Moens, 2008).

CHROMOSOME NUMBER: Meiotic parthenogenetic species. The haploid chromosome number was n = 18 (van der Beek & Karssen, 1997).

POLYTOMOUS KEY CODES: *Female*: A32, B43, C3, D6; *Male*: A423, B3, C23, D1, E1, F2; *J2*: A32, B3, C3, D32, E23, F4.

BIOCHEMICAL AND MOLECULAR CHARACTERISATION: All populations of this species give esterase S1 type and a malate dehydrogenase Nla type (Esbenshade & Triantaphyllou, 1990; Karssen & van Hoenselaar, 1998; Humphreys-Pereira & Elling, 2013). Zijlstra (1995, 1997) developed a relatively simple ITS PCR-RFLP method to differentiate *M. chitwoodi*, *M. fallax*, *M. hapla*, *M. incognita* and *M. javanica* (Gamel *et al.*, 2014). Zijlstra (2000) also proposed a highly sensitive PCR method using species-specific SCAR primers. Wishart *et al.* (2002) developed a sensitive PCR method based on species-specific primers designed from ribosomal IGS regions (Anon., 2009). Overview of diagnostic protocols is given by Anon. (2016a). MtDNA and rRNA genes for *M. chitwoodi* were sequenced by several authors. Humphreys-Pereira & Elling (2013) revealed a high level of mitochondrial heteroplasmy in *M. chitwoodi*.

RELATIONSHIPS (DIAGNOSIS): The species belongs to Molecular group III. *Meloidogyne chitwoodi* can be distinguished from *M. fallax* and other species by the ITS rRNA and *COI* gene sequence. *Meloidogyne chitwoodi* is very similar to *M. fallax* and differs from this species by shorter male stylet length (16-19 *vs* 19-21 μm) and shorter J2 tail length (39-51 *vs* 46-57 μm) and hyaline region (8-14 *vs* 12-16 μm) (Anon., 2009).

18. *Meloidogyne chosenia* Eroshenko & Lebedeva, 1992
(Fig. 65)

COMMON NAME: Willow root-knot nematode.

This species was described from Kamchatka, Russia, from willow and other plants.

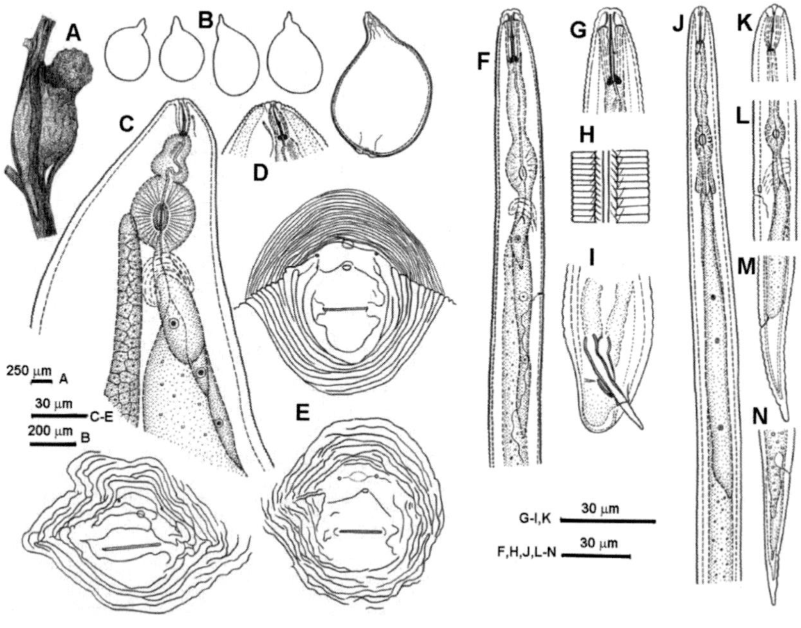

Fig. 65. Meloidogyne chosenia. *A:* Chosenia *root gall; B: Entire female; C, D: Female anterior region; E: Perineal pattern; F: Male anterior region; G: Male labial region; H: Male lateral field; I: Male tail; J: Second-stage juvenile (J2) anterior region; K: J2 labial region; L: J2 median bulb; M, N: J2 tail. (After Eroshenko & Lebedeva, 1992.)*

MEASUREMENTS *(AFTER EROSHENKO & LEBEDEVA, 1992)*

- *Holotype female*: L = 504 μm; W = 465 μm; neck length = 173 μm; stylet = 16 μm; vulval slit = 25 μm; vulva-anus distance = 23 μm; interphasmid distance = 27 μm.
- *Paratype females* (n = 20): L = 487 (388-648) μm; W = 438 (288-547) μm; neck length = 185 (144-216) μm; stylet = 17 (16.0-19.2) μm; stylet knob height = 2.1 μm; stylet knob width = 5.5 μm; DGO = 3-4 μm; median bulb length = 34 μm; median bulb width = 34 μm; anterior end to excretory pore = 9-15 μm; vulval slit = 27 (21-30) μm; vulva-anus distance = 20 (18-24) μm; interphasmid distance = 24 (18-29) μm; EP/ST = 0.4.
- *Males*: L = 973 (860-1050) μm; a = 36 (30-41); b = 9.6 (9-10); c = 98 (95-100); stylet = 19 (17.5-21.0) μm; DGO = 3.5-4.2 μm; pharynx = 93 μm; anterior end to pharyngeal gland base = 176 μm; tail = 10.5 (8-12) μm; spicules = 26 (22-32) μm; gubernaculum = 6 (5-7) μm; tail terminus to phasmids = 7-11 μm.
- *J2*: L = 359 (316-388) μm; a = 21 (18-23); b = 7 (6-7); c = 7.4 (6.3-8.6); c′ = 3.6 (3.2-4.3); stylet = 12 (10.6-13.3) μm; DGO = 3.2 (2.8-4.8) μm; pharynx = 77 μm; anterior end to pharyngeal gland base = 208 μm, anterior end to excretory pore = 67-78 μm; tail = 50 (40-58) μm.

DESCRIPTION

Female

Body without neck rounded, sometimes elongated with rounded posterior end. Cuticular annuli fine, 1.6 μm wide. First 6-10 annuli in anterior end wider than posterior ones. Stylet slender with oval knobs. Procorpus short and wide. Perineal pattern with wave-like cuticle ridges. Cuticle ridges interrupted in area of lateral field.

Male

Labial region with two annuli. Lateral field with four incisures with areolation and anastomoses. Median bulb oval. Hemizonid located 2-3 annuli anterior to excretory pore. Tail rounded with smooth cuticle.

J2

Labial region with labial disc. Procorpus wide. Median bulb elongated. Hemizonid located anterior to excretory pore. Tail conical with smooth terminus. Hyaline region short.

TYPE PLANT HOST: Willow, *Chosenia arbutifolia* (Pall.) A. Skvorts.

OTHER PLANTS: *Filabialendula camtschatica* (Pall.) Maxim., *Urtica platyphylla* Wedd., *Elymus repens* (L.) Gould.

TYPE LOCALITY: Nikolaevka, Yelizovo district, Kamchatka Krai, Russia.

DISTRIBUTION: *Asia*: Russia (Kamchatka).

CHROMOSOME NUMBER: No available data.

POLYTOMOUS KEY CODES: *Female*: A32, B21, C4, D3; *Male*: A4, B32, C32, D1, E1, F1; J2: A3, B23, C3, D3, E32, F43.

BIOCHEMICAL AND MOLECULAR CHARACTERISATION: No available data.

RELATIONSHIPS (DIAGNOSIS): This species has a small EP/ST ratio (0.4), which could be important for a preliminary identification. This species differs from the similar species *M. aquatilis* by the absence of terminal projection in females, a longer stylet in males, and smooth labial region in females and J2; from *M. caraganae* by longer stylet and shorter spicules in males and long thin tail in J2; from *M. carolinensis* by perineal pattern structure, shorter body length and longer J2 tail; and from *M. cruciani* by a more anterior position of the female excretory pore.

19. *Meloidogyne christiei* Golden & Kaplan, 1986

(Figs 66-68)

COMMON NAME: Turkey oak root-knot nematode.

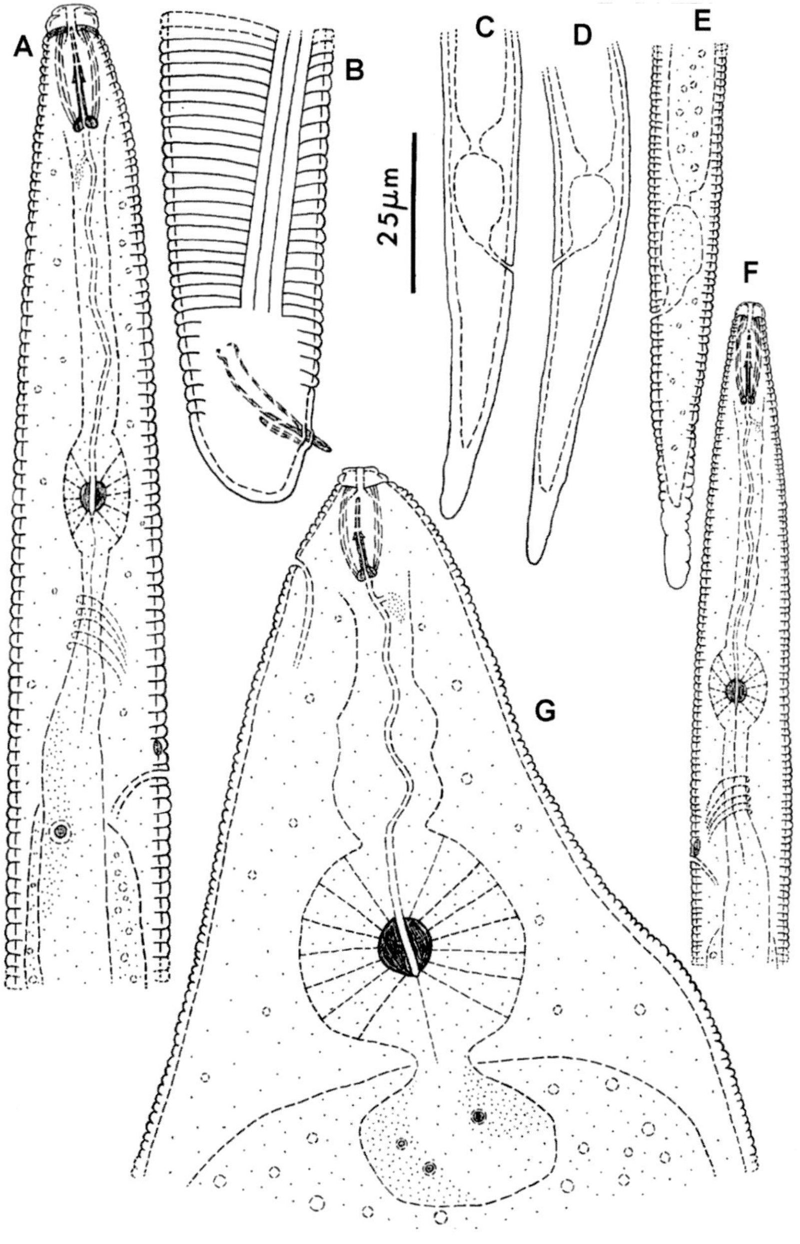

Fig. 66. Meloidogyne christiei. *A: Male anterior region; B: Male tail; C-E: Second-stage juvenile (J2) tail; F: J2 anterior region; G: Female anterior region. (After Golden & Kaplan, 1986.)*

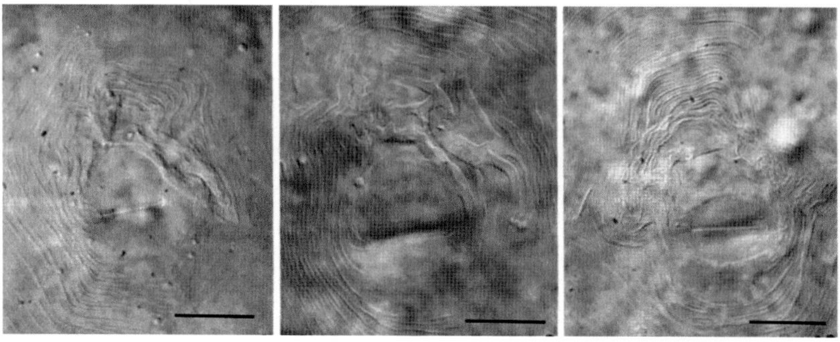

Fig. 67. Meloidogyne christiei. *LM. Perineal patterns. (Scale bars: 25 μm.) (After Golden & Kaplan, 1986.)*

Meloidogyne christiei was first reported infecting turkey oak, *Quercus laevis* Walt., in 1986 in Florida, USA (Golden & Kaplan, 1986), the only locality where it is known to occur. The unusual appearance of the galls and the mono-specific host range of this nematode indicate that this nematode-plant relationship is highly specialised (Kaplan & Koevenig, 1989).

MEASUREMENTS (*AFTER GOLDEN & KAPLAN, 1986*)

- *Holotype female*: L = 681 μm; W = 508 μm; a = 1.3; stylet = 14.2 μm; DGO = 3.5 μm; anterior end to excretory pore = 16.8 μm, to median bulb valve = 68 μm; vulval slit = 25 μm; distance from vulval slit to anus = 21 μm.
- *Females* (n = 50): L = 637 ± 69 (523-779) μm; W = 470 ± 70 (352-623) μm; a = 1.4 ± 0.2 (1.1-2.0); stylet = 13.9 ± 0.7 (13.0-15.3) μm; DGO = 3.8 ± 0.6 (3.0-4.7) μm; anterior end to excretory pore = 17 ± 4.7 (9-28) μm, to median bulb valve = 64 ± 8.8 (53-88) μm; vulval slit length = 23 ± 2.2 (18-26) μm; distance from vulval slit to anus = 22 ± 2.4 (18-28) μm; EP/ST = 1.2.
- *Allotype male*: L = 1235 μm; a = 31; b = 5.8; c = 176; stylet = 17.1 μm; DGO = 3.5 μm; anterior end to median bulb valve = 64 μm; spicules = 25.4 μm; gubernaculum = 8.2 μm; tail = 7 μm.
- *Males* (n = 27): L = 1227 ± 98 (1019-1495) μm; a = 38 ± 3.9 (29-48); b = 6 ± 0.6 (5.2-7.1); c = 160 ± 43 (121-308); stylet = 17.8 ± 0.5 (17.1-18.9) μm; DGO = 4.1 ± 0.5 (3.5-5.3) μm; anterior end to median bulb valve = 71.4 ± 4.5 (64-78 μm); spicules = 25 ± 2.5 (24-

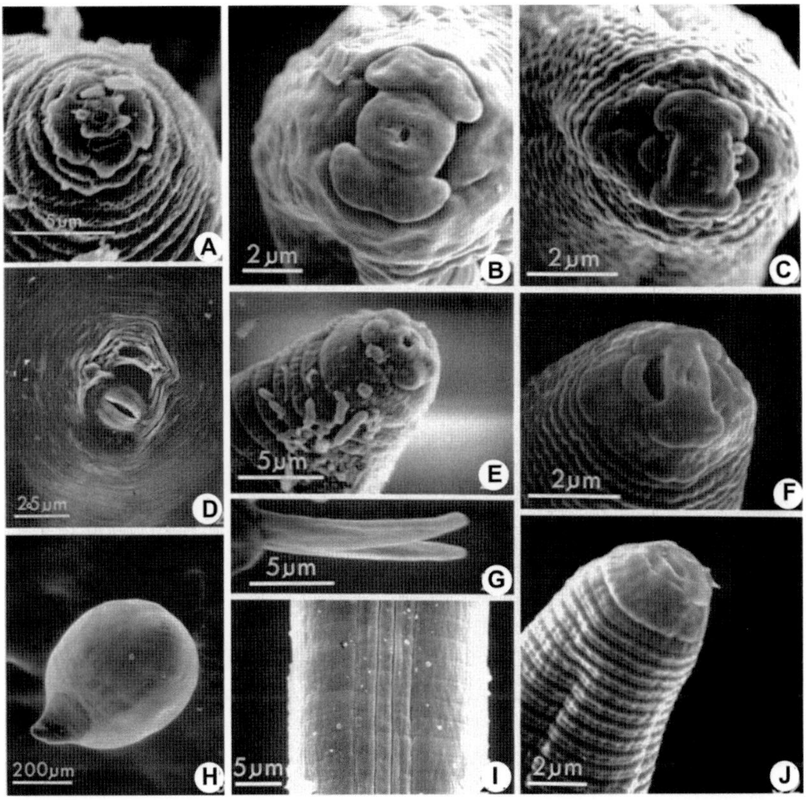

Fig. 68. Meloidogyne christiei. *LM and SEM. A: Female* en face *view; B, E: Male* en face *view; D: Perineal pattern; C, F: Second-stage juvenile (J2)* en face *view; H: Entire female; G: Spicules; I: Male lateral field; J: J2 labial region. (After Golden & Kaplan, 1986.)*

26) μm; gubernaculum = 7.3 ± 0.6 (6.5-8.9) μm; tail = 8.2 ± 1.6 (3.5-10.0) μm.

- *J2* (n = 57): L = 427 ± 21 (374-468) μm; a = 26.5 ± 2.2 (21-30); b = 2.2 ± 0.2 (1.9-2.5); c = 10 ± 0.5 (9-11); stylet = 11.4 ± 0.3 (10.6-11.8) μm; DGO = 4 ± 0.3 (3.5-4.7) μm; anterior end to median bulb valve = 57 ± 2.3 (52-62 μm); W = 16 ± 1.4 (14-21) μm; anterior end to pharyngeal gland base = 196 ± 18 (171-226) μm; tail = 42 ± 2.8 (35-50) μm; hyaline region = 12 ± 1.3 (9-14) μm; b = 4 ± 0.7 (2.8-6.2).
- *Eggs* (n = 30): L = 114 ± 7.2 (101-124) μm: W = 60 ± 5 (47-68) μm; L/W = 1.9 ± 0.2 (1.7-2.3).

DESCRIPTION

Female

Body pearly white, globular to pear-shaped, often with slight posterior protuberance, neck distinct situated anteriorly, commonly off-centre from a median plane with terminal vulva. Labial framework weak, offset from neck but variable in shape, usually bearing one annulus, labial disc having four projections or prongs. Excretory pore distinct, generally located at level near base of retracted stylet. Stylet small, strong, dorsally curved, knobs rounded, sloping posteriorly. Perineal pattern usually with high, squarish arch, striae widely spaced, coarse, broken, tending to diverge at various angles, especially in and above anal area. Beneath pattern and vulval slit a large circular area of exceptionally dense, prominent vaginal muscles usually present. Commonly no egg sac occurring outside gall on root; instead eggs deposited in a tubular, coiled egg sac within gall.

Male

Body slender, vermiform, tapering slightly at both extremities. Labial region slightly offset, with large labial disc, and a prominent post labial annulus. Body annuli distinct. Lateral field with four incisures, forming three bands, centre one slightly smaller, not areolated. Spicules rather short, arcuate, with rounded tips. Tail rounded.

J2

Body small, vermiform, tapering at both extremities but much more so posteriorly. Labial region essentially not offset, with weak labial framework, labial disc and large post labial annulus without striations. Cuticular annulation very fine, measuring *ca* 1 μm wide at mid-body. Lateral field prominent, about one-third of body diam., with four incisures, not areolated. Phasmids small, located in anterior half of tail. Rectum inflated. Tail tapering to a bluntly rounded terminus.

TYPE PLANT HOST: The turkey oak, *Quercus laevis* Walt.

OTHER PLANTS: In glasshouse host range studies *M. christiei* did not produce galls or reproduce on the North Carolina differential host plants and *Citrus limon* (L.) Osbeck.

TYPE LOCALITY: Sanlando Park in Altamonte Springs, Florida, USA.

DISTRIBUTION: *North America*: USA (Florida).

SYMPTOMS: The galls caused by this nematode are distinctive and are even suggestive of nitrogen-fixing nodules in appearance. Galls isolated from roots were 2.4 $\pm$ 0.3 mm in diam. and were typically spheroid and rigid. They occur as discrete nodules, commonly singly but sometimes as a cluster on the side of the root and without adjacent swelling. Young galls appeared tan or orange and became dark brown and hardened with age (Golden & Kaplan, 1986; Kaplan & Koevenig, 1989).

CHROMOSOME NUMBER: No available data.

POLYTOMOUS KEY CODES: *Female*: A23, B32, C3, D6; *Male*: A34, B34, C3, D1, E2, F2; *J2*: A23, B32, C3, D32, E2, F3.

BIOCHEMICAL AND MOLECULAR CHARACTERISATION: The topotype population was characterised using isozyme profiles and ribosomal (18S rRNA, D2-D3 of 28S rRNA and ITS rRNA) and mitochondrial (*COII*-16S rRNA) gene sequences presented by Brito *et al.* (2015). The phenotype N1a detected from a single egg-laying female of *M. christiei* showed one very strong band of malate dehydrogenase activity; however, no esterase activity was identified from a macerate of one or even 20 females per well. *Meloidogyne christiei* formed a separate lineage within *Meloidogyne* and its relationships with any of main *Meloidogyne* clades were not resolved.

RELATIONSHIPS (DIAGNOSIS): The species belongs to Molecular group V and is clearly molecularly differentiated from all other *Meloidogyne* species. *Meloidogyne christiei* is distinctive by the shape of the perineal pattern, the presence of four projections on the female labial disc, and the deposition of eggs in a tubular coiled manner. The perineal pattern is similar to *M. ardenensis* and *M. artiellia*. *Meloidogyne christiei* differs from *M. artiellia* by longer J2 tail length (35-50 *vs* 16-29 μm), and from *M. ardenensis* by shorter female stylet length (13.0-15.3 *vs* 15-19 μm).

20. *Meloidogyne citri* Zhang, Gao & Weng, 1990

(Figs 69, 70)

COMMON NAME: Asian citrus root-knot nematode.

This nematode was found in a citrus orchard in Xiasha, Shuinan, Shunchang county, Fujian province, China, during the spring of 1988 and was later described as a new species. The citrus orchard was planted in 1976 and was heavily infected by this RKN.

MEASUREMENTS (*AFTER ZHANG* ET AL., *1990*)

- *Holotype female*: L = 840 μm; W = 620 μm; stylet = 15 μm; anterior end to excretory pore = 37.5 μm; DGO = 5.0 μm.
- *Paratype females* (n = 20): L = 837 ± 16.7 μm; W = 630 ± 20.8 μm; L/W ratio = 1.3 ± 0.06; neck length = 169.5 ± 0.9 μm; stylet = 15.3 ± 0.2 μm; DGO = 4.0 ± 0.3 μm; stylet knob height = 2.5 ± 0.03 μm; stylet knob width = 4.5 ± 0.2 μm; anterior end to excretory pore = 36.5 ± 0.9 μm, to median bulb valve = 85.5 ± 2.8 μm; median bulb length = 59.4 ± 2.5 μm; median bulb width = 45.4 ± 0.6 μm; vulval slit = 33 ± 0.9 μm; vulva-anus distance = 21.4 ± 1.0 μm; interphasmid distance = 20.3 ± 1.0 μm; EP/ST = 2.4.
- *Allotype male*: L = 2020 μm; W = 45 μm, stylet = 25 μm; stylet knob height = 3.0 μm; stylet knob width = 5.5 μm; DGO = 5.0 μm; anterior end to median valve = 108 μm and to excretory pore = 177 μm; spicules = 35 μm; testis length = 1162 μm, T = 57.5%.
- *Paratype males* (n = 20): L = 1965.5 ± 36.2 μm; W = 43.5 ± 0.7 μm; labial region height = 6.9 (6.0-7.5) μm, labial region diam. = 14.7 (14-15) μm; DGO = 5.0 ± 0.1 μm; stylet = 25.1 ± 0.3 μm; stylet knob height = 3.0 ± 0.1 μm; stylet knob width = 6.1 ± 0.6 μm; anterior end to excretory pore = 202.9 ± 3.7 μm, to median bulb valve = 101.2 ± 1.2 μm; median bulb length = 19.0 ± 0.6 μm; median bulb width = 13 ± 0.3 μm; tail = 15 ± 0.4 μm; spicules = 38.6 ± 0.9 μm; gubernaculum = 6.0-7.0 μm; testis length = 1067.5 ± 42.6 μm; a = 45.4 ± 0.9; T = 54.7 ± 1.7%.
- *Paratype J2* (n = 20): L = 465.1 ± 6.6 μm; W = 17.4 ± 0.1 μm; DGO = 3.7 ± 0.1 μm; stylet = 11.5 ± 0.2 μm; stylet knob height = 2.0 ± 0.03 μm; stylet knob width = 2.7 ± 0.1 μm; anterior end to excretory pore = 96.1 ± 1.0 μm, to median bulb valve = 67.4 ±

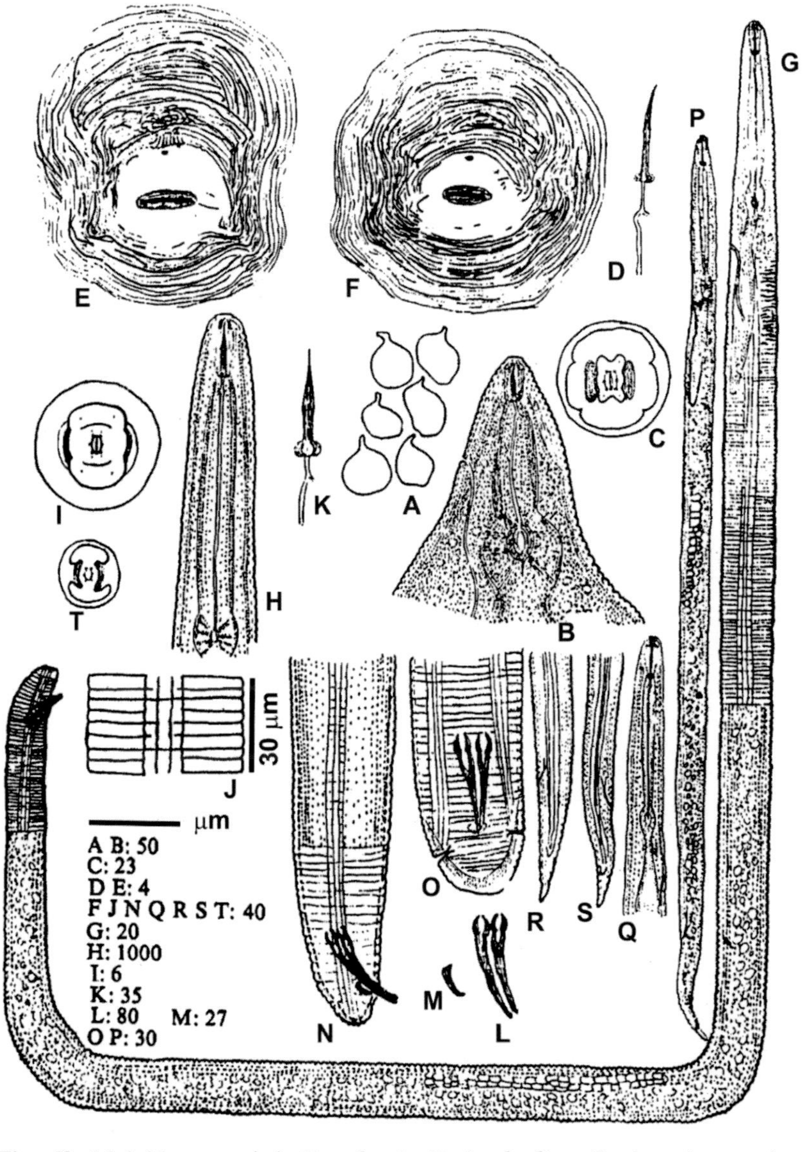

Fig. 69. Meloidogyne citri. *Female. A: Entire bodies; B: Anterior portion; C: Face view of head; D: Stylet; E, F: Perineal patterns; Male. G: Outlines of entire specimens; H: Anterior portion; I: Face view of head; J: Body annulation; K: Stylet; L: Spicules; M: Gubernaculum; N: Tail; O: Tail (ventral); Second-stage juvenile. P: Outlines of entire specimens; Q: Anterior portion; R-S: Tail; T: Face view of head. (After Zhang* et al., *1990.)*

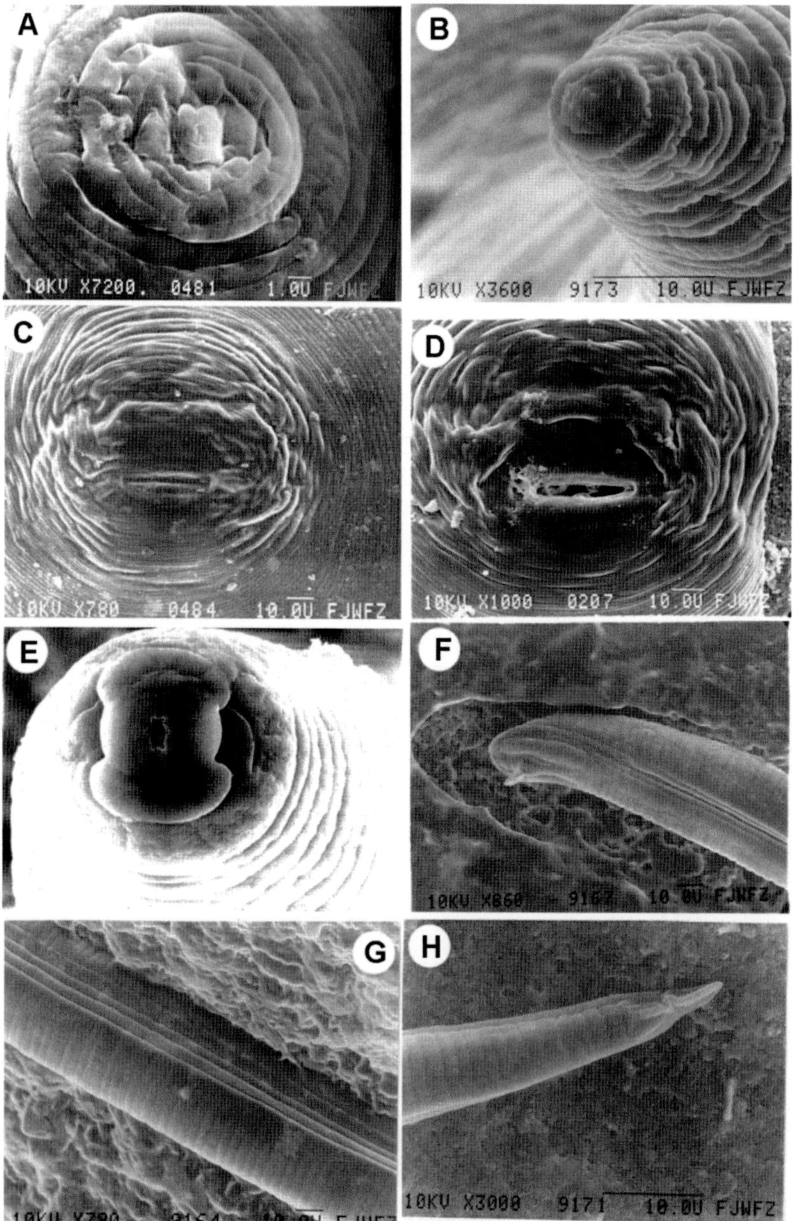

Fig. 70. Meloidogyne citri. *SEM. A, B: Female* en face *view; C, D: Perineal pattern; E: Male* en face *view; F: Male tail; G: Second-stage juvenile (J2) lateral field; H: J2 tail. (After Zhang* et al.*, 1990.)*

0.9 μm; median bulb length = 13.9 $\pm$ 0.5 μm; median bulb width = 9.7 $\pm$ 0.3 μm; tail = 53.8 $\pm$ 0.5 μm; body diam. at anus level = 12.4 $\pm$ 0.1 μm; hyaline region = 16.1 $\pm$ 0.7 μm; a = 27 $\pm$ 0.2; tail/hyaline region = 3.4 $\pm$ 0.1.
- *Eggs* (n = 20): L 98.2 $\pm$ 1.2 (90-110) μm; W = 48.7 $\pm$ 0.7 (40-50) μm; L/W = 2.0 $\pm$ 0.04 (1.8-2.5).

DESCRIPTION

Female

Female white, globular or pear-shaped, labial region prominent with slight posterior protuberance. Labial disc elevated. Median labials wider than labial disc. Lateral labials large, fused with median labials. Amphidial apertures large, slit-like. Two labial annuli. Labial framework moderate, vestibule and vestibule extension prominent. Stylet delicate, stylet cone distinctly curved dorsally, stylet shaft cylindrical. Stylet knobs transversely ovoid, offset from stylet shaft. Perineal pattern round, dorsal arch low and flat, angular striae occurring on both lateral regions of dorsal arch. No lateral lines. Short horizontal striae and longitudinal striae present around anus. Perivulval region free of striae. Fine dense striae found on vulva. Inner cuticle thick, raised with distributed coarse striae with fine short striae present. Striae cheek-shaped in one or both lateral regions of vulva. Striae smooth, continuous ventrally.

Male

Male slender. Labial region not offset from body, labial framework moderately developed. Stylet large, knobs transversely ovoid and offset from shaft. Stylet cone straight, stylet shaft tapering posteriorly. Median labials and labial disc elevated. Labial disc round, elevated above median labials with similar or slightly narrower width. Stoma slit-like. Prestoma hexagonal, surrounded by six inner sensilla. Median labials half-moon-shaped. No lateral labials present. Amphidial aperture slit like. Lateral field slanting at level of seventh annulus, encircling tail. Lateral areolated with four incisures. One testis, directed anteriorly and outstretched with variation in size. Spicule well developed, arcuate with round base and blunt end. Gubernaculum crescentic with wider base. One phasmid at level of cloacal aperture, other one immediately posterior to cloacal aperture. Tail blunt.

J2

Slender, vermiform. Labial region flat, not offset from body. Stylet delicate, stylet cone straight, stylet knobs offset from stylet shaft. Labial disc and median labials fused, dumbbell-shaped in face view. Amphidial apertures slit like. Lateral field elevated with four incisures. Tail tapering posteriorly with 1-2 irregular annulations. Hyaline region clearly defined. Tail blunt. Rectum not dilated.

TYPE PLANT HOST: Unshu mikan *Citrus unshiu* Marc. (Yu.Tanaka *ex* Swingle).

OTHER PLANT: Trifoliate orange (*Poncirus trifoliata* L.) (Vovlas & Inserra, 1996).

TYPE LOCALITY: Xiasha, Shuinan, Shunchang county, Fujian province, China.

DISTRIBUTION: *Asia*: China.

SYMPTOMS: *Citrus unshiu* roots infected by *M. citri* showed swollen root tips on axes. The host response did not differ from those reported by other RKN.

CHROMOSOME NUMBER: No available data.

POLYTOMOUS KEY CODES: *Female*: A2, B2, C2, D3; *Male*: A1, B1, C1, D1, E1, F1; *J2*: A2, B2, C3, D1, E2, F3.

BIOCHEMICAL AND MOLECULAR CHARACTERISATION: No available data.

RELATIONSHIPS (DIAGNOSIS): *Meloidogyne citri* is similar to *M. fujianensis* and *M. donhaiensis*, but the female of *M. citri* is larger with a slight posterior protuberance, the stylet cone is curved dorsally, and the dorsal arch perineal pattern is low and flat with cheek-shaped striae present in the lateral region of the vulva; the J2 of *M. donhaiensis* has shorter hyaline region.

21. *Meloidogyne coffeicola* Lordello & Zamith, 1960
(Fig. 71)

COMMON NAME: Coffee root-knot nematode.

This species is considered to be a destructive pest of *Coffea arabica* in several states of Brazil (Lordello & Zamith, 1960; Castro *et al.*, 2004).

MEASUREMENTS *(AFTER LORDELLO & ZAMITH, 1960)*

- *Females*: L = 992-1348 μm; W = 310-387 μm; stylet = 15.3-17.6 μm; DGO = 3.8-4.6 μm; median bulb length = 33.6-41.3 μm; median bulb width = 32-42.8 μm; EP/ST = 1.8 (estimated from drawings).

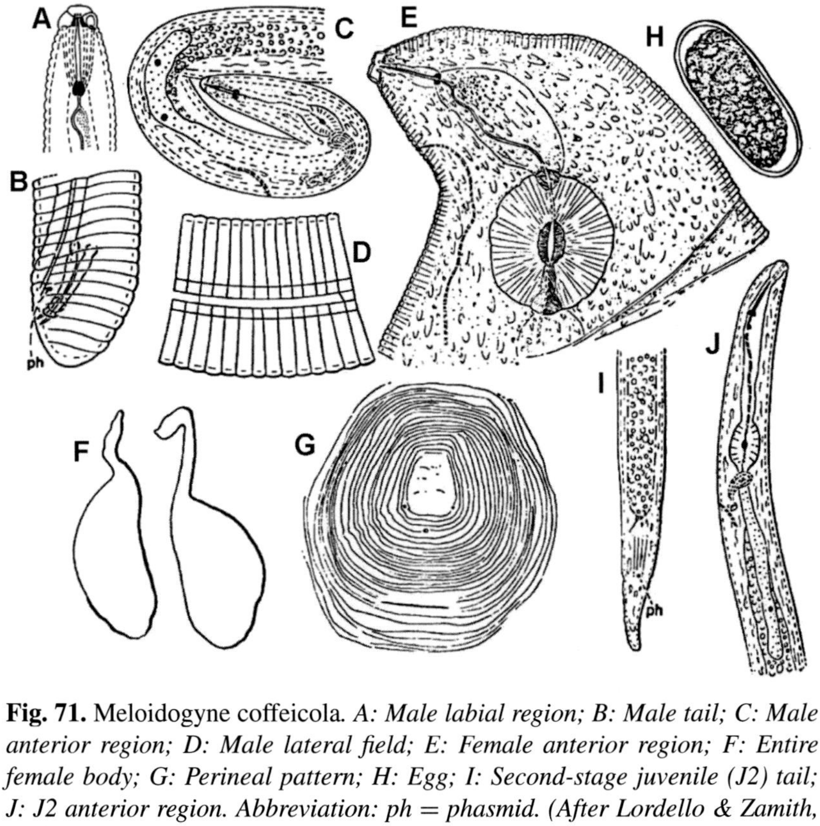

Fig. 71. Meloidogyne coffeicola. *A: Male labial region; B: Male tail; C: Male anterior region; D: Male lateral field; E: Female anterior region; F: Entire female body; G: Perineal pattern; H: Egg; I: Second-stage juvenile (J2) tail; J: J2 anterior region. Abbreviation: ph = phasmid. (After Lordello & Zamith, 1960.)*

- *Males*: L = 1279.0-1595.8 μm; W = 38.3-44.4 μm; stylet = 23-26 μm; DGO = 3.8-4.6 μm; median bulb length = 16.8-23 μm; median bulb width = 9.2-13.8 μm; tail = 18.4-24.5 μm; spicules = 20-29 μm; gubernaculum = 6.0-9.2 μm; a = 32.3-40.1; b = 8.6-14.7; c = 58.3-80.2.
- *J2*: L = 336.6-423.8 μm; W = 15.3-16.8 μm; stylet = 9.2-10.7 μm; DGO = 3.1-3.8 μm; median bulb length = 12.3-13.8 μm; median bulb width = 9.2 μm; tail = 29.1-33.6 μm; a = 22-25.2; b = 5.0-5.9; c = 9.5-13.9.
- *Eggs*: L = 92-101 μm; W = 52-55 μm.

DESCRIPTION *(AFTER WHITEHEAD (1968), MODIFIED FROM LORDELLO & ZAMITH, 1960)*

Female

Body pyroid, brownish, with long neck. Stylet knobs rounded. Excretory pore about 13 annuli posterior to labial region. Posterior cuticular pattern striae closely spaced, very faint, smooth to slightly wavy dorsad with distinct tail whorl and phasmids close to tail tip, which is marked by a few broken striae.

Male

Labial region rather cupolate, offset, with one annulus posterior to labial cap. Lateral field with four main incisures mid-body, outer bands, and occasionally inner band, areolated. Stylet with longitudinally ovoid knobs, not prominent. Phasmids anterior to cloacal aperture.

J2

Labial region slightly offset with one annulus posterior to labial cap, stylet knobs tending to ovoid, weak. Four incisures in lateral field, outer bands and sometimes inner bands areolated. Phasmids small, mid-tail. Tail bluntly rounded.

TYPE LOCALITY: Terra Boa, Paranh, Brazil.

TYPE PLANT HOST: *Coffea arabica*.

OTHER HOSTS: The hosts include arabica (*C. arabica*), robusta (*C. robusta* L.) and Liberian coffee (*C. liberica* Hiern) (Anon., 2010b). The nematode has also been detected on native plants such as *Eupatorium pauciflorum* Kunth and *Psychotria nitidula* Cham. & Schltdl. (Lordello & Lordello, 1972; Jaehn *et al.*, 1980).

DISTRIBUTION: *South America*: Brazil (Minas Gerais, Paraná and São Paulo) (Anon., 2010b).

SYMPTOMS: The nematode does not induce galls, but instead causes cracks in the root tissue. Coffee trees show symptoms of decline, such as peeled and rough roots and fewer side rootlets. Nematode damage often leads the plants to show symptoms of defoliation and chlorosis. For many years, *M. coffeicola* was considered the species with the highest damage potential among all coffee-parasitic nematodes in Brazil. Presently, this nematode is rarely found parasitising coffee because in the infested areas coffee has been replaced by other crops. *Meloidogyne coffeicola* has a low survival rate in soil and a low capacity to infect coffee seedlings and young trees (Campos & Villain, 2005).

CHROMOSOME NUMBER: No available data.

POLYTOMOUS KEY CODES: *Female*: A1, B21, C3, D2; *Male*: A32, B12, C32, D1, E1, F2; *J2*: A32, B3, C4, D2, E21, F4.

BIOCHEMICAL AND MOLECULAR CHARACTERISATION: *Meloidogyne coffeicola* is characterised by its esterase phenotype (Est C2) (Carneiro *et al.*, 2000; Castro *et al.*, 2004). The 18S rRNA gene sequence of this species was provided by Tomalova *et al.* (2012).

RELATIONSHIPS (DIAGNOSIS): The species belongs to Molecular group X and is clearly molecularly differentiated from all other *Meloidogyne* species. *Meloidogyne coffeicola* is similar to *M. africana* and differs from this species by larger body size (992-1348 *vs* 400-1040 μm).

22. *Meloidogyne cruciani* García-Martinez, Taylor & Smart, 1982
(Fig. 72)

COMMON NAME: Crucian root-knot nematode.

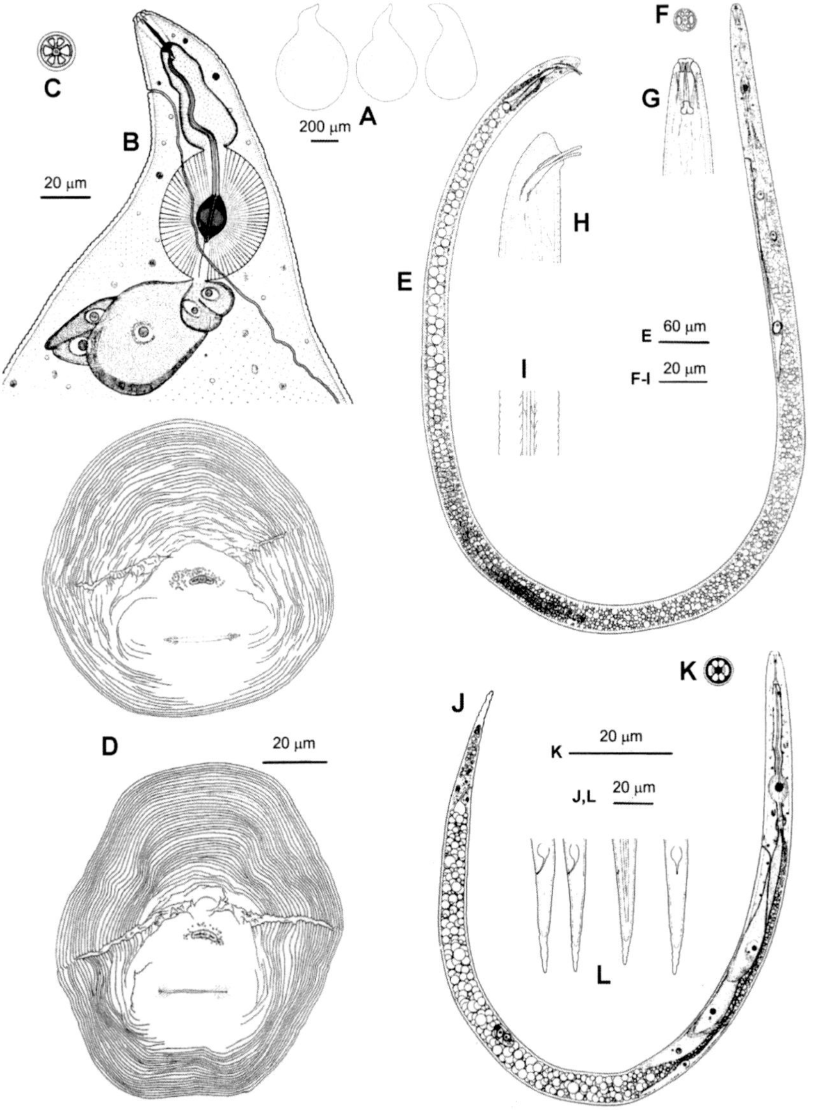

Fig. 72. Meloidogyne cruciani. *A: Entire female; B: Female anterior region; C: Female* en face *view showing labial framework; D: Perineal pattern; E: Entire male; F: Male* en face *view showing labial framework; G: Male labial region; H: Male tail; I: Male lateral field; J: Entire second-stage juvenile (J2); K: J2* En face *view showing labial framework; L: J2 tail. (After García-Martinez* et al., *1982.)*

During a nematode survey conducted in the U.S. Virgin Islands, a RKN morphologically close to *M. javanica* was found in the Agricultural Community Gardens, University of the Virgin Islands, St Croix. This RKN population was studied and described at the University of Florida, Gainesville, Florida, USA, where the nematode was kept in a glasshouse under rigorous phytosanitary conditions for regulatory purposes.

MEASUREMENTS (*AFTER GARCÍA-MARTINEZ* ET AL., *1982*)

- *Holotype female*: L = 949.9 μm; W = 536.4 μm; a = 1.8; stylet = 12.7 μm; stylet knob height = 2.7 μm; stylet knob width = 4.4 μm; DGO = 4.1 μm; anterior and to excretory pore = 31.7 μm, to median bulb valve = 93.3 μm; vulval slit length = 23.8 μm; vulval slit to anus = 20.0 μm; interphasmid distance = 31.7 μm.
- *Paratype females* (n = 21): L = 787.5 ± 86.2 (426-1121.8) μm; W = 505.1 ± 61.5 (315.7-770) μm; a = 1.5 ± 0.1 (1.2-2.1); stylet = 14.2 ± 0.6 (11.4-16.2) μm; stylet knob height = 2.4 ± 0.1 (2.1-2.9) μm; stylet knob width = 4.5 ± 0.2 (3.8-5.1) μm; DGO = 3.9 ± 0.2 (3.2-5.1) μm; anterior end to excretory pore = 32.2 ± 2.2 (25.7-45.1) μm; anterior end to median bulb valve = 78.3 ± 3.4 (65.4-93.3 μm); vulval slit length = 23.2 ± 0.7 (20.0-25.7) μm; vulval slit to anus = 18.1 ± 0.6 (15.9-20.3) μm; interphasmid distance = 30.3 ± 1.2 (25.7-36.8) μm; EP/ST = 2.2 (estimated from drawings).
- *Allotype male*: L = 1440 μm; W = 29.9 μm; stylet = 22.8 μm; stylet knob height = 3.5 μm; stylet knob width = 4.8 μm; DGO = 4.9 μm; anterior end to median bulb valve = 97.4 μm; anterior end to excretory pore = 139.2 μm; anterior end of testis to posterior end = 933.8 μm; spicules = 28.7 μm; gubernaculum = 8.1 μm; phasmid to posterior end = 18.4 μm; a = 48.0; c = 151.3; c′ = 0.4; o = 21.5; T% = 64.9.
- *Paratype males* (n = 25): L = 1378.8 ± 52.3 (1160-1620) μm; W = 33.8 ± 2.4 (23.2-46.3) μm; stylet = 22 ± 0.5 (19.4-24.1) μm; stylet knob height = 3.3 ± 0.1 (2.9-3.8) μm; stylet knob width = 5.2 ± 0.2 (4.1-6.0) μm; DGO = 4.9 ± 0.4 (3.2-7.9) μm; anterior end to median bulb valve = 88.9 ± 4.1 (66.7-118.1) μm; anterior end to excretory pore = 149.2 ± 5.8 (127.3-189.5) μm; anterior end of testis to posterior end = 823.4 ± 60.1 (489.4-1127) μm; spicules = 31.3 ± 0.9 (28.7-38.0) μm; gubernaculum = 8.8 ± 0.5 (6.7-11.1) μm; phasmid to posterior end = 16.7 ± 1.4 (8.3-22.9) μm; a = 43.2 ± 3.7

(31.9-71.2); c = 132.6 ± 13.1 (89.2-238.2); c′ = 0.5 ± 0.1 (0.3-0.7); o = 22.6 ± 1.9 (14.3-36.8); T = 60.1 ± 4.7 (39.8-79.7)%.

- *Paratype J2* (n = 20): L = 435.3 ± 8.7 (418.6-479.8) μm; W = 17.2 ± 0.5 (14.6-18.7) μm; stylet = 10.6 ± 0.2 (9.8-12.1) μm; stylet knob height = 1.4 ± 0.1 (1.1-1.6) μm; stylet knob width = 2.3 ± 0.1 (2.1-2.7) μm; DGO = 3.5 ± 0.1 (3.2-3.9) μm; anterior end to median bulb valve = 57.8 ± 1.4 (51.7-61.9) μm; anterior end to cardia = 76.3 ± 2.1 (69.5-86.7) μm; anterior end to excretory pore = 88.1 ± 3.4 (74.6-103.2) μm; distance from posterior end of glands to anterior end = 202 ± 6.4 (190.4-250.4) μm; genital primordium to posterior end = 163.5 ± 4.3 (148.2-175.5) μm; phasmid to posterior end = 39.2 ± 1.2 (34.9-43.8) μm; tail = 46.6 ± 1.3 (41.3-51.7) μm; width at anus level = 11.2 ± 0.4 (9.8-13.0) μm; a = 25.4 ± 0.8 (22.9-29.8); b = 5.8 ± 0.2 (5.0-7.1); b′ = 2.2 ± 0.1 (1.8-2.4); c = 9.4 ± 0.3 (8.6-10.5); c′ = 4.2 ± 0.1 (3.7-4.6); o = 33.1 ± 0.9 (28.9-37.9).

Description

Female

White, pear-shaped to globular, without prominent posterior protuberance. Neck tapering, curving gently. Labial region offset slightly with labial cap and one or two labial annuli. Labial or cephalic sensilla not observed. Amphidial openings oval, inconspicuous. Labial framework with lateral sectors larger than ventral or dorsal sectors. Stylet robust, with rounded knobs. Excretory pore about one stylet length from base of stylet knobs, variable in exact position; excretory duct easily seen throughout anterior region terminating in a uninucleate gland. Pharyngeal lumen between base of stylet knobs and valve of median bulb well sclerotised with an average width of 2.4 μm. Prominent median bulb with strongly sclerotised valve. Pharyngeal glands consisting of five distinct nucleated lobes. One lobe always larger than other four. Perineal pattern with subcuticular punctations (stippling) almost surrounding anus on lateral and posterior sides. Striae deep, wavy, sometimes broken. Lateral field fairly deep with distinct phasmids. Phasmidial ducts often visible. Vulval lips faintly serrated, margins with very fine striae. Tail terminus indistinct.

Male

Body long, vermiform, tapering at both ends. Labial region offset with two annuli and distinct labial cap. Labial sensilla not observed.

Labial framework with lateral sectors larger than ventral or dorsal sectors, ends of framework slightly forked when viewed laterally. Stylet robust, with rounded knobs sharply tapering into stylet shaft. Amphidial glands prominent posterior to stylet knobs. Cephalids not observed. Median bulb poorly developed, slightly larger than procorpus with well-sclerotised valve. Pharyngeal glands consisting of three distinct nucleated lobes. Hemizonid 3.5 annuli anterior to excretory pore. Excretory duct long, terminating in sac-like unicellular gland. Lateral fields begin anteriorly as two lateral lines near stylet knobs and become four near median bulb. There may be three less pronounced lines between four main ones. There is an anastamosis of lateral lines in posterior end of body. Body twisting anteriorly to spicules. Testis usually one, occasionally two. Spicules slightly arcuate, tips rounded. Gubernaculum with fine serrations on cuneus.

J2

Body vermiform, tapering slightly anteriorly and much more posteriorly. Labial region offset slightly with one annulus, labial cap with weakly visible labial framework, lateral sectors larger than ventral or dorsal sectors. Labial or labial sensilla not observed. Stylet robust, rounded knobs slanting posteriorly. Cephalids not observed. Amphidial glands prominent, posterior to stylet knobs. Pharynx extremely long, posterior extremity of glands averaging 46.4% of total body length. Median bulb well developed with well-sclerotised valve. Pharyngeal glands contained in three distinct nucleated lobes, each with distinct chromocentre. Excretory pore position variable, always posterior to pharyngo-intestinal valve. Hemizonid 2-4 annuli anterior to excretory pore. Excretory duct long, terminating in sac-like unicellular gland. Lateral fields originating as two lines one stylet length posterior to base of knobs, becoming four near median bulb. Two inner lateral lines terminating near phasmids and outer two terminating near tail terminus. Genital primordium easily seen in two-cell stage. Rectum dilated. Phasmids small and difficult to see, one anal body diam. posterior to anus level. Tail gradually tapering, with annuli disappearing near hyaline area. Tail terminus notched, with smooth, bluntly conoid tip.

TYPE PLANT HOST: Tomato, *Solanum lycopersicum* L.

OTHER PLANTS: In a laboratory test, the nematode infected roots of watermelon (*Citrullus lanatus* (Thunb.) Matsum. and Nakai), sweet potato (*Ipomoea batatas* (L.) Lam.), tobacco (*Nicotiana tabacum* L.), corn (*Zea mays*), pepper (*Capsicum annuum*) and cabbage (*Brassica oleracea*) (García-Martinez *et al.*, 1982). This nematode was found on eggplant (*Solanum melongena* L.) in Argentina (Doucet & Pinochet, 1992) and on banana in French Guiana and Guadeloupe (Cofcewicz *et al.*, 2005).

TYPE LOCALITY: Agricultural Community Gardens, University of the Virgin Islands, St Croix, USA Virgin Islands.

DISTRIBUTION: *North America and Caribbean*: USA Virgin Islands, Guadeloupe (Cofcewicz *et al.*, 2005), *South America*: Argentina (Doucet & Pinochet, 1992), French Guiana (Cofcewicz *et al.*, 2005).

CHROMOSOME NUMBER: $2n = 42\text{-}48$ (Esbenshade & Triantaphyllou, 1985a, b).

POLYTOMOUS KEY CODES: *Female*: A213, B324, C2, D5; *Male*: A32, B213, C21, D1, E2, F2; *J2*: A2, B32, C3, D3, E2, F3.

BIOCHEMICAL AND MOLECULAR CHARACTERISATION: Esbenshade & Triantaphyllou (1985b) characterised the species-specific esterase phenotype as M3a. The 18S rRNA gene sequence was provided by Tomalova *et al.* (2012). Based on RAPD data, *M. javanica* and *M. cruciani* were considered to be related species (Cofcewicz *et al.*, 2005).

RELATIONSHIPS (DIAGNOSIS): The species belongs to Molecular group I, the Incognita group. *Meloidogyne cruciani* differs from other species by its perineal pattern with punctations around the anus. Other species reported to have punctations in the perineal area are *M. hapla* and *M. haplanaria*, but the punctations of *M. hapla* are in the area of the tail terminus, and randomly within the pattern area. This species is similar to *M. pisi* and *M. duytsi*, and can be separated from the former in having a different esterase pattern and from the latter only by shorter J2 tail length (41.3-51.7 *vs* 65.1-76.5 μm).

23. *Meloidogyne cynariensis* Pham, 1990
(Fig. 73)

COMMON NAME: Artichoke root-knot nematode.

This nematode was found during a nematological survey in agricultural fields from artichoke roots, *Cynara scolymus* L. in Dalat, Tây Nguyen, Lam Dong province, Vietnam.

MEASUREMENTS *(AFTER PHAM, 1990)*

- *Holotype female*: L = 784 μm; W = 428 μm; a = 1.8; stylet 15 μm, DGO = 4.5 μm; stylet knob width = 5 μm, anterior end to excretory pore = 17 μm; vulval slit = 32 μm; vulva-anus distance = 27 μm.
- *Paratype females* (n = 14): L = 787 (686-843) μm; W = 402 (333-431) μm; a = 1.6 (1.5-1.9); stylet = 15.8 (14.5-17.0) μm; DGO = 4.4 (4.0-5.0) μm; stylet knob width = 4.5-5.0 μm; anterior end to the excretory pore = 16-19 μm; vulval slit = 29 (27-33) μm; vulva-anus distance = 29.3 (26-31) μm; EP/ST = 1.1.
- *Paratype J2* (n = 2): L = 378, 408 μm; a = 29.7, 31.3; b = 4.7, 4.9; c = 6.6, 6.8; c′ = 4.8; stylet = 11.5, 12.5 μm; DGO = 3, 3.4 μm; labial region diam. = 5-6 μm; labial region height = 2 μm; tail = 55 μm; hyaline region = 12 μm.

DESCRIPTION

Female

Body white, elongated pear-shaped or bag-like with long neck. Terminal mound practically non-existent. Labial region not offset from body contours, its anterior end hemispherical. Stylet knobs round and posteriorly sloping. Metenchium slightly curved dorsally. Excretory pore opening perpendicularly at level of stylet knobs. Median bulb almost round (42-43 × 41-45 μm) with muscular walls. Anal-vulval plate almost round with rather dense striae and dorsal side usually lower than ventral, flattened, but slightly elevated. Lateral fields unclear and formed mainly due to anastomoses of dorsal and ventral annulation, or more rarely, due to broken striae in this region. Anal-vulval plate sometimes asymmetrical with one lateral field finer than the other or absent. Perineum with 1-2 lines, vulval lip region containing no striae.

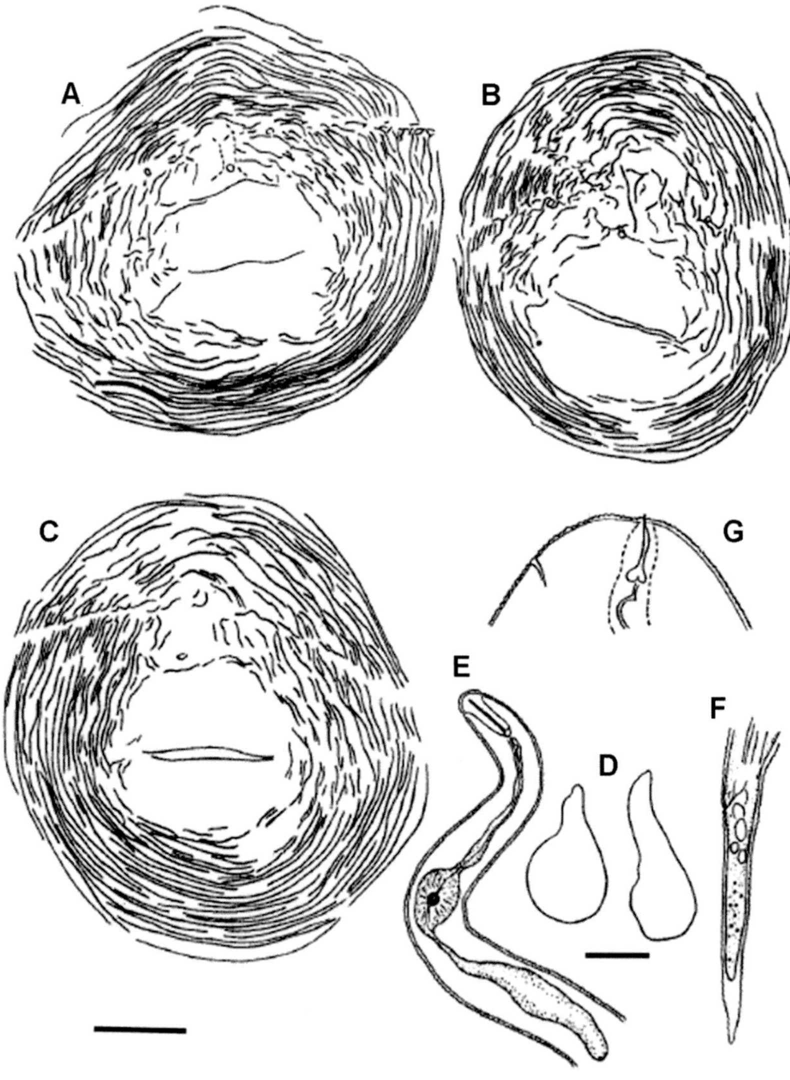

Fig. 73. Meloidogyne cynariensis. *A-C: Perineal pattern; D: Entire female; E: Anterior region of second-stage juvenile (J2); F: J2 tail; G: Female anterior region. (Scale bars: A-C, E-G = 20 μm; D = 300 μm.) (After Pham-Thanh-Binh, 1990.)*

Zone I is in most cases free from striae; if present, then always consisting of broken and erratic line fragments. In ventral sector (Zone II) striae relatively smooth, almost unbroken, usually dense, even though some breaks can be found. Lines in lateral sector (Zone III) often broken and disorganised. Zone IV (dorsal) rather flattened, tail region weakly developed, usually with a few striae. Phasmids well developed, separated by 29-36 μm.

Male

Not found.

J2

Body slender, narrowing towards posterior end. Labial region with three annuli. Lateral field with four distinct incisures. Stylet slender with visible knobs. Procorpus narrow and transforming into a large median bulb of irregular shape with a developed valve. Pharyngeal glands long. Tail long and annulated.

TYPE PLANT HOST: Artichoke plant, *Cynara scolymus* L.

OTHER PLANTS: No other hosts were reported.

TYPE LOCALITY: Dalat, Tây Nguyen, Lam Dong province, Vietnam.

DISTRIBUTION: *Asia*: Vietnam.

SYMPTOMS: Galls single, weakly protruding from surface of the main root. They do not cause strong deformation, but lateral roots are greatly thickened.

CHROMOSOME NUMBER: No available data.

POLYTOMOUS KEY CODES: *Female*: A2, B2, C3, D5; *Male*: -; *J2*: A32, B2, C3, D1, E3, F1.

BIOCHEMICAL AND MOLECULAR CHARACTERISATION: No available data.

RELATIONSHIPS (DIAGNOSIS): *Meloidogyne cynariensis* differs from the majority of species in the shape of the anal-vulval plate, which is practically round. Morphologically, this species is difficult to distinguish from *M. javanica* or *M. maritima*.

24. *Meloidogyne daklakensis* Trinh, Le, Nguyen, Nguyen, Liebanas & Nguyen, 2018

(Figs 74-76)

COMMON NAME: Dak Lak coffee root-knot nematode.

Trinh *et al.* (2018) described this species from roots of coffee in Dak Lak Province, Vietnam.

MEASUREMENTS (*AFTER TRINH ET AL., 2018*)

- *Holotype female*: L = 666 μm; W = 355 μm; stylet = 14.6 μm; DGO = 3.6 μm; median bulb width = 27 μm; a = 1.5; EP/ST = 1.2.
- *Paratype females* (n = 10): L = 645 ± 54 (548-709) μm; W = 378 ± 45 (322-462) μm; stylet = 14.6 ± 0.5 (14-15) μm; DGO = 5.2 ± 0.2 (4.5-5.5) μm; median bulb width = 28 ± 1.4 (26-30) μm; excretory pore to anterior end = 19 ± 1.2 (17.6-20.8) μm; a = 1.7 ± 0.3 (1.4-2.6); EP/ST = 1.3.
- *Paratype males* (n = 20): L = 1228 ± 98 (1085-1365) μm; W = 31 ± 3 (27-34) μm; stylet = 19 ± 1.2 (17-20) μm; stylet knob width = 4.0 ± 0.16 (3.9-4.1) μm; stylet knob height = 2.5 ± 0.5 (2.1-3.12) μm; DGO = 5.3 ± 1 (4.1-7.2) μm; excretory pore to anterior end = 136 ± 17 (107-161) μm; spicules = 24 ± 3.5 (18-29) μm; a = 39.1 ± 2.5 (35.3-42.3); b = 10.8 ± 0.8 (9.4-12.1); b′ = 5.9 ± 0.4 (6.1-6.3).
- *Paratype J2* (n = 20): L = 333 ± 27 (280-373) μm; W = 12 ± 1 (11.4-14.6) μm; stylet = 14 ± 1 (11.4-15.6) μm; DGO = 3.7 ± 0.8 (2.1-5.2) μm; excretory pore to anterior end = 66 ± 6 (55-78) μm; tail length = 50 ± 23 (32-54) μm; hyaline region = 12 ± 2 (9.3-16.6) μm; a = 26.9 ± 2.4 (23.3-31.5); b = 4.5 ± 0.5 (3.4-5.4); b′ = 2.9 ± 0.3 (2.2-3.4); c = 7.4 ± 1 (5.8-10.8); c′ = 4.8 ± 0.6 (3.9-6.2).

DESCRIPTION

Female

Body pearly white, pear-shaped, anterior body portion commonly off-centre from median plane. Anterior end pointed, posterior end varying from rounded to slightly flattened. Neck usually short, not curved. Stoma

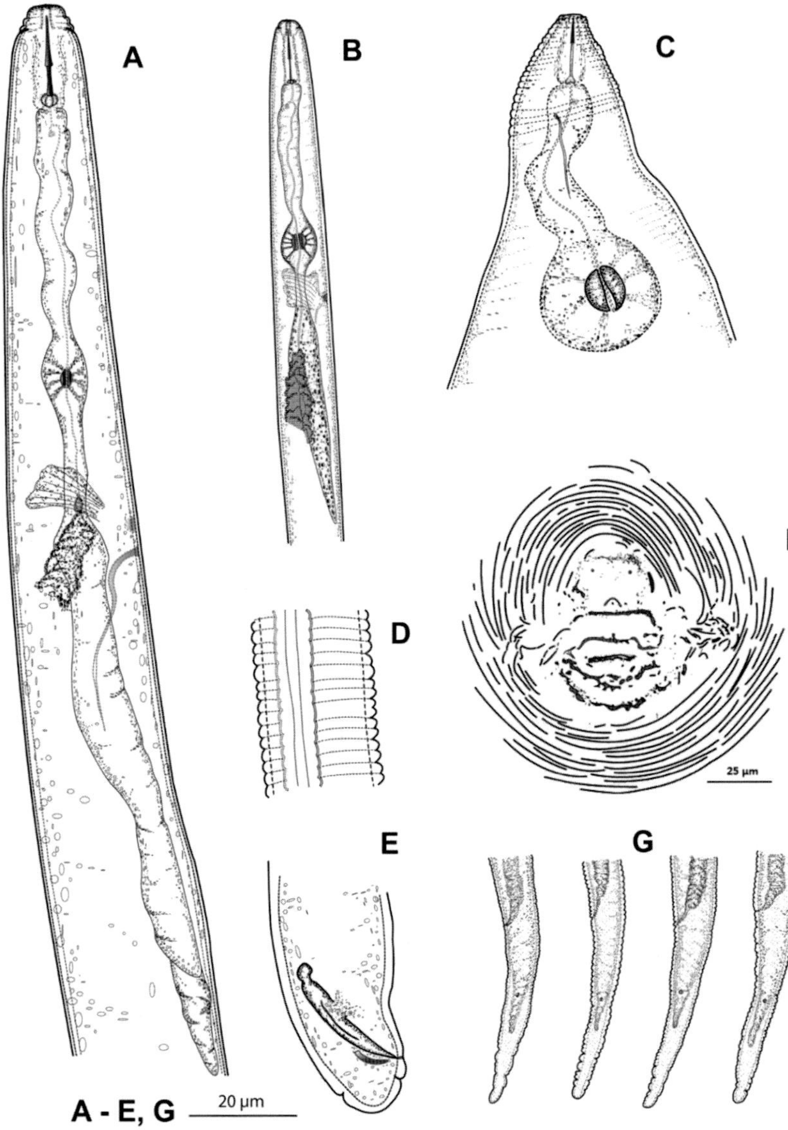

Fig. 74. Meloidogyne daklakensis. *A: Male anterior region; B: Anterior region of second-stage juvenile; C: Female anterior region; D: Male lateral field; E: Male tail; F: Perineal pattern; G: Second-stage juvenile tail. (After Trinh* et al., *2018.)*

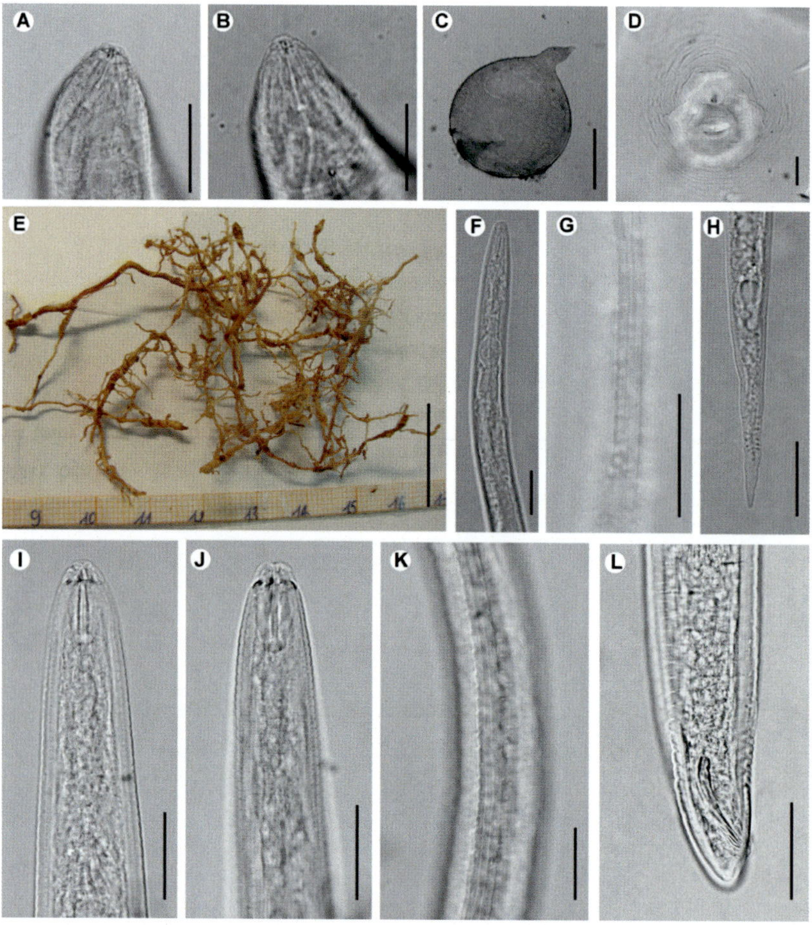

Fig. 75. Meloidogyne daklakensis. *LM. A, B: Female anterior region; C: Entire female; D: Perineal pattern; E: Infected coffee roots; F: Anterior region of second-stage juvenile (J2); G: J2 lateral field; H: J2 tail; I, J: Male anterior region; K: Male lateral field; L: Male tail. (Scale bars: A, B, D, F, G-L = 25 μm; C = 200 μm; E = 2000 μm.) (After Trinh* et al.*, 2018.)*

slit-like, labial region sclerotised, slightly offset from rest of body, bearing one annulus, labial disc not fused with median labials, elevated, lateral sectors enlarged, extension prominent, amphidial openings elongate. Stylet short, cone base triangular and wider than shaft. Stylet tip normally straight (sometimes curved dorsally), shaft cylindrical, same diam. throughout its length. Stylet knobs three, oval and sloping posteriorly.

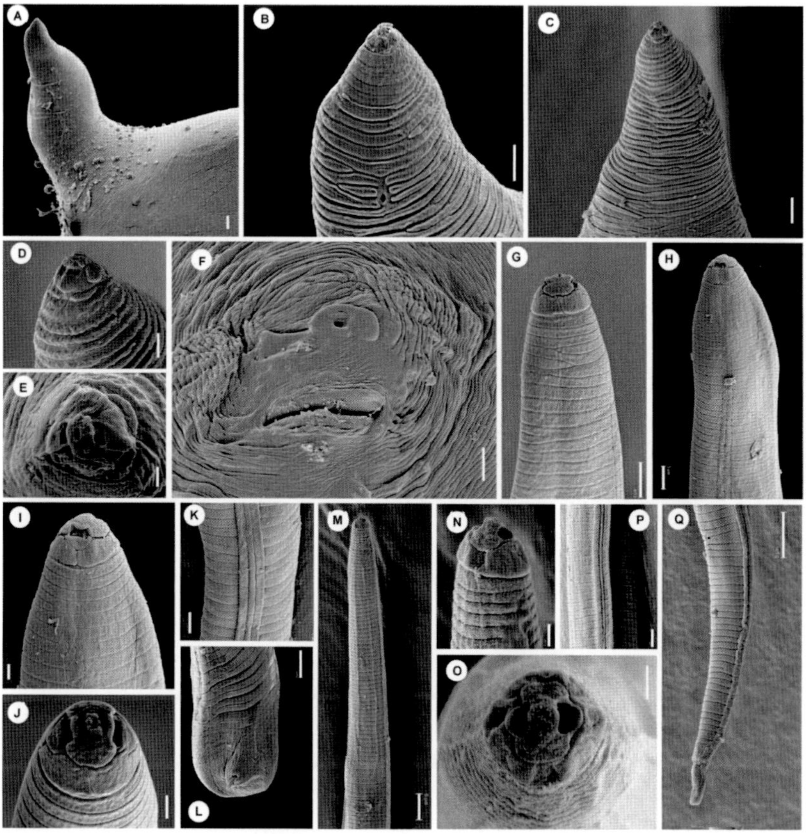

Fig. 76. Meloidogyne daklakensis. *SEM. A-D: Female anterior region; E: Female* en face *view; F: Perineal pattern; G-I: Male anterior region; J: Male* en face *view; K: Male lateral field; L: Male tail; M: Second-stage juvenile (J2) Anterior region; N: J2 labial region; O: J2 lateral field; P: J2* en face *view; Q: J2 tail. (Scale bars: A-F = 10 μm; G-I, Q, M = 5 μm; L, N, P = 1 μm; O = 2 μm.) (After Trinh* et al.*, 2018.)*

Pharyngeal lumen of procorpus wide, often showing rounded protuberances. Excretory pore situated far posterior to stylet knobs and within range from level of procorpus to median bulb. Median bulb large and rounded, valve apparatus oval-shaped and sclerotised. Perineal patterns rounded to oval, striae smooth, lateral lines reduced, dorsal arch low, rounded, enclosing distinct vulva, anus and tail tip. Phasmids indistinct. Vulval slit centrally located in unstriated area and longer than vulva-anus distance.

Male

Body vermiform, anterior end tapering and posterior region bluntly rounded. Labial region slightly offset from body with a high labial region cap consisting of a large labial annulus. Median labials and labial disc bow-tie-shaped. Lateral labials large and triangular, slightly lower than labial disc and median labials. Amphidial openings appearing as long and large slits located between labial disc and lateral labials. Post-labial annuli divided into four lobes by longitudinal slits. Stylet robust and straight, one lateral knob projecting, other two sloping posteriorly. Procorpus distinctly outlined, three times longer than median bulb. Median bulb ovoid, with a strong valve apparatus. Excretory duct curved. Lateral field with four incisures forming three bands, areolated over entire body. Testis one, occupying 58% of body cavity. Spicules slightly curved ventrally with bluntly rounded terminus. Gubernaculum short, crescentic.

J2

Body slender, tapering to an elongated tail. Labial region narrower than body, weak and slightly offset from body, labial disc offset, stoma slit-like, median labials and labial disc bow-tie-shaped, labial disc rounded, amphidial openings enlarged, covered lateral labials. Stylet slender, cone weakly expanded at junction with shaft, knobs rounded. Median bulb broadly oval, valve large and heavily sclerotised. Pharyngo-intestinal junction at level of nerve ring. Excretory pore slightly posterior to nerve ring, opening just posterior to hemizonid. Lateral field with four incisures, not areolated and prominent. Phasmids small, distinct. Tail elongate conoid.

TYPE PLANT HOST: Robusta coffee, *Coffea canephora* Pierre *ex* A. Froehner.

OTHER PLANTS: No other hosts were reported.

TYPE LOCALITY: Western Highland: Dak Lak Province, Vietnam.

DISTRIBUTION: *Asia*: Vietnam. It has only been reported from the type locality (Trinh *et al.*, 2018).

SYMPTOMS: The coffee roots infected with *M. daklakensis* had small galls (1-2 mm diam.).

CHROMOSOME NUMBER: No available data.

POLYTOMOUS KEY CODES: *Female*: A23, B23, C3, D3; *Male*: A34, B34, C32, D1, E2, F2; *J2*: A3, B213, C34, D312, E32, F32.

BIOCHEMICAL AND MOLECULAR CHARACTERISATION: No biochemical data are available for *M. daklakensis*. However, it is molecularly well characterised by ribosomal (ITS, 18S and D2-D3 segments of 28S rRNA) and mitochondrial (*COI*, *COII*-16S) gene (Trinh *et al.*, 2018).

RELATIONSHIPS (DIAGNOSIS): The species belongs to Molecular group X and is clearly molecularly differentiated from all other *Meloidogyne* species. Morphologically, *M. daklakensis* is similar to *M. marylandi*, *M. naasi*, *M. mali* and *M. baetica* and can be distinguished as follows: from *M. marylandi* and *M. baetica* by having a smaller J2 body length 280-373 *vs* 324-489 and 394-422 μm, respectively, and from *M. naasi* and *M. mali* by having a perineal pattern with indistinct phasmids *vs* large phasmids in the females.

25. *Meloidogyne donghaiensis* Zheng, Lin & Zheng, 1990
(Fig. 77)

COMMON NAME: Donghai citrus root-knot nematode.

This nematode was found in 1989 during a survey of citrus plantations. The infected citrus trees showed yellowing leaves, sparse foliage, small fruits and defoliated branches.

MEASUREMENTS *(AFTER ZHENG ET AL., 1990)*

- *Paratype females* (n = 30): L = 659 (565-845) μm; W = 617 (430-740) μm; neck length = 158 (105-225) μm; stylet = 12.2 (10.0-14.6) μm; DGO = 4.8 (4.0-7.3) μm; anterior end to median bulb = 85 (76-123) μm; interphasmid distance = 19 (13-28) μm; vulval slit = 24 (18-32) μm; vulva-anus distance = 21 (16-25) μm; vulval slit to interphasmid distance = 27 (25-33) μm; EP/ST = 3.0 (1.9-4.1).

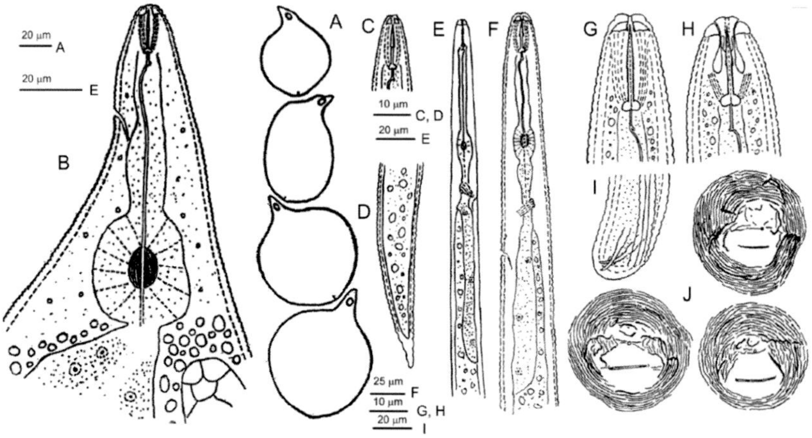

Fig. 77. Meloidogyne donghaiensis. *A: Entire female; B: Female anterior region; C: Second-stage juvenile (J2) labial region; D: J2 tail; E: Anterior region of J2; F: Male anterior region; G, H: Male labial region; I: Male tail; J: Perineal pattern. (After Zheng* et al., *1990.)*

- *Paratype males* (n = 26): L = 1910 (1465-2170) μm; W = 47 (39-50) μm; stylet = 22.2 (22.1-24.5) μm; labial region height = 6.3 (5.5-6.4) μm; labial region diam. = 12.6 (11.6-13.5) μm; DGO = 4.3 (2.5-6.1) μm; spicules = 35 (31-45) μm; gubernaculum = 9 (8-10) μm.
- *Paratype J2* (n = 30): L = 466 (420-485) μm; W = 19 (17-20) μm stylet = 13.3 (12.3-15.3) μm; a = 25.9 (21.4-26.9); b$'$ = 6.4 (5.1-7.0); c = 9.2 (7.6-10.2); c$'$ = 4.2 (3.0-5.0); DGO = 4.0 (3.0-6.0) μm; anterior end to excretory pore = 104 (87.0-105.5) μm.
- *Eggs* (n = 40): L = 94 (91-102) μm; W = 46 (42-50) μm; L/W = 2.1 (1.7-2.3).

DESCRIPTION

Female

Mature female body pyriform, with short neck and distinct annulation. Median bulb rounded. Perineal patterns usually rounded, dorsal arch low, round. Phasmids adjacent to tail terminus. Short striae connected with ends of vulval slit. Lateral field not distinct, sometimes marked by irregular coarse or indistinct striae interrupting pattern. Some coarse striae sometimes parallel with other fine striae intersecting.

Male

Labial region flattened. Stylet knob rounded, posteriorly sloping. Pharyngeal glands overlapping intestine ventrally. Lateral field areolated with four incisures. Tail terminus rounded. Tail length shorter than body diam. at anus level. Phasmids located posterior to anus.

J2

Labial region flattened or slightly rounded, not offset. Hemizonid anterior to excretory pore. Lateral field with four incisures. Tail conoid, terminus rounded without annulation.

TYPE PLANT HOST: Mandarin orange, *Citrus reticulata* Blanco.

OTHER PLANTS: No other hosts were reported.

TYPE LOCALITY: Shenhu town, Jinjiang county, Fujian province, China.

DISTRIBUTION: *Asia*: China.

CHROMOSOME NUMBER: No available data.

POLYTOMOUS KEY CODES: *Female*: A21, B324, C213, D3; *Male*: A123, B21, C12, D1, E1, F1; *J2*: A2, B21, C3, D4, E23, F34.

BIOCHEMICAL AND MOLECULAR CHARACTERISATION: No available data.

RELATIONSHIPS (DIAGNOSIS): This species is most similar to *M. arenaria*, *M. africana* and *M. ottersoni* and some others. It differs from *M. arenaria* in the perineal pattern and from *M. ottersoni* by the presence of some irregular coarse and fine striae in the perineal pattern, and by the striae forming a circle at the tail tip. Additionally, it differs from *M. arenaria* in the J2 stylet length and c′ ratio and from *M. ottersoni* in c′ ratio. It differs from *M. africana* by the absence of vulval lip striae, presence of some coarse striae in the perineal pattern, and male stylet length. Molecular data are required from paratypes in order clearly to separate this from other species.

26. *Meloidogyne dunensis* Palomares-Rius, Vovlas, Troccoli, Liebanas, Landa & Castillo, 2007

(Figs 78-81)

COMMON NAME: Dune root-knot nematode.

Nematode surveys in the Mediterranean coastal sand dunes in central eastern Spain revealed high infection rates of European sea rocket (*Cakile maritima* Scop.) feeder roots by a RKN described as *M. dunensis*.

MEASUREMENTS *(AFTER PALOMARES-RIUS* ET AL., *2007)*

- *Holotype female*: L = 734 μm; W = 520 μm; a = 1.4; stylet = 14 μm; DGO = 4.3 μm; excretory pore from anterior end = 23 μm; vulval slit length = 21 μm; vulva to anus = 14 μm; EP/ST = 1.6.
- *Female paratypes* (n = 12): L = 549 $\pm$ 86 (395-620) μm; W = 400 $\pm$ 80 (380-542) μm; neck length = 124 $\pm$ 5.7 (118-132) μm; a = 1.3 $\pm$ 0.3 (1.0-1.6); stylet = 14 $\pm$ 1.8 (13-16) μm; excretory pore distance 23 $\pm$ 1.3 (19-24) μm; vulval slit = 20 $\pm$ 1.2 (18-22) μm; vulva-anus distance = 15 $\pm$ 0.8 (13-16) μm; EP/ST = 1.6 $\pm$ 0.2 (1.2-1.8).
- *Paratype males* (n = 12): L = 1371 $\pm$ 172.2 (1196-1813) μm; labial region diam. = 11.3 $\pm$ 1.2 (8.0-12.5) μm; labial region height = 5.7 $\pm$ 0.5 (4.5-6.0) μm; stylet = 20 $\pm$ 1.6 (16-22) μm; stylet conus = 11 $\pm$ 0.8 (9.5-11.5) μm; stylet knob width = 4.5 $\pm$ 0.3 (4-5) μm; DGO = 3.5 $\pm$ 0.7 (2.5-4.5) μm; O = 17.4 $\pm$ 3.7 (12.5-24.5)%; anterior end to centre of median bulb = 82 $\pm$ 5.5 (74.5-93.0) μm; median bulb height = 21.5 $\pm$ 2.6 (19.5-26.5) μm; median bulb width = 12.0 $\pm$ 1.7 (9.0-14.5) μm; isthmus length = 22.5 $\pm$ 6.9 (12.0-32.5) μm; pharynx (to cardia) = 124 $\pm$ 11.5 (108-143) μm; pharynx. (to end of gland lobe) = 248 $\pm$ 63.8 (190-407) μm; pharyngeal overlap = 124.5 $\pm$ 62.6 (59.5-286) μm; anterior end to excretory pore = 146 $\pm$ 12.2 (121.5-166) μm; max body diam. = 45 $\pm$ 4.0 (36-49) μm; annuli width = 3.0 $\pm$ 0.4 (2.2-3.7) μm; testis length = 697 $\pm$ 183.4 (486-1057) μm; T = 50 $\pm$ 12.1 (31-67)%; tail = 5.7 $\pm$ 1.3 (3.5-8.0) μm; anal body diam. = 17.5 $\pm$ 1.9 (14.0-19.5) μm; spicules = 35.5 $\pm$ 2.9 (29-38) μm; gubernaculum = 8.5 $\pm$ 1.3 (6.0-10.5) μm; a = 33.4 $\pm$ 4.0 (26.8-41.6); b = 11.6 $\pm$ 1.4 (9.8-14.9); b$'$ = 5.9 $\pm$ 1.1 (4.4-8.0); c = 280.6 $\pm$ 59.5 (186.6-385.3); c$'$ = 0.3 $\pm$ 0.1 (0.2-0.5).

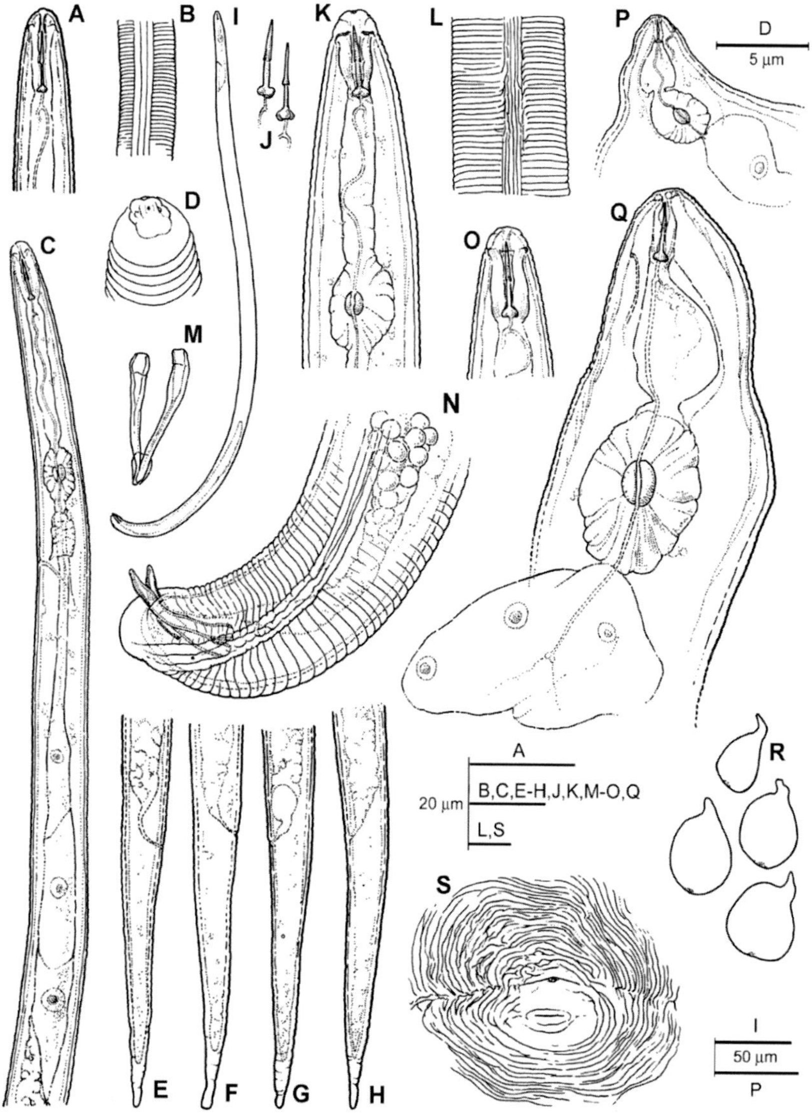

Fig. 78. Meloidogyne dunensis. *A: Second-stage juvenile (J2) labial region; B: J2 lateral field; C: Anterior region of J2; D: J2* En face *view; E-H: J2 tail; I: Entire male; J: Male stylet; K, O: Male anterior end; L: Male lateral field; M: Detail of spicules; N: Male tail; P, Q: Female anterior region; R: Entire females; S: Perineal pattern. (After Palomares-Rius* et al., *2007.)*

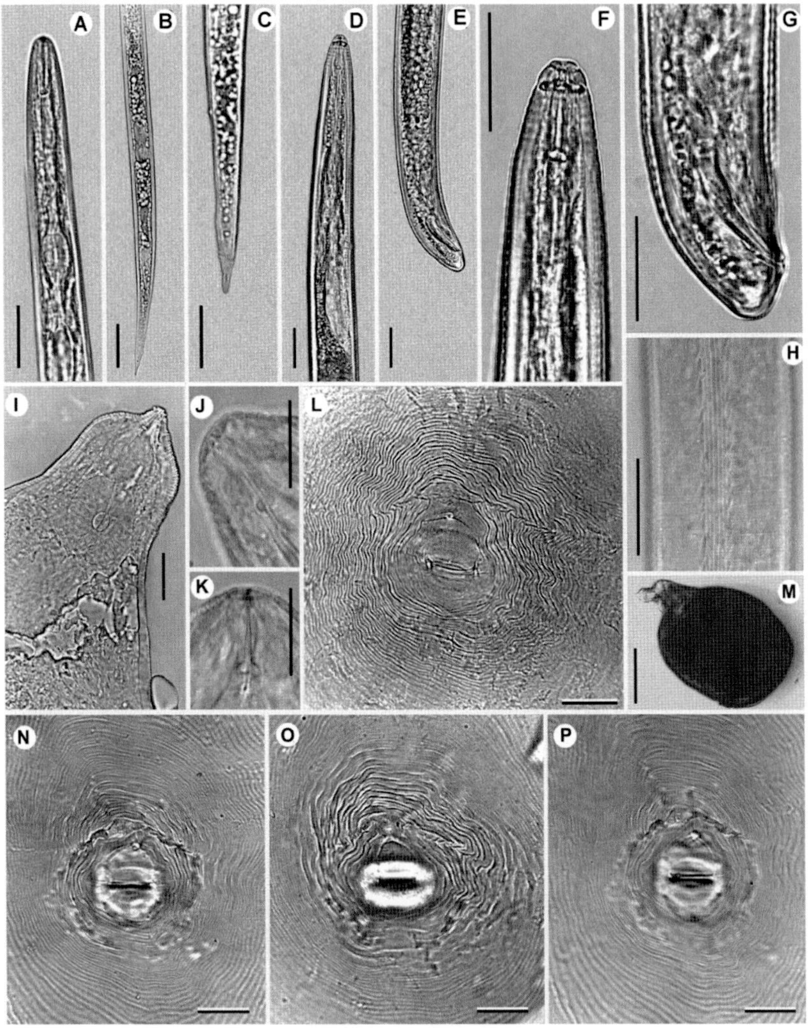

Fig. 79. Meloidogyne dunensis. *Light micrographs. A: Second-stage juvenile (J2) anterior region; B, C: J2 Tail region; D: Male anterior region; E, G: Male tail; F: Male labial region; H: Male lateral field; I: Female anterior region; J, K: Female labial region; L, N-P: Perineal pattern; M: Entire female. (Scale bars: A-I, L, N-P = 25 μm; J, K = 15 μm; M = 250 μm.) (After Palomares-Rius* et al., *2007.)*

- *Paratype J2* (n = 15): L = 446 ± 23 (417-483) μm; labial region diam. = 6.0 ± 0.3 (5.5-6.0) μm; labial region height 2.5 ± 0.2 (2.5-

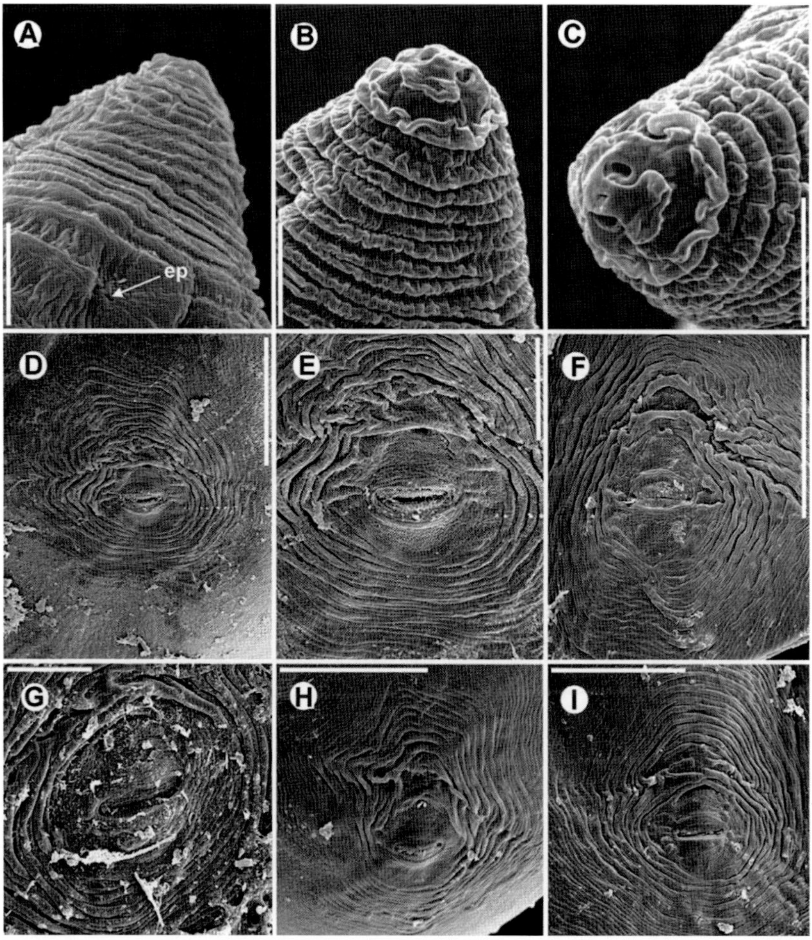

Fig. 80. Meloidogyne dunensis. *SEM. A-C: Female labial region; D-I: Perineal pattern. (ep = excretory pore). (Scale bars: A-C = 5 μm; D, F, I = 50 μm; E, G, H = 20 μm.) (After Palomares-Rius* et al., *2007.)*

3.0) μm; stylet = 11.5 ± 0.6 (11.0-12.5) μm; stylet conus = 6.0 ± 0.5 (5.0-6.5) μm; stylet knob width = 2.0 ± 0.3 (2-3) μm; DGO = 2.5 ± 0.4 (1.5-3.0) μm; O = 21.0 ± 3.8 (14-27)%; anterior end to centre of median bulb = 58 ± 3.4 (53-56) μm; median bulb height = 13 ± 0.9 (12.5-15.0) μm; median bulb width = 9 ± 0.6 (8-10) μm; pharynx (to cardia) = 83 ± 4.3 (77-90) μm; pharynx (to end of gland lobe) = 201 ± 23.7 (152-232) μm; pharyngeal overlap = 115 ± 23.5 (65-146) μm; anterior end to excretory pore = 85 ± 4.5 (79-94) μm;

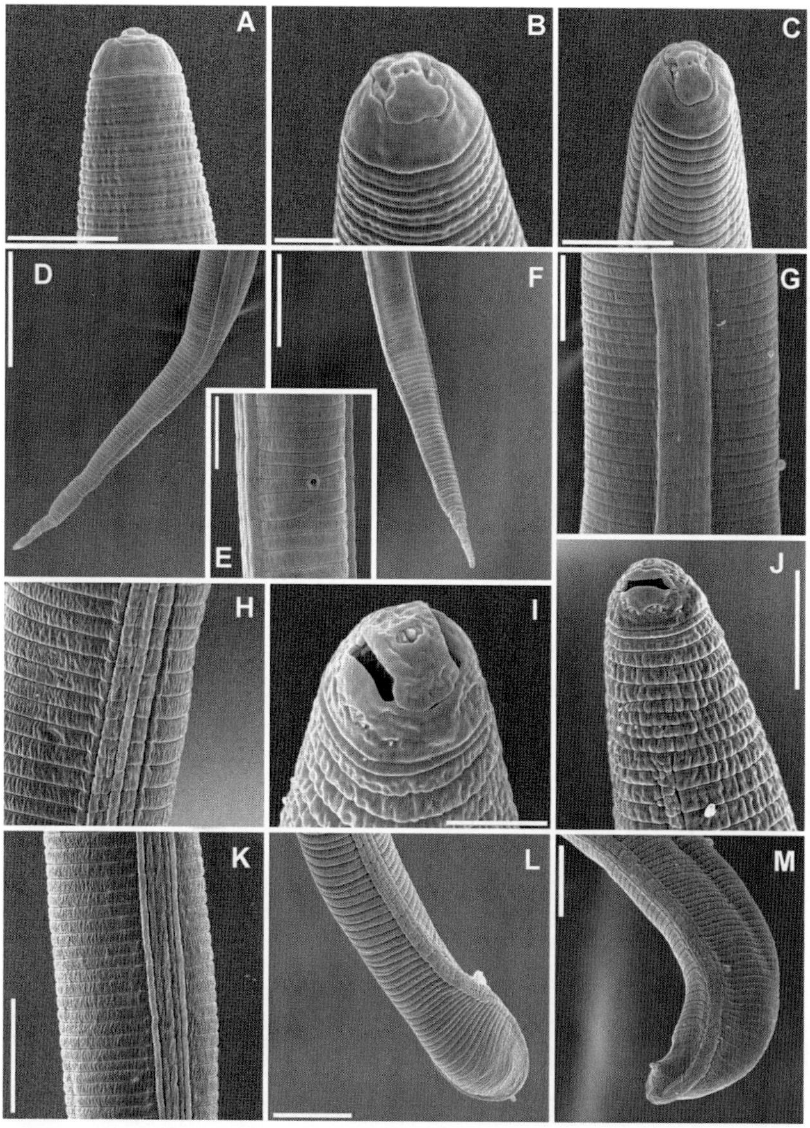

Fig. 81. Meloidogyne dunensis. *SEM. A, B: Second-stage juvenile (J2) labial region; D, F: J2 tail; E: J2 anal region; G, H: J2 lateral field; I, J: Male labial region; K: Male lateral field; L, M: Male tail. (Scale bars: A, C, E, G, I = 5 μm; B = 2 μm; D, F, L, M = 20 μm; H, J, K = 10 μm.) (After Palomares-Rius* et al., *2007.)*

max body diam. = 15 ± 1.1 (13-17) μm; annuli width = 1.1 ± 0.1 (1.0-1.2) μm; lateral field width = 5.0 ± 0.6 (3.5-5.5) μm; anterior end to gonad primordium = 255 ± 44.2 (184-318) μm; tail = 68 ± 7.8 (54-82) μm; anal body diam. = 11 ± 0.8 (9.5-12.5) μm; hyaline region = 14.0 ± 1.9 (9.5-16.5) μm; a = 29.5 ± 2.9 (25.6-34.5); b = 5.3 ± 0.3 (4.9-6.0); b' = 2.2 ± 0.3 (2.0-3.0); c = 6.7 ± 0.8 (5.1-8.3); c' = 6.3 ± 0.5 (5.1-6.8).

- *Embryonated eggs* (n = 30): L = 98 ± 2.8 (94-103) μm; W = 40 ± 1.3 (38-42) μm, L/W = 2.4 ± 0.1 (2.2-2.5).

DESCRIPTION

Female

Body completely or partially enclosed by galled tissue, distinctly annulated, pearly white, pear-shaped, sometimes globose. Neck region distinct. Labial region offset from body, often twisted. Labial cap distinct, variable in shape, labial disc elevated, labial framework weakly sclerotised. Stylet cone slightly curved ventrally, shaft cylindrical, knobs rounded and posteriorly sloping in most specimens. Excretory pore located slightly posterior to level of stylet knobs. Pharyngeal gland with a large mononucleate dorsal lobe and two subventral gland lobes. Perineal pattern rounded-oval, typically formed of numerous fine dorsal and ventral striae and ridges, lateral fields clearly visible. Dorsal arch enclosing usually fine but distinct phasmids. Vulval slit centrally located in unstriated area, slightly wider than vulva-anus distance. Anus fold clearly visible. A large egg sac commonly occurs outside of root gall, containing up to 400 eggs.

Male

Body vermiform, tapering anteriorly; tail rounded, with twisted posterior body portion. Labial region slightly offset, with large labial annulus and a prominent post-labial annulus. Prominent slit-like amphidial openings between labial disc and lateral labials. Labial framework strongly sclerotised; vestibule extension distinct. Stylet with straight cone and shaft. Stylet knobs rounded and posteriorly sloping. Lateral field composed of four incisures anteriorly and posteriorly. Six distinct incisures observed at mid-body, forming five equidistant bands, in 20% of the specimens. In SEM (face view), labial disc high and narrower than labial region, continuous with median labials. Lateral labials absent. Stomatal opening slit-like, located in large elongate prestoma. Amphidial apertures large, elongated, slit-like between labial disc and lateral sectors of labial region. Testis single, long, monorchic, occupying 31-67% of body cavity. Tail usually curved ventrally, short, with bluntly rounded tip and finely annulated. Phasmids small and located at level of cloacal aperture.

J2

Body vermiform, rather long, tapering at both ends with very long, narrow tail. Anterior end angular, labial region continuous with body contour. In SEM view, slit-like stoma located in oval-shaped prestoma, surrounded by six pore-like openings of inner labial sensilla. Median labials and labial disc dumbbell-shaped in face view. Labial disc rounded, raised slightly above median labials. Lateral labials small and oval-rounded, lower than labial disc and median labials. Elongate amphidial apertures located between labial disc and lateral labials. Labial region not annulated. Body annuli distinct but fine. Lateral fields beginning near level of procorpus as two lines. Near median bulb third line beginning and quickly splitting to make four lines, running entire length of body until ending near hyaline region, irregularly areolated. Stylet delicate, cone straight, narrow, sharply pointed. Shaft becoming slightly wider posteriorly. Knobs large, rounded, separate from each other, posteriorly directed. Procorpus faintly outlined, median bulb oval-shaped with enlarged lumen lining. Isthmus not clearly defined, pharyngo-intestinal junction at excretory pore level, or slightly anterior. Gland lobe variable in length, with three equally sized nuclei and overlapping intestine ventrally. Excretory pore distinct, at level with posterior third of isthmus or mid-way. Hemizonid distinct, located 1-2 annuli anteriorly to excretory pore, extending for two additional body annuli. Tail thin, conoid, annuli diminishing in size, become more irregular posteriorly. Hyaline region clearly defined, tail tip finely rounded, rarely clavate. Rectum dilated. A few fat droplets may occur in hyaline region. Phasmids small, difficult to observe, located posterior to anus, at mid-tail level.

TYPE PLANT HOST: European sea rocket, *Cakile maritima* Scop.

OTHER PLANTS: Tomato and chickpea.

TYPE LOCALITY: Mediterranean coastal dunes in Cullera, Valencia, central eastern Spain.

DISTRIBUTION: *Europe*: Spain.

SYMPTOMS: Root galls induced on the roots were variable in size, large (almost three times the root diam.) and located commonly along the root axis but rarely on the root tip. Numerous lateral roots arising from galled root portions were also galled.

CHROMOSOME NUMBER: No available data.

POLYTOMOUS KEY CODES: *Female*: A32, B32, C3, D3; *Male*: A312, B324, C12, D1, E2, F2; *J2*: A2, B23, C213, D213, E3, F12.

BIOCHEMICAL AND MOLECULAR CHARACTERISATION: The isozyme electrophoretic analysis of five specimen groups of young egg-laying females of *M. dunensis* revealed one very slow, weak VS1 Est band after prolonged staining and a N1c Mdh phenotype with two very weak bands that did not occur in the Est and Mdh phenotypes of *M. javanica*, which showed J3 and N1 phenotypes, respectively. The 18S, ITS1-5.8S-ITS2 and D2-D3 regions of 28S rRNA gene sequences of this species were provided by Palomares-Rius *et al.* (2007).

RELATIONSHIPS (DIAGNOSIS): The species belongs to Molecular group II and has a sister relationships with *M. duytsi*. *Meloidogyne dunensis* differs from *M. duytsi* by perineal pattern structure, smaller female EP/ST ratio (1.2-1.8 *vs* 2.4-3.4) and longer spicules (29-38 *vs* 24.0-27.2 μm).

27. *Meloidogyne duytsi* Karssen, van Aelst & van der Putten, 1998
(Figs 82-84)

COMMON NAME: Duyts's root-knot nematode.

This species was described from samples collected from foredunes near Oostvoorne, The Netherlands.

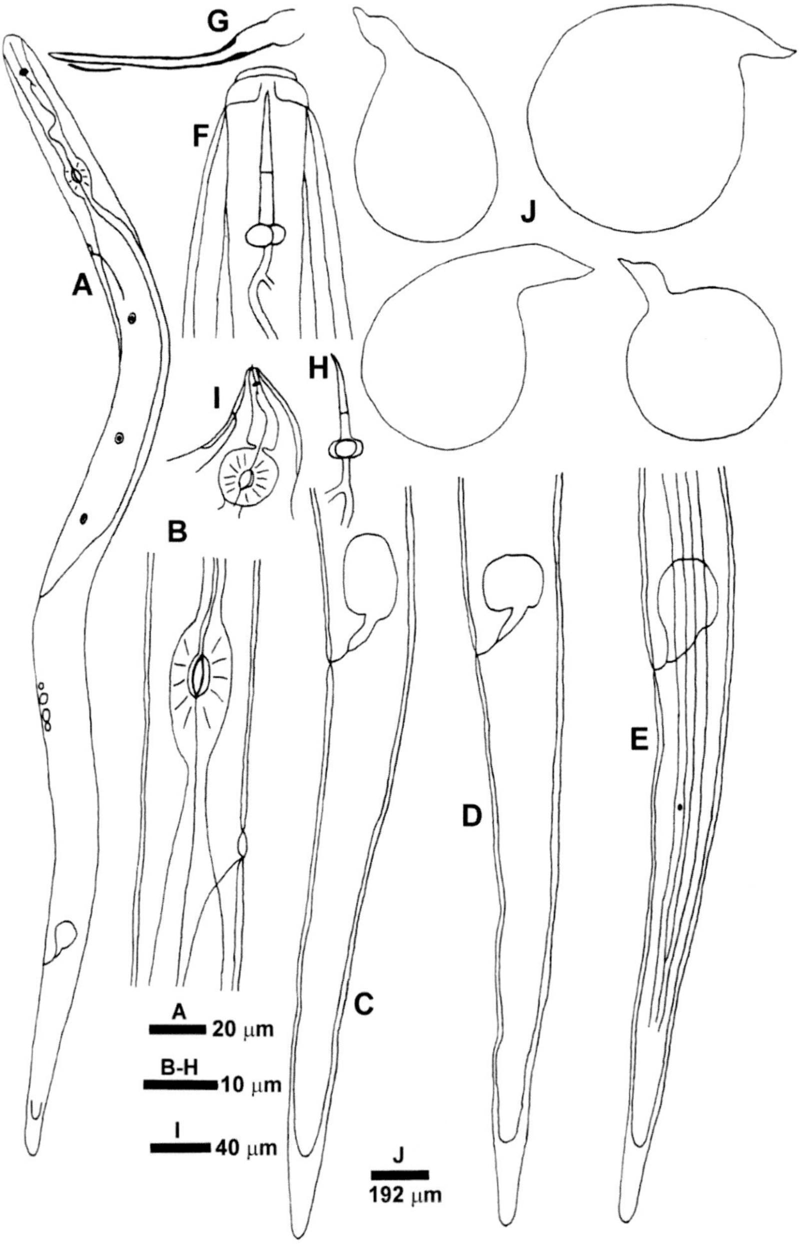

Fig. 82. Meloidogyne duytsi. *A: Entire second-stage juvenile (J2); B: J2 median bulb; C-E: J2 tail; F: Male labial region; G: Spicules; H: Female stylet; I: Female labial region; J: Entire female. (After Karssen* et al., *1998b.)*

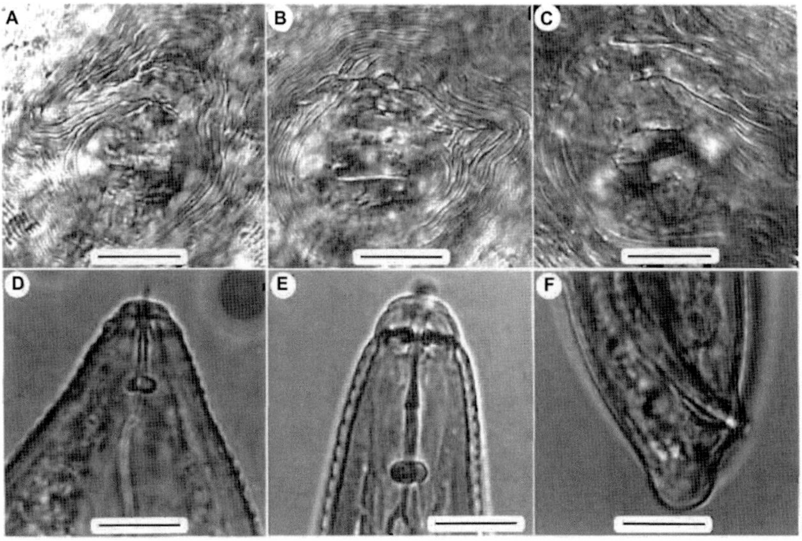

Fig. 83. Meloidogyne duytsi. *LM. A-C: Perineal pattern; D: Female labial region; E: Male labial region; F: Male tail. (Scale bars: A-C = 25 μm; D-F = 10 μm.) (After Karssen* et al., *1998b.)*

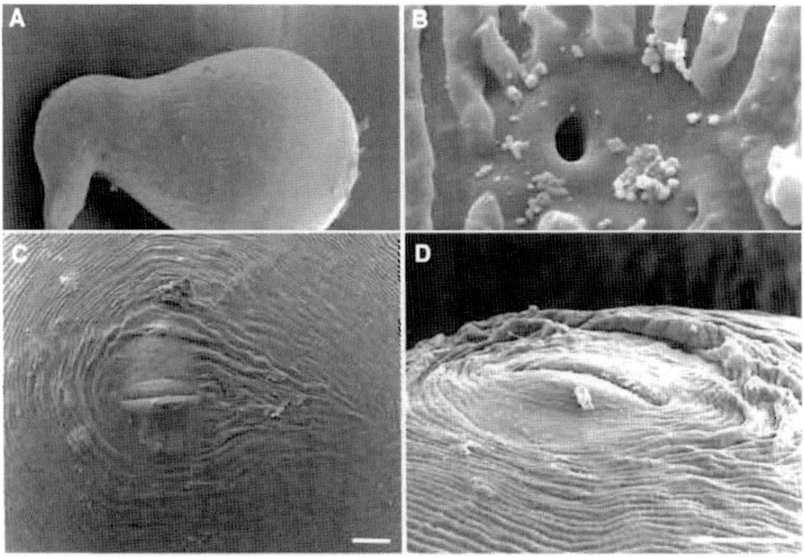

Fig. 84. Meloidogyne duytsi. *SEM. A: Body; B: Excretory pore; C: Perineal pattern; D: Perineal pattern (side view). (After Karssen* et al., *1998b.)*

MEASUREMENTS (*AFTER* KARSSEN ET AL., *1998B*)

- *Females* (n = 30): L = 865.1 ± 108.3 (560-960) μm; W = 591 ± 115.9 (368-800) μm, stylet = 13.3 ± 0.3 (12.6-13.9) μm; DGO = 3.8 ± 0.4 (3.2-4.4) μm; excretory pore to anterior end = 37.5 ± 4.7 (30.3-47.4) μm; vulval slit = 23.2 ± 2.2 (19.6-25.3) μm; vulva-anus distance = 21.5 ± 2.5 (15.8-26.5) μm; a = 1.5 ± 0.2 (1.2-1.8) interphasmid distance 20.2 ± 2.7 μm; EP/ST = 2.8 (estimated from drawings).
- *Males* (n = 30): L = 1316 ± 173.6 (960-1680) μm; W = 34.4 ± 2.7 (27.2-36.7) μm; stylet = 19.8 ± 0.3 (19.0-20.2) μm; DGO = 4.0 ± 0.3 (3.8-5.1) μm; anterior end to median bulb = 74.1 ± 5.0 (63.2-87.2) μm; excretory pore to anterior end = 137.5 ± 11.8 (117.0-155.5) μm; tail = 13.2 ± 1.5 (10.1-15.8) μm; spicules = 25.9 ± 0.9 (24.0-27.2) μm; gubernaculum = 7.1 ± 0.4 (6.3-7.6) μm; testis = 692.2 ± 114.1 (379.2-853.2); a = 38.2 ± 3.6 (30.3-45.8); c = 100.6 ± 15.7 (76.2-130.8); T = 52.6 ± 6.6 (38.7-65.8)%; (excretory pore/L) × 100 = 10.5 ± 1.0 (8.5-13.5).
- *J2* (n = 30): L = 423.6 ± 13.7 (403.2-454.4) μm; W = 17.1 ± 0.6 (15.8-17.7) μm; stylet = 11.1 ± 0.4 (10.7-12.0) μm; DGO = 3.6 ± 0.3 (3.2-3.8) μm; anterior end to median bulb = 55.2 ± 2.0 (49.3-59.4) μm; anterior end to excretory pore = 79.4 ± 86.6 (69.5-86.6) μm; tail = 70.4 ± 3.1 (65.1-76.5) μm; hyaline region = 11.3 ± 1.0 (9.5-13.3) μm; a = 24.9 ± 1.1 (23.2-27.3); c = 6.0 ± 0.2 (5.6-6.6); c′ = 5.7 ± 0.3 (5.2-6.4).
- *Eggs* (n = 30): L = 106.2 ± 5.4 (99.2-116.8) μm; W = 47.9 ± 2.0 (44.8-52.4) μm; L/W = 2.2 ± 0.2 (2.0-2.5).

DESCRIPTION

Female

Body relatively large, annulated, pearly white, globular-shaped, neck region distinct and projecting sideways from body axis, posterior protuberance not observed. Labial region offset from body. Labial cap distinct but rather variable in shape, labial disc slightly elevated. Labial framework weakly sclerotised. Stylet cone slightly curved dorsally, shaft cylindrical. Knobs large, transversely ovoid, offset from shaft. Excretory pore located halfway between median bulb and anterior end. No vesicles observed near lumen lining of median bulb. Pharyngeal glands rather variable in shape and size. Perineal pattern asymmetrical, dorsal arch relatively low, with coarse striae, one lateral wing, variable in size, present in most patterns. Tail terminus distinct and without punctations, with indistinct lateral lines. Phasmids small, located just anterior to covered anus. Ventral pattern region angular-shaped with fine striae.

Male

Body vermiform, annulated and twisted. Four incisures present in lateral field. Outer bands regularly areolated. Labial region offset from body, one relatively high post-labial annulus present, transverse incisures not observed. Labial disc rounded, slightly elevated, fused with submedian labials into a labial cap. Submedian labials rounded, lateral edges slightly wider than labial disc. Prestoma hexagonal in shape with six inner sensilla. Four labial sensilla present on submedian labials and marked by small cuticular depressions. Slit-like amphidial openings present between labial cap and relatively small lateral labials. Labial framework moderately sclerotised. Vestibule extension indistinct. Stylet with straight cone and cylindrical shaft, large transversely ovoid knobs, offset from shaft. Pharynx with slender procorpus and oval-shaped median bulb. Pharyngeal gland lobe ventrally overlapping intestine, variable in length, two subventral gland nuclei present. Testis long, monorchic, usually with outstretched germinal zone. Tail twisted, relatively short, conical with rounded tip. Spicules slender, curved ventrally. Two pores present on spicule tip. Phasmids located posterior to cloacal aperture.

J2

Body vermiform, moderately long and annulated. Lateral field with four incisures, weakly areolated. Labial region truncate and slightly offset from body. Labial framework weakly sclerotised, vestibule extension indistinct. Stylet moderately long, cone straight, shaft cylindrical. Knobs rounded to transversely ovoid. Median bulb ovoid, triradiate lumen with moderately sclerotised lining. Pharyngeal gland lobe relatively long, well developed, ventral overlap of intestine difficult to observe, three gland nuclei present. Hemizonid anterior, adjacent to excretory pore. Tail slightly curved ventrally, relatively long, gradually tapering until distinct short Hyaline region, rectum usually inflated. Phasmids posterior to anus, small, located in ventral incisure of lateral field. One or two cuticular tail constrictions present.

TYPE PLANT HOST: Sand twitch, *Elymus famus* (Viv.) Melderis.

OTHER PLANTS: *Ammophila arenaria*. In a glasshouse test, *Triticum aestivum* 'Minaret' was also infected.

TYPE LOCALITY: Coastal foredunes near Oostvoorne, The Netherlands.

DISTRIBUTION: *Europe*: Belgium, UK, France, The Netherlands (Karssen & van Hoensellar, 1998), Poland (Kornobis, 2001).

SYMPTOMS: Nematodes do not induce or induce only small galls on roots, which makes it difficult to recognise root infection.

CHROMOSOME NUMBER: No available data.

POLYTOMOUS KEY CODES: *Female*: A123, B3, C2, D5; *Male*: A324, B3, C32, D1, E1, F2; *J2*: A2, B32, C12, D32, E3, F21.

BIOCHEMICAL AND MOLECULAR CHARACTERISATION: *Meloidogyne duytsi* is characterised by a unique MDH pattern and one very slow EST band (Table 8). According to the enzyme phenotype coding of Esbenshade & Triantaphyllou (1985a), the patterns are named N2 (MDH) and VS1 (EST), respectively. The 18S rRNA gene sequence of this species was deposited in GenBank by Helder *et al.* (2014).

RELATIONSHIPS (DIAGNOSIS): The species belongs to Molecular group II and has a sister relationship with *M. dunensis*. *Meloidogyne duytsi* differs from *M. dunensis* by perineal pattern structure, larger female EP/ST ratio in and shorter spicules.

28. *Meloidogyne enterolobii* Yang & Eisenback, 1983
(Figs 85, 86)

COMMON NAME: Pacara earpod tree, or guava root-knot nematode.

Fig. 85. Meloidogyne enterolobii. *A: Female anterior region; B: Entire female; C: Perineal pattern; D: Male anterior region; E, G: Male labial region; F: Male lateral field; H: Male tail; I: Second-stage juvenile (J2) anterior region; J, K: J2 labial region; L: J2 lateral field; M-O: J2 Tail. (After Yang & Eisenback, 1983.)*

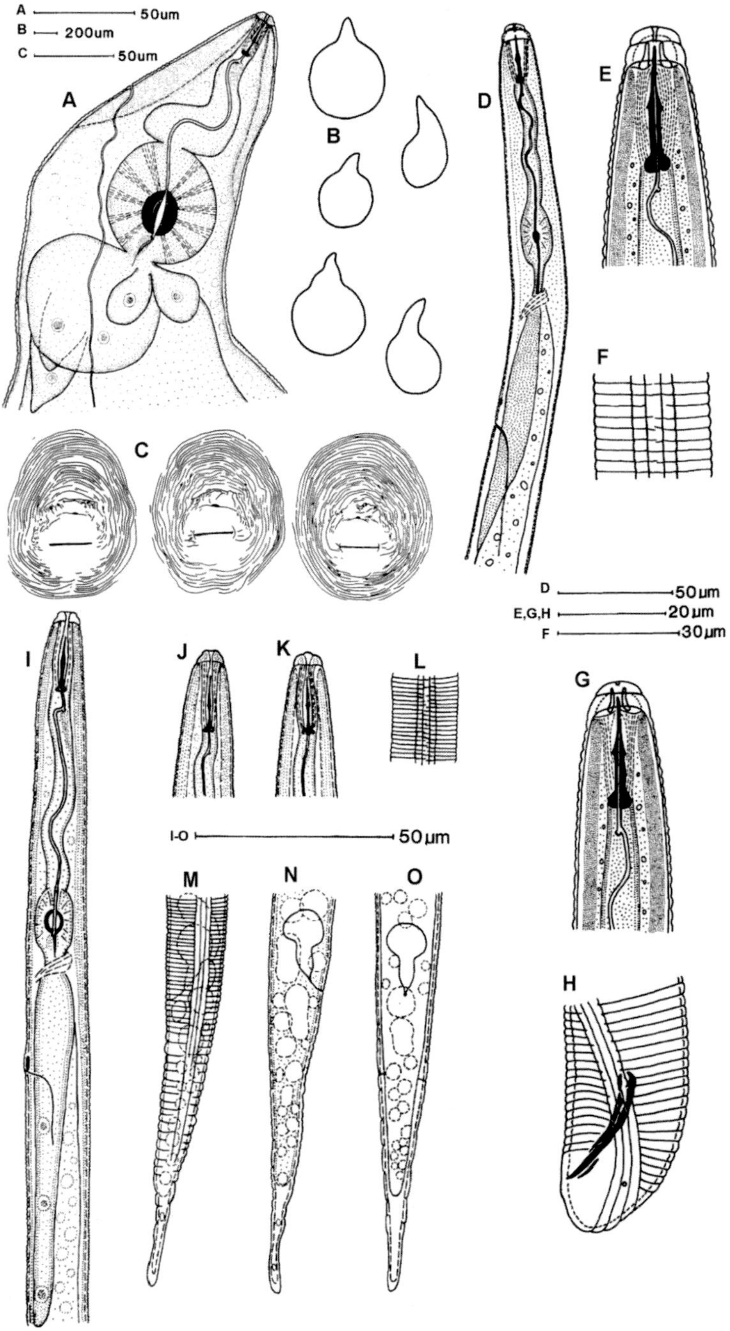
A 50um
B 200um
C 50um
A
B
C
D
E
F
D 50 µm
E,G,H 20 µm
F 30 µm
I
J
K
L
I-O 50 µm
M
N
O
G
H

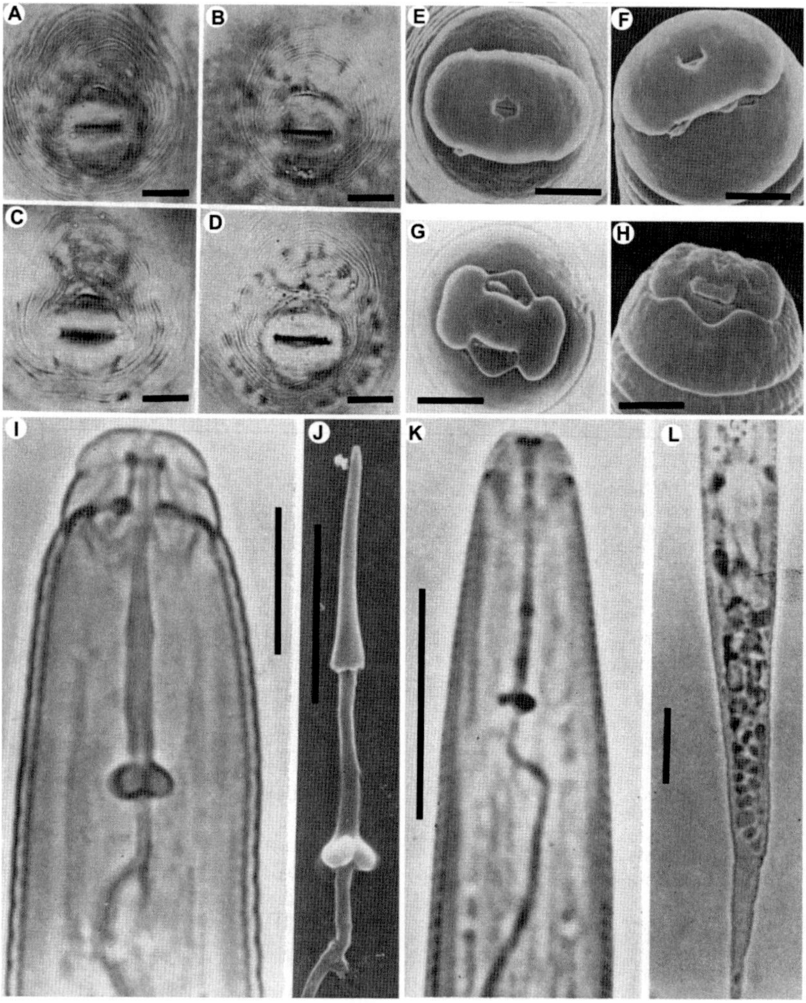

Fig. 86. Meloidogyne enterolobii. *LM and SEM. A-D: Perineal pattern; E, F, I: Male labial region; G, H: Second-stage juvenile (J2)* en face *view; J: Excised male stylet; K: J2 labial region; L: J2 tail. (Scale bars: A-D = 25 μm; E-H = 5 μm; I-L = 10 μm.) (After Yang & Eisenback, 1983.)*

Meloidogyne enterolobii was described by Yang & Eisenback (1983) from roots of pacara earpod trees (*Enterolobium contortisiliquum*) on Hainan Island in China. Later Rammah & Hirschmann (1988) described *M. mayaguensis* from roots of eggplant (*Solanum melongena*) from Puerto Rico and indicated that this species resembles *M. enterolobii*, but shows distinct morphological features and a unique malate dehydrogenase pattern (Karssen *et al.*, 2012). Meng *et al.* (2004) and Xu *et al.* (2004) conducted comprehensive studies on the characterisation of *Meloidogyne* species from China, with isozymes and mtDNA and included *M. enterolobii* populations from Hainan Island isolated from common guava. Based on observations on morphology, host range, mtDNA data and esterase phenotype of the two nominal species, these authors suggested that *M. mayaguensis* could be conspecific with *M. enterolobii*. Karssen *et al.* (2012) re-studied the holotype and paratypes of both species and confirmed *M. mayaguensis* as a junior synonym for *M. enterolobii*.

Meloidogyne enterolobii is a tropical or subtropical nematode and has a broad host range, including cultivated plants and weeds. It is now considered to be one of the most important RKN species as it displays wide virulence and can develop on crop genotypes carrying several RKN resistance genes in economically important crops, including tomato, soybean, cowpea and peppers, and has a higher pathogenicity and reproductive potential than either *M. incognita* or *M. arenaria* (Anon., 2014). Consequently, this species was added to the European and Mediterranean Plant Protection Organization A2 list of pests recommended for regulation as quarantine pests.

MEASUREMENTS

After Yang & Eisenback (1983):

- *Holotype female*: L = 667.2 μm; W = 414.6 μm; stylet = 13.4 μm; DGO = 3.7 μm.
- *Allotype male*: L = 1496.4 μm; W = 37 μm; stylet = 23.6 μm; DGO = 4.9 μm; anterior end to excretory pore = 165.4 μm; tail = 14.2 μm; spicules = 28.3 μm; testis = 880 μm; a = 40; c = 105.4; T = 53.5%.
- *Eggs* (n = 20): L = 95.5 $\pm$ 5.3 (85.7-103.6) μm; W = 38.2 $\pm$ 2.3 (33.6-41.4) μm; L/W = 2.5 $\pm$ 0.2 (2.1-3.0).

After Rammah & Hirschmann (1988):

- *Eggs* (n = 50): L = 98.3 $\pm$ 48 (88.5-111.2) μm; W = 44 $\pm$ 5.9 (37.9-50.4) μm.

See Table 21.

Table 21. *Morphometrics of females, males and second-stage juveniles (J2) of* Meloidogyne enterolobii. *All measurements are in μm and in the form: mean ± s.d. (range).*

Character	China, Yang & Eisenback (1983)	Puerto Rico, Rammah & Hirschmann (1988)	USA, Florida, Broward County, Brito *et al.* (2004a)	USA, Florida, Palm Beach County, Brito *et al.* (2004a)	Venezuela, Perichi & Crozzoli (2010)
Female (n)	20	35	14	15	20
L	735 ± 92.8 (541.3-926.3)	651.2 ± 52.7 (518.4-769.5)	-	-	693.7 ± 51.7 (620.8-800.0)
a	1.3 ± 0.2 (1.0-1.9)	1.3 ± 0.1 (1.1-1.6)	-	-	1.5 ± 0.2 (1.2-2.0)
W	606.8 ± 120.5 (375.7-809.7)	501 ± 44.2 (413.1-599.4)	-	-	490 ± 66.4 (339.2-640.0)
Neck length	218.4 ± 74.2 (114.3-466.8)	170.8 ± 73.1 (81.0-526.5)	-	-	200.9 ± 49.8 (128-352)
Stylet	15.1 ± 1.4 (13.2-18.0)	15.8 ± 0.8 (13.8-16.8)	14.3 ± 0.4 (13.9-15.0)	14.5 ± 0.3 (13.7-15.0)	14.1 ± 1.3 (12.8-16.1)
DGO	4.9 ± 0.8 (3.7-6.2)	4.8 ± 0.8 (3.5-6.7)	4.3 ± 0.3 (3.9-4.9)	4.4 ± 0.5 (3.4-5.3)	5.3 ± 0.9 (3.2-6.4)
Ant. end to excretory pore	62.9 ± 10.5 (42.3-80.6)	48.2 ± 13.6 (25.9-86.6)	-	-	64.3 ± 6.8 (56.0-75.8)
Interphasmid distance	30.7 ± 4.8 (22.2-42.0)	23.2 ± 2.5 (18.1-29.6)	-	-	-
Distance from vulva centre to anus	22.2 ± 1.8 (19.7-26.6)	18.4 ± 1.5 (12.7-21.1)	-	-	19.7 ± 2.0 (16.1-24.2)
Vulval slit	28.7 ± 1.9 (25.3-32.4)	26.1 ± 1.9 (20.9-30.4)	26.5 ± 1.6 (23.5-29.4)	26 ± 2.0 (22.0-30.0)	24.8 ± 2.8 (21.0-29.0)
EP/ST	-	-	-	-	4.5 ± 0.7 (3.1-5.6)
Male (n)	20	30	20	20	20
L	1599.8 ± 159.9 (1348-1913.3)	1503 ± 141.9 (1175-1742)	995.5 ± 97 (856.5-1140.5)	1049 ± 183.6 (782-1397)	1581.6 ± 261.8 (865.3-1802.9)
a	37.9 ± 3.1 (34.1-45.5)	39.9 ± 3.9 (31.1-49.6)	36.8 ± 2.6 (31.5-40.9)	35.2 ± 5.2 (21.9-42.9)	41.3 ± 5.9 (30.0-47.5)
c	131.6 ± 24.2 (72.0-173.4)	105.7 ± 10 (85.8-124.3)	89.1 ± 10 (73.2-102.7)	108.3 ± 19.6 (76.4-128.9)	98.5 ± 17.8 (66.1-124.1)
T	51.0 ± 0.1 (38.4-63.6)	-	-	-	-
W	42.3 ± 3.6 (37.0-48.3)	37.8 ± 3.1 (32.2-44.4)	27 ± 1.8 (24.1-31.5)	31.1 ± 2.8 (26.4-34.3)	38.0 ± 4.8 (28.8-44.2)
Stylet	23.4 ± 1.0 (21.2-25.5)	22.9 ± 1.6 (20.7-24.6)	19.7 ± 0.8 (17.5-20.8)	19.4 ± 1.0 (17.5-21.0)	23.1 ± 1.9 (19.6-26.2)
DGO	4.7 ± 0.4 (3.7-5.3)	4.1 ± 0.4 (3.3-5.0)	4.6 ± 0.4 (3.9-5.0)	3.8 ± 0.5 (3.0-4.8)	6.3 ± 0.5 (4.9-6.7)
Ant. end to median bulb	-	92.1 ± 4.3 (84.8-102)	86 ± 3.8 (78.0-93.1)	80.2 ± 4.7 (71.5-87.7)	101 ± 8.6 (85.2-116.3)
Ant. end to excretory pore	178.2 ± 11.2 (159.7-206.2)	166.4 ± 8.8 (147.2-180.8)	138.3 ± 14.8 (117.5-183.0)	135 ± 15.8 (119-174)	174.7 ± 30.2 (136.0-229.5)

Table 21. *(Continued.)*

Character	China, Yang & Eisenback (1983)	Puerto Rico, Rammah & Hirschmann (1988)	USA, Florida, Broward County, Brito *et al.* (2004a)	USA, Florida, Palm Beach County, Brito *et al.* (2004a)	Venezuela, Perichi & Crozzoli (2010)
Tail	12.5 ± 2.2 (8.6-20.2)	14.3 ± 1.1 (11.3-16.3)	11.2 ± 1.0 (9.8-13.5)	10.2 ± 1.4 (7.8-11.7)	16.1 ± 1.5 (13.1-18.0)
Spicules	30.4 ± 1.2 (27.3-32.1)	28.3 ± 1.5 (24.4-31.3)	26 ± 1.6 (23.5-29.4)	27.3 ± 1.2 (25.4-29.4)	26.8 ± 3.4 (21.3-31.4)
Gubernaculum	6.2 ± 1.0 (4.8-8.0)	7.1 ± 0.6 (6.1-9.3)	6.9 ± 0.4 (6.1-7.7)	6.9 ± 0.7 (5.8-8.3)	7.0 ± 0.8 (5.9-8.2)
J2 (n)	30	30	20	20	20
L	436.6 ± 16.6 (405-472.9)	453.6 ± 28.4 (390.4-528.0)	461.1 ± 15.9 (433-481)	449.2 ± 26.4 (421-483)	473.2 ± 19.8 (439.9-511.3)
a	28.6 ± 1.9 (24-32.5)	30.9 ± 1.9 (26.4-34.7)	30.8 ± 1.2 (28.3-32.5)	28.8 ± 1.5 (27.0-31.2)	31.1 ± 1.8 (27.9-34.0)
b	-	-	5.9 ± 0.3 (5.2-6.3)	6.0 ± 0.2 (5.6-6.4)	-
c	7.8 ± 0.7 (6.8-10.1)	8.3 ± 0.4 (7.0-9.2)	8.2 ± 0.4 (7.6-8.6)	8.5 ± 0.4 (7.8-9.1)	8.9 ± 0.8 (7.8-10.7)
W	15.3 ± 0.9 (13.9-17.8)	14.7 ± 0.5 (13.8-15.8)	15 ± 0.4 (14.5-16.1)	15.6 ± 0.6 (14.7-16.1)	15.2 ± 0.9 (13.1-16.4)
Stylet	11.7 ± 0.5 (10.8-13.0)	11.6 ± 0.3 (11.1-12.2)	10.9 ± 0.3 (10.4-11.5)	10.8 ± 0.3 (10.2-11.4)	10.4 ± 0.7 (9.2-11.6)
DGO	3.4 ± 0.3 (2.8-4.3)	3.9 ± 0.2 (3.3-4.3)	3.8 ± 0.3 (2.9-4.1)	3.4 ± 0.3 (2.9-3.9)	3.4 ± 0.4 (2.6-3.9)
Ant. end to median bulb	-	58.2 ± 1.8 (55.2-62.9)	59.2 ± 1.7 (56.5-61.7)	58.2 ± 3.0 (52.9-63.7)	-
Ant. end to excretory pore	91.7 ± 3.3 (84.0-98.6)	87.6 ± 3.3 (79.9-97.9)	92.4 ± 4.0 (88.2-98.0)	92 ± 19 (85.2-97.0)	96.6 ± 4.0 (89.8-103.5)
Tail	56.4 ± 4.5 (41.5-63.4)	54.4 ± 3.6 (49.2-62.9)	56.4 ± 2.9 (50.5-61.2)	53.1 ± 3.1 (47.0-59.7)	52.5 ± 5.0 (43.0-64.2)
Hyaline region	-	-	11.1 ± 2.6 (5.0-14.7)	11.4 ± 1.9 (8.5-14.7)	-

Description

Female

Body white, pear-shaped to globular, variable in size, with prominent neck variable in size, without posterior protuberance. Labial region not distinctly offset from neck. Labial disc and median labials fused to form labial cap. Hexaradiate labial framework distinct but weak; vestibule and vestibule extension prominent. Cephalids and hemizonid not observed. Position of excretory pore variable, often near median bulb. Cuticular body annulations becoming progressively finer posteriorly. Stylet slender; conical portion slightly curved dorsally, tapering toward tip; cylindrical shaft, posterior end often enlarged. Knobs offset from shaft, distinct from each other, and divided longitudinally by groove so that each knob appears as two. Subventral gland orifices branched, located immediately posterior to enlarged lumen lining of median bulb; subventral gland ampulla small but distinct. Pharyngeal gland comprised of one large uninucleate dorsal pharyngeal gland lobe, two small, nucleated subventral pharyngeal gland lobes usually posterior to dorsal gland lobe but usually oval-shaped, with coarse and smooth striae, dorsal arch moderately high to high, often rounded, nearly square in some specimens. Lateral lines not distinct. Perivulval region generally free of striae, striae may occur on lateral sides of vulva. Striae on ventral area of pattern generally finer and smoother. Tail tip visible, phasmidial ducts large.

Male

Body, vermiform, tapering at both ends. Tail end more rounded than anterior end, twisting through 90° in heat-killed specimens. In lateral view, labial cap high and rounded, labial region only slightly offset from body. Hexaradiate labial framework moderately developed, vestibule and extension distinct. In SEM, stoma slit-like, prestoma hexagonal, surrounded by pit-like openings of six inner labial sensilla. Labial disc and median labials fused, forming elongate labial cap and labial disc slightly elevated above median labials. Four labial sensilla marked on median labials by shallow cuticular depressions. Amphid openings slit-like. Lateral labials absent. Labial region not annulated. Body annuli distinct. Lateral field beginning near level of stylet knobs as two incisures; two additional incisures starting near level of median bulb. Lateral field areolated, encircling tail. Stylet robust, cone straight, pointed, opening located several micrometers from tip. Shaft cylindrical, knobs large, rounded, distinctly offset from shaft, in some specimens each knob divided longitudinally by a groove so that each knob appears as two, but not as pronounced as in female. Procorpus distinct. Median bulb elongate, oval-shaped with enlarged cuticular lumen lining. Pharyngo-intestinal junction indistinct, at level of nerve ring. Gland lobe variable in length, with two nuclei. Excretory pore far from anterior end, terminal duct long. Hemizonid 2-4 annuli anterior to excretory pore. One or two testes, usually outstretched. Spicules arcuate, with rounded base, single tip. Gubernaculum short and simple. Tail short and rounded. Phasmids small, pore-like, at level of cloacal aperture.

J2

Body translucent white, vermiform, rather long, tapering at both ends with very long, narrow tail. Anterior end truncate. Labial region only slightly offset from body. Vestibule and extension more developed than remainder of hexaradiate labial framework. In SEM, stoma slit-like, located in oval-shaped prestoma, surrounded by six pore-like openings of inner labial sensilla. Median labials and labial disc dumbbell-shaped in face view. Labial disc rounded, slightly raised above median labials. Lateral labials large and triangular, lower than labial disc and median labials. Posterior edge of one or both lateral labial may fuse with labial region in some specimens. Elongate amphidial apertures located between labial disc and lateral labials. Labial region not annulated. Body annuli distinct but fine. Lateral field beginning near level of procorpus as two lines. Near median bulb third line beginning and quickly splitting to make four lines, running entire length of body before gradually decreasing to two lines ending near Hyaline region, irregularly areolated. In LM stylet delicate. Cone straight, narrow, sharply pointed, shaft becomes slightly wider posteriorly, knobs large, rounded, separate from each other, offset from shaft. Procorpus faintly outlined, median bulb oval-shaped with enlarged lumen lining, isthmus not clearly defined, pharyngo-intestinal junction difficult to observe. Gland lobe variable in length, with three equally sized nuclei, overlapping intestine ventrally. Excretory pore distinct, hemizonid 1-2 annuli anterior to excretory pore, 3-5 annuli long. Cuticle slightly raised over hemizonid. Tail very thin, annuli increasing in size, becoming more irregular posteriorly. Hyaline region clearly defined. Tail tip broad, bluntly rounded. Rectum dilated. A few fat droplets may occur in hyaline terminus. Phasmids small, difficult to observe, located posterior to anus.

TYPE PLANT HOST: Pacara earpod tree, *Enterolobium contortisiliquum* (Vell.) Morong.

OTHER PLANTS: *Meloidogyne enterolobii* is polyphagous and has many host plants, including cultivated crops, ornamental plants and weeds. It attacks herbaceous as well as woody plants. The principal hosts are *Phaseolus vulgaris* (bean), *Coffea arabica* (coffee), *Gossypium hirsutum* (cotton), *Solanum melongena* (eggplant), *Psidium guajava* (guava), *S. quitoense* Lam. (naranjilla), *Carica papaya* L. (papaya), *Capsicum an-*

nuum (pepper), *S. tuberosum* (potato), *Glycine max* (soybean), *Ipomoea batatas* (sweet potato), *Nicotiana tabacum* (tobacco), *S. lycopersicum* (tomato) and *Citrullus lanatus* (watermelon) (Anon., 2014). Many weed species from more than 12 botanical families are reported as hosts of this nematode (Souza *et al.*, 2006; Rich *et al.*, 2008; Freitas *et al.*, 2017; Bellé *et al.*, 2019a; Moore *et al.*, 2020).

HOST PREFERENCES: Florida, North Carolina and West African *M. enterolobii* populations (Fargette, 1987b; Brito *et al.*, 2004; Ye *et al.*, 2013) showed a similar host preference to that of *M. incognita* race 4 (Taylor & Sasser, 1978). However, populations of *M. enterolobii* from Puerto Rico and Venezuela showed the same host range as *M. incognita* race 2, which does not reproduce on cotton (Rammah & Hirschmann, 1988; Perichi & Crozzoli, 2010).

TYPE LOCALITY: Hainan Island, Hainan Province, China.

DISTRIBUTION: *Africa*: Democratic Republic of Congo, Burkina Faso, Ivory Coast, Kenya, Malawi, Nigeria, Niger, Senegal, South Africa, Togo (Willers, 1997; Anon., 2014; Chitambo *et al.*, 2016; Kolombia *et al.*, 2016); *North America*: USA (Florida, North Carolina, South Carolina, Lousiana) (Brito *et al.*, 2004a, b; Ye *et al.*, 2013; Rutter *et al.*, 2019), Mexico (Ramirez-Suarez *et al.*, 2014, 2016; Carrillo-Fasio *et al.*, 2019); *Asia*: China (Yang & Eisenback, 1983), Vietnam (Iwahori *et al.*, 2009); Thailand (Jindapunnapat *et al.*, 2013), India (Poornima *et al.*, 2016; Ravichandra, 2019). *Central America and Caribbean*: Costa Rica (Humphreys *et al.*, 2012), Cuba (Decker & Rodriguez Fuentes, 1989), Martinique (Carneiro *et al.*, 2000), Puerto Rico (Rammah & Hirschmann, 1988), Trinidad and Tobago; *South America*: Brazil, Venezuela (Anon., 2014); *Europe*: France (Blok *et al.*, 2002) (eradicated), Switzerland (Kiewnick *et al.*, 2008), Portugal (Santos *et al.*, 2019). It was intercepted in The Netherlands, Germany and the UK several times, associated with many ornamental plants imported from Asia, South America and Africa (Anon., 2008; Kiewnick *et al.*, 2008).

PATHOGENICITY: *Meloidogyne enterolobii* is considered as very damaging due to its wide host range, high reproduction rate and induction of large galls. Compared with other RKN species, *M. enterolobii* displays virulence against several sources of RKN-resistance genes and therefore is considered particularly aggressive (Anon., 2014).

CHROMOSOME NUMBER: Reproduction is by mitotic parthenogenesis, the somatic chromosome number is 2n = 44-46 (Yang & Eisenback, 1983; Rammah & Hirschmann, 1988).

POLYTOMOUS KEY CODES: *Female*: A213, B213, C1, D4; *Male*: A213, B213, C2, D1, E1, F2; *J2*: A2, B23, C32, D1, E32, F3.

BIOCHEMICAL AND MOLECULAR CHARACTERISATION: The esterase phenotype En2, M2 (= VS1-S1) or Hi2 with two major bands is characteristic of *M. enterolobii* (Yang & Eisenback, 1983; Esbenshade & Triantaphyllou, 1985a, b; Rammah & Hirschmann, 1988; Fargette *et al.*, 1996; Carneiro *et al.*, 2001; Brito *et al.*, 2004a; Xu *et al.*, 2004) (Fig. 12). Occasionally, one of these bands resolves into two minor bands (Hi4) (Carneiro *et al.*, 2000; Santos *et al.*, 2019). One very strong band and two other weak bands of Mdh activity were observed in isolates (Fig. 12). However, these two minor bands require a large amount of homogenates from several females for their detection (Brito *et al.*, 2004).

Several molecular techniques have been developed for identification of *M. enterolobii*, including sequencing of the IGS rRNA gene (Blok *et al.*, 1997), intergeneric mitochondrial region between the *COII* and lRNA genes (Blok *et al.*, 2002; Brito *et al.*, 2004a; Xu *et al.*, 2004), ITS rRNA (Brito *et al.*, 2004a), *COI*, *COII*, partial 28S rRNA and 18S rRNA genes (Kiewnick *et al.*, 2014), conventional PCR with specific primers (Long *et al.*, 2006; Tigano *et al.*, 2010), loop-mediated isothermal amplification (LAMP) (Niu *et al.*, 2012; He *et al.*, 2013), real-time PCR with specific primers (Kiewnick *et al.*, 2015; Braun-Kiewnick *et al.*, 2016). Recently, recombinase polymerase amplification assays were developed targeting the IGS rRNA gene of *M. enterolobii* (Subbotin, 2019).

RELATIONSHIPS (DIAGNOSIS): The species belongs to Molecular group I and is clearly molecularly differentiated from all other *Meloidogyne* species. *Meloidogyne enterolobii* closely resembles other tropical RKN such as *M. incognita*, *M. arenaria* and *M. javanica*. It differs from *M. incognita* by the following morphological characteristics. In the female, the stylet knobs are rounded, slightly sloping posteriad and divided longitudinally by distinct grooves so that each knob appears as two. The DGO distance is longer in *M. enterolobii* (3.7-6.2 μm) than in *M. incognita* (2-4 μm). The perineal pattern is usually oval-shaped with the dorsal arch being moderately high to high and often rounded. The J2 of some populations can be separated from *M. incognita* by their body length. It differs from *M. arenaria* by J2 body length and from *M. javanica* by male stylet length and J2 body length. The J2 of *M. enterolobii* can be separated from those of *M. incognita* and other *Meloidogyne* species by the very thin and relatively long tail with its clearly defined hyaline region, the posterior part of the hyaline region running straight and parallel (Anon., 2011, 2014).

29. *Meloidogyne ethiopica* Whitehead, 1968
(Figs 87, 88)

COMMON NAME: Ethiopian root-knot nematode.

The species was first described from a single egg mass culture on tomato from the Mlalo region, Lushoto District, Tanga Province, Tanzania. In the original description, the perineal pattern of *M. ethiopica* was characterised as variants of *M. arenaria* and *M. incognita*, even within specimens from the same egg mass (Whitehead, 1968). The species was forgotten for many years, mainly due to difficulties in its accurate identification using the perineal region as major character. The first report of *M. ethiopica* in Brazil was made by Carneiro *et al.* (2003b) in kiwi plants. Later, this species was reported for the first time in Chile and was re-described as *M. ethiopica* by comparing it with a population from Kenya using biochemical and molecular methods (Carneiro *et al.*, 2004a; Monteiro *et al.*, 2017). *Meloidogyne ethiopica* has been also detected in several European countries: Slovenia, Italy and Greece (Širca *et al.*, 2004) and also in Turkey. However, further analysis revealed that all these reports should be considered as belonging to *M. luci* (Gerič Stare *et al.*, 2017).

Meloidogyne brasilensis was described by Charchar & Eisenback (2002) in Brazil from tomato and pea. Morphological and morphometric studies of their description showed important similarities in major characters as well as some general variability in others. Characterisation of esterase isozyme phenotypes of populations of *M. ethiopica* and *M.*

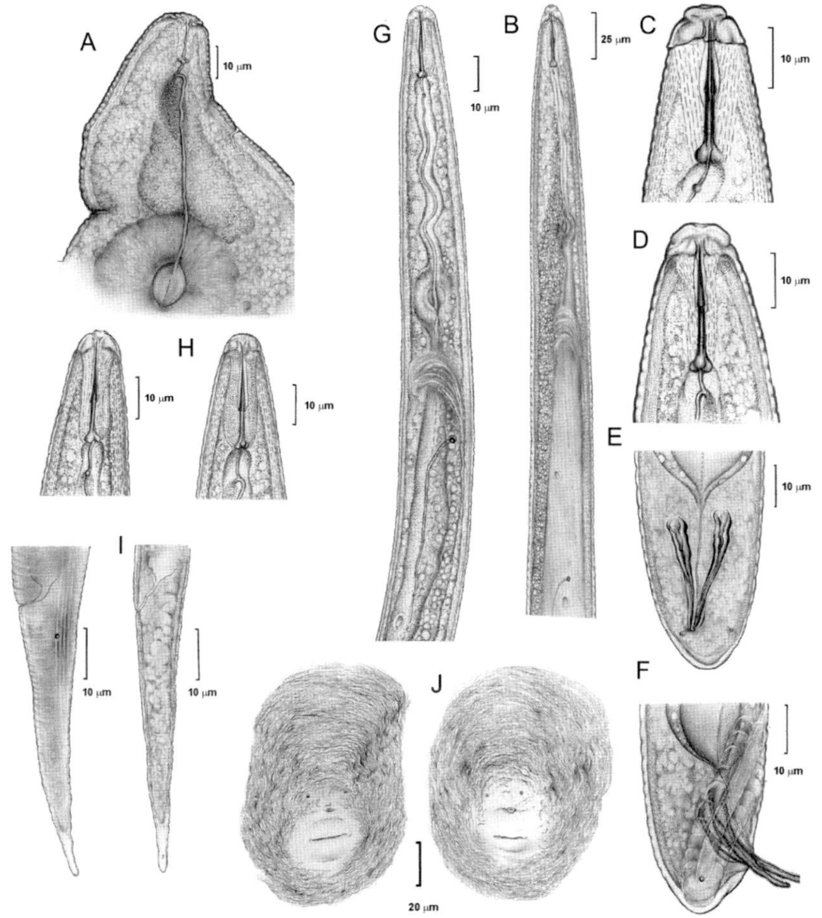

Fig. 87. Meloidogyne ethiopica. *A: Female anterior region; B: Male anterior region; C, D: Male labial region; E, F: Male tail; G: Second-stage juvenile (J2) anterior region; H: J2 labial region; I: J2 tail; J: Perineal pattern. (After Carneiro* et al.*, 2004a.)*

brasilensis resulted in a similar phenotype. Based on the results of molecular analysis of sequences of the ITS1-5.8S-ITS2 and D2-D3 of 28S rRNA gene, RAPD and AFLP data, *M. brasilensis* was therefore considered by Monteiro *et al.* (2017) as a junior synonym of *M. ethiopica*

MEASUREMENTS

See Table 22.

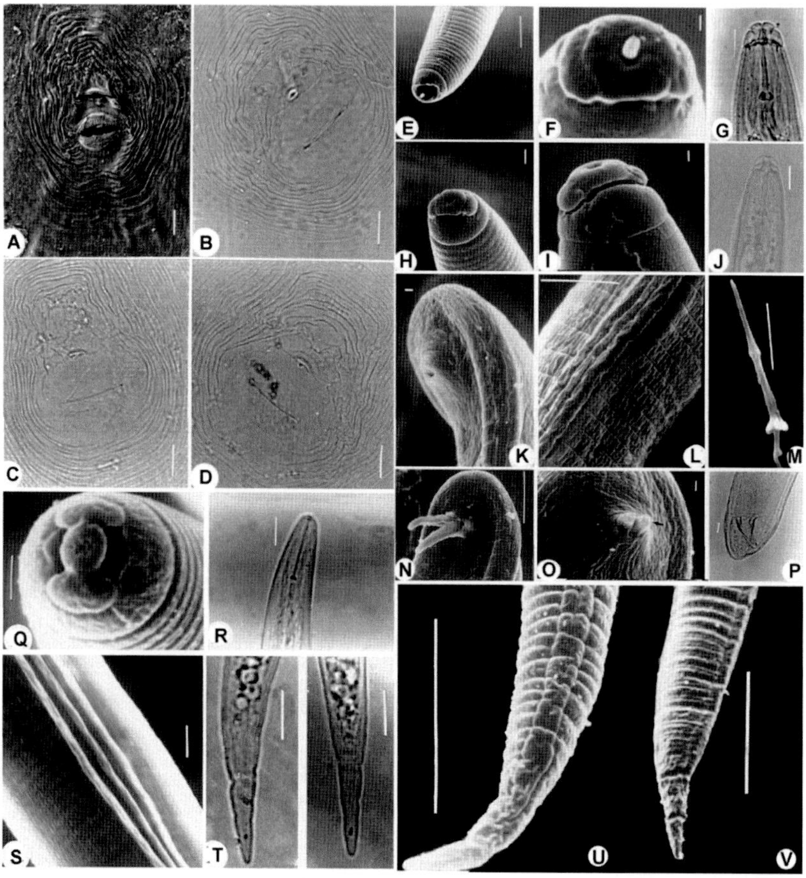

Fig. 88. Meloidogyne ethiopica. *LM and SEM. A-D: Perineal pattern; E-J: Male labial region; K, N-P: Male tail; L: Male lateral field; M: Male excised stylet; Q: Second-stage juvenile (J2)* en face *view; R: J2 labial region; S: J2 lateral field; T-V: J2 tail. (Scale bars: A-D = 10 μm; E, G, J, L-N = 10 μm; F, I, K, O, P = 1 μm; H = 2 μm; Q = 1 μm; R = 5 μm; S = 2 μm; T = 10 μm; U, V = 20 μm.) (After Carneiro* et al.*, 2004a.)*

DESCRIPTION (*AFTER WHITEHEAD, 1968; CARNEIRO* ET AL., *2004A; AND MONTEIRO* ET AL., *2017*)

Female

Body pyroid, neck long and usually well marked off from rest of body. Posterior end of body without protuberance. Labial region offset, usually marked by annulations. Labial framework weakly sclerotised. Labial

Table 22. *Morphometrics of females, males and second-stage juveniles (J2) of* Meloidogyne ethiopica. *All measurements are in μm and in the form: mean ± s.d. (range).*

Character	Tanzania Whitehead (1968)	Brazil Charchar & Eisenback (2002) (= *M. brasilensis*)	Brazil Carneiro *et al.* (2004a)
Female (n)	20	30	30
L	599 ± 72 (459-723)	763 ± 94 (601-959)	700 ± 13 (594-798)
a	-	1.5 ± 0.1 (1.3-1.6)	-
W	-	508 ± 77 (387-728)	252 ± 12 (120-282)
Stylet	13 (11-15)	14.3 ± 0.9 (12.6-16.4)	13.5 ± 0.1 (12-15)
Stylet knob width	2 (2-3)	3.7 ± 0.5 (2.9-4.6)	3.4 ± 0.1 (3.0-5.0)
DGO	4 (3-5)	4.0 ± 0.6 (2.9-5.0)	3.8 ± 0.1 (3.0-5.0)
Ant. end to excretory pore	-	50 ± 11 (35-76)	65.3 ± 2.1 (41-79)
Vulva length	-	23.9 ± 1.8 (19.3-27.1)	26.2 ± 0.7 (25.4-26.9)
Vulva to anus distance	-	19.9 ± 1.9 (15.5-23.2)	19.2 ± 0.5 (17.3-21.0)
Interphasmid distance	-	23.7 ± 2.5 (18.1-29.0)	20.4 ± 0.3 (19-23)
EP/ST*	3.4		
Male (n)	25	30	30
L	1556 ± 274 (831-2101)	1885 ± 414 (883-2489)	1171 ± 48 (890-1500)
a	47.4 ± 8.0 (34.0-61.9)	44.8 ± 4.9 (35.8-55.1)	27.7 ± 0.8 (24.8-31.0)
c	103 ± 24.6 (61-168)	109 ± 23. 2 (52.5-167.7)	-
W	-	42 ± 8.5 (23-57)	48 ± 0.8 (32-59)
Stylet	21.4 ± 1.9 (14.4-24.1)	22.8 ± 1.6 (18.9-25.2)	24.8 ± 0.6 (23-27)

Table 22. *(Continued.)*

Character	Tanzania Whitehead (1968)	Brazil Charchar & Eisenback (2002) (= *M. brasilensis*)	Brazil Carneiro *et al.* (2004a)
Stylet knob width	4.3 ± 0.4 (3.6-5.0)	3.9 ± 0.4 (2.9-4.6)	3.3 ± 0.1 (3.0-4.0)
DGO	-	2.7 ± 0.6 (1.7-4.2)	2.5 ± 0.1 (2.0-3.5)
Ant. end to excretory pore	-	186 ± 30.5 (127-250)	200 ± 3.1 (187-215)
Tail	-	17.3 ± 2.1 (13.4-22.7)	13.4 ± 0.5 (10.2-17.2)
Spicules	33.4 ± 2.2 (28.8-36.0)	31.6 ± 3.3 (25.2-36.1)	39 ± 0.6 (34-42)
Gubernaculum	9.0 ± 1.0 (7.2-10.1)	7.4 ± 0.8 (5.9-8.4)	-
J2 (n)	25	30	30
L	410 ± 12 (383-432)	434 ± 34.5 (322-473)	468 ± 3 (326-510)
a	32.2 ± 1.2 (29.1-34.7)	25.1 ± 1.7 (20.7-27.7)	24 ± 0.27 (21.3-28.2)
c	8.8 ± 0.4 (8.1-9.8)	8.2 ± 0.2 (7.5-8.6)	4.8 ± 0.1 (3.9-6.4)
W	-	17 ± 1.7 (15.6-20.8)	20 ± 0.3 (15-22)
Stylet	10 ± 0.5 (9.1-10.9)	11 ± 0.7 (9.7-12.2)	12.2 ± 0.1 (11-14)
Stylet knob width	-	1.6 ± 0.2 (1.3-2.1)	-
DGO	-	3.0 ± 0.3 (2.5-3.4)	2.6 ± 0.1 (2.0-3.0)
Ant. end to excretory pore	-	88 ± 8.0 (73-108)	93 ± 0.9 (75-106)
Tail	47 ± 2 (41-52)	53 ± 3.9 (43-58)	62 ± 0.6 (52-72)
Hyaline region	-	13.1 ± 1.6 (10-16)	13.5 ± 0.2 (12-15)

* Estimated from drawing.

disc raised, separated from lateral and median labials. Stylet robust, cone generally slightly curved dorsally and gradually increasing in diam. posteriorly. Shaft gradually widening posteriorly to near junction with stylet knobs. Knobs rounded, tapering gradually into shaft. Anterior part of stylet may be curved dorsad, stylet knobs small, tending to be rounded. Posterior cuticular pattern varying from *M. arenaria* type to *M. incognita* type, *i.e.*, striae fairly wide apart, dorsal arch low, rounded to high trapezoidal shape, some striae bifurcating close to lateral line.

Male

Body vermiform, tapering anteriorly, bluntly rounded posteriorly. Labial cap high, rounded, continuous with body contour. In LM, labial framework strongly developed, vestibule and extension distinct. Stylet robust, large, cone straight, pointed and as long as shaft, gradually increasing in diam. posteriorly, shaft cylindrical widening slightly near junction with knobs. Knobs smooth, rounded pear-shaped and posteriorly sloping. Labial region slightly offset, sometimes marked by incomplete annulation. Labial cap with distinct labial disc, amphidial opening elongated slits. In SEM, labial disc high, almost circular to hexagonal, distinctly separated from fused median labials. Median labials crescentic with distinct lateral indentation at junction with labial disc. Diam. of median labials smaller than labial disc. Labial sensilla obscure. Lateral labials absent. Stoma opening slit-like, situated posterior to large, hexagonal prestoma. Inner labial sensilla indistinct, opening into prestomatal cavity. Lateral field with four incisures, areolated. Procorpus well defined. Median bulb oval, with large valve. Pharyngo-intestinal junction obscure and at nerve ring level. Gland lobe variable in length, with only two visible nuclei. Caecum extending to level of median bulb. Excretory pore position variable, terminal excretory duct long. Hemizonid 2-4 annuli anterior to excretory pore. Most males sex reversed with two testes, sometimes one testis in normally developed males. Testis outstretched or distally reflexed. Spicules thick-walled, with strongly ridged arcuate shaft, head cylindrical, offset. Blade tip slightly curved ventrally, with two distal pores. Gubernaculum distinct. Tail short, phasmids at level of cloacal aperture.

J2

Body vermiform, tapering more posteriorly than anteriorly, tail region distinctly narrowed. Body annuli distinct, increasing in size and becoming irregular in posterior tail region. Labial region truncate without annulation, not offset from body. Labial cap low, narrower than labial region. Stoma slit-like (SEM), located in ovoid prestomatal cavity, surrounded by pit-like openings of six inner labial sensilla. Labial disc rounded, distinctly raised. Outer margins of median labials crescentic. Median labials and labial disc dumbbell shaped. Lateral labials fused at 90° with median labials, lower than median labials, margins rounded to slightly triangular, fused or not with labial region. Labial sensilla indistinct in LM, labial framework weak, hexaradiate. Vestibule and vestibule extension more distinct than remainder of framework. Stylet cone gradually increasing in diam., shaft cylindrical to tapering posteriorly. Knobs rounded and offset from shaft. Orifice of dorsal pharyngeal gland branched into three, ampulla indistinct. Procorpus faint, median bulb ovoid with prominent valve; isthmus indistinct, triradiate lining strongly sclerotised, subventral gland orifices branched, located immediately posterior to enlarged lumen of median bulb. Pharyngo-intestinal junction indistinct, at level of nerve ring. Gland lobe of variable length, with three nuclei. Excretory pore distinct. Hemizonid 2-4 annuli anterior to excretory pore. Tail slender, ending in rounded or narrowly pointed tip, annuli occasionally of irregular dimension in posterior region. Hyaline region distinct, rectal dilation large. Lateral fields with four incisures extending entire length of nematode and narrowing in tail region to three areolated incisures. Areolation not always present over entire length of lateral fields. Phasmids in ventral incisure of lateral field, always posterior to anus.

TYPE PLANT HOST: Tomato, *Solanum lycopersicum* L.

OTHER PLANTS: *Meloidogyne ethiopica* can affect numerous plant species. In Brazil, this nematode was found from *Actinidia deliciosa* (Carneiro *et al.*, 2003b), *Glycine max* (soybean) (Castro *et al.*, 2003), *Polymnia sonchifolia* (yacon) (Carneiro *et al.*, 2004a), *Nicotiana tabacum* (tobacco) (Gomes *et al.*, 2005), sugarcane (*Saccharum* spp.), *Phaseolus vulgaris* L. (common bean), melon (*Cucumis melo* L.) (Bellé *et al.*, 2017a, b) and the weed *Oxalis corniculata* (Bellé *et al.*, 2016a). In Chile, it was reported from grapevine (*Vitis vinifera*) (Carneiro *et al.*, 2007) and in Peru from asparagus (*Asparagus officinalis*) (Murga-Gutierrez *et al.*, 2012). In Africa, *M. ethiopica* was found on lettuce (*Lactuca sativa* L.), soybean, cabbage, tobacco, pepper, macadamia, pineapple, carrot, potato, sisal (*Agave sisalana* Perrine) and the weeds *Ageratum conyzoides* L., *Datura stramonium* L. and *S. nigrum* L. (Carneiro *et al.*, 2004a; Onkendi *et al.*, 2014). Bellé *et al.* (2019b) reported many weed species that may be excellent hosts for *M. ethiopica*.

TYPE LOCALITY: Mlalo region, Lushoto District, Tanga Province, Tanzania.

DISTRIBUTION: *South America*: Brazil, Chile, Peru; *Africa*: Ethiopia, Kenya, Mozambique, South Africa, Tanzania, Zimbabwe (EPPO, 2017b).

PATHOGENICITY: *Meloidogyne ethiopica* damages plants by affecting the development of their root system, which is distorted by small and large multiple galls and a lack of fine roots. Affected plants can also show above-ground symptoms such as stunting and wilting. In Brazil and Chile, *M. ethiopica* is considered as a damaging species on kiwi and grapevine as infestations lead to a reduction of plant growth, fruit size and quality. Recently, Santos *et al.* (2020) demonstrated that resistant tomato (*Mi*-1.2 gene) limits the reproduction of *M. ethiopica* and may have potential to be employed as an alternative to nematicides.

CHROMOSOME NUMBER: The reproduction of *M. ethiopica* populations from Brazil and Chile was by mitotic parthenogenesis, the somatic chromosome number being 36-38. Mandefro & Dagne (2000) reported chromosome numbers ranging from 36 to 44 in populations of *M. ethiopica* from Ethiopia (Carneiro *et al.*, 2004a).

POLYTOMOUS KEY CODES: *Female*: A32, B324, C2, D3; *Male*: A34, B2134, C21, D1, E1, F1; *J2*: A23, B3, C3, D213, E23, F2.

BIOCHEMICAL AND MOLECULAR CHARACTERISATION: Species-specific esterase phenotypes (E3) with three bands (Rm: 0.9, 1.05, 1.20) were observed in the three isolates from Brazil, Chile and Kenya, and a malate dehydrogenase phenotype N1 (Rm 1.0) as shared with several species of *Meloidogyne* (Carneiro *et al.*, 2004a). The ITS, 18S and D2-D3 of 28S rRNA, intergenenic *COII*-16S region of mtDNA gene sequences are provided by Tiago *et al.* (2006), Carneiro *et al.* (2014a). PCR with specific primers were developed for this species by Correa *et al.* (2014) (Fig. 15) and for species of the Ethiopica group (Gerič Stare *et al.*, 2019).

RELATIONSHIPS (DIAGNOSIS): The species belongs to Molecular group I (Fig. 18), the Ethiopica group (*M. ethiopica*, *M. luci*, *M. hispanica* and *M. inornata*) proposed by Gerič Stare *et al.* (2019). It is very difficult to distinguish these species based on morphological characteristics alone although they can be differentiated by their EST profiles (Gerič Stare *et al.*, 2019).

30. *Meloidogyne fallax* Karssen, 1996a
(Fig. 89)

COMMON NAME: False Columbia root-knot nematode.

This species was detected for the first time in 1992 in a field plot near Baexem, The Netherlands, and was initially considered as a new race of *M. chitwoodi* (Karssen, 1994, 1995; van Meggelen *et al.*, 1994), although later described as a new species (Karssen, 1996a). The host range, which includes several major agronomic crops such as tomato, carrot, wheat and barley, is similar to that of *M. chitwoodi*. Its most important agronomic host is potato, where it can lead to total yield losses due to quality defects and quarantine issues (Anon., 2006).

MEASUREMENTS *(AFTER KARSSEN, 1996A)*

- *Paratype females* (n = 30): L = 491.3 ± 74.9 (404.1-720.3) μm; W = 361.6 ± 57.7 (256.2-464.1) μm; stylet = 14.5 ± 0.4 (13.9-15.2) μm; stylet knob height = 2.3 ± 0.3 (2.0-2.5) μm; stylet knob width = 4.2 ± 0.3 (3.8-4.4) μm; DGO = 4.3 ± 0.5 (3.8-6.3) μm; anterior end to excretory pore = 22.5 ± 5.3 (12.6-32.9) μm; vulval slit length = 24.7 ± 1.8 (20.2-28.4) μm; vulva-anus distance = 15.9 ± 1.8 (12.6-19.0) μm; a = 1.4 ± 0.3 (0.9-2.0); EP/ST = 1.6 (estimated from drawings).
- *Paratype males* (n = 30): L = 1171 ± 193.6 (736.2-1520.1) μm; W = 30.6 ± 2.1 (27.2-43.8) μm; labial region height = 4.6 ± 0.3 (4.4-

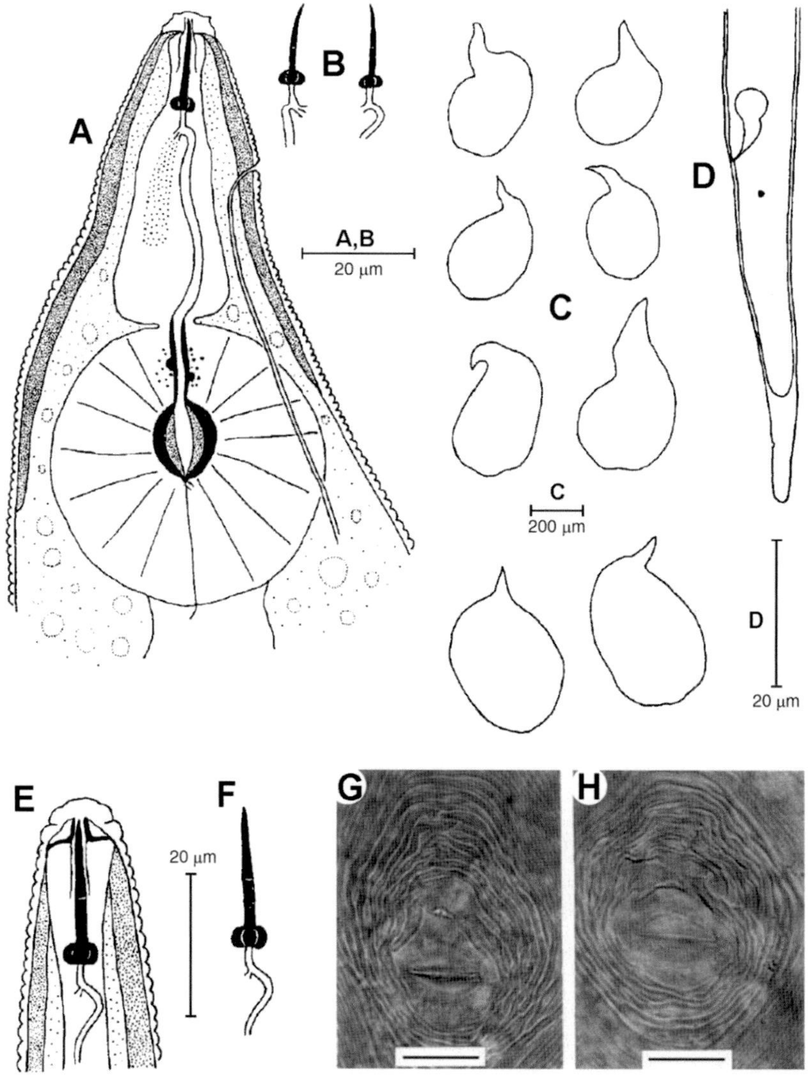

Fig. 89. Meloidogyne fallax. *A: Female anterior region; B: Female stylet; C: Entire female; D: Second-stage juvenile tail; E: Male labial region; F: Male stylet; LM. G, H: Perineal pattern. (Scale bars: G, H = 25 µm.) (After Karssen, 1996.)*

5.1) μm; labial region diam. = 10.7 ± 0.7 (9.5-12.0) μm; stylet = 19.6 ± 0.8 (18.9-20.9) μm; stylet knob height = 3.0 ± 0.3 (2.5-3.2) μm; stylet knob width = 4.9 ± 0.4 (3.8-5.1) μm; DGO = 4.4 ±

0.7 (3.2-5.7) μm; anterior end to median bulb valve = 65.4 ± 4.1 (58.8-72.7) μm; anterior end to excretory pore = 120.9 ± 11.4 (94.8-139.9) μm; tail = 9.2 ± 1.4 (7.6-12.1) μm; spicules = 26.6 ± 2.0 (22.1-29.7) μm; gubernaculum = 7.7 ± 0.5 (7.0-8.5) μm; a = 38.2 ± 6.8 (21.2-53.5); c = 127.8 ± 28.5 (82.7-201.7); T = 42.4 ± 8.4 (24.4-62.1)%.

- *Paratype J2* (n = 30): L = 403.2 ± 15.2 (381.4-435.2) μm; W = 14.3 ± 0.7 (13.3-16.4) μm; labial region height = 2.7 ± 0.4 (1.9-3.2) μm; labial region diam. = 5.5 ± 0.3 (5.1-6.3) μm; stylet = 10.8 ± 0.4 (10.1-11.4) μm; stylet knob height = 1.5 ± 0.3 (1.3-1.9) μm; stylet knob width = 2.3 ± 0.3 (1.9-2.5) μm; DGO = 3.5 ± 0.3 (3.2-3.8) μm; anterior end to median bulb = 48 ± 3.5 (44.2-54.4) μm; anterior end to excretory pore = 69.1 ± 3.4 (63.2-77.1) μm; tail = 49.3 ± 2.2 (46.1-55.6) μm; hyaline region = 13.5 ± 1.0 (12.2-15.8) μm; a = 28.1 ± 1.7 (23.8-40.4); c = 8.2 ± 0.5 (6.9-8.6).
- *Eggs* (n = 30): L = 94.4 ± 3.4 (89.7-103.6) μm; W = 38.9 ± 3.2 (34.1-44.2) μm; L/W = 2.4 ± 0.2 (2.1-2.9).

DESCRIPTION

Female

Body annulated, pearly white, globular to pear-shaped, with slight posterior protuberance and distinct neck region projecting from body axis at an angle of up to 90° to one side. Labial region offset from body, marked with one or two annuli. Labial cap distinct but variable in shape, labial disc slightly elevated. Labial framework weakly sclerotised; vestibule extension distinct. Stylet cone dorsally curved and shaft cylindrical. knobs large, rounded to transversely ovoid, slightly sloping posteriorly from shaft. Excretory pore located between anterior end and median bulb level. One or two large vesicles and several smaller ones located along lumen lining. Pharyngeal glands variable in size and shape. Perineal pattern ovoid to oval-shaped, sometimes rectangular; dorsal arch ranging from low to moderately high, with coarse striae. Tail terminus indistinct without punctations. Phasmids small and difficult to observe. Perivulval area devoid of striae. Lateral lines indistinct, appearing as a weak indentation under SEM, increasing towards tail terminus region and resulting in a relatively large area without striae. Ventral pattern region oval to angular-shaped; striae moderately coarse.

Male

Body vermiform, slightly tapering anteriorly, bluntly rounded posteriorly. Cuticle with distinct transverse striae. Lateral field with four incisures, outer bands irregularly areolated, a fifth, broken, longitudinal incisure rarely present near mid-body. Labial region slightly offset, with a single post-labial annulus usually partly subdivided by a transverse incisure. Labial disc rounded, elevated and fused with median labials. Median labials crescentic with raised edges at lateral sides. Four small labial sensilla marked by cuticular depressions on median labials. Amphidial openings appearing as elongated slits between labial disc and medium-sized lateral labials. Labial framework moderately sclerotised, vestibule extension distinct. Stylet cone straight; shaft cylindrical, knobs large and rounded, offset from shaft. Pharynx with slender procorpus, median bulb oval-shaped with pronounced valve. Ventrally overlapping pharyngeal gland lobe of variable length. Hemizonid, 2-3 μm long, 2-4 annuli anterior to excretory pore. Testis usually long, monorchic, with reflexed or outstretched germinal zone. Tail short and twisted. Spicules slender, curved ventrally; gubernaculum slightly crescentic. Phasmids located anterior to cloacal aperture.

J2

Body moderately long, vermiform, tapering at both ends but posteriorly more so than anteriorly. Body annuli small but distinct. Lateral field with four incisures, not areolated. Labial region truncate, slightly offset from body. Labial cap low and narrower than labial region. Labial framework weakly sclerotised, vestibule extension distinct. Stylet slender and moderately long, cone straight, shaft cylindrical, knobs distinct, rounded, offset from shaft. Pharynx with faintly outlined procorpus and oval-shaped median bulb with distinct valve. Pharyngeal gland lobe variable in length, overlapping intestine ventrally. Hemizonid distinct, at level of excretory pore. Moderately sized tail, gradually tapering to hyaline region, with inflated proctodeum. Phasmids difficult to observe, small, slightly posterior to anus. A rounded hypodermis marking anterior position of smooth hyaline region, tail terminus ending in a broadly rounded tip. Terminus generally marked by faint cuticular constrictions.

TYPE PLANT HOST: Tomato, *Solanum lycopersicum* L.

OTHER PLANTS: *Meloidogyne fallax* can parasitise a wide range of other dicotyledonous plant species and some monocotyledons, including economically important crops such as carrot (*Daucus carota*), black salsify (*Scorzonera hispanica*) and strawberry (*Fragaria* × *ananassa* L.). *Oenothera erythrosepala*, *Phacelia tanacetifolia*, *Hemerocallis* 'Rajah' and *Dicentra spectabilis* are considered good hosts for *M. fallax* (Goossens, 1995; Brinkman *et al.*, 1996). Hairy nightshade (*Solanum physalifolium*) and white clover (*Trifolium repens*) were infected by this nematode in New Zealand (Shah *et al.*, 2010; Rohan *et al.*, 2016).

TYPE LOCALITY: Arable land, 1.6 km north of Baexem, province of Limburg, The Netherlands.

DISTRIBUTION: *Europe*: France, Belgium, Germany, The Netherlands, Switzerland, Ireland, UK (Wesemael *et al.*, 2011; Topalović *et al.*, 2017); *Oceania*: Australia (Nobbs *et al.*, 2001), New Zealand (Marshall *et al.*, 2001); *Africa*: South Africa (Fourie *et al.*, 2001) (not considered as reliable); *North America*: USA (single location in San Francisco County, California (Nischwitz *et al.*, 2013).

CHROMOSOME NUMBER: Meiotic parthenogenetic species. The haploid chromosome number is n = 18 (van der Beek & Karssen, 1997).

POLYTOMOUS KEY CODES: *Female*: A32, B23, C3, D6; *Male*: A34, B32, C23, D1, E1, F2; *J2*: A23, B3, C3, D21, E32, F2.

BIOCHEMICAL AND MOLECULAR CHARACTERISATION: All known *M. fallax* populations share the same rare malate dehydrogenase Nlb type and a 'null' esterase type, prolonged esterase staining, for hours, revealed a very weak three-banded pattern named F3 (van der Beek & Karssen, 1997).

Several molecular diagnostic techniques have been developed for the identification of *M. fallax* including PCR-ITS-RFLP (Zijlstra *et al.*, 1997), PCR with species-specific SCAR primers (Zijlstra, 2000), size discrimination of IGS PCR products (Wishart *et al.*, 2002; Holterman *et al.*, 2012; Nischwitz *et al.*, 2013), sequencing of the ITS rRNA (Karssen *et al.*, 2004; Nischwitz *et al.*, 2013), 18S rRNA (Holterman *et al.*, 2009), D2-D3 of 28S rRNA (Shah *et al.*, 2010; Nischwitz *et al.*, 2013), *hsp90*

(Nischwitz *et al.*, 2013), *COI* (Hodgetts *et al.*, 2016) and ITS-based real time TaqMan PCR (Zijlstra & van Hoof, 2006; de Haan *et al.*, 2014).

RELATIONSHIPS (DIAGNOSIS): The species belongs to Molecular group III. *Meloidogyne fallax* is very similar to *M. chitwoodi* and differs from this species by the longer male stylet and longer J2 tail and hyaline region (Anon., 2009). *Meloidogyne fallax* can be distinguished from *M. chitwoodi* and other species by the ITS rRNA and *COI* gene sequence.

31. *Meloidogyne fanzhiensis* Chen, Peng & Zheng, 1990

(Figs 90, 91)

COMMON NAME: Fanzhi potato root-knot nematode.

This species was found in Fanzhi county, Shanxi province, China, in 1985.

MEASUREMENTS *(AFTER CHEN* ET AL., *1990)*

- *Holotype female*: L = 487 μm; W = 466 μm; neck length = 228 μm; stylet = 12 μm; stylet knob height = 2 μm; stylet knob width = 3 μm; vulval slit length = 33 μm; vulva to anus = 27 μm.
- *Paratype females* (n = 40): L = 532.4 (404.6-892.8) μm; neck length = 227.5 (115.8-264.6) μm; stylet = 11.1 (10.5-12.6) μm; DGO = 4.4 (3.1-5.2) μm; anterior end to excretory pore = 34.6 (22.0-44.5) μm; vulval slit length = 26.7 (19.9-33.5) μm; vulva to anus 23.9 (19.9-30.3) μm; interphasmid distance = 21.3 (18.8-25.1) μm; a = 1.3 (0.9-2.2); EP/ST = 3.1 (estimated from drawings).
- *Allotype male*: L = 985.6 μm; a = 35; stylet = 11.5 μm; spicules = 26.1 μm; gubernaculum = 8.4 μm.
- *Paratype males* (n = 20): L = 1312.7 (925.3-1898.6) μm; stylet = 12.2 (10.5-13.6) μm; DGO = 5.4 (4.2-7.3) μm; anterior end to excretory pore = 131.1 (116.1-175.6) μm; spicules = 25.4 (20.9-29.3) μm; gubernaculum = 10.3 (8.46-12.6) μm; a = 47.9 (37.5-57.8); b′ = 15.6 (12.6-18.6); b = 141.0 (140.2-167.7); c = 4.5 (3.7-5.5).
- *Paratype J2* (n = 40): L = 438.3 (385.4-487.8) μm; stylet = 9.4 (8.4-10.5) μm; DGO = 3.6 (3.1-4.2) μm; anterior end to excretory pore = 76.7 (66.9-83.6) μm; tail = 26.2 (18.8-32.4) μm; hyaline region =

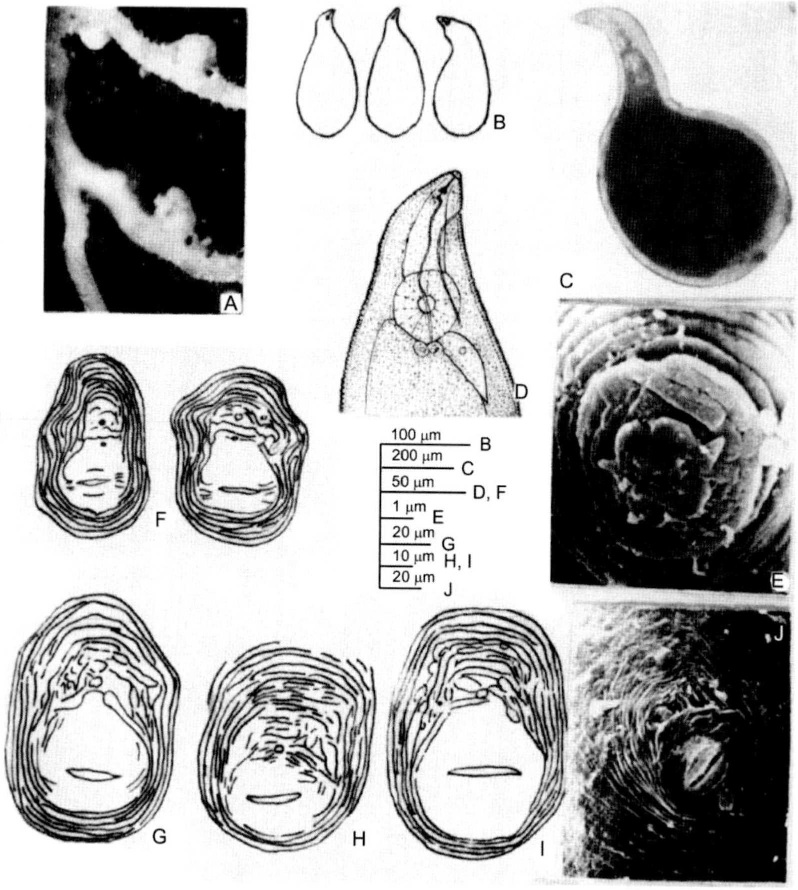

Fig. 90. Meloidogyne fanzhiensis. *A: Females on roots; B: Female bodies; C: Female; D: Anterior end of female; E:* En face *view of female; F-J: Perineal patterns. (After Chen* et al., *1990.)*

10.3 (9.6-11.5) μm; a = 27.4 (25.0-30.3); b′ = 4.5 (3.7-5.4); b = 7.2 (6.0-9.2); c = 17.1 (11.3-21.2).

DESCRIPTION

Female

Female pear-shaped, white, neck angled to side. Labial cap low and flat, smoothly connected to body. SEM showing labial disc and median labials as regular-shaped, median labials forming pairs of sublabi-

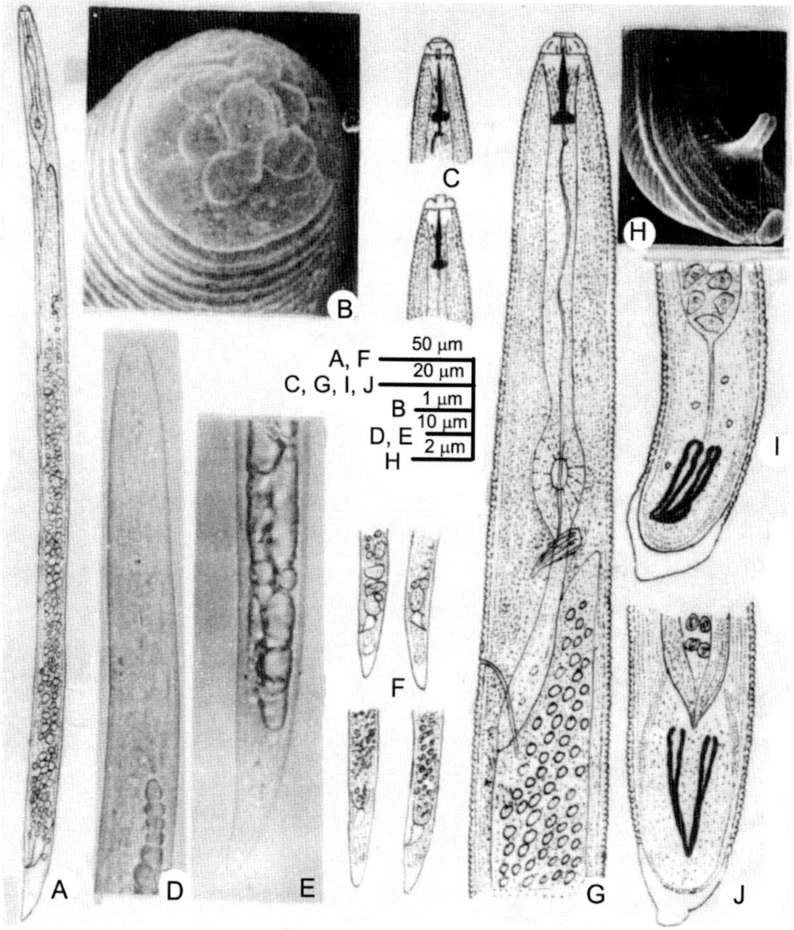

Fig. 91. Meloidogyne fanzhiensis. *A: Second-stage juvenile (J2); B:* En face *view of J2; C, D: J2 anterior region; E: J2 posterior region; F: J2 tail; G: Male anterior region; H-J: Male tail. (After Chen* et al., *1990.)*

als. Lateral labials small, amphid openings clear. Prestoma round, surrounded by six inner labial sensilla. Labial cap weak, stylet medium, basal knobs clearly offset from shaft. Excretory pore located about two stylet lengths posterior to anterior end. Pharyngeal gland overlapping intestine ventrally. Tail not protuberant. Perineal pattern oval, dorsal arch slightly high, striae smooth and round, dorsal striae fine and wavy at tail region, anal area with circular striae. In other areas, striae widely spaced and few at vulva region. Striae at lateral field curving to form whorl-shaped lines, few striae at ventral area, smooth and weak.

Male

Vermiform, labial cap high, narrow at body connection. Contour flattened or slightly rounded in lateral view. Four lines in lateral fields. Amphidial apertures clear in dorsal and ventral views. SEM showing pre-stoma to be oval and surrounded by six inner labial sensilla. Labial disc high, median labials forming pairs of sublabials. Amphid openings slit-like. Stylet short, knobs round, distinctly offset from shaft. Cone about 1.5 times as long as shaft. Anterior end to excretory pore about 1/10 body length. Hemizonid located 2 μm anterior to excretory pore. Spicules blunt.

J2

Body small, vermiform, labial region not offset, labial cap small, flat or slightly rounded in lateral view. Amphidial apertures observed in dorsal and ventral view. SEM showing labial disc to be round, median labials forming pairs of sublabials. Lateral labials large, half-moon shaped. Amphidial aperture a long slit. Stylet knobs rounded. Excretory pore located posterior to median bulb, about five times as long as labial region to stylet knobs distance, hemizonid 1-3 μm posterior to excretory pore. Tail short, bluntly rounded.

TYPE PLANT HOST: Potato, *Solanum tuberosum* L.

OTHER PLANTS: No other hosts were reported.

TYPE LOCALITY: A potato field in Fanzhi county, Shanxi province, China.

DISTRIBUTION: *Asia*: China (Shanxi province).

SYMPTOMS: This species does not produce typical galls on potato roots – the mature female bodies swell and form egg masses outside of the root with the labial region of the nematode attached to the root, which is different to other root-knot species.

CHROMOSOME NUMBER: No available data.

POLYTOMOUS KEY CODES: *Female*: A312, B43, C2, D6; *Male*: A3124, B4, C32, D1, E-, F2; *J2*: A23, B3, C4, D3, E12, F4.

BIOCHEMICAL AND MOLECULAR CHARACTERISATION: No available data.

RELATIONSHIPS (DIAGNOSIS): The labial regions of this species (female, male and J2) are distinctly different and the perineal pattern is also characteristic. The J2 of this species is similar to *M. ardenensis* and *M. propora* in having a blunt tail and the hemizonid located posterior to the excretory pore, but differs in the EP/ST ratio in both species and also by a different perineal pattern to *M. propora*.

32. *Meloidogyne floridensis* Handoo, Nyczepir, Esmenjaud, van der Beek, Castagnone-Sereno, Carta, Skantar & Higgins, 2004

(Figs 92-94)

COMMON NAME: Peach root-knot nematode.

Meloidogyne floridensis was first detected by R.H. Sharpe in 1966 in Gainesville, Florida, USA, where it parasitised the RKN-resistant 'Nemaguard' and 'Okinawa' peach (*Prunus persica* (L.) Batsch) rootstocks. This nematode has been referred to as: *i*) Nemaguard type RKN; *ii*) a new nematode; and *iii*) a biotype of RKN (Sharpe *et al.*, 1969; Sherman *et al.*, 1981; Young & Sherman, 1977). In 1982, this nematode was initially characterised as *M. incognita* race 3 (Sherman & Lyrene, 1981, 1983). The peach RKN was described by Handoo *et al.* (2004). It has been found infecting different crops and weed species in several counties of Florida (Brito *et al.*, 2015). In April-August 2018, samples of galled roots with rhizosphere soil were collected from almond orchards in Atwater, Merced County and Bakersfield, Kern County, California. Almond trees (*Prunus dulcis*) grafted on 'Hansen 536' and 'Brights Hybrid®5' (peach-almond hybrid) rootstocks showed strong symptoms of growth decline. Extracted RKN were identified by both morphological and molecular methods as *M. floridensis*. This detection marked the

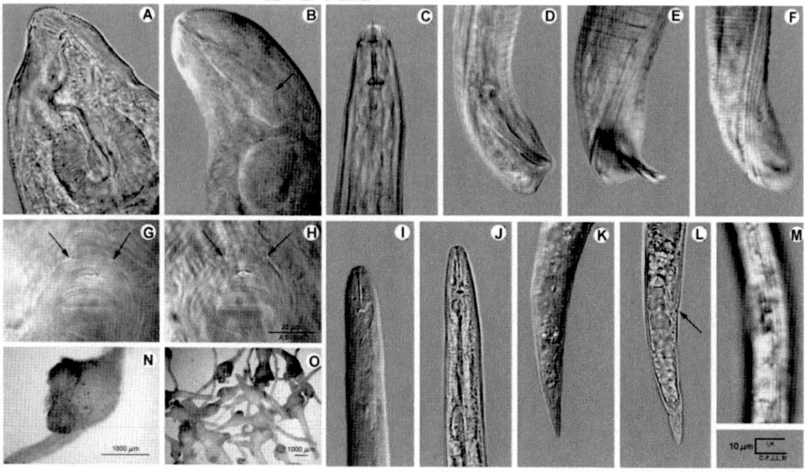

Fig. 92. Meloidogyne floridensis. *LM. A, B: Female anterior region, and excretory pore (arrow); C: Male anterior region; D-F: Male posterior regions showing spicules and lateral field, respectively; 58.2 $\pm$ G, H: Perineal patterns with large phasmids (arrows); I, J: Second-stage juvenile (J2) anterior region; K, L: J2 posterior regions; M: J2 lateral field; N, O: Gall on peach roots. (After Handoo* et al.*, 2004.)*

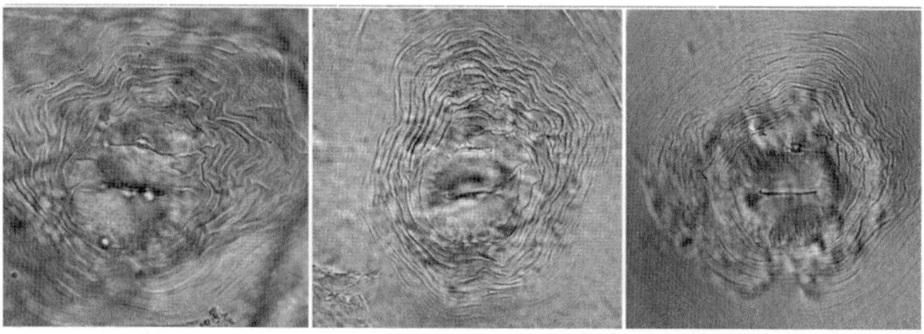

Fig. 93. Meloidogyne floridensis. *Perineal pattern (after Westphal* et al.*, 2019).*

first report of this species in California and the first outside of Florida (Westphal *et al.*, 2019).

Measurements (*after Handoo et al., 2004*)

See Table 23.

- *Holotype female*: L = 700 μm; W = 515 μm; neck length = 148 μm; neck greatest width 75 μm; stylet = 14.5 μm; stylet knob width =

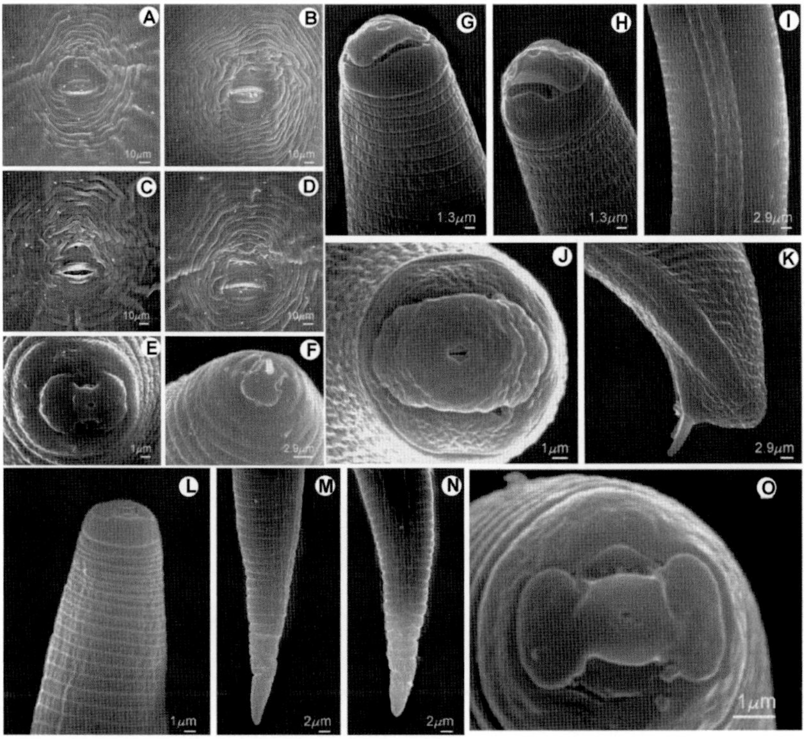

Fig. 94. Meloidogyne floridensis. *SEM. A-D: Perineal pattern; E, F: Female* en face *view; G, H: Male labial region; I: Male lateral field; J: Male* en face *view; K: Male tail; L: Second-stage juvenile (J2) labial region; M, N: J2 tail; O = J2* en face *view. (After Handoo* et al.*, 2004.)*

5 μm; stylet knob height 2.5 μm; DGO = 6 μm; anterior end to excretory pore = 40 μm; cuticle thickness at neck = 3 μm; cuticle thickness at mid-body 7 μm; vulval slit length = 25 μm; distance from vulval slit to anus = 17 μm.

- *Eggs* (n = 25): L = 86.5 ± 4.4 (80-95) μm; W = 44 ± 2.6 (40-50) μm; L/W = 1.9 ± 0.1 (1.7-2.2).

Description *(after Handoo* et al.*, 2004)*

Female

Body pearly white, variable in size, round to pear-shaped with relatively distinct variable-size neck sometimes bent at various angles

Table 23. *Morphometrics of females, males and second-stage juveniles (J2) of* Meloidogyne floridensis. *All measurements are in μm and in the form: mean ± s.d. (range).*

Character	Peach, Gainesville, Florida, USA, Handoo *et al.* (2004) Paratypes	Peach, Gainesville, Florida, USA Stanley *et al.* (2009)	Tomato, Indian River County, Florida, USA, Stanley *et al.* (2009)	Tomato, Hendry County, Florida, USA, Stanley *et al.* (2009)	Cucumber, Hendry County, Florida, USA, Stanley *et al.* (2009)	Almond, Kern County, California, USA Westphal *et al.* (2019)
Female (n)	25	20	20	20	20	-
L	697 ± 96.8 (525-890)	-	-	-	-	-
W	491 ± 87 (356-648)	-	-	-	-	-
Stylet	14.7 ± 0.7 (13.0-16.0)	14.1 ± 0.9 (12.7-16.6)	14.3 ± 0.7 (13-15.6)	13.8 ± 1.3 (10.8-15.7)	14.7 ± 0.7 (13.5-16.1)	-
Stylet knob width	5.0 ± 0.3 (4.0-5.5)	-	-	-	-	-
DGO	4.6 ± 0.9 (3.5-6.0)	3.1 ± 0.4 (2.5-3.9)	4.6 ± 0.7 (3.9-5.9)	3.8 ± 0.5 (2.9-4.7)	3.9 ± 0.2 (3.5-4.4)	-
Ant. end to excretory pore	35 ± 11.3 (17.5-50)	-	-	-	-	-
Vulva length	26 ± 3.0 (21.0-30.0)	25.6 ± 3.1 (21.6-31.3)	22.8 ± 1.6 (21.0-25.9)	22.5 ± 1.8 (19.6-25.5)	23.4 ± 1.3 (21.5-26.4)	-
Vulva to anus distance	19 ± 2 (15-25)	-	-	-	-	-
EP/ST*	2.4	-	-	-	-	-

Table 23. *(Continued.)*

Character	Peach, Gainesville, Florida, USA, Handoo *et al.* (2004) Paratypes	Peach, Gainesville, Florida, USA Stanley *et al.* (2009)	Tomato, Indian River County, Florida, USA, Stanley *et al.* (2009)	Tomato, Hendry County, Florida, USA, Stanley *et al.* (2009)	Cucumber, Hendry County, Florida, USA, Stanley *et al.* (2009)	Almond, Kern County, California, USA Westphal *et al.* (2019)
Male (n)	25	20	20	20	20	5
L	1162 ± 313.4 (564-1742)	1514 ± 326 (793-2038)	1477.8 ± 255.6 (993-1875)	1547 ± 206.5 (1072-1867)	1203 ± 297.6 (838-1847)	1219 ± 411 (675-1725)
a	40.5 ± 7.9 (31.7-56.0)	44.7 ± 7.9 (26.9-58.7)	43.2 ± 6.4 (30.7-56.3)	43.9 ± 5.8 (34.7-52.4)	36.5 ± 7.2 (27.9-54.0)	40.7 ± 10 (25.7-52.8)
b	6.4 ± 1.2 (4.5-8.6)	12.8 ± 2.2 (8.5-16.3)	12.3 ± 2.2 (8.4-16.4)	13.5 ± 2.1 (9.7-17.9)	9.07 ± 2.0 (6.6-14.1)	7.4 ± 2.4 (5.4-10.6)
c	124 ± 35.8 (81-217)	132.4 ± 29.4 (72-174)	114.2 ± 24.2 (84.1-179)	120 ± 15.9 (89.3-153)	118.8 ± 21.9 (75.2-154)	159.3 ± 18.7 (146-172)
W	28.8 ± 6.6 (17-40)	33.2 ± 3.9 (23.5-41.2)	34.2 ± 3.5 (28.4-41.0)	35.7 ± 2.7 (28.4-39.2)	32.9 ± 2.9 (27.4-39.2)	-
Stylet	20.2 ± 1.9 (17-23)	21.2 ± 1.7 (18-24)	21.4 ± 1.7 (17.6-24.5)	21.9 ± 0.7 (20.6-22.8)	22.1 ± 1.2 (20.6-24.5)	20 ± 3.1 (17.5-23.8)
Stylet knob width	5.1 ± 0.3 (5.0-6.0)	5.1 ± 0.4 (4.4-6.0)	5.8 ± 0.3 (4.9-6.3)	5.3 ± 0.3 (4.9-5.7)	5.1 ± 0.5 (4.4-5.9)	-
DGO	3.0 ± 0.5 (2.5-3.5)	3.2 ± 0.5 (2.4-4.4)	2.8 ± 0.3 (2.4-3.4)	3.6 ± 0.5 (2.5-4.4)	3.0 ± 0.5 (2.5-4.4)	-

Table 23. *(Continued.)*

Character	Peach, Gainesville, Florida, USA, Handoo *et al.* (2004) Paratypes	Peach, Gainesville, Florida, USA Stanley *et al.* (2009)	Tomato, Indian River County, Florida, USA, Stanley *et al.* (2009)	Tomato, Hendry County, Florida, USA, Stanley *et al.* (2009)	Cucumber, Hendry County, Florida, USA, Stanley *et al.* (2009)	Almond, Kern County, California, USA Westphal *et al.* (2019)
Ant. end to median bulb	88 ± 16 (63.0-112)	90.1 ± 9.2 (73.5-111)	89.8 ± 9.3 (68.6-106)	91.2 ± 7.1 (71.3-102)	91.6 ± 6.9 (81.3-112)	88.8 ± 12.3 (80.0-97.5)
Ant. end to excretory pore	134 ± 28.8 (90-180)	162.8 ± 30.8 (105-209)	151 ± 20.2 (119-183)	175.6 ± 19.9 (132-212)	155 ± 25.8 (122-226)	122.5 ± 9.0 (112.5-130)
Tail	-	11.4 ± 1.7 (8.8-15)	13.1 ± 2.0 (9.8-18.6)	13.2 ± 1.3 (10.8-15.6)	10.1 ± 1.6 (7.8-13.7)	-
Spicules	27.8 ± 3.4 (23-35)	30.7 ± 2.6 (26.4-34.3)	28.4 ± 2.8 (21.5-34.3)	30.7 ± 2.5 (25.4-35.3)	30 ± 1.9 (26.5-33.3)	31.8 ± 4.8 (27.5-38.8)
Gubernaculum	7.7 ± 1.2 (5-10)	8.6 ± 0.9 (6.9-9.8)	7.6 ± 1.1 (5.9-9.3)	7.7 ± 1.0 (5.9-9.8)	7.8 ± 1.0 (5.9-9.8)	6.3
J2 (n)	25	20	20	20	20	25
L	355 ± 17.7 (310-390)	384 ± 14.9 (348-482)	372 ± 14.8 (338-393)	387 ± 18.2 (352-417)	370 ± 15.0 (335-392)	374 ± 12.5 (357-405)
a	28 ± 1.7 (25-32)	26.0 ± 1.2 (23.0-28.0)	25.0 ± 0.8 (24.0-26.0)	26.7 ± 1.2 (24.5-29.0)	26.0 ± 1.2 (23.0-28.0)	26.1 ± 1.2 (22.8-28.0)
b	2.8 ± 0.5 (2.2-3.9)	3.8 ± 0.2 (3.5-4.1)	4.8 ± 0.2 (4.4-5.2)	4.9 ± 0.5 (4.2-5.6)	4.8 ± 0.2 (4.3-5.0)	4.5 ± 0.3 (3.9-5.6)
c	-	9.3 ± 0.6 (8.1-11.2)	8.5 ± 0.4 (8.0-9.3)	8.9 ± 0.6 (7.7-10.2)	8.5 ± 0.3 (7.8-9.1)	8.7 ± 0.7 (7.3-10.9)

Table 23. *(Continued.)*

Character	Peach, Gainesville, Florida, USA, Handoo *et al.* (2004) Paratypes	Peach, Gainesville, Florida, USA Stanley *et al.* (2009)	Tomato, Indian River County, Florida, USA, Stanley *et al.* (2009)	Tomato, Hendry County, Florida, USA, Stanley *et al.* (2009)	Cucumber, Hendry County, Florida, USA, Stanley *et al.* (2009)	Almond, Kern County, California, USA Westphal *et al.* (2019)
W	12.8 ± 0.4 (12-13.5)	14.9 ± 0.4 (14.0-16.0)	14.7 ± 0.43 (13.7-15.6)	14.5 ± 0.5 (13.7-14.8)	14.5 ± 0.4 (13.2-15.2)	-
Stylet	10.1 ± 0.3 (10-11)	10.9 ± 0.1 (10.0-10.5)	10.1 ± 0.4 (9.8-11.3)	10.2 ± 0.3 (9.8-10.8)	10.7 ± 0.3 (10.2-11.4)	14.1 ± 0.6 (13.0-15.0)
DGO	2.6 ± 0.2 (2.5-3)	2.9 ± 0.2 (2.5-3.8)	3.2 ± 0.3 (2.9-3.9)	3.3 ± 0.3 (2.9-3.9)	3.2 ± 0.4 (2.6-3.9)	3.3 ± 0.7 (2.0-4.8)
Ant. end to median bulb	51.4 ± 2.3 (46-55)	52.0 ± 2.1 (48.0-55.0)	54.5 ± 2.4 (50.0-59.0)	55.1 ± 2.9 (48.2-61.2)	55.7 ± 22 (51.4-59.7)	55.1 ± 3.4 (50.8-67.6)
Ant. end to excretory pore	71.4 ± 4.9 (65-83)	80.9 ± 3.8 (74.5-86.0)	79.5 ± 2.5 (74.0-85.0)	83.7 ± 3.8 (76.4-89.1)	82.1 ± 3.4 (75.4-92.2)	-
Tail	39.4 ± 2.3 (35-42.5)	41.1 ± 2.8 (34.0-45.0)	44.0 ± 2.3 (39.0-48.0)	43.3 ± 3.1 (38.2-48.0)	43.4 ± 2.4 (38.2-48.0)	42.8 ± 3.3 (34.0-51.0)
Hyaline region	9.7 ± 1 (8-12)	10.1 ± 1.1 (8.5-12.0)	8.6 ± 1.1 (5.9-9.8)	8.5 ± 1.4 (5.8-10.7)	10.8 ± 0.9 (8.8-11.8)	8.4 ± 1.2 (5.2-10.4)

* Estimated from drawing.

to body. Labial framework weak, hexaradiate, lateral sectors slightly enlarged, vestibule and extension prominent. Cephalids not observed. Labial region not offset, with labial disc, labial region with one annulus. SEM observations revealed labial disc fused with median labials, dumbbell-shaped, lateral labials indistinct and amphidial openings oval, located between labial disc and lateral labials. Stylet strong, with rounded, broad to posteriorly sloping knobs, cone and shaft straight. Excretory pore distinct, generally located 2-3 stylet lengths posterior to stylet base. Pharynx well developed with elongate cylindrical procorpus, large, rounded median bulb with heavily sclerotised valve. Body cuticle thick at mid-body, thinner near anterior end of neck. Perineal pattern with high to narrowly rounded or ovoid arch, with coarse broken to network-like striae in and above anal area, faint lateral lines interrupting transverse striae, and smooth wavy lines in the outer field, perivulval region without striae, vulva and anus sunken. Phasmids large and distinct with a conspicuous phasmidial canal.

Male

Body cylindrical, vermiform, length variable with both long and short forms, tapering anteriorly, bluntly rounded to clavate posteriorly. Labial region slightly offset, rounded to slightly truncate, without annulation. In SEM, labial disc not raised, continuous with median labials, median labials extending some distance into labial region, lateral labials absent, prestoma hexagonal, surrounded by six inner labial sensilla, stomatal opening slit-like, located in large hexagonal prestoma, amphidial openings appearing as long slits. Body cuticle with transverse annulation. Lateral field with four incisures, encircling tail, outer fields areolated. Stylet robust, cone straight, pointed, knobs large, rounded, sloping posteriorly. Hemizonid prominent, about two annuli long, located one annulus anterior to excretory pore. Excretory pore variable in position, usually near middle of basal pharyngeal bulb, more posteriorly in some specimens. SEM examination of spicules confirmed non-dentate tip of spicules. Spicules arcuate, tips rounded, gubernaculum distinct, short, simple. Tail short, rounded to conoid or clavate.

J2

Body small, vermiform, tapering at both extremities but more so posteriorly. Labial region truncate, slightly offset with labial disc, labial framework weak. SEM observations confirming absence of striations on labial region and on large post-labial annulus. In SEM, stoma slit-like, located in rounded prestoma, surrounded by six pore-like openings of inner labial sensilla, median labials and labial disc dumbbell-shaped in face view, labial disc slightly rounded, raised above crescentic median labials, lateral labials large and triangular, lower than labial disc and median labials, amphidial openings appearing as long slits located between labial disc and lateral labials. Stylet delicate, with small rounded knobs. Cuticular annulations fine, distinct. Lateral field prominent, with four incisures, some areolation, especially in anterior and posterior portion. Excretory pore usually near middle of basal pharyngeal bulb. Hemizonid prominent, about two annuli long, 1-2 annuli anterior to excretory pore. Phasmids indistinct. Rectum inflated. Tail short, tapering to a bluntly rounded terminus.

TYPE PLANT HOST: Peach roots, *Prunus persica* (L.) Batsch.

OTHER PLANTS: Results of the North Carolina differential host test showed that only watermelon (*Citrullus lanatus*) and tomato (*Solanum lycopersicum*) were good hosts for *M. floridensis*, whereas tobacco (*Nicotiana tabacum*), pepper (*Capsicum annuum*), and peanut (*Arachis hypogaea*) were non-hosts. Cotton (*Gossypium hirsutum*) was slightly infected and therefore rated as a poor host (Handoo *et al.*, 2004). Brito *et al.* (2015) listed the following crops as hosts: basil (*Ocimum basilicum*), common bean (*Phaseolus vulgaris*), corn (*Zea mays*), crimson clover (*Trifolium incarnatum*), cucumber (*Cucumis sativus*), dill (*Anethum graveolens* L., Sp. Pl.), eggplant (*S. melongena*), gourd (*Cucurbita pepo* L.), green bean (*P. vulgaris*), lima bean (*P. lunatus* L.), mustard (*Brassica juncea* (L.) Czern.), snapbean (*Phaseolus* sp.), squash (*Cucurbita moschata* (Duchesne *ex* Lam.) Duchesne *ex* Poir.), sugarbeet (*Beta vulgaris* L.), and vetch (*Vicia sativa*). Weed species reported under glasshouse conditions as hosts include: amaranth (*Amaranthus spinosus* L.), American pokeweed (*Phytolacca americana* L.), barnyard grass (*Echinochloa muricata* (P.Beauv.) Fernald), cyprusvine (*Ipomoea quamoclit* L.), dichondra (*Dichondra repens* J.R.Forst. & G.Forst.), English watercress (*Nasturtium officinale* W.T.Aiton), molinillo (*Leonotis nepetaefolia* (L.) R.Br.), morning glory (*Ipomoea triloba* L. and *I. violacea* L.), redroot pigweed (*Amaranthus retroflexus* L.), spurge nettle (*Cnidoscolus stimulosus* (Michx.) Engelm. & Gray), velvet leaf (*Abutilon theophrasti* Medik.), wild mustard (*Sinapis arvensis* L.), wild cucumber (*Cucumis anguria* L.) and zebrina (*Tradescantia zebrina* Heynh. *ex* Bosse) (Brito *et al.*, 2015). In California, the nematode parasitises almond trees (*Prunus dulcis* (Mill.) D.A.Webb) grafted on 'Hansen 536' and 'Brights Hybrid®5' (peach-almond hybrid) and grapevine.

TYPE LOCALITY: University of Florida, Experiment Station farm, Gainesville, Florida, USA.

DISTRIBUTION: *North America*: USA (California, Florida, Georgia, South Carolina) (Handoo *et al.*, 2004; Reighard *et al.*, 2019; Westphal *et al.*, 2019; Marquez *et al.*, 2020).

CHROMOSOME NUMBER: Meiotic parthenogenetic species. The haploid chromosome number is $n = 18$ and possibly sometimes 19 or 20 (Handoo *et al.*, 2004).

POLYTOMOUS KEY CODES: *Female*: A213, B23, C2, D6; *Male*: A324, B324, C213, D1, E1, F1; *J2*: A3, B3, C3, D34, E23, F43.

BIOCHEMICAL AND MOLECULAR CHARACTERISATION: Carneiro *et al.* (2000) have included this species in a comparative study of enzyme phenotypes comprising major RKN species (*Meloidogyne* sp.) and have described its esterase (EST), malate dehydrogenase (MDH), superoxide dismutase (SOD), and glutamate-oxaloacetate transaminase (GOT) phenotypes. The EST phenotype of *M. floridensis* is quite different from the other RKN species tested. This atypical and unique pattern is characterised by the presence of three bands, where the central band is located at the same position as the upper band for *M. javanica*. Other enzymes exhibit banding patterns that correspond to phenotypes already reported. In particular, the MDH phenotype for *M. floridensis* is identical to *M. javanica* and *M. incognita* (N1 type according to Esbenshade & Triantaphyllou, 1990) (Handoo *et al.*, 2004).

Handoo *et al.* (2004) showed that the IGS of RNA gene sequence of *M. floridensis* was different from *M. arenaria*, *M. incognita* and *M. javanica*. PCR products using C2F3 and 1108 primer set were 1.2 kb for *M. floridensis* and digestion with *Hin*fI yielded two unique fragments *ca* 770 bp and 370 bp, which distinguish this species from *M. arenaria* and others (Smith *et al.*, 2015). Sequence of *nad5* fragment of mtDNA allows the clear differentiation of *M. floridensis* from other species (Janssen *et al.*, 2016).

RELATIONSHIPS (DIAGNOSIS): The species belongs to Molecular group I, the Incognita group. This species is similar *M. incognita* and differs from *M. incognita* J2 in the shape of the labial region and tail (smooth labial region *vs* two clear annuli, shorter tail length (34-51 μm) with a bluntly rounded terminus *vs* longer tail 52 (42-62 μm) tapering steadily to subacute terminus); in the female EP/ST ratio 2.4 (1.6-3.7) *vs* 1.4; the nature of the female perineal pattern (high narrowly rounded or ovoid arch with coarse to broken network-like striae in and above anal area, sunken vulva and anus and large and distinct phasmids with conspicuous phasmidial canal *vs* distinct high dorsal arch with smooth to wavy striae, no sunken vulva and anus, and small, indistinct phasmids and phasmidial canals difficult to observe); and in having shorter males with their stylet and spicules relatively shorter. In the host range test, *M. floridensis* reproduced abundantly on 'Nemaguard' and 'Guardian' peach *vs* 'Nemaguard' rootstock being resistant to *M. incognita* (Handoo *et al.*, 2004).

33. *Meloidogyne fujianensis* Pan, 1985
(Fig. 95)

COMMON NAME: Fujian citrus root-knot nematode.

This species was described from mandarin orange in Fujian Province, China, and was found infesting 30% of surveyed citrus orchards in Nanjing district (Pan, 1985).

MEASUREMENTS *(AFTER PAN, 1985; PAN ET AL., 1988; PAN & LIN, 1998)*

- *Holotype female*: Vulval slit length = 22.4 μm, anus to vulva distance = 16.8 μm, interphasmid distance = 19.6 μm.
- *Paratype females* (n = 12): L = 697 (557-875) μm, W = 539 (410-684) μm; neck length = 177 (128-232) μm, neck width = 12.3 (8.0-20.7) μm; stylet = 12.4 (11.4-12.6) μm; DGO = 4.4 (3.6-5.4) μm; median bulb length = 34.6 (34.2-35.0) μm; median bulb width = 29.2 (28.0-30.4 μm). Vulval plates (n = 3): vulval slit length = 23.8 (22.4-25.2) μm; anus to vulva distance = 17.5 (16.1-19.6) μm; interphasmid

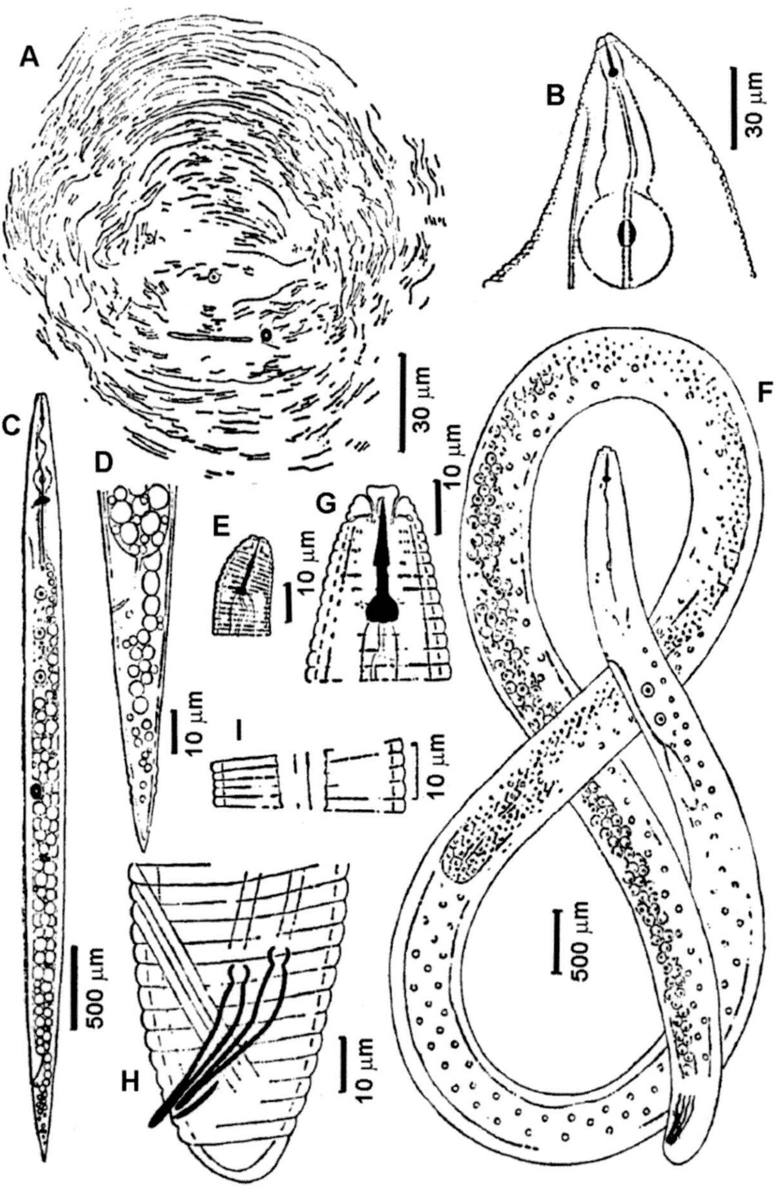

Fig. 95. Meloidogyne fujianensis. *A: Perineal pattern; B: Female anterior region; C: Entire second-stage juvenile (J2); D: J2 tail; E: J2 labial region; F: Entire male; G: Male labial region; H: Male tail; I: Male lateral field. (After Pan, 1985.)*

distance = 20.1 (15.4-25.2) μm; EP/ST = 1.6 (estimated from drawings).

- *Paratype males* (n = 14): L = 1727 (1361-2263) μm; W = 36-57 μm; a = 40 (32-46); b′ = 18 (13-22); stylet = 23 (22-25) μm; stylet knob height = 4 μm; stylet knob width = 5-7 μm; DGO = 6 (4-8) μm; pharyngeal gland length = 133-160 μm; anterior end to excretory pore = 144-171 μm; spicules = 35 (31-39) μm; gubernaculum = 8 (7-9) μm.
- *Paratype J2* (n = 10): L = 415 (348-464) μm; W = 18 (15-22) μm; a = 24 (20-30); b′ = 7.4 (6.0-9.0; c = 8.6 (7.8-10.1); c′ = 3.7 (3.0-4.2); stylet = 13 (13-14) μm; stylet knob height = 1 μm; stylet knob width = 2 μm; DGO = 4 μm; anterior end to median bulb = 70 μm; anterior end to hemizonid = 74 (63-86) μm; body diam. at anus = 13-14 μm; tail = 48 (42-56) μm.
- *Eggs* (n = 20): L = 94 (84-110) μm; W = 43 (38-46) μm, L/W = 2.2.

Description

Female

Body pear-shaped. Labial disc separated from median labials. Labial disc X-shaped in *en face* view. Neck distinct, coarse annulations in neck and tail zone, prominence absent. Stylet slender, stylet knobs small and sloping. Median bulb valve large. Excretory pore situated at 2-4 stylet lengths from anterior end, but anterior to median bulb. Perineal pattern oval, arch moderately high, lateral lines absent. Phasmids small, usually invisible, a small swelling with a central pit outside right-hand margin of vulva, dorsal striae smooth, lateral striae wavy, subventral striae flattened and sparser than that of dorsolateral.

Male

Body cylindrical. Median bulb weak, procorpus narrow, isthmus distinct. Nerve ring just posterior to median bulb. Gland lobe enveloped subventrally. Hemizonid three annuli anterior to excretory pore. Lateral field comprising about 20-25% of body diam., with four incisures. Tail rounded, annuli absent. One testis, very long, may be reflexed. Spicule curved ventrally.

J2

Body slender, annuli 1 μm wide. Median labials and labial disc dumbbell-shaped in face view. Stylet slender. Gland lobe ventrally overlapping. Tail conical with sharply pointed terminus, no annuli.

TYPE PLANT HOST: Mandarin orange, *Citrus reticulata* Blanco.

OTHER PLANT: *Imperata cylindrica* (L.) Breauv. (Pan *et al.*, 1994, 1999).

TYPE LOCALITY: Nanjing County, Fujian province, China.

DISTRIBUTION: *Asia*: China.

SYMPTOMS: Galls of nematode (0.35 × 0.2 mm) were found on the end on infected citrus roots. More secondary roots were produced around galls while infected roots were more coarse. The symptoms above ground were similar to those caused by water or nutrient deficiency. Highly infected citrus failed to set fruit or produced small fruit of unacceptable quality.

CHROMOSOME NUMBER: No available data.

POLYTOMOUS KEY CODES: *Female*: A213, B34, C3, D3; *Male*: A213, B21, C12, D1, E2, F1; *J2*: A23, B2, C3, D-, E2, F3.

BIOCHEMICAL AND MOLECULAR CHARACTERISATION: No available data.

RELATIONSHIPS (DIAGNOSIS): Females of *M. fujianensis* are characterised by a small swelling with a central pit outside the right edge of the vulva. It is similar to *M. aquatilis* and *M. ottersoni* based on females and juveniles in which the majority of the ranges are similar, but with the exception of a longer J2 stylet in *M. fujianensis* than in *M. aquatilis* (13-14 *vs* 10-12 μm) and smaller female in *M. ottersoni* (390-520 *vs* 557-875 μm).

34. *Meloidogyne graminicola* Golden & Birchfield, 1965
(Figs 96, 97)

COMMON NAME: Rice root-knot nematode.

The rice RKN is recognised as one of the important pests of rice. Golden & Birchfield (1965) described this species from roots of

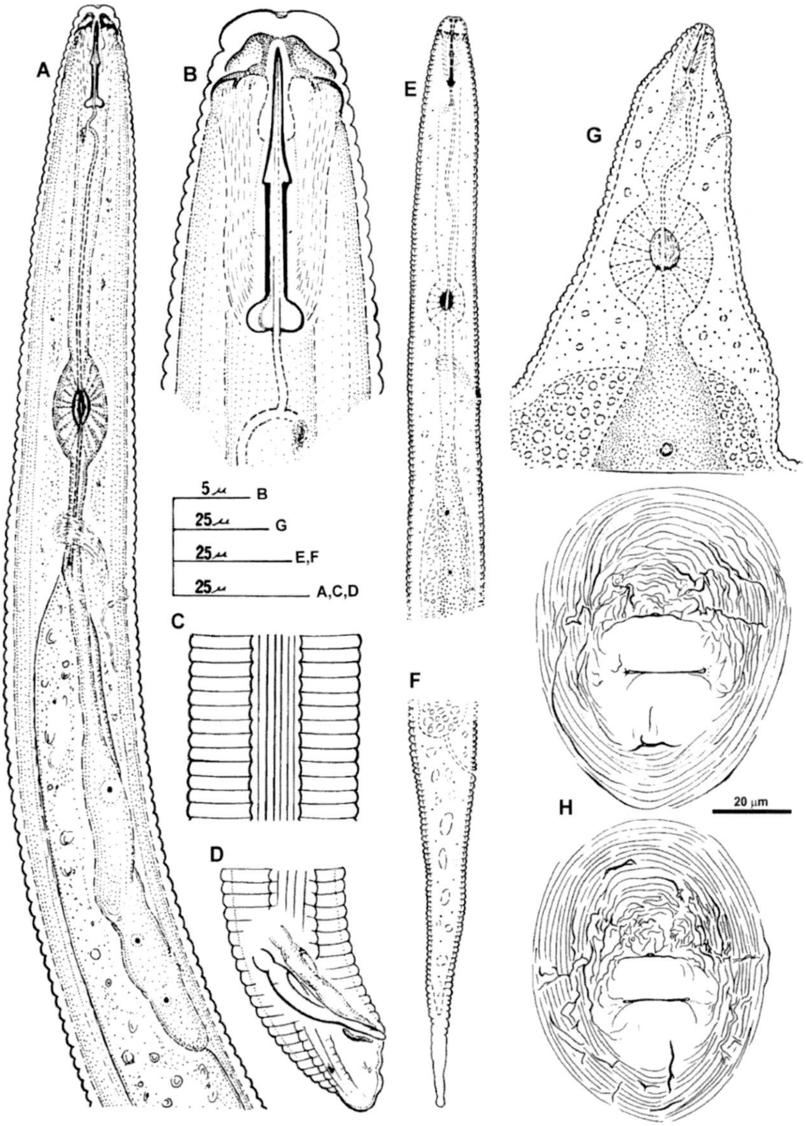

Fig. 96. Meloidogyne graminicola. *A: Male anterior region; B: Male labial region; C: Male lateral field; D: Male tail; E: Second-stage juvenile (J2) anterior region; F: J2 tail; G: Female anterior region; H: Perineal pattern. (After Golden & Birchfield, 1965.)*

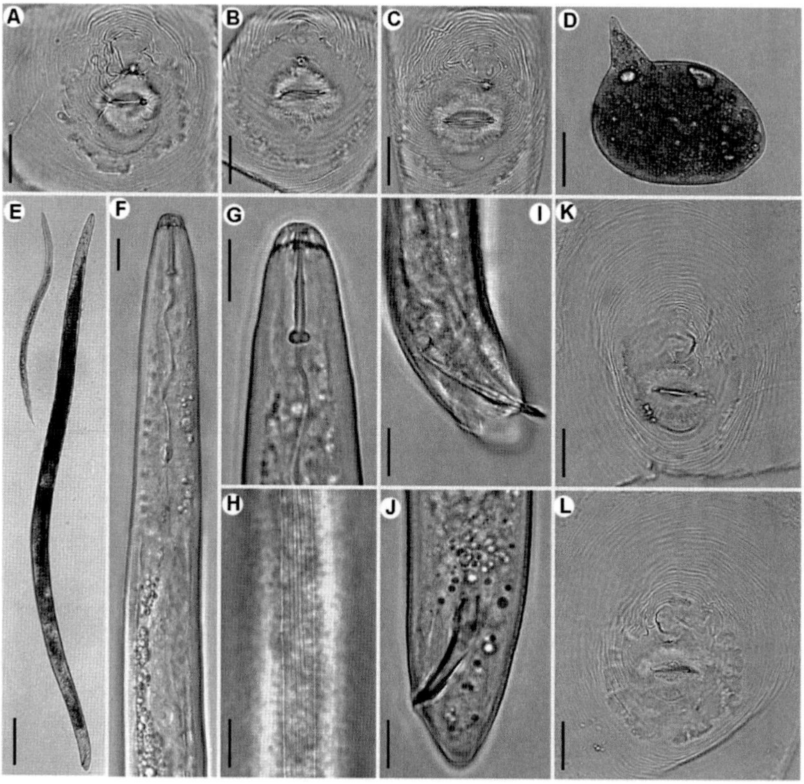

Fig. 97. Meloidogyne graminicola. *LM. A-C, K, L: Female perineal pattern; D: Entire female; E: Entire male and second-stage juvenile; F: Male anterior region; G: Male labial region; H: Male lateral field; I, J: Male tail. (Scale bars: A-C, K, L = 20 μm; D, E = 100 μm; F-J = 10 μm.) (After Fanelli* et al.*, 2017.)*

barnyard grass in Louisiana, USA. This nematode causes production losses on a large scale (ranging from 11 to 80%) in irrigated rice systems in Asia and the Americas (Plowright & Bridge, 1990; Soriano *et al.*, 2000). Descriptions of different stages of this nematodes were provided by several authors (Golden & Birchfield, 1965; Liao & Feng, 1995; Zhao *et al.*, 2001; Harpreet & Rajni, 2012; Salalia *et al.*, 2017; Song *et al.*, 2017b; Tian *et al.*, 2018; Luo *et al.*, 2020; Soares *et al.*, 2020).

Comparison of morphological and morphometric descriptions of *M. graminicola* and *M. hainanensis* syn. n. described by Liao & Feng (1995) from rice in Hainan province, China, revealed similarities in the major characters between species. Since the esterase phenotype of *M. hainanensis* syn. n. is also coincident with that of *M. graminicola*, we propose here the synonymisation of *M. hainanensis* syn. n. with *M. graminicola*. *Meloidogyne lini* syn. n., described by Yang, Hu & Zhu (Yang *et al.*, 1988a), is also considered as a junior synonym of *M. graminicola*, due to high similarities in morphometric and morphological characters.

MEASUREMENTS

See Table 24.

DESCRIPTION

Female

Pearly white, globular to pear-shaped with small neck, cuticle distinctly annulated but often marked with irregular punctations. Labial region smooth, anteriorly flattened, not distinctly offset from neck, with inconspicuous framework. Stylet slender and delicate, knobs rounded with posteriorly sloping anterior margins. Excretory pore conspicuous, anterior to median pharyngeal bulb, more than one stylet length posterior to stylet knobs and 7-16 annuli posterior to labial region. Procorpus elongate cylindrical, median pharyngeal bulb large, situated in the posterior part of neck, highly muscular, rounded to hemispheroid, with strongly cuticularised valve in middle isthmus short and narrow, three pharyngeal glands, each with a prominent nucleus, extending ventrally and ventrolaterally over intestine. Nerve ring obscure. Ovaries two, well developed, convoluted, filling body cavity and overlying intestine, uterus with several eggs. Six large radially arranged, uninucleate rectal glands with prominent nuclei, surrounding rectum. Perineal pattern dorsoventrally oval, sometimes almost circular, dorsal arch low with smooth striae, tail tip marked with prominent, coarse, fairly well separated and disorganised striae, forming an irregular tail whorl, sometimes a few lines converging at either end of vulva. Lateral field obscure or absent. A few well-marked, irregular, short, zigzag striae, distinct from rest and interrupting general pattern, distinguishing it from other species. Phasmids minute, rather close together; distance between phasmids about two-thirds length of vulval slit. Distance from anus to vulva about 2.5-3.0 times distance between anus and phasmid level.

Table 24. *Morphometrics of females, males and second-stage juveniles (J2) of* Meloidogyne graminicola. *All measurements are in μm and in the form: mean ± s.d. (range).*

Character	Holotype, Allotype Louisiana, USA, Golden & Birchfield (1965)	Paratypes, Louisiana, USA, Golden & Birchfield (1965)	Anand, India, Salalia *et al.* (2017)	Vercelli, Italy, Fanelli *et al.* (2017)	Hunan, China, Song *et al.* (2017b)	Hainan, China, Liao & Feng (1995) (= *M. hainanensis* syn. n.)	Hainan, China, Zhao *et al.* (2001), *Allium fistulosum*	Jiangsu, China, Feng *et al.* (2017)	Zhejiang, China, Tian *et al.* (2018)	Brazil, Est: G1, Soares *et al.* (2020)	Brazil, Est: G2, Soares *et al.* (2020)	Brazil, Est: G3, Soares *et al.* (2020)
Female (n)	1	20	12	-	20	45	-	20	7	20	20	20
L	634	573 (445-765)	476.9 (398.8-568.3)	-	619.1 ± 79.9 (479-743)	684 (499-857)	-	585.2	598.9 (499.1-818.7)	534 (340-740)	556 (450-660)	518 (400-620)
W	215	419 (275-520)	299.1 (209.4-378.9)	-	463 ± 90.9 (243-526)	409 (281-624)	-	438.7	354.5 (277-455.5)	379 (180-510)	355 (280-440)	326 (170-420)
Stylet	11.5	11.0 (10.5-11.5)	11.8 (11.0-13.0)	-	12.8 ± 1.2 (10.5-14.8)	12.1 (9.9-14.3)	12.1 (10.8-14.0)	11.3	10.2 (8.1-12.6)	11 (11-13)	12.2 (11-13)	13.0 (12-14)
DGO	-	3.2 (3.0-4.0)	4.6 (4.0-5.0)	-	4.1 ± 0.4 (3.5-5.1)	5.2 (2.9-6.5)	4.3 (3.7-4.7)	3.9	3.7 (2.9-4.9)	3.8 (3-5)	3.6 (3-4)	4.5 (4-5)
Median bulb length	-	10-12	-	-	-	39.0 (33.8-46.8)	-	-	-	-	-	-
Median bulb width	-	20-23	-	-	-	34.4 (28.6-39.0)		-	-	-	-	-
EP/ST	-	2.9*	1.88 (1.0-2.7)	-	-	-	-	-	-	-	-	-
Vulval slit length	-	-	25.9 (22.0-31.0)	-	-	-	22.0- (18.8-27.5)	23.0	20.7 (17.3-25.5)	-	24.3 (18-27)	27.0 (22-32)
Vulval slit to anus distance	-	-	17.8 (15.0-22.0)	-	-	-	-	-	-	-	18.2 (16-22)	15.5 (11-18)
Interphasmid distance	-	-	14.7 (11.0-20.0)	-	-	-	-	-	-	-	17.5 (14-24)	17.0 (13.5-24.0)
Male (n)	1	20	10	10	20	30	5	15	17	20	20	20

Table 24. *(Continued.)*

Character	Holotype, Allotype Louisiana, USA, Golden & Birchfield (1965)	Paratypes, Louisiana, USA, Golden & Birchfield (1965)	Anand, India, Salalia *et al.* (2017)	Vercelli, Italy, Fanelli *et al.* (2017)	Hunan, China, Song *et al.* (2017b)	Hainan, China, Liao & Feng (1995) (= *M. hainanensis* syn. n.)	Hainan, China, Zhao *et al.* (2001), *Allium fistulosum*	Jiangsu, China, Feng *et al.* (2017)	Zhejiang, China, Tian *et al.* (2018)	Brazil, Est: G1, Soares *et al.* (2020)	Brazil, Est: G2, Soares *et al.* (2020)	Brazil, Est: G3, Soares *et al.* (2020)
L	1156	1222 (1020-1428)	1261.2 (1116.6-1445.7)	1264 ± 105.3 (1052-1420)	1475.8 ± 169.7 (1246-1832)	2023 (1612-2472)	1295 (1000-1515)	1302.9	1270 (1043.4-1553.4)	1349 (980-1610)	1122 (990-1500)	1303 (1100-1540)
a	-	41.3 (41.1-42.5)	41.7 (36.8-51.6)	42.1 ± 3.2 (37.6-47.3)	-	34.9 (27.8-41.3)	-	39.6	41.5 (38.5-53.3)	36.6 (24.5-48)	29.7 (22-36)	41.3 (31.5-50.0)
c	-	117.4 (72.8-215.0)	126.5 (108.8-150.7)	129 ± 31.2 (94.1-180)	-	151.5 (100.0-212.5)	-	125.5	137.7 (129.1-146.3)	141 (89-187)	119 (76-136)	132 (87-177)
W	-	29.8 (24.0-35.0)	30.4 (28.0-33.0)	30.1 ± 1.7 (27.3-31.7)	39.0 ± 5.0 (30.6-48.4)	58.4 (44.2-72.8)	-	32.6	25.9 (21.3-31.5)	37.3 (30-40)	38.1 (28-50)	31.8 (28-40)
Stylet	16.8	16.8 (16.0-17.5)	18.1 (17.0-19.0)	15.9 ± 0.5 (15.5-16.9)	19.3 ± 0.9 (17.9-20.6)	17.2 (15.2-18.9)	16.4 (16.0-17.2)	17.3	17.2 (15.2-18.9)	17.1 (15-20)	16.6 (15-20)	17.5 (16-19)
Stylet knob width	-	3.5-4.0	-	-	-	3.8 (3.1-4.7)	-	-	-	4.3 (3-5)	3.9 (3.0-4.5)	4.5 (3-5)
DGO	-	3.3 (3.0-4.0)	3.0 (3.0-3.0)	4.3 ± 0.8 (3.5-5.8)	3.8 ± 0.5 (2.9-4.6)	7.1 (5.2-7.8)	3.3 (3.0-3.8)	3.3	3.6 (3.2-4.0)	4.0 (3-5)	4.4 (4-5)	4.0 (3.5-5)
Ant. end to excretory pore	-	-	111.2 (101.0-130.0)	122.5 ± 13.6 (107-153)	-	142.9 (124.8-163.8)	-	-	-	112 (58-150)	124 (90-142)	98 (85-120)
Spicules	-	28.1 (27.5-29.0)	28.2 (26.5-30.0)	25.9 ± 1.1 (23.5-26.9)	30.7 ± 2.4 (27.2-36.1)	33.5 (28.6-36.4)	-	27.2	21.1 (20.1-21.9)	30.5 (20-36)	24.6 (19-29)	28.9 (25-32)
Gubernaculum	-	6.1 (5.5-7.0)	6.7 (6.0-7.0)	6.5 ± 0.6 (5.5-7.0)	-	11.0 (10.4-11.7)		-	-	8.8 (7-11)	7.0 (5-9)	6.0 (5-9)
Tail	11.2	11.1 (6.0-15.0)	10 (9.0-11.0)	10.2 ± 2.2 (7.2-13.5)	11.3 ± 2.3 (8.4-16.6)	13.5 (10.4-18.2)	-	10.8	9.2 (8.1-10.3)	9.7 (8-12)	11.1 (9-13)	10.0 (8-14)
J2 (n)	-	20	15	10	20	30	25	20	20	20	20	20

Table 24. *(Continued.)*

Character	Holotype, Allotype Louisiana, USA, Golden & Birchfield (1965)	Paratypes, Louisiana, USA, Golden & Birchfield (1965)	Anand, India, Salalia *et al.* (2017)	Vercelli, Italy, Fanelli *et al.* (2017)	Hunan, China, Song *et al.* (2017b)	Hainan, China, Liao & Feng (1995) (= *M. hainanensis* syn. n.)	Hainan, China, Zhao *et al.* (2001), *Allium fistulosum*	Jiangsu, China, Feng *et al.* (2017)	Zhejiang, China, Tian *et al.* (2018)	Brazil, Est: G1, Soares *et al.* (2020)	Brazil, Est: G2, Soares *et al.* (2020)	Brazil, Est: G3, Soares *et al.* (2020)
L	-	441 (415-484)	484.6 (408.8-568.3)	441 ± 22.3 (416-485)	483.0 ± 22.4 (427-515)	482.0 (442.0-535.6)	456.4 (410.0-510.0)	447.4	456.7 (403-509)	450 (410-570)	502 (470-520)	430 (400-490)
a	-	24.8 (22.3-27.3)	32.4 (27.3-37.9)	28.0 ± 1.9 (25.8-30.9)	-	32.6 (28.3-37.2)		27.7	28.6 (23.0-32.9)	25.1 (20.5-30.7)	28.4 (21-38)	23.5 (19-28)
b	-	3.2 (2.9-4.0)	5.9 (5.2-6.3)	5.6 ± 0.2 (5.3-5.7)	-	-	-	-	-	-	-	-
c	-	6.2 (5.5-6.7)	6.4 (6.1-6.8)	6.3 ± 0.5 (5.8-7.1)	-	7.3 (6.4-8.2)	-	-	6.2 (5.5-6.7)	7.6 (5.9-9.3)	6.7 (5.5-7.6)	6.0 (5.1-7.1)
c′	-	4.6 (4.0-5.0)	7.1 (5.4-8.5)	6.4 ± 0.4 (5.6-7.2)		-	-	-	-	-	-	-
W	-	12.0	14.9 (14.0-16.0)	15.8 ± 0.5 (15.5-16.5)	17.5 ± 1.6 (15.5-20.0)	14.8 (13.0-15.6)	-	16.3	16.1 (12.9-19.1)	18.3 (15-20)	18.1 (13-24)	18.4 (15-23)
Stylet	-	11.4 (11.0-12.0)	11.4 (11.0-12.0)	10.6 ± 0.6 (10.0-11.8)	14.0 ± 0.5 (13.2-15.5)	12.8 (10.4-15.1)	-	13.2	12.1 (10.6-13.1)	10.9 (10-12)	11.6 (11-13)	11.3 (10-12)
DGO	-	2.8 (2.8-3.4)	3.1 (3.0-4.0)	2.9 ± 0.3 (2.6-3.6)	3.4 ± 0.4 (2.9-4.3)	5.2 (4.7-5.2)	-	4.0	2.6 (2.1-2.8)	2.9 (2-4)	2.7 (2-3)	3.0
Ant. end to excretory pore	-	-	81.9 (70.0-95.0)	74 ± 4.0 (70.8-80)	-	80.9 (67.6-93.6)	-	-	-	79 (66-91)	72 (65-80)	68 (59-83)
Tail	-	70.9 (67.0-76.0)	75.3 (64.0-92.0)	70.0 ± 5.7 (60.0-78.5)	73.7 ± 4.0 (68.4-83.1)	66.8 (57.2-75.4)	72.9 (60.0-85.0)	71.6	70.2 (61.2-79.8)	60 (50-73)	73 (66-80)	72 (66-80)
Hyaline region	-	17.9 (14.0-21.0)	21.6 (16.0-30.0)	21.0 ± 1.1 (19.5-23.0)	20.2 ± 2.7 (15.0-25.0)	15.3 (10.4-18.2)	22.1 (12.5-27.5)	19.9	19.5 (16.5-22.6)	21.1 (20-24)	19.3 (16-22)	18.6 (14-24)

* Estimated from drawing.

Male

Body cylindrical, vermiform, tapering more towards anterior than posterior extremity. Cuticle prominently annulated. Annuli *ca* 2.0-2.5 μm apart near mid-body. Labial region continuous with body or slightly offset by a constriction, nearly flat anteriorly, consisting of a prominent labial annulus followed by one or sometimes two wide post-labials. Labial framework conspicuously sclerotised. Stylet strong with rounded posteriorly-sloping knobs; anterior conical part of stylet forming about 50% of entire length. Anterior and posterior cephalids at about second and seventh annuli posterior to labial region. Excretory pore distinct, 51-64 annuli posterior to labial region (*ca* 0-7 annuli posterior to nerve ring). Hemizonid 1-2 annuli wide, 1-3 annuli anterior to excretory pore. Hemizonion a few annuli posterior to excretory pore but inconspicuous. Procorpus elongate, cylindrical, wider than isthmus. Median pharyngeal bulb hemispheroid to fusiform with strongly cuticularised valve in middle. Isthmus, a narrow tube, encircled by nerve ring near middle, three pharyngeal glands forming a compact lobe overlapping intestine ventrally and ventrolaterally. Lateral field 7.7 μm (6.5-9.5) wide or about 25% of body diam., marked with four incisures in young and eight in large and old specimens, near mid-body. Outer incisures crenate and outer bands areolated at extremities. Testis single, outstretched, sometimes reflexed anteriorly. Spicules arcuate or slightly bent ventrally near middle. Gubernaculum rod-shaped. Tail with smooth terminus. Phasmids small, post-cloacal, located near middle of tail.

J2

Body cylindrical, vermiform, tapering towards posterior extremity. Cuticle finely marked with distinct transverse striae, about 1 μm apart near mid-body. Labial region continuous with body, weakly sclerotised, marked with three faint post-labial annuli. Stylet delicate with posteriorly-sloping rounded knobs. Excretory pore at level of nerve ring or slightly posterior. Hemizonid just anterior to excretory pore. Median pharyngeal bulb rounded, almost spherical, with prominent refractive valve. Lateral field with three incisures, occupying one-quarter to one-third of body diam. near middle. Outer incisures finely crenate. Tail long, with irregularly annulated posterior hyaline region, and 4-5 times as long as anal body diam. Tail terminus rounded, often slightly clavate.

TYPE PLANT HOST: Barnyard grass, *Echinochloa colona* (L.) Link.

OTHER PLANTS: *Meloidogyne graminicola* has been reported to infect over 100 plant species, including cereals and grasses. It has been found parasitising roots of several crops and weed species including rice (*Oryza sativa*), onion (*Allium cepa* L.), oat (*Avena sativa* L.), wheat (*Triticum aestivum*), barley (*Hordeum vulgare*), tomato (*Solanum lycopersicum*), bean (*Phaseolus vulgaris*), soybean (*Glycine max*), corn (*Zea mays*), *Alopecurus carolinianus* Walt., *Allium fistulosum* L., *Brachiaria mutica* (Forssk.) Stapf, *Cyperus compressus* L., *C. imbricatus* Retz., *C. rotundus* L., *C. procerus* Rottb., *C. pulcherrimus* Willd. *ex* Kunth, *Eleusine indica* (L.) Gaertn., *Fuirena glomerata* Lam., *Fimbristylis miliacea* (L.) Vahl, *Monochoria vaginalis* (Burm.f.) C.Presl *ex* Kunth, *Musa nana* Auth., *Panicum repens* L., *Paspalum scrobiculatum* L., *Poa annua* L., *Ranunculus pusillus* Poir., *Scirpus articulatus* L., *Sphaeranthus senegalensis* Candolle, *Sphenoclea zeylanica* Gaertner (Birchfield, 1964, 1965; Buangsuwon *et al.*, 1971; Mulk, 1976; Yik & Birchfield, 1979; MacGowan, 1989; MacGowan & Langdon, 1989; Zhao *et al.*, 2001; Gergon *et al.*, 2002; Vaish *et al.*, 2012; Negretti *et al.*, 2014; Zhou *et al.*, 2015; Bellé *et al.*, 2019c, d). Some weed hosts may play a role as significant reservoirs of this rice parasite, particularly on the edge of fields or near paths, and this could be important for the management of this nematode species (Yik & Birchfield, 1979; Bellafiore *et al.*, 2015). This nematode was recently reported as a pest of soybean (*G. max*) in China (Long *et al.*, 2017). *Meloidogyne graminicola* is able to infect and complete its life cycle in the model plant *Nicotiana benthamiana* Domin, although experiments demonstrated a lower susceptibility compared to rice (Naalden *et al.*, 2018).

HOST RACES: Two pathotypes of *M. graminicola* showing variability in host plant specificity have been identified, with one pathotype that can infect and reproduce in rice 'BR11', whereas the other pathotype was not able to reproduce on this rice cultivar although the first pathotype can reproduce on wheat (Pokharel *et al.*, 2010). No Vietnamese *M. graminicola* populations were able to reproduce on tomato, green bean, soybean or corn (Bellafiore *et al.*, 2015). By contrast, these four plants have been reported as hosts for other *M. graminicola* populations (MacGowan & Langdon, 1989; Pokharel *et al.*, 2010). These data suggest that *M. graminicola* consists of more than one race showing variable ability to parasitise some crops (Mantelin *et al.*, 2017).

TYPE LOCALITY: Field at Baton Rouge, Louisiana, USA.

DISTRIBUTION: *North America*: USA (Florida, Georgia, Louisiana, Mississippi) (Birchfield, 1964, 1965; Anon., 2016b, 2018a), *Asia*: Bangladesh, China, India, Indonesia, Laos, Myanmar, Nepal, Pakistan, Philippines, South Korea, Sri Lanka, Thailand, Vietnam (Golden & Birchfield, 1968; Manser, 1968; Rao, 1970; Roy, 1973; Nugaliyadde *et al.*, 2001; Soriano & Reversat, 2003; Pokharel, 2009; Harpreet & Rajni, 2012; Bellafiore *et al.*, 2015; Jabbar *et al.*, 2015; Zhou *et al.*, 2015; Katsuta *et al.*, 2016; Long *et al.*, 2017; Salalia *et al.*, 2017; Song *et al.*, 2017b; Tian *et al.*, 2017, 2018; Wang *et al.*, 2017; Anon., 2018a; Mwamula *et al.*, 2021); *South America*: Brazil, Colombia, Ecuador, Suriname; *Africa*: South Africa and Madagascar (Kleynhans, 1991; Sperandio & Monteiro, 1991; Bastidas & Montealegre, 1994; Chapuis *et al.*, 2016); *Europe*: in lowland and upland rice fields of northern Italy (Fanelli *et al.*, 2017).

PATHOGENICITY: *Meloidogyne graminicola* is a major constraint to rice production in the world and is widely distributed in all rice-growing agroecosystems, including upland, lowland, deep water and irrigated rice (Le *et al.*, 2009; Mantelin *et al.*, 2017) and, in particular, S.E. Asia and the USA (Pankaj & Prasad, 2010). Symptoms in above-ground parts include tip drying, chlorotic leaves, and emerging leaves and panicles crinkled and with reduced tiller height (Patnaik & Padhi, 1987). Below-ground galls of *M. graminicola* on rice are elongate and usually located just posterior to the root tip, affected roots assuming a characteristic hook-shape, as well as abnormal development with small slender lateral roots producing a hairy root system (Patnaik & Padhi, 1987; Mantelin *et al.*, 2017). In India, it causes yield losses of 17-30% (Rao *et al.*, 1977; Prasad *et al.*, 1986). The impact of *M. graminicola* on growth and yield of lowland rainfed rice was assessed in north-western Bangladesh (Padgham *et al.*, 2004) and was recognised as an important rice parasite in Western and South Vietnam (Dang-Ngoc *et al.*, 1982; Bellafiore *et al.*, 2015). In the Philippines, growth and yield of Yellow Granex onion was reduced by 7-82% when plants were inoculated with 50-10 000 J2 of *M.*

graminicola (Gergon *et al.*, 2002). Numerous varieties and selections of rice have been tested, nearly all of which have proved to be susceptible. Some sources of resistance to *M. graminicola* have been identified in African rice species (*O. glaberrima* and *O. longistaminata*), as well as in a few Asian rice cultivars (Mantelin *et al.*, 2017). This nematode is also considered a potential pest of wheat in Nepal, India, Pakistan and Bangladesh.

CHROMOSOME NUMBER: *Meloidogyne graminicola* reproduces by facultative meiotic parthenogenesis and has n = 18 (Triantaphyllou, 1985b).

POLYTOMOUS KEY CODES: *Female*: A32, B4, C2, D3; *Male*: A34, B43, C2, D2, E2, F2; *J2*: A2, B32, C12, D12, E3, F3.

BIOCHEMICAL AND MOLECULAR CHARACTERISATION: *Meloidogyne graminicola* showed an esterase G1 (VS1) phenotype with one slow band with a large drawn-out area of enzymatic activity (Esbenshade & Triantaphyllou, 1985a, b; Negretti *et al.*, 2017). Two other esterase variants (Est G2 = R2, G3 = R3) were also revealed from rice-flooded areas in Brazil (Soares *et al.*, 2020). The malate dehydrogenase N1a phenotype is similar to that of *M. chitwoodi* and *M. salasi* (Esbenshade & Triantaphyllou, 1985a, b; Negretti *et al.*, 2017).

Sequences of 18S rRNA, D2-D3 of 28S rRNA, ITS and some mtDNA genes of samples, which originated from different countries and hosts, are published for this species. McClure *et al.* (2012) and Fanelli *et al.* (2017) showed the presence of two haplotypes in the mtDNA region between the *COII* and 16*S* rRNA genes for *M. graminicola* associated with two geographic origins: Asia *vs* North America and Europe.

A set of species-specific primers were proposed by Bellafiore *et al.* (2015), Htay *et al.* (2016) and Mattos *et al.* (2019) for identification of *M. graminicola* (Table 8). However, Negretti *et al.* (2017) and Soares *et al.* (2020) questioned its specificity and showed that these molecular markers are not species-specific for *M. graminicola* only and can give a false positive result with *M. oryzae* and *M. ottersoni*. Katsuta *et al.* (2016) developed real-time PCR for quantification of *M. graminicola* in soil with primers designed based on specificity in the ITS rRNA gene, but these primers were not widely tested. Fanelli *et al.* (2017) developed the ITS-PCR-RFLP assay for identification of this species. The mitochondrial genome of *M. graminicola* was also recently sequenced (Sun *et al.*, 2014).

RELATIONSHIPS (DIAGNOSIS): The species belongs to Molecular group III, the Graminicola group (*M. aegracyperi*, *M. graminicola*, *M. salasi* and *M. trifoliophila*). This species is morphologically similar to *M. trifoliophila*. Presently available biochemical and molecular data does not allow these species to be distinguished from one another. The host ranges of *M. trifoliophila* and *M. graminicola* also overlap. These species can be differentiated from each other mainly by female characters. Females of *M. graminicola* differs from *M. trifoliophila* by perineal pattern: dorsoventrally elongated with prominent ridges and angled striae in the dorsal arch *vs* round, striae smooth, dorsal arch without prominent interruptions or ridges, and by the female excretory pore located at more than one stylet length *vs* one stylet length or less posterior to the stylet knobs.

35. *Meloidogyne graminis* (Sledge & Golden, 1964) Whitehead, 1968

(Figs 98, 99)

COMMON NAME: Grass root-knot nematode.

Meloidogyne graminis was first identified in 1959 as a parasite of turfgrass in Florida by E.B. Sledge (Sledge & Golden, 1964). Nematode infection was associated with decline of St Augustine grass, which became chlorotic and died.

MEASUREMENTS

See Table 25.

DESCRIPTION *(AFTER SLEDGE & GOLDEN, 1964)*

Female

Body white and oval with protruding neck usually situated well to one side of median plane through vulva. Cuticle finely annulated and quite thick, measuring 21.5 (17.0-32.5) μm at thickest point on body. Labial region bearing no annuli and not distinctly offset from neck.

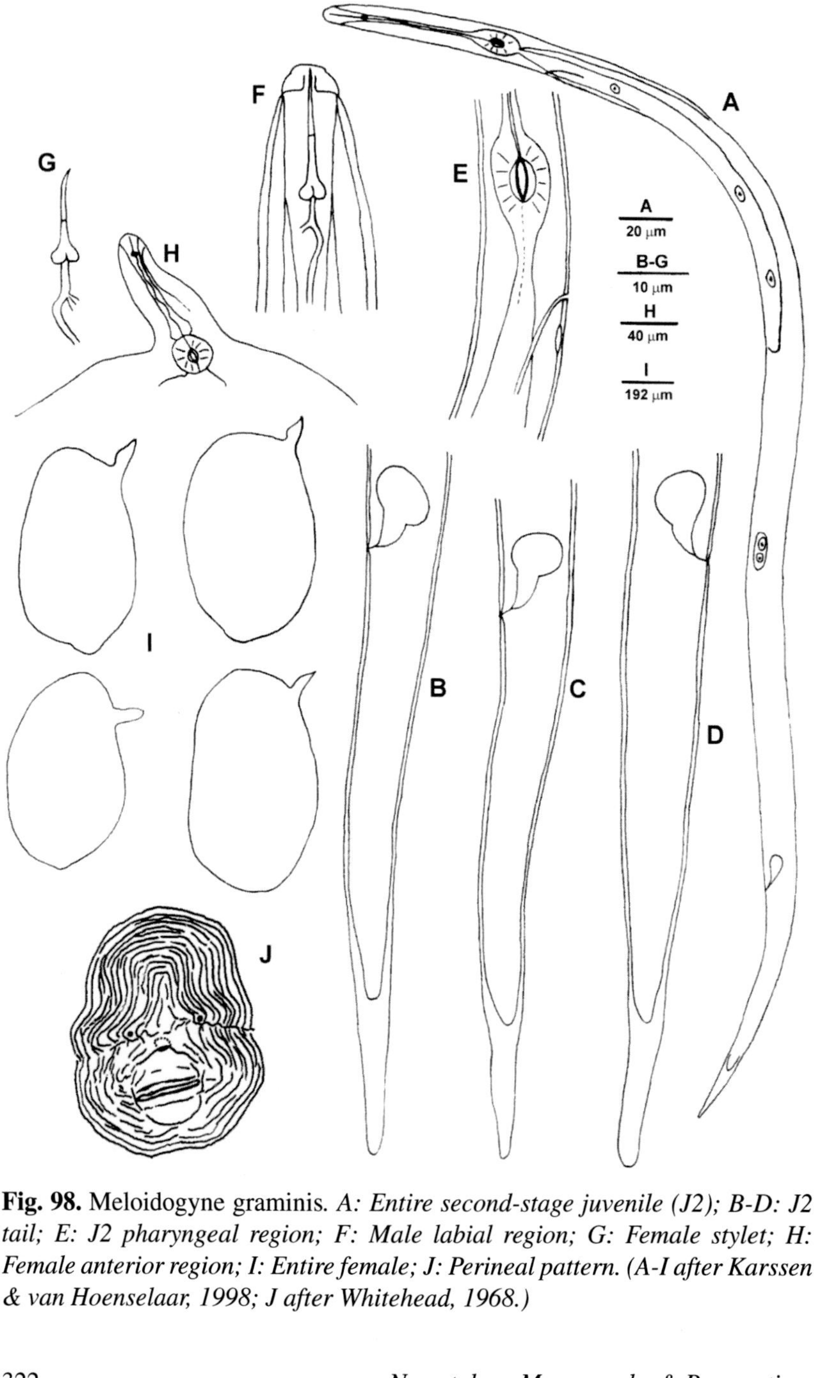

Fig. 98. Meloidogyne graminis. *A: Entire second-stage juvenile (J2); B-D: J2 tail; E: J2 pharyngeal region; F: Male labial region; G: Female stylet; H: Female anterior region; I: Entire female; J: Perineal pattern. (A-I after Karssen & van Hoenselaar, 1998; J after Whitehead, 1968.)*

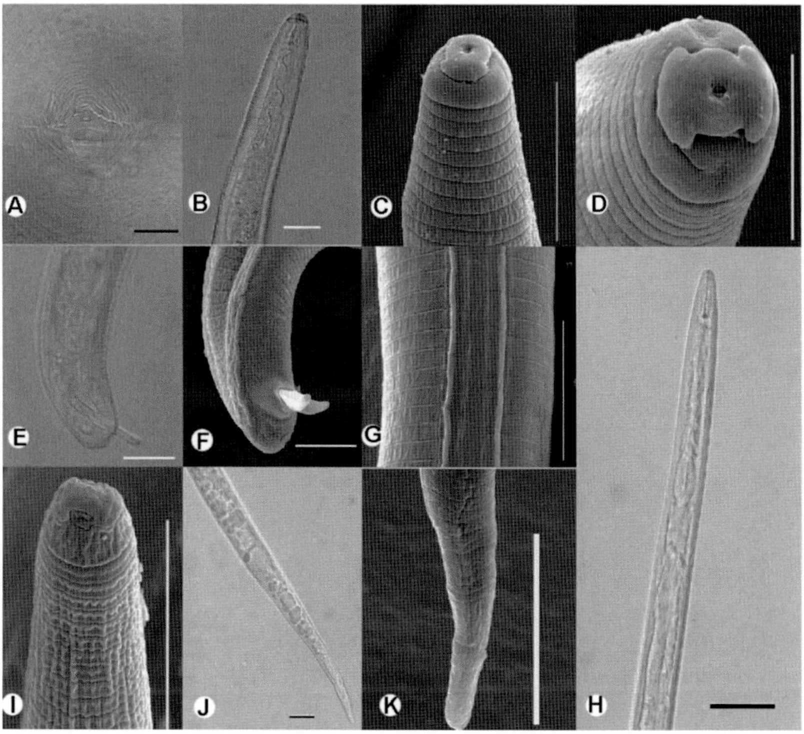

Fig. 99. Meloidogyne graminis. *LM and SEM. A: Perineal pattern; B, C: Male anterior region; D: Male* en face *view; E, F: Male tail; G: Male lateral field; H: Second-stage juvenile (J2) anterior region; I: J2* en face *view; J, K: J2 tail. (Scale bars: A, B, E, H, J = 20 μm; C, F, K = 10 μm; D, G, I = 5 μm.) (After Azevedo de Oliveira* et al., *2018.)*

Labial region variable in exact shape but apparently with circumoral elevation, labial framework indistinct. Cephalids not observed. Stylet knobs rounded posteriorly. Pharynx well developed with elongate, cylindrical procorpus and large spherical median bulb provided with heavily sclerotised valve. Pharyngeal glands with three prominent nuclei. Junction of pharynx and intestine obscure. Excretory pore distinct; generally located less than one stylet length posterior to labial region. Ovaries two, becoming indistinguishable when eggs packing uterus and enlarging, eventually filling body cavity. Vulva and anus situated posteriorly on a slight but distinct button-like protrusion of body. Perineal pattern coarse, with rather high arch, and lateral lines not completely interrupting transverse striations, tail terminus area free of striae.

Table 25. *Morphometrics of females, males and second-stage juveniles (J2) of* Meloidogyne graminis. *All measurements are in μm and in the form: mean ± s.d. (range).*

Character	Holotype, Allotype Florida, USA, Sledge & Golden (1964)	Paratypes, Florida, USA, Sledge & Golden (1964)	Paratypes, Karssen & van Hoenselaar (1998)	Gainesville, Florida, USA, Whitehead (1968)	Brazil, Oliveira *et al.* (2018)
Female (n)	-	20	6	7	-
L	816	726 (586-841)	-	658 ± 91 (541-797)	-
W	490	472 (280-680)	-	453 ± 58 (358-527)	-
Stylet	12.5	12.5 (11.7-13.4)	12.8 ± 0.4 (12.5-13.5)	12.0 (10.0-16.0)	-
DGO	-	3.7 (3.5-4.5)	4.1 ± 0.6 (3.5-5.0)	3.0 (2.0-4.0)	-
Ant. end to excretory pore	-	-	11.4 ± 1.6 (8.0-13.0)	-	-
EP/ST	-	-	0.9	-	-
Male (n)	1	20	3	24	30
L	1420	1512 (1275-1734)	1248 ± 96 (1152-1344)	1442 ± 159 (1115-1743)	1270 ± 240 (900-1705)
a	41.2	43.5 (37.4-50.4)	36 ± 0.5 (33.5-36.5)	43 ± 4.94 (35.6-50.8)	34.8 ± 6 (23.1-45.4)
b	7.1	-	-		-
c	220	187.3 (131.7-273.9)	-	144 ± 24.9 (112-189)	-
c′	-	-	-		1.1 ± 0.1 (0.8-1.4)
W	-	34.9 (31.0-42.0)	34.8 ± 3.2 (32.0-38.0)	-	36.6 ± 5.0 (25.0-47.5)
Stylet	18.5	18.3 (18.0-19.0)	18.1 ± 0.4 (18.0-18.5)	19.6 ± 1.1 (17.3-21.2)	15.5 ± 1.0 (13.0-17.0)
DGO	-	2.5 (2.0-3.0)	2.7 ± 0.4 (2.5-3.2)	3.0 ± 0.7 (2.2-4.7)	3.4 ± 1 (2.5-5.0)
Median bulb length	-	-	-	21.6 ± 2.6 (18.0-25.9)	17.0 ± 3.0 (13.0-22.0)
Median bulb width	-	-	-	10.7 ± 1.6 (6.8-13.7)	10.9 ± 1 (10.0-14.0)
Ant. end to excretory pore	-	-	-	-	132.9 ± 15 (100-153)
Spicules	-	28.3 (28.0-29.0)	27.4 ± 3.6 (25.0-32.0)	26.5 ± 2.5 (21.2-30.2)	30.7 ± 2 (27.0-34.0)

Table 25. *(Continued.)*

Character	Holotype, Allotype Florida, USA, Sledge & Golden (1964)	Paratypes, Florida, USA, Sledge & Golden (1964)	Paratypes, Karssen & van Hoenselaar (1998)	Gainesville, Florida, USA, Whitehead (1968)	Brazil, Oliveira *et al.* (2018)
Gubernaculum	-	8.1 (8.0-8.5)	8.0 ± 0.4 (7.5-8.5)	7.5 ± 0.5 (6.8-7.9)	6.3 ± 1 (4-8)
Tail	-	8.5 (6.0-11.0)	-		26.3 ± 5 (20.0-37.5)
J2 (n)		20	10	24	40
L	-	475 (420-510)	463 ± 31.4 (403-506)	439 ± 18 (409-473)	413.7 ± 22.6 (365-475)
a	-	31.7 (28.8-34.0)	33.9 ± 1.9 (31.5-37.3)	29.1 ± 2.5 (24.3-33.8)	25.8 ± 2.5 (21.1-31.6)
b	-	2.3 (2.1-3.0)	-	2.2 ± 0.2 (1.9-2.5)	-
c	-	6.1 (5.7-6.8)	6.1 ± 0.3 (5.7-6.4)	6.2 ± 0.5 (5.6-7.7)	6 ± 0.4 (5.5-7.7)
c′	-	-	7.7 ± 0.5 (6.8-8.3)		6.3 ± 0.7 (4.7-8.1)
W	-	15 (14.5-16.0)	15 (14.5-16.0)	-	16 ± 1.2 (13.0-18.0)
Stylet	-	12.6 (12.0-13.5)	12.4 ± 0.4 (12.0-13.5)	10.7 ± 0.5 (9.9-11.9)	10.3 ± 0.9 (9.0-12.0)
DGO	-	2.5 (2.0-3.0)	2.5 ± 0.3 (2.0-3.5)	2.1 ± 0.4 (1.5-2.8)	3.2 ± 0.5 (2.0-4.0)
Median bulb length	-	-	-	11.9 ± 1.1 (10.1-13.7)	12.5 ± 1.4 (10.0-15.0)
Median bulb width	-	-	-	7.8 ± 1.0 (6.1-10.1)	8.4 ± 1.5 (7.0-8.0)
Ant. end to excretory pore	-	-	-	-	71.9 ± 5.9 (62-90)
Tail	-	78.3 (68-88)	76.0 ± 4.9 (66-82)	71.0 ± 6.0 (56-81)	68.5 ± 4.5 (52-76)
Hyaline region	-	18.5 (14.0-22.5)	18.9 ± 1.8 (16.0-21.5)		13.2 ± 1.8 (10.0-17.0)

Male

Body cylindroid, vermiform, tapering gradually at either end. Labial region slightly offset from body, cap relatively small, labial disc not elevated, lateral labials absent. Cuticular annulation distinct. Annuli *ca* 2.5 μm wide in middle region of body, becoming smaller toward both ends of body. Lateral field 8.0 (7.0-8.5) μm wide, with four lines, not areolated except in extreme anterior portion. Labial framework prominent. Stylet stout, with knobs rounded posteriorly but not so much anteriorly. Median bulb elongate with well-developed sclerotised valve. Length of pharynx 212 (202-224) μm and from centre of median bulb to base of stylet 74 (68-81) μm. Hemizonid prominent, located on first two annuli anterior to excretory pore. Hemizonion small but distinct, located about eight annuli posterior to excretory pore. Spicules arcuate. Phasmids about 4 μm from tail tip.

J2

Body cylindrical, vermiform, tapering considerably toward posterior end. Labial region not offset from body and bearing no visible annuli. Cuticular annulation of body well marked. Lateral field 4.5 (4.0-5.0) μm wide with four lines. Labial framework indistinct. Stylet weak with rounded knobs. Median bulb elongate with prominent sclerotised valve. Length of pharynx 200 (179-224) μm and from centre of median bulb to base of stylet 43.3 (39-47) μm. Hemizonid located *ca* 4-5 μm posterior to excretory pore. Tail long, with dilated rectum. Terminus rounded.

TYPE PLANT HOST: St Augustine grass, *Stenotaphrum secundatum* (Walt.) Kuntze.

OTHER PLANTS: *Meloidogyne graminis* parasitise grasses and cereals: *Agrostis stolonifera* L., *Cynodon transvaalensis* Davy, *C. dactylon*, *Dactylis glomerata* L., *Digitaria sanguinalis* (L.) Scop., *Hemarthria altissima* (Poir.) Stapf & CE. Hubb., *Eremochloa ophiuroides* (Munro) Hack., *Festuca elatior* L., *Hordeum vulgare*, *Paspalum notatum* Flugge, *Poa pratensis* L., *Sorghum bicolor* (L.) Moench, *Triticum aestivum* L., *Zea mays*, *Zoysia japonica* Steud. (MacGowan, 1984; Jepson, 1987; Karssen & van Hoenselaar, 1998; Sanchez *et al.*, 2018). Studies at the University of Florida have also shown that the common agricultural weeds yellow nutsedge (*Cyperus esculentus*) and purple nutsedge (*C. rotundus*) are both hosts to *Meloidogyne graminis*.

TYPE LOCALITY: Lawn around FDACS, Division of Plant Industry, Cowperthwaite Buildg., 9 Winter Haven, Florida, USA.

DISTRIBUTION: *North America*: USA (Alabama, Arizona, California, Florida, Georgia, Hawaii, Kansas, Maryland, Nevada, New England, North Carolina, South Carolina, Tennessee, Texas and Virginia) (Dickerson, 1966; MacGowan, 1984; McClure *et al.*, 2012; Zeng *et al.*, 2012; Ye *et al.*, 2015, 2019); *South America*: Venezuela (Perichi *et al.*, 2006), Brazil (Oliveira *et al.*, 2018); *Asia*: China (Zhuo *et al.*, 2011), Malaysia (Bojang *et al.*, 2019). *Europe*: reported from coastal dunes in Germany (Sturhan, 1976) and The Netherlands, although detailed analysis of the slides indicated the presence of *M. maritima* and not *M. graminis* (Karssen & van Hoenselaar, 1998). Karssen & van Hoenselaar (1998) also believed that there is no indication that this species is present elsewhere in Europe.

PATHOGENICITY: *Meloidogyne graminis* is damaging to many turf and forage grasses. This nematode cause large circular areas of dead or dying grasses, which may show chlorosis at the margins (Grisham *et al.*, 1974). *Meloidogyne graminis* was pathogenic to bermudagrass (*Cynodon dactylon*) (Heald, 1969), the most widely used turfgrass on golf courses worldwide. On turf, *M. graminis* proliferates in the upper soil profile (Laughlin & Williams, 1971). Research at the University of Florida has revealed that on golf greens the majority of *M. graminis* infect roots growing in the thatch and upper inch (= 2.54 cm) of soil. *Meloidogyne graminis* was also discovered in roots of declining creeping bentgrass (*Agrostis stolonifera* 'Penn A-4') greens on a golf course in Indian Wells, CA (Ploeg, pers.comm.).

CHROMOSOME NUMBER: *Meloidogyne graminis* reproduces by facultative meiotic parthenogenesis and has n = 18 (Triantaphyllou, 1985b).

POLYTOMOUS KEY CODES: *Female*: A23, B34, C4, D5; *Male*: A32, B3, C2, D1, E1, F2; *J2*: A21, B2, C12, D12, E3, F1.

BIOCHEMICAL AND MOLECULAR CHARACTERISATION: Esbenshade & Triantaphyllou (1987) described the esterase (VS1 type) and malate dehydrogenase (N4 type) patterns for *M. graminis*. A profile with single EST band with a very slow migration for this species was also published by Brito *et al.* (2010), Zhuo *et al.* (2011) and Oliveira *et al.* (2018). The D2-D3 of 28S rRNA, ITS rRNA gene sequences and sequence of mtDNA region between the *COII* and 16S rRNA genes can be used for identification of *M. graminis* from other RKN. Ye *et al.* (2015) developed PCR with species-specific primers for *M. graminis* diagnostics. The *COII*-16S rRNA-PCR-RFLP with the restriction enzymes *Dra*I and *Ssp*I discriminates *M. graminis* from *M. marylandi* (McClure *et al.*, 2012). McClure *et al.* (2012) distinguished two *COII*-16S rRNA haplotypes (A and B) of *M. graminis*.

RELATIONSHIPS (DIAGNOSIS): The species belongs to Molecular group IV and is clearly molecularly differentiated from all other *Meloidogyne* species. *Meloidogyne graminis* is close to *M. sasseri* and other members of the *graminis* group (Jepson, 1987). From *M. sasseri* it can be differentiated by shorter J2 body length (420-510 *vs* 470-650 μm), shorter stylet (12-13.5 *vs* 13-14.5 μm), and shorter tail (68-88 *vs* 83-115) μm. The perineal pattern of this species also resembles *M. hispanica* and *M. sasseri*, but females differ in EP/ST ratio (0.9 *vs* 0.8-4.8, 0.5-2.2, respectively) and several morphometrics in male and J2.

36. *Meloidogyne hapla* Chitwood, 1949
(Figs 100, 101)

COMMON NAME: Northern root-knot nematode.

Northern RKN is one of the four commonest species worldwide. Chitwood (1949) described this species from roots of potato in Long Island, New York, USA. Morphological and morphometric descriptions of this species were given by Whitehead (1968), Orton Williams (1974), Eisenback & Hirschmann (1979a), Eisenback & Triantaphyllou (1991), Eisenback (1993), Karssen & van Hoenselaar (1998), Hunt & Handoo (2009) and others.

MEASUREMENTS

See Table 26.

- *Eggs* (n = 20): L = 78 (71-91) μm; W = 31 (26-40) μm.

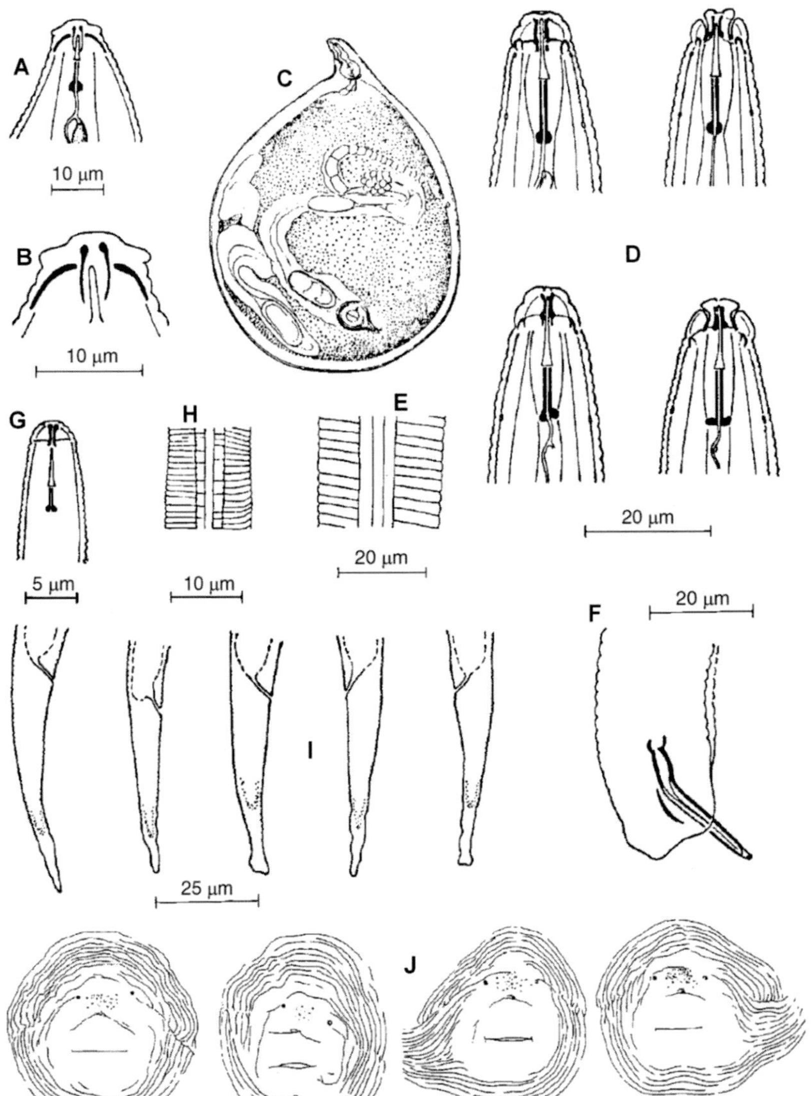

Fig. 100. Meloidogyne hapla. *A, B: Female labial region; C: Entire female; D: Male labial region; E: Male lateral field; F: Male tail; G: Second-stage juvenile (J2) labial region; H: J2 lateral field; I: J2 tail; J: Perineal pattern. (After Whitehead, 1968.)*

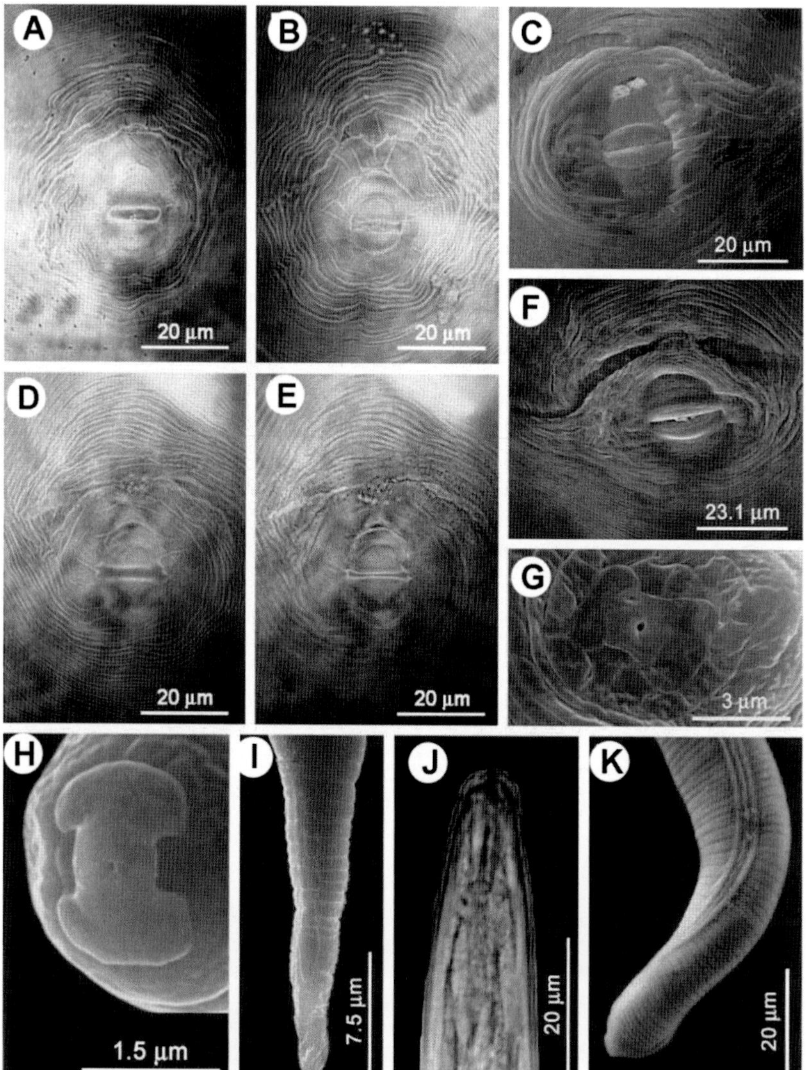

Fig. 101. Meloidogyne hapla. *LM and SEM. A-F: Perineal pattern; G: Female* en face *view; H: Second-stage juvenile (J2)* en face *view; I: J2 tail; J: Male labial region; K: Male tail, lateral field. (After Handoo* et al., *2005.)*

Table 26. *Morphometrics of females, males and second-stage juveniles (J2) of* Meloidogyne hapla. *All measurements are in μm and in the form: mean ± s.d. (range).*

Character	New York, USA Whitehead (1968) Paratypes	Hawaii, USA Handoo *et al.* (2005a)	HRBB100, Ethiopia Meressa *et al.* (2015)	Iran Sohrabi *et al.* (2015)
Female (n)	20	25	45	-
L	612 (419-845)	-	529 ± 42.4 (486-574)	-
a	-	1.7 ± 0.2 (1.4-2.1)	-	-
W	430 (311-561)	342 ± 65 (225-495)	-	-
Stylet	11.0 (10.0-13.0)	-	14.7 ± 0.9 (14-16)	-
Stylet knob width	-	3.0 ± 0.5 (2.5-3.5)	3.5 ± 0.5 (3.0-4.0)	-
Stylet knob height	-	2.5 ± 0.1 (2.5-3.0)	-	-
DGO	5.0 (4.0-6.0)	5.2 ± 0.5 (5.0-6.5)	5.2 ± 0.4 (5.0-6.0)	-
Median bulb length	36 (31-43)	-	-	-
Median bulb width	31 (26-37)	-	-	-
Median bulb valve width	10 (9.0-11.0)	-	-	-
Median bulb valve length	12 (10.0-13.0)	-	-	-
Ant. end to excretory pore	-	36.4 ± 8.0 (25.0-50.0)	33.4 ± 5.8 (28.0-41.0)	-
EP/ST	-	2.9 ± 0.7 (1.9-4.0)	2.3 ± 0.3 (1.9-2.6)	-
Male (n)	25	-	-	-
L	1139 (791-1432)	-	-	-
a	41.7 (33.3-47.0)	-	-	-

Table 26. *(Continued.)*

Character	New York, USA Whitehead (1968) Paratypes	Hawaii, USA Handoo *et al.* (2005a)	HRBB100, Ethiopia Meressa *et al.* (2015)	Iran Sohrabi *et al.* (2015)
b	15.5 (12.8-19.2)	10.1 ± 1.5 (7.6-11.8)	-	-
c	118 (73-283)	-	-	-
Stylet	20.0 (17.0-23.0)	-	-	-
Stylet knob height	-	2.7 ± 0.2 (2.5-3.0)	-	-
DGO	2.9 (2.5-3.5)	-	-	-
Spicules	25.7 (22.0-28.0)	23.3 ± 2.5 (20.0-25.0)	-	-
Gubernaculum	8.2 (7.0-8.5)	7.2 ± 0.3 (7.0-7.5)	-	-
J2 (n)	20	25	45	12
L	337 (312-355)	323 ± 18.4 (284-355)	360.7 ± 14.7 (345-375)	394 ± 39.3 (348-450)
a	23.9 (20.1-26.6)	30.2 ± 2.6 (24.1-35.5)	30.7 ± 1.1 (29.3-31.9)	30.9 ± 4 (24.4-37.6)
b	-	4.7 ± 0.3 (4.3-5.2)	-	4.6 ± 0.44 (4.0-5.1)
c	7.9 (7.3-10.2)	7.7 ± 0.6 (6.8-9.4)	-	8.0 ± 1 (6.2-10.3)
c′	4.4 (3.7-4.7)	5.4 ± 0.6 (4.5-6.0)	-	5.3 ± 0.8 (3.5-6.3)
W	-	10.8 ± 0.9 (10.0-12.0)	-	-
Stylet	9.7 (8.0-11.0)	-	12.9 ± 0.4 (12.2-13.3)	12.1 ± 0.8 (11.0-13.0)
DGO	-	2.5 ± 0.0 (2.5-2.5)	2.5 ± 0.2 (2.2-2.7)	-
Ant. end to excretory pore	-	-	71.1 ± 2.1 (68.8-74)	-

Table 26. *(Continued.)*

Character	New York, USA Whitehead (1968) Paratypes	Hawaii, USA Handoo *et al.* (2005a)	HRBB100, Ethiopia Meressa *et al.* (2015)	Iran Sohrabi *et al.* (2015)
Tail	43.0 (33.0-48.0)	42.6 ± 4.2 (30.0-47.5)	47.4 ± 3.1 (42.4-50.4)	50 ± 5.6 (42.0-57.0)
Hyaline region	-	10.9 ± 2.1 (5.0-15.0)	-	15 ± 1.8 (12.0-18.0)

DESCRIPTION

Female

Body pyroid with short neck. Cuticle becoming thicker in posterior half of body, sometimes considerably. Labial region with two annuli posterior to labial-cap. Stylet knobs rounded, inconspicuous. Excretory pore 14-20 annuli posterior to labial region, hemizonid just posterior to pore. Perineal cuticular pattern roughly circular, composed of closely spaced smooth or slightly wavy striae. Dorsal arch low. Lateral field may be unmarked, may be marked only by slight irregularities in striae, or dorsal and ventral striae may meet at a slight angle along field. Some forking of striae at lateral field may also occur. In some cases, ventral striae may extending laterally on one or both sides to form 'wings' in which the dorsal striae meet almost at right angles. Tail with few striae but distinct punctations forming a stippled area between anus and tail terminus. Stippling may be more diffuse over inner part of pattern. Phasmids fairly widely spaced. For photomicrographs of patterns, see Taylor *et al.* (1955). SEM studies showed that morphology of cytological races A and B of *M. hapla* is similar for some characters and different for others, *viz.*, female stylet morphology is considered the most stable character examined, whereas male stylet morphology is correlated with cytological race (Eisenback, 1993).

Male

Numerous in some populations, absent in others. Labial region not offset, a truncate cone to hemispherical in outline. Usually only one annulus posterior to labial-cap. Stylet slender, stylet knobs rounded, not offset. Anterior cephalid on second body annulus, posterior cephalid just anterior to level of relaxed stylet. Hemizonid 45-58 annuli posterior to labial region, 0-4 annuli anterior to excretory pore. Lateral field with four incisures. Tail terminus bluntly rounded, phasmids at about cloacal aperture level. One or two testes. Spicules slightly curved, with small sharp processes projecting from spicule wall at junction of head region and shaft in to spicule head region. Gubernaculum crescentic, proximal end thicker than distal end. Thorne (1961) mentions presence in some populations, especially on strawberries, of miniature males, which may be only half the usual length.

J2

Labial region not offset, a truncated cone shape. Labial cap followed by three annuli on sublateral labial region sectors, one annulus on lateral labial region sectors. Stylet knobs rounded. Lateral field with four incisures, outer bands irregularly cross-striated. Rectum not dilated. Tail tip variable, subacute or sometimes bifid. SEM studies by Eisenback & Hirschmann (1979) showed non-annulated labial region with median labials and labial disc (dumbbell-shaped) in same contour and forming a smooth continuous structure.

TYPE PLANT HOST: Potato, *Solanum tuberosum* L.

OTHER PLANTS: *Meloidogyne hapla* attacks over 550 plant host species. In cooler climates nearly all vegetables of economic importance are hosts of this nematode, as well as crops such as clover, lucerne, sugarbeet, fruits (strawberry), ornamentals (carnation, freesia, gypsophila, rose, statice), groundnut, soybean, pyrethrum, *etc.* (Orton Williams, 1974; Jepson, 1987; Meressa *et al.*, 2014, 2015). Barley or Sudan grass are non-hosts and can be used in rotations to reduce *M. hapla* infections (Bélair, 1996; Viaene & Abawi, 1998).

TYPE LOCALITY: A field at Bridgehampton, Long Island, New York, USA.

DISTRIBUTION: *Meloidogyne hapla* generally occurs in the cooler regions of the world, but has also been recorded worldwide from temperate regions to tropical and subtropical regions at higher altitudes (more than 1000 m a.s.l.). It is widely distributed in agricultural areas, but also occurs in natural habitats like coastal dunes, woods and riverbanks (Karssen & van Hoenselaar, 1998). It has been reported in: *Australia*, *North America* (Canada, USA), and several countries of *Africa*, *Asia*, *South America* and *Europe*, including the Mediterranean Basin and Russia (Orton Williams, 1974; Karssen & van Hoenselaar, 1998; LaMondia, 2002; Conceição *et al.*, 2009; Vovlas *et al.*, 2010; Wesemael *et al.*, 2011; Pedroche *et al.*, 2013; Villain *et al.*, 2013; Onkendi *et al.*, 2014; Meressa *et al.*, 2015; Panahi & Barooti, 2015; Sohrabi *et al.*, 2015; Ali *et al.*, 2016; Akyazi *et al.*, 2017).

SYMPTOMS: *Meloidogyne hapla* induces small galls that have numerous secondary roots emerging. Heavily galled root systems generally appear thick and matted; however, some populations may produce galls that are similar to those of the other three most common species (Eisenback & Triantaphyllou, 1991). Plants infected by *M. hapla* were stunted, low yielding, with shortened growth life and galled roots.

PATHOGENICITY: *Meloidogyne hapla* causes important economic losses worldwide for several horticultural, vegetable crops and pastures, including carrots, lettuce, lucerne, onion, potato, rose, sugarbeet, strawberry, and white clover, with a damage potential of 41-84% of crop yield under pot or microplot conditions. (Townshend & Potter, 1978; Griffin *et al.*, 1982; Inserra *et al.*, 1983; Viaene & Abawi, 1996; LaMondia, 2002; Wobalem & Viaene, 2005; Gugino *et al.*, 2006; Conceição *et al.*, 2009; Pang *et al.*, 2009; Zasada *et al.*, 2012; Karnkowski *et al.*, 2013; Meressa *et al.*, 2016).

CYTOLOGICAL RACES AND CHROMOSOME NUMBER: Triantaphyllou (1966, 1984) reported that this species occurs as two cytological races, A and B, including several cytological forms. After extensive cytological examination, *M. hapla* was considered as the most biologically complex nematode species known to date (Triantaphyllou & Hirschmann, 1980). *Meloidogyne hapla* cytological race A reproduces by facultative meiotic parthenogenesis and has $n = 13\text{-}17$, polyploids $n = 28, 34$; and cytological race B reproduces by obligatory mitotic parthenogenesis with $2n = 30\text{-}32, 43\text{-}48$ (Triantaphyllou, 1966, 1985b; Eisenback & Triantaphyllou, 1991; Liu & Williamson, 2006).

The J2 of race A populations are 357-467 (413) μm long, whereas populations of race B are longer, 410-517 (474) μm. The tail length of race A populations is 46-58 (53) μm and in race B populations 54-

69 (62) μm. The hyaline region is 12-19 (16) μm long (Eisenback & Triantaphyllou, 1991) The shaft of the stylet of males from populations of race A gradually increased in width posteriorly, whereas the shaft was cylindrical in males of race B (Eisenback, 1993).

HOST RACES: Populations of *M. hapla* reproduce on peanut, pepper, NC95 tobacco, and tomato, but do not reproduce on cotton or watermelon (Hartman & Sasser, 1985). Populations of race A and race B varied in their ability to reproduce on *Tagetes patula* (Eisenback, 1987). Race B populations readily reproduced on *T. patula* but race A populations did not (Eisenback & Triantaphyllou, 1991).

POLYTOMOUS KEY CODES: *Female*: A23, B43, C213, D1; *Male*: A34, B324, C32, D1, E1, F1; *J2*: A3, B3, C34, D324, E32, F3.

BIOCHEMICAL AND MOLECULAR CHARACTERISATION: The majority of *M. hapla* samples showed the H1 esterase phenotype. The malate dehydrogenase H1 phenotype is unique for this species (Esbenshade & Triantaphyllou, 1985a, b; Carneiro *et al.*, 2000) (Fig. 12).

Sequences of ribosomal genes: 18S rRNA, ITS rRNA, the D2-D3 of 28S rRNA, IGS rRNA and mtDNA: *COII*-16S rRNA fragment (Fig. 13), *COI* and *COII* gene clearly differentiate *M. hapla* from all other RKN. Several specific primers were designed for diagnostics of this species (Table 10). Peng *et al.* (2017) developed loop-mediated isothermal amplification methods for identification of *M. hapla*. Several authors proposed real-time PCR, LAMP and RPA assays for detection of *M. hapla* in root galls and in soil (Table 11).

RELATIONSHIPS (DIAGNOSIS): The species belongs to Molecular group II and is clearly molecularly differentiated from all other *Meloidogyne* species. *Meloidogyne hapla* can be distinguished from all other *Meloidogyne* spp. by morphology of the female perineal pattern, which is rounded with a low dorsal arch and, especially, by the distinct punctations between the anus and tail terminus. This species is close to *M. haplanaria* from which it can be differentiated by perineal pattern, shape, morphology of the labial region and stylet, shape and constitution of the median bulb, and position of the excretory pore in females (Eisenback *et al.*, 2003).

37. *Meloidogyne haplanaria* Eisenback, Bernard, Starr, Lee & Tomaszewski, 2003
(Figs 102-105)

COMMON NAME: Texas peanut root-knot nematode.

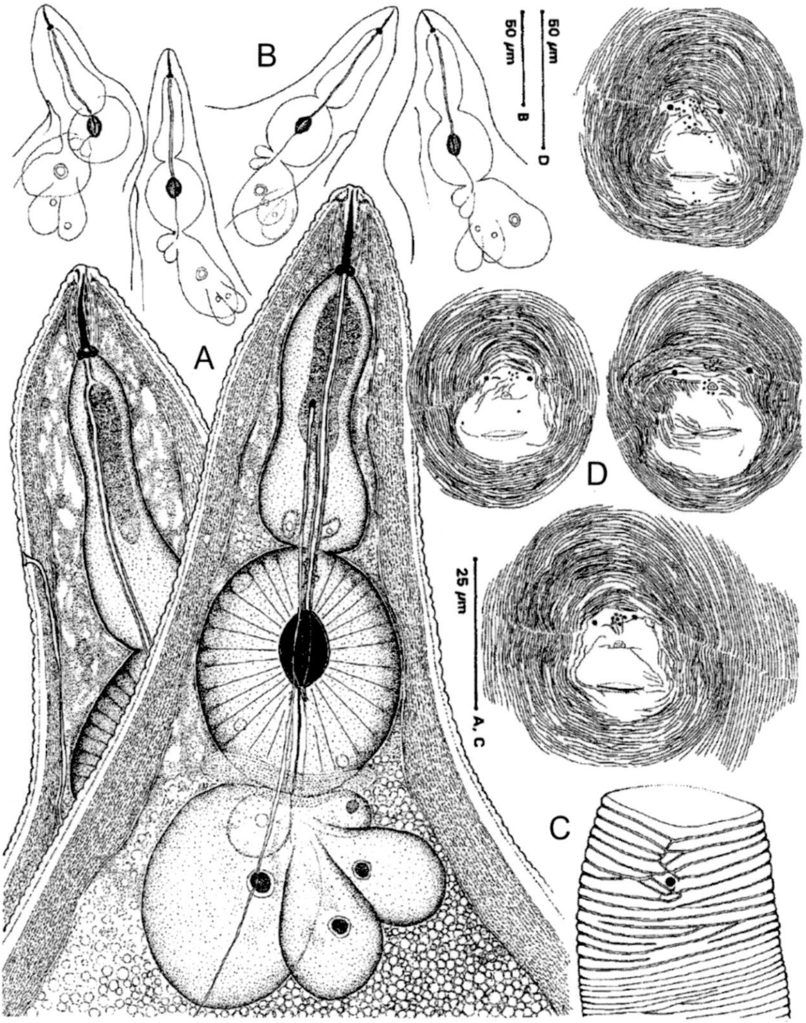

Fig. 102. Meloidogyne haplanaria. *A, B: Female anterior region; C: Excretory pore; D: Perineal pattern. (After Eisenback* et al., *2003.)*

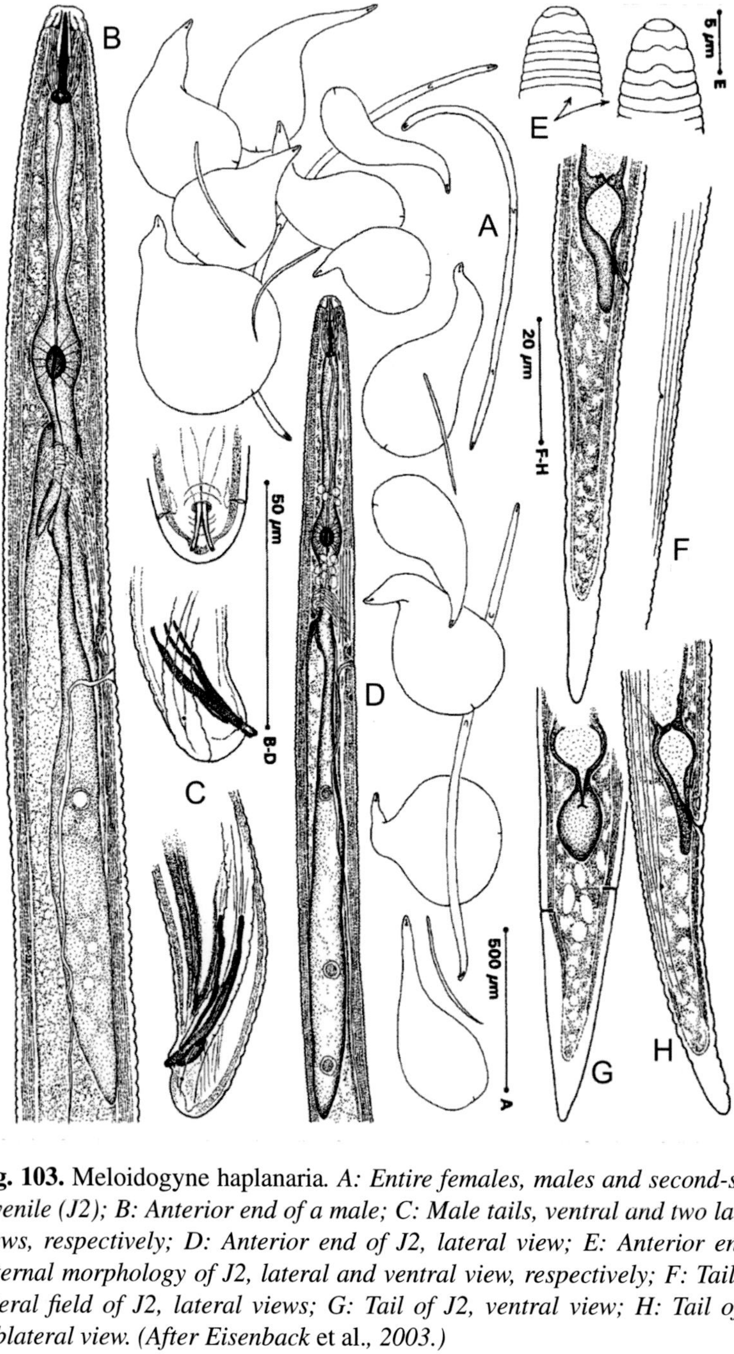

Fig. 103. Meloidogyne haplanaria. *A: Entire females, males and second-stage juvenile (J2); B: Anterior end of a male; C: Male tails, ventral and two lateral views, respectively; D: Anterior end of J2, lateral view; E: Anterior end of external morphology of J2, lateral and ventral view, respectively; F: Tail and lateral field of J2, lateral views; G: Tail of J2, ventral view; H: Tail of J2, sublateral view. (After Eisenback* et al.*, 2003.)*

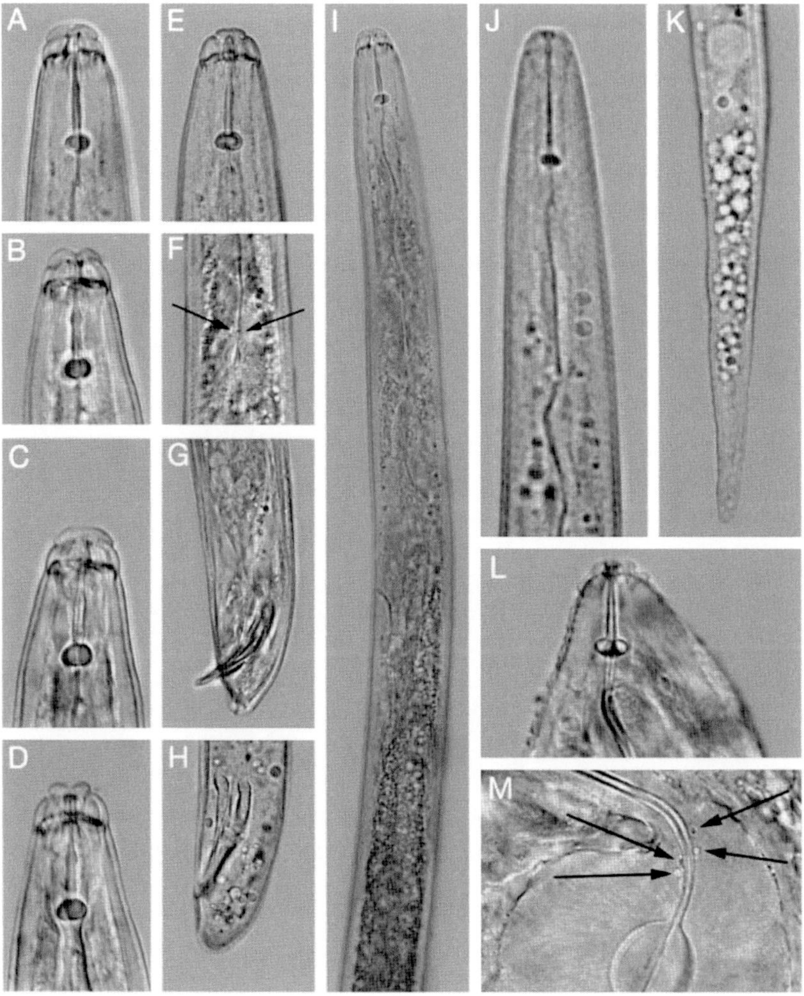

Fig. 104. Meloidogyne haplanaria. *LM. A-C: Anterior end of male (lateral views); D-E: Anterior end of male (dorsal views); F: Median bulb of male showing vesicles near lumen lining (arrows); G-H: LM of male tail (lateral and nearly lateral views, respectively); I: LM of anterior end of male showing indistinct outline of pharynx; J: LM of anterior end of second-stage juvenile (J2); K: LM of tail of J2; L: Anterior end of female; M: Anterior portion of the median bulb showing vesicles near lumen lining (arrows). (After Eisenback* et al., *2003.)*

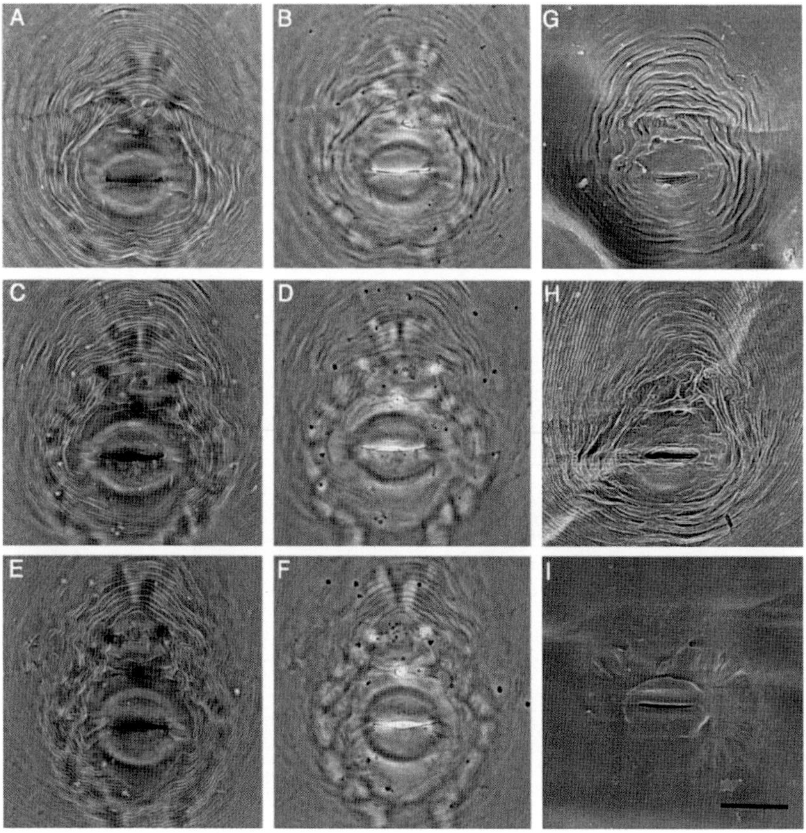

Fig. 105. Meloidogyne haplanaria. *LM and SEM. Perineal patterns. A, C, E: Light micrographs of surface details; B, D, F: Light micrographs of details below surface of pattern shown in adjacent photograph; G-H: SEM of outside surface; I: SEM of internal surface. (Scale bar: 25 μm.) (After Eisenback* et al., *2003.)*

Eisenback *et al.* (2003) described this species from roots of peanut in Collingsworth, Texas, USA. The current distribution of *M. haplanaria* in peanut fields is limited to Texas and Arkansas, USA. This nematode has also been reported in Florida, which poses a potential threat to peanut and tomato production in that state (Joseph, 2016).

MEASUREMENTS

See Table 27.

Table 27. *Morphometrics of females, males and second-stage juveniles (J2) of* Meloidogyne haplanaria. *All measurements are in μm and in the form: mean ± s.d. (range).*

Character	Holotype-Allotype, Texas, USA, Eisenback *et al.* (2003)	Paratypes, Texas, USA, Eisenback *et al.* (2003)	Florida, USA, Joseph *et al.* (2016)
Female (n)	-	30	-
L	921	936 ± 158.8 (675-1475)	-
a	-	1.6 ± 0.3 (1.0-2.2)	-
W	576	588 ± 80.3 (450-775)	-
Stylet	14.5	14.3 ± 0.9 (13.0-16.0)	-
Stylet knob width	-	2.4 ± 0.5 (1.8-2.6)	-
Stylet knob height	-	4.7 ± 0.9 (3.7-5.4)	-
DGO	5.5	5.4 ± 0.8 (5.0-6.0)	-
Ant. end to excretory pore	70	69.1 ± 24.3 (31.0-118.0)	-
Vulval slit	-	23.3 ± 4.3 (16.4-25.5)	
Vulval slit to anus distance	-	19.6 ± 4.5 (17.3-28.9)	
Interphasmid distance	-	18.8 ± 3.7 (18.6-25.5)	
EP/ST	4.8	4.8	
Male (n)	1	30	-
L	1873	1847 ± 49 (1350-2425)	-
a	41.4	43.8 ± 4.8 (32.8-54.6)	-
c	102.9	102 ± 21.3 (54.6-159.7)	-
T	44	44.2 ± 5.8 (30.0-56.0)	-

Table 27. *(Continued.)*

Character	Holotype-Allotype, Texas, USA, Eisenback *et al.* (2003)	Paratypes, Texas, USA, Eisenback *et al.* (2003)	Florida, USA, Joseph *et al.* (2016)
W	-	38.1 ± 4.4 (24.5-45.5)	-
Stylet	19	18.1 ± 1.6 (16.5-22.0)	-
Stylet knob width	-	5.5 ± 0.8 (4.5-6.3)	-
Stylet knob height	-	3.2 ± 0.5 (2.7-3.5)	-
DGO	6.0	5.2 ± 0.9 (3.7-6.4)	-
Ant. end to excretory pore	163.5	167.4 ± 33.3 (150.0-180.9)	-
Spicules	-	38.5 ± 3.3 (36.0-42.0)	-
Gubernaculum	7.5	7.5	-
Tail	-	10.8 ± 2.5 (7.0-17.0)	-
J2 (n)	-	30	20
L	-	419 ± 21.1 (365-480)	448.2 ± 2.3 (435.2-470.5)
a	-	22 ± 2.0 (18.5-27.8)	23.5 ± 0.3 (21.2-27.0)
c	-	7.3 ± 1.2 (5.5-9.9)	7.3 ± 0.0 (7.0-7.7)
W	-	19.1 ± 1.6 (16.5-23.0)	19 ± 0.2 (16.2-20.7)
Stylet	-	10.4 ± 1.8 (9.0-12.0)	12.8 ± 0.3 (11.9-15.4)
Stylet knob width	-	-	2.3 ± 0.0 (2.2-2.3)
Stylet knob height	-	-	1.8 ± 0.0 (1.7-1.9)
DGO	-	2.9 ± 0.6 (2.0-3.5)	2.9 ± 0.0 (2.9-3.0)
Ant. end to excretory pore	-	82.7 ± 21.5 (74.5-105.5)	91.5 ± 0.4 (86.9-95.2)

Table 27. *(Continued.)*

Character	Holotype-Allotype, Texas, USA, Eisenback *et al.* (2003)	Paratypes, Texas, USA, Eisenback *et al.* (2003)	Florida, USA, Joseph *et al.* (2016)
Tail	-	65 ± 10.7 (58.0-74.0)	61 ± 0.2 (59.5-62.5)
Hyaline region	-	14.6 ± 3.1 (11.0-16.0)	16.9 ± 0.2 (13.6-17.8)

- *Eggs* (n = 30): L = 98.8 (93-110) μm; W = 53.4 (51-57) μm; L/W = 1.9 (1.7-2.0).

Description

Female

Body translucent white, variable in size, pear-shaped with short neck, posteriorly rounded, without tail protuberance. In SEM, stoma slit-like, located in ovoid prestoma, surrounded by pit-like openings of six inner labial sensilla. Labial disc fused with median labials, dumbbell-shaped in face view. Median labials crescentic. Lateral labials large, triangular, separated from median labials and labial region. Labial region not offset from regular body annuli. In LM, labial framework distinct, hexaradiate, lateral sectors enlarged. Vestibule and extensions prominent. Cephalids and hemizonid not observed. Distance of excretory pore to labial region variable, located in most specimens mid-way between anterior end and median bulb, terminal excretory duct very long. Stylet long and robust, cone of same size as shaft, tip straight or slightly curved dorsally, widening gradually posteriorly, junction of cone and shaft uneven. Shaft cylindrical and same width throughout, or widening slightly near junction with knobs; knobs broad laterally, offset from shaft, distinctly separated from each other, knobs very slightly indented anteriorly. Dorsal gland orifice branching into three ducts, dorsal gland ampulla large., subventral gland orifices branched, located posteriorly to enlarged triradiate lumen lining of median bulb. Pharyngeal lumen lining with small rounded vesicles anterior to triradiate lumen lining. Pharyngeal glands large, trilobed, dorsal lobe largest, uninucleate, two subventral nucleated lobes variable in size, shape, and position, located posterior to dorsal gland lobe. Pharyngo-intestinal cells two, small, rounded, nucleated, located between median bulb and intestine. Two ovaries and six rectal glands characteristic of genus. Perineal patterns extremely rounded to oval-shaped. Dorsal arch high and rounded except for striae near vulva, which are low with rounded shoulders. Lateral field with distinctly forked striae. Ventral striae varying from wavy to coarse. Tail tip area well defined, free of striae, often with a few to several subcuticular punctations. Subcuticular punctations located randomly within pattern area. Perivulval region not striated, rarely striae occur near lateral edges of vulva. Vulva located in depression, surrounded by wide cuticular ridge. Phasmidial ducts distinct, phasmid surface structure often obscured by striae in SEM. Anus distinct, surrounded by a thick cuticular layer.

Male

Body translucent white, vermiform; body tapering anteriorly, bluntly rounded posteriorly, tail twisting through 90° in heat-killed specimens. Labial cap high in lateral view, extending posteriorly onto distinctly offset labial region. Labial region high in lateral view, tapering posteriorly, distinctly offset from body. Hexaradiate labial framework well sclerotised, vestibule and extension distinct. Prestoma large, hexagonal. Stoma slit-like, located in large, hexagonal prestomatal cavity, surrounded by pore-like openings of six inner labial sensilla. In SEM, labial disc rounded, very large, often separated from median labials by a shallow groove. Median labials very wide, outer margins crescentic, posteriorly sloping. Labial disc and median labials may or may not be fused to form elongate and wide labial structure extending posteriorly onto labial region. Four labial sensilla marked on median labials by shallow, elongated, ovoid, depressions. Amphidial apertures large, elongated, slit-like between labial disc and lateral sectors of labial region. Lateral labials absent. Labial region smooth, annulation absent. Body annuli large, distinct. Lateral field with four incisures, two beginning near level of stylet knobs and two near level of median bulb, lateral field areolated, encircling tail. Stylet robust, large, cone straight, pointed, gradually increasing in diam. posteriorly, opening located several micrometers from stylet tip, cone of same size as shaft. Shaft cylindrical, posterior end wider than anterior end. Knobs large, wide, rounded, offset from shaft. Dorsal gland duct branched into three ducts, gland ampulla indistinct. Procorpus indistinctly outlined, indistinct median bulb elongated, oval-shaped with valve enlarged, triradiate cuticular lumen lining, subventral pharyngeal gland orifices branched, located posteriorly to median bulb. Pharyngo-intestinal junction indistinct. Gland lobe variable in length, with indistinct nuclei rarely visible. Excretory pore distinct, variable in position (150-181 μm), terminal duct long. Hemizonid located anterior to excretory pore. Intestinal caecum short, extending anteriorly on dorsal side to base of median bulb. Usually one testis, rarely two, outstretched, or reflexed anteriorly. Spicules long, slender, slightly arcuate with single tip, short labial region, wide vellum, and indistinct shaft. Gubernaculum distinct, crescentic. Tail short and rounded. Phasmids slit-like, opening near level of cloacal aperture.

J2

Body translucent white, long, slender, tapering anteriorly but more so posteriorly. Body annuli distinct, increasing in size and becoming irregular in posterior tail region. Lateral field starting approximately at middle of procorpus and extending to near phasmids, with four incisures, areolated in some specimens. Stoma slit-like, located in oval-shaped prestomatal depression, surrounded by pore-like openings of six inner labial sensilla. Labial cap high, narrower than labial region. Labial disc elongated, rounded, completely fused with median labials. Median labials with outer margins crescentic, smooth. Median labials and labial disc dumbbell-shaped. Lateral labials distinct, lower than median labials, margins crescentic. Labial region smooth without annulation. Amphidial apertures elongate, located between labial disc and lateral labials. Labial region high, distinctly offset from body. Hexaradiate framework weakly sclerotised in LM, vestibule and vestibule extension distinct. Stylet moderately long but delicate, stylet cone sharply pointed, gradually increasing in width posteriorly, shaft cylindrical, may widen slightly posteriorly, knobs rounded. DGO moderately long, orifice branching into three ducts, ampulla poorly defined. Procorpus faintly outlined, median bulb ovoid with distinct valve, subventral pharyngeal gland orifices posterior to valve, ampulla distinct. Pharyngo-intestinal junction indistinct, at level of nerve ring. Pharyngeal gland lobe variable in length with three small nuclei of same size. Excretory pore distinct, variable in position, terminal duct very long. Hemizonid distinct, located anteriorly to excretory pore. Tail slender, ending in slightly round tip; tail annuli larger and irregular posteriorly. Hyaline tail end long, variable in size. Rectal dilation large. Phasmids small, indistinct, located at one edge of lateral field, posterior to anus.

TYPE PLANT HOST: Peanut, *Arachis hypogaea* L.

OTHER PLANTS: *Meloidogyne haplanaria* reproduced on tomato and peanut, but reproduction was low on corn and wheat. Cotton (*Gossypium hirsutum*), tobacco (*Nicotiana tabacum*), pepper (*Capsicum frutescens* L.), and watermelon (*Citrullus vulgaris* Schard.) were non-hosts (Eisenback *et al.*, 2003). Common bean, garden pea, radish and soybean were confirmed as hosts, and cowpea and eggplant were poor hosts in glasshouse tests (Bendezu *et al.*, 2004). This species was found in rhizosphere soil of Indian hawthorn, okra, ash, oak, cherry laurel, maple, tomato, willow, rivercane, elm, bermudagrass, and birch in Arkansas (Khanal *et al.*, 2016a; Ye *et al.*, 2019). *Meloidogyne haplanaria* has been shown to overcome the resistance in tomato conferred by the *Mi* gene (Bendezu *et al.*, 2004). Recently, it has been reported in Florida in *Mi*-resistant tomato (Joseph *et al.*, 2016).

TYPE LOCALITY: Collingsworth, Texas, USA, in a commercial field located 6.4 km north of the intersection of Highway 204 and road FM 1547 (community of Quail) on the east side of the highway.

DISTRIBUTION: *North America*: USA (Florida, Texas, Arkansas). *Meloidogyne haplanaria* is distributed throughout the peanut production areas (Eisenback *et al.*, 2003; Joseph *et al.*, 2016).

PATHOGENICITY: This nematode is a major pest of peanut in Texas (Eisenback *et al.*, 2003). The *M. haplanaria* infestation in tomato fields in Naples, Florida produced heavy galling and severe yield reduction (Joseph *et al.*, 2016). Infected plants showed typical RKN symptoms, such as stunting and extensive root galls.

CHROMOSOME NUMBER: No available data.

POLYTOMOUS KEY CODES: *Female*: A12, B32, C1, D3; *Male*: A123, B324, C1, D1, E2, F2; *J2*: A23, B32, C21, D213, E32, F3.

BIOCHEMICAL AND MOLECULAR CHARACTERISATION: The isozyme phenotype of *M. haplanaria* is characterised by a single MDH isozyme (Rf = 0.44) and a single isozyme of esterase activity at Rf = 0.61. The esterase isozyme was further characterised by a low intensity of staining relative to that of other species and that of MDH (Eisenback *et al.*, 2003).

PCR amplifications with primer set C2F3/1108 located in the mtDNA *COII* and 16S mtDNA, respectively, produced a 540 bp amplicon from *M. haplanaria*. Amplification of the TRNAH/MRH106 fragment following digestion with *Mnl*I and *Hinf*I allowed the differentiation of *M. haplanaria* from other RKN (Joseph *et al.*, 2016). Phylogenetic analysis based on the *COII*-16S rRNA gene fragment and *COI* showed that *M. haplanaria* is related to *M. enterolobii* (Joseph *et al.*, 2016; Álvarez-Ortega *et al.*, 2019).

RELATIONSHIPS (DIAGNOSIS): This species belongs to Molecular group I. *Meloidogyne haplanaria* can be distinguished from the four most common species of RKN (*M. arenaria*, *M. hapla*, *M. incognita*, and *M. javanica*) (Eisenback *et al.*, 1981) by the unique form of the perineal pattern, shape and morphology of the labial region and stylet, shape and constitution of the median bulb, and position of the female excretory pore. *Meloidogyne haplanaria* can be easily diagnosed and separated from the most common RKN by an inability to infect cotton, tobacco, pepper, and watermelon. Pepper is a good host for *M. arenaria* and *M. hapla* (Hartman & Sasser, 1985) but not *M. haplanaria*. Sequences of mtDNA *COII* and 16S gene fragment and *COI* gene can differentiate *M. haplanaria* from other species.

38. *Meloidogyne hispanica* Hirschmann, 1986
(Figs 106-110)

COMMON NAME: Seville root-knot nematode.

This nematode, isolated from peach rootstock, (*Prunus persica silvestris* Batsch), in Seville, Spain, was studied for the first time by Dalmasso & Bergé (1978) and later described as *M. hispanica* by Hirschmann (1986). Subsequently, Seville RKN was detected in other areas of the country on beet (*Beta vulgaris* L.) and grapevines and found also to reproduce on wheat (*Triticum aestivum* L.) in glasshouse conditions (Karssen & van Hoenselaar, 1998; Castillo *et al.*, 2009b). This species has a worldwide distribution and has been reported infecting economically important crops.

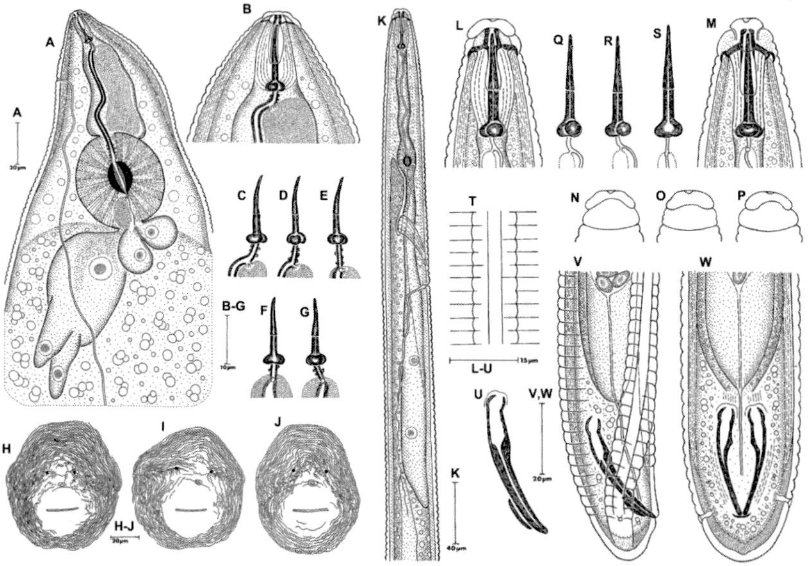

Fig. 106. Meloidogyne hispanica. *A: Female anterior region; B: Female labial region; C-G: Female stylets; H-J: Perineal pattern; K: Male anterior region; L, M-P: Male labial region; Q-S: Male stylet; T: Male lateral field; U: Spicules; V, W: Male tail. (After Hirschmann, 1986.)*

Measurements

See Table 28.

- *Eggs* (n = 50): L = 91.5 (80-105) μm; W = 42.4 (37-52) μm; L/W = 2.2 (1.7-2.6).

Description

Female

Body ivory coloured, ovoid to globular, characteristically with long neck, posteriorly rounded, without tail protuberance. Labial region not offset from body, not annulated. Labial cap broad, squarish. First 8-10 annuli posterior to labial region smaller than remaining body annuli. Hexaradiate labial framework weakly sclerotised, lateral sectors slightly enlarged, vestibule and vestibule extension distinctly sclerotised. Stylet small, easily dislodged posteriorly. Stylet cone slightly curved dorsally, widening gradually posteriorly. Shaft of same width throughout, or widening slightly near junction with knobs. Stylet knobs

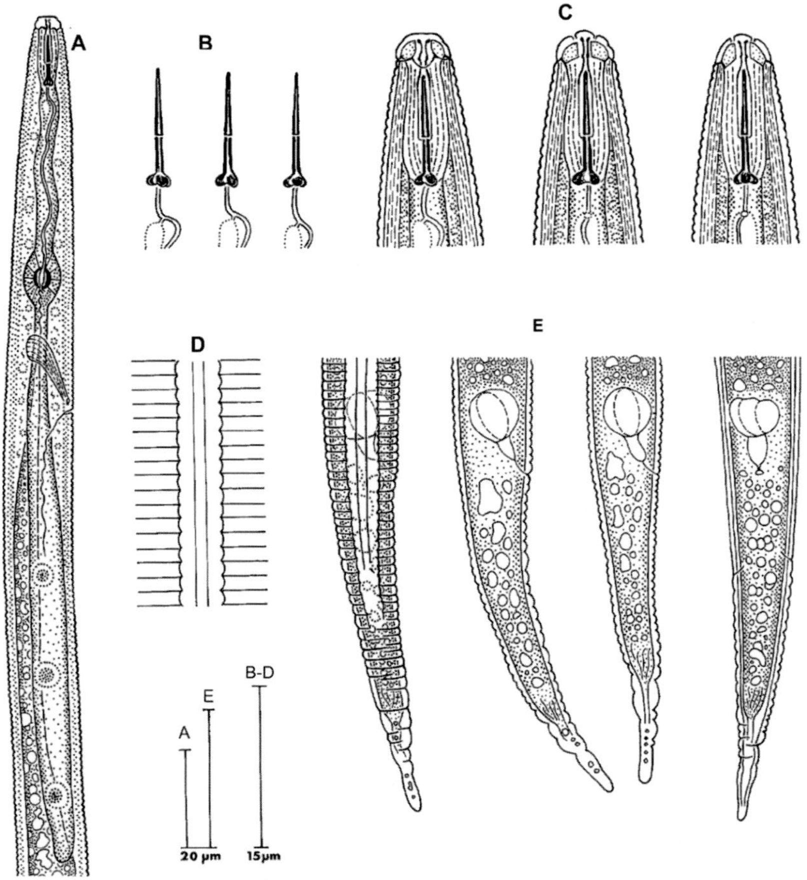

Fig. 107. Meloidogyne hispanica. *A: Second-stage juvenile (J2) anterior region; B: J2 stylet; C: J2 labial region; D: J2 lateral field; E: J2 tail. (After Hirschmann, 1986.)*

broad with indented anterior margins, distinctly separate, usually offset from shaft, sometimes slightly sloping posteriorly. Distance between stylet base and dorsal pharyngeal gland orifice short to moderately long (2.8-4.0 μm), gland orifice branching into three ducts, dorsal gland ampulla large, subventral gland orifices branched, located posteriorly to enlarged triradiate lumen lining of median bulb valve. Pharyngeal lumen lining with small knots anterior to DGO and median bulb valve. Pharyngeal glands large, trilobed, dorsal lobe largest, two subventral lobes variable in size, shape, and position, located posteriorly to dorsal

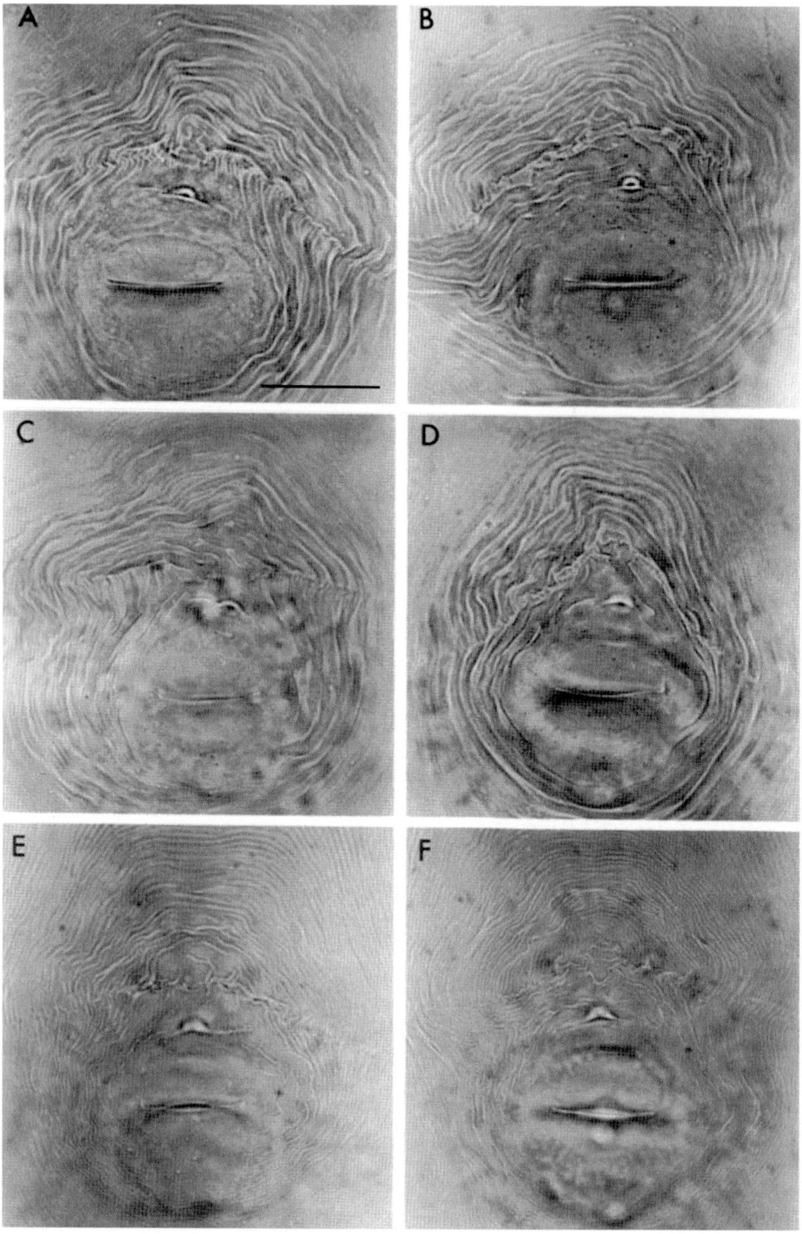

Fig. 108. Meloidogyne hispanica. *LM. A-F: Perineal pattern. (Scale bar: 20 μm.) (After Hirschmann, 1986.)*

gland lobe. Pharyngo-intestinal cells two, near junction of median bulb and intestine. Excretory pore position variable, close to labial region, in most specimens at level of DGO; terminal excretory duct very long. Perineal patterns oval-shaped to rectangular. Dorsal arch generally low, some patterns with higher squarish arch. Dorsal striae varying from fine and wavy to coarse. Ventral pattern area generally with fine, smooth striae, may form ventral wings on one side. Lateral lines distinctly forked, frequently spaced widely, with fringe-like striae between lines. Tail tip area well defined, free of striae. Perivulval region not striated, rarely striae near lateral edges of vulva. Vulval edges slightly crenate. Phasmidial ducts distinct, no phasmid surface structure apparent in SEM.

Male

Generally large nematodes, rather robust, body tapering anteriorly, bluntly rounded posteriorly, tail twisting through 90° in heat-killed specimens. Body annuli large, distinct. Lateral field with four incisures, rarely with a fifth central, broken incisure, faint areolation in outer fields throughout; on tail region areolation extending into central field. Lateral field bulging slightly near level of stylet base. Labial cap in lateral view high and rounded, extending posteriorly on to distinctly offset labial region. In SEM (face view), labial disc very large, wider than long axis of median labials, elongated, and slightly raised. Median labials narrow, crescentic with smooth outer margins. Distinct indentations frequently seen at junction of labial disc and median labials. Labial disc and median labials fused to form elongate labial structures extending posteriorly on to labial region. Occasionally, slight dorsoventral asymmetry of labial structures. Lateral labials generally indicated or fully formed. Labial region usually smooth but may have up to three incomplete annulations. Prestoma large, hexagonal. Stoma opening slit-like, located in large, hexagonal prestomatal cavity. Six inner labial sensilla small pits, opening at edge of, or into, prestomatal cavity. Labial sensilla distinct. Amphidial apertures large, elongate slits. Labial framework well sclerotised, vestibule and vestibule extension distinct. Stylet robust, large; cone straight, pointed, gradually increasing in diam. posteriorly, stylet opening marked by slight protuberance several micrometers from stylet tip, shaft cylindrical, knobs large, rounded, slightly offset from shaft, rarely sloping posteriorly. Dorsal gland duct branched into three ducts, gland ampulla poorly defined. Procorpus distinctly outlined, median bulb elongate, oval-shaped with large valve. Subventral gland openings branched,

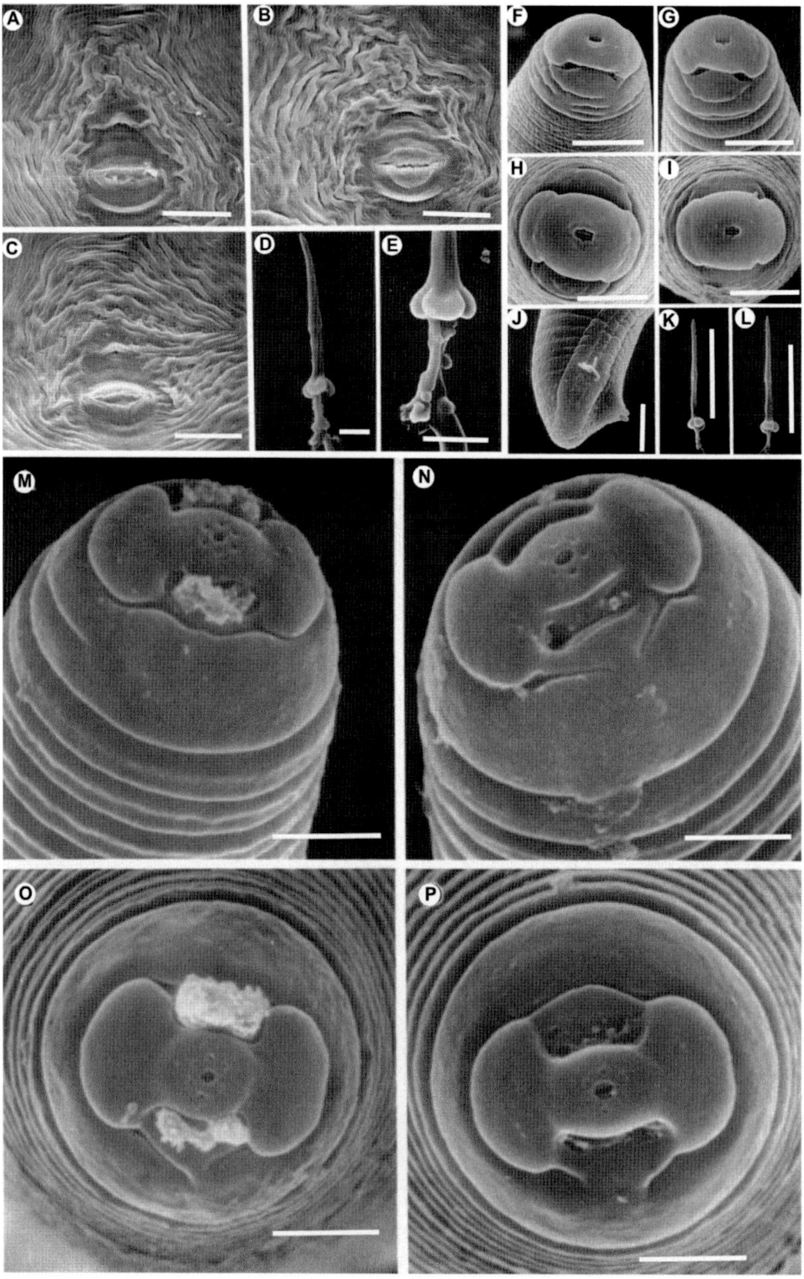
A
B
C
D
E
F
G
H
I
J
K
L
M
N
O
P

Fig. 109. Meloidogyne hispanica. *SEM. A-C: Perineal pattern; D, E: Female stylet; F-I: Male labial region; J: Male tail; K, L: Male stylet; M-P: Male lip region; N-O: Second-stage juvenile labial region. (Scale bars: A-C = 20 μm; D, E = 5 μm; F-I = 10 μm; K-L = 20 μm; M-P = 5 μm.) (After Hirschmann, 1986.)*

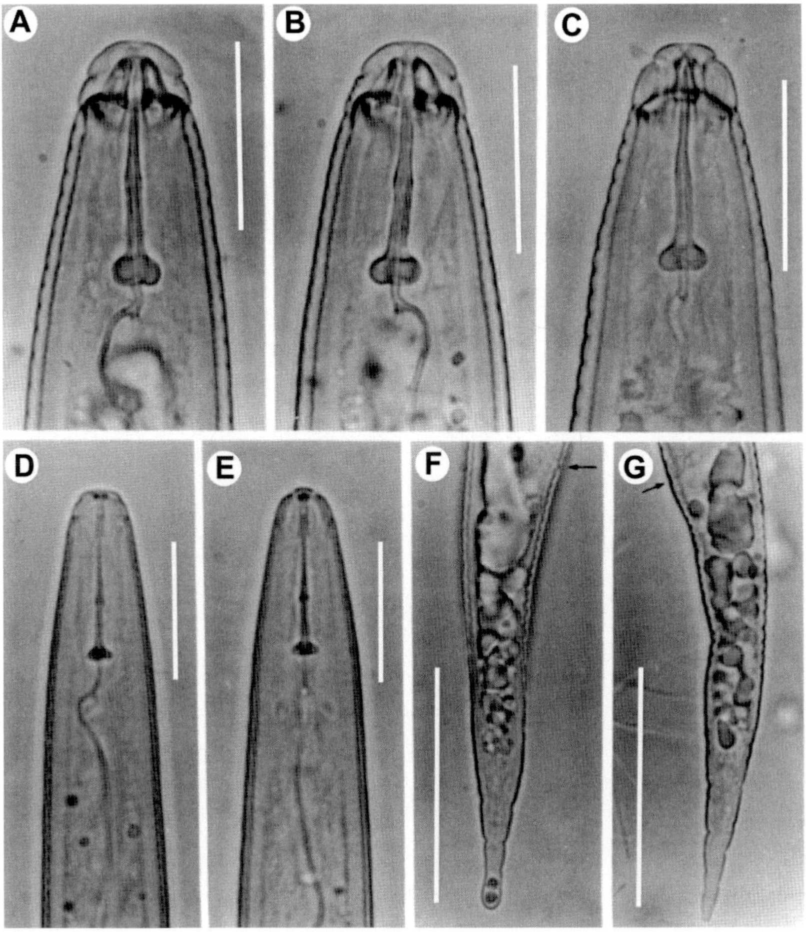

Fig. 110. Meloidogyne hispanica. *LM. A-C: Male labial region; D, E: Second-stage juvenile (J2) labial region; F, G: J2 tail (anus arrowed). (Scale bars: A-C = 20 μm; D, E = 10 μm; F, G = 20 μm.) (After Hirschmann, 1986.)*

Table 28. *Morphometrics of females, males and second-stage juveniles (J2) of* Meloidogyne hispanica. *All measurements are in μm and in the form: mean ± s.d. (range).*

Character	Holotype, Allotype, Seville, Spain, Hirschmann (1986)	Paratypes, Seville, Spain, Hirschmann (1986)	Portugal, Maleita *et al.* (2012c)
Female (n)	-	30	-
L	995	830 ± 150.4 (570-1180)	723.3 ± 79.2 (570-920)
a	1.6	1.7 ± 0.3 (0.9-2.2)	1.5 ± 0.2 (1.2-2.0)
W	620	503 ± 108.9 (330-740)	503.8 ± 62.6 (365-660)
Stylet	13.9	14.1 ± 0.3 (13.5-14.5)	14.7 ± 1.3 (11.1-19.0)
Stylet knob width	-	4.6 ± 0.2 (4.1-5.1)	4.3 ± 0.3 (3.2-5.0)
Stylet knob height	-	2.5 ± 0.2 (2.1-2.8)	2.5 ± 0.28 (1.8-3.7)
DGO	4.0	3.2 ± 0.3 (3.0-4.0)	4.8 ± 0.9 (2.8-6.8)
Median bulb length	-	48.8 ± 4.8 (39.5-63.2)	-
Median bulb width	-	45.5 ± 3.9 (39.5-52.6)	-
Median bulb valve width	-	12.6 ± 0.5 (11.5-13.8)	-
Median bulb valve length	-	15.6 ± 1.4 (14.1-120.5)	-
Ant. end to excretory pore	33.2	30.5 ± 12.3 (12.0-71.0)	35.2 ± 10.3 (13.2-77.9)
Vulval slit	-	23.5 ± 1.2 (20.0-25.4)	25.8 ± 3.1 (14.3-34.3)
Vulval slit to anus distance	-	19 ± 1.3 (17.2-22.6)	-
Interphasmid distance	-	22.1 ± 2.8 (16.7-28.1)	24.2 ± 3.6 (15.7-35.7)
EP/ST	2.4	2.4 ± 0.7 (0.8-4.8)	-

Table 28. *(Continued.)*

Character	Holotype, Allotype, Seville, Spain, Hirschmann (1986)	Paratypes, Seville, Spain, Hirschmann (1986)	Portugal, Maleita *et al.* (2012c)
Male (n)	1	30	-
L	1770	1678 ± 168.4 (1341-1990)	1846 ± 170.1 (1400-2357)
a	50.3	41.2 ± 6.1 (31.4-61.4)	44.7 ± 4.3 (36.5-55.4)
c	128.3	128.1 ± 19 (98.2-172.6)	176.7 ± 36.9 (104.7-350.4)
T	45	48.7 ± 8.2 (33.0-62.0)	51.3 ± 8.0 (28.3-69.7)
W	-	41.1 ± 4.4 (32.5-47.5)	41.6 ± 4.8 (32.5-58.6)
Labial region height	-	12.8 ± 0.5 (11.8-13.7)	5.1 ± 1.2 (2.4-9.2)
Labial region diam.	-	7.6 ± 0.3 (7.0-8.1)	12.6 ± 0.6 (11.0-14.5)
Stylet	23.7	23.5 ± 0.6 (22.0-24.0)	23 ± 1.0 (20.0-24.7)
Stylet knob width	-	5.6 ± 0.3 (5.1-6.1)	5.1 ± 0.3 (4.2-5.8)
Stylet knob height	-	3.2 ± 0.2 (2.9-3.5)	3.2 ± 0.3 (2.4-4.0)
DGO	2.8	2.5 ± 0.5 (1.4-3.6)	3.4 ± 0.8 (2.1-7.9)
Ant. end to excretory pore	117.8	181.5 ± 20.3 (149.0-254.0)	185.4 ± 18.0 (147.5-240.0)
Spicules	31.8	32.1 ± 0.8 (31.0-34.0)	34.7 ± 2.1 (28.0-41.1)
Gubernaculum	7.9	8.3 ± 0.5 (7.5-9.0)	8.1 ± 0.8 (5.8-11.8)
Tail	-	13.3 ± 1.7 (11.0-16.2)	10.8 ± 1.9 (5.8-16.6)
J2 (n)	-	50	-
L	-	392.6 ± 18.7 (356-441)	418.4 ± 28.8 (320-514)
a	-	27.1 ± 1.2 (24.6-30.9)	28.7 ± 2.5 (22.4-35.7)

Table 28. *(Continued.)*

Character	Holotype, Allotype, Seville, Spain, Hirschmann (1986)	Paratypes, Seville, Spain, Hirschmann (1986)	Portugal, Maleita *et al.* (2012c)
c	-	8.5 ± 0.4 (7.7-9.4)	8.7 ± 0.9 (5.8-12.8)
W	-	14.5 ± 0.5 (13.5-16.0)	14.6 ± 1.2 (13.3-17.8)
Stylet	-	11.1 ± 0.3 (10.4-11.9)	10.6 ± 0.8 (9.2-13.0)
Stylet knob width	-	2.6 ± 0.1 (2.3-2.8)	2.1 ± 0.2 (1.6-2.9)
Stylet knob height	-	1.4 ± 0.1 (1.2-1.6)	1.2 ± 0.2 (1.0-1.6)
DGO	-	2.8 ± 0.3 (2.0-3.5)	3.6 ± 0.6 (2.4-5.3)
Median bulb valve width	-	4.1 ± 0.1 (3.8-4.4)	-
Median bulb valve length	-	4.6 ± 0.1 (4.3-4.8)	-
Labial end to excretory pore	-	80.1 ± 2.7 (74.0-86.0)	79.8 ± 3.8 (61.6-86.8)
Tail	-	46.4 ± 2.8 (41.0-53.5)	48.1 ± 3.8 (33.2-56.8)

located posteriorly to median bulb valve. Pharyngo-intestinal junction distinct. Gland lobe variable in length, two equal-sized nuclei variable in position. Intestinal caecum extending anteriorly on dorsal side to level of median bulb. Excretory pore variable in position, terminal duct long, curved, ending in excretory cell with large nucleus near right lateral chord. Hemizonid 1-5 annuli anterior to excretory pore. Usually one testis, rarely two, outstretched, or reflexed anteriorly. Sperm large, rounded, granular. Spicules long, slender, slightly arcuate with single tip. Gubernaculum crescentic. Tail short, elongate conoid. Phasmids at level of cloacal aperture, with slit-like openings in SEM.

J2

Body annuli distinct, increasing in size and becoming irregular in posterior tail region. Lateral field starting approximately at middle of procorpus and extending to near phasmid, 4.6-5.3 μm wide, with four incisures, not areolated, outer lines crenate. Labial region truncate, distinctly offset from body. Labial cap low, narrower than labial region. In SEM labial disc rounded, distinctly raised. Outer margins of median labials crescentic to rounded, smooth. Median labials and labial disc dumbbell-shaped. Lateral labials fused at right angle with median labials, lower than median labials, margins rounded to triangular, may fuse with labial region. Labial region smooth, occasionally with 1-2 short, broken annulations. Prestoma oval, stoma slit-like, large. Inner labial sensilla pit-like, large, opening on labial disc, symmetrically arranged around prestoma. Area around prestomatal opening not recessed. Labial sensilla indistinct. Labial framework weakly sclerotised. Vestibule and vestibule extension distinct. Stylet moderately sized, but delicate. Stylet cone sharply pointed, increasing in width gradually posteriorly, shaft cylindrical, may widen slightly posteriorly, knobs robust, distinctly separated, rounded, sloping posteriorly. Dorsal gland ampulla poorly defined. Procorpus faintly outlined, median bulb ovoid with prominent valve; isthmus not clearly outlined. Pharyngo-intestinal junction indistinct, at level of nerve ring. Pharyngeal gland lobe variable in length with three distinct nuclei about equal in size. Hemizonid 1-2 annuli anterior to excretory pore. Tail slender, ending in bluntly rounded tip. Posterior tail region, with large annuli of variable size, frequently appearing knobbly. Hyaline region indistinct. Rectal dilation large, filled with matrix material. Phasmids obscure, located a short distance posterior to anal opening.

TYPE PLANT HOST: Peach rootstock, *Prunus persica* (L.) Batsch.

OTHER PLANTS: The host range of *M. hispanica* includes the families Alliaceae, Apiaceae, Asteraceae, Brassicaceae, Caryophyllaceae, Chenopodiaceae, Cucurbitaceae, Fabaceae, Poaceae, and Solanaceae. *Meloidogyne hispanica* reproduced on tomato (*Solanum lycopersicum*) and has been detected on beet (*Beta vulgaris* L.), *Amaranthus* sp., banana, carnation (*Dianthus caryophyllum* L.), cucumber, fig tree (*Ficus carica* L.), granadilla (*Passiflora ligularis* A.Juss.), grapevine (*Vitis vinifera* L.), geranium (*Pelargonium notatum* (L.) L'Herit), sugarcane (*Saccharum officinarum* L.), snapdragon (*Antirrhinum majus* L.), squash (*Cucurbita moschata*), sunflower (*Helianthus annuus* L.), sweet pepper (*Capsicum annuum* L.), white mulberry (*Morus alba* L.), and also found to reproduce on wheat (*Triticum aestivum*) in glasshouse conditions (Esbenshade & Triantaphyllou, 1985a, b; Fargette, 1987a; Kleynhans, 1993; Hugall *et al.*, 1994; Karssen & van Hoenselaar, 1998; Trudgill *et al.*, 2000; Carneiro *et al.*, 2004b; Karssen, 2004; Cofcewicz *et al.*, 2005; Chaves *et al.*, 2007; Castillo *et al.*, 2009b; van der Wurff *et al.*, 2010; Tzortzakakis *et al.*, 2014; Shokoohi *et al.*, 2016). Melon (*Cucumis sativus* L.) 'Gazver' and 'Jazzer', bean (*Phaseolus vulgaris*) 'Foicinha' and 'Rajado', corn (*Zea mays*) 'Belgrano' and 'PR35P12', tomato (*Solanum lycopersicum*) 'Roma', 'Rutgers' and 'Sinatra', and potato 'Baraka' and 'Diana', were rated as susceptible to *M. hispanica* (Maleita *et al.*, 2005). Furthermore, lettuce (*Lactuca sativa* L.) 'Apulia' and 'Esperie' and tomato 'Viriato F1' exhibited galls but did not support nematode reproduction while the pepper 'Galileo' was resistant (Maleita *et al.*, 2005). Maleita *et al.* (2012b) also evaluated the reproduction of *M. hispanica* in 63 species/cultivars in pot assays. Cultivars of aubergine, bean, beetroot, broccoli, carnation, corn, cucumber, French garlic, lettuce, melon, onion, parsley, pea, potato, spinach, tobacco, and two of cabbage were susceptible (Maleita *et al.*, 2012b). Cabbage 'Bacalan', cauliflower 'Temporão' and pepper 'Zafiro R2' were poor hosts, and pepper 'Aurelio' and 'Solero' were resistant (Maleita *et al.*, 2012b).

TYPE LOCALITY: Peach orchard, Seville province (Andalusia), Spain.

DISTRIBUTION: This species has a worldwide distribution and has been reported infecting economically important crops in *Africa*: Burkina Faso, Malawi, South Africa (Kleynhans, 1991, 1993; Trudgill *et al.*, 2000; Onkendi *et al.*, 2014); *Asia*: China (Fu *et al.*, 2011), Iran (Shokoohi *et al.*, 2016), Korea (Esbenshade & Triantaphyllou, 1985a); *Europe*: Greece, The Netherlands, Portugal, Spain (Hirschmann, 1986; Karssen, 2004; Abrantes *et al.*, 2008; Castillo *et al.*, 2009b; van der Wurff *et al.*, 2010; Tzortzakakis *et al.*, 2014); *North America*: USA (South Carolina) (Handoo *et al.*, unpubl.); *Central America and Caribbean*: Costa Rica, Martinique; *South America*: Brazil, Ecuador, French Guiana (Janati *et al.*, 1982; Carneiro *et al.*, 2004b; Cofcewicz *et al.*, 2005); and *Oceania*: Australia, Fiji (Esbenshade & Triantaphyllou, 1985a, b).

PATHOGENICITY: The pathogenicity of *M. hispanica* was assessed on 25 tomato genotypes, 60 days after inoculation with 5000 eggs based on root gall index and reproduction factor (Rf) by Maleita *et al.* (2011). All the tomato genotypes were susceptible (excellent or good hosts), except the genotype 'Rapit', considered as resistant (poor host). Significant differences in reproduction were detected between the *Mi* allelic conditions and genotypes within *Mi* allelic conditions. The increasing number of *Mi* alleles (0, 1 or 2) was associated with decreasing Rf, which suggests a possible dosage effect of the *Mi* gene (Maleita *et al.*, 2011). Maleita *et al.* (2012a) studied the thermal requirements for *M. hispanica* under growth chamber conditions in tomato and constant temperatures (10-35°C), showing that no egg development occurred at 10 or 35°C. An increase in invasion of tomato roots by *M. hispanica* J2 was correlated with an increase in temperature. Resistant tomato (*Mi*-1.2 gene) limited reproduction of *M. hispanica* at 20, 25, and 30°C, but not at 35°C, indicating that these high temperatures blocked the resistance mechanism provided by the *Mi*-1.2 gene (Maleita *et al.*, 2012a).

Gene silencing using RNAi is a powerful tool for functional analysis of nematode genes and can provide a new strategy for the management of RKN. The transcript of the venom allergen-like protein gene (*Mhi-vap*-1) of *M. hispanica* is localised in the subventral pharyngeal gland cells of the J2 and the gene is highly transcribed in this developmental nematode stage. The purpose of this study was to assess whether the silencing of the *Mhi-vap*-1 gene could affect nematode attraction to roots, penetration, development and reproduction in tomato plants. Duarte *et al.* (2017) determined by quantitative RT-PCR analysis the relative expression of the *Mhi-vap*-1 gene in J2 incubated with a soaking solution. Silencing of the *Mhi-vap*-1 gene interfered with the completion of the nematode life cycle and caused a reduction in nematode attraction to roots, penetration and infection of plants. A small difference in the number of females and galls formed was also observed, which caused a small decrease in the nematode reproduction factor. The use of RNAi silencing of the *Meloidogyne* effector gene *Mhi-vap*-1 showed that this gene is important for the plant-nematode interaction during the early events of infection and could be a target gene for anti-nematode strategies (Duarte *et al.*, 2017).

CHROMOSOME NUMBER: Cytologically, *M. hispanica* is similar to the diploid race of *M. arenaria* and differs from the diploid race of *M. incognita* in that prophase I is not prolonged and the chromosomes do not clump together; 2n = 33-36 (Triantaphyllou, 1985b).

POLYTOMOUS KEY CODES: *Female*: A213, B32, C2134, D5; *Male*: A213, B2, C2, D1, E2, F2; *J2*: A32, B32, C3, D2, E23, F3.

BIOCHEMICAL AND MOLECULAR CHARACTERISATION: Biochemically, *M. hispanica* has a characteristic esterase phenotype (two bands-Hi2 or four bands of esterase activity-two major bands and two minor and fainter bands-S2-M1 or Hi4 (Rm: 0.32, 0.35, 0.38 and 0.41) (Fig. 12), different from that of other *Meloidogyne* species (Maleita *et al.*, 2012c; Santos *et al.*, 2019); and a N1 malate dehydrogenase phenotype (Esbenshade & Triantaphyllou, 1985a, b).

Molecularly this nematode can be identified by PCR-RFLP analysis and sequencing of the mtDNA region from *COII* and 16S rRNA genes (Maleita *et al.*, 2012c). The 18S, ITS and D2-D3 of 28S rRNA gene sequences were not able to discriminate *M. hispanica* reliably from the other RKN species.

RELATIONSHIPS (DIAGNOSIS): The species belongs to Molecular group I (Fig. 18) and is similar to species of the Ethiopica group, but can be separated by perineal pattern. This species can also be differentiated from other species by the esterase phenotype and the *COII* and 16S rRNA gene sequences.

39. *Meloidogyne ichinohei* Araki, 1992
(Figs 111, 112)

COMMON NAME: Japanese root-knot nematode.

Araki (1992a) described this species in roots of *Iris laevigata* Fisch. (rabbit ear iris), cultivated in a paddy field from Kyushu, Japan.

MEASUREMENTS *(AFTER ARAKI, 1992A)*

- *Holotype female*: L = 857.5 μm; W = 396.5 μm; stylet = 13.0 μm; DGO = 4.1 μm; excretory pore to anterior end = 49.9 μm; a = 2.2; EP/ST = 3.8.
- *Allotype male*: L = 1451 μm; W = 44.4 μm; a = 32.7; c = 105.1; stylet = 16.6 μm; DGO = 6.9 μm; median bulb to anterior end = 71.4 μm; excretory pore to anterior region = 140.8 μm; spicules = 32.9 μm; gubernaculum = 9.3 μm.

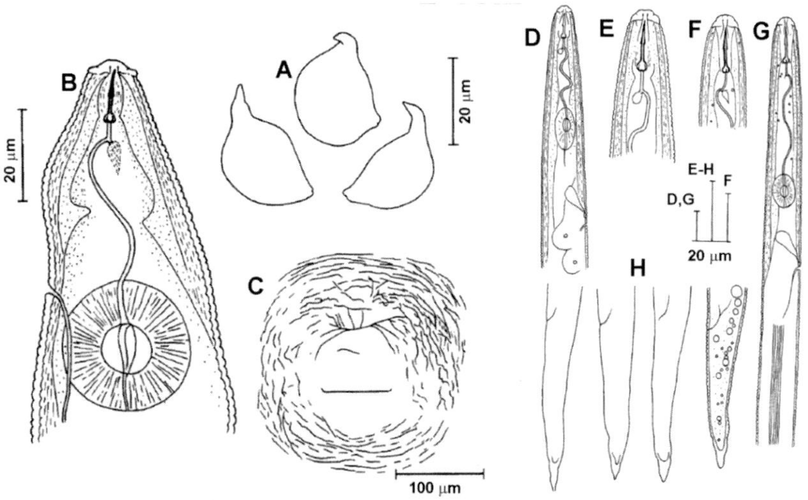

Fig. 111. Meloidogyne ichinohei. *A: Entire female; B: Female anterior region; C: Perineal pattern; D: Male anterior region; E: Male labial region; F: Second-stage juvenile (J2) labial region; G: J2 anterior region; H: J2 tail. (After Araki, 1992a.)*

- *Paratype females* (n = 10): L = 796.1 ± 114.5 (619.8-1019) μm; W = 393.2 ± 93.3 (300-575) μm; stylet = 12.3 ± 0.8 (11.0-13.6) μm; stylet knob height = 2.1 ± 0.2 (1.8-2.4) μm; stylet knob width = 4.1 ± 0.4 (3.3-4.6) μm; DGO = 5.4 ± 0.9 (3.8-6.8) μm; excretory pore to anterior end = 48.7 ± 8.3 (36.7-62.0) μm; a = 2.1 ± 4.5 (1.5-2.6); interphasmid distance = 14 μm; vulval slit length = 34.2 ± 6.2 (25.5-46.1) μm; vulva-anus distance = 25.1 ± 4.5 (14.1-29.6) μm; EP/ST = 3.6 ± 0.6 (2.9-4.7).
- *Paratype males* (n = 2): L = 1516 (1451-1581) μm; stylet = 17.0 (16.6-17.4) μm; stylet knob height = 3.1-3.2 μm; stylet knob width = 4.0 (3.8-4.2) μm; DGO = 6.2 (6.1-6.9) μm; excretory pore to anterior end = 135.9 (131.0-140.8) μm; spicules = 33.3 (32.9-33.7) μm; gubernaculum = 8.9 (8.4-9.3) μm; a = 35.5 (32.7-38.3); c = 113.9 (113.3-114.5).
- *Paratype J2* (n = 90): L = 469.6 ± 20.0 (413-524) μm; W = 15.0 ± 0.5 (14.1-16.0) μm; stylet = 11.3 ± 0.7 (9.7-12.9) μm; DGO = 5.2 ± 0.6 (3.6-6.4) μm; excretory pore to anterior end = 72.8-100.0 μm; tail = 54.2 ± 6.0 (37.5-69.6) μm; hyaline region = 7.7 ± 2.8 (2.9-

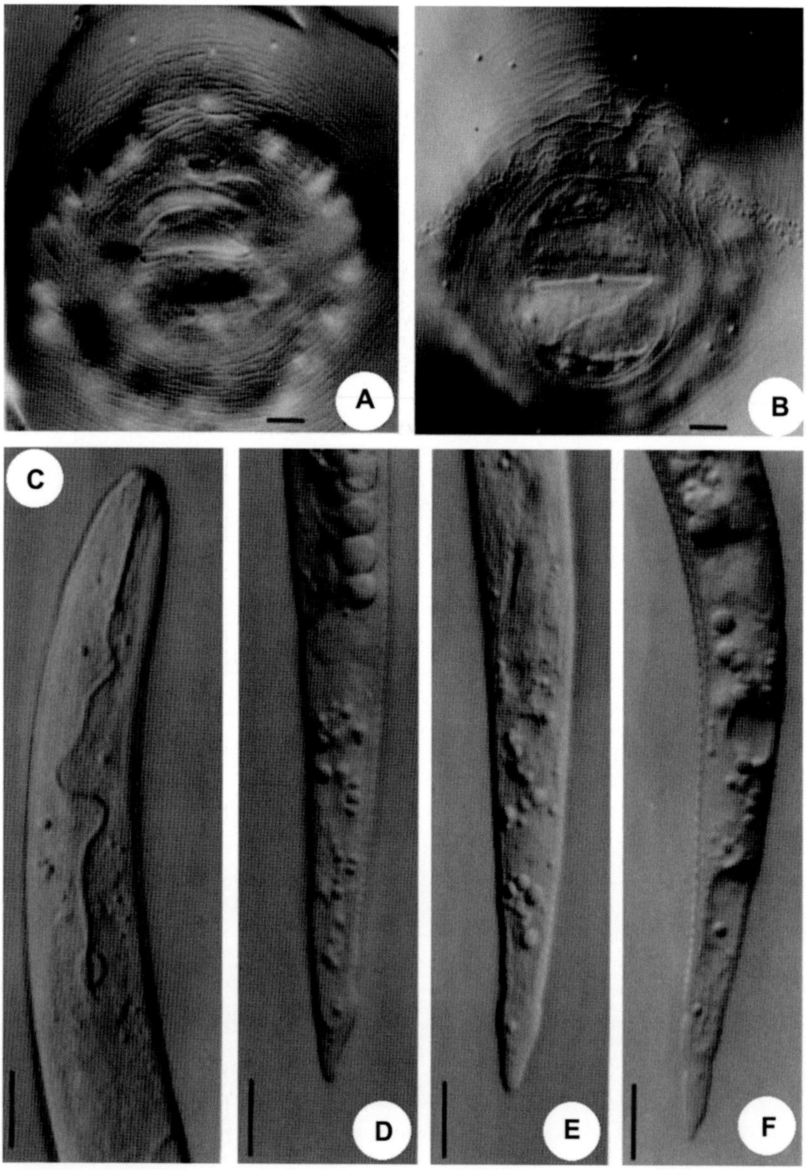

Fig. 112. Meloidogyne ichinohei. *LM. A, B: Perineal pattern; C: Second-stage juvenile (J2) anterior region; D-F: J2 tail. (Scale bar = 5 μm.) (After Araki, 1992a.)*

17.7) μm; a = 31.3 ± 1.7 (27.1-34.9); c = 8.8 ± 0.8 (7.0-12.1); c′ = 5.3 ± 0.7 (3.7-6.8).
- *Eggs* (n = 50): L = 23.4 (19.0-29.0) μm; W = 11.2 (10.0-12.0) μm; L/W = 2.1 (1.6-2.6).

DESCRIPTION

Female

Body globular to ellipsoidal, pearly white, variable in size; neck prominent, always situated on ventral side of body, posterior end of body with prominent and undulate posterior protuberance. Body cuticle annulated. Labial region offset, usually marked by an annulus. Labial cap distinct, labial framework weakly sclerotised, vestibule and vestibule extension distinct. Stylet relatively short and delicate, stylet cone dorsally curved, shaft cylindrical. Stylet knobs separated, offset from shaft, transversely ovoid. Excretory pore located about four times stylet length posterior to stylet knobs. Perineal patterns difficult to prepare due to prominent posterior protuberance. Area around anus usually covered by fold of tail tip. Shape of perineal pattern rounded, striae extremely faint, broken and somewhat spaced. Vulval slit like, without surrounding striae, phasmids small, obscure. Dorsal arch low and rounded, lateral lines absent, sometimes with lines of particles instead. Ventral pattern region rounded. Spermatheca with 18-30 cells forming irregular lobes (Bert *et al.*, 2002).

Male

Rare, body vermiform, tapering anteriorly, bluntly rounded posteriorly. Cuticle with distinct annulations. Lateral field with seven or eight lateral lines, not areolated. Labial region offset, labial cap prominent, narrower than labial region. Labial framework moderately sclerotised. Stylet moderately developed, cone straight, slightly longer than shaft, tip pointed, base of cone broadened near junction with shaft. Shaft cylindrical, slightly broadened posteriorly. Knobs offset from shaft, rounded, sloping posteriorly. Median bulb oval, with large valve. Hemizonid anterior to or just at level of excretory pore. Testis one, directed anteriorly. Spicules identical. Labial region rounded, offset. Blade arcuate, tapering towards tip.

J2

Body vermiform, tapering at both ends, more so posteriorly. Body annulation distinct, becoming larger and irregular in posterior tail region. Lateral field with six lateral lines, not areolated. Labial region slightly offset, no annulation seen in light microscopy. Labial cap high, narrower than labial region. Labial framework weakly sclerotised; vestibule and vestibule extension distinct. Stylet slender, stylet cone straight, pointed, gradually increasing in width posteriorly, shaft cylindrical, may widen slightly posteriorly. Knob small, slightly separate, rounded. Procorpus faintly outlined. Median bulb oval, with prominent valve. Pharyngo-intestinal junction indistinct, at level of nerve ring. Gland lobe variable in length, with three nuclei. Hemizonid just anterior to excretory pore, sometimes appearing posterior depending on angle of observation. Excretory pore located at 16.3-20.5% of body length. Tail tapering gradually towards end, usually abruptly narrowing at level of hyaline region, hyaline region distinct, tip rounded, shape of hyaline tail triangular, but tail tip variable even from a cohort of one female, sometimes protruding to finely rounded tip with various intermediate forms observed. Rectum not dilated. Phasmid obscure, 1/2-2/3 tail length posterior to anal opening.

Eggs

Morphology similar to that of eggs of other *Meloidogyne* species. Eggshell without markings in light microscopy.

TYPE PLANT HOST: Rabbit ear iris, *Iris laevigata* Fisch.

OTHER PLANTS: No other hosts were reported.

TYPE LOCALITY: Hisayama, Fukuoka, Kyushu, Japan.

DISTRIBUTION: *Asia*: Japan.

PATHOGENICITY: No damage has been reported on type host plant.

CHROMOSOME NUMBER: No available data.

POLYTOMOUS KEY CODES: *Female*: A21, B34, C12, D1; *Male*: A32, B43, C2, D2, E1, F2; *J2*: A21, B32, C312, D4123, E23, F213.

BIOCHEMICAL AND MOLECULAR CHARACTERISATION: Biochemically, *M. ichinohei* has a characteristic esterase phenotype (IC2), which is different from that of other *Meloidogyne* species, and a malate dehydrogenase N1 phenotype (Janssen *et al.*, 2017). This nematode was characterised by *COI* (Janssen *et al.*, 2017) and rRNA gene sequences (18S, ITS and D2-D3 of 28S rRNA) (De Ley *et al.*, 1999, 2002; Holterman *et al.*, 2008; Rybarczyk-Mydlowska *et al.*, 2014).

RELATIONSHIPS (DIAGNOSIS): The species belongs to Molecular group X and is clearly molecularly differentiated from all other *Meloidogyne* species (Fig. 18). *Meloidogyne ichinohei* has a distinct perineal pattern. *Meloidogyne ichinohei* is close to *M. kralli* but differs from it in having a setoff labial region, more posteriorly situated excretory pore (48.7 *vs* 15.8 μm), fewer and shorter striae in the perineal pattern of the female, longer DGO (5.2 *vs* 4.4 μm), non-dilated rectum, shorter hyaline region (7.7 *vs* 17.4 μm), and blunter J2 tail tip.

40. *Meloidogyne incognita* (Kofoid & White, 1919) Chitwood, 1949
(Figs 113-115)

COMMON NAME: Southern root-knot nematode.

Meloidogyne incognita is considered to be one of the major *Meloidogyne* species due to its worldwide economic importance. It is commonly encountered in tropical regions and in temperate regions restricted to glasshouses.

Kofoid & White (1919) described *Oxyuris incognita* from eggs isolated from the faeces of soldiers at Camp Travis, Texas and elsewhere in the USA. Sandground (1923) proved by experimentation that these eggs were those of *Heterodera radicicola* (as *Meloidogyne* spp. were then called), ingested in infected plant material. Chitwood (1949) considered that his RKN from carrot, Texas, was conspecific with *O. incognita* calling it *Meloidogyne incognita*. Descriptions of *M. incognita* are given by many authors (Chitwood, 1949; Whitehead, 1968; Eisenback, 1982; Eisenback & Hirschmann, 1979, 1980, 1981;

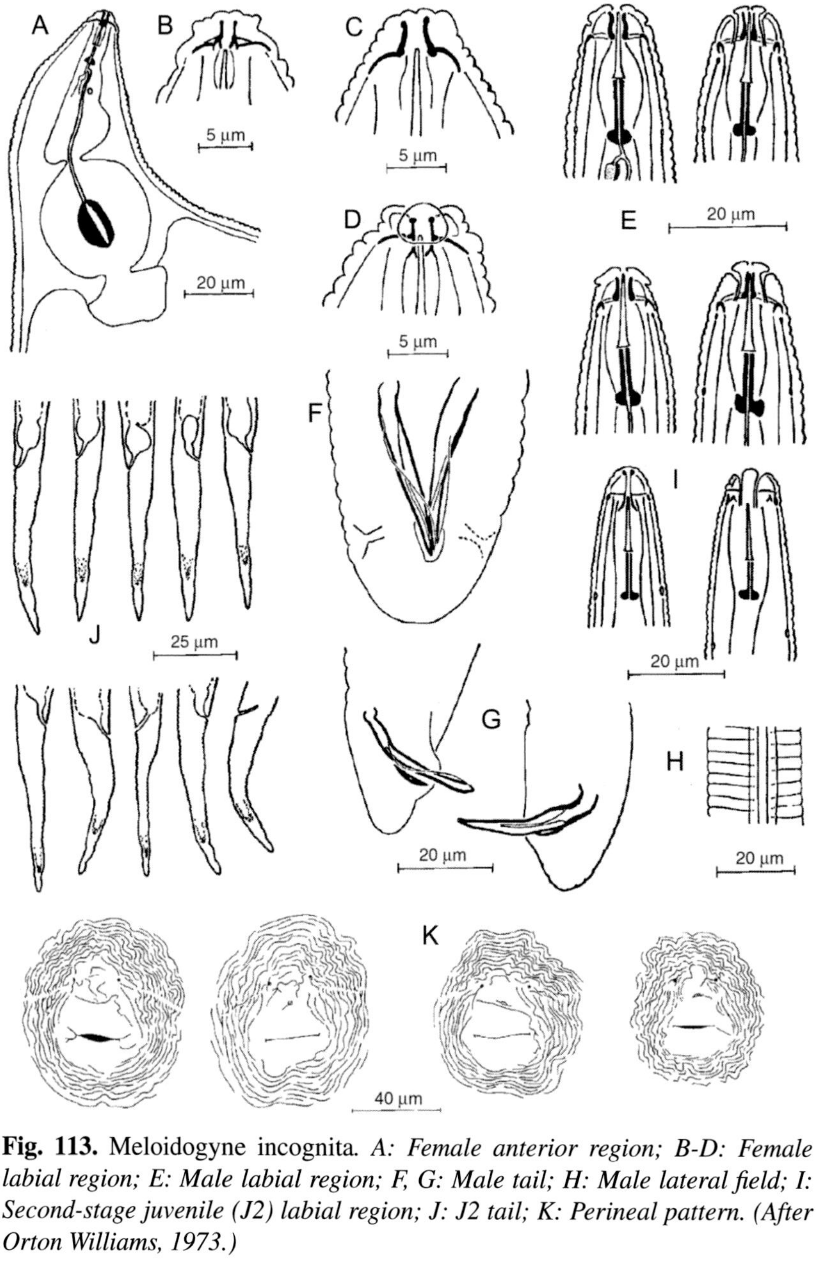

Fig. 113. Meloidogyne incognita. *A: Female anterior region; B-D: Female labial region; E: Male labial region; F, G: Male tail; H: Male lateral field; I: Second-stage juvenile (J2) labial region; J: J2 tail; K: Perineal pattern. (After Orton Williams, 1973.)*

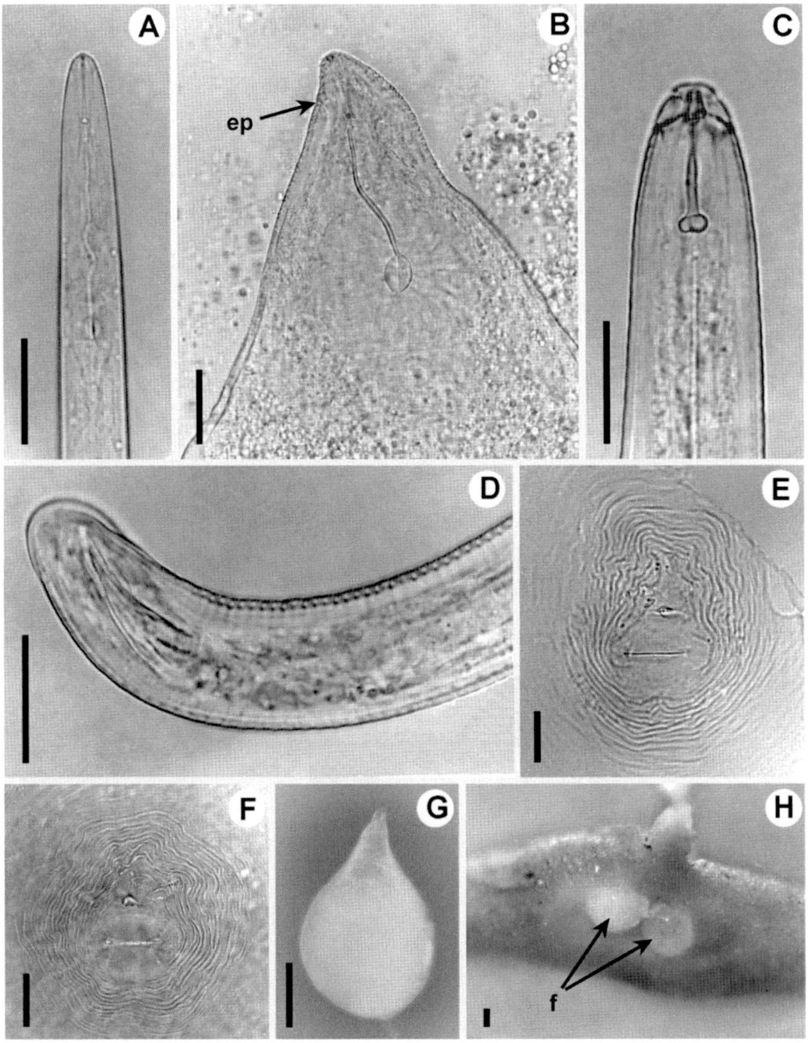

Fig. 114. Meloidogyne incognita. *LM. A: Second-stage juvenile anterior region; B: Female anterior region showing excretory pore (ep); C: Male anterior region; D: Male tail; E, F: Perineal pattern; G: Entire female; H: Root gall on celery with females (f). (Scale bars: A-F = 20 μm; G = 200 μm; H = 150 μm.) (Courtesy of N. Vovlas.)*

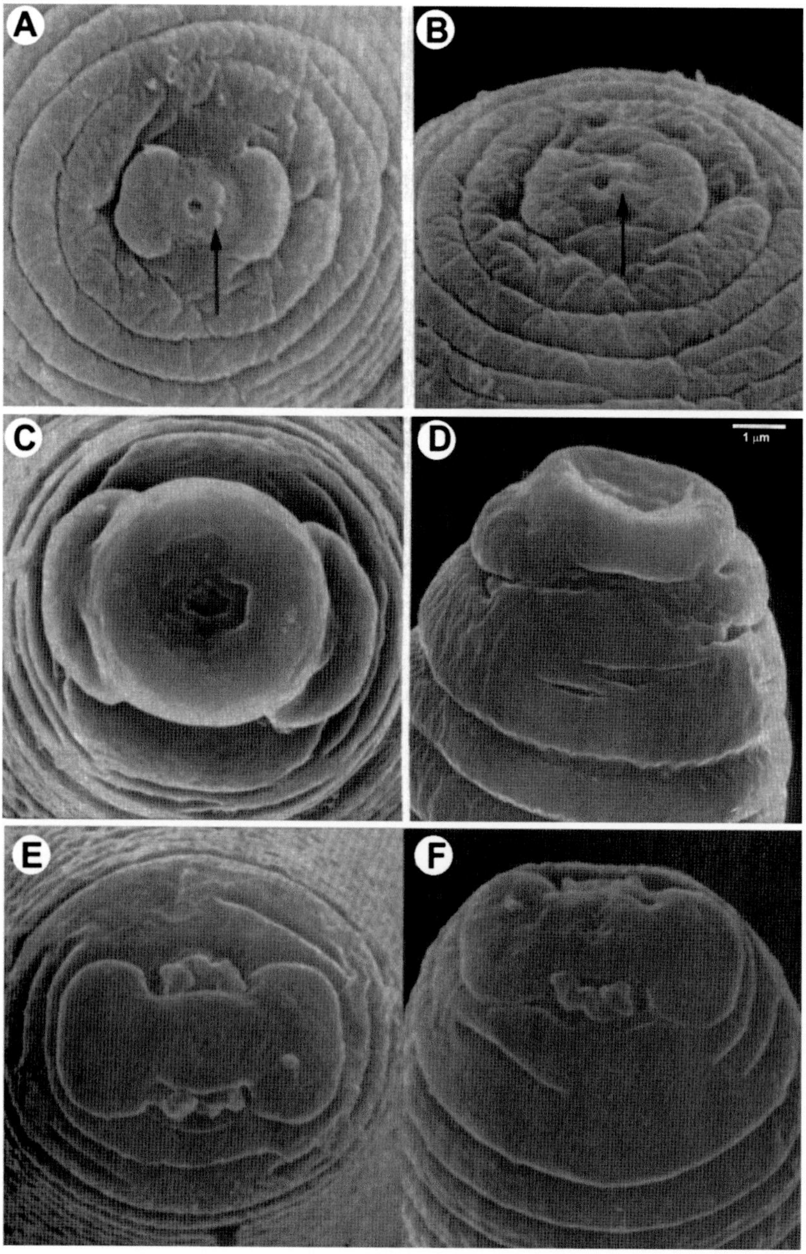
A
B
C
D
1 μm
E
F

Fig. 115. Meloidogyne incognita. *SEM. A, B: Female labial region; C, D: Male labial region; E, F: Second-stage juvenile labial region. Prestoma arrowed. (Scale bars: A-F = 1 μm.) (After Eisenback & Hirschmann, 1979b, 1980; Eisenback* et al.*, 1980.)*

Eisenback *et al.*, 1981; Jepson, 1987; Brito *et al.*, 2004; Kaur & Attri, 2013; Sumita *et al.*, 2018 and others). Recently, Monteiro *et al.* (2019), based on detailed morphological, biochemical and molecular analyses, synonymised *M. polycephannulata* with *M. incognita*.

Measurements

See Table 29.
Eggs (n = 20): L = 77 (63-90) μm; W = 32 (24-37) μm.

Description

Female

Body spherical with projecting neck, no posterior protuberance. Labial region with two or occasionally three annuli posterior to labial cap. Cuticle thickening abruptly at base of retracted stylet. Stylet knobs rounded or drawn out laterally. Excretory pore at level of, or posterior to, stylet knobs, 10-20 annuli posterior to labial region. Perineal pattern probably most variable of all known species. Typical perineal pattern oval to rounded, 'Incognita' type with striae closely spaced, very wavy to zigzag especially dorsally and laterally. Dorsal arch high, rounded. Lateral field not clear, sometimes marked by breaks in striae, broken ends often forked, pattern merging into body striae. 'Acrita' type with striae smoother, more widely spaced (or with coarse widely spaced striae separated by fine closely spaced striae visible for short distances). Dorsal arch variable, may be flattened at top or trapezoid. Striae often forked along a 'lateral line'. Limits of pattern more or less well defined. Jepson (1987) noticed that one of the most characteristic feature of the pattern is the ∧-shaped formed by striae in region just dorsal to tail. Aberrant patterns also occur, *e.g.*, Dropkin (1953) "Type E". For photomicrographs of patterns see Taylor *et al.* (1955), Triantaphyllou & Sasser (1960) and Whitehead (1968). Phasmids not prominent and about same distance apart as width of vulva.

Table 29. *Morphometrics of females, males and second-stage juveniles (J2) of* Meloidogyne incognita. *All measurements are in μm and in the form: mean ± s.d. (range).*

Character	Paratypes, Texas, USA, Chitwood (1949)	Whitehead (1968)	Venezuela, Perichi & Crozzoli (2010)	Passion fruit, India Khan *et al.* (2017)	Tomato, Brazil Monteiro *et al.* (2019) = *M. polycephannulata*
Female (n)	-	20	20	20	30
L	510-690	609 (500-723)	775.2 ± 72.8 (664-896)	629 ± 51.6 (510-770)	864 ± 23.2 (700-1100)
a	-	-	1.3 ± 0.1 (1.0-1.5)	-	-
W	300-430	415 (331-520)	616.5 ± 48.9 (516-709)	415 ± 27.4 (350-460)	684 ± 14.2 (595-800)
Stylet	15.0-16.0	14.0 (13.0-16.0)	15.3 ± 1.4 (13.0-17.7)	15.9 ± 0.8 (15.2-18.1)	12.9 ± 0.2 (12.0-14.0)
DGO	-	3.0 (2.0-4.0)	3.9 ± 0.8 (3.2-4.8)	3.9 ± 0.4 (3.3-4.8)	3.6 ± 0.1 (2.5-4.2)
Median bulb length	-	46.0 (37.0-63.0)	-	38.3 ± 2.6 (33.3-42.8)	-
Median bulb width	-	39.0 (31.0-49.0)	-	34.4 ± 2.7 (26.6-47.5)	-
Median bulb valve width	-	12 (11-13)	-	-	-
Median bulb valve length	-	14 (13-16)	-	-	-
Ant. end to excretory pore	-	-	28.0 ± 3.8 (19.3-33.9)	34.9 ± 7.8 (26.6-47.5)	43 ± 2.3 (21-60)
Vulval slit		-	25.0 ± 2.3 (19.3-27.4)	-	-
EP/ST	-	1.4	1.9 ± 0.4 (1.2-2.5)	-	-
Male (n)	-	14	20	4	30
L	-	1583 (1108-1953)	1488 ± 287 (935-2031)	1830 ± 177.7 (1600-2020)	1851 ± 36.1 (1500-2200)
a	-	46.3 (31.4-55.4)	47.8 ± 8.0 (31.0-60.5)	45.9 ± 3.2 (42.1-48.8)	47 ± 0.7 (39.5-52.4)
b′	-	17.4 (13.8-20.5)	-	19.1 ± 0.4 (18.7-19.6)	-
c	-	146 (97-255)	-	-	49.9 ± 0.9 (41.9-56.6)
W	-	-	31.2 ± 3.8 (22.6-37.1)	39.9 ± 3.0 (36.1-42.8)	39.3 ± 0.4 (35-42)

Table 29. *(Continued.)*

Character	Paratypes, Texas, USA, Chitwood (1949)	Whitehead (1968)	Venezuela, Perichi & Crozzoli (2010)	Passion fruit, India Khan *et al.* (2017)	Tomato, Brazil Monteiro *et al.* (2019) = *M. polycephannulata*
Stylet	-	25.0 (23.0-27.0)	22.9 ± 2.0 (19.0-25.8)	22.6 ± 0.9 (21.9-23.8)	23.9 ± 0.3 (22-26)
Stylet knob width	-	-	-	4.9 ± 0.2 (4.8-5.3)	4.9 ± 0.1 (4.5-5.5)
Stylet knob height	-	-	-	3.0 ± 0.2 (2.9-3.3)	3.0 ± 0.1 (2.5-3.5)
DGO	-	2.1 (1.4-2.5)	3.5 ± 1.1 (1.6-4.8)	2.4 ± 0.4 (1.9-2.9)	3.0 ± 0.1 (2.3-3.5)
Median bulb length	-	20 (14.4-25.2)	-	-	-
Median bulb width	-	11.2 (8.6-15.8)	-	-	-
Median bulb valve length	-	7.2 (5.8-9.0)	-	-	-
Ant. end to excretory pore	-	-	152.7 ± 18.6 (124.1-185.4)	160.1 ± 12.5 (142-172)	132 ± 15 (117-200)
Spicules	-	35.2 (28.8-40.3)	28.7 ± 3.9 (21.0-35.5)	30.6 ± 0.9 (29.5-31.4)	37.0 ± 0.9 (30.0-46.0)
Gubernaculum	-	11.2 (9.4-13.7)	7.6 ± 1.1 (6.4-9.7)	-	9.1 ± 0.3 (7.0-11.0)
J2 (n)	-	25	20	20	30
L	360-393	371 (337-403)	453 ± 64.8 (380-593)	410.5 ± 21.5 (365-456)	415 ± 4.4 (330-470)
a	-	28.3 (24.9-31.5)	34.0 ± 4.4 (26.7-41.8)	30.8 ± 1.9 (28.0-34.2)	20.7 ± 0.3 (16.5-23.7)
b	-	2.4 (2.0-3.1)	-	3.0 ± 0.3 (2.7-3.6)	-
b′	-	7.1 (6.4-8.4)	-	7.4 ± 0.5 (6.7-8.5)	-
c	-	8.1 (6.9-10.6)	8.8 ± 1.4 (7.0-12.3)	8.3 ± 0.5 (7.5-9.6)	7.9 ± 0.1 (6.3-9.5)
c′	-	-	-	5.3 ± 0.3 (4.9-6.1)	-
W	-	-	13.4 ± 1.2 (11.6-16.1)	13.4 ± 0.6 (12.5-14.3)	20.1 ± 0.2 (18.0-22.0)
Stylet	-	10.5 (9.6-11.7)	10.4 ± 1.0 (9.0-12.9)	10.4 ± 0.3 (10.0-11.0)	10.7 ± 0.2 (9.0-17.0)

Table 29. *(Continued.)*

Character	Paratypes, Texas, USA, Chitwood (1949)	Whitehead (1968)	Venezuela, Perichi & Crozzoli (2010)	Passion fruit, India Khan *et al.* (2017)	Tomato, Brazil Monteiro *et al.* (2019) = *M. polycephannulata*
DGO	-	-	2.5 ± 0.3 (1.9-2.6)	3.2 ± 0.2 (2.9-3.3)	2.3 ± 0.1 (1.8-3.0)
Median bulb length	-	11.3 (10.1-12.9)	-	8.4 ± 0.9 (8.1-8.6)	-
Median bulb valve width	-	7.3 (5.8-8.3)	-	-	-
Median bulb valve length	-	5.2 (3.6-6.5)	-	-	-
Ant. end to excretory pore	-	-	84.1 ± 11.6 (66.1-100)	82.1 ± 3.9 (79.3-86.5)	77.0 ± 1.6 (58.0-91.0)
Tail	-	46.0 (38.0-55.0)	52 ± 6.9 (40.0-61.9)	49.5 ± 2.2 (46.5-53.2)	53 ± 7.8 (42.0-60.0)
Hyaline region	-	-	-	11.9 ± 1.1 (10.5-13.3)	13.9 ± 0.2 (11.0-16.0)

Male

Labial region not offset, high truncate cone-shaped, clearly annulated. Labial cap with stepped outline in lateral view, annuli number posterior to labial cap very variable, usually 1-3 on sublateral labial sectors and 1-5 on lateral labial sectors. SEM studies showing a high labial cap formed by a large round labial disc raised above median labials and centrally concave. Labial cap as wide as labial region. Labial region generally bearing 2-4 incomplete annuli, although some populations may have complete annulations. Labial region not offset. Conus of stylet longer than shaft, stylet knobs prominent, usually of greater width than length, with flat, concave or 'toothed' anterior margins. Excretory pore at level of posterior end of isthmus, hemizonid usually 0-5 annuli more anterior. Lateral field with four incisures, outer bands areolated, inner band rarely cross-striated except at posterior end. Testes one or two. Tail bluntly rounded, terminus unstriated. Phasmids at cloacal aperture level or just anterior. Spicules slightly curved, gubernaculum crescentic.

J2

Labial region not offset, truncated cone shape in lateral view, subhemispherical in dorsoventral view. Labial-cap wide followed by two clear annuli on sublateral labial sectors (lateral view), three annuli on lateral labial sectors (dorsal or ventral view). Stylet knobs prominent, rounded. Hemizonid three annuli long, located just anterior to excretory pore. Lateral field with four incisures, outer bands cross striated. Rectum dilated. Tail tapering to subacute terminus, striae coarsening posteriorly.

TYPE PLANT HOST: Carrot, *Daucus carota* L. Isolated from the faeces of soldiers.

OTHER PLANTS: *Meloidogyne incognita* is extremely polyphagous, attacking both monocotyledons and dicotyledons (Orton Williams, 1973). Many vegetables (*e.g.*, beans, *Brassica* spp., carrot, *Cucurbita* spp., lettuce, okra, pea, *Phaseolus* spp., tomato), grasses and pasture legumes (*e.g.*, bermudagrass, clovers, lucerne), shrubs and trees (*e.g.*, tea, *Prunus* spp., *Vitis* spp.), cereals, ornamentals, and crops such as sugarcane, potato, sweet potato, tobacco, and turfgrasses are attacked. Over 700 species and varieties are listed as hosts (Goodey *et al.*, 1965; Orton Williams, 1973; Rich *et al.*, 2008; Ye *et al.*, 2015; Bačić *et al.*, 2016; Karuri *et al.*, 2017). According to the North Carolina differential host test, *M. incognita* is made up of six host races. Populations of all four races reproduce on pepper, watermelon and tomato, but they vary in their response to resistant tobacco and cotton. Race 1 populations do not reproduce on tobacco or cotton, race 2 populations reproduce on tobacco but not on cotton, race 3 populations do not reproduce on tobacco but do reproduce on cotton, and race 4 populations reproduce on tobacco and cotton (Hartman & Sasser, 1985) (Table 3).

Meloidogyne incognita also parasitises many weeds, which act as reservoirs of the nematode, including: *Abutilon theophrasti* Medik., *Achyranthes aspera* L., *Ageratum conzyoides* L., *Amaranthus blitoides* S.Wats., *A. graecizans* L., *A. retroflexus* L., *A. spinosus* L., *Ampelamus laevis* (Michx.) Krings, *Anchusa azurea* Mill., *Borreria hispida* K. Schum., *Brachiaria reptans* (Trin.) Griseb., *Celosia argentea* L., *Chenopodium album* L. Bosc *ex* Moq., *Cleome viscosa* L., *Cnidoscolus stimulosus* (Michx.) Engelm. & Gray, *Cucumis anguria* L., *Cyperus esculentus* L., *C. rotundus* L., *Datura stramonium* L., *Dichondra repens* J.R.Forst. & G.Forst., *Eleusine indica* (L.) Gaertn., *Emilia sonchifolia* (L.) DC. *ex* Wight, *Euphorbia hirta* L., *Ipomoea hederacea* Jacq., *I. triloba* L., *Lactuca runcinata* L., *L. scariola* L., *Leonotis nepetaefolia*

(L.) R.Br., *Lucas aspera* (Willd.) Linn., *Malva neglecta* Wallr., *Ocimum canum* Sims., *Phytolacca americana* L., *Plantago lanceolata* L., *Polygonum aviculare* L., *P. lanceolatum* Gand., *P. persicaria* L., *Portulaca oleraceae* L., *Rumex dentatus* L., *Setaria viridis* (L.) P.Beauv., *Sida acuta* Brum. F., *Sinapis alba* L., *Solanum dulcamara* L., *S. nigrum* L. (Kaur *et al.*, 2007; Rich *et al.*, 2008; Castillo *et al.*, 2008; Anwar *et al.*, 2009; Singh *et al.*, 2010; Gharabadiyan *et al.*, 2012), moth plant (*Araujia sericifera* Brot.) (D'Errico *et al.* (2014), and dill (*Anethum graveolens* L.), growing in the glasshouse vegetable production area in Yalova, Turkey (Kepenekci & Dura, 2017).

TYPE LOCALITY: A field at El Paso, Texas, USA.

DISTRIBUTION: *Meloidogyne incognita* is one of the four commonest species worldwide although restricted to protected cultivation in temperate regions. Distribution includes Africa, Australia, Central and South America, India, Japan, Malaysia, USA and glasshouses in Northern Europe, Canada and Russia (Orton Williams, 1973; Guzman-Plazola *et al.*, 2006; Wesemael *et al.*, 2011; Kaur & Attri, 2013; Onkendi *et al.*, 2014; Karuri *et al.*, 2017; Khan *et al.*, 2017; Zhao *et al.*, 2017). The geographic distribution of *M. incognita* is largely dependent on environmental factors such as temperature and moisture (Sasser, 1977).This species occurs over a wider geographic range than any other species, from approximately 40°N latitude to 33°S (Taylor *et al.*, 1982).

PATHOGENICITY: *Meloidogyne incognita* typically incites large, usually irregular galls, and causes important economic losses worldwide. *Meloidogyne incognita* tends to form large and irregular galls some distance from the root tip (Hunt & Handoo, 2009). The damage potential of *M. incognita* is difficult to ascertain in cases where crops are suffering from inadequate nutrition or soil moisture or simultaneous attack by other soil-borne pathogens or pests (*i.e.*, fungi, bacteria, viruses, insects or other nematodes). The damage potential, considered as the maximum percentage of yield loss, has been studied in different crops after inoculation with eggs or J2 (at different inoculum densities) by several researchers (Wesemael *et al.*, 2011). This damage potential was estimated from 100% in artichoke (Di Vito & Zaccheo, 1991), cantaloupe (Di Vito *et al.*, 1983), or common bean (Di Vito *et al.*, 2004a), to 50%

in parsley (Aguirre *et al.*, 2003), or 45% in grapevine (Sasanelli *et al.*, 2006).

CHROMOSOME NUMBER: Populations of *M. incognita* reproduce exclusively by mitotic parthenogenesis and occur as two cytological forms: the triploid (3n = 41-46), which is the most common, and the diploid form (2n = 32-38) (Triantaphyllou, 1981, 1985b; Eisenback & Triantaphyllou, 1991).

POLYTOMOUS KEY CODES: *Female*: A23, B32, C3, D6; *Male*: A2134, B12, C12, D1, E2, F1; *J2*: A32, B32, C3, D32, E32, F23.

HOST RACES: Six host races have been recognised and accepted in this species using the North Carolina differential host test (Table 3). Populations of all races reproduce on tomato and watermelon, but vary in their response to tobacco or cotton or pepper.

BIOCHEMICAL AND MOLECULAR CHARACTERISATION: *Meloidogyne incognita* populations showed I1, I2 and S2 esterase types and the malate dehydrogenase N1 type phenotype (Esbenshade & Triantaphyllou, 1985a, b; Carneiro *et al.*, 2000; Santos *et al.*, 2012) (Fig. 12). The malate dehydrogenase phenotype N1 is similar to that of *M. javanica*, *M. exigua* and some populations of *M. arenaria* (Esbenshade & Triantaphyllou, 1985a, b).

Several molecular markers and PCR with species-specific primers have been developed for the accurate identification of *M. incognita* (Zijlstra, 1997; Adam *et al.*, 2007; Niu *et al.*, 2011) (Table 11). This nematode can be identified by *nad5* (Janssen *et al.*, 2016) gene sequence.

RELATIONSHIPS (DIAGNOSIS): The species belongs to Molecular group I, the Incognita group. *Meloidogyne incognita* can be distinguished from all other *Meloidogyne* spp. by several morphological and molecular characteristics, including: female cuticle thickening abruptly at base of relaxed stylet, typical perineal pattern with dorsal arch high and rounded, striae closely spaced, very wavy to zigzag, lateral field not clear, sometimes marked by breaks in striae, male labial region with a high truncated cone shape, stylet knobs prominent with flat, concave or toothed anterior margins.

41. *Meloidogyne indica* Whitehead, 1968
(Figs 116-118)

COMMON NAME: Indian citrus root-knot nematode.

Whitehead (1968) described this species from roots of citrus in New Delhi, India.

MEASUREMENTS

See Table 30.

- *Eggs* (n = 30): L = 77 (71-88) μm; W = 30 (26-35) μm.

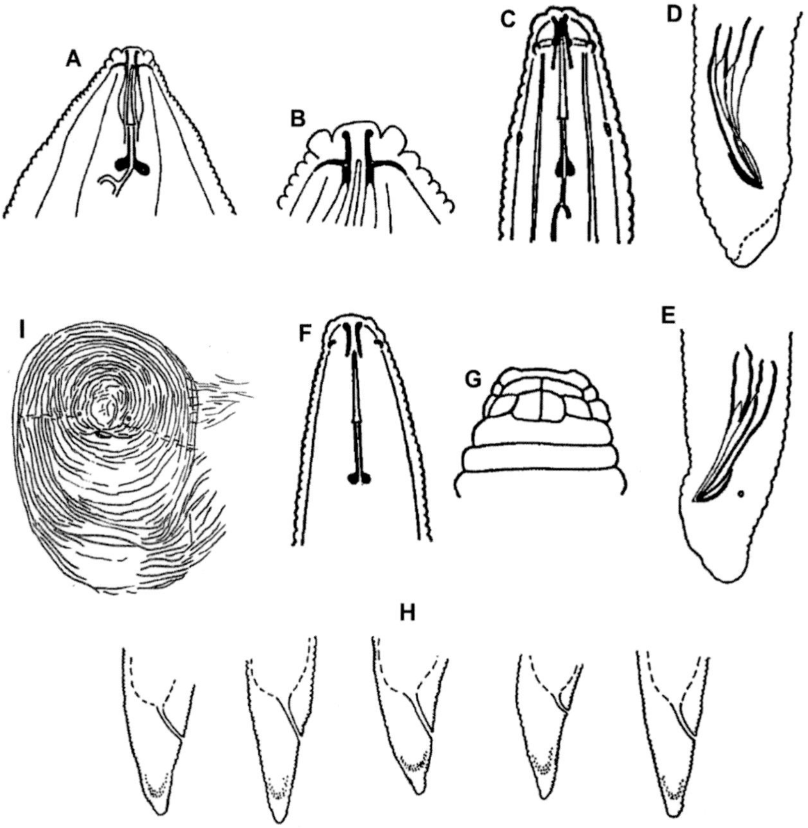

Fig. 116. Meloidogyne indica. *A, B: Female labial region; C: Male labial region; D, E: Male tail; F, G: Second-stage juvenile (J2) labial region; H: J2 tail; I: Perineal pattern. (After Whitehead, 1968.)*

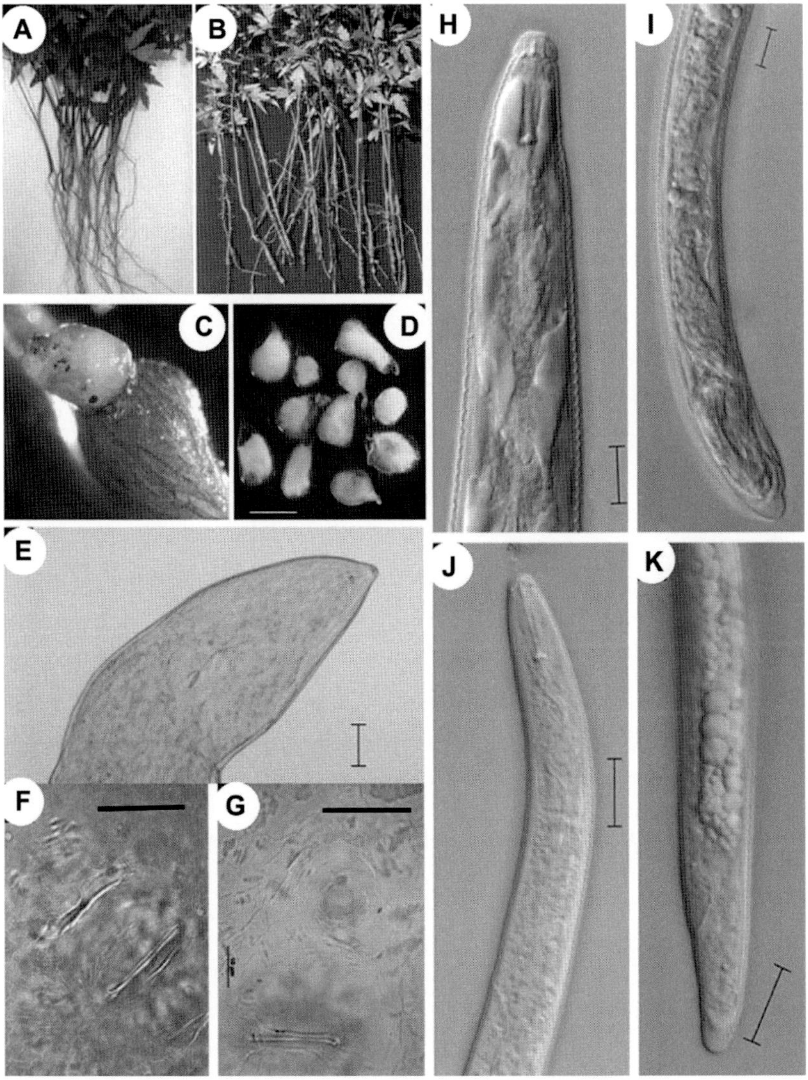

Fig. 117. Meloidogyne indica. *LM. A, B: Healthy and infected neem seedlings; C: Root gall; D: Entire female; E: Female anterior region; F, G: Perineal pattern; H: Male anterior region; I: Male tail region; J: Second-stage juvenile (J2) anterior end; K: J2 tail. (Scale bars: D = 550 μm; E = 100 μm; F, G = 20 μm; H, I = 10 μm; J, K = 20 μm.) (After Phani* et al., *2018.)*

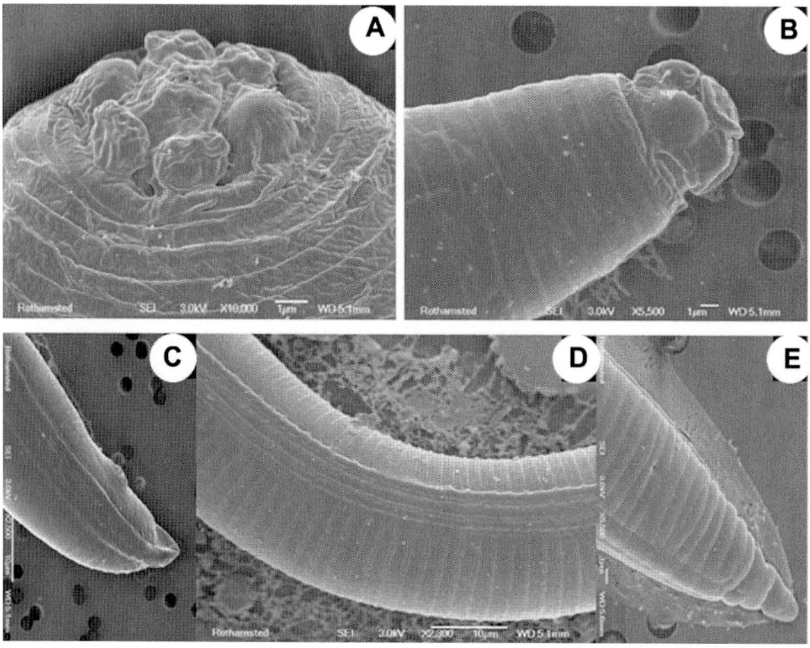

Fig. 118. Meloidogyne indica. *SEM. A: Female labial region; B: Male labial region; C: Male tail; D: Male lateral field; E: Second-stage juvenile tail. (After Phani* et al.*, 2018.)*

DESCRIPTION

Female

Body saccate with fairly thick cuticle, neck short, labial region divided into six sectors, laterals only slightly larger than subdorsals and subventrals, laterals characteristically subdivided and subdorsals and subventrals each with a 'papilla', labial region with one annulus posterior to labial-cap. Excretory pore about 20 annuli (16-22) from anterior end, anterior margins of stylet knobs either concave or posteriorly sloping. Posterior cuticular pattern faint, composed of closely spaced, usually smooth striae forming a distinct tail whorl with low, rounded dorsal arch, phasmids not as close to tail terminus and lateral field usually absent.

Male

Labial region hemispherical or truncated cone-shaped, not offset, with two annuli posterior to labial cap, anterior annulus longer than posterior, appearing 'tiled' in incident illumination; basal plate fairly thick, an-

Table 30. *Morphometrics of females, males and second-stage juveniles (J2) of* Meloidogyne indica. *All measurements are in μm and in the form: mean ± s.d. (range).*

Character	Paratypes, India, Whitehead (1968)	Bt cotton, India, Khan *et al.* (2018)	Citrus, India, Khan *et al.* (2018)	Neem and citrus, India, Phani *et al.* (2018)
Female (n)	8	-	-	30
L	-	683 ± 138.2 (370-900)	-	653 ± 92.2 (450-790)
a	-	1.6 ± 0.3 (1.2-2.2)	-	1.6 ± 0.3 (1.4-2.1)
W	-	418 ± 70.6 (300-520)	-	408 ± 75 (325-550)
Stylet	14 (12-16)	13.1 ± 0.6 (12.4-15.2)	-	13.7 ± 0.4 (13.0-14.0)
Stylet knob width	-	1.0 ± 0.2 (1.0-1.4)	-	-
Stylet knob height	-	3.8 ± 0.2 (3.3-4.3)	-	-
DGO	3.0 (2.0-4.0)	2.9 ± 0.3 (2.4-3.8)	-	2.9 ± 0.3 (2.5-3.7)
Median bulb length	38 (31-43)	31 ± 3.1 (25.6-36.1)	-	-
Median bulb width	39 (33-46)	28.6 ± 3.9 (23.8-36.1)	-	-
Median bulb valve width	9 (7-11)	8.4 ± 0.9 (6.7-9.5)	-	-
Median bulb valve length	14 (13-16)	12.6 ± 0.9 (11.4-15.2)	-	-
Ant. end to excretory pore	-	22.7 ± 6.4 (14.3-36.1)	-	-
Vulval slit	-	21 ± 3.0 (13.3-25.7)	-	-
Vulval slit to anus distance	-	21.8 ± 3.6 (14.3-27.6)	-	-
Interphasmid distance	-	24.2 ± 3.6 (17.1-30.4)	-	-
EP/ST	-	1.7	-	-

Table 30. *(Continued.)*

Character	Paratypes, India, Whitehead (1968)	Bt cotton, India, Khan *et al.* (2018)	Citrus, India, Khan *et al.* (2018)	Neem and citrus, India, Phani *et al.* (2018)
Male (n)	-	-	-	15
L	-	1140 ± 171.4 (820-1400)	1266 ± 99.1 (1130-1400)	1253 ± 80 (1180-1380)
a	-	44.1 ± 6.2 (34.2-55.8)	40.5 ± 6.0 (32.2-46.6)	28 ± 4.0 (24.5-34.7)
b′	-	16.2 ± 2.4 (11.5-20.7)	16.7 ± 1.9 (13.7-18.9)	-
W	-	26 ± 3.1 (20.9-31.4)	31.7 ± 4.3 (26.6-35.2)	28 ± 4.0 (24.5-35.0)
Stylet	-	16.6 ± 0.6 (16.2-18.0)	16.4 ± 0.6 (15.7-17.1)	16.3 ± 0.4 (16.0-17.0)
Stylet knob width	-	1.8 ± 0.2 (1.4-1.9)	1.8 ± 0.2 (1.4-1.9)	-
Stylet knob height	-	3.0 ± 0.5 (1.9-3.8)	3.0 ± 0.2 (2.9-3.3)	-
DGO	-	3.3 ± 0.4 (2.9-3.8)	3.6 ± 0.5 (2.9-4.3)	3.1 ± 0.1 (2.9-3.3)
Ant. end to excretory pore	-	108.2 ± 14.4 (81.7-130.1)	111.9 ± 16.6 (87.4-126.4)	-
Spicules	-	25.8 ± 2.5 (21.9-31.4)	27.4 ± 1.2 (25.7-28.5)	26.0 ± 0.6 (25.9-27.5)
J2 (n)	20	-	-	20
L	414 ± 4.5 (381-448)	462.8 ± 32.5 (400-535)	474.3 ± 41.8 (390-525)	484 ± 31.5 (430-520)
a	-	27.6 ± 3.5 (22-34.2)	26 ± 2.4 (23.3-30.3)	26.7 ± 1.9 (23.3-30.3)
b	-	3.7 ± 0.3 (3.2-4.2)	3.6 ± 0.3 (3.0-4.1)	3.5 ± 0.1 (3.1-3.7)
b′	-	9.1 ± 0.7 (7.8-10.2)	9.3 ± 1.2 (6.7-10.7)	-
c	24.9 ± 1.4 (21.2-31.0)	24.1 ± 1.3 (21.8-26.6)	35.4 ± 6.0 (25.7-47.4)	26.2 ± 1.2 (24.2-27.7)
c′	-	1.8 ± 0.1 (1.5-2.0)	1.5 ± 0.2 (1.1-1.8)	1.6 ± 0.1 (1.5-1.9)

Table 30. *(Continued.)*

Character	Paratypes, India, Whitehead (1968)	Bt cotton, India, Khan *et al.* (2018)	Citrus, India, Khan *et al.* (2018)	Neem and citrus, India, Phani *et al.* (2018)
W	-	17.1 ± 3.0 (13.3-21.9)	18.4 ± 2.2 (15.2-21.9)	18.0 ± 1.5 (15.2-21.9)
Stylet	12.0 ± 0.9 (10-14)	13.2 ± 0.6 (12.4-14.3)	12.3 ± 0.3 (11.9-13.3)	13.8 ± 0.1 (13.5-14.0)
DGO	-	2.8 ± 0.2 (2.4-3.3)	3.0 ± 0.3 (2.4-3.8)	2.8 ± 0.2 (2.5-3.0)
Median bulb length	-	12.5 ± 0.9 (11.4-14.3)	11.9 ± 1.1 (10.5-14.3)	-
Median bulb width	-	8.7 ± 0.9 (7.6-10.5)	9.2 ± 0.9 (8.1-10.9)	-
Ant. end to excretory pore	-	80.9 ± 5.2 (73.2-91.2)	79.7 ± 7.3 (65.6-92.2)	-
Tail	16.8 ± 1.9 (13-20)	19.2 ± 1.6 (17.1-21.9)	13.7 ± 1.9 (9.5-16.2)	18 ± 0.6 (17.5-19.5)
Hyaline region	-	4.7 ± 0.4 (3.8-5.7)	4.2 ± 0.6 (2.9-5.2)	-

terior part of stylet longer than shaft, stylet knobs small and rounded triangular in shape, anterior margins well swept back, posterior cephalid about seven annuli from anterior end, pharynx overlapping intestine ventrally, excretory pore near anterior end of posterior pharyngeal region and 2-5 annuli posterior to hemizonid, hemizonid one annulus long, one testis; lateral field with four incisures for greater length of body, outer bands fully areolated, inner band occasionally cross-striated towards posterior end of body where lateral field fully areolated, tail rather long, apparently concave ventrally, striae passing around terminus, phasmids opposite cloacal aperture.

J2

Labial region not offset, a low truncated cone, labial cap outline slightly posterior to general contour of labial region, two, sometimes three, annuli posterior to labial cap, labial annuli apparently marked by longitudinal striae, distal sclerotisation of labial region thicker than proximal, cephalids not seen, stylet knobs fairly prominent rounded, anterior margins often posteriorly sloping, rectum not dilated, tail conoid, terminus usually blunt, unstriated.

TYPE PLANT HOST: Lime, *Citrus aurantifolia* (Christm.) Swingle.

OTHER PLANTS: *Meloidogyne indica* also infects *Citrus* spp. such as sweet orange (*Citrus sinensis* Osbeck), morinda (*Morinda officianalis* F.C.How) (Zhang & Weng, 1991), and *Amaranthus* sp., watermelon, cowpea (Patel *et al.*, 2003), medicinal plants (Lin *et al.*, 2004), cotton (Khan *et al.*, 2018), and neem or Indian lilac (*Azadirachta indica* A. Juss.) (Phani *et al.*, 2018).

TYPE LOCALITY: New Delhi, India.

DISTRIBUTION: *Asia*: India (Whitehead, 1968; Patel *et al.*, 1999; Khan *et al.*, 2018; Phani *et al.*, 2018) and China (Fujian Province) (Zhang & Weng, 1991).

PATHOGENICITY: Nematode-infected citrus plants exhibited symptoms very similar to dieback or decline of older plants. *Citrus* plants exhibited a sick appearance with yellowing of leaves, drying of twigs and yielding no fruit (Khan *et al.*, 2018).

CHROMOSOME NUMBER: No available data.

POLYTOMOUS KEY CODES: *Female*: A213, B32, C3, D2; *Male*: A34, B43, C32, D1, E2, F1; *J2*: A23, B23, C4, D4, E1, F4.

BIOCHEMICAL AND MOLECULAR CHARACTERISATION: Esterase enzyme activity was not detected (Khan *et al.*, 2018). Molecularly this nematode can be identified by the unique sequences of 28S rRNA, ITS rRNA and *COI* genes, which are different to all other *Meloidogyne* spp. (Phani *et al.*, 2018).

RELATIONSHIPS (DIAGNOSIS): The species belongs to Molecular group XI and is clearly molecularly differentiated from all other *Meloidogyne* species. It has a sister relationship with *M. nataliei*. It is similar to *M. ovalis* from which it can be differentiated by the continuous striae (female group 2 in Jepson, 1987) *vs* a round/oval pattern without marked lateral lines (group 3, Jepson, 1987) in *M. ovalis*.

42. *Meloidogyne inornata* Lordello, 1956
(Figs 119-123)

COMMON NAME: Brazilian root-knot nematode.

Lordello (1956) described this species from roots of soybean from Estaçao Experimental Central do Instituto Agronômico, Campinas, State of São Paulo, Brazil. This nematode was reported from tobacco (Figueiredo, 1958), yacon (Carneiro *et al.*, 2000, 2008), and common bean (Machado *et al.*, 2013; Correia *et al.*, 2016). Whitehead (1968) and Hewlett & Tarjan (1983) considered *M. inornata* to be closely related to *M. incognita*. Although *M. inornata* was synonymised by Jepson (1987) and Eisenback & Triantaphyllou (1991) to *M. incognita* based on morphological features, a comprehensive study conducted by Carneiro *et al.* (2008) confirmed the validity of *M. inornata*.

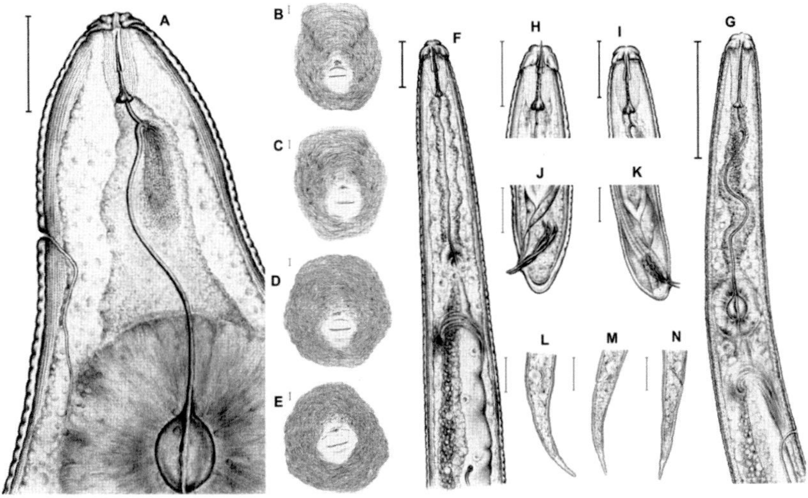

Fig. 119. Meloidogyne inornata. *A: Female anterior region; B-E: Perineal pattern; F, G: Male anterior region; H, I: Male labial region; J, K: Male tail; L-N: Second-stage juvenile tail. (Scale bars = 20 μm.) (After Carneiro* et al.*, 2008.)*

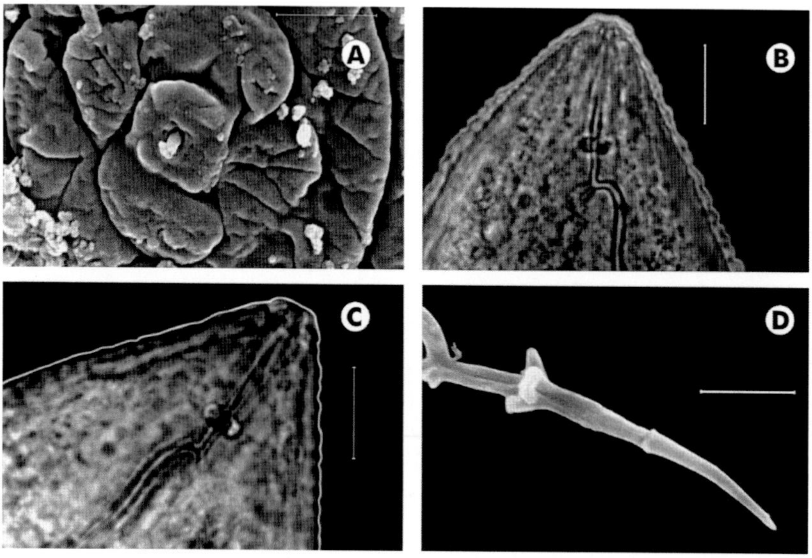

Fig. 120. Meloidogyne inornata. *Female. SEM and LM. A: Anterior region, face view; B, C: Anterior region; D: Excised stylet. (Scale bars: A = 2 μm; B, C = 10 μm; D = 5 μm.) (After Carneiro* et al., *2008.)*

MEASUREMENTS

See Table 31.

- *Eggs*: L = 90.0-112.5 μm; W = 37.5-60.0 μm.

DESCRIPTION (*AFTER CARNEIRO* ET AL., *2008*)

Female

Body completely enclosed by gall tissue. Body translucent white, variable in size, round or more or less flattened posteriorly. Neck prominent, bent at various angles to body. Body cuticle distinctly annulated, annuli smaller in anterior neck region. Labial region offset from body, sometimes annulated. Stoma slit-like, located in ovoid prestomatal cavity, surrounded by pit-like openings of six inner labial sensilla. Labial disc and median labials dumbbell-shaped. Labial disc raised, separated from lateral and median labials. Median labials never dividing into labial pairs. Lateral labials large, triangular, fused laterally with labial region for short distance. Amphidial apertures located between labial disc and lateral labials. In LM, labial framework weakly

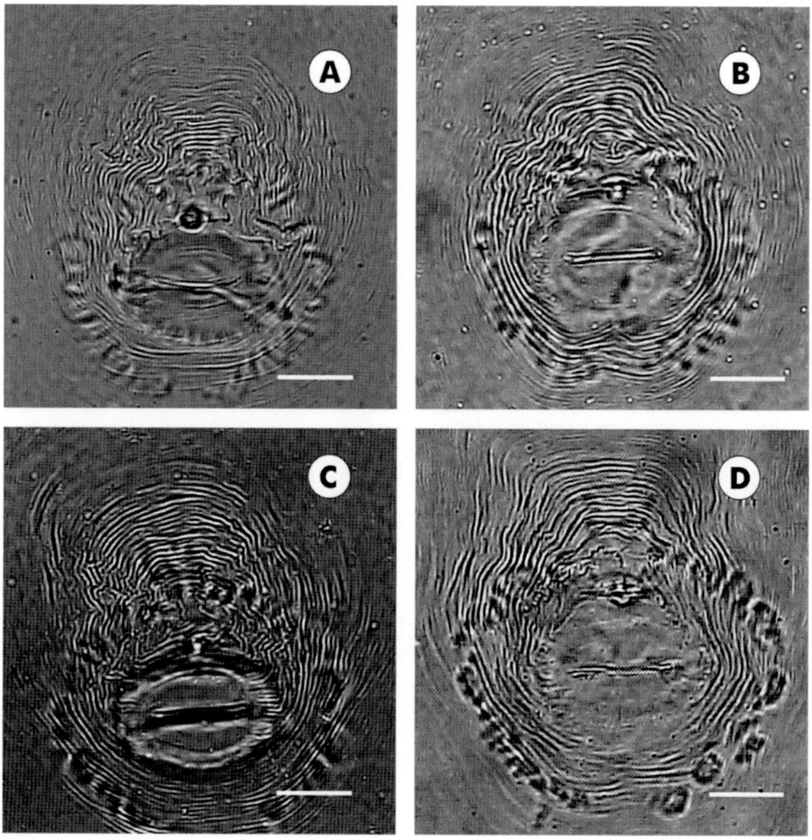

Fig. 121. Meloidogyne inornata. *LM. A-D: Perineal pattern. (Scale bars = 30 μm.) (After Carneiro* et al., *2008.)*

developed, hexaradiate, lateral sectors slightly enlarged, vestibule and extension prominent. Stylet robust, easily dislodged posteriorly. Stylet cone generally slightly curved dorsally with well-developed knobs. Shaft slightly wider near junction with knobs. Knobs well developed, distinctly separated, oval-shaped, posteriorly sloping. Excretory pore located between dorsal pharyngeal gland orifice and median bulb, neck gradually increasing in diam. posteriorly. Subventral gland orifices branched, located immediately posterior to enlarged lumen of median bulb. Pharyngeal glands with one large dorsal lobe with two nuclei, two small nucleated subventral gland lobes of variable in shape, position and size, usually located posterior to dorsal gland lobe. Perineal cuticular patterns variable, oval to squarish in shape, striae coarse near centre

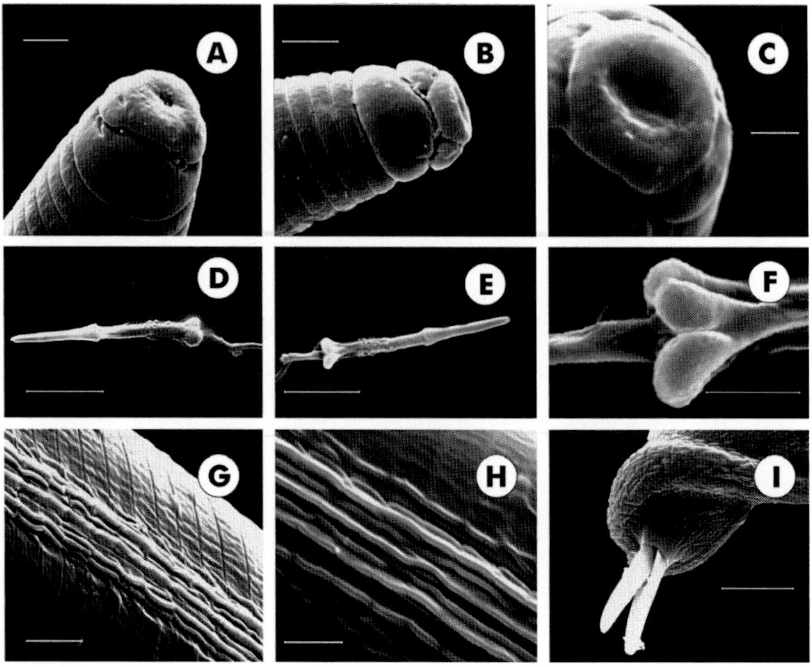

Fig. 122. Meloidogyne inornata. *SEM. A-C: Male labial region; D-F: Male excised stylet; G, H: Male lateral field; I: Tail with spicules. (Scale bars: A, B = 5 μm; C, F = 2 μm; D, E, I = 5 μm; G, H = 1 μm.) (After Carneiro* et al., *2008.)*

and finer outside. Striae smooth to wavy, widely separated, usually continuous without distinct whorl. Dorsal arch moderately high to high, rounded to squarish, never forming 'shoulders'. Lateral field without distinct incisures, sometimes faintly appearing as a discontinuous linear depression in LM. Perivulval region generally free of striae. Phasmids small, opposite, located posterior to anus. Fold over anus present.

Male

Body vermiform, tapering anteriorly, bluntly rounded posteriorly, twisting through 90°; tail arcuate. Body annuli large, distinct. In lateral view labial cap high and rounded, labial region offset from body, without annulation. In SEM, stoma slit-like, located in ovoid to hexagonal cavity surrounded by pit-like openings of six inner labial sensilla. Large rounded labial disc distinctly raised above median labials and possibly centrally concave. In some specimens labial disc not

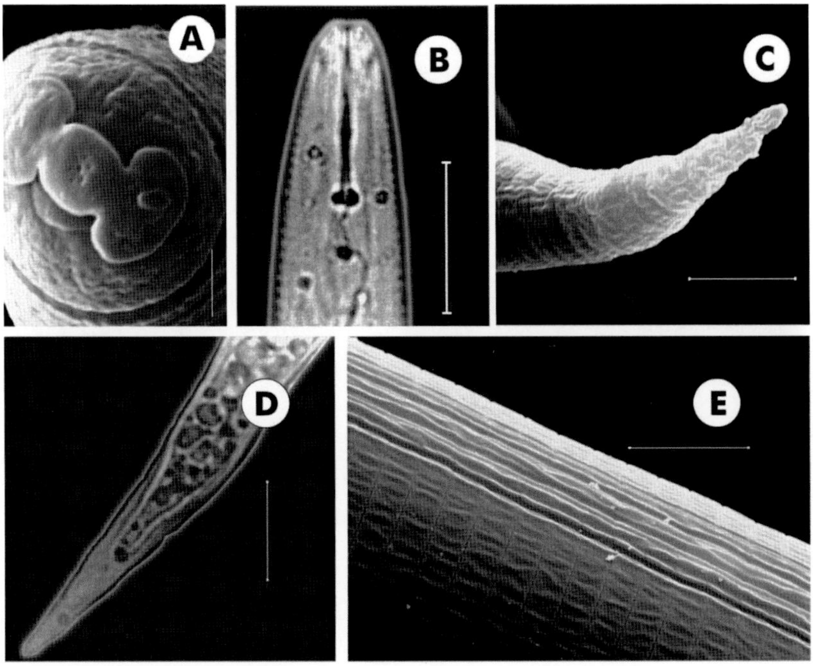

Fig. 123. Meloidogyne inornata. *SEM and LM. A: Second-stage juvenile (J2)* en face *view; B: J2 labial region; C, D: J2 tail; E: J2 lateral field. (Scale bars: A = 1 μm; B, D, E = 10 μm; C = 5 μm.) (After Carneiro* et al.*, 2008.)*

Table 31. *Morphometrics of females, males and second-stage juveniles (J2) of* Meloidogyne inornata. *All measurements are in μm and in the form: mean ± s.d. (range).*

Character	Paratypes, Brazil, Lordello (1956)	Neotype, Carneiro *et al.* (2008)	Brazil, Carneiro *et al.* (2008)	Chile, Gerič Stare *et al.* (2019)
Female (n)	-	1	30	20
L	542-980	645	606 ± 15 (594-781)	789.2 ± 135.0 (676.2-1012.5)
a	1.5	-	1.3 ± 0.8 (1.1-1.4)	1.6 ± 0.3 (1.4-2.0)
W	448-620	435	517 ± 12 (437-675)	509.5 ± 39.0 (448.8-555.3)

Table 31. *(Continued.)*

Character	Paratypes, Brazil, Lordello (1956)	Neotype, Carneiro *et al.* (2008)	Brazil, Carneiro *et al.* (2008)	Chile, Gerič Stare *et al.* (2019)
Stylet	15.0-16.5	16	15.3 ± 0.2 (15.0-17.0)	15.2 ± 0.6 (14.7-16.1)
Stylet knob width	-	-	4.0 ± 0.5 (4.0-4.5)	3.8 ± 0.2 (3.5-4.1)
Stylet knob height	2.5-3.3	-	2.8 ± 0.2 (2.5-3.5)	2.5 ± 0.3 (2.0-2.9)
DGO	3.0-5.0	4.0	3.9 ± 0.1 (3.5-4.5)	4.1 ± 0.6 (3.2-4.6)
Ant. end to median bulb			104 ± 11 (90-120)	95.3 ± 18.4 (77.6-122.0)
Anterior end to excretory pore	-	29.0	36.9 ± 2.8 (25.0-53.0)	49.1 ± 4.7 (42.3-55.1)
Vulva length			24.4 ± 0.8 (22.5-26.2)	24.8 ± 1.1 (23.2-25.9)
Vulva-anus distance			19.0 ± 0.5 (17.5-22.5)	20.5 ± 1.7 (18.6-22.8)
Interphasmid distance			24.5 ± 0.7 (22.5-26.3)	23.5 ± 1.7 (21.6-25.3)
EP/ST	-	1.8	2.4	-
Male (n)			30	20
L	1207-1995	-	1594 ± 58 (1101-2063)	1181.8 ± 216.5 (881.2-1423)
a	24.2-51.7	-	27.7 ± 0.8 (24.4-51.0)	35 ± 4.6 (31.1-42.7)
c	145.5-198.7	-	152.2 ± 12.2 (104.5-187.2)	91.9 ± 17.9 (64.3-117.6)
W	33.0-50.0	-	47.0 ± 0.9 (32.0-51.0)	33.7 ± 4.0 (28.3-38.9)
Labial region height	5.0-7.0	-	-	-
Stylet	20.0-25.0	-	21.7 ± 0.6 (20.0-25.0)	22.3 ± 1.5 (20.8-24.9)
Stylet knob width	-	-	4.8 ± 0.6 (4.0-4.5)	4.0 ± 0.6 (3.2-4.9)
Stylet knob height	2.5-3.3	-	3.0 ± 0.5 (2.5-3.5)	3.2 ± 0.3 (3.0-3.7)
DGO	4.0-5.0	-	4.5 ± 0.2 (4.0-5.0)	3.4 ± 0.5 (2.8-4.2)

Table 31. *(Continued.)*

Character	Paratypes, Brazil, Lordello (1956)	Neotype, Carneiro *et al.* (2008)	Brazil, Carneiro *et al.* (2008)	Chile, Gerič Stare *et al.* (2019)
Median bulb length	18-28	-	-	-
Median bulb width	10.0-11.6	-	-	-
Ant. end to median bulb			84 ± 3.2 (75-120)	95.7 ± 13.7 (74.5-116.2)
Ant. end to excretory pore	-	-	167.3 ± 3.1 (135-200)	154.9 ± 14.6 (134.5-172)
Tail	10.0-13.0		13.5 ± 0.5 (10.0-15.0)	12.9 ± 1.2 (11.2-14.5)
Spicules	26.5-33.0	-	33 ± 0.6 (26.0-38.0)	31.2 ± 4.3 (27-37.2)
Gubernaculum	-	-	6.7	
J2 (n)	-		30	20
L	375-420	-	418 ± 3 (394-487)	421.6 ± 16.4 (401-446.2)
a	28.0-36.0	-	23 ± 0.3 (17.9-28.7)	27.2 ± 2.1 (25.3-31.1)
c	11.6-12.7	-	10.2 ± 0.4 (6.7-13.9)	8.5 ± 0.6 (7.7-9.4)
c′	-	-	-	4.8 ± 0.7 (3.9-5.9)
W	11.6-15.0	-	19.3 ± 0.6 (17.0-22.0)	15.6 ± 1.2 (13.9-16.8)
Stylet	10.0-13.0	-	11.5 ± 0.1 (10.0-13.0)	12.6 ± 0.9 (11.4-13.9)
DGO	-	-	3.0 ± 0.1 (2.5-3.5)	3.3 ± 0.4 (2.9-4.0)
Median bulb length	11.6-13.3	-	-	-
Median bulb width	6.6-10.0	-	-	-
Ant. end to median bulb	-		122 ± 2.8 (102-134)*	58.4 ± 2.7 (55.7-61.5)
Ant. end to excretory pore	-	-	158 ± 0.9 (152-164)*	77.5 ± 12.2 (60.2-93.5)
Tail	33.0-36.5	-	49.4 ± 0.6 (35.0-58.0)	50.1 ± 5.0 (42.6-55.8)
Hyaline region	-	-	13.9 ± 0.3 (10.0-15.0)	11.1 ± 1.6 (9.5-13.7)

* Possible incorrect measurements noticed by Gerič Stare *et al.* (2019).

so rounded or raised, more continuous with median labials. Median labials crescentic with distinct lateral indentation at junction with labial disc. Labial sensilla not clearly demarcated on median labials. Lateral labials absent. Amphidial apertures elongated slits between labial disc and lateral sides of labial region. In LM, labial framework moderately developed, hexaradiate, lateral sectors slightly enlarged. Vestibule extensions distinct. Stylet robust, large, cone straight, pointed, as long as shaft, slightly increasing in diam. posteriorly at junction of shaft. Shaft cylindrical with several small projections, widening slightly near junction with knobs. Knobs rounded, distinctly indented, posteriorly sloping, merging gradually with shaft. Dorsal pharyngeal gland orifice branching into three ducts, ampulla distinct. Amphids very distinct, often producing exudates. Procorpus distinct, median bulb ovoid, triradiate, lining of enlarged lumen of median bulb thinner than in female. Pharyngo-intestinal junction at level of nerve ring, indistinct. Three nuclei in gland lobe, lobe variable in length. Intestinal caecum extending anteriorly, sometimes approaching level of dorsal pharyngeal gland orifice. Excretory pore distinct, 6-10 annuli posterior to hemizonid. Lateral field starting just posterior to stylet knob level (12 annuli posterior to labial region) as two incisures and extending to tail tip. Composed of a variable number of crenate incisures in different parts of body. Areolated lateral field showing four or five, even more, incisures in SEM. Even external incisures of field may be made up of interrupted lines. Usually, sex reversed males (with two testes) are common. Testes outstretched or anteriorly reflexed. Spicules arcuate, gubernaculum distinct. Tail short, phasmids distinct, at cloacal aperture level.

J2

Body vermiform, clearly annulated, tapering more posteriorly than anteriorly. Anterior end truncate, labial region only slightly offset from body. Vestibule and extension more developed than remainder of hexaradiate labial framework. In SEM, prestoma opening rounded, surrounded by small, pore-like openings of six inner labial sensilla. Median labials and labial disc dumbbell-shaped in face view. Lateral labials small and triangular, lower than labial disc and median labials. Posterior edge of one or both lateral labials sometimes fusing with labial region. Labial disc oval, slightly raised above median labials. Median labials with crescentic to rounded margins. Labial sensilla not visible externally. Elongated amphidial apertures located between labial disc and lateral labials. Labial region not annulated. Lateral field beginning at procorpus level as two lines, composed of 4-6 straight or undulate incisures near median bulb. In LM, stylet delicate, cone straight, narrow, sharply pointed; shaft becoming slightly wider posteriorly, knobs small, oval-shaped, posteriorly sloping. Procorpus faintly outlined, median bulb oval-shaped with enlarged lumen lining, isthmus clearly defined, pharyngo-intestinal junction difficult to observe. Gland lobe variable in length, overlapping intestine ventrally. Excretory pore distinct; hemizonid anterior to or adjacent with excretory pore. Tail conoid with rounded, unstriated, terminus. Hyaline region clearly defined. Rectum dilated. Phasmids small, difficult to observe, located posterior to anus.

TYPE PLANT HOST: Soybean, *Glycine max* (L.) Merr.

OTHER PLANTS: *Meloidogyne inornata* has been reported to infect *Anthurium andreanum* Lind., *G. max* and *Nicotiana tabacum*, yacon (*Smallanthus sonchifolius* (Poepp. (H.Rob.)) (Carneiro *et al.*, 2008; Camara *et al.*, 2019) and common bean (*Phaseolus vulgaris*) (Machado *et al.*, 2013; Correia *et al.*, 2016). The population of *M. inornata* from Capão Bonito reproduced on tomato, tobacco and watermelon (The North Carolina differential host test). No reproduction occurred on cotton, pepper, or on peanut. *Meloidogyne inornata* has the same differential host response as *M. javanica* (Carneiro *et al.*, 2008).

TYPE LOCALITY: Estaçao Experimental Central do Instituto Agronômico, Campinas, State of São Paulo, Brazil.

DISTRIBUTION: *South America*: Brazil, Chile (Gerič Stare *et al.*, 2019).

PATHOGENICITY AND SYMPTOMS: Plants of yacon in commercial farms located in the district of Alto Norte, Muniz Freire, Espírito Santo, Brazil, showed uniformly intense leaf yellowing (leaf chlorosis) symptoms distributed on the leaf surface, especially on the lower leaves, small size (dwarfism), wilting at the hottest hours of the day, and galls throughout the plant root system.

CHROMOSOME NUMBER AND REPRODUCTION MODE: Mitotic parthenogenesis, 3n = 54-58 chromosomes (Carneiro *et al.*, 2008).

POLYTOMOUS KEY CODES: *Female*: A23, B2, C2, D2; *Male*: A2134, B213, C213, D1, E2, F2; *J2*: A23, B23, C3, D23, E213, F3.

BIOCHEMICAL AND MOLECULAR CHARACTERISATION: *Meloidogyne inornata* showed a species-specific esterase (EST) phenotype (In3 = Y3) with three bands (Rm: 0.80, 1.10, 1.30) and MDH phenotype N1 (Rm: 1.0) The esterase phenotype is unique and is the most useful character to differentiate *M. inornata* from *M. incognita*, *M. ethiopica* and all other species (Carneiro *et al.*, 2000, 2008). Molecularly this nematode belongs to the Ethiopica group and showed a sister relationship with *M. ethiopica* based on the analysis of 18S rRNA, mtDNA and other gene sequences (Tigano *et al.*, 2005; Janssen *et al.*, 2016; Álvarez-Ortega *et al.*, 2019). *Meloidogyne inornata* (*Meloidogyne* sp. 3) differed from other species in a PCR-RAPD study (Randig *et al.*, 2002).

RELATIONSHIPS (DIAGNOSIS): This species belongs to Molecular group I and the Ethiopica group. *Meloidogyne inornata* can be differentiated from *M. incognita* by the labial region of males and J2 showing only one wide, post labial annulus instead of three, and the male stylet is robust with small projections surrounding the shaft *vs* no projections in *M. incognita* (Eisenback & Triantaphyllou, 1991). The location of the DGO of female, male and J2 *M. inornata* is further posterior to the stylet knobs and the excretory pore in *M. inornata* is located *ca* 2.5 stylet lengths posterior to the anterior extremity *vs ca* one in *M. incognita*. Taken in combination, these characters allow *M. inornata* to be differentiated from *M. incognita* (Carneiro *et al.*, 2008). *Meloidogyne inornata* can be differentiated from *M. ethiopica* by various features, including perineal pattern of the *M. incognita* type *vs* varying from *M. incognita* to *M. arenaria* type and longer female stylet length. The J2 tail length of *M. inornata* is shorter at 33-58 *vs* 41-72 μm. The presence of large phasmids in female *M. ethiopica* (Golden, 1992) is another diagnostic character for differentiating *M. inornata* from *M. ethiopica* (Carneiro *et al.*, 2008).

43. *Meloidogyne izalcoensis* Carneiro, Almeida, Gomes & Hernandez, 2005a

(Figs 124-128)

COMMON NAME: El Salvador coffee root-knot nematode.

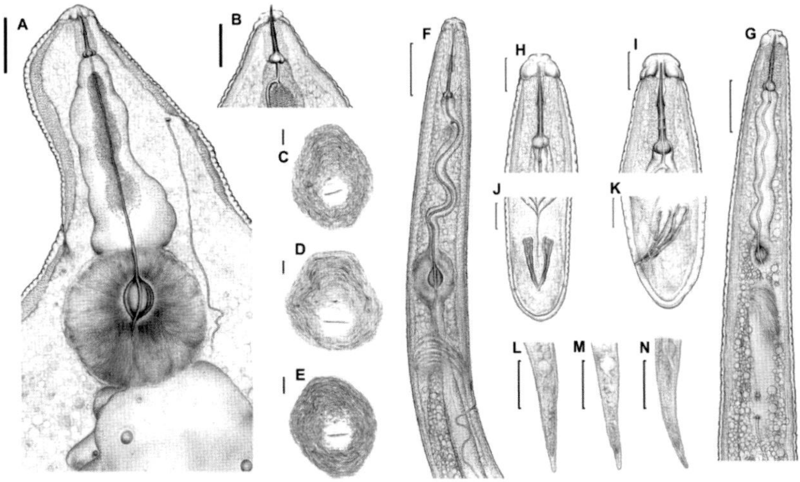

Fig. 124. Meloidogyne izalcoensis. *A: Female anterior region; B: Female labial region; C-E: Perineal pattern; F, G: Male anterior region; H, I: Male labial region; J, K: Male tail; L-N: Second-stage juvenile tail. (Scale bars: A-E = 20 μm; F, G = 10 μm; H-N = 20 μm.) (After Carneiro* et al., *2005a.)*

This nematode was characterised by Hernandez *et al.* (2004a) and described as a new species by Carneiro *et al.* (2005a) from roots of coffee from Izalco, Sonsonate Department, El Salvador. The diagnosis of *M. izalcoensis* is difficult due to the similarity of the perineal pattern with that of *M. incognita* and *M. paranaensis*.

MEASUREMENTS

See Table 32.

DESCRIPTION

Female

Completely enclosed by gall tissue. Body translucent to white, variable in size, elongated, ovoid to pear-shaped. Neck sometimes prominent, bent at various angles to body. Body cuticle distinctly annulated, annuli smaller in anterior neck region. Labial region offset from body, sometimes annulated. Stoma slit-like, located in ovoid prestomatal cavity, surrounded by pit-like openings of six inner labial sensilla. Labial disc and median labials dumbbell-shaped. Labial disc

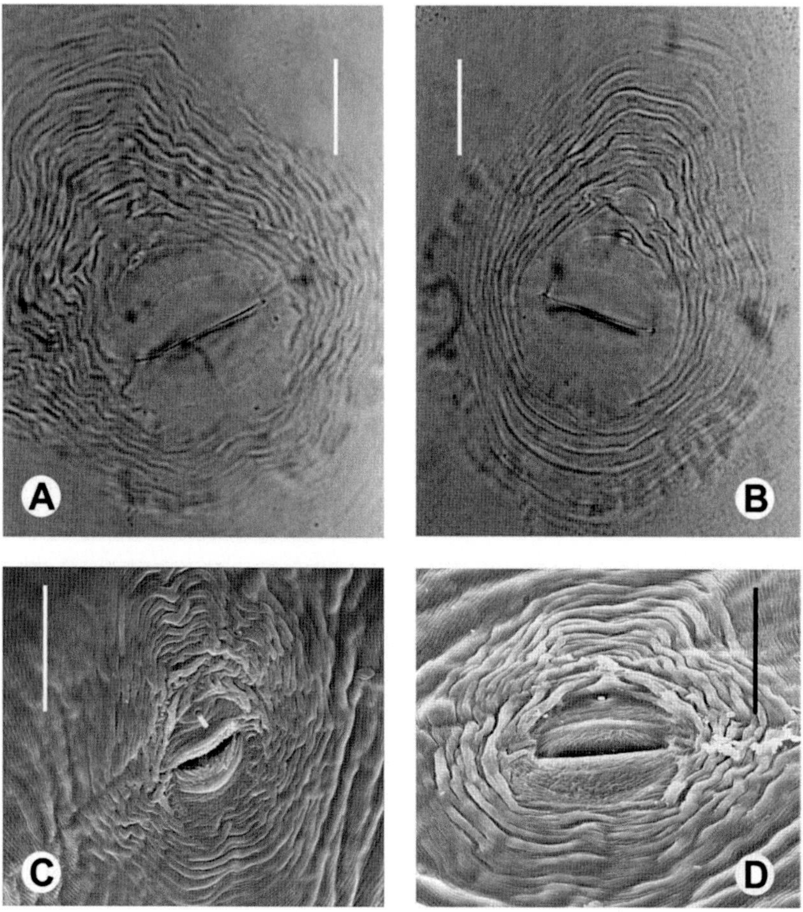

Fig. 125. Meloidogyne izalcoensis. *SEM and LM. Perineal patterns. (Scale bars: A, B = 30 μm; C, D = 20 μm.) (After Carneiro* et al., *2005a.)*

with two prominent elevations or bumps on ventral side, these being slightly raised above median labials. Median labials never dividing into labial pairs. Lateral labials large, fused laterally with labial region for short distance. Amphidial apertures oval, located between labial disc and lateral labials. Under LM, labial framework weakly developed, hexaradiate, lateral sectors slightly enlarged, vestibule and extension prominent. Stylet robust, easily dislodged posteriorly. Stylet cone generally slightly curved dorsally, widening slightly posteriorly near junction with shaft. Shaft gradually widening posteriorly near junction

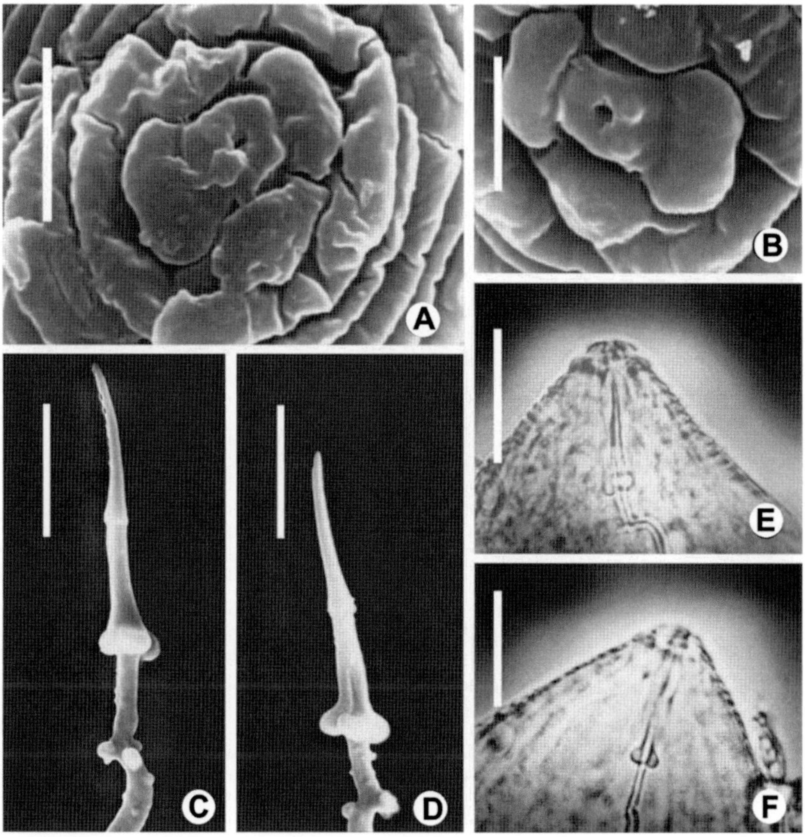

Fig. 126. Meloidogyne izalcoensis. *SEM and LM. Female. A, B: Anterior region; C, D: Excised stylets; E, F: Anterior region. (Scale bars: A, B = 2 μm; C, D = 5 μm; E, F = 11 μm.) (After Carneiro* et al.*, 2005a.)*

with knobs. Knobs well developed, distinctly separated, oval-shaped and posteriorly sloping. Excretory pore located *ca* 1.5-2.5 stylet lengths posterior to stylet, neck gradually increasing in diam. posteriorly. DGO *ca* one shaft length, orifice branching into three ducts, ampulla large. Subventral gland orifices branched, located immediately posterior to enlarged lumen of median bulb. Pharyngeal glands with one large dorsal lobe with two nuclei, subventral gland lobes two, small, nucleated, variable in shape, position and size, usually located posterior to dorsal gland lobe. Perineal cuticular patterns variable, oval to squarish in shape, striae coarse near centre and finer towards outside of pattern. Striae

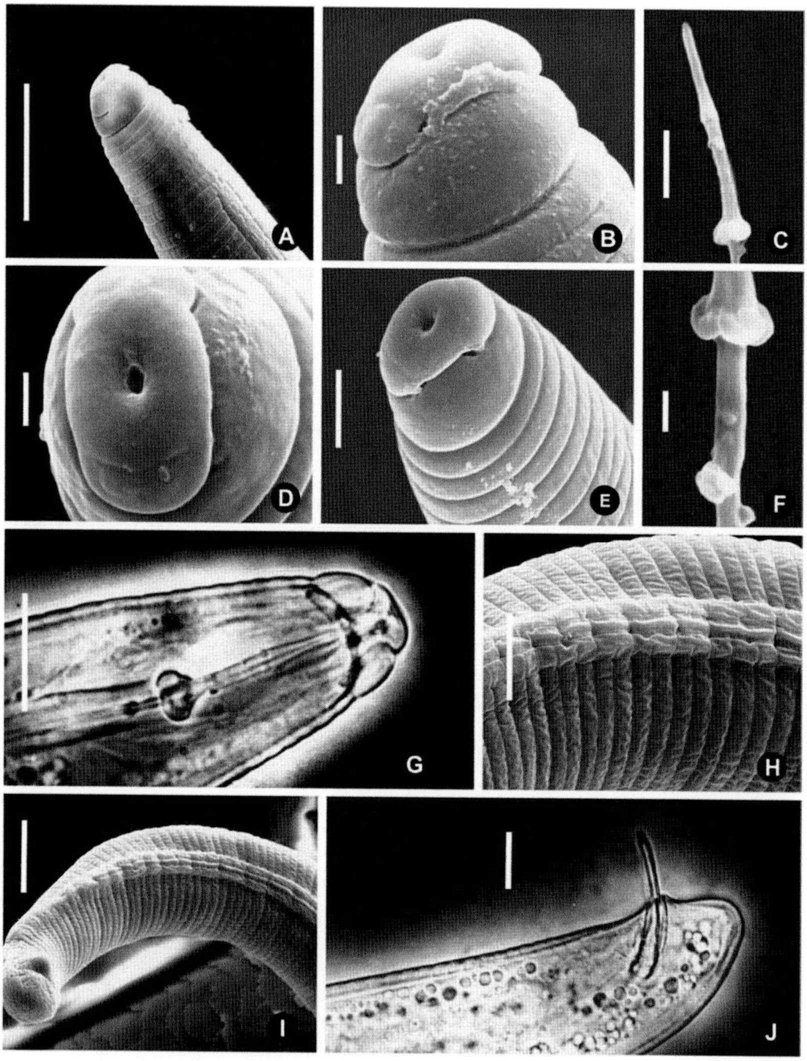

Fig. 127. Meloidogyne izalcoensis. *Male. A, B, E: SEM of anterior region; D: SEM of labial region,* en face *view; C, F: SEM of excised stylets; G: LM of head end in dorsal view; H: SEM of lateral fields; I, J: SEM and LM of tail. (Scale bars: A, I = 20 μm; B, D, F = 2 μm; C, H, J = 10 μm; E = 5 μm; G = 8 μm.) (After Carneiro* et al., *2005a.)*

smooth to wavy, widely separated, usually continuous, distinct whorl lacking. Dorsal arch moderately high to high, rounded to squarish, never

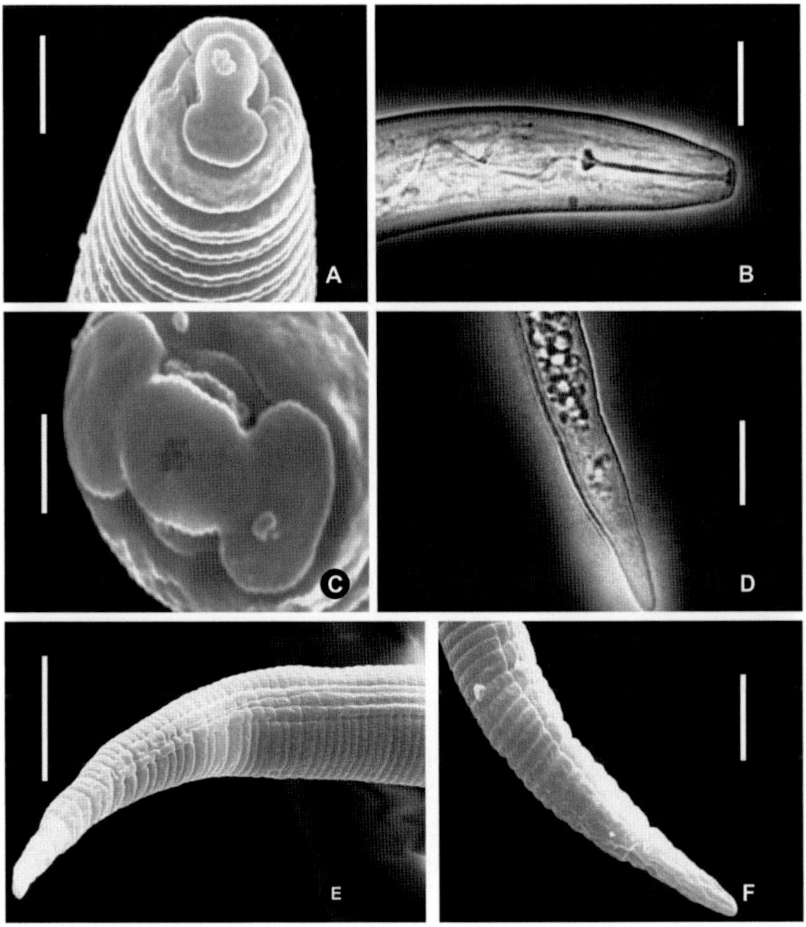

Fig. 128. Meloidogyne izalcoensis. *Second-stage juvenile. A, B: SEM and LM micrographs of anterior region; C: SEM of labial region, face view; D: LM of tail; E, F: SEM of tails showing lateral fields. (Scale bars: A = 2 μm; B, D = 12 μm; C = 1 μm; E = 10 μm; F = 5 μm.) (After Carneiro* et al.*, 2005a.)*

forming 'shoulders'. Lateral lines without distinct incisures, sometimes appearing as a faint, discontinuous, linear depression under LM and SEM. Perivulval region generally free of striae; striae rarely present on lateral sides of vulva. Tail terminus sometimes visible. Phasmids small, located immediately on both side, and posterior to anus. Fold over anus present.

Table 32. *Morphometrics of females, males and second-stage juveniles (J2) of* Meloidogyne izalcoensis. *All measurements are in μm and in the form: mean ± s.d. (range).*

Character	Holotype, allotype, El Salvador, Carneiro *et al.* (2005a)	Paratypes, El Salvador, Carneiro *et al.* (2005a)	Africa, Jorge *et al.* (2016)
Female (n)	-	30	-
L	770	794 ± 114 (650-1190)	-
a	1.6	1.5 ± 0.8 (1.3-2.3)	-
W	480	576 ± 134 (400-940)	-
Stylet	16.0	15.6 ± 0.4 (15.0-16.0)	15.0-16.0
DGO	5.5	5.4 ± 0.6 (4.5-6.0)	4.5-6.0
Ant. end to excretory pore	49	45 ± 23 (20-68)	-
Vulval slit	-	29.0 ± 0.5 (25.0-34.0)	
Vulval slit to anus distance	-	20.0 ± 0.4 (19.1-25.2)	
Interphasmid distance	-	32.0 ± 0.6 (26.0-40.0)	
EP/ST	3.1	2.9	-
Male (n)	1	30	-
L	1550	1707 ± 147 (1300-1870)	-
a	46.9	38.1 ± 6.8 (31.4-48.0)	-
c	96.8	113 ± 27 (76-140)	
W	33.0	45.0 ± 5.0 (33.0-50.0)	-
Stylet	25.0	25.0 ± 0.1 (23.0-26.0)	23.0-26.0
DGO	5.0	5.0 ± 0.5 (4.0-7.0)	4.0-7.0

Table 32. *(Continued.)*

Character	Holotype, allotype, El Salvador, Carneiro *et al.* (2005a)	Paratypes, El Salvador, Carneiro *et al.* (2005a)	Africa, Jorge *et al.* (2016)
Ant. end to excretory pore	162	174 ± 0.2 (145-211)	-
Spicules	28.0	31.4 ± 4.0 (27.0-35.0)	-
Gubernaculum	12.0	10.2 ± 0.9 (8-12)	-
Tail	16.0	15.3 ± 2.2 (13-20)	-
J2 (n)	-	30	-
L	-	417 ± 13.5 (400-427)	-
a	-	25.9 ± 0.5 (22.3-26.1)	-
c	-	9.0 ± 0.6 (7.9-10.1)	-
W	-	17.0 ± 1.5 (15.0-20.0)	-
Stylet	-	12.8 ± 0.2 (12.0-13.0)	12.0-13.0
DGO	-	3.3 ± 0.2 (3.0-4.0)	-
Ant. end to excretory pore	-	79.5 ± 3.0 (75.0-83.0)	-
Tail	-	45.5 ± 2.5 (45.0-48.0)	45.0-48.0
Hyaline region	-	12.5 ± 1.4 (10.0-14.0)	-

Male

Body vermiform, tapering anteriorly, bluntly rounded posteriorly twisting through 90°; tail arcuate. Body annuli large, distinct. In lateral view, labial cap high and rounded, labial region offset from body, without annulation. Under SEM, stoma slit-like, located in ovoid to hexagonal cavity surrounded by pit-like openings of six inner labial sensilla. Labial disc rounded. Median labials divided into labial pairs, fused with labial disc forming elongate labial cap with small depressions. Labial sensilla not clearly demarcated on median labials. Lateral labials absent. Amphidial apertures elongate slits located between labial disc and lateral sides of labial region. Under LM, labial framework moderately developed, hexaradiate, lateral sectors slightly enlarged. Vestibule extensions distinct. Stylet robust, opening several microns from stylet tip, cone pointed, smaller than shaft, slightly increasing in diam. posteriorly, junction of cone and shaft uneven. Shaft cylindrical sometimes with small projections, widening slightly near junction with knobs. Knobs rounded, reniform, distinctly indented, posteriorly sloping, transversely elongated, merging gradually with shaft. Dorsal pharyngeal gland orifice branching into three ducts, ampulla distinct. Amphids very distinct. Procorpus distinct, median bulb ovoid, triradiate, lining of enlarged lumen of median bulb thinner than in female. Pharyngo-intestinal junction at level of nerve ring, indistinct. Three nuclei in gland lobe, lobe variable in length. Intestinal caecum extending anteriorly, sometimes approaching level of dorsal pharyngeal gland orifice. Excretory pore distinct, located 6-10 annuli posterior to hemizonid. Lateral field areolated with four incisures beginning near level of stylet knobs or 11 annuli posterior to labial region as two incisures; two additional incisures starting near level of median bulb. Sex reversed males (with two testes) usually common compared to normal males (one testis). Testes outstretched or anteriorly reflexed. Spicules arcuate, gubernaculum distinct. Tail short, distinct phasmids at cloacal aperture level.

J2

Body vermiform, clearly annulated, tapering more posteriorly than anteriorly. Anterior end truncated, labial region only slightly offset from body. Vestibule and extension more developed than remainder of hexaradiate labial framework. Under SEM, prestoma opening rounded, surrounded by small, pore-like openings of six inner labial sensilla. Median labials and labial disc dumbbell-shaped in face view. Lateral labials small, triangular, lower than labial disc and median labials. Posterior edge of one or both lateral labials occasionally fused with labial region. Labial disc oval, slightly raised above median labials. Median labials with crescentic to rounded margins. Labial sensilla not visible externally. Elongated amphidial apertures located between labial disc and lateral labials. Labial region not annulated, body annuli distinct but fine. Lateral field beginning near level of procorpus as two lines, third line beginning near median bulb and quickly dividing, making four lines running entire length of body before gradually decreasing to two lines that end near hyaline region, lateral field irregularly areolated. Under LM, stylet delicate, cone straight, narrow, sharply pointed, shaft becoming slightly wider posteriorly, knobs small, oval-shaped and posteriorly sloping. Procorpus faintly outlined, median bulb oval with enlarged lumen lining, isthmus clearly defined, pharyngo-intestinal junction difficult to observe. Gland lobe variable in length, overlapping intestine ventrally. Excretory pore distinct, hemizonid anterior, adjacent to excretory pore. Tail conoid with rounded unstriated terminus. Hyaline region not clearly defined. Rectum dilated. Phasmids small, difficult to observe, located posterior to anus.

TYPE PLANT HOST: Coffee, *Coffea arabica* L.

OTHER PLANTS: This species was detected on banana (*Musa* sp.) from Vietnam (Carneiro *et al.*, 2005a), cabbage (*Brassica oleracea* var. *capitata*) and purple sage (*Salvia dorrii* (Kellogg) Abrams) from Benin, tomato (*Solanum lycopersicum*) and sweet pepper (*Capsicum annuum*) from Tanzania (Jorge Junior *et al.*, 2016; Santos *et al.*, 2018a). This nematode reproduced on tomato 'Rutgers', tobacco 'NC95', pepper 'California Wonder', watermelon 'Charleston Gray', while no reproduction occurred on cotton 'Deltapine 61' or peanut 'Florunner' (Carneiro *et al.*, 2005a).

TYPE LOCALITY: Izalco, Sonsonate Department, El Salvador.

DISTRIBUTION: *South America*: El Salvador, Brazil, *Asia*: Vietnam, *Africa*: Benin, Kenya, Tanzania (Carneiro *et al.*, 2005a; Jorge Junior *et al.*, 2016; Santos *et al.*, 2018a; Stefanelo *et al.*, 2019).

PATHOGENICITY: *Meloidogyne izalcoensis* causes typical relatively small galls, mostly on the extremities of new roots. Egg masses are produced outside the roots in large quantities (Carneiro *et al.*, 2005a). On the 'Mundo Novo' and 'Catuaí Amarelo' cultivars of *Coffea arabica* and 'Timor Hybrids' (*C. arabica* × *C. canephora*) there is a tendency to form lateral roots in the gall region. Necrotic areas can also be seen on the galled roots and these may be aggravated by secondary infections so that some sections of the root die. The formation of cracks on infected roots was not observed in 1-year-old plants in glasshouse conditions. The root symptoms caused by this species on coffee are completely different from those caused by *M. paranaensis*, *M. incognita*, *M. arabicida* and *M. exigua*. *Meloidogyne izalcoensis* causes severe root destruction, frequently killing coffee trees in El Salvador (Carneiro *et al.*, 2005a).

CHROMOSOME NUMBER: Reproduction is by mitotic parthenogenesis, 2n = 44-48 (Carneiro *et al.*, 2005a).

POLYTOMOUS KEY CODES: *Female*: A21, B2, C2, D6; *Male*: A213, B12, C213, D1, E2, F2; *J2*: A2, B2, C3, D23, E23, F3.

BIOCHEMICAL AND MOLECULAR CHARACTERISATION: *Meloidogyne izalcoensis* showed a species-specific esterase (EST) phenotype (Iz4) with four bands (Rm: 0.86, 0.96, 1.24, 1.30). This esterase phenotype (Est I4) is unique and is the most useful character to differentiate *M. izalcoensis* from all other species (Carneiro *et al.*, 2005a; Santos *et al.*, 2018a; Stefanelo *et al.*, 2018). MDH phenotype N1 (Rm: 15.2). Molecularly this nematode can be identified by species-specific primers developed by Correa *et al.* (2013). PCR-RFLP of the *COII*-16S rRNA mitochondrial gene region enable a clear diagnostic differentiation between *M. izalcoensis* and *M. arabicida* (Carneiro *et al.*, 2005b; Humphreys-Pereira *et al.*, 2014; Pagan *et al.*, 2015) (Fig. 13).

RELATIONSHIPS (DIAGNOSIS): *Meloidogyne izalcoensis* belongs to Molecular group I, the Incognita group. The perineal pattern of *M. izalcoensis* is similar to *M. incognita* and *M. paranaensis*. In the female the stylet cone is slightly curved dorsally and the knobs are not anteriorly indented, whereas in *M. incognita* the stylet cone is distinctly curved dorsally and the knobs are anteriorly indented. DGO in *M. izalcoensis* is longer at 4.5-6.0 *vs* 2.0-4.8 μm in *M. incognita*. The J2 labial region of *M. izalcoensis* is not annulated whereas in *M. incognita* it bears 2-4 incomplete annuli (Chitwood, 1949; Eisenback & Triantaphyllou, 1991; Carneiro *et al.*, 2005b). The EP/ST ratio can also separate this two species from *M. izalcoensis*.

44. *Meloidogyne javanica* (Treub, 1885) Chitwood, 1949
(Figs 129-131)

COMMON NAME: Javanese root-knot nematode.

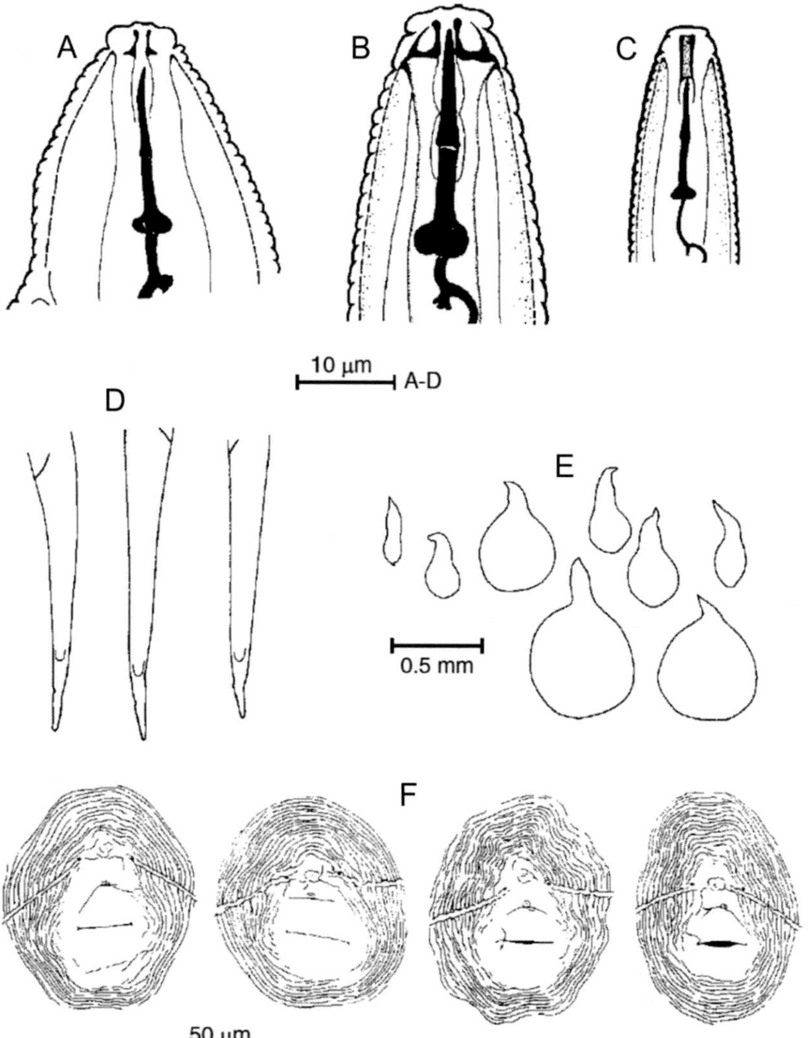

Fig. 129. Meloidogyne javanica. *A: Female labial region; B: Male labial region; C: Second-stage juvenile (J2) labial region; D: J2 tail; E: Entire female; F: Perineal pattern. (After Orton Williams, 1972.)*

Meloidogyne javanica is one of the most common and important RKN species in tropical and subtropical regions. It has a wide host range and is considered a major agricultural pest (Taylor & Sasser, 1978). This nema-

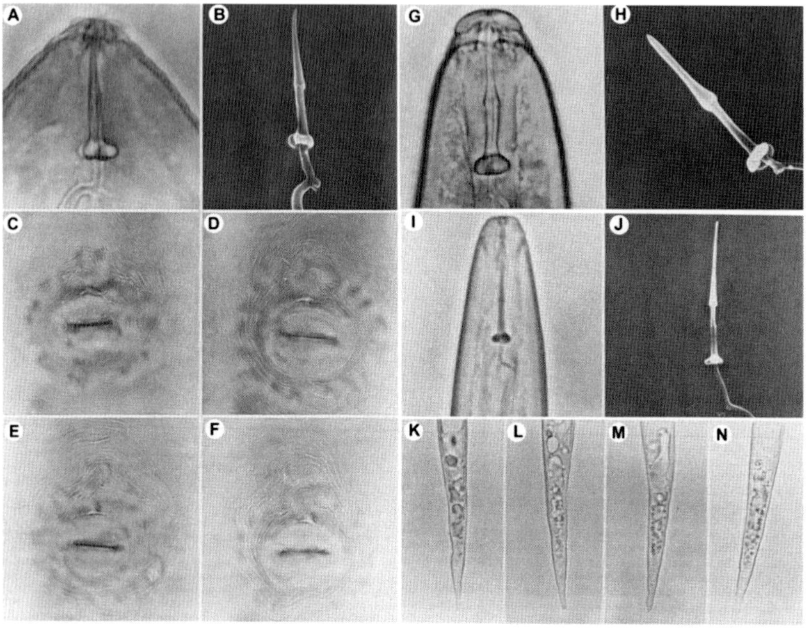

Fig. 130. Meloidogyne javanica. *LM. A: Female labial region; B: Female excised stylet; C-F: Perineal pattern; G: Male labial region; H: Male excised stylet; I: Second-stage juvenile (J2) labial region; J: J2 excised stylet; K-N: J2 tail. (After Eisenback & Hirschmann, 1979, 1980; Eisenback* et al.*, 1980.)*

tode species was collected in sugarcane by J. van der Vecht, Head of the Institute for Plant Diseases and Pests, Buitenzorg, Java, Indonesia, and described by Treub (1885). The descriptions of *M. javanica* are given by many authors (Chitwood, 1949; Whitehead, 1968; Orton Williams, 1972; Eisenback & Hirschmann, 1979, 1980, 1981; Eisenback *et al.*, 1981; Eisenback, 1982; Jepson, 1987; Rammah & Hirschmann, 1990a; Carneiro *et al.*, 1998; Latha *et al.*, 1998; Sahoo & Ganguly, 2000; Singh *et al.*, 2018).

Liu & Zhang (2001) described a new species, *M. dimocarpus*, with similar morphological and morphometric data to *M. javanica*, with the exception of continuous striae crossing between the vulva and anus. Since no additional biochemical or molecular data were provided, we consider this difference a minor intraspecific variability and regarded *M. dimocarpus* n. syn. as a junior synonym of *M. javanica*.

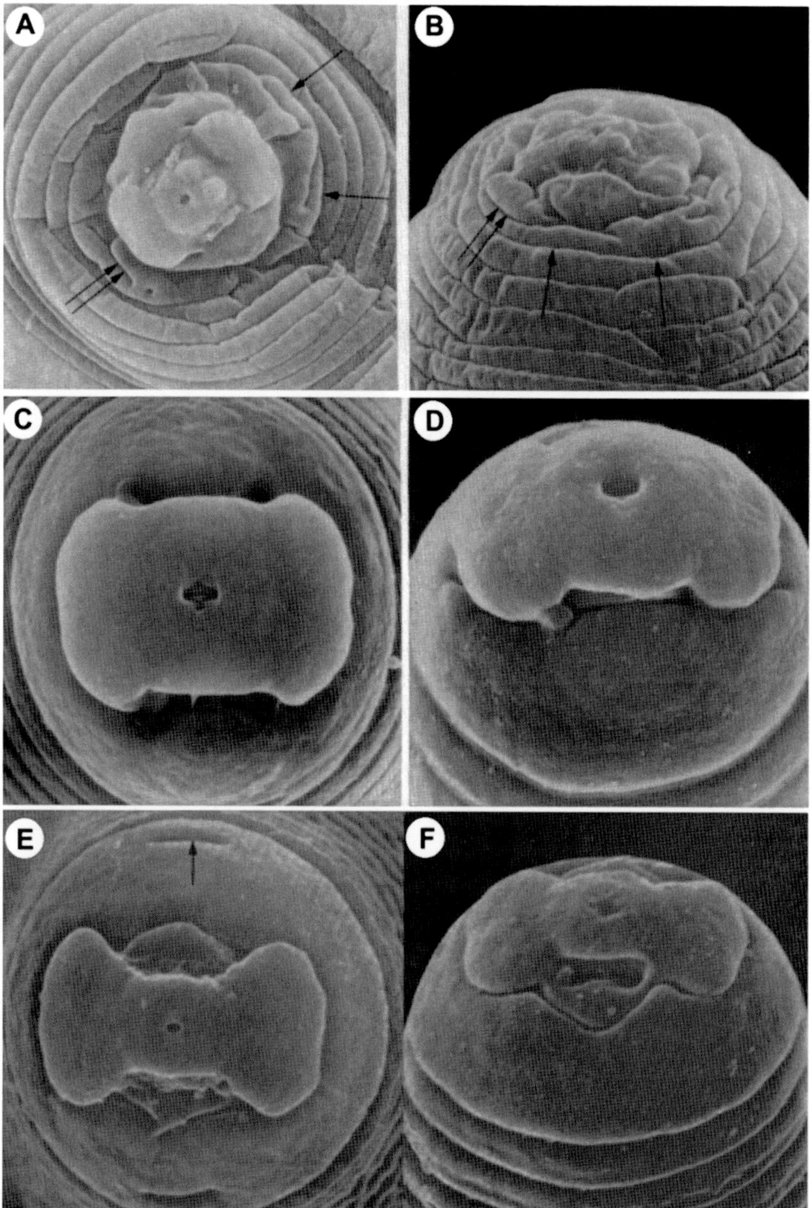

Fig. 131. Meloidogyne javanica. *SEM. A, B: Female labial region; C, D: Male labial region; E, F: Second-stage juvenile labial region. (After Eisenback & Hirschmann, 1979, 1980; Eisenback* et al., *1980.)*

MEASUREMENTS

See Table 33.

- *Eggs* (n = 20): 81 (71-89) × 30 (27-35) μm.

DESCRIPTION

Female

Body pear-shaped, nearly spherical in adult with projecting neck tapering to labial region, posterior end rounded, or with slightly protruding perineal region. Labial region with one annulus posterior to labial cap, slightly wider than first body annulus. Stylet slender, dorsally curved, knobs rounded. Shaft cylindrical, broadening near base.

Table 33. *Morphometrics of females, males and second-stage juveniles (J2) of* Meloidogyne javanica. *All measurements are in μm and in the form: mean ± s.d. (range).*

Character	Paratypes, Java, Indonesia, Chitwood (1949)	Whitehead (1968)	Georgia, USA Rammah & Hirschmann (1990a)	Pallabialalem, India Latha *et al.* (1998)	Delhi, India Sahoo & Ganguly (2000)
Female (n)	-	20	25	20	20
L	545-800	657 ± 67 (541-804)	786 ± 23.8 (599.4-1012.5)	758 ± 12.2 (678-850)	582 ± 31.2 (490-692)
a	-	-	-	-	1.3 ± 0.1 (0.8-1.7)
b	-	-	-	-	9.8 ± 1.2 (6.6-13.8)
W	300-545	431 ± 63 (311-581)	576.4 ± 14 (486-729)	567 ± 14.1 (471-678)	340.2 ± 29 (262-475)
Stylet	16.0	15.0 (14.0-18.0)	15.9 ± 0.2 (14.8-17.8)	15.9 ± 0.2 (14.1-17.2)	15.4 ± 0.7 (13.0-18.0)
Stylet knob width	4.0-5.0	4.0 (2.0-5.0)	4.9 ± 0.1 (4.4-5.5)	-	-
Stylet knob height	2.0	-	2.1 ± 0.03 (1.9-2.4)	-	-
DGO	3.0-4.0	3.0 (2.0-5.0)	3.2 ± 0.1 (2.2-4.5)	3.1 ± 0.1 (2.0-4.1)	4.2 ± 0.2 (3.5-5.0)
Median bulb length	-	42 (38-46)	-	-	44 ± 1.4 (40-50)

Table 33. *(Continued.)*

Character	Paratypes, Java, Indonesia, Chitwood (1949)	Whitehead (1968)	Georgia, USA Rammah & Hirschmann (1990a)	Pallabialalem, India Latha *et al.* (1998)	Delhi, India Sahoo & Ganguly (2000)
Median bulb width	-	35 (31-44)	-	-	39 ± 2.0 (35-50)
Median bulb valve width	-	-	-	-	10.4 ± 0.2 (10.0-11.0)
Median bulb valve length	-	11.0 (10.0-13.0)	-	-	14.7 ± 0.4 (14.0-16.0)
EP/ST	2.8	1.8*	-	-	-
Vulval slit length	-	-	25.4 ± 0.5 (21.5-28.1)	-	-
Interphasmid distance	-	-	27.9 ± 0.5 (24.1-34.2)	-	-
Male (n)	-	25	25	20	-
L	940-1440	1131 ± 120 (757-1297)	1444.1 ± 34.9 (1093.5-1782)	1917.9 ± 50.7 (1421-2295)	-
a	26-42	37.5 ± 5.2 (17.5-42.9)	33.0 ± 0.5 (27.9-40.1)	-	-
b	7-13	15.7 ± 1.2 (13-18.4)		-	-
c	-	91 ± 23.6 (50-144)		-	-
W	-	-	43.8 ± 0.6 (37.0-49.4)	-	-
Labial region height	-	6.6 ± 0.4 (5.8-7.6)	-	-	-
Labial region diam.	-	-	-	-	-
Stylet	20.0-21.0	21.2 ± 1.5 (20.0-23.0)	21.8 ± 0.2 (19.5-23.3)	19.6 ± 0.3 (17.0-21.0)	-
Stylet knob width	5.0	4.3 ± 0.4 (3.6-5.4)	5.3 ± 0.1 (4.6-5.8)	-	-
Stylet knob height	3.0-3.5	-	2.5 ± 0.03 (2.2-2.9)	-	-
DGO	3.0	2.9 ± 0.8 (2.2-4.7)	2.8 ± 0.1 (2.2-3.7)	2.9 ± 0.1 (2.2-3.6)	-
Median bulb length	-	19.7 ± 2.1 (15.1-23.7)	-	-	-
Median bulb width	-	10.9 ± 0.9 (9.4-12.9)	-	-	-

Table 33. *(Continued.)*

Character	Paratypes, Java, Indonesia, Chitwood (1949)	Whitehead (1968)	Georgia, USA Rammah & Hirschmann (1990a)	Pallabialalem, India Latha *et al.* (1998)	Delhi, India Sahoo & Ganguly (2000)
Spicules	30.0-31.0	26.7 ± 3.4 (20.9-31.7)	31.9 ± 0.4 (29.6-35.3)	-	-
Gubernaculum	-	8.4 ± 0.6 (7.2-9.4)	8.2 ± 0.2 (7.4-9.4)	-	-
Tail	-	-	-	-	-
J2 (n)	-	25	25	20	20
L	340-400	417 ± 22 (387-459)	429.7 ± 3.0 (406.4-470.4)	458 ± 6.8 (402-515)	339.8 ± 3.7 (325-350)
a	24.0-26.0	30.6 ± 2.1 (27.1-35.9)	27.5 ± 0.3 (25.0-30.3)	-	27.5 ± 0.5 (25.4-28.5)
b	8.0	2.4 ± 0.3 (2.1-3.4)	-	-	2.3 ± 0.1 (1.9-2.7)
b′	-	7.5 ± 0.2 (7.1-8.0)	7.2 ± 0.04 (6.6-7.6)	-	7.5 ± 0.2 (6.3-7.9)
c	5.8-6.6	8.5 ± 0.7 (7.3-11.1)	7.7 ± 0.1 (7.0-8.1)	-	8.9 ± 0.3 (7.7-10.0)
c′	-	-	-	-	5.7 ± 0.2 (5.3-6.3)
W	-	-	15.6 ± 0.1 (14.8-16.9)	-	12.4 ± 0.2 (12.0-13.0)
Stylet	10.0	-	11.5 ± 0.1 (10.7-11.9)	15 ± 0.1 (14-16)	12.8 ± 0.3 (12.0-14.0)
DGO	4.0	-	3.7 ± 0.1 (3.2-4.1)	3.3 ± 0.1 (3.0-4.0)	3.7 ± 0.2 (3.0-4.5)
Median bulb length	-	12.1 ± 0.7 (10.8-13.7)	-	-	10.7 ± 0.4 (10.0-12.0)
Median bulb width	-	6.6 ± 0.5 (5.4-7.6)	-	-	6.1 ± 0.2 (5.5-7.0)
Median bulb valve length	-	4.3 ± 0.5 (3.2-5.0)	-	-	-
Ant. end to excretory pore	-	-	83.9 ± 0.4 (80.8-86.7)	-	71.9 ± 2.2 (67.0-80.0)
Tail	-	49 ± 4.0 (36.0-56.0)	56.1 ± 0.5 (51.8-60.8)	55.3 ± 0.8 (50.5-61.5)	38.1 ± 1.1 (35.0-42.0)
Hyaline region	-	-	-	-	6.8 ± 0.7 (3.0-9.0)

* Estimated from drawing.

Large transverse elongate knobs offset from shaft and often marked anteriorly by a shallow indentation. Perineal pattern round or oval to pear-shaped, dorsal arch varying from rounded, to moderate height, sometimes flattened dorsally. Striae smooth to wavy, tail tip often marked by an irregular whorl. Typically lateral fields well defined by a double incisure bordering a definite break in striae clearly dividing pattern into dorsal and ventral sectors. Lateral fields may be visible for some distance from tail but do not extend into neck region, or may be poorly marked. Perineal patterns with distinct lateral lines. Phasmids usually distinct.

Male

Labial region not offset, outline in lateral view characteristic with first annulus projecting anterolaterally, well separated by constriction from a wider basal annulus, which may be subdivided into two annuli either completely or partially (giving appearance of one annulus on one side of body and two on other). In dorsal or ventral view, labial region outline rounded, amphidial apertures distinct. In face view, lateral edges of oral disc nearly parallel with each other and overlap amphidial opening. Median labials rounded, lateral edges slightly wider than labial disc. Stylet unique for species and useful for separating *M. javanica* males from other species with similar labial region shape. Stylet knobs transversely elongated, distinctly offset from shaft; conus bluntly pointed. Valves of pharyngeal bulb large. Pharyngeal gland overlapping intestine for some distance ventrally. Excretory pore distinct, duct visible for some way along body. Hemizonid 0-4 annuli anterior to pore. One testis or two. Intersexes common, displaying female characters to varying degrees ranging from a slight swelling anterior to cloacal aperture to possession of a well-developed vulva. Lateral field usually with four incisures ending at terminus or just anterior. Bands between incisures plain or outer ones sometimes areolated to some degree. Torsion of posterior region of body occurring to some extent making true lateral views of tail hard to obtain. Tail shape somewhat variable, digitate in lateral view, bluntly rounded in ventral. Terminus not clearly striated. Phasmids located in lateral field at cloacal level or slightly anterior. Spicules slightly curved, with ventral flange distally, tips pointed. Gubernaculum thin, crescentic. Some populations may produce intersexes appearing as normal males, but with a rudimentary vagina in various states of development (Eisenback & Triantaphyllou, 1991).

J2

Labial region not offset, in lateral view truncate cone shape with three annuli posterior to labial cap, more rounded in dorsal and ventral view with one annulus posterior to labial cap. Labial disc and median labials often dumbbell-shaped, but in some specimens may be bowtie-shaped (Eisenback & Triantaphyllou, 1991). Stylet knobs rounded, not prominent. Hemizonid usually three annuli long, located immediately anterior to excretory pore. Lateral field with four incisures, outer bands not clearly cross-striated. Rectum dilated, tail tapering to a subacute or finely rounded terminus, annuli coarsening posteriorly.

TYPE PLANT HOST: Sugarcane, *Saccharum officinarum* L.

OTHER PLANTS: *Meloidogyne javanica* is extremely polyphagous, attacking both monocotyledons and dicotyledons and with a long list of over 770 host species or varieties (Goodey *et al.*, 1965), many of economic importance such as tea, tobacco, potato, grapevine, tomato, many legumes, vegetables, fruit trees, cereals and ornamentals, sugarcane and yam (Orton Williams, 1972; Onkendi *et al.*, 2014; Vovlas *et al.*, 2015; Kolombia *et al.*, 2017; Medina *et al.*, 2017). Groundnut is not a good host for *M. javanica*. In fact, most populations of *M. javanica* in the USA do not reproduce well on groundnut but populations of this species from India and northern Africa generally reproduce well on groundnut. *Meloidogyne javanica* also parasitises many weeds that act as reservoirs of these nematodes similarly to *M. incognita* (Kaur *et al.*, 2007; Castillo *et al.*, 2008; Rich *et al.*, 2008; Anwar *et al.*, 2009).

TYPE LOCALITY: Cheribon or Buitenzorg, Java, Indonesia.

DISTRIBUTION: *Meloidogyne javanica* is widely distributed in warm and tropical climates where it is often the dominant RKN at higher altitudes. Range includes Africa, Asia, Australia, Brazil, Central and South America, Ceylon, Colombia, Cyprus, India, Israel, Malaysia, Pakistan, Spain, Trinidad, USA and glasshouses in northern Europe (Orton Williams, 1972; Wesemael *et al.*, 2011; Onkendi *et al.*, 2014; Meza *et al.*, 2016; Song *et al.*, 2017a; Zhao *et al.*, 2017). The latitude range for *M. javanica* is from approximately 33°N to 33°S, about 3° less than that of *M. incognita* (Taylor *et al.*, 1982). This species is predominant in areas with a distinct dry season with less than 5 mm of precipitation per month for 3 or more successive months.

PATHOGENICITY: Galls produced by *M. javanica* are similar to those of *M. incognita* or *M. arenaria* and many other species. Damage potential, considered as the max. percentage of yield loss, has been studied in different crops after inoculation with eggs or J2 (at different inoculum densities) by several researchers (Wesemael *et al.*, 2011). This damage potential was estimated from 100% in common bean (Di Vito *et al.*, 2007), tomato and pepper (Mekete *et al.*, 2003), rice (Di Vito *et al.*, 1996a), or sunflower (Di Vito *et al.*, 1996b), to 50% in olive (Sasanelli *et al.*, 2002), or 40% in potato (Vovlas *et al.*, 2005).

Meloidogyne javanica is the most severe pest of crop plants in Central Africa (Daulton & Curtis, 1964) and the predominant species in tobacco-growing areas of Malawi and Zimbabwe where losses have been estimated at 18-25 million pounds weight of cured leaf per annum, worth £2 750 000 to £3 750 000 (Daulton, 1963).

CHROMOSOME NUMBER: *Meloidogyne javanica* reproduces exclusively by mitotic parthenogenesis, chromosome numbers of 2n = 42-48 (Triantaphyllou, 1985b).

POLYTOMOUS KEY CODES: *Female*: A23, B213, C3, D5; *Male*: A34, B23, C23, D1, E2, F1; *J2*: A23, B2, C3, D43, E23, F21.

HOST RACE: Host races of *M. javanica* have not been widely recognised (Table 3), although a few populations vary from the normal host response in the standard differential host test (Hartman & Sasser, 1985). No morphological differences had been detected between populations of the three host races (Rammah & Hirschmann, 1990a). Usually, populations of *M. javanica* reproduce on watermelon, tomato and *M. incognita*-resistant tobacco, but do not reproduce on cotton, pepper, or peanut (Hartman & Sasser, 1985).

BIOCHEMICAL AND MOLECULAR CHARACTERISATION: *Meloidogyne javanica* showed the J3 esterase (Fig. 12) and a N1 malate dehydrogenase phenotype (Esbenshade & Triantaphyllou, 1985a, b). One population of *M. javanica* from Florida, isolated from tobacco (*Nicotiana tabacum*), showed esterase J2a phenotype (Brito *et al.*, 2008), and another from Cerrado native plants in Brazil (Silva *et al.*, 2014), but J2a was not stable when the nematode isolate was cultured on tomato. One population from Bangladesh and one from Korea showed the N3 phenotype, similar to a few populations of *M. arenaria* (Eisenback & Triantaphyllou, 1991).

Meloidogyne javanica can be identified using PCR by species-specific primers (Table 10). Several molecular markers and techniques have been developed for the accurate identification of *M. javanica* (Zijlstra, 1997; Adam *et al.*, 2007; Niu *et al.*, 2011; Kiewnick *et al.*, 2013; Janssen *et al.*, 2016, 2017).

RELATIONSHIPS (DIAGNOSIS): The species belongs to Molecular group I, the Incognita group. *Meloidogyne javanica* can be distinguished from all other *Meloidogyne* spp. by several morphological and molecular characteristics. Useful diagnostic characters include the morphology of the female perineal pattern that has distinct lateral lines clearly delineating the dorsal and ventral regions of the pattern, dorsal arch moderately high and narrow; the shape of the male labial region, and EP/ST = 2.8. This species is close to *M. arenaria* and *M. incognita* from which it can be differentiated by perineal pattern and the EP/ST ratio (1.8-2.8 *vs* 2.4, 1.2-2.5, respectively).

45. *Meloidogyne jianyangensis* Baojun, Hu, Chen & Zhu, 1990
(Fig. 132)

COMMON NAME: Jianyang citrus root-knot nematode.

Baojun *et al.* (1990) described this species from roots of mandarin orange in Jianyang County, Sichuan province, China.

MEASUREMENTS (*AFTER BAOJUN* ET AL., *1990*)

- *Holotype female*: L = 504.6 μm; W = 414.6 μm; stylet = 14.6 μm; DGO = 3.2 μm; excretory pore to anterior end = 18.4 μm; EP/ST = 1.3.
- *Allotype male*: L = 1343.6 μm; W = 51.0 μm; stylet = 19.2 μm; DGO = 2.6 μm; excretory pore to anterior end = 114.8 μm;

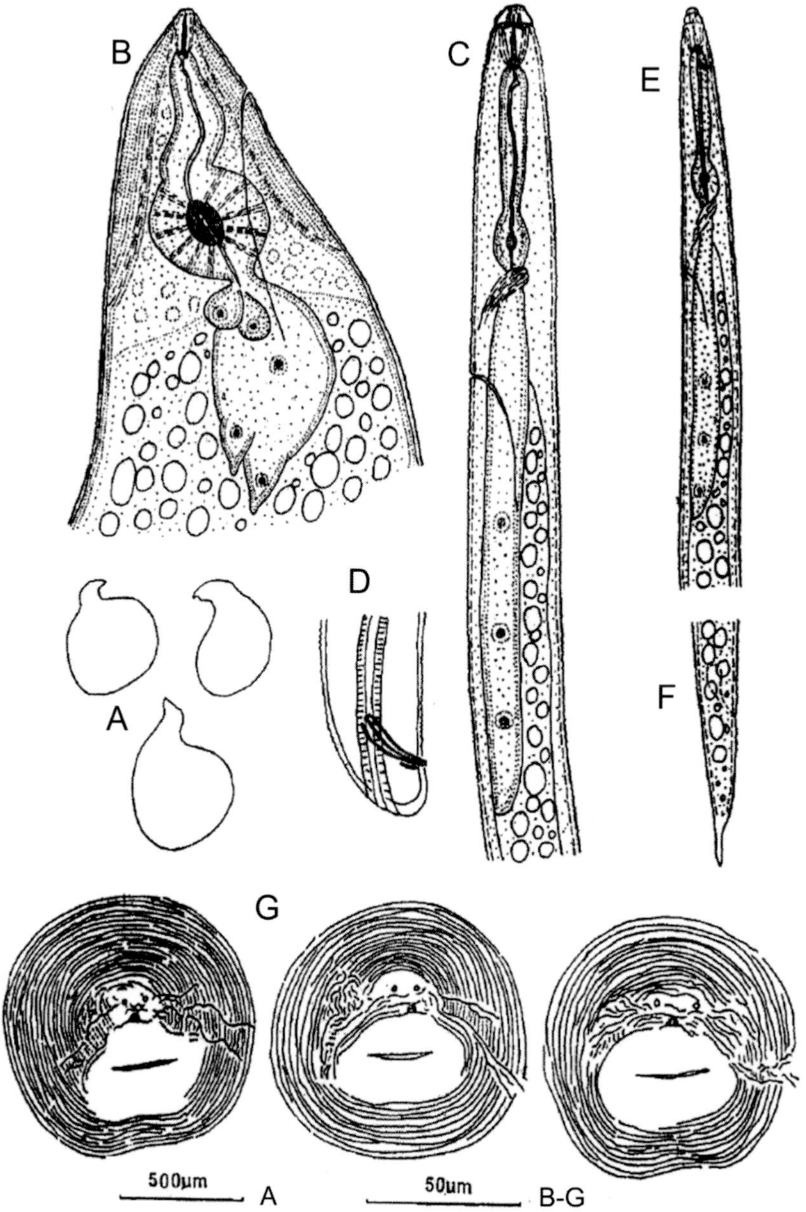

Fig. 132. Meloidogyne jianyangensis. *A: Entire female; B: Female anterior region; C: Male anterior region; D: Male tail; E: Second-stage juvenile (J2) anterior region; F: J2 tail; G: Perineal pattern. (After Yang* et al., *1990.)*

spicules = 31.1 μm; gubernaculum = 4.0 μm; tail = 15.0 μm; a = 26.4; b = 17.1; c = 80.9; T = 57.7%.

- *Paratype females* (n = 30): L = 510 $\pm$ 64.5 (394-677) μm; W = 360 $\pm$ 66.2 (254-513) μm; stylet = 13.8 $\pm$ 1.0 (11.8-15.6) μm; stylet knob width = 3.6 $\pm$ 0.2 (3.1-4.3) μm; stylet knob height = 1.9 $\pm$ 0.1 (1.7-2.1) μm; DGO = 3.6 $\pm$ 0.3 (2.9-3.9) μm; anterior end to median bulb = 85.1 $\pm$ 13.8 (64.8-133.6) μm; anterior end to excretory pore = 25.1 $\pm$ 4.6 (18.4-36.1) μm; interphasmid distance = 12.9 $\pm$ 2.0 (9.8-18.4) μm; vulval slit length = 26.3 $\pm$ 2.4 (19.6-29.4) μm; anus-vulva distance = 15.8 $\pm$ 2.0 (8.7-18.2) μm; a = 1.5 $\pm$ 0.3 (1.0-2.2); EP/ST = 1.3 (estimated from drawings).
- *Paratype males* (n = 30): L = 1477 $\pm$ 175 (1084-1728) μm; W = 51.0 $\pm$ 9.1 (37.6-67.0) μm; stylet = 21.8 $\pm$ 1.8 (19.1-26.0) μm; stylet knob width = 4.8 $\pm$ 0.4 (4.3-5.8) μm; stylet knob height = 3.0 $\pm$ 0.2 (2.3-3.4) μm; DGO = 3.0 $\pm$ 0.3 (2.5-3.5) μm; excretory pore to anterior end = 110.1 $\pm$ 10.8 (92.7-145.7) μm; spicules = 30.7 $\pm$ 2.0 (26.0-34.0) μm; gubernaculum = 9.1 $\pm$ 1.0 (7.0-11.0) μm; tail = 15.0 $\pm$ 2.8 (8.5-20.0) μm; a = 29.6 $\pm$ 4.8 (18.5-43.0); c = 103.2 $\pm$ 30.7 (60.9-191.6); T = 68.0 $\pm$ 14.8 (44.4-90.5)%.
- *Paratype J2* (n = 30): L = 423.2 $\pm$ 25.0 (387.6-483.3) μm; W = 18.4 $\pm$ 1.4 (16.0-22.6) μm; stylet = 15.1 $\pm$ 1.0 (13.0-16.8) μm; stylet knob width = 2.8 $\pm$ 0.4 (2.4-3.6) μm; stylet knob height = 1.5 $\pm$ 0.1 (1.3-1.8) μm; DGO = 2.5 $\pm$ 0.4 (2.1-3.7) μm; excretory pore to anterior end = 83.1 $\pm$ 4.9 (73.3-97.1) μm; tail = 48.7 $\pm$ 6.5 (41.0-62.4) μm; hyaline region = 7.8 $\pm$ 1.1 (5.8-10.7) μm; a = 23.1 $\pm$ 1.7 (17.7-27.2); b = 6.4 $\pm$ 0.5 (5.5-7.4); c = 8.8 (6.8-10.2).
- *Eggs*: L = 104.0 (92.0-115.0) μm; W = 45.5 (39.0-61.0) μm.

DESCRIPTION

Female

Mature female creamy to yellow, globular with prominent neck. Body length from end of neck to end of body bigger than body diam. Variable in size. Labial region not distinctly offset from neck, no annuli observed, framework slightly sclerotised, body cuticle around neck annulated, posterior of body finely annulated with posterior protuberance, excretory pore usually apparent, located *ca* twice two stylet lengths posterior, stylet slender, conical portions hardly curved dorsally. Knobs offset from shaft. Subventral gland orifices located ventrally posterior to median bulb. Two pharyngo-intestinal cells located between median bulb and intestine. Perineal pattern rounded with smooth and fine striae. Lateral lines distinct and continuous. Striae occurring on lateral sides of tail tip. Number of striae variable. Spine-like striae stretch radially from tail terminus to lateral regions. Striae among the radial striae could be broken and look like punctations. One semicircular ridge surrounding perivulval region dorsally but can be discontinuous and appear as many dots. Perivulval region free of striae. Phasmids located dorsally to anus, distance between phasmids short.

Male

Vermiform body transparent or translucent and of variable size, tapering at both ends. Tail end more rounded than anterior end. Body annulated with four incisures. Areolation present except for two inner incisures. Labial cap high with no annulation. Labial disc and median labials fused and labial disc slightly elevated above median labials. Median labials indented inward slightly in middle. Stoma slit-like, lateral labial absent. Amphid slit-like, posterior to labial disc. Stylet straight, pointed, knobs distinctly offset from shaft. Procorpus cylindrical, median bulb oval. Subventral gland orifices located ventrally posterior to median bulb. Pharynx overlapping intestine. Opening of excretory pore near hemizonid. Hemizonid indistinct. Spicule and gubernaculum accurate. Phasmid small.

J2

Body vermiform, tapering at both ends. Body annulated with four incisures. Labial cap high without annulation. Labial region offset from body. Labial disc and median labials fused, dumbbell-shaped. Labial disc slightly elevated above median labials. Lateral disc present. Amphidial apertures slit-like, located between labial disc and lateral labials. Stylet straight, knobs distinctly offset from shaft. Pharynx overlapping intestine. Hemizonid anterior to excretory pore. Excretory duct distinct, hemizonid obscure. Intestine enlarged slightly, tapering at end with pointed tail. Phasmid obscure.

TYPE PLANT HOST: Mandarin orange, *Citrus reticulata* Blanco.

OTHER PLANTS: No other hosts were reported.

TYPE LOCALITY: Jianyang County, Sichuan Province, China.

DISTRIBUTION: *Asia*: China.

PATHOGENICITY: The damage caused to mandarin orange in China has not been assessed.

CHROMOSOME NUMBER: No available data.

POLYTOMOUS KEY CODES: *Female*: A32, B32, C3, D1; *Male*: A324, B213, C23, D1, E2, F2; *J2*: A23, B12, C32, D43, E2, F1.

BIOCHEMICAL AND MOLECULAR CHARACTERISATION: This species has three major bands of esterase activity at Rm = 0.41, 0.45 and 0.48, but no gel picture was provided by the authors in the original description. No molecular data of this species are available.

RELATIONSHIPS (DIAGNOSIS): The perineal pattern of *M. jianyangensis* is characteristic and it is round with spine-like striae present from the tail terminus to the lateral region. No shoulder-like striae are present in the arch. In the male, the knobs are distinctly offset from the shaft. *Meloidogyne jianyangensis* is similar to *M. arenaria* but differs from the latter species in several morphological characteristics. In the female, *M. jianyangensis* has a shorter stylet (11.0-14.0 *vs* 13.4-17.8 μm). It is similar to *M. exigua*, from which differs in longer males (1084-1728 *vs* 660-1450 μm) and longer J2 (388-483 *vs* 281-450 μm) with longer stylet (13-17 *vs* 8.6-14.0 μm).

46. *Meloidogyne jinanensis* Zhang & Su, 1986
(Fig. 133)

COMMON NAME: Jinan root-knot nematode.

Zhang & Su (1986) described this species from roots of bonfire salvia (*Salvia splendens* Sellow *ex* Schult.) in Tami Aghu Park, Jinan City, Shandong Province, China.

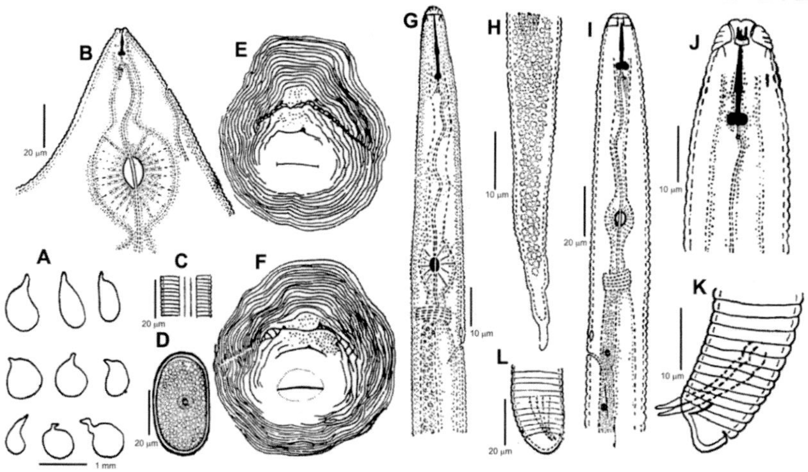

Fig. 133. Meloidogyne jinanensis. *A: Entire female; B: Female anterior region; C: Second-stage juvenile (J2) lateral field; D: Egg; E, F: Perineal pattern; G: J2 anterior region; H: J2 tail; I: Male anterior region; J: Male labial region; K, L: Male tail. (After Zhang & Su, 1986.)*

MEASUREMENTS *(AFTER ZHANG & SU, 1986)*

- *Paratype females* (n = 37): L = 527 (344-760) μm; W = 385 (272-560) μm; labial region height = 3.5 (2.5-4.0) μm; labial region diam. = 10.2 (7.0-21) μm; stylet = 13.6 (11.5-14.5) μm; stylet knob height = 2.7 (2.5-3.5) μm; stylet knob width = 4.5 (4.0-5.0) μm; DGO = 4.8 (4.0-7.0) μm; excretory pore to anterior end = 47.2 (21.0-71.0) μm; median bulb length = 46.7 (40.0-52.5) μm; median bulb width = 42.7 (38.5-52.5) μm; a = 1.4 (1.0-2.1); EP/ST = 3.5; vulva width = 25 (22.5-27.0) μm, interphasmid distance = 22 (19-27) μm; distance from anus to vulva = 15.6 (11.0-17.5) μm.
- *Paratype males* (n = 30): L = 1299 (880-1552) μm; W = 33.7 (28.0-48.0) μm; stylet = 19.6 (17.5-24.0) μm; stylet knob height = 3.0 (2.5-3.5) μm; stylet knob width = 12.3 (10.5-14.5) μm; labial region diam. = 11.7 (10.0-13.5) μm; distance from base of stylet knob to anterior end = 24.0 (22.0-26.5) μm; tail = 12.0 (8.0-16.0) μm; diam. at anus level = 19.5 (15.1-32.0) μm; excretory pore to anterior end = 142 (120-161) μm; spicules = 28.3 (21.0-35.5) μm; gubernaculum = 7.2 (4.0-10.0) μm; a = 37.3 (31.0-50.6); b = 15.6 (11.7-20.0); c = 100.5 (63.4-162.0).

- *Paratype J2* (n = 30): L = 430 (381-530) μm; W = 16.5 (14.0-20.0) μm; stylet = 12.0 (10.0-14.5) μm; stylet knob height = 1.7 (1.5-2.5) μm; stylet knob width = 2.9 (2.0-4.5) μm; DGO = 4.0 (3.0-5.5) μm; labial region height = 3.0 (2.5-5.5) μm; labial region diam. = 5.5 (5.0-7.0) μm; distance from base of stylet knobs to anterior end = 15.7 (14.0-17.5) μm; excretory pore to anterior end = 83.4 (63.0-96.0) μm; anterior end to the centre of median bulb = 59.2 (53.5-66.5) μm; median bulb height = 15 (11.0-24.0) μm; median bulb width = 10.7 (8.0-16.0) μm; tail = 53.5 (47.5-59.5) μm; anal body diam. = 10.8 (10.0-12.0) μm; a = 26.3 (20.7-31.5); b = 7.5 (5.7-8.8); c = 8.1 (6.8-10.8); c′ = 5.0 (4.2-5.6).
- *Eggs* (n = 30): L = 92.8 (80.0-105.0) μm; W = 42.3 (32.0-51.0) μm; L/W = 2.2 (1.7-3.2).

DESCRIPTION

Female

Labial region with two or three annuli. Excretory pore located ventrally, 3-5 times stylet length from anterior end (equal to 2.7 times length from anterior end to base of stylet knobs). Hemizonid 3.0 μm long, 1-2 annuli anterior to excretory pore. Perineal pattern circular or oval outline, dorsal arch moderate, usually trapezoid, often many short striae inside two inner lateral lines, a curved line present between two inner lateral lines and no or few striae above curved line. Punctate region between tail tip and anus, sometimes punctate region spreading on both sides of vulva, anus flap distinct, striae of perineal pattern smooth, continuous, slightly wavy.

Male

Body vermiform, anterior portion narrower than median and posterior portion, tail end rounded. Labial cap 2.5 (1.6-3.5) μm high, labial region 8.1 (6.9-9.6) μm wide, labial region with three annuli in dorsoventral view, opening of amphids distinct. Stylet stronger than that of adult female and juvenile. Hemizonid 3.5 μm long, 1-2 annuli anterior to excretory pore. Four lateral lines in lateral field. Spicules isometric and isomorphic, gubernaculum crescentic, located dorsal to spicules. Phasmids distinct, posterior to cloacal aperture and 7.0 (3.5-13.0) μm from tail tip.

J2

Labial region with two or three annuli in lateral view. Opening of amphids distinct. Hemizonid 3.8 μm long, 1-2 annuli anterior to excretory pore. Tail long, cone-shaped, tail end rounded with no annuli, tail length moderate, rectum not dilated.

TYPE PLANT HOST: Bonfire salvia, *Salvia splendens* Sellow *ex* Schult.

OTHER PLANTS: No other hosts were reported.

TYPE LOCALITY: Tami Aghu Park, Jinan City, Shandong Province, China.

DISTRIBUTION: *Asia*: China.

PATHOGENICITY: The damage caused by the Jinan root-knot nematode in China has not been assessed.

CHROMOSOME NUMBER: No available data.

POLYTOMOUS KEY CODES: *Female*: A32, B324, C1, D5; *Male*: A324, B32, C213, D1, E-, F2; *J2*: A213, B213, C3, D2, E32, F32.

BIOCHEMICAL AND MOLECULAR CHARACTERISATION: No available data.

RELATIONSHIPS (DIAGNOSIS): This species is very similar to *M. microcephala* in female and J2 morphology but has a punctate perineal pattern in *M. jinanensis*. *Meloidogyne jinanensis* can be separated from *M. ardenensis* by perineal pattern form.

47. *Meloidogyne kikuyensis* De Grisse, 1961
(Figs 134-137)

COMMON NAME: Kikuyu grass root-knot nematode.

De Grisse (1961) described this species from roots of Kikuyu grass near the headquarters of the East African Agriculture and Forestry Research Organisation, Muguga, Kenya. Descriptions of this species were also given by Eisenback & Spaull (1988) and Eisenback & Vieira (2020).

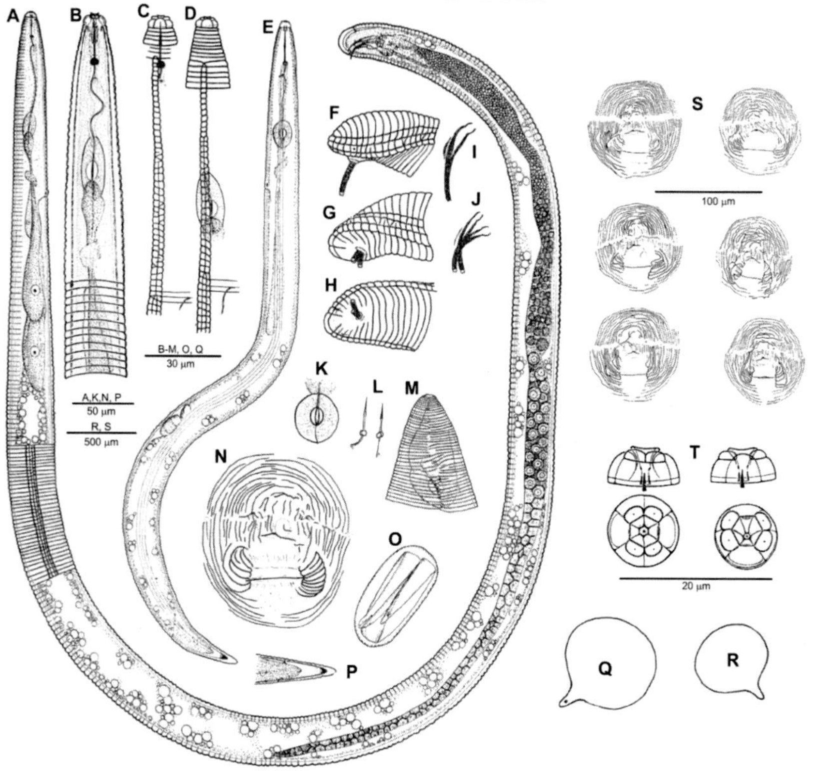

Fig. 134. Meloidogyne kikuyensis. *A: Entire male; B: Male anterior region; C, D: Male labial region; E: Entire second-stage juvenile (J2); F-H: Male tail; I, J: Spicules; K: Female median bulb; L: Female stylet; M: Female anterior region; N: Perineal pattern; O: Egg; P: J2 tail; Q, R: Entire female; S: Perineal pattern; T: Male labial region. (After De Grisse, 1960.)*

MEASUREMENTS *(AFTER DE GRISSE, 1961)*

- *Holotype male*: L = 1650 μm; stylet = 18.5 μm; a = 41.3; b = 5.5; c = 118.0.
- *Paratype females* (n = 29): L = 720 (580-880) μm; W = 620 (480-750) μm; stylet = 15.0 (13.5-16.0) μm; DGO = 4 (3.5-5.5) μm; a = 1.2 (1.1-1.3); EP/ST = 2.1 (estimated from drawing); interphasmid distance = 18 μm (14-24); vulva width = 26 μm (21-29); vulva to interphasmid line = 26 μm (17-31); anus to interphasmid line = 9 μm (6-11).

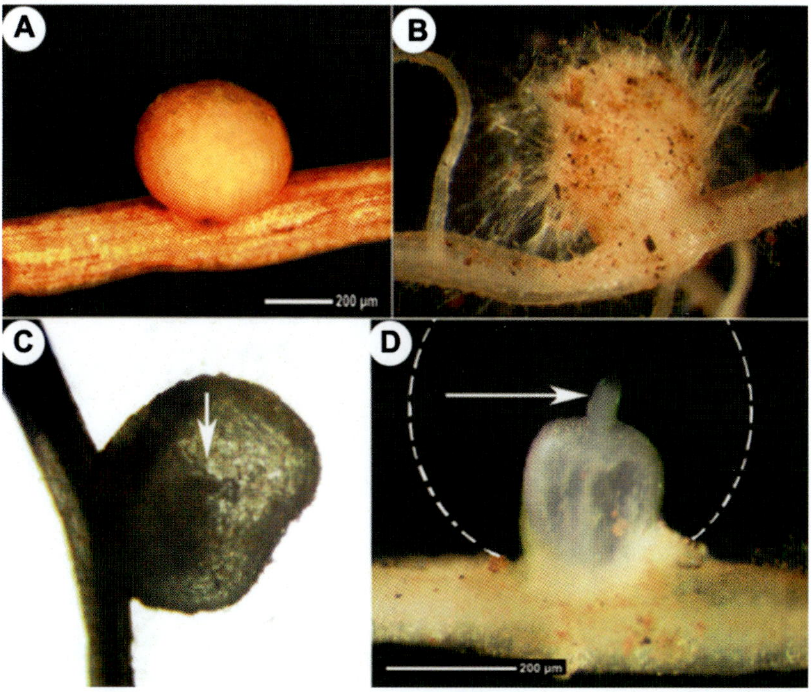

Fig. 135. Meloidogyne kikuyensis. *A-D: Sugar cane roots infected by the nematode showing galls, developing second-stage juvenile (arrowed) and xylem vessels. (After Eisenback & Dodge, 2012.)*

- *Paratype males* (n = 45): L = 1220 (810-1650) μm; stylet = 18.0 (17.0-20.0) μm; DGO = 5 (4.5-6.0) μm; spicules = 33 (31.0-35.0) μm; gubernaculum = 9.0 (8.0-11.0) μm; a = 39.4 (28.6-44.5); b = 5.6 (3.6-7.9); c = 104.5 (70.5-149.0); T = 51 (31-68)%.
- *Paratype J2* (n = 31): L = 320 (290-360) μm; stylet = 13.0 (12.0-15.0) μm; DGO = 4 (3.5-5.0) μm; excretory pore to anterior end = 66 (57-69) μm; median bulb length = 46 (43-49) μm; a = 20.4 (17.1-23.2); b = 6.7 (6.3-7.0); c = 11 (10.0-12.3).
- *Eggs* (n = 65): L = 96.0 (83.0-108.0) μm; W = 45.0 (42.0-49.0) μm.

DESCRIPTION

Female

Body subspherical. Body contour at posterior end smooth. Labial region a truncated cone with two annuli and a labial cap. Stylet slender

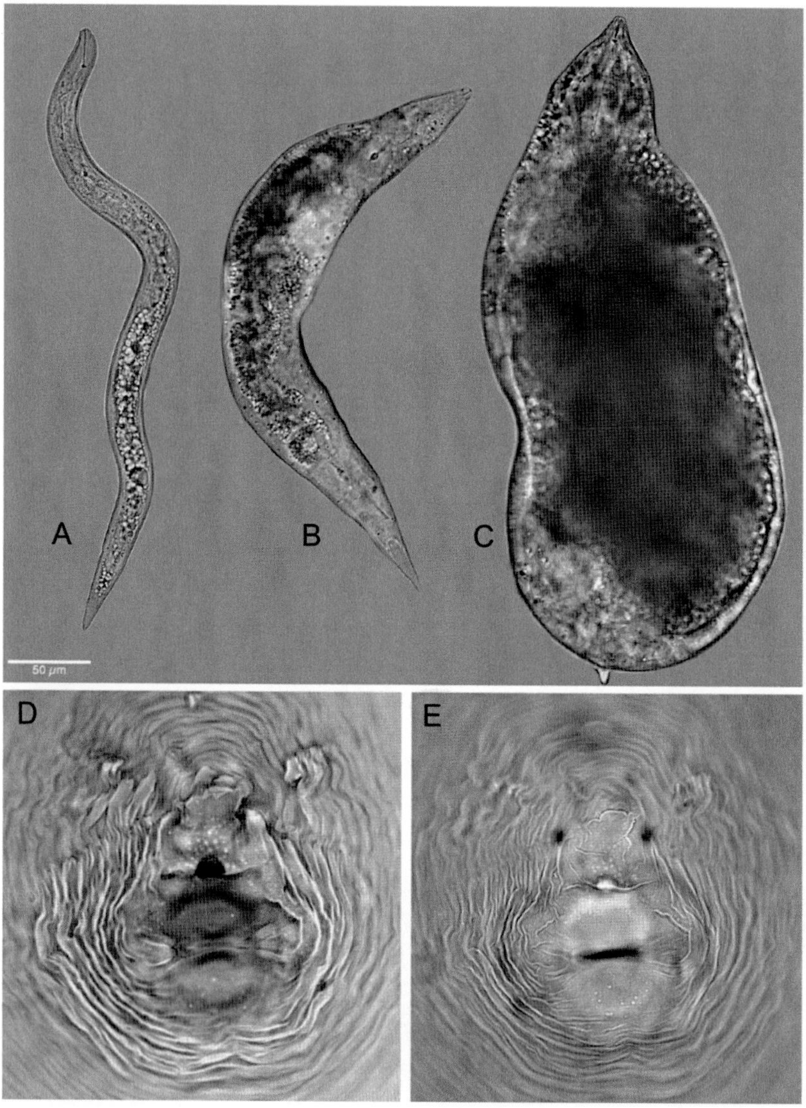

Fig. 136. Meloidogyne kikuyensis *LM. A: Pre-infective migratory second-stage juvenile (J2); B: Slightly swollen post-infective sedentary J2; C: Swollen post-infective J2; D, E: Perineal pattern. (After Eisenback & Vieira, 2020.)*

with rounded basal knobs; anterior (prorhabdion) generally shorter than posterior section. Procorpus thick, swollen, median bulb large, ovate, with crescentic valves, anterior lumen of median bulb wider than lumen

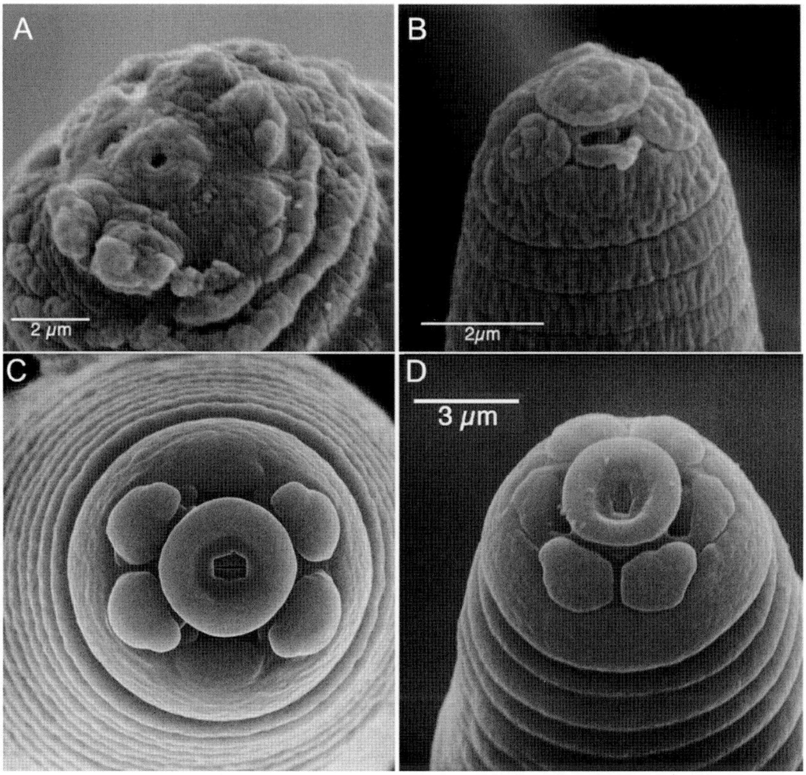

Fig. 137. Meloidogyne kikuyensis. *SEM. A: Female anterior end; B: Anterior end of second-stage juvenile; C: Face view showing male with a distinct labial disk and six lips; D: Lateral view of male labial disk. (After Eisenback & Vieira, 2020.)*

posterior to valves, wall showing two small tooth-like structures at different levels, halfway along anterior lumen. Excretory pore 13 (9-19) annuli posterior to base of labial region. Spermatheca made up of 28-34 large rounded cells with undulating borders, oviduct containing 6-8 cells. Lateral field showing as a slight break in annuli at sides of body anteriorly and as a zigzag line in mid-body region. Posterior cuticular pattern characterised by a cheek-like structure on each side of vulva marked by prominent striae directed towards vulva. Vulva beneath general level of pattern. Immediately anterior to anus, rectum is connected to a characteristic sinus. No distinct tail whorl, tail terminus marked by irregular striae only. Anterior parts of lateral field in form of clear spaces unmarked by incisures, in some examples, posterior regions of lateral fields containing short striae, lateral field rarely crossed by striae. In total, 10-12 horizontal striae may occur between the anus and the tail remnant, or the area may be free of striae.

Male

Body transversely annulated, annuli 3-4 μm wide at mid-body. Lateral field commencing at about eighth body annulus (7-14), marked by two incisures, at 20th annulus (19-31) a third incisure appearing in lateral field, splitting into two at about 46th annulus (45-49) to give four incisures in lateral field for rest of its length. Lateral field then one quarter body diam. across and transversely striated over rest of its length, outer bands regularly, inner band irregularly areolated. Lateral field continuing around tail, which is always twisted. Labial region with hexagonal labial cap, partly covering large lateral and smaller subdorsal and subventral labials. Amphid apertures on lateral labials, slit-like near inner margins. Four small papillae close to mouth; subdorsal and subventral labials each with a single papilla located in middle of labial cuticle. Labial region with two post labial annuli, anterior large, posterior only about 1 μm long. Anterior part of stylet 9.0 (7.5-10.0) μm long, surrounded by tri-partite tubular guide. Posterior part of stylet surrounded by thin muscular tube. Anterior cephalids situated near second and third body annuli, posterior cephalids near eighth body annulus. Median bulb of pharynx elongate, twice as long as wide. Two pharyngeal glands open close to posterior end of median bulb valves, forming a swelling partly covering median bulb, dorsal pharyngeal gland small, lying on left side of body. Nerve ring encircling short isthmus one bulb length posterior to median bulb. Excretory pore 51 (45-56) annuli posterior to labial region, opposite nerve ring. Hemizonid one or two annuli or immediately anterior to excretory pore. Phasmids situated from closely posterior to cloacal aperture to within about 6 μm of tail tip, generally not opposed. Spicules equal, curved inner margins thicker than outer, spicule tips two-layered with hyaline extremities. Gubernaculum one-third spicule length with 'layering' apparent in some cases. Testis 51 (31-68)% of body length, always single and not observed reflexed. *Vas deferens* full of spermatocytes and spermatozoa, ending in ejaculatory duct passing between spicules. Glandular structures found anterior to spicule heads, in some cases with ducts passing into spicules.

J2

Body fusiform and finely annulated. Labial region formed by slight constriction, with two annuli and a labial cap. Stylet long and slender with rounded basal knobs. Anterior part of stylet slightly longer than posterior, surrounded by a thin guiding tube. Nerve ring slightly more than halfway between median pharyngeal bulb and excretory pore. Excretory pore situated on annuli 42-52, posterior to labial region. Hemizonid anterior to excretory pore. Anterior end of lateral field not observed, middle incisure appearing one bulb length before median pharyngeal bulb, four incisures from level of excretory pore. Lateral field 4-6 μm wide. Opposite anus, two inner incisures replaced by a single middle one on which, in two cases, phasmids appeared to lie. Phasmids not opposed, located one-quarter tail length posterior to anus. Tail 2.6 anal body diam. long. Tail tip normally smooth but sometimes annulated. Genital primordium situated at 66-68% of body length.

TYPE PLANT HOST: Kikuyu grass, *Pennisetum clandestinum* Höchst.

OTHER PLANTS: The nematode also parasitises cowpea (*Vigna unguiculata* (L.) Walp.) in Kenya (De Guiran & Netscher, 1970), sugarcane in South Africa (Eisenback & Spaull, 1988; Triantaphyllou, 1990; Kleynhans, 1991), and coffee in Kenya (Campos & Villain, 2005). Eisenback & Vieira (2020) reported that no galls were found on cowpea in the host test experiment.

TYPE LOCALITY: Near the headquarters of the East African Agriculture and Forestry Research Organisation, Muguga, Kenya.

DISTRIBUTION: *Africa*: Kenya, South Africa and Tanzania (Triantaphyllou, 1990; Campos & Villain, 2005; Onkendi *et al.*, 2014).

SYMPTOMS: The galls induced by *M. kikuyensis* are unique and more complex than those caused by most root-knot nematode species. The vascular tissues that supply the giant cells with nutrients occur at a right angle to the vascular cylinder in the main root. Unlike most species of root-knot nematodes, feeding cells of *M. kikuyensis* appear to be formed by the dissolution of cell walls that contribute to the makeup of the enlarged giant cells (Dodge, 2014). The females were completely enclosed by the gall tissue. The unique galls that form on one side of the root resemble the nitrogen-fixing nodules produced on legumes in response to infection by *Rhizobium* and related bacteria (Fig. 133) (Eisenback & Dodge, 2012).

PATHOGENICITY: The damage caused to Kikuyu grass has not been assessed.

CHROMOSOME NUMBER: Reproduction appears to be by amphimixis. The cytogenetic study demonstrated that both males and females have chromosomes only n = 7 (Triantaphyllou, 1990).

POLYTOMOUS KEY CODES: *Female*: A231, B23, C2, D5; *Male*: A324, B34, C21, D1, E2, F2; *J2*: A3, B21, C4, D4, E2, F4.

BIOCHEMICAL AND MOLECULAR CHARACTERISATION: No biochemical data for this species are available. Eisenback & Vieira (2020) provided sequences of the *COII*-16S rRNA gene fragment and ITS1 rRNA gene.

RELATIONSHIPS (DIAGNOSIS): *Meloidogyne kikuyensis* differs from *M. africana* and *M. pisi* in J2 ratio c′; additionally, from *M. africana* by the complete cross-annulation of the lateral field of the male and by the possession of no more than four incisures. The J2 of *M. africana* are also longer and have the phasmids close to the tail terminus.

48. *Meloidogyne konaensis* Eisenback, Bernard & Schmitt, 1994
(Figs 138-141)

COMMON NAME: Kona root-knot nematode.

This nematode was first found in 1991 in several coffee plantations on the island of Hawaii, USA. Eisenback *et al.* (1994) described a new species from a population isolated from soils cultivated with coffee (*Coffea arabica*) in Kona Island, Hawaii, USA, and then reared on tomato plants for the species description. According to the description, the perineal pattern of *M. konaensis* overlapped with the perineal pattern characteristics of other species such as *M. arenaria* and *M. incognita*. Based on the results of morphological, molecular and pathological tests, Monteiro *et al.* (2016) concluded that coffee is not a host of *M. konaensis*

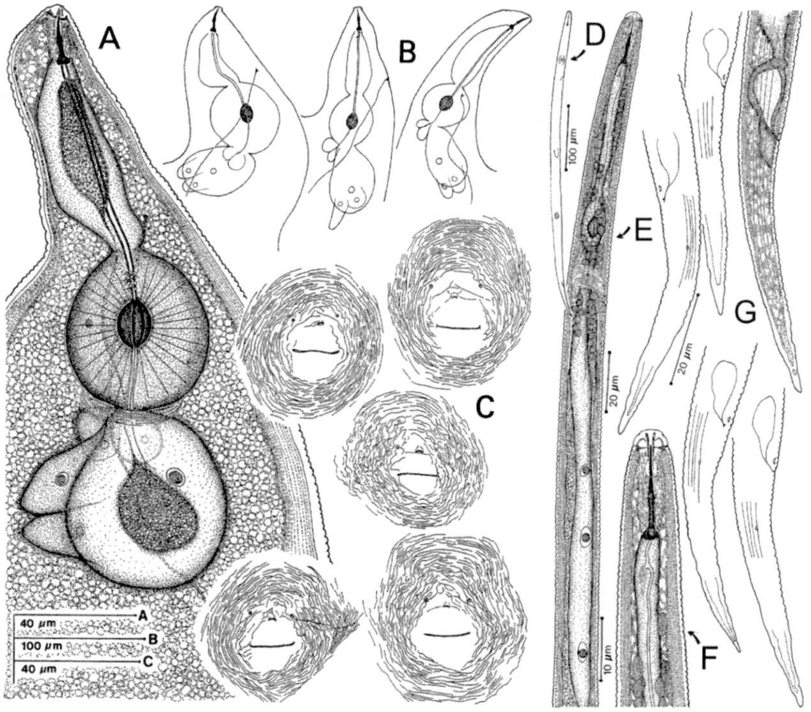

Fig. 138. Meloidogyne konaensis. *A, B: Female anterior region; C: Perineal pattern; D: Entire second-stage juvenile (J2); E, F: J2 anterior region; G: J2 posterior region. (After Eisenback* et al., *1994.)*

as previously reported in the original description of this species and that a mixture of two species (*M. konaensis* and *M. paranaensis*) occurred in Hawaii.

MEASUREMENTS

See Table 34.

Eggs (n = 30): L = 96.7 (89-104) μm; W = 48.0 (42-56) μm; L/W = 2.0 (1.7-2.4).

DESCRIPTION

Female

Body translucent white, variable in size, pear-shaped, sometimes with elongate neck 0.5-1.0 times body length minus neck. Neck prominent,

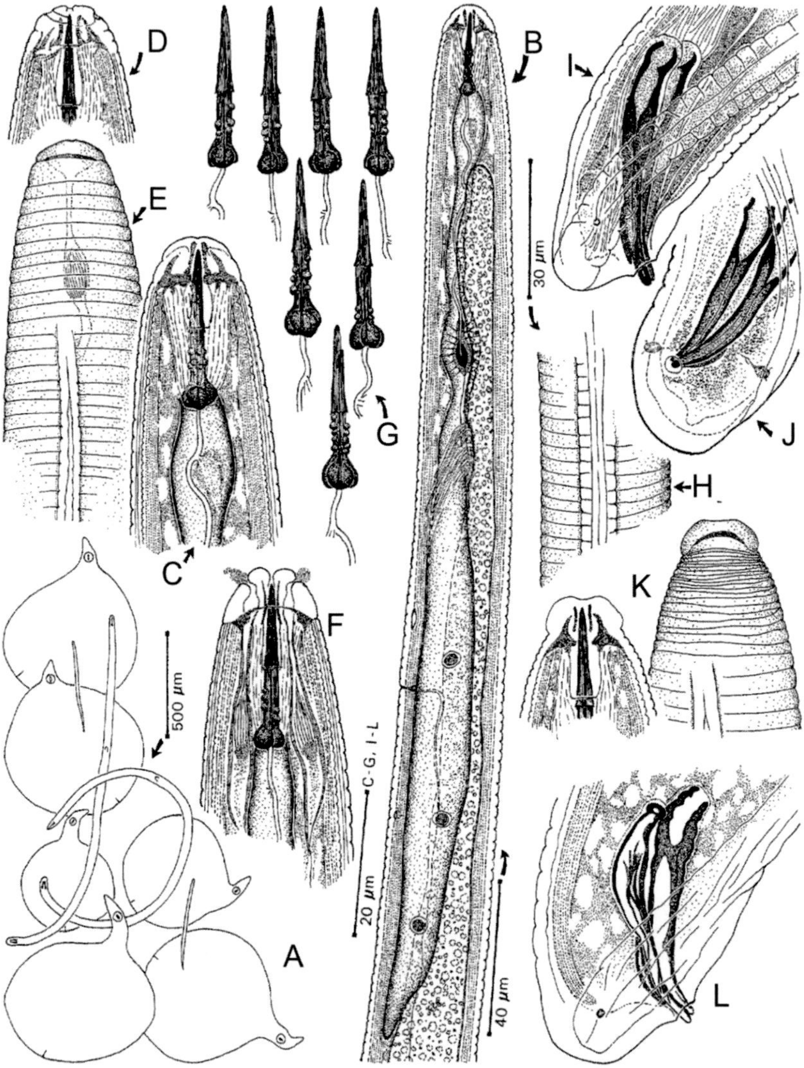

Fig. 139. Meloidogyne konaensis. *A: Entire females, males and second-stage juveniles; B: Male anterior region; C-F, K: Male labial region; H: Male lateral field; I, J, L: Male tail. (After Eisenback* et al., *1994.)*

sometimes bent at various angles to body. In LM, labial framework weak, hexaradiate, lateral sectors slightly enlarged, vestibule and extension prominent. Excretory pore 2-3 stylet lengths posterior to stylet base. Stylet strong, cone slightly curved dorsally, shaft enlarged posteriorly,

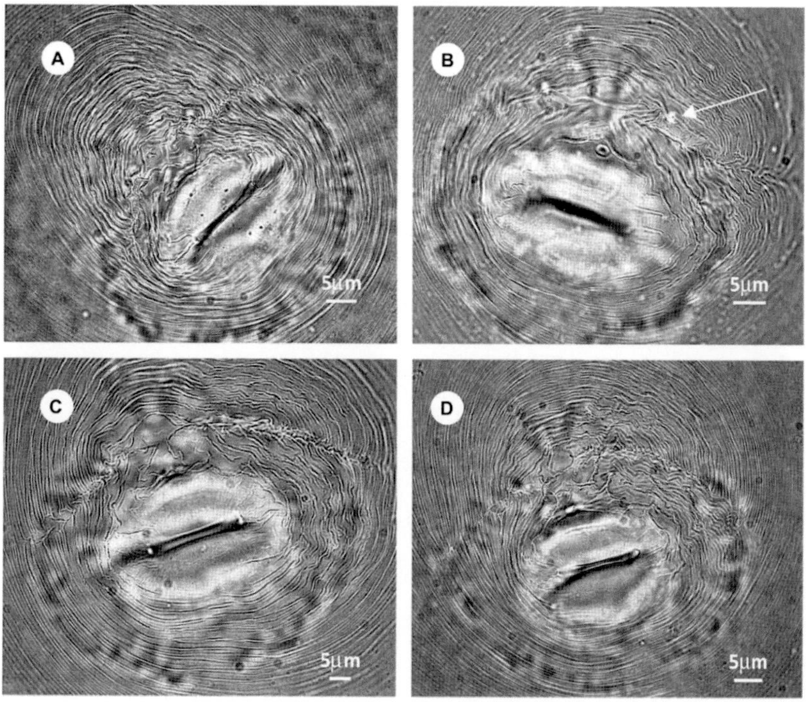

Fig. 140. Meloidogyne konaensis. *A-D: LM photos showing typical variation for perineal patterns of* M. konaensis *from Brazil (arrow in 'B' points to phasmid). (After Monteiro* et al., *2016.)*

three large knobs tapering onto shaft. DGO at *ca* one shaft-length, orifice branching into three ducts, ampulla large. Subventral gland orifices branched, located immediately posterior to enlarged lumen of median bulb. Vesicles present on lumen lining in anterior portion of median bulb. Pharyngeal gland with one large dorsal lobe with one nucleus, two small, nucleated subventral gland lobes, variable in shape, position, and size, usually posterior to dorsal gland lobe, two small, rounded pharyngo-intestinal cells with nuclei attached dorsally to median bulb. In SEM, stoma slit-like, located in ovoid prestomatal cavity, surrounded by pit-like openings of six inner labial sensilla. Labial disc often rectangular, fused with median labials. Median labials divided into distinct labial pairs, lateral labials large and triangular, labial annulus without annulation. Perineal patterns variable in shape, striae coarse, sometimes continuous, smooth to wavy. Dorsal arch rounded

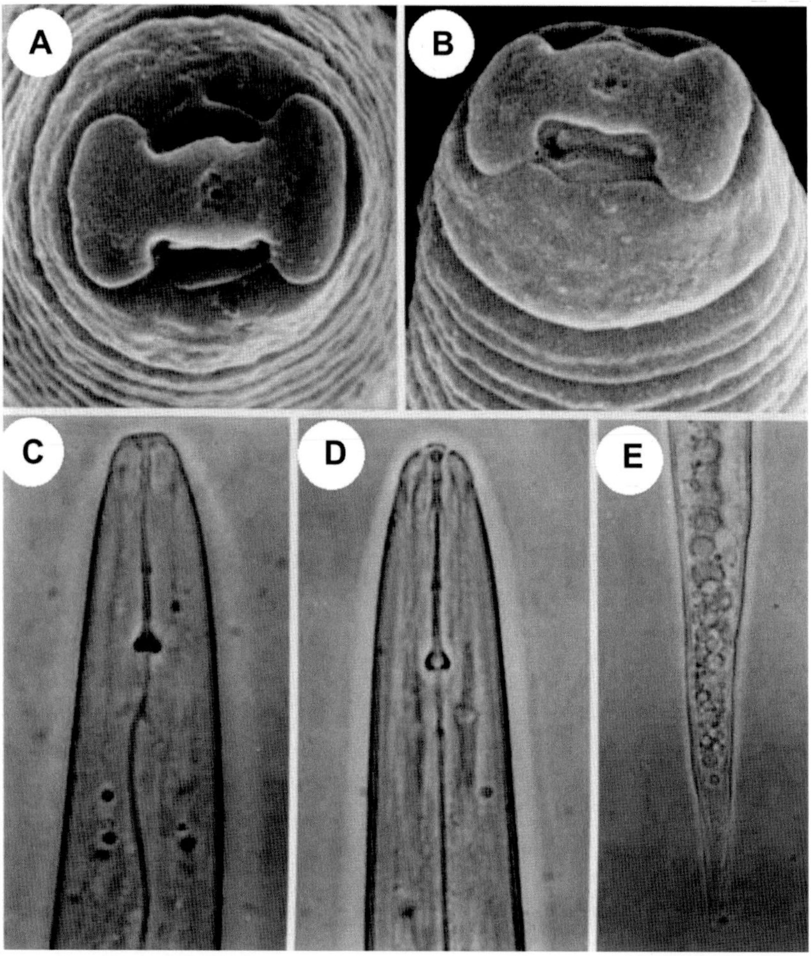

Fig. 141. Meloidogyne konaensis *second-stage juveniles. A, B: SEM of anterior end, face and lateral views, respectively; C, D: LM of anterior end, lateral and ventral views, respectively; E: Tail, lateral view. (After Eisenback* et al., *1994.)*

to squarish. Perivulval region free of striae. Phasmids small, directly opposite and posterior to anus, surface structure not apparent in SEM. Egg morphology typical for genus. Egg masses usually protruding from root tissues, a few sometimes enclosed within root tissues.

Male

Body vermiform, tapering anteriorly, bluntly rounded posteriorly, tail arcuate, twisting through 90°. Labial cap high, rounded, tapering

Table 34. *Morphometrics of females, males and second-stage juveniles (J2) of* Meloidogyne konaensis. *All measurements are in μm and in the form: mean ± s.d. (range).*

Character	Holotype, Allotype, Hawaii, USA, Eisenback *et al.* (1994)	Paratypes, Hawaii, USA, Eisenback *et al.* (1994)	Brazil Monteiro *et al.* (2016)
Female (n)	-	30	30
L	990	992 ± 240 (532-1510)	860 ± 198 (650-1200)
a	2.1	2.2 ± 0.5 (1.4-3.8)	-
W	470	470 ± 106 (320-723)	501 ± 191 (350-770)
Stylet	16.0	16.0 ± 0.9 (14.3-18.5)	16.0 ± 0.4 (14.0-20.0)
Stylet knob width	-	4.4 ± 0.3 (3.8-5.0)	3.6 ± 0.1 (3.0-5.0)
Stylet knob height	-	2.8 ± 0.5 (1.7-3.8)	2.2 ± 0.1 (1.5-3.0)
DGO	5.0	5.0 ± 0.8 (3.4-6.7)	5.0 ± 0.2 (4.0-7.0)
Ant. end to excretory pore	47.0	47.6 ± 13.3 (27.2-87.3)	46.0 ± 20.6 (29-75)
Vulval slit		30.5 ± 2.5 (25.2-35.3)	23.1 ± 0.2 (20.0-25.0)
Vulval slit to anus distance		18.9 ± 2.7 (11.8-24.4)	18.6 ± 0.3 (14.0-21.0)
Interphasmid distance		28.8 ± 5.1 (19.3-38.6)	31.0 ± 0.5 (22-36)
EP/ST	2.9	3.0	2.9
Male (n)		30	30
L	1500	1522 ± 167 (1149-1872)	1813 ± 640 (1060-2050)
a	40.5	41.2 ± 3.8 (34.5-51.1)	-
W	37.0	37.0 ± 3.7 (29.1-43.7)	-
Stylet	22.0	22.1 ± 0.9 (20.2-24.4)	23.2 ± 0.4 (20.0-24.0)

Table 34. *(Continued.)*

Character	Holotype, Allotype, Hawaii, USA, Eisenback *et al.* (1994)	Paratypes, Hawaii, USA, Eisenback *et al.* (1994)	Brazil Monteiro *et al.* (2016)
Stylet knob width	-	3.7 ± 0.5 (3.4-4.2)	4.9 ± 0.1 (4.0-5.0)
Stylet knob height	-	4.6 ± 0.4 (3.4-5.0)	3.2 ± 0.1 (3.0-4.0)
DGO	7.0	6.9 ± 0.7 (5.9-8.4)	5.8 ± 1.0 (5.0-9.0)
Ant. end to excretory pore	155.0	155.3 ± 12.2 (134-178)	162 ± 14.2 (146-179)
Spicules	27.0	26.9 ± 1.9 (21.8-29.4)	34 ± 0.5 (30.0-36.0)
Gubernaculum		-	9.8 ± 0.2 (8.0-11.0)
Tail	12.0	12.0 ± 1.9 (8.4-13.4)	13.6 ± 0.4 (10.0-16.0)
J2 **(n)**	-	30	30
L	-	502 ± 15.9 (468-530)	474 ± 65 (395-550)
a	-	27.8 ± 2.9 (20.2-33.2)	-
b	-	6.6 ± 0.4 (6.2-7.4)	-
c	-	8.1 ± 0.6 (6.1-8.8)	-
c′	-	5.6 ± 0.6 (4.7-6.9)	-
W	-	18.2 ± 1.4 (16.8-19.3)	14.9 ± 0.5 (10.0-21.0)
Stylet	-	13.4 ± 0.6 (12.6-14.3)	14.5 ± 0.2 (13.0-15.0)
Stylet knob width	-	-	2.5 ± 0.1 (2.0-3.0)
Stylet knob height	-	-	1.5 ± 0.04 (1.0-2.0)
DGO	-	4.6 ± 0.5 (4.2-5.9)	4.8 ± 0.2 (4.0-6.0)

Table 34. *(Continued.)*

Character	Holotype, Allotype, Hawaii, USA, Eisenback *et al.* (1994)	Paratypes, Hawaii, USA, Eisenback *et al.* (1994)	Brazil Monteiro *et al.* (2016)
Ant. end to excretory pore	-	95.6 ± 5.8 (89-111)	87 ± 21.4 (65-120)
Tail	-	58 ± 6.6 (48.7-73.1)	56 ± 17.1 (35.0-76.0)
Hyaline region	-	-	16 ± 0.5 (13.0-19.0)

posteriorly, labial region distinct from first body annulus, usually without annulation. In LM, labial framework moderately developed, hexaradiate, lateral sectors slightly enlarged. Vestibule and extension distinct. In SEM, stoma slit-like, located in ovoid to hexagonal prestomatal cavity surrounded by pit-like openings of six inner labial sensilla. Labial disc rounded. Median labials often divided into labial pairs, fused with labial disc forming elongate labial cap. Labial sensilla not clearly demarcated on median labials. Lateral labials absent. Amphidial apertures elongate slits between labial disc and lateral sectors of labial region. Labial region not annulated. Stylet morphology distinct. Stylet opening several microns from stylet tip, cone pointed, gradually increasing in diam. posteriorly, junction of cone and shaft uneven. Shaft cylindrical often slightly wider near middle with numerous large projections. Knobs broadly elongate, tapering onto shaft, indented slightly anteriorly, rounded posteriorly. DGO branched into three ducts, ampulla distinct. Amphids very distinct, often producing exudates. Procorpus distinct, median bulb ovoid, triradiate lining of enlarged lumen of median bulb thinner than in female. Subventral gland orifices posterior to lining of median bulb, branched. Pharyngo-intestinal junction at level of nerve ring, indistinct. Three nuclei in gland lobe, lobe variable in length. Intestinal caecum extending anteriorly, sometimes approaching level of dorsal pharyngeal gland orifice, displacing anterior pharynx ventrad. Excretory pore distinct, 6-10 annuli posterior to hemizonid. Areolated lateral field beginning near level of stylet base usually with four incisures, in some areas one additional central incisure present. Usually two testes (sex reversed males), sometimes one testis (normal males), outstretched or anteriorly reflexed. Spicules arcuate, gubernaculum distinct. Tail short, distinct phasmids at level of cloacal aperture.

J2

Body vermiform, tapering more posteriorly than anteriorly. In LM, labial framework weak, hexaradiate. Vestibule and vestibule extension more distinct than rest of framework. In SEM, stoma slit-like located in ovoid prestomatal cavity, surrounded by pit-like openings of six inner labial sensilla. Labial disc, median labials, and lateral labials fused into one structure. Labial disc elevated above stoma. Median labials with rounded margin, labial sensilla indistinct. Amphidial apertures between labial disc and elongate lateral labials may fuse with labial region. Labial region smooth, body annuli distinct. Stylet cone increasing in width gradually, shaft cylindrical to tapering posteriorly, knobs rounded and offset from shaft. DGO long (4-6 μm), orifice branching into ducts, ampulla indistinct. Median bulb ovoid, triradiate lining strongly sclerotised, subventral gland orifices branched, located immediately posterior to enlarged lumen of median bulb. Pharyngo-intestinal junction indistinct, at level of nerve ring. Gland lobe of variable length with three nuclei. Excretory pore distinct; hemizonid 2-4 annuli anterior to excretory pore. Areolated lateral field with four incisures, beginning as two near base of stylet. Tail often distinctly curved ventrally, annulations larger posteriorly. Hyaline region distinct. Phasmids small, in ventral incisure of lateral field, always posterior to anus.

TYPE PLANT HOST: Rhizosphere of *Coffea arabica* L.

OTHER PLANTS: *Meloidogyne konaensis* reproduced on tomato. Zhang & Schmitt (1994) studied the host suitability of 32 species of monocotyledonous and dicotyledonous plants. Good hosts included barley, bean ('Hawaiian Wonder'), cabbage, cantaloupe, carrot, corn, cucumber, eggplant, *Gardenia* sp., hilograss, lettuce, nutsedge, pea, pepper, pumpkin, radish, tomato, watermelon and wheat. Poor or non-host were bean ('Nanoa Wonder'), broccoli, chrysanthemum, cotton, ginger, pineapple, oat, peanut, taro, soybean, sunn hemp, water chestnut and sweet potato (Zhang & Schmitt, 1994). *Meloidogyne konaensis* was detected in Brazil parasitising three crops and one weed species: cabbage (*Brassica capi-*

tata L.), papaya (*Carica papaya*), noni (*Morinda citrifolia* L.) and canapum (*Physalis angulata* L.) (Silva & Santos, 2012; Silva, 2014).

TYPE LOCALITY: Kona Experiment Station at Kealakekua, Hawaii County, Hawaii, USA.

DISTRIBUTION: *Oceania*: USA (Hawaii), *South America*: Brazil (Monteiro *et al.*, 2016).

PATHOGENICITY: Eisenback *et al.* (1994) reported that in the Kona area on the island of Hawaii (USA), *M. konaensis* caused the death of coffee trees in some commercial plantations. This species has not been found on any other crop or in the forested area of the island. However, pathogenicity tests by Monteiro *et al.* (2016) indicated that coffee is not a host of *M. konaensis* as previously reported in the original description of this species and is considered a minor species.

CHROMOSOME NUMBER: Reproduction of *M. konaensis* is by mitotic parthenogenesis, and the two populations presented a karyotype 2n = 40-44.

POLYTOMOUS KEY CODES: *Female*: A123, B213, C2, D3; *Male*: A312, B213, C23, D1, E2, F1; *J2*: A12, B21, C312, D12, E32, F213.

BIOCHEMICAL AND MOLECULAR CHARACTERISATION: Biochemically, *M. konaensis* was originally described as possessing an F1 esterase phenotype (Eisenback *et al.*, 1994). Sipes *et al.* (2005) reported that three different esterase phenotypes (F1, I1 and F1-I1) were detected even in single-egg-mass lines derived from an F1 female. Three isolates of *M. konaensis* differing in esterase phenotype (F1, I1, and F1-I1) did not differ morphologically, and all of them had N1 malate dehydrogenase phenotypes yet differed in their parasitic ability (Sipes *et al.*, 2005). Only the F1 isolate parasitised *C. arabica*. The F1-I1 isolate had greater reproduction on tomato and cucumber than either the I1 or F1 isolate (Sipes *et al.*, 2005). Monteiro *et al.* (2016) considered that *M. konaensis* had esterase pattern-K3 (= M3-F1), which was unique and species-specific with three major bands Rm 1.0, 1.17, 1.27 and a secondary band Rm 1.10, whereas *M. paranaensis* had esterase pattern P1 (= F1).

Molecularly this nematode can be identified by 18S rRNA, D2-D3 of 28S rRNA and *COI* gene sequences and can be separated from *M. paranaensis* (Tomalova *et al.*, 2012; Monteiro *et al.*, 2016; Powers *et al.*, 2018).

RELATIONSHIPS (DIAGNOSIS): The species belongs to Molecular group I, the Incognita group. Males of *M. konaensis* have knobs that are not offset, are posteriorly sloping and have 6-12 large projections surrounding the shaft, whereas in *M. paranaensis* (Carneiro *et al.*, 1996a), the knobs are transversely elongated, broad and offset from the shaft, with sometimes one or two projections surrounding the shaft. This character represents the most useful criterion for differentiating *M. konaensis* from *M. paranaensis* (Carneiro *et al.*, 1996a; Monteiro *et al.*, 2016).

49. *Meloidogyne kongi* Yang, Wang & Feng, 1988
(Fig. 142)

COMMON NAME: Chinese citrus root-knot nematode.

Yang *et al.* (1988b) described this species from roots of *Citrus* sp. from Guangxi, China.

MEASUREMENTS (*AFTER YANG ET AL., 1988B*)

- *Holotype female*: L = 816.7 μm; W = 564.7 μm; stylet = 14.8 μm; DGO = 4.9 μm; median bulb to anterior end = 93.7 μm; excretory pore to anterior end = 40.5 μm; EP/ST = 2.7.
- *Allotype male*: L = 1687.7 μm; W = 48.8 μm; stylet = 22 μm; DGO = 6.1 μm; excretory pore to anterior end = 129.7 μm; spicules = 34.7 μm; gubernaculum = 10.7 μm; tail = 15.3 μm; a = 34.4; b = 13.5; c = 110.4; T = 63.4%.
- *Paratype females* (n = 30): L = 715.4 ± 104.9 (611-820) μm; W = 491.7 ± 83.4 (408-575) μm; stylet = 14.5 ± 0.8 (13.7-15.3) μm; stylet knob width = 4.7 ± 0.3 (4.5-5.0) μm; stylet knob height = 2.3 ± 0.1 (2.2-2.4) μm; DGO = 5.1 ± 0.7 (4.5-5.8) μm; excretory pore to anterior end = 35.6 ± 9.8 (25.9-45.4) μm; a = 1.5 ± 0.2 (1.3-1.7); EP/ST = 2.5.
- *Paratype males* (n = 30): L = 1701 ± 178.2 (1523-1879) μm; W = 46.5 ± 4.9 (41.6-51.5) μm; labial region height = 6.3 ± 0.5 (5.9-

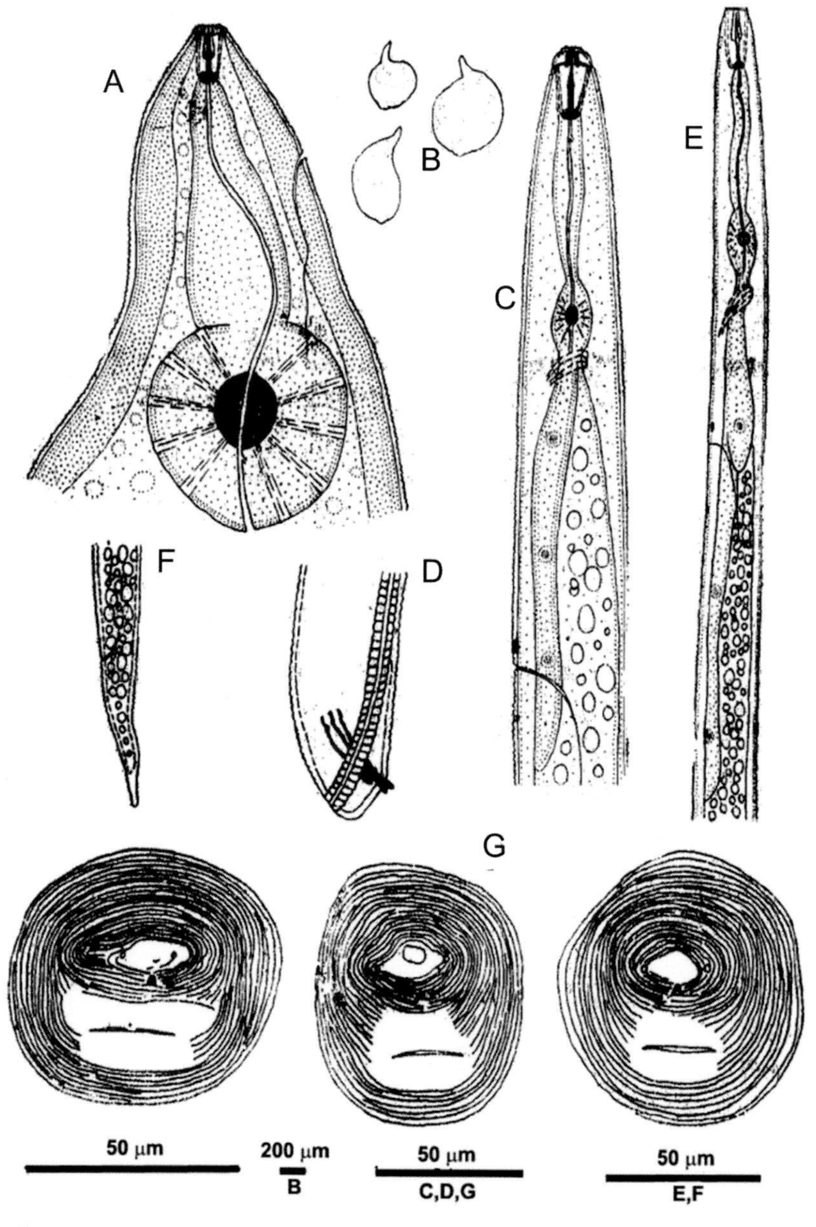

Fig. 142. Meloidogyne kongi. *A: Anterior end of female; B: Females; C: Anterior end of male; D: Tail of male; E: Anterior end of second-stage juvenile (J2); F: Tail of J2; G: Perineal patterns. (After Yang* et al., *1988.)*

6.8) μm; stylet = 23.0 ± 1.0 (22.0-24.0) μm; stylet knob width = 6.1 ± 0.3 (5.8-6.4) μm; stylet knob height = 3.4 ± 0.3 (3.2-3.7) μm; DGO = 6.6 ± 0.8 (5.8-7.5) μm; excretory pore to anterior end = 176.7 ± 21.5 (130-198) μm; spicules = 35.7 ± 2.8 (32.9-38.4) μm; gubernaculum = 9.2 ± 1.0 (8.1-10.7) μm; tail = 15.7 ± 1.8 (13.9-17.5) μm; a = 36.8 ± 3.9 (32.9-40.7); b = 12.4 ± 1.5 (10.9-13.9); c = 109.3 ± 14.8 (94.5-124.1); T = 60.5 ± 7.9 (40.0-74.3)%.
- *Paratype J2* (n = 30): L = 414.5 ± 21.9 (360-448) μm; W = 16.9 ± 0.7 (15.4-17.9); stylet = 12.8 ± 0.6 (12.0-14.2) μm; stylet knob width = 3.0 ± 0.1 (2.8-3.3) μm; stylet knob height = 1.7 ± 0.1 (1.6-1.9) μm; DGO = 4.9 ± 0.5 (3.9-5.8) μm; excretory pore to anterior end = 86.3 ± 4.4 (76-98) μm; tail = 43.4 ± 5.0 (35.3-52.7) μm; hyaline region = 10.7 ± 2.3 (6.7-12.4) μm; a = 24.4 ± 1.1 (22.4-26.7); b = 4.0 ± 0.4 (3.4-4.9); c = 9.7 ± 1.4 (7.7-12.4).
- *Eggs*: L = 93.3 (82.3-108.0) μm; W = 43.0 (40.0-46.0) μm.

DESCRIPTION

Female

Creamy to yellow, with prominent neck, pear-shaped, variable in size with distinct protuberance. Annulation on neck visible, annulation on body obscure. Labial region offset from body. No annulation found on labial region. Labials not elevated, labial framework delicate and weak. Excretory pore to anterior end 2-3 times stylet length. Pharynx well developed. Stylet straight, pointed. Cone-shaft junction enlarged. Stylet knobs large, distinct from shaft. Median bulb rounded with well-developed valve. Two subventral pharyngeal lobes opening posteriorly below median bulb valve. Pharyngeal lobes overlapping anterior portion of intestine. Perineal patterns rounded to oval-shaped. Dorsal arch rounded, moderately high to low. Striae smooth to fine, continuous dorsally and free of short striae. Tip of tail visible, circular striae around tail and anus. Vulval slit-like. Striae visible in lateral region of vulva, striae may occur in prevulval region or around vulva. Phasmids long, round, within circular striae around tail tip.

Male

Vermiform, tapering anteriorly, bluntly rounded posteriorly. Body annulation distinct. Lateral field with four incisures areolated except for two central incisures. Labial cap high. Labial region offset from body without annulation. Labial framework well developed. Labial disc and median labials fused. Labial disc rounded, slightly elevated above median labials. Lateral labials semi-circular, prestoma slit-like, slightly depressed with six depressed sensilla. Amphid apertures large, slit-like, between labial disc and median labials. Stylet pointed. Base of cone slightly wider than shaft, which is cylindrical. Stylet knobs well developed, offset from stylet base. Median bulb oval-shaped, with well-developed valve. Subventral pharyngeal glands opening posterior to median bulb valve. Pharynx overlapping anterior portion of intestine. Nerve ring posterior to median bulb. Excretory pore located 2-3 annuli posterior to hemizonid. Hemizonid visible. Spicules arcuate, gubernaculum one, phasmids obscure.

J2

Vermiform, tapering at both ends. Anterior end blunter than tail end, annulation on body distinct. Lateral field with four incisures, areolated except for two central incisures. Excretory pore 3-4 annuli posterior to hemizonid. Labial disc and median labials fused, dumbbell-shaped. Labial disc slightly elevated above median labials. Median labials elongate. No lateral labials. Amphidial apertures slit-like, located below median labials. Prestoma circular. Stylet pointed. Three stylet knobs, offset from shaft. Median bulb oval-shaped, with valve. Subventral pharyngeal gland opening posterior to median bulb valve. Pharyngeal gland lobes overlapping anterior portion of intestine. Tail tapering with blunt end. Rectum obscure.

TYPE PLANT HOST: Roots of *Citrus* sp.

OTHER PLANTS: The North Carolina differential host test indicated that *M. kongi* could not reproduce on all of the test hosts except hot pepper.

TYPE LOCALITY: Guangxi, China.

DISTRIBUTION: *Asia*: China.

PATHOGENICITY: The damage caused to *Citrus* sp. in China has not been assessed.

CHROMOSOME NUMBER: No available data.

POLYTOMOUS KEY CODES: *Female*: A2, B23, C2, D2; *Male*: A213, B2, C12, D1, E2, F2; *J2*: A23, B21, C3, D324, E23, F3.

BIOCHEMICAL AND MOLECULAR CHARACTERISATION: No available data.

RELATIONSHIPS (DIAGNOSIS): *Meloidogyne kongi* is close to *M. arenaria* and *M. enterolobii*. From *M. arenaria* it can be separated by an overview of the perineal pattern, which is oval-shaped and has continuous striae from the dorsal to ventral region *vs* meeting at an angle at the lateral lines as in *M. arenaria*. The continuous striae, no short striae in the lateral region, circular striae around the tail tip and anus, and large phasmids are characters that can be used to separate *M. kongi* from *M. enterolobii*.

50. *Meloidogyne kralli* Jepson, 1983
(Figs 143, 144)

COMMON NAME: Krall root-knot nematode.

Jepson (1983b) described this species from roots of sedge (*Carex acuta* L.) Elva River near Elva, Tartu district, Estonia. Karssen & van Hoenselaar (1998) also provided measurements of the paratypes.

MEASUREMENTS *(AFTER JEPSON, 1983B)*

- *Holotype J2*: L = 442 μm; stylet = 11.0 μm; DGO = 4.0 μm; excretory pore to anterior region = 78.5 μm; tail = 63.0 μm; hyaline region = 17.0 μm.
- *Paratype females* (n = 15): L = 463 $\pm$ 60 (351-564) μm; W = 306 $\pm$ 69 (218-487) μm; stylet = 13.3 $\pm$ 0.3 (12.5-14.5) μm; DGO = 4.4 $\pm$ 0.2 (3.5-4.5) μm; excretory pore to anterior end = 15.8 $\pm$ 2.1 (10.0-24.5) μm; EP/ST = 0.8.
- *Paratype males* (n = 6): L = 1076 $\pm$ 77 (947-1143) μm; stylet = 18.8 $\pm$ 0.9 (18.0-20.0) μm; DGO = 4.4 $\pm$ 1.0 (3.5-6.0) μm; excretory pore to anterior end = 127.4 $\pm$ 4.8 (120-133.5) μm; spicules = 26.3 $\pm$

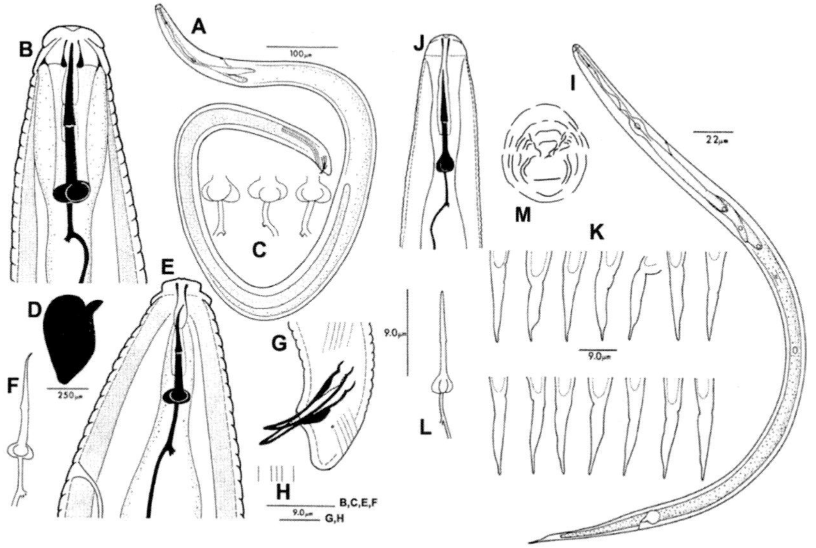

Fig. 143. Meloidogyne kralli. *A: Entire male; B: Male labial region; C: Male stylet knobs; D: Entire female; E: Female anterior region; F: Female stylet; G: Male tail; H: Male lateral field; I: Entire second-stage juvenile (J2); J: J2 labial region; K: J2 tail; L: J2 stylet. (After Jepson, 1983b.) M: Perineal pattern. (After Jepson, 1987.)*

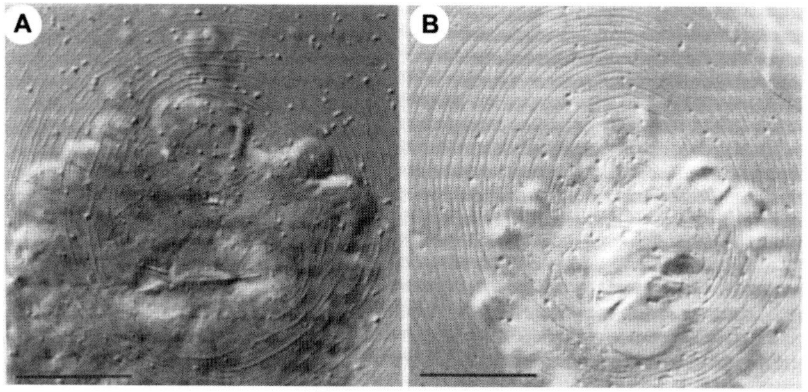

Fig. 144. Meloidogyne kralli. *A, B: Perineal patterns. (Scale bars: A, B = 25 μm.) (After Karssen & van Hoenselaar, 1998.)*

2.1 (22.5-28.0) μm; a = 31.7 ± 5.0 (25.7-39.4); c = 116.1 ± 9.4 (104.3-127.1).

- *Paratype J2* (n = 30): L = 439 ± 6.0 (408-476) μm; stylet = 10.8 ± 0.1 (10.5-11.5) μm; DGO = 4.1 ± 0.2 (3.0-4.5) μm; excretory pore to anterior end = 78 ± 2.1 (64-88) μm; tail = 68 ± 1.6 (61-78) μm; hyaline region = 17.4 ± 0.6 (14.5-21.0) μm; a = 31.0 ± 1.3 (26.7-35.3); b = 6.5 ± 0.2 (5.9-7.1); c = 6.5 ± 0.2 (5.6-7.1); c′ = 7.0 ± 0.4 (5.5-8.9).

DESCRIPTION

Female

Body pearly white, globular to elongate, with a distinct neck and labial region not offset, vulva on a posterior protuberance. Stylet cone forming about half of stylet length with anterior dorsal curvature, knobs offset from shaft, rounded and transversely ovoid in lateral view. Excretory pore variable in position, from anterior to stylet knobs to posterior to dorsal pharyngeal gland orifice. Perineal pattern with faint, fine striae. Overall pattern circular to ovoid in anteroposterior direction, with a low arch with same contour as other striae. Striae single, straight and widely spaced, circular and broken in tail area, vertical striae just anterior to tail tip, usually a fold over anus. Lateral field faintly visible at junction of posterior and anterior striae. A characteristic posterolaterally directed irregular double incisure on either side of tail region. Perivulval region free from striae, often bounded laterally by more distinct anterolateral striae. Phasmids small and less widely spaced than the width of vulva.

Male

Cuticle with transverse annulations. Lateral field with 4-8 incisures, frequently five, not areolated. Labial region truncate, not offset. Labial cap flat with slight depression at stoma and rounded median labials. Labial cap almost as broad across as single post-labial annulus at its apex and base. Post-labial annulus straight sided and slightly tapering anteriorly. Labial annulus shallower than post-labial annulus. Stylet cone forming about half of stylet length, knobs offset from shaft, rounded and somewhat transversely ovoid in lateral view. Hemizonid one or two annuli anterior to excretory pore. Single testis. Tail morphology typical for genus including spicules, gubernaculum and tail shape, phasmids at level of cloacal aperture.

J2

Cuticle with transverse annulation. Lateral field with four main incisures but in some individuals 5-8 incisures may be visible. Lateral field not areolated. Labial region not offset, with labial cap and one postlabial annulus. Stylet cone forming about half total length, knobs prominent, rounded and sloping posteriorly. Hemizonid directly anterior to excretory pore. Tail shape distinct, hyaline region tapering gradually and then more sharply towards terminus, terminal portion very narrow ending in a finely rounded tip. Phasmids distinct at about one-third of hyaline region from tail tip.

TYPE PLANT HOST: Acute sedge, *Carex acuta* L.

OTHER PLANTS: This nematode has been also reported in Cyperaceae under natural conditions: *Carex vesicaria* L., *C. riparia* Curt., *C. pseudocyperus* L. and *Scirpus sylvaticus* L. In the glasshouse it also reproduced on *Festuca pratensis* Huds. and *Hordeum vulgare* L. (Jepson, 1983b; Wesemael *et al.*, 2011).

TYPE LOCALITY: Elva River near Elva, Tartu district, Estonia.

DISTRIBUTION: *Europe*: Estonia, Latvia, Lithuania, Poland, Russia (Pskov and Leningrad regions), Slovakia, Switzerland, UK. *Meloidogyne kralli* occurs in sandy, peat and silt soils, usually in wet places on the banks of rivers and shores of lakes (Jepson, 1983b; Brzeski, 1998; Kornobis, 2001; Karssen & Grunder, 2002; Karssen & van Hoenselaar, 2002; Wesemael *et al.*, 2011; Renko & Murin, 2013).

PATHOGENICITY: The damage caused to host plants has not been assessed.

CHROMOSOME NUMBER: No available data.

POLYTOMOUS KEY CODES: *Female*: A3, B32, C4, D3; *Male*: A43, B3, C32, D1, E2, F1; *J2*: A2, B32, C21, D12, E3, F12.

BIOCHEMICAL AND MOLECULAR CHARACTERISATION: A Nlc type malate dehydrogenase pattern and a very weak multiple banding esterase pattern were detected (Karssen & van Hoenselaar, 1998). A unique 18S rRNA gene sequence of this species was published by van Megen *et al.* (2009).

RELATIONSHIPS (DIAGNOSIS): This species belongs to Molecular group III and related to *M. naasi*. *Meloidogyne kralli* most closely morphologically resembles two species that also parasitise Graminaceae and Cyperaceae: *M. graminicola* and *M. sewelli*. *Meloidogyne kralli* is distinguished from *M. sewelli* by the J2 DGO at 3.0-4.5 *vs* 7-8 μm, respectively. *Meloidogyne kralli* and *M. sewelli* have very similar J2 tapering tails, each with a sharply tapering terminus and narrow terminal portion ending in a finely rounded tip, although that of *M. sewelli* is more slender. The tail of *M. graminicola* is different in shape with a more or less clavate terminus. *Meloidogyne kralli* differs from *M. sewelli* in tail length (61-78 *vs* 68-78 μm) but overlaps with the range of mean lengths from different populations of *M. graminicola* (60-85 μm). The perineal pattern of *M. kralli* most closely resembles that of *M. graminicola* in overall shape and presence of the posterolaterally directed double incisure on either side of the tail region, but the striae of *M. kralli* are much less distinct, finer and more widely spaced. The pattern of *M. sewelli* is very different from the other two species, having a characteristic circle of striae between the phasmids and the rest of the pattern with fine but mostly indistinct striae.

51. *Meloidogyne lopezi* Humphreys-Pereira, Flores-Chaves, Gomez, Salazar, Gomez-Alpízar & Elling, 2014

(Figs 145-147)

COMMON NAME: Puntarenas coffee root-knot nematode.

Humphreys-Pereira *et al.* (2014) described this species from roots of coffee (*Coffea arabica* 'Catuaí') collected from Santa Marta, Volcán de Buenos Aires, Puntarenas province in southern Costa Rica.

MEASUREMENTS (*AFTER HUMPHREYS-PEREIRA* ET AL., *2014*)

- *Holotype female*: L = 758 μm; W = 570 μm; stylet = 16.4 μm; stylet knob width = 4.6 μm; stylet knob height = 2.2 μm; DGO = 3.6 μm; excretory pore to anterior end = 35.0 μm; a = 1.3; EP/ST = 2.1.

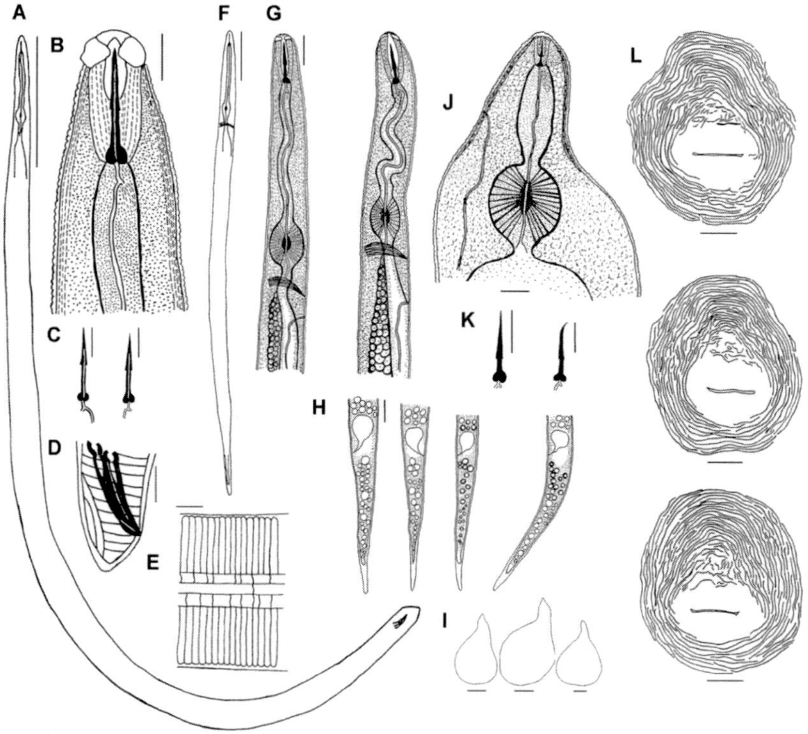

Fig. 145. Meloidogyne lopezi. *A: Entire male; B: Male anterior region; C: Male stylet; D: Male tail region; E: Male lateral field; F: Entire second-stage juvenile (J2); G: J2 anterior region; H: J2 tail region; I: Entire female; J: Female anterior region; K: Female stylet; L: Perineal pattern. (Scale bars: A, I = 200 μm; B-E, K = 10 μm; F = 50 μm; G, H = 11 μm; J = 18 μm; L = 20 μm.) (After Humphreys-Pereira* et al.*, 2014.)*

- *Paratype females* (n = 30): L = 833 ± 80 (717-956) μm; W = 521 ± 68.2 (368-598) μm; stylet = 18.0 ± 1.9 (15.4-23.0) μm; stylet knob width = 4.3 ± 0.4 (3.4-5.3) μm; stylet knob height = 3.2 ± 0.5 (2.2-4.3) μm; DGO = 4.5 ± 0.9 (2.9-6.2) μm; excretory pore to anterior end = 39 ± 9.8 (22-59) μm; interphasmid distance = 26.0 ± 3.5 (20.2-32.0) μm; vulval slit length = 26.0 ± 2.5 (21.4-30.8) μm; vulva-anus distance = 20.1 ± 2.0 (16.6-24.9) μm; a = 1.6 ± 0.2 (1.3-2.0); EP/ST = 2.2 ± 0.6 (1.4-3.8).
- *Paratype males* (n = 30): L = 1963 ± 195.2 (1616-2340) μm; W = 43 ± 3.2 (37-47) μm; labial region height = 6.2 ± 0.5 (5.5-7.2) μm; labial region diam. = 13.3 ± 0.7 (11.97-14.6) μm; stylet = 23.4 ±

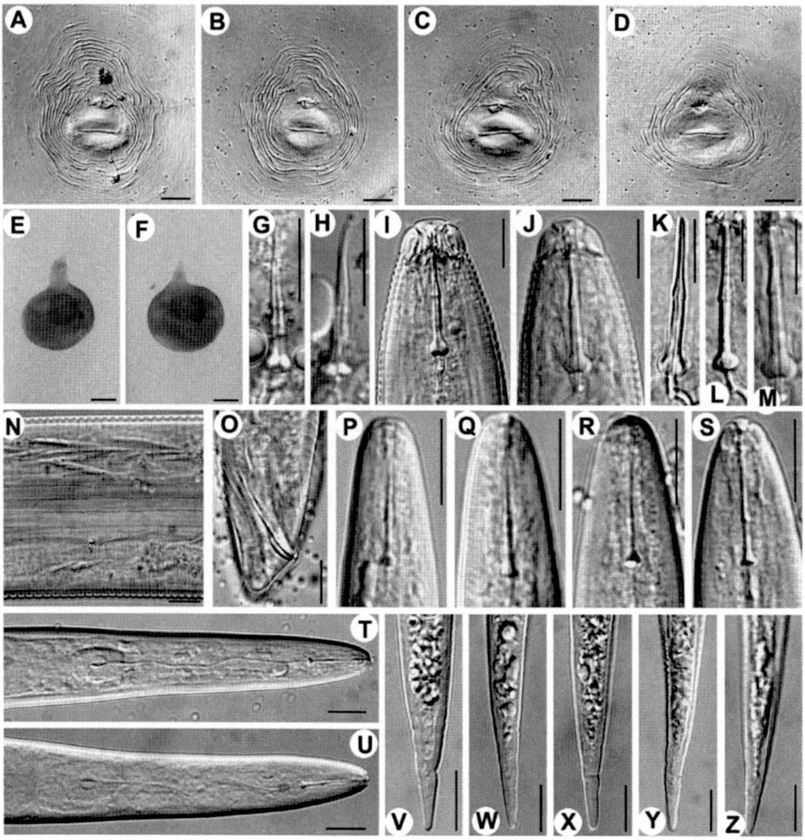

Fig. 146. Meloidogyne lopezi. *LM A-D: Perineal pattern; E, F: Entire female; G, H: Female stylet; I, J: Male anterior region; K-M: Male stylet; N: Male lateral field; O: Spicules and gubernaculum; P-S: Second-stage juvenile (J2) stylet; T, U: J2 labial region; V-Z: J2 tail. (Scale bars: A-D = 20 μm; E, F = 200 μm; G-Z = 10 μm.) (After Humphreys-Pereira* et al., *2014.)*

1.0 (21.5-24.5) μm; stylet knob width = 5.0 ± 0.39 (4.3-6.2) μm; stylet knob height = 3.8 ± 0.39 (3.1-4.31) μm; DGO = 2.5 ± 0.4 (2.0-3.5) μm; excretory pore to anterior end = 156 ± 22.6 (128-194) μm; tail = 17.4 ± 1.4 (14.4-19.2) μm; spicules = 33.9 ± 2.8 (29.0-38.0) μm; gubernaculum = 9.2 ± 0.7 (7.7-10.5) μm; a = 46.2 ± 2.8 (42.2-50.0); c = 113 ± 5.5 (105-124); T = 47.8 ± 8 (31.9-61.9)%.

- *Paratype J2* (n = 30): L = 532 ± 23.5 (478-574) μm; W = 18 ± 2.0 (15.0-22.0) μm; labial region height = 3.4 ± 0.3 (2.7-3.8) μm;

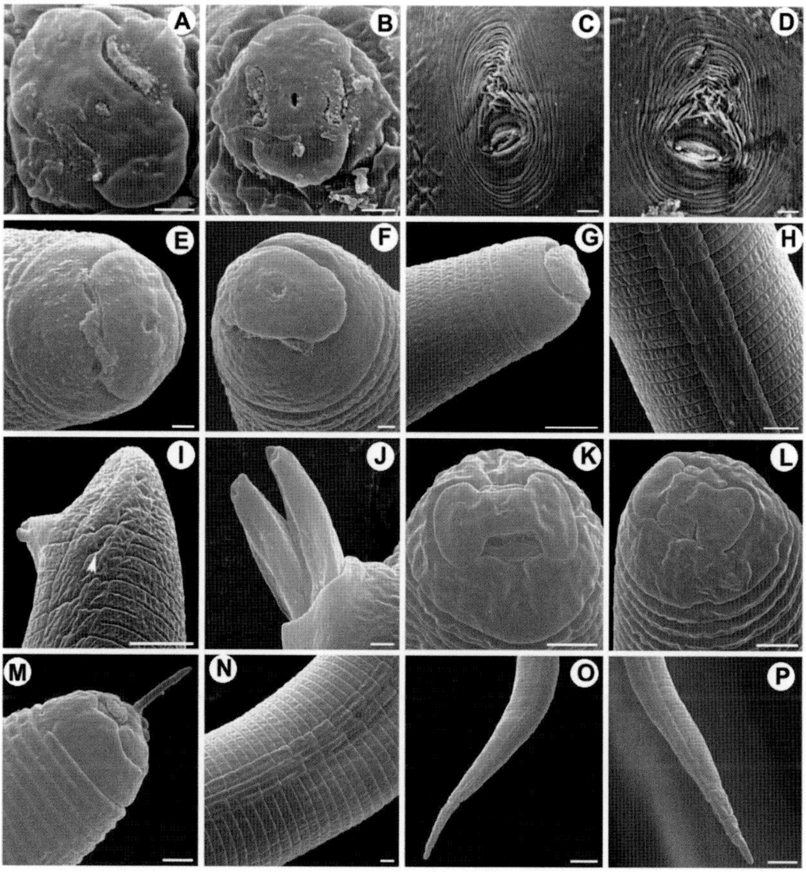

Fig. 147. Meloidogyne lopezi. *SEM. A, B: Female* en face *view; C, D: Perineal pattern; E-G: Male labial region; H: Male lateral field; I: Male tail showing spicules and phasmid (arrowhead); J: Pores in tip of male spicules; K, L: Second-stage juvenile (J2) labial region; M: J2 labial region; N: J2 lateral field; O, P: J2 tail. (Scale bars: A, B, E, F, J-N = 1 μm; G-I, O, P = 5 μm; C, D = 10 μm.) (After Humphreys-Pereira* et al., *2014.)*

labial region diam. = 6.4 ± 0.3 (5.8-6.9) μm; stylet = 11.3 ± 0.9 (9.0-13.0) μm; stylet knob width = 2.2 ± 0.2 (1.92-2.63) μm; stylet knob height = 1.6 ± 0.2 (1.2-1.92) μm; DGO = 3.0 ± 0.3 (2.5-3.5) μm; excretory pore to anterior end = 100 ± 3.1 (95-109) μm; tail = 58.5 ± 2.9 (53.2-62.3) μm; hyaline region = 11.0 ± 0.9 (9.8-12.9) μm; a = 30.2 ± 3.7 (24.4-36.7); c = 9.1 ± 0.6 (8.4-10.5).

DESCRIPTION

Female

Pearly white body, varying in shape from ovoid to saccate and with a variable neck diam. and length. Anterior end pointed, posterior end varying from rounded to slightly flattened. Labial region slightly offset from rest of body and sclerotised. Labial disc and median labials fused. In labial disc, six inner labial sensilla surrounding ovoid prestoma. Stoma slit-like. Lateral labials large, triangular, separated from labial region. Amphidial apertures elongated, located between labial disc and lateral labials. Stylet short, cone base triangular and wider than shaft. Stylet tip normally straight, but can be curved (10% of specimens). Shaft cylindrical, of same diam. throughout its length. Stylet knobs three, oval and sloping posteriad. Pharyngeal lumen of procorpus wide, often showing rounded protuberances. Excretory pore located at level of anterior end of procorpus. Excretory duct long and curved, very visible and bulky near excretory pore. Median bulb rounded or oval, with oval-shaped and sclerotised valve apparatus. Perineal patterns of females ovoid, with moderately high dorsal arches that are mostly rounded and sometimes squarish. Dorsal striae variable, mainly coarse and smooth, but can be fine or wavy. Weak lateral lines frequently present on both sides of perineal patterns, appearing as depressions and not incisures. Punctations and striae absent in perineum.

Male

Body vermiform, variable in length, anterior end tapering and posterior region bluntly rounded. Labial region offset from body and with a high labial region cap. Labial framework strong and sclerotised. Large stoma-like slit located in a hexagonal prestoma and surrounded by six inner labial sensilla. Labial disc and median labials fused, forming an elongated labial region cap. Lateral labials absent. Amphidial apertures elongated, located between labial disc and labial region. Labial region high and lacking annulation. Lateral field consisting of four incisures with areolations along body but with only a few actually crossing central field. Stylet robust and straight. Cone pointed, larger than shaft and slightly wider at base near junction with shaft. Stylet knobs large, rounded and sloping posteriad. Procorpus distinctly outlined, three times larger than median bulb. Median bulb ovoid, with a strong valve apparatus. Excretory duct curved. Excretory pore distinct and usually located six annuli posterior to hemizonid. Normally only one testis present, extending anteriorly. Spicules of variable length, arcuate and with two pores clearly visible at tip. Gubernaculum distinct. Phasmids with slit-like openings at level of cloacal aperture and located in central lateral field.

J2

Body vermiform, tapering more towards posterior than anterior end. Labial region narrower than body and slightly offset. Labial cap slightly elevated. Labial framework weakly developed. Labial disc and median labials fused. In labial disc, a stoma-like slit located in an ovoid prestoma and surrounded by six inner labial sensilla. Labial region smooth. Sometimes anteriormost 1-2 body annuli bordering labial region incomplete. Lateral labials small and triangular, sometimes elongated. Labial sensilla not visible. Elongated amphidial apertures located between labial disc and lateral labials. Body annulated from anterior end to terminus. Lateral field consisting of four incisures, with areolations along body but with only a few extending across field. Stylet weak, knobs rounded and sloping posteriad. Ampulla not well developed. Procorpus 2.0-2.5 times length of median bulb. Median bulb ovoid with sclerotised valve apparatus in centre. Hemizonid located anterior to excretory pore, extending for *ca* two body annuli. Excretory pore located posterior to nerve ring. Excretory duct curved and discernible when reaching intestine. Rectum dilated. Tail conoid, terminus rounded. Hypodermal terminus mostly rounded and in some cases pointed.

TYPE PLANT HOST: Coffee, *Coffea arabica* L., 'Catuaí'.

OTHER PLANTS: No other hosts were reported.

TYPE LOCALITY: Santa Marta, Volcán de Buenos Aires, Puntarenas province in southern Costa Rica.

DISTRIBUTION: *Central America*: Costa Rica.

SYMPTOMS: Coffee roots infected with *M. lopezi* had relatively small galls (1-2 mm diam.) when compared to other coffee-associated *Meloidogyne* spp. Each gall was characterised by a single female exposing an egg mass on the outside of the gall. No cracking of the root surface or corky roots, as typically caused by *M. arabicida*, were observed.

CHROMOSOME NUMBER: No available data.

POLYTOMOUS KEY CODES: *Female*: A21, B12, C213, D6; *Male*: A12, B21, C21, D1, E2, F2; *J2*: A12, B32, C32, D32, E23, F2.

BIOCHEMICAL AND MOLECULAR CHARACTERISATION: Biochemically *M. lopezi* has a characteristic esterase phenotype L2 consisting of a minor band close in size to the largest band in *M. javanica* phenotype J3 and two major bands similar in size to the smallest band in *M. javanica*. The Mdh phenotype of *M. lopezi* had three bands, two major and one minor. The largest band of the *M. lopezi* Mdh phenotype was similar in size to the major band of the *M. javanica* N1 Mdh phenotype. The Mdh phenotype of *M. lopezi* was similar to the N3 phenotype described by Esbenshade & Triantaphyllou (1985a).

Molecularly this nematode can be identified by *COII* and the 16S rRNA mtDNA genes, D2-D3 fragment of 28S and 18S rRNA gene (Humphreys-Pereira *et al.*, 2014). Phylogenetic analyses based on *COII* showed that *M. lopezi* is a sister species to *M. arabicida*, whereas phylogeny based on 18S rRNA gene showed a clade with *M. arabicida*, *M. izalcoensis*, *M. paranaensis* and *M. konaensis* (Humphreys-Pereira *et al.*, 2014). PCR-RFLP based on the region between the *COII* and the *16S rRNA* mtDNA genes proved to be a reliable method to differentiate *M. lopezi* from other tropical and coffee-parasitising RKN such as *M. paranaensis*, *M. incognita*, *M. javanica*, *M. izalcoensis* and *M. arabicida* (Humphreys-Pereira *et al.*, 2014) (Fig. 13).

RELATIONSHIPS (DIAGNOSIS): This species belongs to Molecular group I, the Incognita group. A phylogenetic analysis based on mtDNA showed that *M. lopezi* n. sp. is closely related to *M. arabicida*, although they differ strongly in morphology and isozyme phenotypes. *Meloidogyne lopezi* can be distinguished from other *Meloidogyne* spp. that infect coffee by several morphological features: females of *M. lopezi* have a greater mean body length of 833 *vs* 681-791 μm in *M. paranaensis*, 773 μm in *M. arabicida* and 794 μm in *M. izalcoensis*, but are smaller than females of *M. konaensis* with 860-992 μm. *Meloidogyne lopezi* females have a greater average stylet length of 18.0 *vs* 12.0 μm in *M. arabicida*, 15.6 μm in *M. izalcoensis*, 16.0 μm in *M. konaensis* and 16.0-17.3 μm in *M. paranaensis* (López & Salazar, 1989; Eisenback *et al.*, 1994; Carneiro *et al.*, 1996a, 2005a). The perineal patterns of *M. lopezi* mainly have moderately high and rounded dorsal arches, a squarish shape, as found in *M. paranaensis* and *M. izalcoensis*, being rare. In contrast to *M. lopezi*, the latter two species also show high dorsal arches similar to *M. incognita*. In J2 of *M. lopezi*, the average stylet length is 11.3 μm, similar to *M. arabicida*, but smaller than the average stylet length of *M. izalcoensis* with 12.8 μm, *M. konaensis* with 13.4-14.5 μm and *M. paranaensis* with 13.5-15 μm.

52. *Meloidogyne luci* Carneiro, Correa, Almeida, Gomes, Mohammad Deimi, Castagnone-Sereno & Karssen, 2014
(Figs 148-151)

COMMON NAME: Luc's root-knot nematode.

Carneiro *et al.* (2014) described this species from roots of lavender (*Lavandula spica* Cav.) plantations, in Caxias do Sul, Rio Grande do Sul state, Brazil. Later, *M. luci* was detected in a potato field near Coimbra, Portugal (Maleita *et al.*, 2018). In 2017, this species was added to the European Plant Protection Organization Alert List (EPPO, 2017b). It has been found parasitising corn and kiwi in Greece, and tomato in Italy and Slovenia (Širca *et al.*, 2004; Conceição *et al.*, 2012; Maleita *et al.*, 2018). In Brazil, Iran, Chile and Turkey, *M. luci* has been found

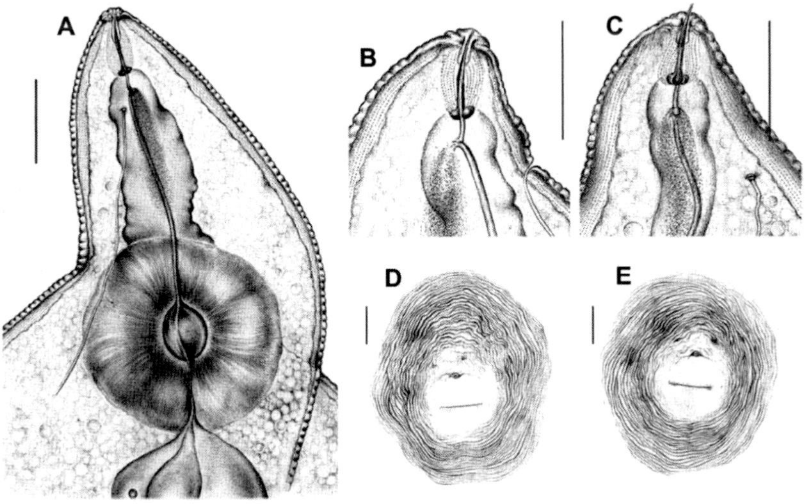

Fig. 148. Meloidogyne luci. *A: Female anterior region; B, C: Female labial region; D, E: Perineal pattern. (Scale bars: A-E = 20 μm.) (After Carneiro* et al., *2014.)*

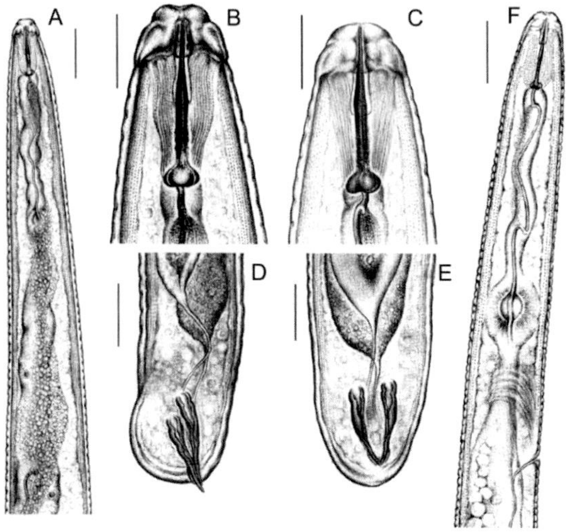

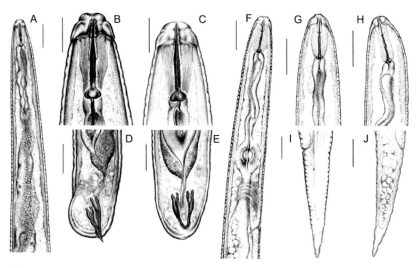

Fig. 149. Meloidogyne luci. *A: Male anterior region; B, C: Male labial region; D, E: Male tail; F-H: Second-stage juvenile (J2) anterior region; I, J: J2 tail. (Scale bars: A, F = 20 μm; B-E, G-J = 10 μm.) (After Carneiro* et al.*, 2014.)*

associated with several important vegetable plants and fruit tree species (Aydinli *et al.*, 2013; Carneiro *et al.*, 2014; Bellé *et al.*, 2016; Janssen *et al.*, 2016; Machado *et al.*, 2016)., *Meloidogyne luci* may have been misidentified as *M. ethiopica* in several surveys (Santos *et al.*, 2019) because of its morphological resemblance to *M. ethiopica* and similar esterase phenotype.

MEASUREMENTS *(AFTER CARNEIRO* ET AL., *2014)*

- *Holotype female*: L = 680 μm; W = 450 μm; stylet = 15.0 μm; stylet knob width = 3.5 μm; stylet knob height = 2.2 μm; DGO = 3.5 μm; anterior end to excretory pore = 23.5 μm; a = 1.5; EP/ST = 1.6.

See Table 35.

DESCRIPTION

Female

Females completely enclosed by gall tissue. Body translucent white, variable in size, elongated, ovoid to pear-shaped. Neck sometimes prominent, bent at various angles to body. Body cuticle distinctly annulated, annuli smaller in anterior neck region. Labial region offset

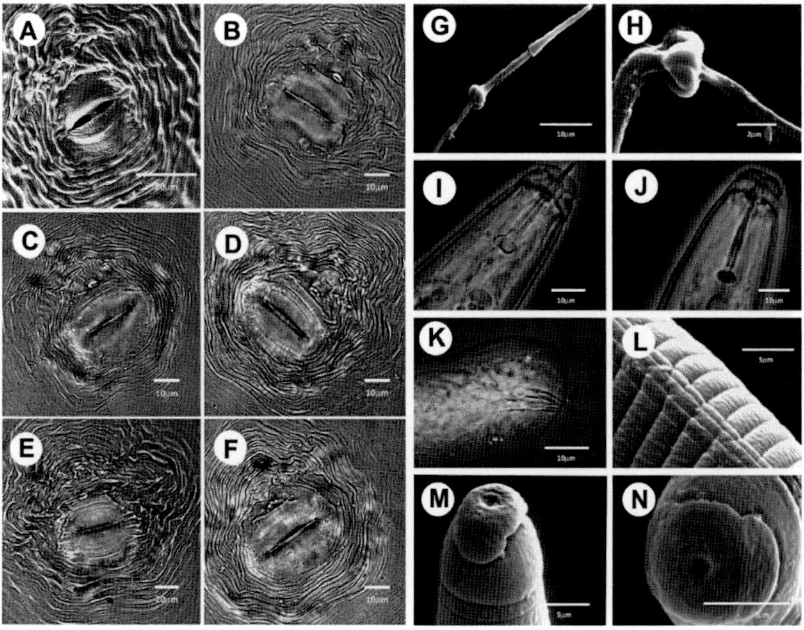

Fig. 150. Meloidogyne luci. *LM and SEM. A-F: Perineal pattern; G, H: Male stylet; I, J, M, N: Male labial region; K: Male tail; L: Male lateral field. (After Carneiro* et al.*, 2014.)*

from body, sometimes annulated. Stoma slit-like, located in ovoid prestomatal cavity, surrounded by pit-like openings of six inner labial sensilla. Labial disc and median labials dumbbell-shaped. Labial disc with two prominent bumps on ventral side slightly raised above median labials. Median labials never dividing into labial pairs. Lateral labials large, fused laterally with labial region for short distance. Amphidial apertures oval-shaped, located between labial disc and lateral labials. In LM, labial framework weakly developed, hexaradiate, lateral sectors slightly enlarged, vestibule and extension prominent. Stylet robust, easily dislodged posteriorly. Stylet cone generally slightly curved dorsally, widening slightly posteriorly near junction with shaft. Shaft gradually wider posteriorly near junction with knobs. Knobs well developed, distinctly separated, reniform, posteriorly sloping. Excretory pore located posterior to stylet at *ca* 0.5-1.0 stylet lengths and neck gradually increasing in diam. posteriorly. DGO *ca* one shaft length long, orifice branching into three ducts, ampulla large. Subventral gland orifices branched, lo-

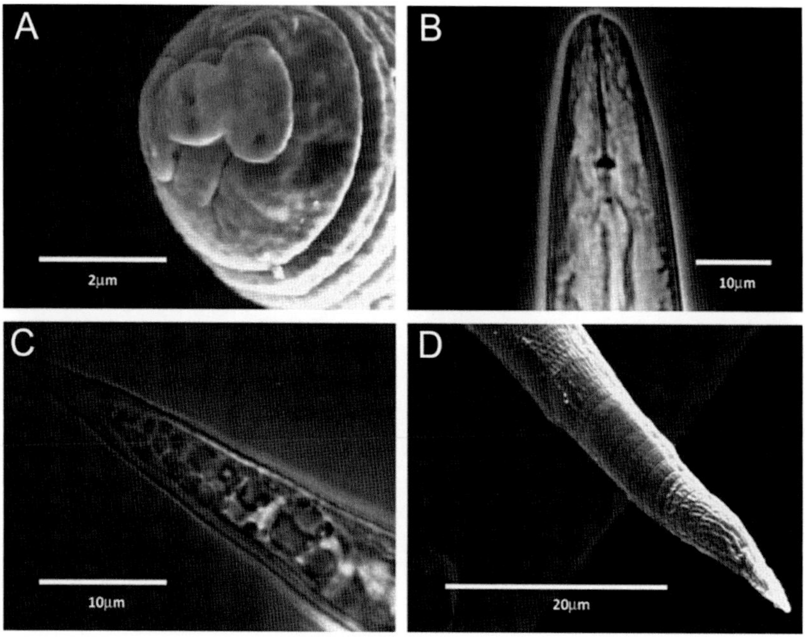

Fig. 151. Meloidogyne luci. *A, B: Second-stage juvenile (J2) labial region; C, D: J2 tail. (After Carneiro* et al.*, 2014.)*

cated immediately posterior to enlarged lumen of median bulb. Pharyngeal glands with one large dorsal lobe with two nuclei, two small nucleated subventral gland lobes, variable in shape, position and size, usually posterior to dorsal gland lobe. Perineal cuticular patterns highly variable, not very useful for identification. Dorsal arch high and squarish to low and round, striae coarse and smooth to wavy, widely separated, usually continuous, without distinct whorl. In lateral areas lines in dorsal arch tending to curve sharply toward tail terminus meeting ventral striae at an angle. Also, these striae becoming forked and wider near lateral areas, although distinct lateral incisures generally absent. If lateral incisures present, extending only a short distance anteriorly. Perivulval region generally free of striae, striae rarely present on lateral regions of vulva. Tail terminus sometimes visible. Phasmids small, directly opposite and posterior to anus. Fold over anus present.

Male

Body vermiform, tapering anteriorly, bluntly rounded posteriorly twisting through 90°, tail arcuate. Body annuli large, distinct. In lateral

Table 35. *Morphometrics of females, males and second-stage juveniles (J2) of* Meloidogyne luci. *All measurements are in μm and in the form: mean ± s.d. (range).*

Character	Brazil, Carneiro *et al.* (2014)	Turkey, Gerič Stare *et al.* (2019)	Slovenia, Gerič Stare *et al.* (2019)
Female (n)	30	130	20
L	704 ± 115 (570-800)	702 ± 94 (555-920)	865 ± 84.4 (785-983)
a	1.43 ± 0.39 (1.04-1.82)	1.4 ± 0.2 (1.0-1.8)	1.7 ± 0.3 (1.3-2.0)
W	503 ± 55 (440-550)	508 ± 78 (314-682)	520 ± 69.9 (433.4-602.6)
Stylet	15.5 ± 0.5 (15.0-16.0)	14.1 ± 1.9 (11.3-18.7)	15.2 ± 1.2 (13.5-16.8)
Stylet knob width	3.6 ± 0.5 (3.2-4.0)	3.7 ± 0.6 (2.4-5.0)	3.6 ± 0.8 (2.4-4.8)
Stylet knob height	2.0 ± 0.1 (1.8-2.3)	2.0 ± 0.3 (1.0-2.7)	1.8 ± 0.5 (1.3-2.8)
DGO	3.2 ± 0.8 (3.0-4.0)	3.7 ± 0.7 (2.6-5.2)	3.3 ± 0.6 (2.3-4.1)
Ant. end to median bulb	-	73.6 ± 9.9 (52.6-101.8)	71.4 ± 13.4 (50.9-89.8)
Ant. end to excretory pore	31.3 ± 18 (18.5-48.5)	38.9 ± 11.6 (19.2-77.1)	46.9 ± 16.9 (28.8-73.8)
Vulva length	23.3 ± 0.5 (20.0-26.0)	25.1 ± 2.5 (16.2-31.2)	25.6 ± 2.7 (20.9-29.4)
Vulva-anus distance	17.4 ± 0.6 (15.0-26.0)	18.3 ± 2.7 (10.8-23.6)	20.1 ± 3.0 (16.3-24.9)
Interphasmid distance	26.3 ± 0.8 (20.0-36.5)	23.1 ± 4.6 (11.5-35.0)	27.8 ± 5.9 (18.2-35.2)
EP/ST	2.0	-	-
Male (n)	30	130	20
L	1602 ± 520 (1090-2130)	1252 ± 308 (706-2038)	1534 ± 171.7 (1340-1940)
a	39.7 ± 9.8 (21.8-57.6)	30.6 ± 5.5 (21.6-49.3)	42.1 ± 7.0 (30.9-52.5)
c	322 ± 72 (73-710)	102 ± 26.5 (56.1-180.5)	82.4 ± 38.9 (30.9-156.5)

Table 35. *(Continued.)*

Character	Brazil, Carneiro *et al.* (2014)	Turkey, Gerič Stare *et al.* (2019)	Slovenia, Gerič Stare *et al.* (2019)
W	43 ± 5.0 (37-50)	41 ± 7.9 (23-56)	37.3 ± 7.2 (29-56)
Stylet	22.1 ± 2.7 (20.8-23)	20.7 ± 1.4 (17.3-23.6)	21.2 ± 1.7 (17.5-23.7)
Stylet knob width	4.2 ± 0.4 (3.8-4.5)	4.0 ± 0.5 (3.1-5.4)	4.2 ± 0.4 (3.3-4.6)
Stylet knob height	2.6 ± 0.3 (2.5-3.0)	2.6 ± 0.2 (2.0-3.3)	2.8 ± 0.6 (2.0-3.7)
DGO	3.5 ± 1.0 (2.5-4.5)	3.7 ± 0.7 (2.2-5.4)	3.2 ± 0.5 (2.4-3.8)
Ant. end to median bulb	-	86.8 ± 12.8 (60.7-114.7)	75.3 ± 8.9 (62.2-89)
Ant. end to excretory pore	199 ± 30.2 (150-217)	162.1 ± 24.6 (108.9-233.4)	150.3 ± 30.2 (111-181)
Tail	9.5 ± 1.0 (3.0-15.0)	12.4 ± 1.9 (9.3-18.1)	13.0 ± 2.1 (9.7-18.0)
Spicules	31.3 ± 4.0 (24.0-35.0)	26.8 ± 2.6 (22.7-36.2)	31.2 ± 3.0 (27.0-37.6)
J2 (n)	30	130	20
L	383 ± 85 (300-470)	383 ± 26 (321-439)	351 ± 28.7 (321-408)
a	25.6 ± 10.5 (15.0-36.1)	23.2 ± 2.2 (17.8-27.6)	23.2 ± 2.1 (19.4-26.8)
c	8.7 ± 2.6 (6.2-11.5)	8.2 ± 1.2 (5.5-12.1)	7.8 ± 0.7 (5.8-8.7)
c′	-	4.3 ± 0.6 (3.0-5.5)	4.2 ± 0.3 (3.6-4.5)
W	16.0 ± 1.5 (13.0-20.0)	16.7 ± 1.8 (13.6-21.0)	15.3 ± 2.1 (12.4-21.0)
Stylet	12.5 ± 0.2 (12.0-13.5)	13.6 ± 0.7 (11.6-15.3)	13.6 ± 0.4 (13.0-14.0)
DGO	2.9 ± 0.5 (2.3-3.3)	3.0 ± 0.4 (2.1-3.8)	2.6 ± 0.4 (2.1-3.4)
Ant. end to median bulb	-	54.3 ± 3.6 (47.5-67.6)	55.2 ± 4.6 (46.2-66.0)

Table 35. *(Continued.)*

Character	Brazil, Carneiro *et al.* (2014)	Turkey, Gerič Stare *et al.* (2019)	Slovenia, Gerič Stare *et al.* (2019)
Ant. end to excretory pore	73.0 ± 10.0 (62.0-82.0)	80.1 ± 8.4 (64.3-103.8)	76.2 ± 10.7 (63.0-92.0)
Tail	44.0 ± 4.5 (40.0-48.5)	47.5 ± 6.1 (30.6-58.6)	45.5 ± 5.0 (37.6-58.0)
Hyaline region	11.7 ± 3.0 (9.0-15.0)	11.3 ± 1.3 (8.0-13.5)	11.1 ± 1.4 (8.9-14.0)

view, labial cap high and rounded, labial region not offset from body, without annulation. In SEM, stoma slit-like, located in ovoid to hexagonal cavity surrounded by pit-like openings of six inner labial sensilla. Labial disc rounded, median labials divided into labial pairs, fused with labial disc forming elongate labial cap with small depressions. Labial sensilla not clearly demarcated on median labials. Lateral labials absent. Amphidial apertures elongated slits between labial disc and lateral sides of labial region. In LM, labial framework moderately developed, hexaradiate, lateral sectors slightly enlarged. Vestibule extensions distinct. Stylet robust, opening 3-4 μm from bluntly pointed tip, cone larger than shaft, slightly increasing in diam. posterior to junction of cone and shaft. Shaft cylindrical, widening slightly near junction with knobs. Knobs small and rounded to heart-shaped, merging gradually into shaft. Dorsal pharyngeal gland orifice branching into three ducts, ampulla distinct. Amphids very distinct, often producing exudates. Procorpus distinct, median bulb ovoid, triradiate, lining of enlarged lumen of median bulb thinner than in female. Pharyngo-intestinal junction at level of nerve ring, indistinct. Three nuclei in gland lobe, lobe variable in length. Intestinal caecum extending anteriorly, sometimes approaching level of dorsal pharyngeal gland orifice. Excretory pore distinct, 6-10 annuli posterior to hemizonid. Lateral field areolated with four incisures beginning near level of stylet knobs or 12 annuli posterior to labial region as two incisures, two additional incisures starting near level of median bulb. Sex reversed males (two testes) common compared to normal males (one testis). Testes outstretched or anteriorly reflexed. Spicules arcuate, gubernaculum distinct. Tail short to very short, indistinct phasmids at cloacal aperture level.

J2

Body vermiform, clearly annulated, tapering more posteriorly than anteriorly. Anterior end truncate, labial region only slightly offset from body. Vestibule and extension more developed than remainder of hexaradiate labial framework. In SEM, prestoma opening rounded, surrounded by small, pore-like openings of six inner labial sensilla. Median labials and labial disc dumbbell-shaped in face view. Lateral labials small, triangular, lower than labial disc and median labials. Posterior edge of one or both lateral labials sometimes fusing with labial region. Labial disc oval, slightly raised above median labials. Median labials with crescentic to rounded margins. Labial sensilla not visible externally. Elongated amphidial apertures located between labial disc and lateral labials. Labial region not annulated, body annuli distinct but fine. Lateral field beginning near level of procorpus as two lines, third line beginning near median bulb and quickly splitting into four lines, running entire length of body before gradually decreasing to two lines ending near hyaline region, irregularly areolated. In LM, stylet delicate, cone straight, narrow, sharply pointed, shaft becoming slightly wider posteriorly, knobs small, oval-shaped, posteriorly sloping. Procorpus faint, median bulb oval-shaped with enlarged lumen lining, isthmus clearly defined, pharyngo-intestinal junction difficult to observe. Gland lobe variable in length, overlapping intestine ventrally. Excretory pore distinct, hemizonid anterior adjacent towards excretory pore. Tail conoid with finely rounded unstriated terminus. Hyaline region not clearly defined. Rectum dilated. Phasmids small, difficult to observe, located posterior to anus.

TYPE PLANT HOST: Lavender, *Lavandula spica* Cav.

OTHER PLANTS: This nematode parasitises numerous crop species (Susič *et al.*, 2020a). *Meloidogyne luci* was found in Brazil infecting cucumber (*Cucumis sativus*) and lettuce (*Lactuca sativa*) in Vargem Bonita (Gama, DF), broccoli (*Brassica oleracea* var. *italica*) in Alexandre Gusmão (Ceilândia, DF), okra (*Abelmoschus esculentus* (L.) Moench) in Ceilândia, DF, green bean (*Phaseolus vulgaris*) in Braslândia, DF, yacon (*Smallanthus sonchifolius*) in Planaltina, DF, and kiwi (*Actinidia deliciosa*) A.Chev) C.F.Liang) in Nova Roma do Sul, RS. In Chile, it was found on grapevine (*Vitis vinifera*) in La Rotunda, Casa Branca, and in Iran on rose (*Rosa* sp.), snapdragon (*Antirrhinum majus*), sedum (*Hylotelephium spectabile* (Boreau) H. Ohba) in Tehran city (Carneiro *et al.*, 2014), soybean and common bean in Brazil (Bellé *et al.*, 2016b; Machado *et al.*, 2016), potato (Maleita *et al.*, 2018). Santos *et al.* (2019) reported this species from the ornamental plant *Cordyline australis* (G.Forst) Hook.f. and the weed *Oxalis corniculata* L. Conçalves *et al.* (2020) reported this species from kiwi and corn and maintained on tomato in Greece.

TYPE LOCALITY: Plantations in Caxias do Sul, Rio Grande do Sul state, Brazil.

DISTRIBUTION: *South America*: Argentina, Brazil, Bolivia, Chile, Ecuador; *Central America*: Guatemala; *Asia*: Iran (Carneiro *et al.*, 2014), Turkey; *Europe*: Greece, Italy, Serbia, Slovenia (Strajnar *et al.*, 2009; Conceição *et al.*, 2012; Maleita *et al.*, 2012c; Aydinli *et al.*, 2013; Aydinli, 2018; Susič *et al.*, 2020a) and Portugal (Maleita *et al.*, 2018; Rusinque *et al.*, 2021). Recent biochemical and molecular studies made by Gerič Stare *et al.* (2017) demonstrated that some European populations identified as *M. ethiopica* are in fact misidentifications of *M. luci*.

DIFFERENTIAL HOSTS: This species reproduced on tomato 'Rutgers', tobacco 'NC95' and on pepper 'California Wonder'. No reproduction occurred on watermelon 'Charleston Gray', cotton 'Deltapine 61', or peanut 'Florunner'.

PATHOGENICITY: The pathogenicity to potato was assessed in 16 commercial cultivars and compared with *M. chitwoodi* (Maleita *et al.*, 2018). All potato cultivars were susceptible to both *Meloidogyne* species with gall indices of 5 and higher, and reproduction factor values ranging from 12.5 to 122.3, suggesting that *M. luci* may constitute a potential threat to potato production (Maleita *et al.*, 2018). Recently, Santos *et al.* (2020) demonstrated that resistant tomato (*Mi*-1.2 gene) limits the reproduction of *M. luci* and may have potential to be employed as an alternative to the use of nematicides.

CHROMOSOME NUMBER: Reproduction is mitotic parthenogenesis; 2n = 42-46 chromosomes.

POLYTOMOUS KEY CODES: *Female*: A23, B2, C2, D2; *Male*: A2134, B2, C213, D1, E2, F1; *J2*: A32, B2, C3, D32, E23, F3.

BIOCHEMICAL AND MOLECULAR CHARACTERISATION: *Meloidogyne luci* showed a species-specific esterase phenotype (L3) with three bands (Rm: 1.05, 1.10, 1.25) and MDH N1 (Rm: 1.0) in all isolates from Brazil and Iran. The SOD and GOT phenotypes were also studied for L3 isolates but were not species-specific and were shared with various *Meloidogyne* spp. (Esbenshade & Triantaphyllou, 1985a; Carneiro *et al.*, 2000).

Carneiro *et al.* (2014) reported that this nematode can be identified by ITS1-5.8S-ITS2 rRNA and D2-D3 fragment of 28S rRNA gene sequences from closely related species including *M. ethiopica* and *M. inornata*. However, these genes show a high similarity with those of *Meloidogyne* spp. from Ethiopica and Incognita groups. Sequence of mtDNA region between *COII* and 16S rRNA genes can be used for reliable identification of this species (Gerič Stare *et al.*, 2017; Santos *et al.*, 2019).

RELATIONSHIPS (DIAGNOSIS): This species belongs to Molecular group I (Fig. 18), the Ethiopica group. The perineal pattern of *M. luci* is similar to *M. ethiopica* and differs from *M. inornata* (closer to *M. incognita*). In the female, the stylet conus is slightly curved dorsally and the reniform knobs are similar to *M. inornata* and different from *M. ethiopica*, the shaft is wide in *M. ethiopica* when compared to *M. inornata* and *M. luci*. In SEM, the male stylet is completely different between the three species: in *M. luci* the knobs are small, rounded to heart-shaped, in *M. ethiopica* reniform, and in *M. inornata* rounded, distinctly indented, sloping posteriorly and like tear drops. Average DGO is intermediate in males of *M. luci* at 3.2-3.7 *vs* 3.4-4.5 μm in *M. inornata* and 2.5-2.7 μm in *M. ethiopica*. In SEM, the male lateral field, with four areolated incisures, is similar in *M. luci* and *M. ethiopica* but completely different to *M. inornata* where it may show four, five, or even more incisures and the external incisures of the field may be made up of interrupted lines. The average male tail length is shorter in *M. luci*, 9.5 *vs* 13.4-17.3 μm in *M. ethiopica* and 12.9-13.5 μm in *M. inornata*. In the J2 of *M. luci*, L = 300-470 *vs* 322-510 μm in *M. ethiopica* and 394-487 μm in *M. inornata*.

53. *Meloidogyne lusitanica* Abrantes & Santos, 1991
(Figs 152-154)

COMMON NAME: Olive root-knot nematode.

Abrantes & Santos (1991) described this species from roots of olive trees in Miranda do Corvo, Portugal.

MEASUREMENTS *(AFTER ABRANTES & SANTOS, 1991)*

- *Holotype female*: L = 695 μm; W = 510 μm; stylet = 19.0 μm; stylet knob width = 4.5 μm; stylet knob height = 2.5 μm; DGO = 5.0 μm; median bulb length = 45 μm; median bulb width = 37 μm; median bulb valve width = 9.0 μm; median bulb valve length = 12.0 μm; excretory pore to anterior end = 50.0 μm; a = 1.4; EP/ST = 2.6.
- *Allotype male*: L = 1240 μm; W = 34.5 μm; stylet = 19.0 μm; labial region diam. = 12.0 μm; labial region height = 3.5 μm; stylet knob width = 4.5 μm; stylet knob height = 3.0 μm; DGO = 4.5 μm; median bulb width = 9.0 μm; median bulb valve length = 5.0 μm; excretory pore to anterior end = 137.5 μm; tail = 11.5 μm; spicules = 34.5 μm; gubernaculum = 10.0 μm; a = 35.9; c = 107.8; T = 60.5%.
- *Paratype females* (n = 30): L = 873 ± 116 (620-1118) μm; W = 534 ± 70 (345-650) μm; stylet = 17.1 ± 0.9 (17.0-19.0) μm; stylet knob width = 5.0 ± 0.1 (4.0-5.5) μm; stylet knob height = 2.5 ± 0.1 (2.0-3.0) μm; DGO = 4.0 ± 0.6 (3.0-5.0) μm; median bulb length = 46.5 ± 0.8 (39.0-58.5) μm; median bulb width = 43.4 ± 0.8 (35.0-56.0) μm; median bulb valve width = 9.8 ± 0.1 (9.0-11.0) μm; median bulb valve length = 12.1 ± 0.1 (10.5-13.5) μm; excretory pore to anterior end = 44.1 ± 8.3 (28.0-60.0) μm; vulval slit length 20.1 ± 2.2 (16.0-26.5); anus to vulva distance 19.3 ± 2.8 (14.5-26.0); interphasmid distance 27.9 ± 4.4 (16.5-35.5); a = 1.6 ± 0.2 (1.3-1.9); EP/ST = 2.6 ± 0.5 (1.6-3.8).
- *Paratype males* (n = 30): L = 1613 ± 240 (960-1.80) μm; W = 43.0 ± 5.50 (34.5-52.0) μm; labial region diam. = 12.7 ± 0.1 (12.0-14.0) μm; labial region height = 4.3 ± 0.1 (3.5-5.0) μm; stylet = 24.5 ± 1.1 (21.0-27.0) μm; stylet knob width = 4.7 ± 0.1 (4.0-5.5) μm; stylet knob height = 3.1 ± 0.1 (2.5-3.5) μm; DGO = 5.0 ± 0.7 (4.0-6.0) μm; median bulb width = 11.2 ± 0.2 (9.0-12.5) μm; median bulb valve width = 4.0 ± 0.1 (3.0-5.5) μm; median bulb valve length = 5.8 ±

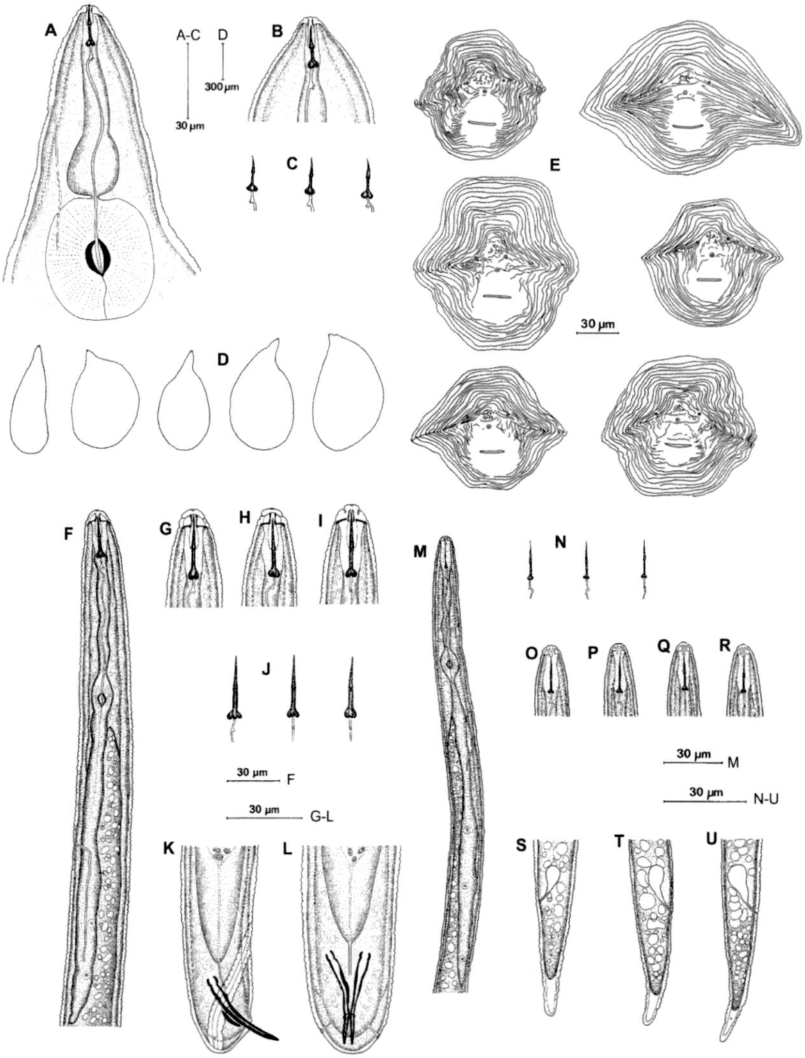

Fig. 152. Meloidogyne lusitanica. *A: Female anterior region; B: Female labial region; C: Female stylet; D: Entire female; E: Perineal pattern; F: Male anterior region; G-I: Male labial region; J: Male stylet; K, L: Male tail; M: Second-stage juvenile (J2) anterior region; N: J2 stylet; O-R: J2 labial region; S-U: J2 tail. (After Abrantes & Santos, 1991.)*

0.1 (4.8-7.5) μm; excretory pore to anterior end = 170 $\pm$ 20.0 (130-207) μm; tail = 9.5 $\pm$ 1.0 (3.0-15.0) μm; spicules = 37.9 $\pm$ 2.9 (32.0-

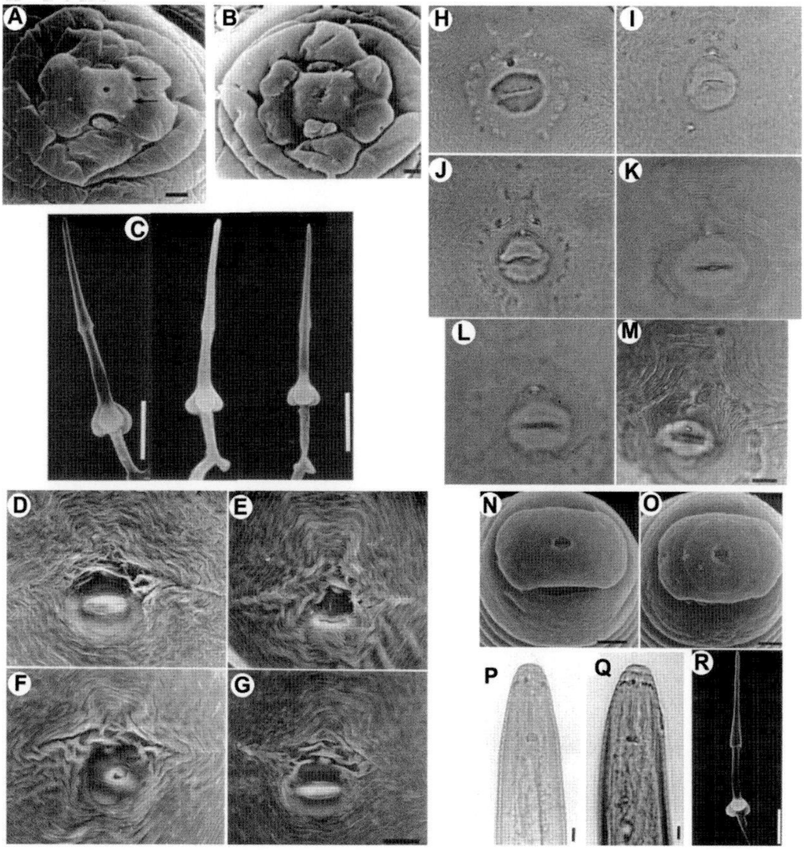

Fig. 153. Meloidogyne lusitanica. *LM and SEM. A, B: Female* en face *view; C: Female stylet; D-M: Perineal pattern; N, O: Male* en face *view; P, Q: Male labial region; R: Male stylet. (Scale bars: A, B = 1 μm; C = 5 μm; D-M = 20 μm; N, O = 2 μm; P, Q = 4 μm; R = 5 μm.) (After Abrantes & Santos, 1991.)*

44.5) μm; gubernaculum = 10.2 ± 0.8 (8.5-12.0) μm; a = 37.7 ± 4.8 (26.3-45.3); c = 130.2 ± 21.4 (76.8-170.5); T = 51.3 ± 7.0 (40.4-70.6)%.

- *Paratype J2* (n = 30): L = 450 ± 35 (390-515) μm; W = 18.8 ± 1.0 (17.0-20.0) μm; labial region diam. = 6.3 ± 0.1 (6.0-6.5) μm; labial region height = 2.0 ± 0.02 (1.8-2.5) μm; stylet = 14.2 ± 0.9 (13.0-16.0) μm; stylet knob width = 2.2 ± 0.04 (2.0-2.5) μm; stylet knob height = 1.3 ± 0.04 (1.0-1.5) μm; DGO = 3.9 ± 0.4 (3.5-4.5) μm;

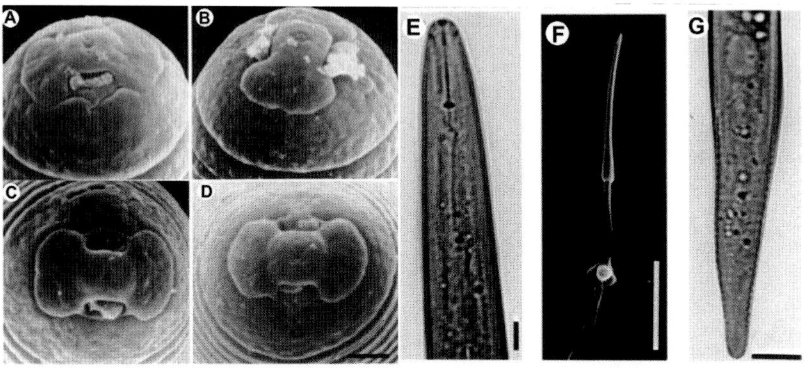

Fig. 154. Meloidogyne lusitanica. *A, B: SEM of second-stage juvenile (J2) heads, lateral views; C, D: SEM of J2 anterior end, face views; E: LM of J2 anterior region; F: SEM of excised J2 stylet; G: LM of J2 tail. (Scale bars: A-C same scale as D = 1 μm; E, F = 5 μm; G = 10 μm.) (After Abrantes & Santos, 1991.)*

median bulb valve width = 3.5 ± 0.03 (3.2-3.8) μm; median bulb valve length = 4.4 ± 0.1 (4.0-4.8) μm; excretory pore to anterior end = 90 ± 5.4 (78-102) μm; tail = 44.1 ± 3.0 (39.0-50.0) μm; hyaline region = 12.0 ± 1.2 (10.0-14.0) μm; a = 24.1 ± 1.8 (20.0-27.4); c = 10.2 ± 0.5 (9.1-11.4); c′ = 3.3 ± 0.3 (3.0-3.7).
- *Eggs* (n = 30): L = 124.4 (108-144) μm; W = 47.2 (37-57) μm; L/W = 2.7 (2.4-2.9).

DESCRIPTION

Female

Females completely enclosed by gall tissue. Body pearly white, variable in size, elongate ovoid to pear-shaped, with short neck, posteriorly rounded, without tail protuberance. Labial region not offset from body, not annulated. Rounded prestoma located centrally on labial disc. Pore-like openings of six inner labial sensilla surrounding prestoma. Labial disc with two bumps on ventral side, slightly raised above median labials. Median labials usually indented medianly, often dividing into labial pairs. Lateral labials large, fused laterally with labial region for short distance. Amphidial openings oval, between labial disc and lateral labials. Labial framework weakly developed. Stylet long, easily dislodged posteriorly. Stylet cone slightly curved dorsally, widening gradually posteriorly, shaft of same width throughout, or widening slightly near junction with knobs, knobs well developed, distinctly separate, pear-shaped. Excretory pore located posterior to stylet about 1.5-2.5 stylet lengths or 16-23 annuli from anterior end. Characteristic cuticular pattern of perineal region trapezoidal in shape, striae coarse, sometimes continuous, smooth to wavy, forming a medium to high, trapezoidal, dorsal arch. Ventral pattern area with fine, smooth or wavy striae. Lateral lines indicated by dorsal and ventral striae meeting at an angle, or by short lateral striae, sometimes striae discontinuous, may fork near lateral lines. Occasionally, some striae extending laterally, forming one or two wings. Tail tip area well defined, marked by few punctations; striae often bending toward vulva. Perivulval region free of striae. Phasmids distinct; surface structure not apparent in SEM.

Male

Body vermiform, tapering anteriorly, bluntly rounded posteriorly. Labial cap in lateral view quite high and rounded, extending posteriorly onto distinctly offset labial region. In SEM (face view), labial disc very large and round, stoma slit-like located in ovoid prestomatal cavity surrounded by pore-like openings of six inner labial sensilla. Median labials with smooth or slightly indented outer margins, fused with labial disc to form a continuous elongate structure with parallel sides. Four labial sensilla marked by cuticular depressions on median labials. Amphidial apertures large, elongate slits between labial disc and lateral sectors of labial region. Lateral labials absent. Labial region usually smooth, but may have one or two incomplete annulations. Body annuli distinct. Lateral field with four incisures beginning near level of stylet base as two incisures. In LM, labial framework moderately developed. Stylet robust, large, cone straight, pointed, gradually increasing in diam. posteriorly, stylet opening marked by very faint protuberance several micrometers from stylet tip, shaft cylindrical, knobs elongate, pear-shaped, slightly offset from shaft. Procorpus and median bulb distinct, median bulb elongate, oval with large valve plates. Hemizonid distinct, three or four annuli anterior to well-defined excretory pore. Testis usually one, sometimes two, generally outstretched. Spicules long, moderately curved ventrally, gubernaculum crescentic. Tail short, terminus not striated. Phasmids posterior to cloacal level.

J2

Body vermiform, clearly annulated, tapering more posteriorly than anteriorly. In SEM, prestoma opening ovoid, surrounded by small, pore-like openings of six inner labial sensilla. Labial disc, median labials and lateral labials fused into one structure. Labial disc rounded, slightly raised above median labials. Median labials with crescentic to rounded margins, sometimes slightly indented, suggesting subdivision into labial pairs, wider than labial disc. Labial sensilla not expressed externally. Amphidial apertures between labial disc and lateral labials. Labial cap narrower than labial region. Labial region smooth, occasionally with one or two short, incomplete annulations. Lateral field marked by four incisures. Labial framework weak. Stylet long, but delicate. Stylet cone sharply pointed, increasing in width gradually posteriorly, shaft cylindrical, knobs distinctly separated, pear-shaped, slightly offset from shaft. Median bulb ovoid with prominent valve plates. Excretory pore distinct, hemizonid one or two annuli anterior to excretory pore. Tail conoid with rounded unstriated terminus, hyaline region distinct. Rectal dilation large. Phasmids small, always posterior to anal level.

TYPE PLANT HOST: Olive, *Olea europaea* L. 'Galega'.

OTHER PLANTS: Among the host differentials commonly used for identification of *Meloidogyne* species, watermelon and tomato were lightly galled but no reproduction occurred. No infection was found on tobacco, cotton, pepper or peanut.

TYPE LOCALITY: A field near Cadaixo, Miranda do Corvo, Portugal.

DISTRIBUTION: *Europe*: Portugal. The geographic distribution of *M. lusitanica* is unknown, although in a survey of Portuguese olive fields this species was found in only two of the 90 fields sampled (Abrantes & Santos, 1991).

SYMPTOMS: Egg masses of *M. lusitanica* are always large and yellowish to reddish in colour. No females were observed to protrude from roots. Histological examination of olive roots infected by this nematode species demonstrated that giant cell formation was induced. Hyperplasia and hypertrophy were common phenomena in the cortical and vascular parenchyma.

PATHOGENICITY: The species may be an important pest of olive, but no pathogenicity test has been performed (Abrantes & Santos, 1991).

CHROMOSOME NUMBER: No available data.

POLYTOMOUS KEY CODES: *Female*: A12, B21, C213, D1; *Male*: A234, B12, C12, D1, E1, F2; *J2*: A213, B12, C3, D32, E2, F43.

BIOCHEMICAL AND MOLECULAR CHARACTERISATION: Pais & Abrantes (1989) described the esterase (P1 type) and malate dehydrogenase (P3 type) patterns for *M. lusitanica*. Karssen & van Hoenselaar (1998) considered these bands as A1 for esterase and N1c for malate dehydrogenase. No molecular data for this nematode are available.

RELATIONSHIPS (DIAGNOSIS): This species has a distinctive perineal pattern. *Meloidogyne lusitanica* is close to *M. megatyla*, *M. partityla* and *M. ardenensis*. However, *M. lusitanica* differs from these species by the presence of distinct punctations above the anus in the perineal pattern, by the absence of deep indentations in the stylet knobs of the female, male, and J2, and by the more posterior position of the excretory pore at 44.1 (28.0-60.0) μm in the female, which in *M. megatyla* is 20.4 (12.0-32.0) μm, and in *M. partityla* is 27.9 (17.0-41.0) μm. The J2 body length is similar to those of *M. megatyla* and *M. partityla*; however, they are distinct in tail or rectum shape.

54. *Meloidogyne mali* Itoh, Ohshima & Ichinohe, 1969
(Figs 155, 156)

COMMON NAME: Apple root-knot nematode.

Itoh *et al.* (1969) described this species from roots of apple root-stock from Toyono, Kamiminochi, Nagano Prefecture, Japan. *Meloidogyne mali* has been added to the EPPO A2 List of pests recommended for regulation as quarantine pests in 2017 (Curto *et al.*, 2017; EPPO, 2017a). Descriptions of this species were also given by Okamoto & Yaegashi

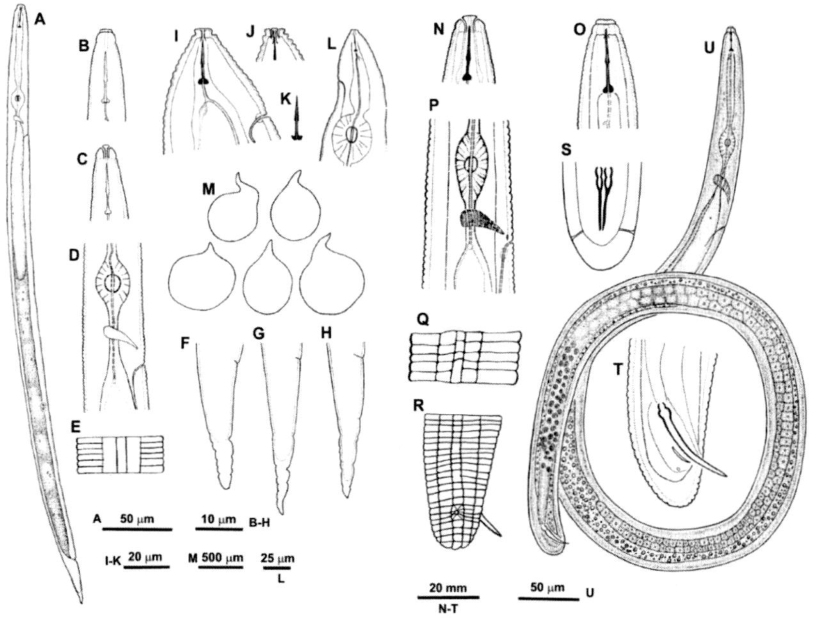

Fig. 155. Meloidogyne mali. *A: Entire second-stage juvenile (J2); B, C: J2 labial region; D: J2 median bulb; E: J2 lateral field; F-H: J2 tail; I, J, L: Female labial region; K: Female stylet; M: Entire female; N, O: Male labial region; P: Male median bulb; Q: Male lateral field; R-T: Male tail; U: Entire male. (After Itoh* et al.*, 1969.)*

(1981a, b), Okamoto *et al.* (1983), Palmisano & Ambrogioni (2000) and Gu *et al.* (2013).

Measurements

See Table 36.

- *Eggs* (n = 50): L = 104 (90-120) μm; W = 42 (30-45) μm; L/W = 2.5.

Description

Female

Mature females completely enclosed in the root tissue and labial region embedded in xylem. Eggs laid within the cortex and later epidermis broken. Body variable in shape, primarily determined by the type of root

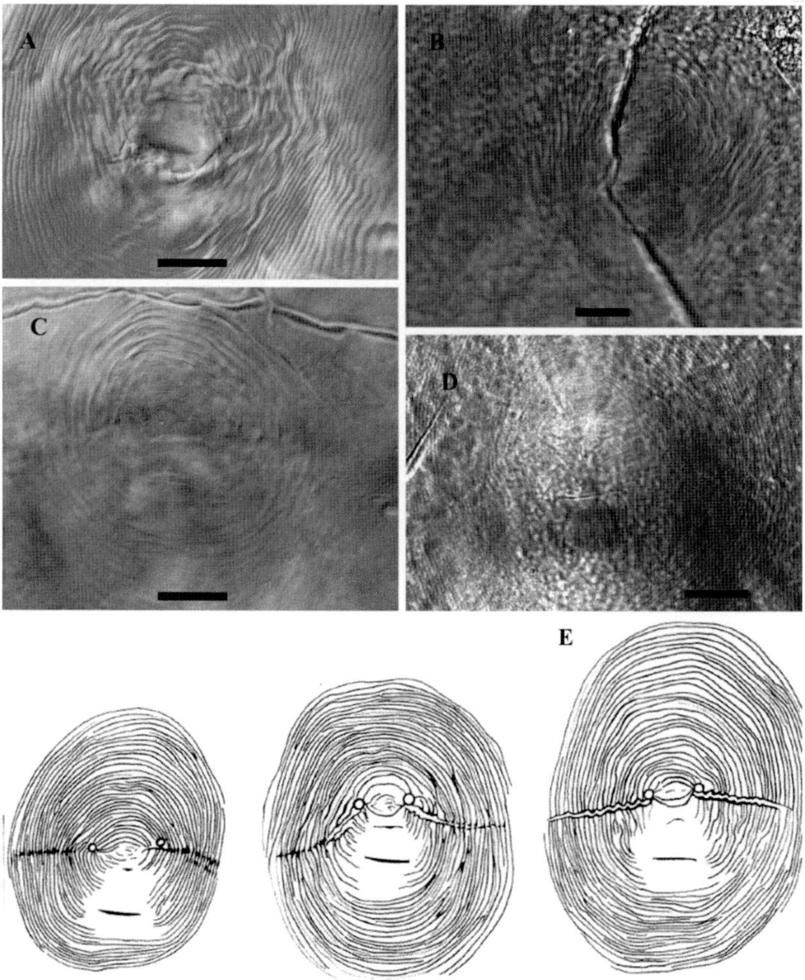

Fig. 156. Meloidogyne mali. *A-E: Perineal pattern. (Scale bars = 10 μm.) (After Itoh* et al.*, 1969; Ahmed* et al.*, 2013.)*

on which the nematode develops; in young and soft tissue of feeder roots they are white and pyriform with long neck, whilst in hard root tissue they are spherical or onion-shaped with short neck. Cuticle in mature female 20 (8-40) μm thick on spherical part of body. Excretory pore located ventrally on 23rd (20-29th) annulus from anterior end. Stylet curved dorsally, with well-developed knobs that tend to slope either posteriorly or anteriorly in the ratio of 16:8 individuals. In dorsoventral

Table 36. *Morphometrics of females, males and second-stage juveniles (J2) of* Meloidogyne mali. *All measurements are in μm and in the form: mean ± s.d. (range).*

Character	Holotype, Allotype, Japan, Itoh *et al.* (1969)	Paratypes, Japan, Itoh *et al.* (1969)	Tokyo, Japan, Okamoto *et al.* (1983)	Tuscany, Italy, Palmisano & Ambrogioni (2000)	Japan, Gu *et al.* (2013)	Japan, Gu *et al.* (2020)
Female (n)	-	25	-	30	-	3
L	690	847 ± 93.4 (684-1044)	-	770.9 ± 140.4 (568-1043)	-	710 ± 32.7 (670-750)
a	-	-	-	1.3 ± 0.3 (1.0-2.4)	-	-
W	500	660 ± 95.8 (540-864)	-	618.3 ± 151.6 (357-1007)	-	-
Stylet	14.0	15 (13-17)	-	14.2 ± 1.0 (12-16)	-	12.0 ± 1.1 (10.9-13.4)
Stylet knob width	-	-	-	3.5 ± 0.7 (2.6-5.2)	-	-
Stylet knob height	-	-	-	1.8 ± 0.6 (1.1-3.9)	-	-
DGO	4.0	5.5 (4.0-7.0)	-	4.6 ± 0.8 (3.5-6.5)	-	3.5 ± 0.7 (2.6-4.4)
Median bulb length	13.0	49 (40-73)	-	42.6 ± 6.5 (33-59)	-	-
Median bulb width	9.0	39 (32-44)	-	40.9 ± 7.0 (31-59)	-	-
Median bulb valve width	13.0	10 (9-11)	-	9.7 ± 1.4 (7.2-12.4)	-	-
Median bulb valve length	9.0	12 (11-13)	-	12.4 ± 1.0 (11.1-14.4)	-	-
Vulval slit	-	18 ± 2.5 (12-24)	-	22.0 ± 2.9 (17.0-28.7)	-	20.7 ± 2.1 (18.3-23.5)
Vulval slit to anus distance	-	17 ± 1.8 (14-22)	-	19.0 ± 1.9 (15.0-22.2)	-	18.6 ± 2.1 (15.6-20.3)
Interphasmid distance	-	22 ± 3.5 (17-29)	-	19.2 ± 3.8 (13.7-28.9)	-	21.4 ± 1.6 (19.6-23.4)
Ant. end to excretory pore	-	-	-	32.3 ± 7.8 (15.7-41.5)	-	-
EP/ST	-	-	-	2.3	-	-
Male (n)	1	25	-	30	2	1
L	1400	1447 (1270-1630)	-	1461.7 ± 189.9 (1053-1776)	1686, 1494	1365
a	-	38.0 (31.0-44.0)	-	39.7 ± 4.0 (32.9-46.3)	44.0, 44.5	34.1
b	-	15 (11-21)	-	13.1 ± 1.1 (11.3-15.1)	15.4, 14.4	4.5
c	-	40.0 (32.0-58.0)	-	128.3 ± 19.2 (87.0-172.9)	-	-

Table 36. *(Continued.)*

Character	Holotype, Allotype, Japan, Itoh *et al.* (1969)	Paratypes, Japan, Itoh *et al.* (1969)	Tokyo, Japan, Okamoto *et al.* (1983)	Tuscany, Italy, Palmisano & Ambrogioni (2000)	Japan, Gu *et al.* (2013)	Japan, Gu *et al.* (2020)
W	36.0	38.0 ± 1.0 (30.0-47.0)	-	36.9 ± 4.3 (26.6-48.4)	38, 33.5	40
Labial region height	-	-	-	5.5 ± 0.5 (4.8-6.1)	-	-
Labial region diam.	-	-	-	10.4 ± 0.5 (9.7-10.9)	-	-
Stylet	19.0	20.0 (18.0-22.0)	-	19.4 ± 1.2 (17.5-23.0)	20.9, 22.2	-
Stylet knob width	-	-	-	3.9 ± 0.4 (3.0-4.8)	-	-
Stylet knob height	-	-	-	2.4 ± 0.1 (2.0-3.0)	-	-
DGO	8.0	8.0 (6.0-13.0)	-	6.3 ± 0.8 (4.8-8.5)	9.4, 7.9	8.9
Median bulb length	-	-	-	19.3 ± 3.0 (15.0-28.0)	-	-
Median bulb width	-	-	-	11.4 ± 1.6 (7.3-16.9)	-	-
Median bulb valve length		-	-	6.0 ± 0.9 (4.2-7.3)	-	-
Median bulb valve width			-	4.2 ± 0.6 (3.6-6.1)	-	-
Ant. end to excretory pore	-	-	-	147.1 ± 18.8 (97-189)	-	156.4
Spicules	35.0	32.0 (28.0-35.0)	-	33.8 ± 1.9 (30.0-37.5)	31, 33	34.9
Gubernaculum	10.0	8.5 (7.0-10.0)	-	9.0 ± 0.9 (7.3-9.8)	9.5, 11.0	9.6
Tail	-	-	-	11.5 ± 1.1 (8.5-13.3)	-	-
T	-	55 (34-65)	-	48.7 ± 9.7 (27.9-70.9)	-	-
J2 (n)	-	25	-	30	20	20
L	-	418 (390-450)	430 (370-490)	412.6 ± 20.6 (373-460)	425 ± 30.1 (362-466)	445 ± 28.3 (401-507)
a	-	28.5 (27.0-31.0)	-	29.5 ± 3.4 (22.3-35.5)	30.4 ± 2.6 (25.2-34.5)	31.9 ± 2.5 (27.4-36.6)
b	-	7.2 (6.0-8.0)	-	5.7 ± 0.3 (5.3-6.4)	8.0 ± 0.6 (6.8-9.2)	-
c	-	13.3 (12.0-15.0)	14.8 (13.6-16.4)	13.3 ± 1.2 (11.5-16.6)	13.2 ± 1.1 (11.6-15.3)	14.8 ± 2.5 (12.1-19.4)
c′	-	3.7 (3.0-5.0)	2.9 (2.6-3.2)	3.7 ± 0.5 (2.5-4.7)	4.2 ± 0.7 (3.1-5.6)	3.8 ± 0.6 (3.0-4.8)
W	-	14.5 (14.0-16.0)	-	14.2 ± 1.8 (12.1-18.2)	14.0 ± 1.1 (13.1-18.1)	14.0 ± 0.6 (13.1-15.4)

Table 36. *(Continued.)*

Character	Holotype, Allotype, Japan, Itoh *et al.* (1969)	Paratypes, Japan, Itoh *et al.* (1969)	Tokyo, Japan, Okamoto *et al.* (1983)	Tuscany, Italy, Palmisano & Ambrogioni (2000)	Japan, Gu *et al.* (2013)	Japan, Gu *et al.* (2020)
Labial region height	-	-	-	3.4 ± 0.3 (3.0-4.2)	2.6 ± 0.3 (2.1-3.2)	-
Labial region diam.	-	-	-	4.8 ± 0.2 (4.2-5.2)	5.1 ± 0.4 (4.1-5.7)	-
Stylet	-	14.0 (12.0-15.0)	-	10.0 ± 0.8 (8.5-11.1)	10.5 ± 0.5 (9.5-11.6)	10.4 ± 0.5 (9.4-11.1)
Stylet knob width	-	-	-	2.4 ± 0.1 (1.8-2.4)	2.1 ± 0.2 (1.8-2.7)	-
Stylet knob height	-	-	-	1.2	1.2 ± 0.2 (0.8-1.6)	-
DGO	-	4.7 (4.0-6.0)	-	3.6 ± 0.4 (3.0-4.8)	4.4 ± 0.6 (3.5-5.5)	3.2 ± 0.5 (2.4-4.0)
Median bulb length	-	-	-	12.6 ± 1.2 (10.4-16.9)	-	-
Median bulb width	-	-	-	7.9 ± 0.7 (6.7-9.7)	-	-
Median bulb valve width	-	-	-	3.6 ± 0.2 (3.0-4.2)	-	-
Median bulb valve length	-	-	-	3.9 ± 0.5 (3.3-4.8)	-	-
Ant. end to excretory pore	-	-	-	72.2 ± 5.1 (60-77)	74.1 ± 4.2 (68.8-82.3)	75.1 ± 4.2 (72.8-83.3)
Tail	-	31 (30-34)	37 (24-31)	31.3 ± 3.1 (24.2-37.5)	32.7 ± 3.0 (29.2-39.3)	30.5 ± 4.5 (23.5-35.8)
Hyaline region	-	-	-	8.2 ± 1.8 (4.8-12.7)	-	9.8 ± 1.6 (6.8-12.0)

view, opening of amphid visible between cap and first annulus. In portion immediately posterior to labial region, neck rapidly expanding to its max. followed by roughly same width until it joins spherical body. Labial skeleton furnished with a cylindrical stylet guide with short 'arms' protruding at level of labial region at front and expanding at base in neck. Perineal region forming a typical pattern of *Meloidogyne*, except for fact that a portion of cuticle surrounding anus and vulva is slightly elevated and, in consequence, vulval slit never distinctly observed at same focus as that of surrounding annulations at high magnification. Perineal pattern oval, made up of smooth striae finely spaced, dorsal arch low and flat, some transverse striae toward both ends of vulva from either side pronounced, and consequently vulval plate small, ventral arch similarly flat, tail terminus forming circular striae. Phasmids large, wider apart than length of vulval slit with measurement of former 22 (17-29) μm and latter 18 (12-24) μm, and a distinct striae always located bending downwards spanning phasmids. Lateral field clearly marked with single or double incisures. Distance of vulva from an imaginary line joining phasmids = 25 (19-31) μm, and closest distance between vulva and anus = 17 (14-22) μm.

Male

General structure typical of genus. When killed by gentle heat, body becoming straight and posteriorly twisted. Face view under SEM showed an oval labial disc and lateral labials absent or obscure (Okamoto *et al.*, 1983). Lateral field about one-third body diam. with four incisures and areolated on tail. Phasmids closely posterior to anus and tail length of 28-44 μm somewhat shorter than anal body diam. Excretory pore located 20 (7-26) annuli posterior to median pharyngeal bulb and one or two annuli posterior to hemizonid. Labial cap with six labials. Amphidial opening slit-like and situated at each lateral labial. In dorsoventral view, a constriction seen on lateral labials one-third from anterior end. Massive finely pointed stylet with rounded basal knobs. All specimens examined possessing single testis, sometimes reflexed, 788 (540-970) μm long. Spicules slightly curved ventrally with a bluntly rounded terminus. Gubernaculum short and crescentic.

J2

Body slender and clearly annulated. Lateral field consisting of four incisures, without transverse striae, and with outer bands crenate and wider than inner one. Labial region offset by a slight constriction, labial annulus smooth, labial cap shallow. Amphid position similar to that of male. Labial disc oval and fused to median labials which are fused to lateral labials in same contour (Okamoto *et al.*, 1983). Stylet very slender, with posteriorly sloping knobs. Median pharyngeal bulb oval. Hemizonid located 21 (15-26) annuli posterior to muscular bulb. Tail conoid, with irregular, rounded, and unstriated terminus. Rectum not swollen. Genital primordium, consisting of a small group of cells, located at about two-thirds distance from anterior end.

TYPE PLANT HOST: Apple, *Malus prunifolia* Borkh.

OTHER PLANTS: *Meloidogyne mali* is a polyphagous RKN that infects a wide range of host plants and is subject to phytosanitary restrictions in many countries. The nematode was found from the following plants in Japan: apple (*Malus pumila* Mill.) and apple root-stocks of *Malus* spp. including *M. sieboldii* Rehd, carrot (*Daucus carota* var. *sativa*), cherry (*Prunus yedoensis* Matsum.), sweet chestnut (*Castanea crenata* Sieb. *et* Zucc.), cucumber (*Cucumis sativus*), Japanese maple tree (*Acer palmatum* Thunb.), eggplant, fig tree (*Ficus carica*), grapevine (*Vitis vinifera*), mulberry (*Morus bombycis* Koidz.), pepper, rose (*Rosa hybrida* Hort.), soybean, tomato, and white clover (*Trifolium repens*) (Itoh *et al.*, 1969). Recently, *M. mali* has been repeatedly detected from Japanese maple and Crape myrtle (*Lagerstroemia indica*) imported from Japan in Ningbo port, China (Gu *et al.*, 2013; Gu & He, 2015). Other hosts included raspberry (*Rubus idaeus* L.), mountain ash (*Sorbus aucuparia* L.), common beech (*Fagus sylvatica* L.), common oak (*Quercus robur* L.), yew (*Taxus baccata* L.), small balsam (*Impatiens parviflora* DC.), dandelion (*Taraxacum officinale* F.H. Wigg.), common nettle (*Urtica dioica* L.), common fern (*Dryopteris filix-mas* (L.) Schott), red avens (*Geum coccineum* Lindl.), narrow buckler fern (*Dryopteris carthusiana* (Vill.) H.P. Fuchs), Herb Robert (*Geranium robertianum* L.) (Ahmed *et al.*, 2013) and elm (*Ulmus chenmoui* W.C.Cheng) (Palmisano & Ambrogioni, 2000).

TYPE LOCALITY: Toyono, Kamiminochi, Nagano Prefecture, Japan.

DISTRIBUTION: *Asia*: Japan; *Europe*: Italy, France, The Netherlands, UK; *North America*: USA (New York) (Itoh *et al.*, 1969; Toida, 1991; Palmisano & Ambrogioni, 2000; Karssen *et al.*, 2008; Gu & He, 2015; Prior *et al.*, 2019). Reports of *M. mali* from citrus and cottonwood in China remain questionable (Gu *et al.*, 2020).

PATHOGENICITY: This nematode has been regarded as one of the most important nematodes damaging apple trees in Northern Japan. Stunting and severe decline of infected trees are reported in Japan in orchards infested by this nematode (Itoh *et al.*, 1969; Nyczepir & Halbrendt, 1993). Mulberry has been damaged by *M. mali* infections in Japan, reducing plant growth and leaf weight by 10-20% (Toida, 1991). It has been also reported from a declining hedge of Manhattan

Euonymus (*Euonymus kiautschovicus* Loes.) in Harrison, New York, USA (Eisenback *et al.*, 2017).

BIOLOGY: *Meloidogyne mali* requires 18-22 weeks to complete one full generation on apple and does so only once in a year (Inagaki, 1978).

SYMPTOMS: *Meloidogyne mali* induces normal size galls (up to 0.5 cm in diam.) on young roots. However, on older roots these galls develop into relatively large galls (1-2 cm in diam.). Above-ground symptoms in trees are only visible when the trees become heavily infested when they show early leaf fall and reduced growth (Anon., 2018b).

CHROMOSOME NUMBER: *Meloidogyne mali* (2n = 22) was found to reproduce by amphimixis (Janssen *et al.*, 2017).

POLYTOMOUS KEY CODES: *Female*: A21, B23, C2, D4; *Male*: A32, B32, C21, D1, E2, F2; *J2*: A23, B21, C43, D423, E12, F34.

NOTE: *Meloidogyne ulmi* Palmisano & Ambrogioni, 2000 was synonymised with *M. mali* based on morphological and morphometric similarities, common hosts, as well as biochemical and molecular similarities at both protein and DNA levels (Holterman *et al.*, 2009; Ahmed *et al.*, 2013). Morphological and morphometric studies of holotypes and paratypes of *M. mali* and *M. ulmi* revealed important similarities in the major characters as well as some general variability in a few others (Ahmed *et al.*, 2013). Host test also showed that, besides the two species being able to parasitise the type hosts of the other, they share some other common hosts. Esterase and malate dehydrogenase isozyme phenotypes of some *M. ulmi* populations gave a perfectly comparable result to that already known for *M. mali* (Ahmed *et al.*, 2013). Finally, phylogenetic studies of their 18S and 28S rDNA sequence data revealed that the two species are not distinguishable at the DNA level (Ahmed *et al.*, 2013).

BIOCHEMICAL AND MOLECULAR CHARACTERISATION: *Meloidogyne mali* showed an esterase phenotype of weak single bands, corresponding to the VS1 type (Esbenshade & Triantaphyllou, 1985a). The MDH phenotype showed that some individuals gave single-banded patterns of the H1 type (Esbenshade & Triantaphyllou, 1985a), while others revealed a three-banded pattern, designated H3 (Ahmed *et al.*, 2013). Usually, the H1 type had two additional weaker bands at the same level as the upper two H3 bands. There was also an additional observation in the types of single bands some of the specimens produced. These single bands were positioned at the same level as the upper H3 band, which are given the name H1a (Ahmed *et al.*, 2013).

Molecularly this nematode can be differentiated from other RKN by ribosomal (18S, ITS and 28S rRNA gene) and mitochondrial (*COI*) gene sequences (Holterman *et al.*, 2009; Ahmed *et al.*, 2013; Gu *et al.*, 2013, 2020; Gu & He, 2015; Eisenback *et al.*, 2017; Janssen *et al.*, 2017). Zhou *et al.* (2017) developed a LAMP technique for quick diagnostics of *M. mali*.

RELATIONSHIPS (DIAGNOSIS): The species belongs to Molecular group VIII and is clearly molecularly differentiated from all other *Meloidogyne* species. *Meloidogyne mali* resembles *M. arenaria* and *M. ardenensis* in the general form of female perineal pattern, but differs in the length of the vulval slit and the distance between the phasmids from the latter two species. The J2 of *M. mali* can be distinguished from other species by shorter body and tail length.

55. *Meloidogyne maritima* Jepson, 1987
(Figs 157-159)

COMMON NAME: Maritime root-knot nematode.

Jepson (1987) described this species from roots of *Ammophila arenaria* (L.) Link on dunes at Perranporth, Cornwall, UK.

MEASUREMENTS

Table 37.

- *Eggs* (n = 30): L = 105.4 (99-122) μm; W = 45.8 (42-50) μm; L/W = 2.3 (2.1-2.7).

Fig. 157. Meloidogyne maritima. *A: Entire second-stage juvenile (J2); B: J2 median bulb; C-E: J2 tail; F: Male labial region; G: Spicules; H: Female stylet; I: Female anterior region; J: Entire female. (After Karssen* et al.*, 1998a.) K: Perineal pattern. (After Jepson, 1987.)*

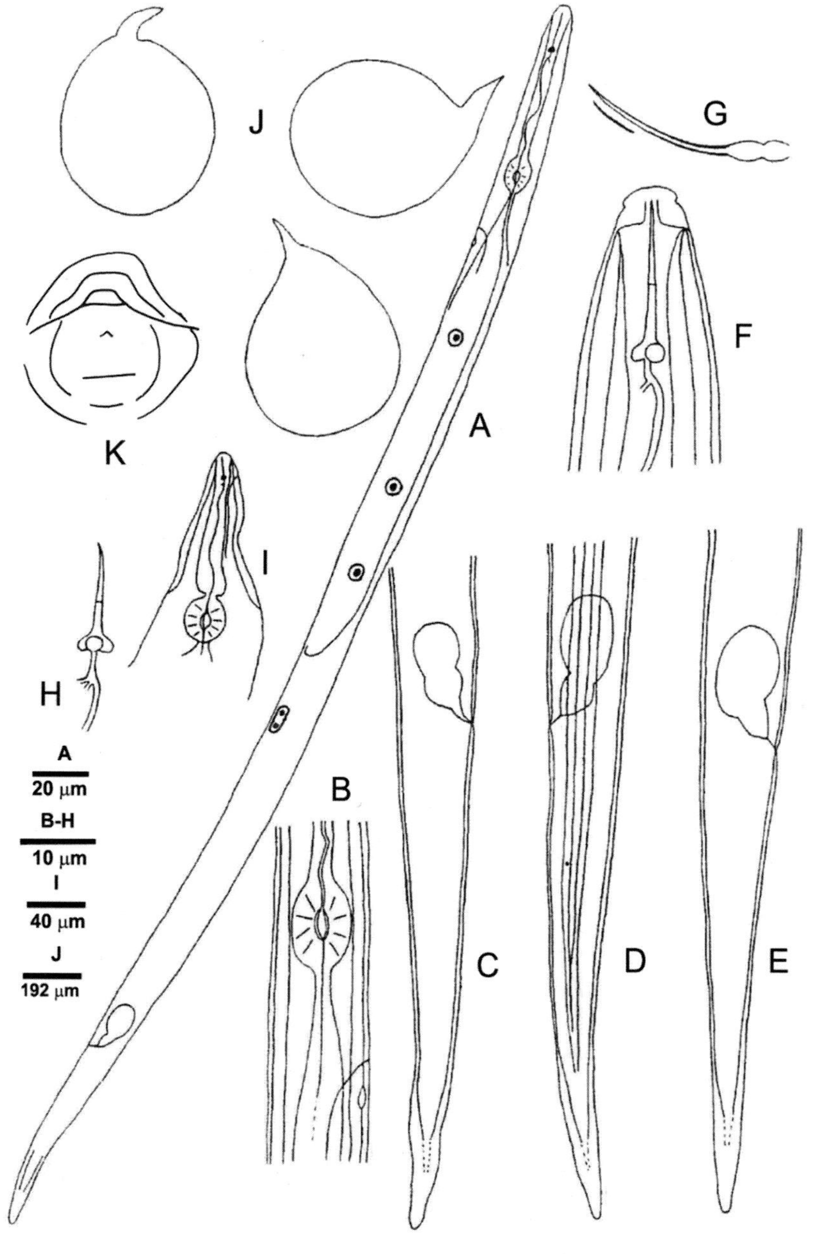
J
G
F
A
K
I
H
A
20 μm
B-H
10 μm
I
40 μm
J
192 μm
B
C
D
E

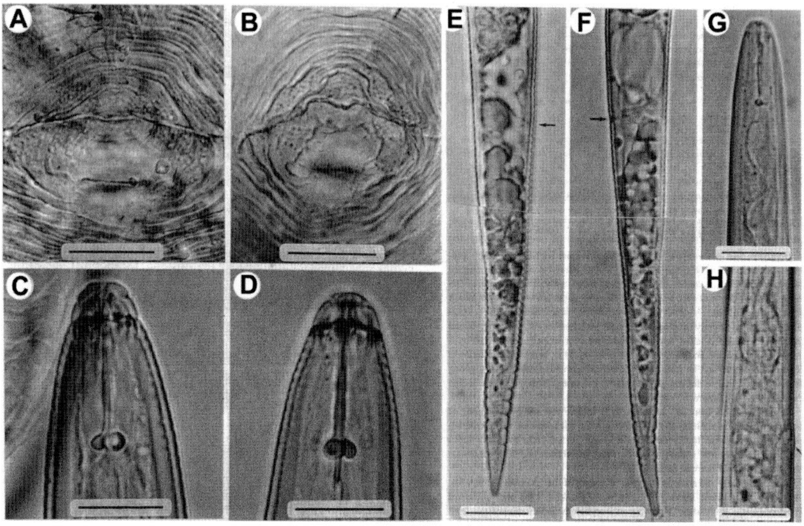

Fig. 158. Meloidogyne maritima. *LM. A, B: Perineal pattern; C, D: Male labial region; E, F: Second-stage juvenile (J2) tail; G: J2 labial region; H: J2 metacorpus. (Scale bars: A, B = 25 μm; C-H = 10 μm.) (After Karssen* et al., *1998a.)*

DESCRIPTION

Female

Body pearly white, globular to elongate with a neck relatively small but distinct, cuticle at the base thickened, with labial region not offset, vulva on a protuberance. Labial disc slightly elevated. Labial framework distinct, weakly sclerotised, vestibule extension distinct. Stylet cone about half stylet length with anterior dorsal curvature, knobs offset from shaft, smooth and rounded to slightly transversely ovoid in lateral view. Excretory pore near level of stylet base. Median bulb with vesicles. Perineal pattern small and rounded, inner part with coarse, widely spaced striae. There is a deep anal fold. Lateral lines strongly marked as folded and twisted junction of dorsal and ventral striae, these continue inwards, dividing and encircling tail region. Dorsal arch rounded, rarely squared, and moderately high with striae forming arch very widely spaced. Perivulval region free of striae, area dorsal to anus free of striae. Phasmids small and closer together than width of vulva.

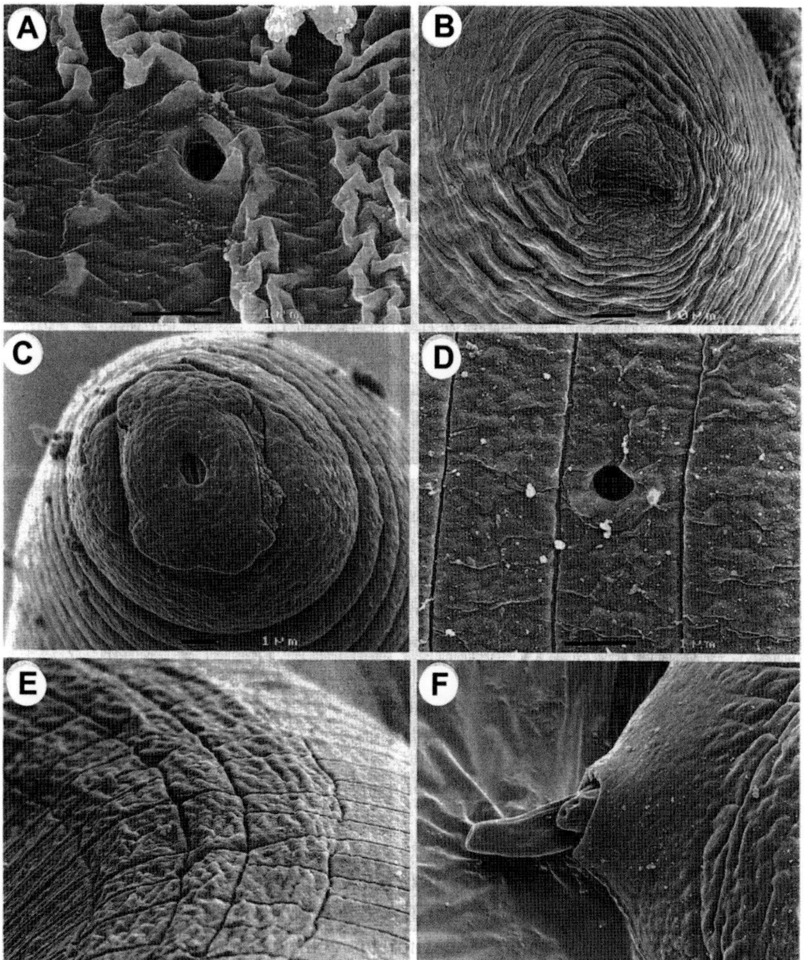

Fig. 159. Meloidogyne maritima. *SEM. A: Female excretory pore; B: Perineal pattern; C: Male* en face *view; D: Male excretory pore; E: Male lateral field; F: Cloacal region with spicule. (Scale bars = 1 μm.) (After Karssen* et al.*, 1998a.)*

Male

Cuticle with transverse annulations (annuli about 2 μm wide). Lateral field with four incisures, middle and outer bands regularly areolated. Labial region not offset, anteriorly tapering with shallowly rounded labial cap, single post labial annulus. Labial disc oval, elevated, fused with median labials. Relatively long slit-like amphidial opening present

Table 37. *Morphometrics of females, males and second-stage juveniles (J2) of* Meloidogyne maritima. *All measurements are in µm and in the form: mean ± s.d. (range).*

Character	Holotype, UK, Jepson (1987)	Paratypes, UK, Jepson (1987)	The Netherlands Karssen *et al.* (1998a)
Female (n)	-	5	30
L	-	631 ± 116.6 (484-785)	748 ± 136 (496-896)
a	-	1.5 ± 0.2 (1.2-1.8)	1.4 ± 0.1 (1.3-1.7)
W	-	432 ± 70.7 (309-484)	530 ± 89 (352-624)
Stylet	-	13.5	14.2 ± 0.5 (14-15)
Stylet knob width	-	3.6	-
Stylet knob height	-	1.8	-
DGO	-	3.5	3.6 ± 0.3 (3.0-4.0)
Median bulb length	-	-	30.7 ± 3.3 (26.5-37.9)
Median bulb width	-	-	28.5 ± 4.6 (22.8-37.9)
Median bulb valve width	-	-	9.1 ± 1.8 (5.1-12.0)
Median bulb valve length	-	12.6	11.3 ± 0.4 (10.7-12.0)
Ant. end to excretory pore	-	23.4	15 ± 2.9 (9.5-19.0)
Vulval slit	-	-	22.2 ± 1.5 (19.2-24.0)
Vulval slit to anus distance	-	-	17.9 ± 2.1 (16.0-22.4)
EP/ST	-	1.7	1.1
Male (n)	-	19	30
L	-	1187 ± 106.7 (1037-1462)	1034 ± 157 (749-1357)
a	-	-	33.3 ± 3.0 (27.6-38.4)

Table 37. *(Continued.)*

Character	Holotype, UK, Jepson (1987)	Paratypes, UK, Jepson (1987)	The Netherlands Karssen *et al.* (1998a)
c	-	-	86 ± 14.4 (66.0-109)
W	-	-	31.1 ± 11.1 (24.0-35.5)
Labial region height	-	-	9.8 ± 0.5 (8.9-10.1)
Labial region diam.	-	-	5.1 ± 0.8 (4.4-7.0)
Stylet	-	21.2 ± 0.8 (18.0-22.5)	20.5 ± 0.7 (20.0-22.0)
Stylet knob width	-	5.3 ± 0.5 (4.5-6.3)	5.1 ± 0.3 (4.4-5.7)
Stylet knob height	-	2.0 ± 0.2 (1.8-2.3)	2.6 ± 0.3 (2.5-3.2)
DGO	-	3.0 ± 0.4 (2.5-3.5)	2.8 ± 0.4 (2.5-3.2)
Median bulb valve length	-	4.8 ± 0.5 (4.1-5.4)	-
Ant. end to excretory pore	-	129.6 ± 4.6 (122-138)	109 ± 12.6 (90-139)
Spicules	-	30.4 ± 2.6 (26.2-34.6)	28.9 ± 0.5 (28-30)
Gubernaculum	-	-	7.3 ± 0.5 (6.0-8.0)
Tail	-	-	12.1 ± 0.7 (11.0-13.0)
T	-	-	49.2 ± 9.5 (25.5-67.2)
J2 (n)	1	20	30
L	400	424.6 ± 23.7 (388-486)	471 ± 21 (442-512)
a	-	-	27.8 ± 1.6 (25.8-31.0)
c	-	-	6.6 ± 0.2 (6.2-7.0)

Table 37. *(Continued.)*

Character	Holotype, UK, Jepson (1987)	Paratypes, UK, Jepson (1987)	The Netherlands Karssen *et al.* (1998a)
c′	-	-	5.8 ± 0.3 (5.2-6.3)
W	-	-	17.0 ± 0.7 (16.0-18.0)
Labial region height	-	-	2.6 ± 0.4 (1.9-3.2)
Labial region diam.	-	-	5.9 ± 0.3 (5.7-6.2)
Stylet	-	-	12.4 ± 0.3 (12.0-13.0)
Stylet knob width	-	2.8 ± 0.2 (2.7-3.2)	2.9 ± 0.4 (2.5-3.2)
Stylet knob height	-	1.4 ± 0.3 (0.9-1.8)	1.6 ± 0.3 (1.3-1.9)
DGO	3.0	3.5 ± 0.3 (2.7-3.6)	2.9 ± 0.3 (2.5-3.5)
Median bulb valve width	-	-	3.1 ± 0.2 (2.5-3.2)
Median bulb valve length	-	-	3.7 ± 0.2 (3.2-3.8)
Ant. end to excretory pore	81.0	76.5 ± 4.1 (65.7-82.8)	81 ± 3.0 (75.0-87.0)
Tail	-	63.1 ± 2.4 (59.4-68.4)	71.5 ± 2.5 (66.0-76.0)
Hyaline region	12.0	13.6 ± 2.0 (9.0-17.1)	-

between labial disc and lateral side of post-labial annulus. Stylet with cone about half total length, knobs offset and projecting anteriorly to make an acute angle with shaft, each knob smooth, transversely ovoid with anterior surface concave. Hemizonid *ca* 4 μm long, anterior to excretory pore. Pharyngeal gland lobe overlapping intestine ventrally, two subventral gland nuclei present. Single testis, variable in length, rarely reflexed. Tail rounded, short and twisted. Spicules slender and curved ventrally, two pores on each spicule tip. Phasmids near level of cloacal aperture.

J2

Cuticle with transverse annulations (about 1 μm). Lateral field with four incisures, weakly areolated. Labial region rounded to truncate, not offset from body, with labial cap and single post labial annulus. Labial framework weakly sclerotised, vestibule extension distinct. Stylet with cone about half total length, knobs rounded and slightly sloping posteriorly, offset from shaft. Hemizonid situated posterior to excretory pore. Tail relatively long, tapering and conical with constriction near terminus, delimiting end portion which is finely rounded, rectum dilated. Phasmids relatively small but distinct, posterior to anus near ventral incisure of lateral field.

TYPE PLANT HOST: Marram grass, *Ammophila arenaria* (L.). Link.

OTHER PLANTS: No other host reported. Since roots of *Festuca* spp. and *Carex* spp. were concomitant with rhizomes of marram grass, these plants need to be verified as hosts.

TYPE LOCALITY: Perranporth, Cornwall, UK.

DISTRIBUTION: *Europe*: Belgium, France, Germany, The Netherlands, UK (Sturhan, 1976; Brinkman, 1985; Karssen *et al.*, 1998a; Karssen, 2002; Wesemael *et al.*, 2011). This nematode has been recorded from coastal sand dunes.

PATHOGENICITY: The damage caused to marram grass has not been assessed.

CHROMOSOME NUMBER: No available data.

POLYTOMOUS KEY CODES: *Female*: A23, B32, C3, D5; *Male*: A34, B23, C213, D1, E2, F1; *J2*: A23, B2, C23, D213, E3, F21.

BIOCHEMICAL AND MOLECULAR CHARACTERISATION: Karssen *et al.* (1998a) described the esterase (VS1-S1 type) and malate dehydrogenase (N1a type) patterns for *M. maritima*. Ribosomal 18S rRNA gene sequences are provided by Holterman *et al.* (2009).

RELATIONSHIPS (DIAGNOSIS): The species belongs to Molecular group VI and is clearly molecularly differentiated from all other *Meloidogyne* species. *Meloidogyne maritima* differs from *M. duytsi* by an anterior hemizonid position and a distinct hyaline region, a non-tapering male labial region with large transversely ovoid knobs, offset from the shaft and a shorter DGO (2.5-3.5 *vs* 3.8-5.1 μm), and a perineal pattern without distinct lateral lines and different malate dehydrogenase and esterase patterns (Karssen *et al.*, 1998a). Morphologically and ecologically it is related to species of the *graminis* group, like *M. graminis*, *M. kralli*, *M. naasi* and *M. sasseri*, which differ in morphology and morphometrics from *M. maritima* (Karssen *et al.*, 1998a).

56. *Meloidogyne marylandi* Jepson & Golden, 1987 in Jepson, 1987
(Figs 160, 161)

COMMON NAME: Maryland grass root-knot nematode.

Jepson & Golden (1987) described this species from roots of Bermuda grass (*Cynodon dactylon*) in College Park, Maryland, USA.

MEASUREMENTS

See Table 38.

- *Eggs* (n = 30): L = 91.6 ± 5.2 (80-100) μm; W = 46.6 ± 4.2 (40-53) μm; L/W = 2.0 ± 0.2 (1.6-2.4).

DESCRIPTION

Female

Body pearly white, old specimens sometimes light brown. Shape globular to elongate, neck distinct, often to one side of a median plane through vulva. Vulva located on small protuberance. Stylet short. Stylet knobs distinct, rounded, usually sloping posteriorly. Labial region not

Fig. 160. Meloidogyne marylandi. *LM. A, B: Female anterior region; C-F: Female labial region; G: Entire second-stage juvenile (J2); H: J2 lateral field; I, J: Mid-body sections showing lateral field; K: J2 anterior region; L, M: J2 tail. (After Golden, 1989.)*

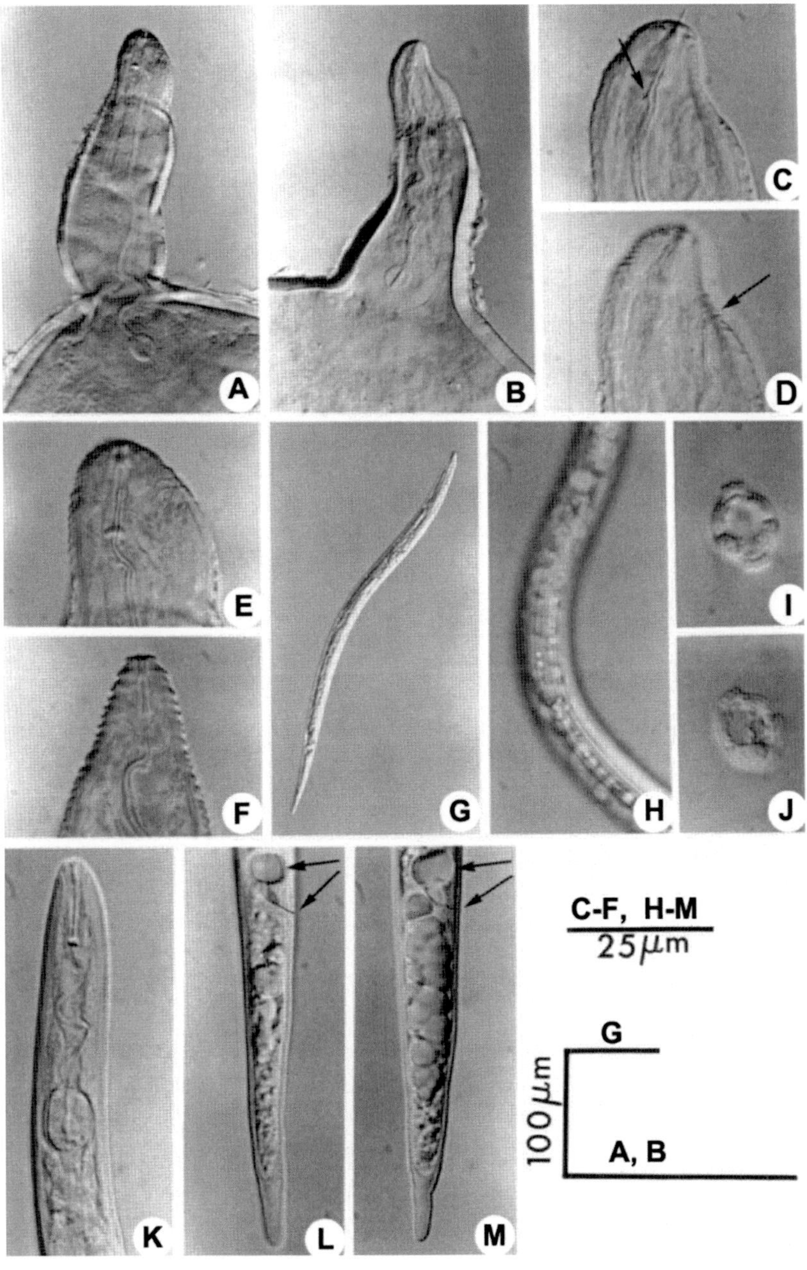
A
B
C
D
E
F
G
H
I
J
K
L
M
C-F, H-M
25μm
G
100μm
A, B

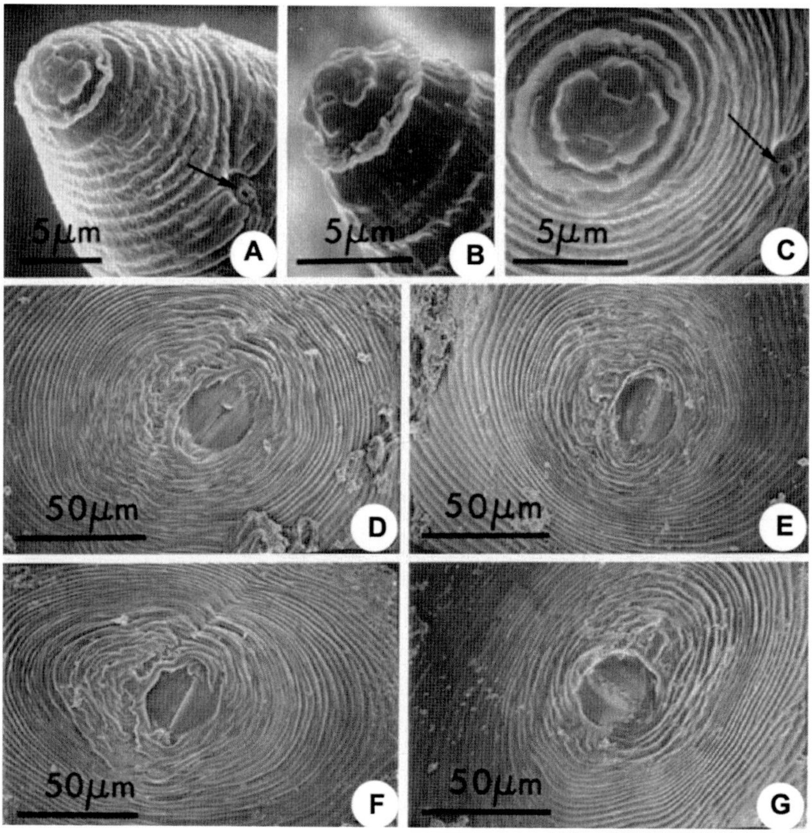

Fig. 161. Meloidogyne marylandi. *SEM. A-C: Female anterior region; D-G: Perineal pattern. Excretory pore arrowed. (After Golden, 1989.)*

offset, variable in shape, labial framework distinct, labial cap usually prominent with one labial annulus. Excretory pore usually near base of retracted stylet. Body cuticle thick, thinner near anterior end of neck. Median bulb prominent, usually near base of neck or sometimes posterior. Perineal pattern ovoid to round. Striae wavy, coarse, well spaced, wavy striae usually forming a low, rounded dorsal arch, some patterns with a higher, squarish dorsal arch. Perivulval region without striae, although occasionally some striae extending to each end of vulva.

Male

Not found.

Table 38. *Morphometrics of females and second-stage juveniles (J2) of* Meloidogyne marylandi. *All measurements are in μm and in the form: mean ± s.d. (range).*

Character	Holotype, Maryland, USA, Jepson & Golden (1987)	Paratypes, Maryland, USA, Jepson & Golden (1987)	Zoysia culture Calverton, Maryland, USA, Golden (1989)	*Panicum crus-galli*, Japan, Araki (1992b)	*Avena strigosa*, Israel, Oka *et al.* (2003)
Female (n)	-	25	30	10	-
L	-	747.4 ± 110.6 (525-923)	566.5 ± 64.3 (427-692)	-	611.4 ± 68.6 (500-721)
a	-	2.5 ± 0.4 (1.8-3.3)	1.8 ± 0.2 (1.5-2.3)	-	-
W	-	297.0 ± 46.8 (208-421)	314.4 ± 42.1 (248-427)	-	362.6 ± 37.5 (280-400)
Stylet	-	14.4 ± 0.3 (14.2-4.8)	13.5 ± 0.4 (12.8-14)	13.9 ± 0.7 (12.8-15.2)	12.5 ± 0.7 (11.8-14.6)
Stylet knob width	-	-	4.2 ± 0.4 (3.5-4.7)	3.9 ± 0.2 (3.6-4.2)	3.4 ± 0.4 (3.1-4.3)
DGO	-	3.9 ± 0.4 (3.5-4.7)	2.5 ± 0.4 (1.8-2.9)	3.9 ± 0.6 (2.6-4.9)	3.3 ± 0.5 (2.6-4.3)
Ant. end to excretory pore	-	13.5 ± 3.5 (8.8-19.5)	-	16.5 ± 3.4 (12.4-24.3)	11.9 ± 1.3 (9.8-13.4)
Vulval slit	-	25.1 (20.7-29.5)	24.3 ± 2.3 (20.6-26.5)	-	23.2 ± 1.7 (19.5-26.8)
Vulval slit to anus distance	-	12.8 (11.8-15.3)	13.7 ± 2.3 (11.8-16.5)	-	12.1 ± 0.8 (10.9-13.4)
EP/ST	-	0.9	-	1.1	1.0
J2 (n)	1	40	37	100	-
L	412	424.6 ± 20.5 (385-489)	370.1 ± 13.7 (324-398)	392.3 ± 22.1 (339-449)	422.9 ± 13.5 (396-446)
a	31.7	28.8 ± 1.5 (25.2-33.1)	24.5 ± 2.4 (20.1-28.4)	25 ± 1.3 (22.3-28.4)	28.8 ± 1.7 (26.2-31.3)
b	2.2	2.2 ± 0.2 (1.9-2.7)	2.0 ± 0.2 (1.8-2.6)	-	-
c	6.9	7.0 ± 0.3 (6.6-7.7)	6.7 ± 0.3 (5.9-7.4)	6.2 ± 0.3 (5.5-7.1)	6.8 ± 0.4 (6.3-7.6)
W	13	14.8 ± 0.6 (13.6-17.1)	15.2 ± 1.7 (13.0-18.3)	15.7 ± 0.4 (14.2-16.6)	14.8 ± 0.8 (13.4-15.9)
Labial region height		2.3 ± 0.1 (1.8-2.4)	5.0 ± 0.3 (4.1-5.3)	-	-

Table 38. *(Continued.)*

Character	Holotype, Maryland, USA, Jepson & Golden (1987)	Paratypes, Maryland, USA, Jepson & Golden (1987)	Zoysia culture Calverton, Maryland, USA, Golden (1989)	*Panicum crus-galli*, Japan, Araki (1992b)	*Avena strigosa*, Israel, Oka *et al.* (2003)
Labial region diam.		5.3 ± 0.1 (4.7-5.4)	2.3 ± 0.2 (1.8-2.4)	-	-
Stylet	12	11.4 ± 0.3 (10.8-11.8)	11.2 ± 0.4 (10.0-11.8)	12.9 ± 0.5 (11.5-14.1)	11 ± 0.4 (10.4-11.9)
DGO	2.5	2.4 ± 0.2 (2.4-3.0)	2.4 ± 0.2 (1.8-2.9)	2.5 ± 0.3 (1.8-3.1)	2.6 ± 0.4 (2.2-2.9)
Ant. end to excretory pore	68	69.2 ± 1.9 (66.1-73.0)	-	-	-
Tail	59.6	60.6 ± 3.4 (52.5-68.4)	54.9 ± 3.6 (46.0-61.9)	63.7 ± 2.7 (54.0-70.5)	63.1 ± 3.1 (60.4-69.3)
Hyaline region	11.8	11.8 ± 1.1 (9.4-13.8)	11.4 ± 0.9 (9.4-12.9)	12.3 ± 1.1 (9.3-14.7)	12.4 ± 0.9 (9.9-13.4)

J2

Body vermiform, tapering at both extremities, but more so posteriorly. Labial region not offset, with labial disc, without annulation. Stylet delicate, with small rounded knobs. Cuticular annulation fine, distinct. Lateral field prominent, with four incisures, some areolation, especially in anterior portion. Excretory pore 3-4 annuli anterior to hemizonid. Rectum dilated. Phasmids small, indistinct, at about 70% of tail length from terminus. Tail tapering gradually, terminus bluntly rounded.

TYPE PLANT HOST: Bermuda grass, *Cynodon dactylon* (L.) Pers.

OTHER PLANTS: It has been reported in *Zoysia* sp. and bermuda and turf-grasses (Oka *et al.*, 2003). Host range tests showed that the turfgrasses *Stenotaphrum secundatum*, *Dactyloctenium australe* Steud., *Paspalum vaginatum* Sw., corn (*Zea mays*) and oat (*Avena sativa* L.) were non-hosts or resistant, whereas Kikuyu grass (*Pennisetum clandestinum* Hochst. *ex* Chiov.), wheat (*Triticum aestivum*), barley (*Hordeum vulgare*), bristle oat (*A. strigosa* Schreb.), Siberian millet (*Echinochloa frumentaceae* Link) and pearl millet (*P. glaucum* (L.) R.Br.) were susceptible to the nematode (Oka *et al.*, 2003). Faske & Starr (2009) recorded that this nematode reproduced on several grass species (Poaceae) including *Buchloe dactyloides* (Nutt.) Engelm. (buffalograss), *E. colona* (jungle rice), *Eragrostis curvula* (Schrad.) Nees (weeping lovegrass), *Paspalum dilatatum* Poir. (dallisgrass), *P. notatum* Flüggé (bahiagrass), *Sorghastrum nutans* (L.) Nash (indiangrass), *Tripsacum dactyloides* L. (eastern gamagrass), and *Zoysia matrella* (zoysiagrass). Mwamula *et al.* (2021) found this species on *Poa pratensis* and *Z. japonica*.

TYPE LOCALITY: College Park, Maryland, USA.

DISTRIBUTION: *Asia*: Japan (Araki, 1992b; Orui, 1998), Israel (Oka *et al.*, 2003, 2004), South Korea (Mwamula *et al.*, 2021); *North America*: USA (Maryland, Florida (Sekora *et al.*, 2012), Arkansas (Elmi *et al.*, 2000; Khanal *et al.*, 2016b), Arizona, California, Hawaii, Nevada, and Utah (McClure *et al.*, 2012), Oklahoma (Walker, 2014), North Carolina, South Carolina (Ye *et al.*, 2015), Texas; *Central America*: Costa Rica (Salazar *et al.*, 2013).

PATHOGENICITY: Damage caused to turfgrass has been recorded under field conditions (McClure *et al.*, 2012).

CHROMOSOME NUMBER: No available data.

POLYTOMOUS KEY CODES: *Female*: A213, B32, C4, D3; *Male*: (not found); *J2*: A23, B32, C23, D32, E3, F32.

BIOCHEMICAL AND MOLECULAR CHARACTERISATION: Oka *et al.* (2003) described the esterase (VS1 type) and malate dehydrogenase (N1c type) patterns for *M. marylandi*.

Molecularly this nematode can be differentiated from other RKN by the D2-D3 expansion segments of 28S and ITS of rRNA (McClure *et al.*, 2012), and the mitochondrial region between the *COII* and the 16S RNA genes (McClure *et al.*, 2012; Ye *et al.*, 2015). PCR-RFLP of *COII* and the 16S RNA gene fragment with restriction enzyme SspI was able to distinguish populations of *M. graminis* from *M. marylandi*, providing a fast and inexpensive method for diagnosis of these nematodes from turf (McClure *et al.*, 2012). The RFLP profile for this species was also given by Orui (1998).

RELATIONSHIPS (DIAGNOSIS): The species belongs to Molecular group IV and is clearly molecularly differentiated from all other *Meloidogyne* species. *Meloidogyne marylandi* differs from *M. graminis*, *M. maritima*, *M. aquatilis* and *M. naasi* by length of tail and hyaline region for J2, perineal pattern and other characters. Males unknown in *M. marylandi* and present in all other four species (Golden, 1989).

57. *Meloidogyne megadora* Whitehead, 1968

(Figs 162, 163)

COMMON NAME: Angolan coffee root-knot nematode.

Whitehead (1968) described this species from roots of robusta coffee at the Coffee Research Station, Amboim, Angola, Africa. *Meloidogyne megadora* has been found in several African countries.

MEASUREMENTS

See Table 39.

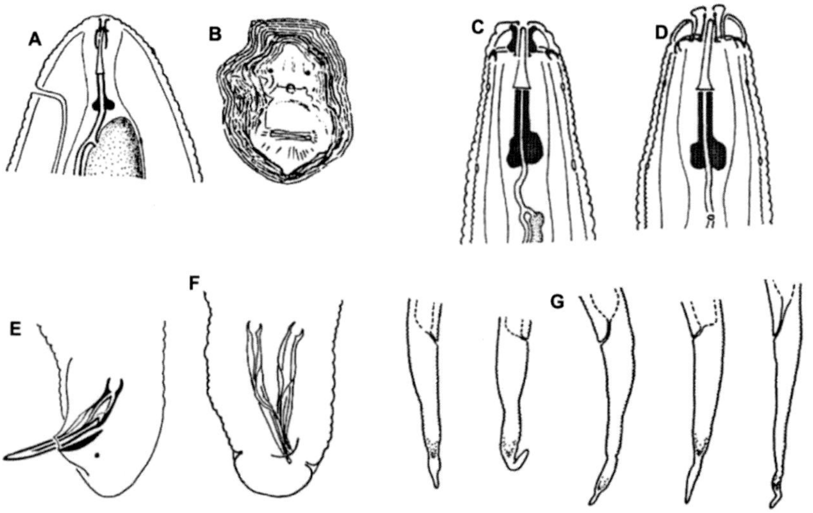

Fig. 162. Meloidogyne megadora. *A: Female labial region; B: Perineal pattern; C, D: Male labial region; E, F: Male tail; G: Second-stage juvenile tail. (After Whitehead, 1968.)*

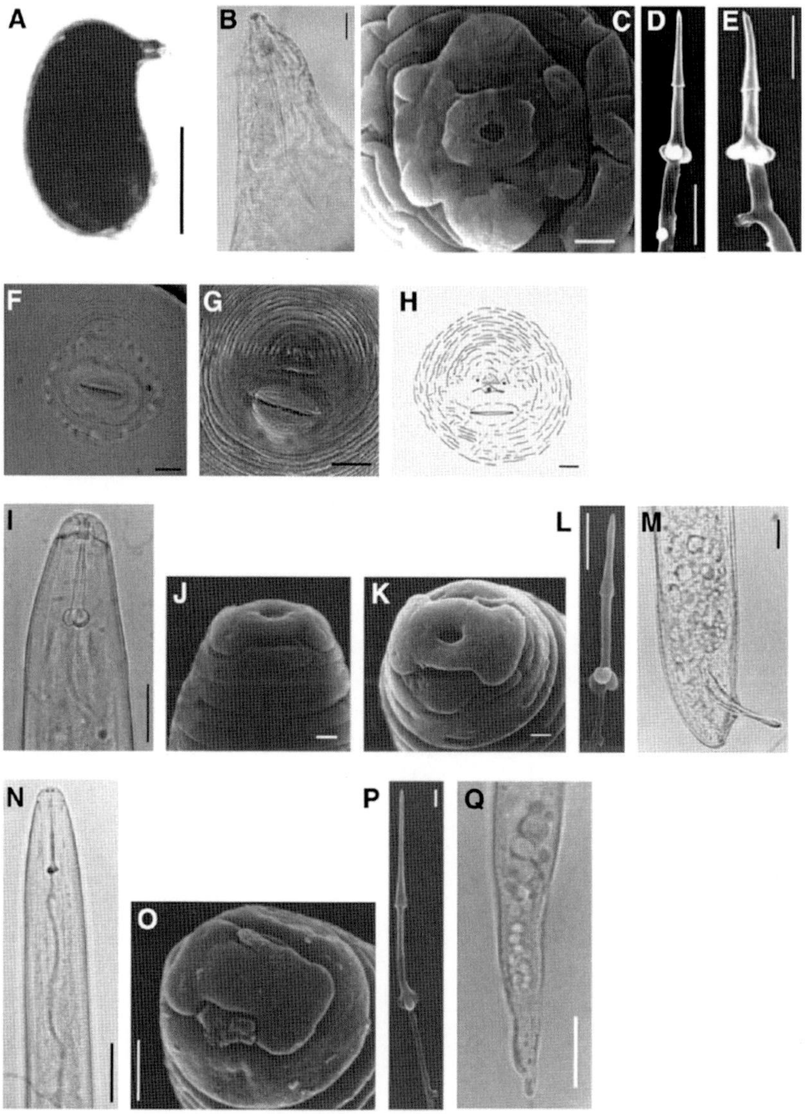

Fig. 163. Meloidogyne megadora. *LM and SEM. A: Entire female; B: Female labial region; C: Female* en face *view; D, E: Female stylet; F-H: Perineal pattern; I-K: Male labial region; L: Male stylet; M: Male tail; N: Second-stage juvenile (J2) anterior region; O: J2* en face *view; P: J2 stylet; Q: J2 tail. (Scale bars: A = 500 μm; B, I, M, N, Q = 10 μm; C, J, K, O, P = 1 μm; D, E, L = 5 μm; F, G, H = 20 μm.) (After Maleita* et al., *2016.)*

Table 39. *Morphometrics of females, males and second-stage juveniles (J2) of* Meloidogyne megadora. *All measurements are in μm and in the form: mean ± s.d. (range).*

Character	Paratypes, Amboim, Angola, Whitehead (1968)	Democratic Republic of São Tomé and Príncipe, Coffee Maleita *et al.* (2016)	Democratic Republic of São Tomé and Príncipe, Bean Maleita *et al.* (2016)
Female (n)	12	20	20
L	683 ± 87 (554-845)	1118 ± 203 (830-1520)	112 ± 104.2 (925-1360)
a	1.5 ± 0.2 (1.1-1.8)	2.0 ± 0.3 (1.6-2.9)	2.0 ± 0.2 (1.7-2.4)
W	-	570 ± 81 (440-750)	568 ± 71.3 (450-720)
Stylet	15.0 (13.0-17.0)	15.4 ± 0.9 (14.0-16.5)	14.7 ± 0.6 (13.0-15.3)
Stylet knob width	-	4.4 ± 0.4 (4.0-5.0)	3.9 ± 0.2 (3.3-4.0)
Stylet knob height	-	2.6 ± 0.4 (2.0-3.0)	2.2 ± 0.2 (2.0-2.8)
DGO	6.0 (4.0-8.0)	5.8 ± 0.6 (5.0-6.5)	5.4 ± 0.6 (4.0-6.5)
Median bulb length	51 (40-63)	-	-
Median bulb width	43 (34-58)	-	-
Median bulb valve width	12.0 (10.0-14.0)	12.9 ± 1.0 (11.5-15.0)	12.7 ± 1.7 (11.0-17.0)
Median bulb valve length	16.0 (14.0-19.0)	16.1 ± 1.6 (12.5-19.0)	15.9 ± 1.5 (13.0-19.0)
Ant. end to excretory pore	-	33.3 ± 4.3 (26.0-40.0)	38.5 ± 5.3 (32.0-55.0)
Vulval slit	-	41.5 ± 6.2 (30.0-51.0)	36.5 ± 6.0 (25.0-52.0)
Vulval slit to anus distance	-	26.6 ± 4.1 (17-33)	24.4 ± 3.2 (18.0-30.0)
Interphasmid distance	-	25.5 ± 4.2 (19.0-34.0)	22 ± 4.7 (15.0-30.0)
EP/ST	-	2.2	-

Table 39. *(Continued.)*

Character	Paratypes, Amboim, Angola, Whitehead (1968)	Democratic Republic of São Tomé and Príncipe, Coffee Maleita *et al.* (2016)	Democratic Republic of São Tomé and Príncipe, Bean Maleita *et al.* (2016)
Male (n)	25	20	20
L	1906 ± 330 (905-2277)	2280 ± 369 (1170-2720)	1864 ± 405.7 (1000-2520)
a	52.8 ± 7.06 (36.9-62.8)	49.3 ± 9.4 (35.5-67.4)	40.1 ± 3.8 (33.2-45.8)
b	22.1 ± 5.9 (11.1-44.5)	-	-
c	137 ± 29.9 (98-218)	161.4 ± 31.5 (97.5-237.0)	126.1 ± 20.7 (93.3-174.2)
W	-	47.1 ± 8.6 (33.0-60.5)	46.3 ± 8.4 (28-59.5)
Labial region height	5.2 ± 0.5 (4.3-6.1)	3.3 ± 0.4 (2.5-4.0)	2.6 ± 0.3 (2.0-3.0)
Labial region diam.	-	10.7 ± 0.9 (8.5-12.5)	10.6 ± 0.9 (8.0-12)
Stylet	20.4 ± 1.1 (18.3-21.9)	21.3 ± 1.1 (18-23)	20.1 ± 1.1 (18.5-21.5)
Stylet knob width	-	5.7 ± 0.4 (5.0-6.0)	5.1 ± 0.4 (4.5-5.5)
Stylet knob height	-	3.7 ± 0.4 (3.0-4.5)	3.3 ± 0.3 (2.5-3.5)
DGO	6.5 ± 1.2 (4.0-8.3)	6.1 ± 1.2 (4.0-8.0)	6.1 ± 0.5 (5.0-7.0)
Median bulb length	21.9 ± 3.9 (14.4-32.4)	-	-
Median bulb width	10.5 ± 1.9 (7.6-15.8)	11.5 ± 1.8 (7.0-14.0)	12.8 ± 1.1 (10.5-15.0)
Median bulb valve length	-	7.0 ± 0.8 (6.0-8.5)	7.0 ± 1.1 (5.0-8.5)
Median bulb valve width	7.3 ± 1.0 (5.0-8.6)	5.2 ± 0.6 (4.0-6.0)	4.8 ± 0.5 (4.0-6.0)
Ant. end to excretory pore	-	159.0 ± 32.8 (89.0-196.0)	149.7 ± 22.6 (82.0-189.0)

Table 39. *(Continued.)*

Character	Paratypes, Amboim, Angola, Whitehead (1968)	Democratic Republic of São Tomé and Príncipe, Coffee Maleita *et al.* (2016)	Democratic Republic of São Tomé and Príncipe, Bean Maleita *et al.* (2016)
Spicules	32.6 ± 2.9 (25.2-36.0)	38.5 ± 3.3 (30.0-43.0)	37.4 ± 2.3 (32.5-43.0)
Gubernaculum	10.6 ± 0.9 (9.4-11.9)	9.9 ± 0.9 (8.0-12.0)	9.0 ± 0.8 (7.5-10.0)
Tail	-	14.4 ± 2.3 (10.0-18.5)	14.8 ± 3.2 (10.0-18.5)
J2 (n)	26	20	20
L	451 ± 27 (413-548)	434 ± 28 (370-485)	471 ± 25.1 (410-510)
a	27.8 ± 2.3 (23.1-32.9)	27.1 ± 2.2 (23.3-32.3)	29.0 ± 2.9 (22.8-33.0)
b	2.5 ± 0.2 (2.8-3.0)	-	-
b′	7.8 ± 0.4 (7.2-8.6)	-	-
c	8.4 ± 0.6 (7.6-11)	7.5 ± 0.8 (5.8-9.5)	8.2 ± 0.9 (6.5-9.6)
c′	5.0 ± 0.5 (4.3-6.2)	-	-
W	-	16.1 ± 0.9 (15.0-18.0)	16.3 ± 1.1 (15.0-18.5)
Labial region height	-	2.1 ± 0.2 (1.8-2.5)	1.9 ± 0.2 (1.5-2.0)
Labial region diam.	-	5.7 ± 0.4 (5.0-6.5)	5.5 ± 0.4 (5.0-6.0)
Stylet	12.0 ± 0.6 (10.7-13.2)	12.5 ± 1.0 (11.0-15.0)	11.5 ± 0.7 (10.3-13.0)
Stylet knob width	-	2.3 ± 0.2 (2.0-2.5)	2.0 ± 0.2 (1.5-2.5)
Stylet knob height	-	1.7 ± 0.2 (1.2-2.0)	1.4 ± 0.2 (1.0-1.5)
DGO	3.9 ± 1.1 (2.3-4.8)	4.1 ± 0.5 (3.5-5.0)	3.6 ± 0.4 (3.0-5.0)

Table 39. *(Continued.)*

Character	Paratypes, Amboim, Angola, Whitehead (1968)	Democratic Republic of São Tomé and Príncipe, Coffee Maleita *et al.* (2016)	Democratic Republic of São Tomé and Príncipe, Bean Maleita *et al.* (2016)
Median bulb length	12.1 ± 1.5 (10.1-15.8)	-	-
Median bulb width	8.0 ± 0.8 (6.8-9.4)	-	-
Median bulb valve length	4.6 ± 0.6 (3.2-6.5)	4.1 ± 0.2 (3.5-4.5)	3.8 ± 0.5 (3.0-5.5)
Median bulb valve width	-	3.7 ± 0.3 (3.0-4.4)	3.4 ± 0.3 (3.0-4.0)
Ant. end to excretory pore	-	80.3 ± 9.5 (65.0-108.0)	83.9 ± 5.0 (73.0-91.0)
Tail	53.0 ± 3.0 (47.0-58.0)	58.0 ± 5.3 (48.5-76.0)	57.9 ± 4.6 (51.0-70.0)
Hyaline region	-	16.7 ± 3.1 (10.6-23.5)	16.4 ± 3.8 (8.0-23.0)

DESCRIPTION

Female

Oval or elongated to pear-shaped body, large and asymmetrical, some females with protuberance and well-defined tail. Neck distinct, sometimes quite long, projecting sideward at an angle to longitudinal body axis. Labial region continuous with body contour. Oval prestoma, six inner labial sensilla surrounding prestoma, labial disc square, raised above median labials, with four bumps, median labials rectangular, in face view, with two bumps and a pair of labial sensilla, large lateral triangular, fused with median labials, amphidial openings oval. Conical part of stylet curved dorsally. Shaft becoming broader posteriorly, knobs posteriorly sloping, transverse elongate, distinctly offset from shaft and deeply indented anteriorly, usually with one groove. Excretory pore located posterior to stylet base, 19-28 annuli from labial region and 10-19 annuli from stylet base. Hemizonid posterior 5-6 annuli to excretory pore. Perineal patterns variable in shape, although some features distinctive. Perineal pattern rounded, composed of fine, wavy striae, often broken. Dorsal arch low and rounded. Cuticle often folded in dorsal region of pattern. Anal fold present. Phasmids fairly close to tail terminus. Vulva wider than interphasmid distance. Some horizontal, fine, broken striations between vulva and anus. Rudimentary lateral field marked by breaks in striae and minute disordered striae within field. Concentric circles of striae sometimes forming a distinct raised tail pattern, some striae directed toward angles of vulva.

Male

Body with anterior end slightly tapering, tail conoid, bluntly rounded, terminus not striated. Labial cap low, truncate, narrower than labial region, labial region not offset, without annulation or with a few (1-4) irregular, incomplete lines. Labial disc round, elevated, fused with median labials, lateral labials absent. Stoma slit-like, located in ovoid prestomatal cavity surrounded by pore-like openings of six inner labial sensilla, four labial sensilla marked by cuticular depressions on median labials, amphidial apertures elongate slits. Labial framework moderately developed. Vestibule and extension distinct. In most males stylet robust, opening about 3.0 μm from stylet tip, cone bluntly pointed at tip, widening in diam. posteriorly, in dwarf males, *ca* 8% of 75 observed, stylet thinner than normal. Procorpus and median bulb distinct, median bulb elongate. Hemizonid distinct, about 3.0 μm long, variable in position: anterior to, or anterior to (one annulus), or posterior to (two annuli) excretory pore. Lateral field usually with four incisures, in some areas one more incisure was present, two outer bands not completely areolated. Testes predominantly one, two occasionally (4%), several with reflexed germinal zone. Spicules long, distal half nearly straight; gubernaculum weakly curved in lateral view. Phasmids posterior to cloacal level.

J2

Labial region offset with 1-3 irregular, incomplete annulations. Labial disc slightly raised above rounded-off median labials, lateral labials elongate, in same contour as labial region, labial disc and median labials dumbbell-shaped in face view. Prestoma opening ovoid, surrounded by small, pore-like openings of six inner labial sensilla. Labial sensilla appearing externally as small, round, cuticular depressions. Labial framework weak. Stylet moderately long. Stylet cone gradually increasing in width posteriorly, shaft cylindrical, slightly wider near middle, knobs elongate and gradually merge with shaft. Lateral field marked by four incisures, weakly areolated. Excretory pore distinct. Hemizonid, 2-3 annuli long, at level of excretory pore, in two specimens more anterior and adjacent to nerve ring. Tail long, with constrictions along length and near terminal part, ending in a narrow, rounded tip, hyaline region long (average 16.0 μm). Tail shape variable, even from a cohort of one female.

TYPE PLANT HOST: Robusta coffee, *Coffea canephora* Pierre.

OTHER PLANTS: *Meloidogyne megadora* is known to infect Rubiaceae, including *Coffea arabica*, *C. congensis* A. Froehner and *C. eugenioides* S. Moore (Whitehead, 1968). The host range of *M. megadora* also includes Musaceae (Zhang & Weng, 1991) and possibly Apiaceae, Asteraceae, Euphorbiaceae, Myrtaceae and Solanaceae (Decker *et al.*, 1980; Yassin & Zeidan, 1982). It has been also reported in banana (*Musa paradisiaca* L. var. *sapientum*) (Zhang & Wen, 1991), carrot and cassava (Eisenback, 1997; Onkendi *et al.*, 2014). Other plants recorded as susceptible include *Cucumis melo*, *C. sativus*, *Ipomoea batatas*, *Musa acuminata* L., *M. sapientum* L., *P. vulgaris*, *Pisum sativum* L., and *Vicia faba* (Almeida *et al.*, 1997; Almeida & Santos, 2002). Some plants including *Beta vulgaris*, *Brassica napus*, *Glycine max*, and *Raphanus sativus* appeared to be resistant/hypersensitive and exhibited galling but did not support reproduction (Almeida *et al.*, 1997). All North Carolina differential plants were non-hosts of *M. megadora* (Maleita *et al.*, 2016).

SYMPTOMS: Host-parasite relationship studies showed that the reaction of coffee (*C. arabica* 'Catuaí amarelo') roots to *M. megadora* could represent an intermediate response between susceptibility and hypersensitivity (Rodrigues *et al.*, 2000). The giant cells induced by *M. megadora* in 'Catuaí amarelo' roots were multinucleate and increased their synthetic activity and were usually associated with poorly developed nematodes and had abundant inclusions of various forms (Rodrigues *et al.*, 2000). However, in Catimor (*Coffea arabica* 'Caturra × Timor Hybrid'), necrosis of injured root cells was induced. Following minor cytoplasmatic changes, the nurse cells showed condensation of chromatin, altered carbohydrate metabolism, development of lysosomes and autophagic vacuoles. This response seems to be a hypersensitive-like response that leads to cell death a few days after nematode infection (Rodrigues *et al.*, 2000).

TYPE LOCALITY: Coffee Research Station, Amboim, Angola, Central West Africa.

DISTRIBUTION: *Africa*: Angola (Whitehead, 1968), Sudan (Decker *et al.*, 1980; Yassin & Zeidan, 1992), Uganda (Whitehead, 1969), Democratic Republic of São Tomé and Príncipe (Lordello & Fazuoli, 1980; Santos *et al.*, 1992; Abrantes *et al.*, 1995; Maleita *et al.*, 2012c); *Asia*: China (Zhang & Wen, 1991; Yassin & Zeidan, 1992).

PATHOGENICITY: This species causes great losses in coffee fields in the Democratic Republic of São Tomé and Príncipe (Abrantes *et al.*, 1995).

CHROMOSOME NUMBER: No available data.

POLYTOMOUS KEY CODES: *Female*: A23, B23, C2, D3; *Male*: A1234, B32, C213, D1, E2, F1; *J2*: A21, B23, C312, D123, E3, F1.

BIOCHEMICAL AND MOLECULAR CHARACTERISATION: Maleita *et al.* (2012c, 2016) described the esterase (Me3 type) pattern with three electrophoretic bands (Relative mobility (Rm) = 0.68, 0.77, and 0.82) and malate dehydrogenase (Me1 type) pattern with one (Rm = 0.17) band for *M. megadora*. Molecularly this nematode can be identified by the unique ITS rRNA gene sequences, which is different to all other *Meloidogyne* spp. described (Maleita *et al.*, 2016).

RELATIONSHIPS (DIAGNOSIS): The species belongs to Molecular group X and is clearly molecularly differentiated from all other *Meloidogyne* species. Jepson (1987) included this species in the *acronea* group (Table 13). This species can be distinguished from *M. africana* by the number of incisures in the male lateral field (4 *vs* 10), and from *M. acronea* by EP/ST ratio. The perineal pattern of this species also resembles *M. donghaiensis*, but differs in stylet length at 13-17 *vs* 10.0-14.6 μm, and EP/ST ratio at 2.2 *vs* 3.0 (1.9-4.1), and several morphometrics in male and J2.

58. *Meloidogyne megatyla* Baldwin & Sasser, 1979
(Figs 164, 165)

COMMON NAME: Pine root-knot nematode.

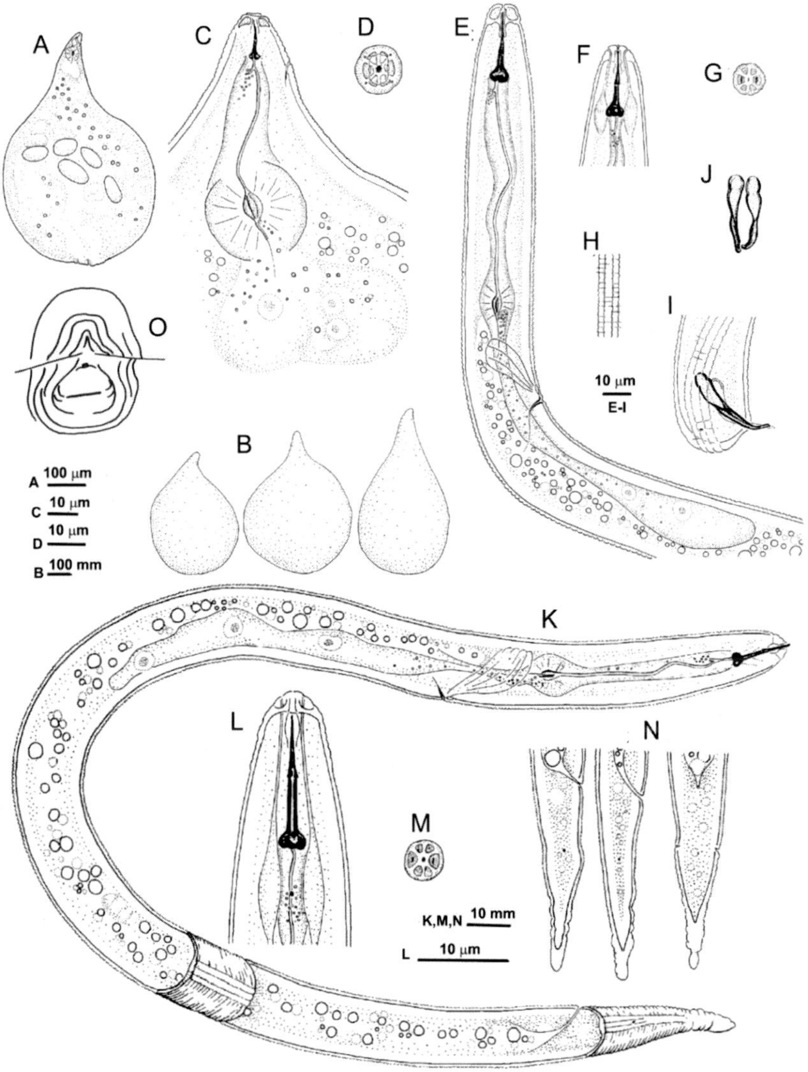

Fig. 164. Meloidogyne megatyla. *A, B: Entire female; C: Female anterior region; D: Female labial framework; E: Male anterior region; F: Male labial region; G: Male labial framework; H: Male lateral field; I: Male tail; J: Spicules; K: Entire second-stage juvenile (J2); L: J2 labial region; M: J2 labial framework; N: J2 tail; O: Perineal pattern. (A-N after Baldwin & Sasser, 1979; O after Jepson, 1987.)*

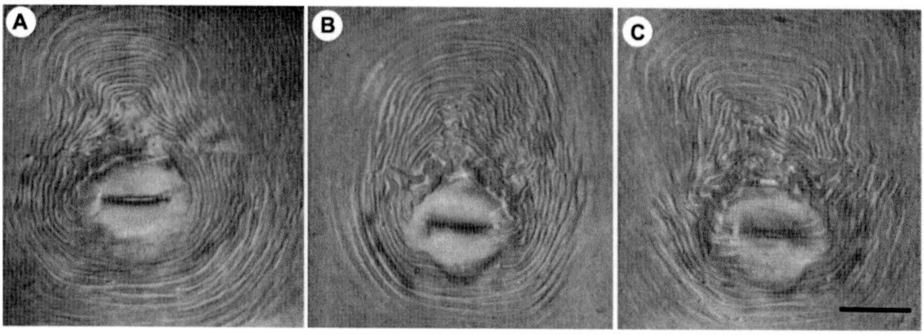

Fig. 165. Meloidogyne megatyla. *LM. A-C: Perineal pattern. (After Baldwin & Sasser, 1979.)*

Baldwin & Sasser (1979) described this species from roots of pine in North Carolina. Additional morphological details of *M. megatyla* were given by Eisenback *et al.* (1985).

MEASUREMENTS *(AFTER BALDWIN & SASSER, 1979)*

- *Holotype female*: L = 540 μm; W = 480 μm; stylet = 15 μm; DGO = 4.5 μm; excretory pore to anterior region = 28 μm.
- *Allotype male*: L = 1367 μm; stylet = 22 μm; DGO = 6.0 μm; excretory pore to anterior region = 145.5 μm; spicules = 30 μm; a = 40.2; b = 14.9; T = 80%.
- *Paratype females* (n = 25): L = 560 ± 31.7 (406-691) μm; W = 358 ± 32.1 (211-491) μm; stylet = 16.4 ± 0.5 (14.0-18.0) μm; DGO = 5.0 ± 0.5 (4.0-7.0) μm; stylet knob height = 2.9 ± 0.2 (2.2-3.5) μm; stylet knob width = 5.0 ± 0.2 (3.8-7.3) μm; median bulb length = 44.4 ± 2.9 (32-56) μm; median bulb width = 42 ± 2.9 (31-58) μm; excretory pore to anterior end = 20.4 ± 2.9 (12-32) μm; vulval slit to anus = 18.9 ± 0.8 (15.2-23.4) μm; interphasmid distance = 17.3 ± 1.1 (12.9-22.3) μm; a = 1.6 ± 0.1 (1.1-2.2); EP/ST = 1.2.
- *Paratype males* (n = 20): L = 1312.9 ± 68.4 (960-1545) μm; W = 35.3 ± 1.3 (30.3-42.4) μm; stylet = 23.9 ± 0.5 (21.7-25.5) μm; DGO = 5.2 ± 0.3 (4.2-6.3) μm; stylet knob height = 4.5 ± 0.2 (3.9-5.1) μm; stylet knob width = 5.2 ± 0.3 (4.2-6.3) μm; median bulb to anterior end = 128.5 ± 7.1 (106.1-149.5) μm; spicules = 33.7 ± 1.4 (29.3-36.9) μm; a = 37.2 ± 1.7 (30.9-42.0); b = 15.1 ± 0.7 (10.9-17.4); T = 63 ± 4.4 (44-82)%.
- *Paratype J2* (n = 23): L = 416.4 ± 7.8 (393-457) μm; W = 16.2 ± 0.3 (14.7-17.9) μm; stylet = 14.6 ± 0.3 (13.8-16.6) μm; DGO = 5.1 ±

0.2 (4.2-5.9) μm; stylet knob height = 2.3 ± 0.04 (2.1-2.5) μm; stylet knob width = 3.0 ± 0.1 (2.5-3.4) μm; excretory pore to anterior end = 83.1 ± 1.8 (76.2-97.3) μm; tail = 39.8 ± 1.3 (31.6-45.1) μm; a = 25.7 ± 0.8 (21.9-28.8); b = 7.1 ± 0.1 (6.7-7.8); c = 10.5 ± 0.4 (9.5-13.5); c′ = 3.3 ± 0.1 (2.5-3.9).

- *Eggs* (n = 30): L = 95.7 ± 1.3 (88.5-102.4) μm; W = 48.3 ± 1.3 (43-57) μm; L/W = 2.2 ± 0.04 (1.7-2.2).

DESCRIPTION

Female

Female white, variable in size, pear-shaped with gradually tapering neck and posteriorly rounded with no tail protuberance. Cuticle with max. thickness about 10.0 μm and finely striated, cephalids not observed. Labial region slightly offset with two or three annuli. Labial cap dorsoventrally elongate, amphidial openings apparently small and oval, six inner labial sensilla and four labial sensilla. Labial framework approximately hexaradiate with slightly enlarged lateral sectors. Stylet dorsally curved with large rounded knobs, offset from shaft, often indented anteriorly. Dorsal gland orifice branching into three ducts, subventral gland orifices immediately posterior to medium bulb valve and also branched. Dorsal gland ampulla very large, gland lobe indistinct. Excretory pore often near level of stylet knobs but position highly variable. Perineal pattern composed of deep, smooth, rarely broken coarse striae superimposed over fine, sometimes wavy striae. Dorsal arch high and broad, flattened dorsally, and coarse striae sometimes diverging anteriorly (outward) as they approach single lateral lines. Striations interrupted but otherwise little altered at lateral lines. Phasmidial ducts faintly visible, little or no phasmid surface structure. Anus covered by cuticular flap, tail terminus indistinct. Vulva generally surrounded by smooth perivulval region.

Male

Body vermiform, tapering and rounded at both ends. Heat-killed specimens curving concavely on ventral side and tail twisting through 90°. Lateral field with four lines, spaced approximately equally throughout most of lateral field, areolation and crenation indistinct anteriorly but very pronounced posteriorly. Anteriorly, outer lines originating at about level of stylet knobs, inner lines beginning very close together or as a single line at about mid-region of procorpus. Occasionally, four lateral lines separated by faint intermediate lines. Labial region usually with three annuli dorsally and two ventrally. Labial disc not pronounced, amphidial openings apparently small and elongate, amphidial cheeks indistinct. Stoma slit-like, prestoma hexagonal, surrounded by pore like openings of six inner labial sensilla. Labial disc and median labials fused, forming one smooth continuous labial cap. Slit-like amphidial apertures between labial disc and lateral sectors on labial region. Lateral labials absent. Labial framework moderately heavy and hexaradiate with lateral sectors only slightly enlarged. Stylet heavy with massive rounded knobs. Amphidial glands broadest near level of knobs. Cephalids not observed. Orifice of dorsal gland duct branched as in female, but gland ampulla small and poorly defined. Median bulb oval and only slightly broader than procorpus, valve large and well developed with branched ducts of subventral gland orifices near its base. Poorly defined pharyngo-intestinal junction at level of nerve ring, intestinal caecum extending anteriorly to level of subventral gland orifices. Gland lobe variable in length, but more than 100 μm long with two distinct nuclei. Excretory pore pronounced with hemizonid located immediately anterior. Spicules arcuate; gubernaculum not clearly observed. One testis or two testes. Tail short with cloacal aperture generally very far posterior, near level of phasmids.

J2

Body vermiform, tapering slightly anteriorly and more pronouncedly posteriorly. Heat-killed specimens straight or curved slightly concavely on ventral side. Lateral field with four lateral lines similar to those of male except that anterior origin may be at level of posterior half of procorpus and lines terminate 4-5 μm from tail tip rather than encircling. Labial region slightly offset, usually with three annuli dorsally and two ventrally. Labial disc not pronounced, amphidial openings apparently small and oval. Large slit-like stoma located in very large ovoid prestoma, surrounded by pore-like openings of six inner labial sensilla. Labial framework hexaradiate with lateral sectors slightly enlarged. Stylet robust with large rounded knobs. Amphidial apertures located between labial disc and median labials. Lateral labials elongate to triangular, often fused with labial region for a short distance. Cephalids not observed. Median bulb about twice diam. of procorpus. Poorly defined pharyngo-intestinal junction posterior to nerve ring, small intestinal caecum. Gland lobe variable in length with two distinct subventral gland nuclei and one less distinct dorsal gland nucleus. Tail gradually tapering, terminus with enlarged annuli, sometimes smooth at tip or forming an elongate swelling. Rectum not dilated.

TYPE PLANT HOST: Loblolly pine, *Pinus taeda* L.

OTHER PLANTS: *Meloidogyne megatyla* reproduced only on pine and no signs of nematode infection were observed on the differential hosts: corn, cotton, peanuts, pepper, strawberry, sweet potato, tobacco, tomato, watermelon. It has not yet been determined whether conifers other than *P. taeda* are susceptible (Baldwin & Sasser, 1979).

TYPE LOCALITY: A property surrounding the residence of Mrs Wallace Baxley, U.S. Highway 701, Elizabethtown, North Carolina, USA.

DISTRIBUTION: *North America*: USA (North Carolina). It has been reported only in the type locality. Some *Meloidogyne* populations with similar stylet knobs have been reported from *P. taeda* in Florida but the identification was not confirmed (Baldwin & Sasser, 1979).

PATHOGENICITY: Galling on pine roots was slight. A single root tip was often infected with several females, even when the overall infection of a plant was light. Reproduction on pine was very slow under glasshouse conditions (apparently a single generation may take more than 10 weeks), and more than 50 eggs were rarely observed in a given egg mass. Males might be produced only under specific conditions; only on one occasion were large numbers recovered from a glasshouse culture. It has not been determined if other conifers are susceptible to this nematode (Baldwin & Sasser, 1979).

CHROMOSOME NUMBER: *Meloidogyne megatyla* (n = 18) was found to reproduce by amphimixis (Goldstein & Triantaphyllou, 1982).

POLYTOMOUS KEY CODES: *Female*: A32, B213, C3, D6; *Male*: A34, B21, C21, D1, E2, F2; *J2*: A23, B12, C34, D3, E21, F43.

BIOCHEMICAL AND MOLECULAR CHARACTERISATION: No available data.

RELATIONSHIPS (DIAGNOSIS): *Meloidogyne megatyla* is close to *M. mali*, *M. lusitanica* and *M. thailandica*. From *M. mali* it differs in J2 tail length (32-45 *vs* 24-39 μm), and stylet knobs of J2 and males are much larger in *M. megatyla* than those for *M. mali* and the perineal pattern is different; from *M. lusitanica* it differs by the stylet knobs of the female, male, and J2, the more anterior position of the excretory pore in the female (20.4 (12-32) *vs* 44 (28-60) μm) and different perineal pattern; and from *M. thailandica* by having J2 with shorter body length (416 (392-457) *vs* 484 (450-540) μm), longer and more robust stylet (14.6 (14.0-17.0) *vs* 10.2 (10.0-11.0) μm), shorter tail (39.7 (32.0-45.0) μm with terminus having enlarged annuli *vs* 61.2 (55.0-65.0) μm with long, gradually tapering terminus without enlarged annuli and swelling), and females having an EP/ST ratio of 1.2 *vs* 2.3.

59. *Meloidogyne mersa* Siddiqi & Booth, 1991
(Figs 166-168)

COMMON NAME: Mangrove root-knot nematode.

Siddiqi & Booth (1991) described this species from roots of *Sonneratia alba* J. Smith. trees in Brunei Darussalam.

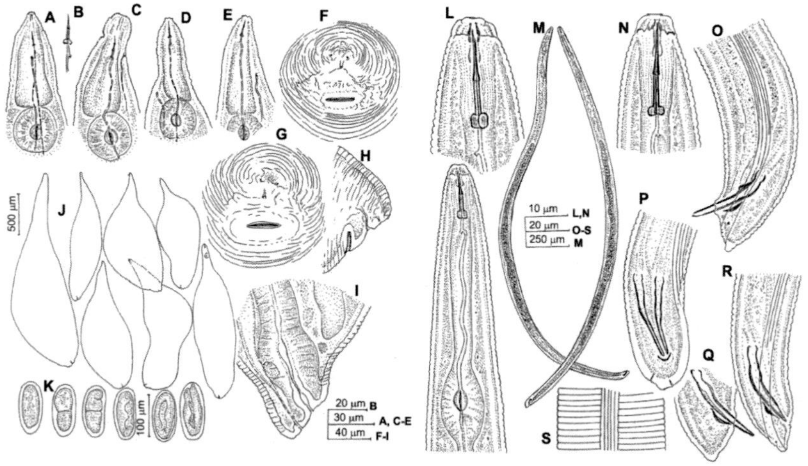

Fig. 166. Meloidogyne mersa. *A, C, D, E: Female anterior region; B: Female stylet; F-I: Perineal pattern; J: Entire female; K: Eggs; L, N: Male labial region; M: Entire male; O-R: Male tail; S: Male lateral field. (After Siddiqi & Booth, 1991.)*

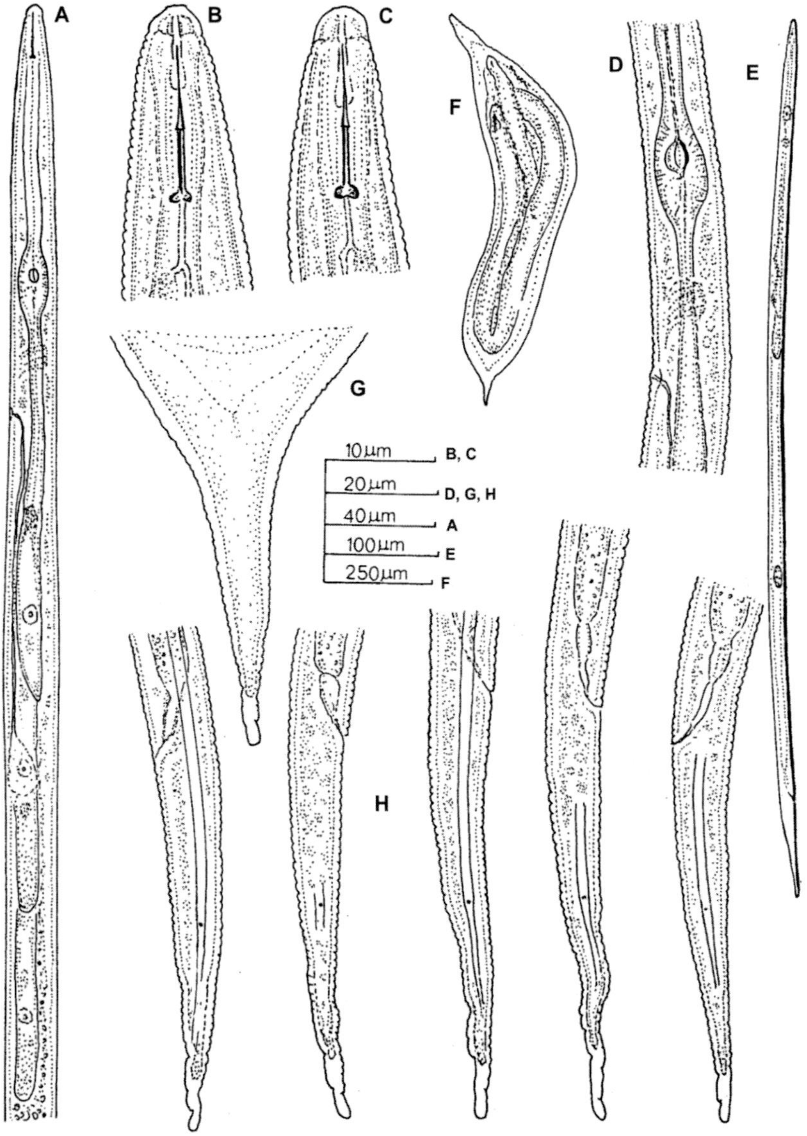

Fig. 167. Meloidogyne mersa. *A: Anterior region of second-stage juvenile (J2); B, C: J2 labial region; D: J2 median bulb region; E: Entire J2; F: Developing male; G: Tail of developing male; H: J2 tail. (After Siddiqi & Booth, 1991.)*

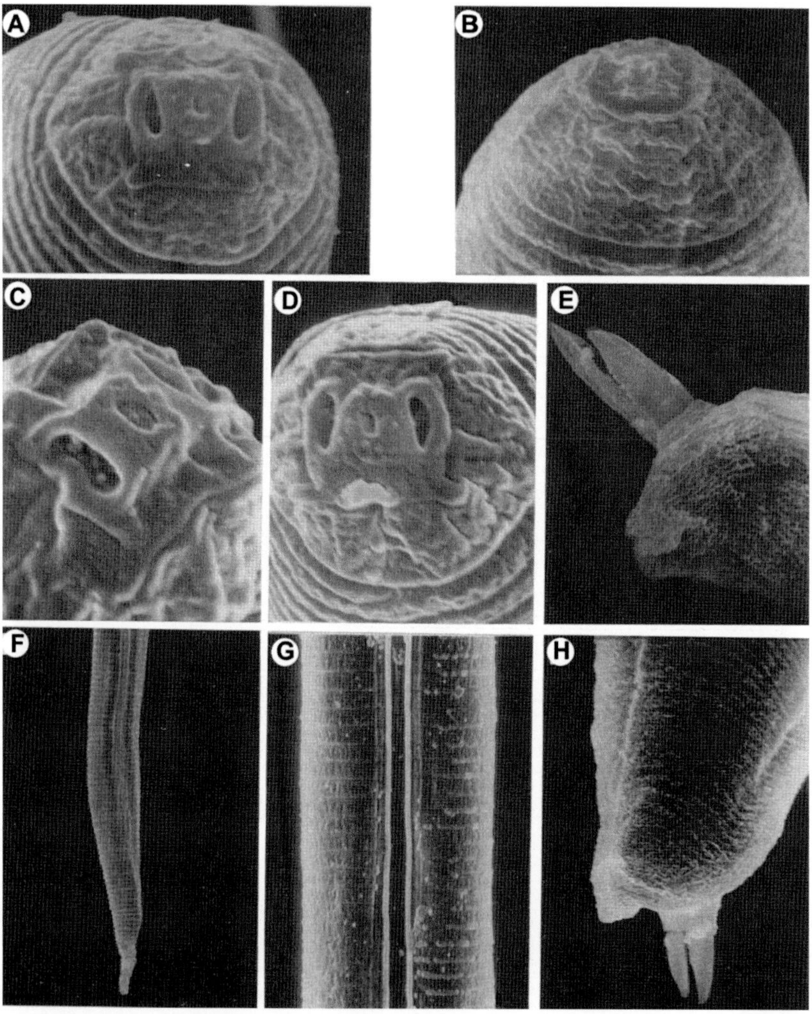

Fig. 168. Meloidogyne mersa. *SEM. A, C, D: Second-stage juvenile (J2) labial region; B: Male labial region; E, H: Male tail; F: J2 tail; G: J2 lateral field. (After Siddiqi & Booth, 1991.)*

MEASUREMENTS *(AFTER SIDDIQI & BOOTH, 1991)*

- *Holotype male*: L = 2500 μm; stylet = 22.5 μm; a = 51; c = 147; T = 72%; spicules = 38 μm; gubernaculum = 11 μm.
- *Paratype males* (n = 100): L = 2400 ± 300 (1880-2890) μm; labial region diam. = 9.5-11.5 μm; labial region height = 5-6 μm;

stylet = 21.5 ± 0.7 (20.0-23.0) μm; stylet knob width = 4.0-5.0 μm; DGO = 7.5 (6.0-9.0) μm; distance from anterior end to median bulb valve = 112 (100-123) μm; median bulb length = 25-32 μm; median bulb width = 6-18 μm; median bulb valve length = 7-9 μm; median bulb valve width = 5-7 μm; excretory pore to anterior end = 186 ± 17 (156-215) μm; spicules = 37.5 ± 1.2 (35.0-39.0) μm; gubernaculum = 11.5 ± 1.4 (10.0-16.0) μm; a = 53 ± 8 (40-66); c = 133 ± 31 (82-180); T = 68 ± 7.6 (55-76)%.
- *Paratype females* (n = 50): L = 1850 ± 370 (1150-2530) μm; W = 550 ± 129 (360-850) μm; L/W = 2.9 ± 0.5 (1.8-4.5); neck length = 200-700 μm; stylet = 14.3 ± 0.5 (13.5-15.0) μm; stylet knob width = 9.5-11.5 μm; stylet knob height = 5.0-6.0 μm; DGO = 7.0 (6.0-8.5) μm; excretory pore to anterior end = 50 ± 15.8 (28-85) μm; median bulb length = 43 (40-46) μm; median bulb width = 40 (37-42) μm; EP/ST = 3.7 (2.6-4.4).
- *Paratype J2* (n = 100): L = 720 ± 60 (610-870) μm; W = 15.8 (14.5-17.5) μm; stylet = 14.5 ± 0.7 (13.0-16.0) μm; DGO = 7.0 (6.0-8.0) μm; median bulb length = 17-20 μm; median bulb width = 9-10 μm; median bulb valve length = 5.0-7.0 μm; median bulb valve width = 5.0-6.0 μm; median bulb to anterior end = 90 (85-96) μm; excretory pore to anterior end = 122 ± 12 (95-135) μm; tail = 70 ± 5.3 (63-81) μm; hyaline region = 10.2 ± 1.4 (8.0-13.0) μm; a = 46 ± 4 (39-58); c = 10.7 ± 0.6 (8.4-11.7); c′ = 6.7 ± 0.6 (5.7-10.5).
- *Eggs*: L = 133 ± 11.5 (112-165) μm; W = 48 ± 5.7 (40-60) μm; L/W = 2.6 ± 0.3 (2.0-3.3).

DESCRIPTION

Female

Body swollen, pearly white to transparent, elongate obese, may be pear- or lemon-shaped, saccate, bilobed, but never globular, with a distinct long tapering, straight to sometimes arcuate, neck and a terminal elevated perineal region. Cuticle 3.5-5.0 μm thick, transversely striated, striae irregular, about 2 μm apart posterior to stylet region, indistinct at mid-body, forming a distinct pattern in perineal region. Labial region offset, discoid, with an elevated oral disc about 5 μm in diam. and one or two annuli, *en face* view showing six equal sectors of lightly sclerotised labial framework. Distance from anterior end of body to excretory pore irregular depending upon length and contraction of anterior region.

Stylet strong, with straight to dorsally arcuate conus 7 (7.0-7.5) μm long and large basal knobs with convex to indented posterior surfaces. Procorpus swollen with max. width at base of 34 μm, offset from median bulb by a deep constriction. Median bulb round, highly muscular, with a large round valvular apparatus at centre. Perineal region distinctly raised. Vulva a conspicuous transverse slit located in a subterminal body depression. Perineal pattern round to longitudinally oval with smooth fine striae and low round dorsal arch, tail region elevated, usually with a whorl of striae and stippled area beneath cuticle, phasmids close together, less than vulval length apart, some broken lines occurring in lateral region. Anus a small pore located subterminally, 25 (19-34) μm posterior to vulva, surface between anus and vulva normally convex but may be flat or concave. Tail elevation 12 (9-15) μm, hemispherical. Vagina thick-walled, anteriorly directed. Two branches of reproductive organs enormously developed, directed anteriorly, ovaries coiled. Up to 125 eggs seen in body.

Male

Body normally ventrally arcuate to C-shaped, with posterior end always twisted by 90-180 degrees. Body annulation distinct, annuli averaging 2.5 (2.3-2.8) μm wide at mid-body and about 2 μm wide posterior to stylet. Lateral field distinct beginning as a line at about halfway from anterior end of body to median pharyngeal bulb, becoming 0.2-0.3 times as wide as body posterior to excretory pore level and ending on tail tip, usually with six incisures, but four or more than six (up to 12) seen on some males. Labial region smooth, lacking annulation, hemispherical or slightly tapering anteriorly, slightly offset from body by a constriction, occasionally continuous, oral disc low, 5.0-6.5 μm in diam., amphidial apertures in form of dorsoventrally elongated oval slits at base of oral disc, no longitudinal indentation on labial surface. Stylet well developed, about two labial region diam. long, conus about half stylet length, with a sharply pointed, solid-appearing tip, basal knobs rounded, sometimes with a median depression near middle. Median pharyngeal bulb well developed, muscular, oval to fusiform, not offset from procorpus or isthmus, with a large valvular apparatus in centre. Pharyngeal glands ventral to intestine. Hemizonid poorly developed, 2-4 annuli long, 0-3 annuli anterior to excretory pore. Testis long, spermatogonia in 1-3 rows, sperm round, 10 (9-11) μm diam. Spicules cephalated, similar, straight to slightly arcuate ventrally, tapering to a blunt terminus. Gubernaculum fixed, with tapering end, very thick near middle. Tail short conoid-rounded, sometimes bluntly rounded or with a knob-like tip, lacking a bursa but cuticle thickened and raised in lateral field region. Phasmids punctiform, mid-way on tail.

J2

Body slender, usually arcuate ventrally. Cuticle striated transversely, striae 1.0-1.5 μm apart at mid-body. Lateral field a narrow band, about 2.0-2.5 μm wide, one-eighth to one-sixth as wide as body, beginning as a line between anterior end of body and median pharyngeal bulb and ending before tail terminus, with two ridges or three or four incisures, in SEM, two ridges but six incisures may be seen, not areolated. Labial region hemispherical to slightly tapering anteriorly, smooth, occasionally with two fine striae, offset by a constriction, 5.5-6.5 μm wide and 2.5-3.5 μm high, oral disc inconspicuous but sometimes terminal region elevated; in SEM, oral disc squarish with narrow lobe-like submedian pseudolabials and six pit- or papilla-like sensilla around oral opening. Labial framework slightly sclerotised. Stoma and stylet guiding tube narrow and anteriorly sclerotised. Stylet slender, conus sharply pointed, solid-appearing anteriorly, 7.0-8.5 μm long or 53 (51-56)% of total stylet length, knobs close together, rounded with anterior surfaces sloping posteriorly, 2-3 μm across. Median pharyngeal bulb oval, usually occupying three-fifths to two-thirds of body diam., with a large oval refractive valvular apparatus. Nerve ring slightly posterior to median bulb. Excretory pore posterior to nerve ring, at or just posterior to hemizonid, which is 3-4 annuli long. Pharyngeal glands elongated with subventrals asymmetrical and extending well posterior to dorsal gland, ventral to intestine, may reach almost to middle of body. Genital primordium two-celled, on right side of intestine, at 64 (62-67)% of body from anterior end. Rectum not dilated or swollen abnormally but sometimes with wide lumen. Anus distinct in lateral view as followed by a slight depression in body surface. Tail posterior to anus subcylindrical then tapering to irregularly indented terminal region having a characteristic ventral indentation before a smooth rounded tip, with 56 (49-68) annuli ventrally. Phasmids punctiform, at about 44 (39-49)% of tail length from anus, 32 (27-37) μm and 42 (37-47) μm from anus and tail tip, respectively.

TYPE PLANT HOST: Mangrove, *Sonneratia alba* J. Smith.

OTHER PLANTS: No other hosts were reported.

TYPE LOCALITY: The intertidal area of Pulau Bedukang, Brunei Bay, Brunei.

DISTRIBUTION: *Asia*: Brunei. It has only been reported from the type locality.

SYMPTOMS: Galling on mangrove tree either ranging in size from 0.5 to 10 mm in diam. on tips of fine feeder roots or occasionally larger in parenchymatous roots in the intertidal region. Newly formed galls were green and had 1-10 females and up to two males, while older and larger galls could contain up to 20 females, ten males and numerous eggs and juveniles (Siddiqi & Booth, 1991). No damage was documented on the type plant host.

CHROMOSOME NUMBER: No available data.

POLYTOMOUS KEY CODES: *Female*: A1, B32, C12, D4; *Male*: A1, B23, C1, D2, E1, F2; *J2*: A1, B12, C12, D324, E23, F12.

BIOCHEMICAL AND MOLECULAR CHARACTERISATION: No available data.

RELATIONSHIPS (DIAGNOSIS): This species was included into the *graminis* group by Jepson (1987). *Meloidogyne mersa* differs from *M. spartinae* by males having an offset, more rounded labial region, more incisures in the lateral field, a longer stylet and a more posteriorly located excretory pore, females being longer and having straight rarely arcuate neck, not situated to one side, and a more posterior excretory pore. From *M. graminis*, *M. mersa* differs by its males having a longer body, stylet and spicules, females being longer and having a longer stylet and a more posterior excretory pore, and J2 having a longer body, a longer stylet and a shorter hyaline region.

60. *Meloidogyne microcephala* Cliff & Hirschmann, 1984
(Figs 169-172)

COMMON NAME: Thailand tobacco root-knot nematode.

Cliff & Hirschmann (1984) described this species from roots of tobacco (*Nicotiana tabacum*) in Thailand.

MEASUREMENTS

See Table 40.

- *Eggs* (n = 30): L = 91.7 (82.5-101.0) μm; W = 39.8 (39.0-44.0) μm.

DESCRIPTION

Female

Body usually small, ivory-coloured, globular to ovoid in shape with distinct neck region and flattened to rounded posteriorly. Labial region small, slightly offset from body. Labial cap low, dumbbell-shaped in face view, in lateral view labial disc slightly raised above median labials. Annulation often present on labial region. Labial framework delicate, with distinct vestibule and vestibule extension. Stylet small, delicate. Cone narrow, curved dorsally, tapering to blunt tip, shaft straight, same width throughout. Stylet knobs relatively large, square to rectangular, sloping slightly posteriorly away from shaft. Anterior surface of knobs occasionally indented. Distance between base of stylet and DGO moderately long. Gland orifice branching, ampulla large. Procorpus and median bulb distinct, median bulb valve prominent. Subventral gland orifices branched and posterior to valve. Pharyngeal gland large, trilobed, dorsal lobe largest. Two small pharyngo-intestinal cells present near junction of median bulb and intestine. Excretory pore usually near level of DGO. Perineal patterns ovoid to rectangular. Dorsal arch usually low and rounded, occasionally high and more squarish. Lateral lines visible, more pronounced near tail tip. Phasmids distinct, close to tail tip. Anus not visible in SEM, covered by cuticular flap. Striations usually continuous. Dorsal striae coarse and wavy, forming characteristic series of cuticular flaps, often in a windmill-shaped pattern, around tail terminus. Ventral striae frequently forked or divided near lateral line, may form a wing on one side of pattern. Edges of vulva usually smooth. Perivulval region with fine striae near lateral edges of vulva.

Male

Generally slender nematodes, tapering to bluntly rounded at either end with labial region narrower than tail end. Tail twisted through 90°.

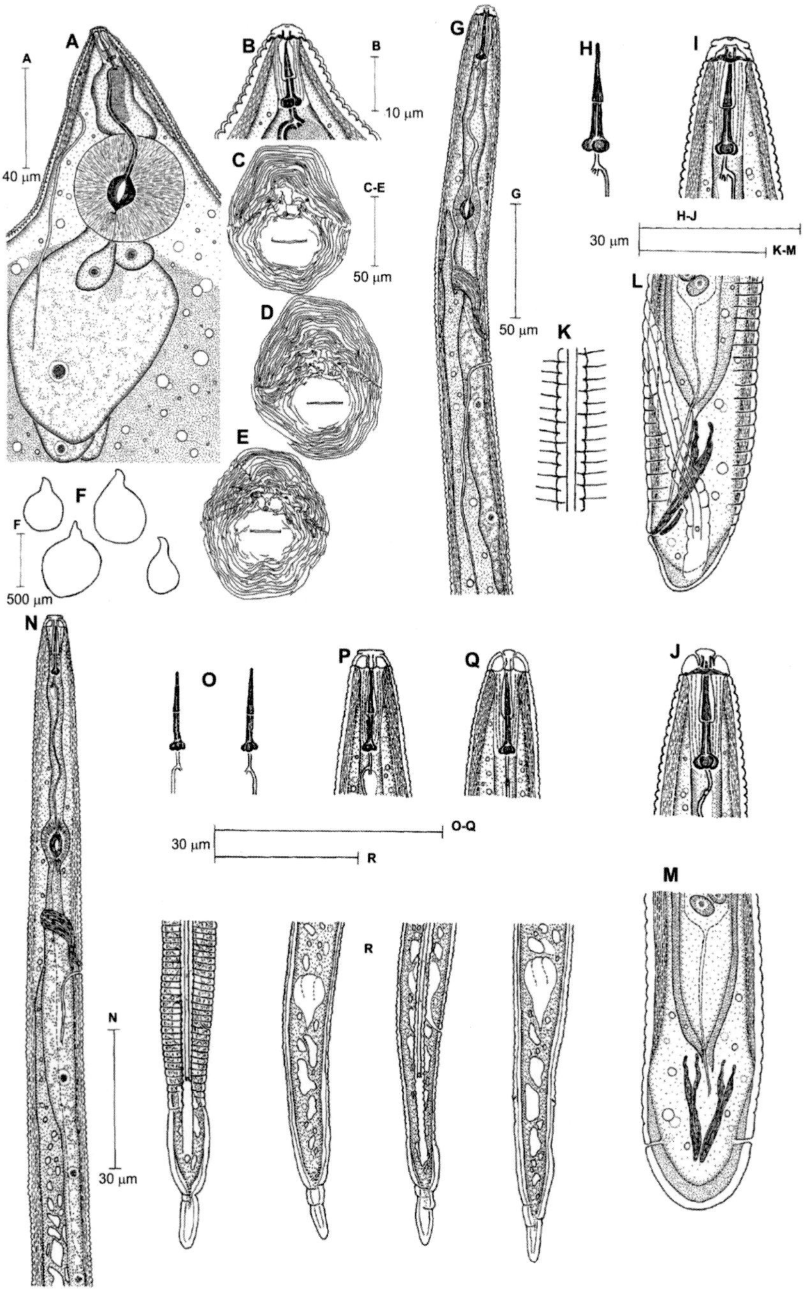
A
B
C
D
E
F
G
H
I
J
K
L
M
N
O
P
Q
R
40 μm
10 μm
C-E
50 μm
500 μm
50 μm
H-J
30 μm
K-M
O-Q
30 μm
30 μm

Fig. 169. Meloidogyne microcephala. *A: Female anterior region; B: Female labial region; C-E: Perineal pattern; F: Entire female; G: Male anterior region; H: Male stylet; I, J: Male labial region; K: Male lateral field; L, M: Male tail; N: Second-stage juvenile (J2) anterior region; O: J2 stylet; P, Q: J2 labial region; R: J2 tail. (Scale bar = 20 μm.) (After Cliff & Hirschmann, 1984.)*

Body annuli pronounced, 2.0-2.5 μm at mid-body. Lateral field with four incisures, incompletely areolated. Labial region small, squarish, distinctly narrower than first body annulus. Labial cap low, not rounded, narrower than labial region in lateral view. In SEM, labial structures small, rectangular in face view. Labial disc not wider than long axis of median labials and slightly raised above median labials. Indentations between median labials and labial disc usually slight. Median labials roughly rectangular, outer margins occasionally irregular, labials not extending posteriorly over labial region. Indication of lateral labials rarely present. Labial region not annulated. Prestoma small, oval to hexagonal in shape. Inner labial sensilla obscure. Labial sensilla usually distinct. Labial framework weakly developed, vestibule and vestibule extension distinct. Stylet slender, delicate, cone pointed, gradually increasing in diam. posteriorly, occasionally slightly wavy, shaft same width throughout or slightly wider anterior to knobs. Stylet knobs small, angular, offset from shaft. Distance from stylet base to DGO moderately long. Procorpus and median bulb distinct, median bulb valve pronounced, subventral gland openings posterior to valve, pharyngeal gland long with two equal-sized nuclei. Pharyngo-intestinal junction indistinct, intestinal caecum present. Excretory pore distinct, terminal excretory duct long. Hemizonid 2-5 annuli anterior to excretory pore. Usually one outstretched testis, rarely two testes. Sperm large, rounded, granular. Spicules slender, blades curved dorsally. Gubernaculum small, crescentic. Tail bluntly rounded. Phasmids distinct.

J2

Body small, slender, tapering at both ends. Labial region truncate. Labial cap small, difficult to discern. In SEM, labial disc small, rounded, slightly elevated above median labials. Median labials large, outer margins indented. Median labials and labial disc fused to form elongate labial cap. Lateral labials triangular, margins may be irregular, frequently fused to labial region. Fusion of lateral and median labials forming

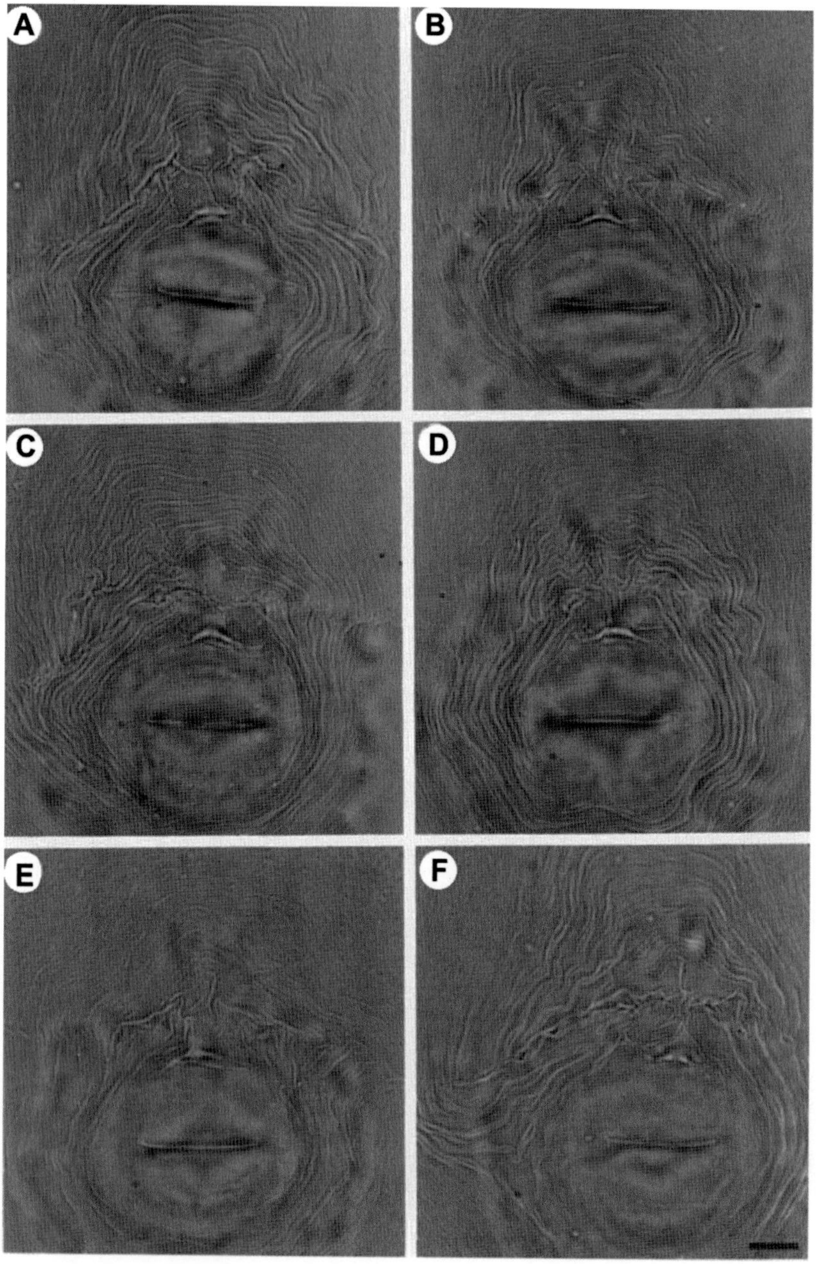

Fig. 170. Meloidogyne microcephala. *LM. A-F: Perineal pattern. (Scale bar = 10 μm.) (After Cliff & Hirschmann, 1984.)*

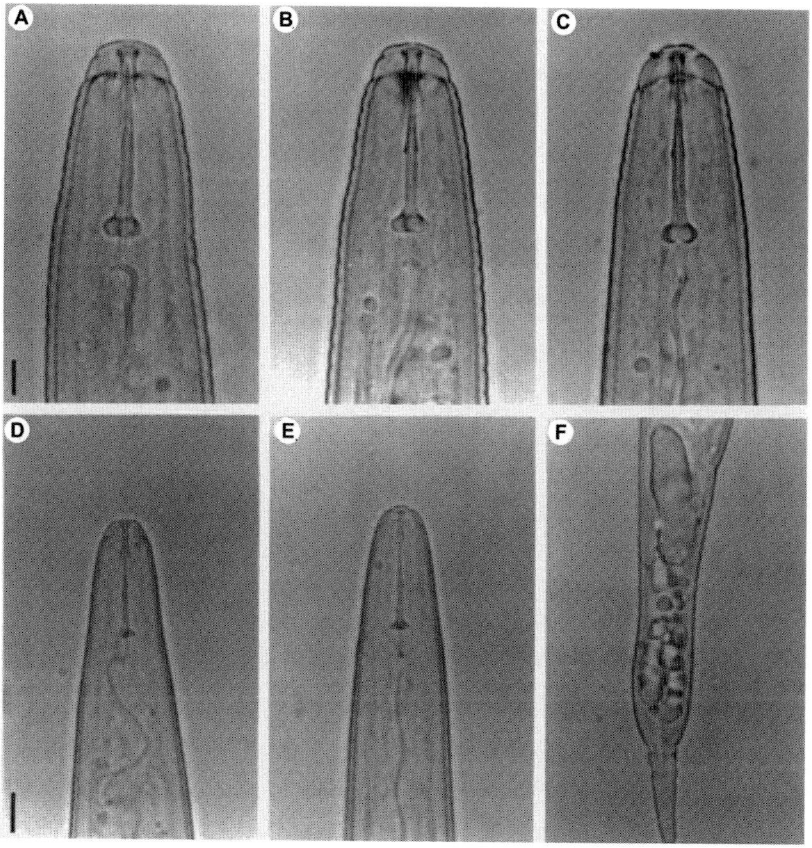

Fig. 171. Meloidogyne microcephala. *LM. A-C: Male labial region; D, E: Second-stage juvenile (J2) labial region; F: J2 tail. (Scale bars: A-F = 4 μm.) (After Cliff & Hirschmann, 1984.)*

right angle with lateral edge of labial disc. Labial region not annulated. Prestoma small, oval-shaped. Inner labial sensilla distinct, opening on labial disc. Prestoma and inner labial sensilla in slight depression on labial disc. Labial sensilla indistinct. Labial framework weakly sclerotised. Vestibule and vestibule extension distinct. Stylet slender and delicate, cone pointed, increasing gradually in width posteriorly, shaft cylindrical, slightly wider anterior to stylet knobs, knobs low, sloping slightly posteriorly. Distance from base of stylet to DGO long. Procorpus slender, median bulb distinct, ovoid with prominent valve. Pharyngo-intestinal junction between nerve ring and excretory pore obscure, pharyngeal

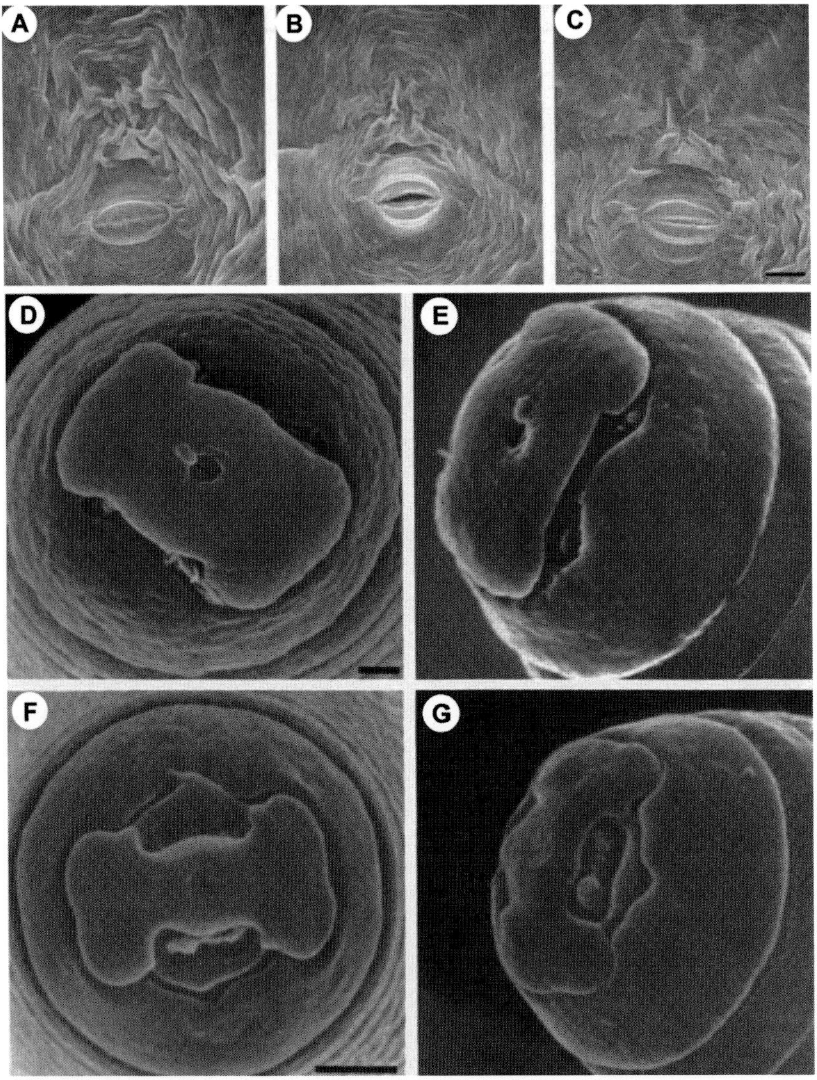

Fig. 172. Meloidogyne microcephala. *SEM. A-C: Perineal pattern; D, E: Male labial region; F, G: Second-stage juvenile labial region. (Scale bars: A-C = 10 μm; D-G = 1 μm.) (After Cliff & Hirschmann, 1984.)*

gland lobe long, with three nuclei. Hemizonid 1-2 annuli anterior to excretory pore. Lateral field with four incisures. Body annuli small and distinct, increasing in size and becoming irregular in posterior tail region. Tail short, plump, hyaline region indistinct. Tail tip shape characteristic,

Table 40. *Morphometrics of females, males and second-stage juveniles (J2) of* Meloidogyne microcephala. *All measurements are in μm and in the form: mean ± standard error of mean (range).*

Character	Holotype, allotype, Thailand, Cliff & Hirschmann (1984)	Paratypes, Diploid Thailand, Cliff & Hirschmann (1984)	Tetraploid Triantaphyllou & Hirschmann (1997)
Female (n)	-	30	30
L	587	601 ± 17.6 (450-840)	795 ± 20.4 (482-960)
a	1.2	1.5 ± 0.03 (1.1-1.9)	1.6 ± 0.04 (1.3-2.0)
W	490	400 ± 14. 8 (255-525)	510 ± 13.4 (370-682)
Stylet	14.2	14.4 ± 0.1 (13.5-15.0)	19.2 ± 0.1 (18.0-20.5)
Stylet knob width	4.4	4.3 ± 0.05 (3.7-4.8)	5.2 ± 0.04 (4.8-5.6)
Stylet knob height	2.0	2.4 ± 0.03 (2.2-2.9)	2.9 ± 0.09 (2.7-3.2)
DGO	3.2	3.7 ± 0.1 (3.0-5.0)	5.7 ± 0.2 (4.0-8.0)
Ant. end to excretory pore	32	51.6 ± 2.1 (32.0-71.0)	-
Vulval slit	-	20.8 ± 0.3 (17.8-23.8)	29.0 ± 0.4 (25.8-33.2)
Vulval slit to anus distance	-	17.9 ± 0.3 (14.3-23.5)	22.4 ± 0.5 (15.8-28.0)
Interphasmid distance	-	23.6 ± 0.5 (17.8-27.9)	31.6 ± 0.7 (21.8-42.7)
EP/ST	2.3	3.8	-
Male (n)	1	30	30
L	1301	1346 ± 25.7 (979-1522)	2041 ± 55.4 (1280-2540)
a	44.3	53.0 ± 0.8 (44.7-60.7)	55.3 ± 1.5 (44.7-60.7)
c	116.7	101.3 ± 2.4 (74.5-133.6)	128.5 ± 4.3 (74.9-177.4)
W	29.4	30.4 ± 0.5 (25.5-35.5)	37.7 ± 1.0 (28.5-50.0)

Table 40. *(Continued.)*

Character	Holotype, allotype, Thailand, Cliff & Hirschmann (1984)	Paratypes, Diploid Thailand, Cliff & Hirschmann (1984)	Tetraploid Triantaphyllou & Hirschmann (1997)
Stylet	21.2	20.6 ± 0.1 (19.0-22.0)	26.2 ± 0.2 (24.0-28.0)
Stylet knob width	3.9	4.2 ± 0.05 (3.7-4.7)	5.8 ± 0.07 (4.5-6.3)
Stylet knob height	2.2	2.4 ± 0.04 (2.0-2.9)	3.6 ± 0.07 (2.8-4.3)
DGO	3.1	3.5 ± 0.05 (3.0-4.0)	5.5 ± 0.2 (3.0-8.0)
Ant. end to excretory pore	147	146 ± 2.6 (118-169)	217 ± 4.1 (150-261)
Spicules	29.4	26.6 ± 0.3 (24.0-29.5)	35.0 ± 0.4 (32.0-39.0)
Gubernaculum	7.4	7.5 ± 0.3 (7.0-8.5)	10.3 ± 0.1 (9.0-11.5)
Tail	11.2	13.0 ± 0.2 (11.0-15.5)	16.2 ± 0.4 (12.5-20.0)
J2 (n)	-	30	30
L	-	456 ± 2.6 (416-472)	552 ± 4.6 (508-607)
a	-	32.4 ± 0.4 (29.0-38.3)	36.2 ± 0.5 (29.2-40.6)
b	-	5.6 ± 0.2 (5.3-5.7)	-
c	-	10.1 ± 0.1 (9.1-13.6)	8.6 ± 0.1 (8.0-9.0)
c′	-	4.5 ± 0.1 (3.4-5.0)	-
W	-	14.2 ± 0.2 (12.0-16.0)	15.5 ± 0.2 (14-17.5)
Stylet	-	9.3 ± 0.04 (9.0-10.0)	12.4 ± 0.1 (12.0-13.0)
DGO	-	3.1 ± 0.1 (2.5-4.0)	3.2 ± 0.1 (3.0-4.0)
Ant. end to excretory pore	-	83.6 ± 0.6 (74.0-89.0)	114 ± 0.5 (106-120)

Table 40. *(Continued.)*

Character	Holotype, allotype, Thailand, Cliff & Hirschmann (1984)	Paratypes, Diploid Thailand, Cliff & Hirschmann (1984)	Tetraploid Triantaphyllou & Hirschmann (1997)
Tail	-	45.4 ± 0.5 (43.0-50.0)	65.1 ± 0.6 (59.0-79.0)
Hyaline region	-	21.0 ± 1.1 (19.5-23.0)	-

tip frequently offset by deep annuli as small finger-like projection from remainder of tail. Rectal dilation large. Phasmids difficult to discern, located short distance posterior to anal opening.

TYPE PLANT HOST: Tobacco, *Nicotiana tabacum* L.

OTHER PLANTS: In differential host tests (Hartman & Sasser, 1985), the responses were similar to those expected for *M. arenaria* host race 2, infective on tobacco, watermelon, tomato and non-infective on peanut, pepper and cotton (Cliff & Hirschmann, 1984).

TYPE LOCALITY: Chiang Mai, Thailand.

DISTRIBUTION: *Asia*: Thailand.

CHROMOSOME NUMBER: Cytologically, *M. microcephala* revealed a chromosome number of 2n = 37, which is consistent for cytological race B of *M. arenaria*. Triantaphyllou & Hirschmann (1997) demonstrated evidence of direct polyploidisation by the propagation in tomato of 20 exceptionally large J2 isolated from a diploid population, resulting in a tetraploid population with 2n = 74; both populations reproduced exclusively by mitotic parthenogenesis. Morphological and morphometric studies revealed that differences between diploid and tetraploid populations involved larger dimensions of the various life stages (Table 39) but similar morphology. The observation suggested that under field conditions, the tetraploid form would probably have failed to survive (Triantaphyllou & Hirschmann, 1997).

POLYTOMOUS KEY CODES: *Female*: A23, B32, C1, D5; *Male*: A34, B23, C23, D1, E2, F1; *J2*: A2, B3, C3, D1, E21, F3, G4.

BIOCHEMICAL AND MOLECULAR CHARACTERISATION: Esbenshade & Triantaphyllou (1985a) reported the A1 phenotype for esterase and N1 for malate dehydrogenase. No molecular data are available on this species.

RELATIONSHIPS (DIAGNOSIS): This species has a large EP/ST ratio at 3.8, which could differentiate it from other species. This species resembles *M. arenaria* in a few characteristics, but differs in knob shape. The J2 stylet length is shorter, 9.0-10.0 *vs* 10.1-15.6 μm in *M. arenaria.*

61. *Meloidogyne microtyla* Mulvey, Townshend & Potter, 1975
(Fig. 173)

COMMON NAME: Creeping red fescue root-knot nematode.

Mulvey *et al.* (1975) described this species from roots of turfgrass at Cambridge (Preston) in South-Western Ontario, Canada.

MEASUREMENTS *(AFTER MULVEY* ET AL., *1975)*

- *Holotype female*: L = 480 μm; W = 430 μm; stylet = 14 μm.
- *Allotype male*: L = 1380 μm; stylet = 19.0 μm; spicules = 29.0 μm; gubernaculum = 8.0 μm; a = 43.1.
- *Paratype females* (n = 20): L = 560 (480-650) μm; W = 460 (410-520) μm; L/W = 1.2; stylet = 14.0 (13.0-15.0) μm; DGO = 5-6 μm; EP/ST* = 1.2 (*estimated from figure).
- *Paratype males* (n = 20): L = 1210 (1060-1400) μm; stylet = 19.0 (18.0-20.0) μm; DGO = 4 μm; pharynx = 212-218 μm; excretory pore to anterior end = 127-138 μm; spicules = 29.0 (28.0-30.0) μm; gubernaculum = 7.0-9.0 μm; a = 43.1 (36.3-49.6).
- *Paratype J2* (n = 20): L = 380 (350-400) μm; stylet = 11.0-12.0 μm; DGO = 2.5-3.5 μm; tail = 40-48 μm; hyaline region = 9-10 μm; a = 28.8 (26.9-30.8); c = 8.2-8.7.
- *Eggs* (n = 20): L = 80-90 μm; W = 38-43 μm.

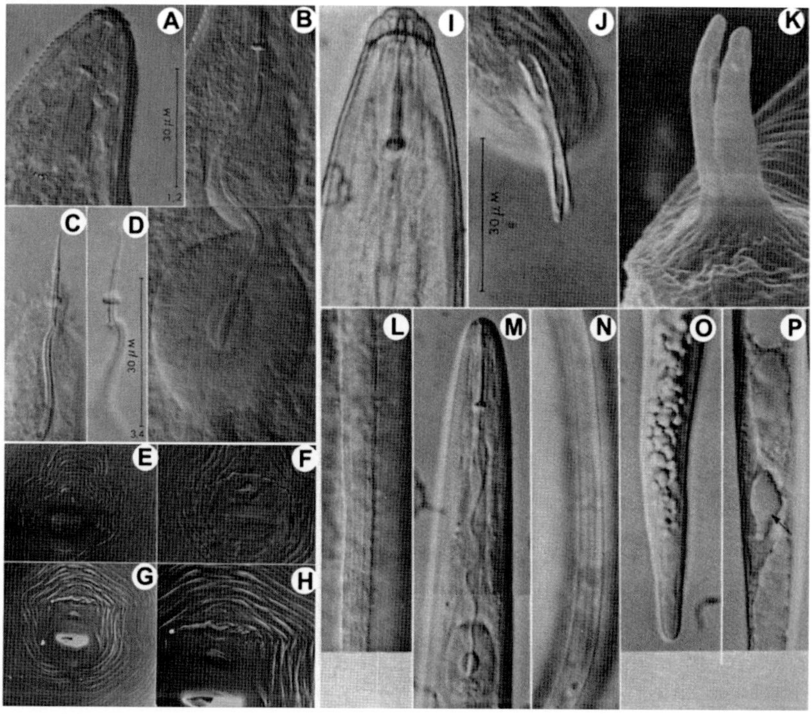

Fig. 173. Meloidogyne microtyla. *LM. A, B: Female anterior region; C, D: Female stylet; E-H: Perineal pattern; I: Male labial region; J, K: Male tail; L: Male lateral field; M: Second-stage juvenile (J2) anterior region; N: J2 lateral field; O: J2 tail; P: J2 anal region. Inflated rectum of J2 arrowed. (After Cliff & Hirschmann, 1984.)*

Description

Female

Body pearly white with relatively long neck, globular to pear-shaped. Neck cuticle 3-4 µm thick, body cuticle 6-7 µm thick. Labial region distinctly offset with two annuli posterior to labial cap. Labial framework present, cephalids not observed. Stylet slender with distinct, rounded knobs. Pharynx very well developed with large cylindrical procorpus and large rounded median bulb provided with a heavily sclerotised valve. Pharyngeal glands obscure. Excretory pore distinct, situated 3-4 annuli posterior to base of retracted stylet. Posterior cuticular pattern not on a protuberance. Lateral incisures absent. Perineal pattern mostly with medium arch, phasmids slightly dorsal to anus and about same distance apart as width of vulval aperture. Striae fairly widely separated, wide and smooth. Lateral area well defined with broken and forked striae. Spermatheca with 18-26 cells and cell boundaries only slightly interlaced (Bert *et al.*, 2002), coincident with hypothesis that strictly amphimictic species have a larger number of spermatheca cells (Bert *et al.*, 2002).

Male

Labial region offset from body, bearing a well-developed labial annulus and a large post-labial annulus. Labial framework prominent, cephalids obscure. Stylet stout with rounded knobs. Cuticular annulation distinct with annuli *ca* 2 μm wide near mid-body. Lateral field distinct with four incisures. Median bulb rounded with well-developed sclerotised valve 68-75 μm from anterior end of body. Hemizonid large, about two annuli anterior to excretory pore duct level. Hemizonid obscure. Spicules arcuate, testis one, occasionally reflexed a short distance from anterior end. Phasmids small, postanal, about mid-way on tail.

J2

Labial region continuous with body, labial framework slight. Labial region bearing three faint post-labial annuli, stylet slender with small basal knobs. Median bulb rounded with prominent sclerotised valve. Labial disc raised above median lips. Rounded shape of labial disc is prominent, prestoma very small. Area around prestoma, surrounded by inner labial sensilla, is slightly depressed. Median lips often deeply indented, forming lip pairs. Large lateral labials triangular, junction of median and lateral labials forming a perpendicular angle with lateral edge of labial disc. Labial sensilla expressed externally, and labial region not annulated (Eisenback & Hirschmann, 1979). Pharynx 150-170 μm long. Hemizonid located slightly anterior and adjacent to excretory pore. Lateral field with four incisures. Tail short, terminus rounded. Rectum dilated in 35 of 50 J2 examined.

TYPE PLANT HOST: Fescue, *Festuca rubra* L., 'Elco'.

OTHER PLANTS: In differential host tests (Hartman & Sasser, 1985) and other host plants, *M. microtyla* was infective on tomato, sugarbeet, oat (*Avena sativa*), brome grass (*Bromus inermis* Leyss), orchard grass (*Dactylis glomerata*), barley (*Hordeum vulgare*), timothy (*Phleum pratense*), rye (*Secale cereale* L.), wheat (*Triticum aestivum*), red clover (*Trifolium pratense*), and white clover (*T. repens*); and non-infective on peanut, pepper, corn (*Zea mays*), birdsfoot trefoil (*Lotus corniculatus* L.), alfalfa (*Medicago sativa*), and carrot (*Daucus carota*) (Mulvey *et al.*, 1975). *Meloidogyne microtyla* reproduced on 62 out of 87 plant species tested in glasshouse conditions (Townshend *et al.*, 1984); graminaceous hosts were best and dicotyledonous weeds and dicotyledonous vegetable species, such as garden pea, celery, and parsley, were poor to intermediate hosts, but capable of maintaining the nematode in the infested area (Townshend *et al.*, 1984).

TYPE LOCALITY: A field at Preston, Ontario, Canada.

DISTRIBUTION: *North America*: Canada. It has been reported only in the type and several localities in Canada (Mulvey *et al.*, 1975).

PATHOGENICITY: *Meloidogyne microtyla* severely damaged an apple cover crop of creeping red fescue (Olthof, 1986). Also, it was pathogenic on creeping red fescue, reducing accumulated top clippings (42%) (Townshend & Potter, 1986). Kentucky bluegrass was not affected by this nematode (Townshend & Potter, 1986). Survival studies (Townshend & Potter, 1986) indicated that nematode populations recovered from soil stored at 5°C increased, whereas those in soil stored at 22°C were reduced to almost undetectable numbers within 98 days after incubation; the reproduction of *M. microtyla* in tomato in soil previously stored at 5°C was nine-fold that of soil from fresh tomato, and 100-fold that of soil stored at 22°C (Townshend & Potter, 1986). The occurrence of a diapause was proposed (Townshend & Potter, 1986).

POLYTOMOUS KEY CODES: *Female*: A32, B32, C3, D5; *Male*: B34, B3, C2, D1, E1, F2; *J2*: A32, B23, C213, D3, E23, F3.

CHROMOSOME NUMBER: *Meloidogyne microtyla* reproduces by amphimixis and has a haploid chromosome number of n = 18-19 (Eisenback & Hirschmann, 1979; Triantaphyllou, 1985b).

BIOCHEMICAL AND MOLECULAR CHARACTERISATION: Esbenshade & Triantaphyllou (1985a) reported M1 phenotype for esterase and H1 for malate dehydrogenase. Tandingan De Ley *et al.* (2002) published a sequence for the 18S rRNA gene.

RELATIONSHIPS (DIAGNOSIS): The species belongs to Molecular group II and is clearly molecularly differentiated from all other *Meloidogyne* species. *Meloidogyne microtyla* can be distinguished by a short female stylet, characteristic perineal pattern with widely spaced striae and in having an area of striae interruption above the anus. This species differs from the most similar species, *M. incognita*, in male spicule length (28-30 *vs* 21-46 μm), and DGO of female *M. microtyla* located 5-6 μm posterior to the stylet base *vs* 2.0-4.8 μm. *Meloidogyne microtyla* perineal patterns differ from that of *M. incognita* in more widely spaced striae and in having an area of striae interruption above the anus, which is consistent in all the patterns examined. From *M. javanica*, it can be distinguished by male lateral field areolation and male labial region shape.

62. *Meloidogyne mingnanica* Zhang, 1993
(Fig. 174)

COMMON NAME: Fujian citrus root-knot nematode.

Zhang (1993) described this species from citrus roots in an orchard in Hubei Longhu, Jinjiang county, south Fujian, China.

MEASUREMENTS *(AFTER ZHANG, 1993)*

- *Holotype female*: L = 900 μm; W = 700 μm; stylet = 13.0 μm; excretory pore to anterior end = 30.0 μm; DGO = 3.2 μm; EP/ST = 2.3.
- *Allotype male*: L = 1940 μm; W = 48 μm; stylet = 23.0 μm; stylet knobs height = 2.5 μm; DGO = 7.5 μm; median bulb to anterior end = 113 μm; excretory pore to anterior end = 195 μm; tail = 15.0 μm; spicules = 40.0 μm; gubernaculum = 10.0 μm, testis length = 880 μm; T = 45.4%.
- *Paratype females* (n = 20): L = 853 $\pm$ 16.3 μm; W = 631 $\pm$ 16.6 μm; stylet = 12.7 $\pm$ 0.15 μm; stylet knob height = 2.5 $\pm$ 0.03 μm; stylet knob width = 4.4 $\pm$ 0.12 μm; DGO = 4.4 $\pm$ 0.13 μm; excretory pore to anterior end = 30 $\pm$ 0.10 μm; L/W = 1.4 $\pm$ 0.03; median bulb to

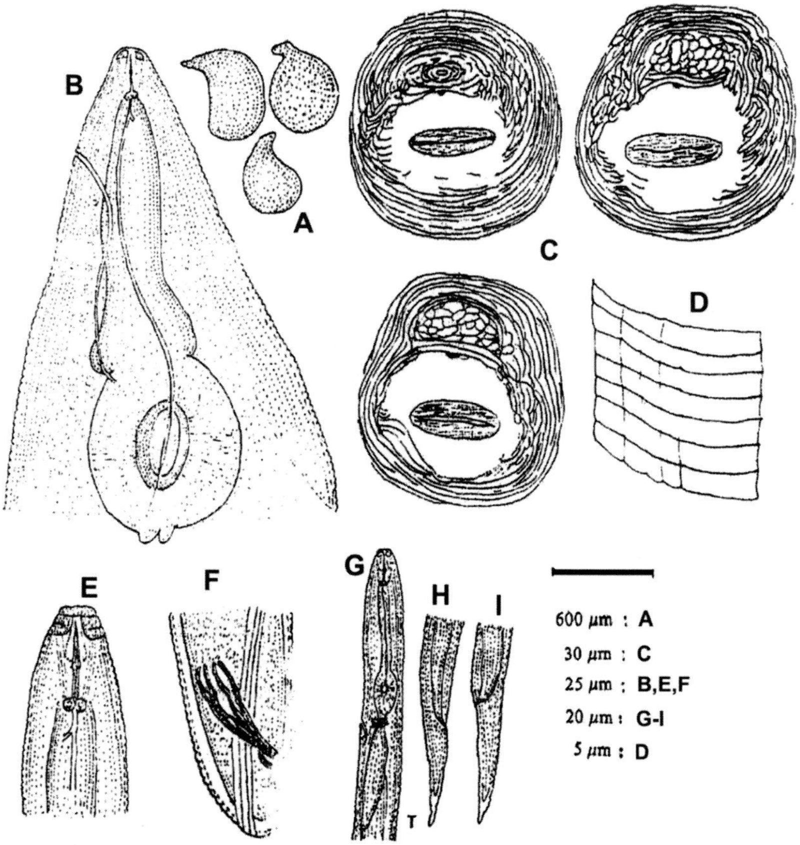

Fig. 174. Meloidogyne mingnanica. *A: Entire female; B: Female anterior region; C: Perineal pattern; D: Female lateral field; E: Male anterior region; F: Male tail; G: Second-stage juvenile (J2) anterior region; H, I: J2 tail. (After Zhang, 1993.)*

anterior end = 82.1 ± 2.7 μm; vulval slit = 34.3 ± 1.0 μm; vulval slit to anus distance = 18.4 ± 0.5 μm; interphasmid distance = 21.3 ± 0.5 μm; EP/ST = 2.4 ± 0.03.

- *Paratype males* (n = 20): L = 1840 ± 31 μm; W = 45.4 ± 0.99 μm; stylet = 22.8 ± 0.18 μm; stylet knobs height = 3.0 ± 0.07 μm; stylet knobs width = 7.2 ± 0.1 μm; DGO = 4.4 ± 0.15 μm; excretory pore to anterior end = 173.7 ± 4.9 μm; median bulb to anterior end = 97 ± 2.0 μm; median bulb length = 23.5 ± 1.3 μm; tail = 17.5 ± 1.4 μm; spicules = 37.5 ± 0.6 μm; T = 48 ± 1.3%; testis = 883.2 ± 26.0 μm.

- *Paratype J2* (n = 20): L = 406 ± 4.31 μm; W = 18.3 ± 0.24 μm; stylet = 9.5 ± 0.21 μm; stylet knobs height = 1.5 ± 0.01 μm; stylet knobs width = 2.6 ± 0.1 μm; DGO = 2.7 ± 0.06 μm; excretory pore to anterior end = 84.2 ± 1.2 μm; median bulb to anterior end = 63.1 ± 0.6 μm; median bulb length = 13.3 ± 0.3 μm; tail = 10.7 ± 0.32 μm; hyaline region = 9.8 ± 0.25 μm.
- *Eggs* (n = 30): L = 111.4 (98-126) μm; W = 47.2 (34.0-57.0) μm; L/W = 2.3 (1.9-3.7).

DESCRIPTION

Female

Body white to slightly yellow, spherical to pear-shaped. Cuticle thick, neck bent ventrally, with a slight protuberance posteriorly. Excretory pore posterior to base of stylet, stylet delicate, cone straight, shaft cylindrical and anterior part of stylet knobs transverse ovoid-shaped. Observation under SEM showed labial disc slightly elevated, median labials indented medianly, overall outline of median and labial disc X-shaped, two anterior portions of a median labial elevated anteriorly with round or blunt ends. Prestoma oval-shaped, stoma slit-like, six inner labial sensilla, labials separated or fused. Perineal pattern 8-shaped, dorsal arch low, square or semi-circle, tail area and perineum area separated by one or two continuous ridge-striae anterior to anus. No striae between anus and vulva. Cuticle of tail terminus irregularly thickened with a whorl or wrinkles. Inner-lateral cuticle ridges raised. Phasmids small and obscure.

Male

Body vermiform, labial region flat, stylet robust with broadly rounded knobs. Spicules well developed, arcuate, gubernaculum crescentic. Observation under SEM showing labial disc rounded, slightly elevated, stoma slit-like, prestoma oval-shaped, six inner labial sensilla. Median labials wider than labial disc, indented or smooth median labials, lateral labials degenerate, amphidial opening large and posterior to labial disc. Incisures starting at sixth body annulus and running to end of tail, initially three lateral lines increasing to four with areolation in middle. Phasmid small, anterior to cloacal aperture.

J2

Body vermiform, flat labial region, stylet straight with distinct broadly rounded stylet knobs. Observation under SEM showing labial disc fused with median labials, slightly elevated, oval-shaped. Prestoma rounded, median labials of variable shape, semicircular, slit-like, or indented to fork-like. Amphidial opening slit-like, between labial disc and lateral labials, lateral labial small, fused at both ends with median labials. Cuticle of incisures thick, incisures starting and ending with three lateral lines. Four lines found in mid-body region, areolation in anterior part of incisures. Tail gradually tapering with blunt end bearing one or two indentations. Rectum not dilated. Hyaline region short.

TYPE PLANT HOST: Unshu mikan, *Citrus unshiu* (Yu.Tanaka *ex* Swingle).

OTHER PLANTS: No other hosts were reported.

TYPE LOCALITY: Citrus orchard at Hubei Longhu, Jinjiang county, Fujian province, China.

DISTRIBUTION: *Asia*: China.

CHROMOSOME NUMBER: No available data.

POLYTOMOUS KEY CODES: *Female*: A1, B3, C2, D1; *Male*: A1, B2, C1, D1, E2, F1; *J2*: A2, B3, C4, D4, E1, F4.

BIOCHEMICAL AND MOLECULAR CHARACTERISATION: No available data.

RELATIONSHIPS (DIAGNOSIS): *Meloidogyne mingnanica* is similar to *M. kongi*. It can be distinguished from *M. kongi* and other *Meloidogyne* spp. by a characteristic 8-shaped perineal pattern with tail area and the perineum area separated by one or two continuous striae above the anus.

63. *Meloidogyne minor* Karssen, Bolk, van Aelst, van den Beld, Kox, Korthals, Molendijk, Zijlstra, van Hoof & Cook, 2004

(Figs 175-178)

Karssen *et al.* (2004) described this species from roots of potato near Zeijerveld, The Netherlands. *Meloidogyne minor* is causing increasing

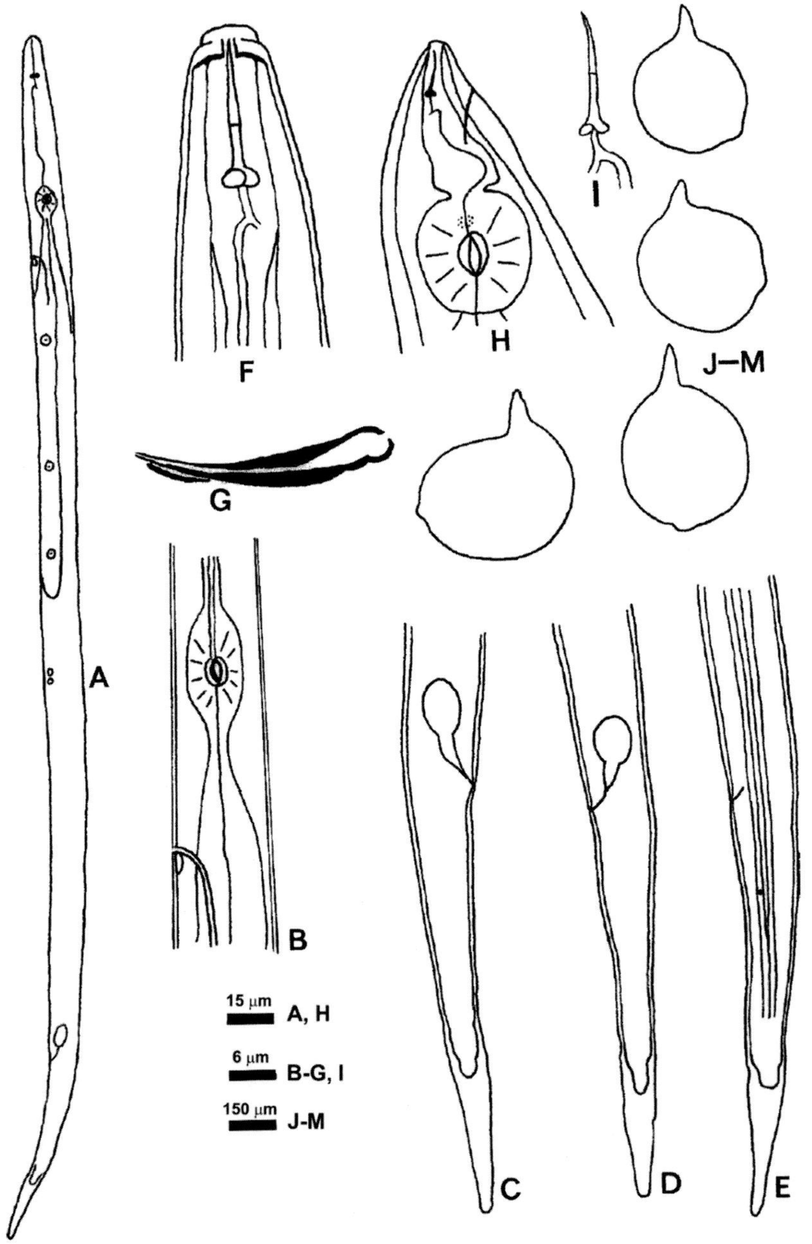

Fig. 175. Meloidogyne minor. *A: Entire second-stage juvenile (J2); B: J2 median bulb; C-E: J2 tail; F: Male labial region; G: Spicules; H: Female labial region; I: Female stylet; J-M: Entire female. (After Karssen* et al.*, 2004.)*

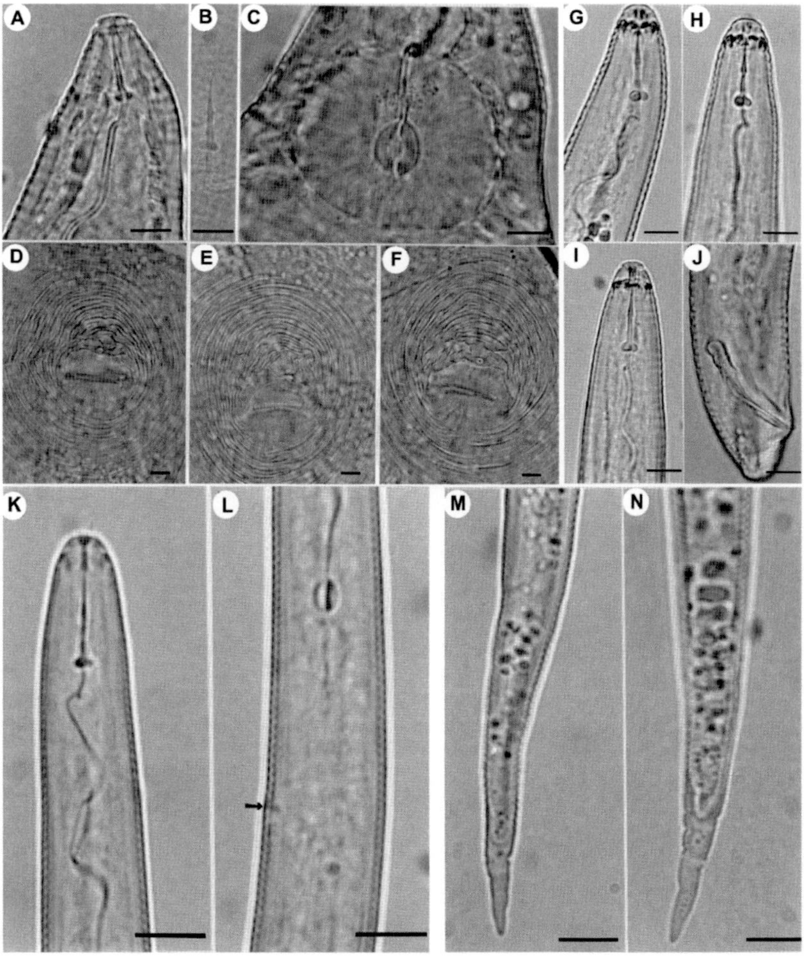

Fig. 176. Meloidogyne minor. *LM. A: Female anterior region; B: Female stylet; C: Female median bulb; D-F: Perineal pattern; G-I: Male labial region; J: Male tail; K: Second-stage juvenile (J2) anterior region; L: J2 median bulb region, excretory pore arrowed; M, N: J2 tail. (Scale bars = 10 μm.) (After Karssen* et al., *2004.)*

concern in temperate agriculture and horticulture. Initially identified from a potato field in The Netherlands that used to be a pasture, *M. minor* has been found primarily on golf courses and sports grounds in The Netherlands, Belgium, UK and Ireland, where it tends to cause yellow patch disease within a few years of new greens have been established

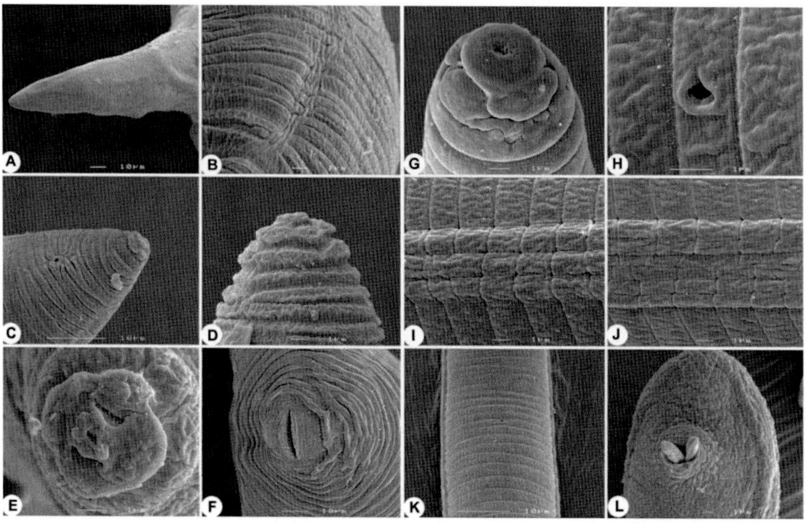

Fig. 177. Meloidogyne minor. *SEM. A, C, D: Female anterior region; B: Female lateral field at neck region; E: Female* en face *view; F: Perineal pattern; G: Male labial region; H: Male excretory pore; I, J: Male lateral field; K: Male excretory pore; L: Male tail. (Scale bars: A, C, F = 10 μm; B, D, E = 1 μm; G-J, L = 1 μm; K = 10 μm.) (After Karssen* et al., *2004.)*

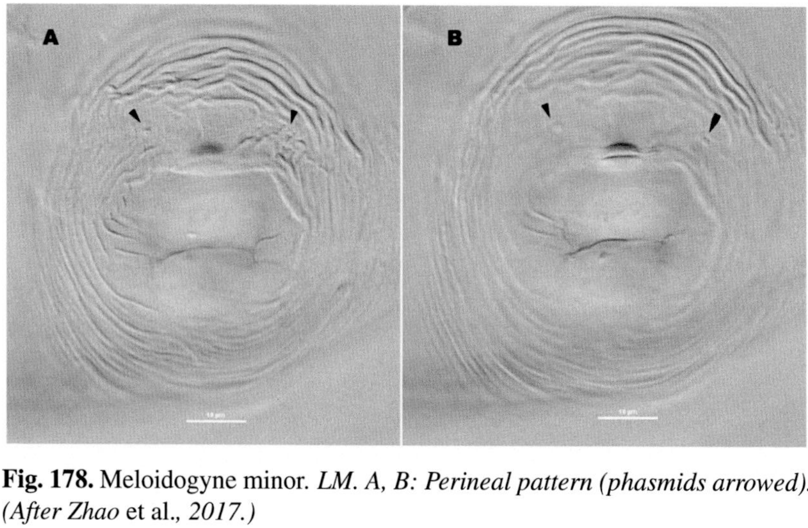

Fig. 178. Meloidogyne minor. *LM. A, B: Perineal pattern (phasmids arrowed). (After Zhao* et al., *2017.)*

(Elling, 2013). *Meloidogyne minor* was also discovered on golf course greens in the state of Washington, USA (Nischwitz *et al.*, 2013).

MEASUREMENTS

See Table 41.

- *Eggs* (n = 25): L = 90.6 (80-103) μm; W = 47.2 (38.5-57.5) μm; L/W = 2.0 (1.6-2.3).

DESCRIPTION

Female

Body relatively small, weakly annulated, pearly white, usually globose, sometimes elongated, neck region distinct, often bent, young females with a slight posterior protuberance. Labial region offset from body. Labial cap distinct, highly variable in shape, labial disc elevated, lateral labials prominent, labial framework weakly sclerotised. Stylet cone slightly curved dorsally, shaft cylindrical, knobs transversely ovoid, slightly sloping posteriad from shaft. Excretory pore located near stylet knob level. Several small vesicles observed near lumen lining of median bulb. Pharyngeal glands variable in size and shape. Perineal pattern small, rounded with fine striae, dorsal arch low with coarse striae, tail remnant area distinct, without punctation, in some patterns weak lateral lines present, phasmids small, usually not visible, located above covered anus. Egg mass *ca* 5-6 times larger than female body size.

Table 41. *Morphometrics of females, males and second-stage juveniles (J2) of* Meloidogyne minor. *All measurements are in μm and in the form: mean ± s.d. (range).*

Character	Holotype, The Netherlands, Karssen *et al.* (2004)	Paratypes, The Netherlands, Karssen *et al.* (2004)	USA Nischwitz *et al.* (2013)	New Zealand Zhao *et al.* (2017)
Female (n)	-	25	-	11
L	570	526 ± 71 (416-608)	-	413 ± 70 (332-534)
a	1.4	1.6 ± 0.3 (1.1-2.3)	-	1.6 ± 0.2 (1.4-1.9)
W	397	339 ± 55 (240-464)	-	258 ± 47 (199-363)

Table 41. *(Continued.)*

Character	Holotype, The Netherlands, Karssen *et al.* (2004)	Paratypes, The Netherlands, Karssen *et al.* (2004)	USA Nischwitz *et al.* (2013)	New Zealand Zhao *et al.* (2017)
Stylet	13.9	14.2 ± 1.1 (12.6-15.2)	-	13.3 ± 0.4 (12.8-13.8)
Stylet knob width	3.2	3.5 ± 0.5 (3.2-3.8)	-	2.5 ± 0.3 (2.1-2.8)
Stylet knob height	1.5	1.7 ± 0.5 (1.3-1.9)	-	1.4 ± 0.3 (1.1-1.7)
DGO	-	4.1 ± 1.2 (3.2-6.3)	-	4.6 ± 0.5 (4.2-5.5)
Median bulb length	32	35.0 ± 7 (27.0-46.0)	-	30.2 ± 6.7 (22.7-41)
Median bulb width	29	31.0 ± 7.3 (22.0-48.0)	-	24.9 ± 4.4 (20.2-31.8)
Median bulb valve width	9.6	8.9 ± 1.2 (7.0-10.1)	-	7.6 ± 0.9 (6.0-8.5)
Median bulb valve length	11.2	11.5 ± 1.6 (9.5-13.3)	-	10.3 ± 1.5 (7.8-12.3)
Ant. end to excretory pore	16.4	18.3 ± 7.8 (13.9-25.9)	-	13.1 ± 1.4 (11.6-14.4)
Vulval slit	24	25.8 ± 2.5 (22.8-29.1)	-	23.8 ± 4.8 (17-29.4)
Vulval slit to anus distance	13.8	15.3 ± 2.5 (12.6-17.1)	-	14.1 ± 2.2 (11.5-15.9)
EP/ST	1.2	1.3	-	1.0
Male (n)	-	25	-	11
L	-	1045 ± 54 (790-1488)	-	1219 ± 271 (818-1689)
a	-	39.0 ± 4.2 (29.8-48.3)	-	29.8 ± 6.7 (19.2-37.7)
b	-	-	-	18.1 ± 4.0 (14.0-22.0)
c	-	101 ± 21.3 (72.4-140)	-	136.5 ± 11.7 (121.1-146.3)
T	-	48.4 ± 12.3 (29.9-73.2)	-	58-60

Table 41. *(Continued.)*

Character	Holotype, The Netherlands, Karssen *et al.* (2004)	Paratypes, The Netherlands, Karssen *et al.* (2004)	USA Nischwitz *et al.* (2013)	New Zealand Zhao *et al.* (2017)
W	-	26.9 ± 4.5 (21.5-31.6)	-	42.0 ± 11 (25.0-52.0)
Labial region height	-	3.9 ± 0.7 (3.2-4.4)	-	4.3 ± 1.0 (3.4-6.1)
Labial region diam.	-	9.6 ± 0.9 (8.9-10.7)	-	10.7 ± 1.2 (9.1-12.3)
Stylet	-	17.8 ± 1.0 (17.1-19.0)	-	17.9 ± 0.8 (16.3-18.8)
Stylet knob width	-	4.2 ± 0.5 (3.8-5.1)	-	3.9 ± 0.3 (3.3-4.3)
Stylet knob height	-	2.0 ± 0.3 (1.9-2.5)	-	2.0 ± 0.3 (1.6-2.6)
DGO	-	3.8 ± 0.4 (3.2-4.4)	-	4.4 ± 0.4 (3.7-4.8)
Median bulb length	-	-	-	23.6 ± 4.3 (20.6-26.6)
Median bulb width	-	9.0 ± 1.7 (7.6-12.0)	-	10.9 ± 1.4 (9.3-12.0)
Median bulb valve length	-	5.0 ± 0.7 (4.4-5.7)	-	6.3 ± 0.6 (5.5-6.9)
Median bulb valve width	-	3.6 ± 0.5 (3.2-3.8)	-	4.0 ± 0.3 (3.5-4.3)
Ant. end to excretory pore	-	114 ± 24.9 (88-137)	-	115.7 ± 31.5 (77.9-160.3)
Spicules	-	25.6 ± 3.4 (22.8-28.4)	-	26.4 ± 2.0 (23.1-28.7)
Gubernaculum	-	6.1 ± 0.6 (5.7-6.3)	-	7.6 ± 0.7 (7.1-8.3)
Tail	-	10.5 ± 2.3 (8.2-12.6)	-	9.2 ± 1.6 (7.4-11.2)
J2 (n)	-	25	21	13
L	-	377 ± 8.0 (310-416)	323 ± 18.8 (300-361)	377 ± 8.0 (370-390)
a	-	28.4 ± 2.0 (23.9-32.4)	27.3 ± 2.0 (23.2-32.5)	22.8 ± 3.4 (17.7-28.7)

Table 41. *(Continued.)*

Character	Holotype, The Netherlands, Karssen *et al.* (2004)	Paratypes, The Netherlands, Karssen *et al.* (2004)	USA Nischwitz *et al.* (2013)	New Zealand Zhao *et al.* (2017)
b	-	-	2.7 ± 0.4 (2.2-3.3)	6.5 ± 0.2 (6.1-6.8)
c	-	7.0 ± 0.3 (6.2-7.6)	6.9 ± 0.4 (6.4-7.6)	6.9 ± 0.4 (6.1-7.8)
c′	-	5.7 ± 0.4 (4.5-6.3)	5.5 ± 0.7 (4.2-6.4)	4.8 ± 1.5 (3.7-5.4)
W	-	13.3 ± 1.3 (12.0-15.8)	11.9 ± 0.8 (10.0-13.0)	17.0 ± 3.0 (13.0-22.0)
Labial region height	-	2.0 ± 0.2 (1.9-2.5)	-	2.6 ± 0.4 (2.0-3.2)
Labial region diam.	-	5.2 ± 0.4 (5.1-5.7)	-	5.8 ± 0.4 (5.2-6.1)
Stylet	-	9.2 ± 0.9 (7.6-10.1)	9.7 ± 0.5 (9.0-10.0)	11.3 ± 1.2 (9.7-12.8)
Stylet knob width	-	1.3 ± 0.2 (1.2-1.4)	-	2.0 ± 0.3 (1.5-2.5)
Stylet knob height	-	1.9 ± 0.3 (1.8-2.0)	-	1.1 ± 0.2 (1.0-1.5)
DGO	-	3.0 ± 0.5 (2.5-3.2)	-	3.4 ± 0.5 (2.7-4.2)
Median bulb width	-	-	-	8.9 ± 0.1 (8.9-9.3)
Median bulb length	-	-	-	14.6 ± 1.0 (13.5-17.1)
Median bulb valve width	-	2.9 ± 0.5 (2.5-3.2)	-	2.8 ± 0.2 (2.7-3.1)
Median bulb valve length	-	3.3 ± 0.3 (3.2-3.8)	-	3.9 ± 0.4 (3.6-4.5)
Ant. end to excretory pore	-	70.5 ± 6.6 (58.1-77.1)	-	72.3 ± 6.2 (64.9-77.9)
Tail	-	54.0 ± 6.2 (49.0-63.0)	47.1 ± 3.7 (40.0-51.5)	54.8 ± 4.3 (52.9-62.6)
Hyaline region	-	16.1 ± 3.9 (12.0-22.1)	14.8 ± 1.5 (12.0-17.5)	15.9 ± 1.1 (14.0-18.0)

Male

Body vermiform, annulated, usually not twisted, tail region curved. Four incisures present in raised lateral field, often with one or two incomplete incisures in middle near mid-body, outer bands irregularly areolated. Labial region not offset from body, one post-labial annulus present, often with one or two incomplete transverse incisures. Labial disc rounded, elevated, fused with anchor-shaped submedian labials. Prestoma hexagonal in shape, surrounded by six inner sensilla. Four labial sensilla present on submedian labials, close to labial disc and marked by small slits. Slit-like amphidial openings present between labial disc and prominent lateral labials. Labial framework strongly sclerotised, vestibule extension distinct. Stylet with straight cone and cylindrical shaft, large transversely ovoid knobs, slightly sloping posteriorly from shaft. Dorsal gland orifice close to stylet knobs. Pharynx with slender procorpus and oval-shaped median bulb. Pharyngeal gland lobe ventrally overlapping intestine, two subventral gland nuclei present. Hemizonid 3.5-4.5 μm long, anterior to excretory pore. Testis very long, monorchic, with outstretched germinal zone. Tail usually curved ventrally, short, conical with bluntly rounded tip. Spicules and gubernaculum slender, curved ventrally, two small pores present on each spicule tip. Phasmids small, located posterior to cloacal aperture.

J2

Body vermiform, relatively short, annulated, anterior part tapering posteriorly to stylet-knob level, posterior part slightly ventrally curved when heat relaxed. Lateral field with four incisures, areolation not visible. Labial region rounded, not offset from body. Labial framework weakly sclerotised, vestibule extension distinct. Stylet small, cone straight, shaft cylindrical, knobs transversely ovoid, slightly sloping posteriorly. Median bulb relatively large, ovoid, triradiate lumen with clear sclerotised lining. Pharyngeal gland lobe relatively long, ventral intestine overlap clearly visible, three gland nuclei present. Hemizonid posterior, adjacent to excretory pore, 2.0-2.5 μm long. Tail straight, sometimes slightly curved ventrally, gradually tapering to finely pointed tail tip, rectum usually weakly dilated. Hyaline region distinct, relatively long and narrow, anterior hypodermal part rounded and relatively narrow, often one or two cuticular constrictions present on tail terminus. Phasmids posterior to anus, at about 33% of tail length, very small, located in ventral incisure of lateral field.

TYPE PLANT HOST: Potato, *Solanum tuberosum* L.

OTHER PLANTS: *Meloidogyne minor* is primarily a pest of creeping bentgrass (*Agrostis stolonifera*) on golf greens and other turfgrasses. *Meloidogyne minor* failed to reproduce on marigold (*Tagetes patula* L.) and corn (*Zea mays*) but reproduced on carrot (*Daucus carota*), phacelia (*Phacelia tanacetifolia* Benth.), alfalfa (*Medicago sativa*), Italian ryegrass (*Lolium multiflorum*), perennial ryegrass (*L. perenne*), oat (*Avena sativa*), lettuce (*Lactuca sativa*), tomato, vetch (*Vicia sativa*) (Karssen *et al.*, 2004). Under field conditions, rye (*Secale cereale* 'Sorum') and sugarbeet (*Beta vulgaris* 'Shakira') were revealed as non-hosts (Thoden *et al.*, 2012).

TYPE LOCALITY: Collected near Zeijerveld, The Netherlands, but species described from a glasshouse culture on the roots of *Solanum lycopersicum*.

DISTRIBUTION: *Europe*: The Netherlands, Belgium, UK, Ireland and Portugal; *North America*: USA (Washington); *Oceania*: New Zealand; *South America*: Chile (Viaene *et al.*, 2007; Wesemael *et al.*, 2011; McClure *et al.*, 2012; Zhao *et al.*, 2017).

PATHOGENICITY: Although *M. minor* was described from tomato roots, it was originally isolated from potato roots. Potato tuber symptoms are more or less comparable with those caused by *M. chitwoodi* and *M. fallax*, *i.e.*, numerous small pimple-like raised areas on the tuber surface with egg-laying females present just below the skin causing small dots of necrotic and brownish tissue. The only difference noticed, compared to the external tuber symptoms of *M. chitwoodi* and *M. fallax*, was that the pimple-like raised areas were more corky with *M. minor*.

Meloidogyne minor appears to be a potentially dangerous plant-parasitic nematode capable of infecting many mono- and dicotyledonous plants. Golf greens of creeping bent grass (*A. stolonifera* var. *stolonifera*) showed yellow patch disease. The patches appear about 10 days after heavy spring rainfall from late May to early June and persisted until

October. On affected greens, the patches appeared in new positions each season (Karssen *et al.*, 2004). The symptoms have been observed in greens constructed with sand/peat (90/10%) sown with creeping bent cultivars. Greens affected include some on established courses and some on greens newly constructed on land formerly used for grass or arable farming. Field trials using plant nutrient solutions and fungicide treatments have not resulted in any improvement in the colour of affected greens (Mark Hunt and Kate Entwhistle, pers. comm.). In a glasshouse experiment, a single treatment of nematicide (oxamyl) applied to cores taken from affected and unaffected turf from three golf courses to control *M. minor* resulted in improved grass growth and colour in affected cores. There were many more *M. minor* J2 in the soil and roots of the yellow cores for all golf courses, densities averaging 3400, 2700 and 1700 J2 (100 g dry soil)$^{-1}$ in yellow cores while unaffected cores from the same greens had <100, 0 and 0 J2 (100 g)$^{-1}$. Gall indices differentiated yellow and green cores and courses in the same way. Pathogenicity studies by Wesemael *et al.* (2014) on potato 'Bintje' determined the tolerance limit in 41 J2 (100 cm^3 soil)$^{-1}$. Based on the nematode development at 22.3°C, Wesemael *et al.* (2014) calculated that *M. minor* requires 606-727 degree days based on 5°C to complete its life cycle on potato. Under field conditions, Thoden *et al.* (2012) tested the potential for damaging potato production in terms of quantity as well as quality in 'Bartina', 'Astérix' and 'Markies'. In all potato cultivars, galls were located at the beginning of lateral roots leading to a thickened root base. Sometimes lateral roots continued to grow, which led to a 'tail' on top of the galls. Tubers also displayed slight symptoms of infection at harvest, showing white dots (young or adult females) under the potato skin but no signs of infection were present on the outside of tubers.

The spatial and temporal distribution of *M. minor* on a creeping bentgrass green in Ireland was determined (Morris *et al.*, 2013). J2 were absent from the soil from November to February, when soil temperatures were below 10°C. Both galls and egg masses were present throughout the year but were more abundant in late summer and early autumn. J2, galls and egg masses were more prevalent in the top 10 cm of soil than deeper zones. The nematode population tended to decrease as distance from the centre of the yellow patches increased. The mean diam. of five sampled patches increased from 23 ± 7 cm in June to 45 ± 2 cm in August (Morris *et al.*, 2013). Egg masses incubated at constant temperatures, J2 hatched between 15 and 25°C, with limited hatch (<1%) at 10 and

30°C. The percentage hatch was lower at 15°C (43%) than at 20-25°C (63-76%). The temperatures at which J2 was active ranged from 4 to 30°C, with greatest activity between 15 and 25°C (Morris *et al.*, 2011).

CHROMOSOME NUMBER: It reproduces by facultative meiotic parthenogenesis, with a haploid chromosome number of n = 17.

POLYTOMOUS KEY CODES: *Female*: A32, B32, C3, D3; *Male*: A43, B34, C32, D1, E2, F1; *J2*: A32, B3, C32, D123, E3, F213.

BIOCHEMICAL AND MOLECULAR CHARACTERISATION: This species is characterised by a N1a malate dehydrogenase (Mdh) pattern with two additional weaker bands after prolonged staining and one very slow weak VS1 esterase (Est) band (Karssen *et al.*, 2004).

Meloidogyne minor can be distinguished from other species by the ITS rRNA, IGS2 rRNA, D2-D3 expansion segments of 28S rRNA, *hsp90* and *COI* gene sequences (McClure *et al.*, 2012; Nischwitz *et al.*, 2013; Hodgetts *et al.*, 2016). Gamel *et al.* (2014) provided PCR-IGS-RFLP profile for *M. minor*. The real-time PCR method for detection of *M. minor* has been also developed (De Weerdt *et al.*, 2011).

RELATIONSHIPS (DIAGNOSIS): The species belongs to Molecular group III and is clearly molecularly differentiated from all other *Meloidogyne* species. Based on morphology, *M. minor* is close to *M. chitwoodi* and *M. microtyla* but differs from them by stylet knob shape, perineal pattern shape, male labial region shape and most J2 characteristics. *Meloidogyne minor* also differs in host range, isozyme patterns from *M. chitwoodi* and *M. microtyla* (Townshend *et al.*, 1984; Esbenshade & Triantaphyllou, 1985b; Ebsary, 1986; De Ley *et al.*, 2002).

64. *Meloidogyne moensi* Le, Nguyen, Nguyen, Liebanas, Nguyen & Trinh, 2019

(Figs 179-181)

COMMON NAME: Moens's coffee root-knot nematode.

Le *et al.* (2019) described this species from roots of coffee in Dak Lak Province, Vietnam.

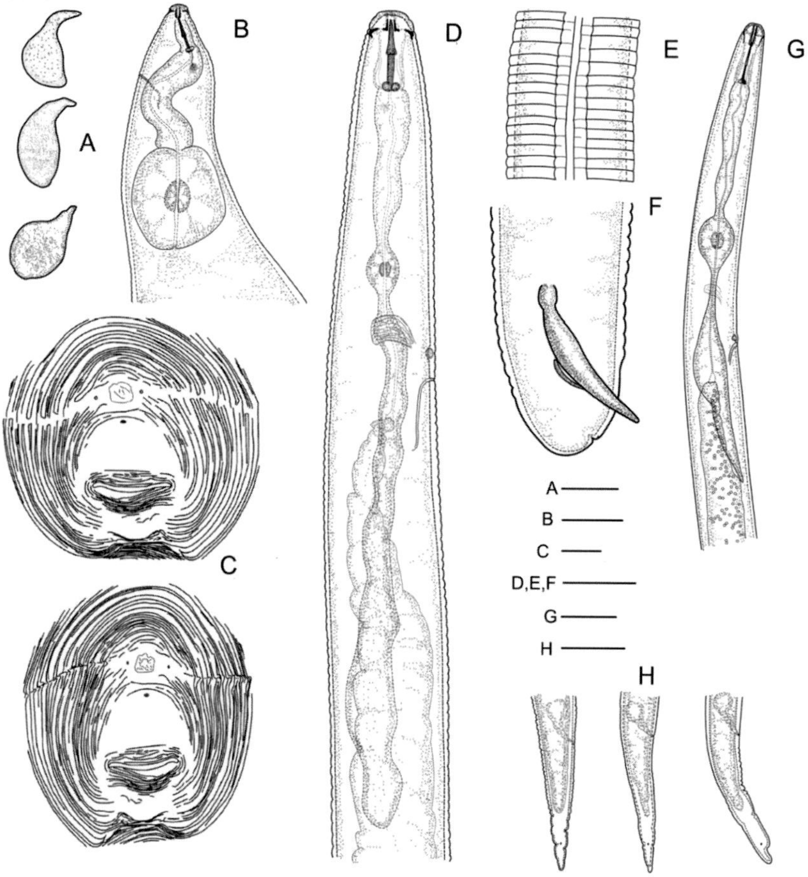

Fig. 179. Meloidogyne moensi. *A: Entire female; B: Female anterior region; C: Perineal pattern; D: Male anterior region; E: Male lateral field; F: Male tail; G: Second-stage juvenile (J2) anterior region; H: J2 tail. (Scale bars: A = 200 μm; B, H = 20 μm; C = 25 μm; D-F = 15 μm; G = 10 μm.) (After Le* et al.*, 2019.)*

MEASUREMENTS *(AFTER LE ET AL., 2019)*

- *Holotype female*: L = 741.4 μm; W = 328 μm; stylet = 14.5 μm; DGO = 5.1 μm; median bulb width = 34 μm; excretory pore to anterior end =34.4 μm; median bulb to anterior end = 82.7 μm; vulval slit = 21.5 μm; vulval slit to anus distance = 11.6 μm; interphasmid distance =13.7 μm; a = 2.3; EP/ST = 2.4.

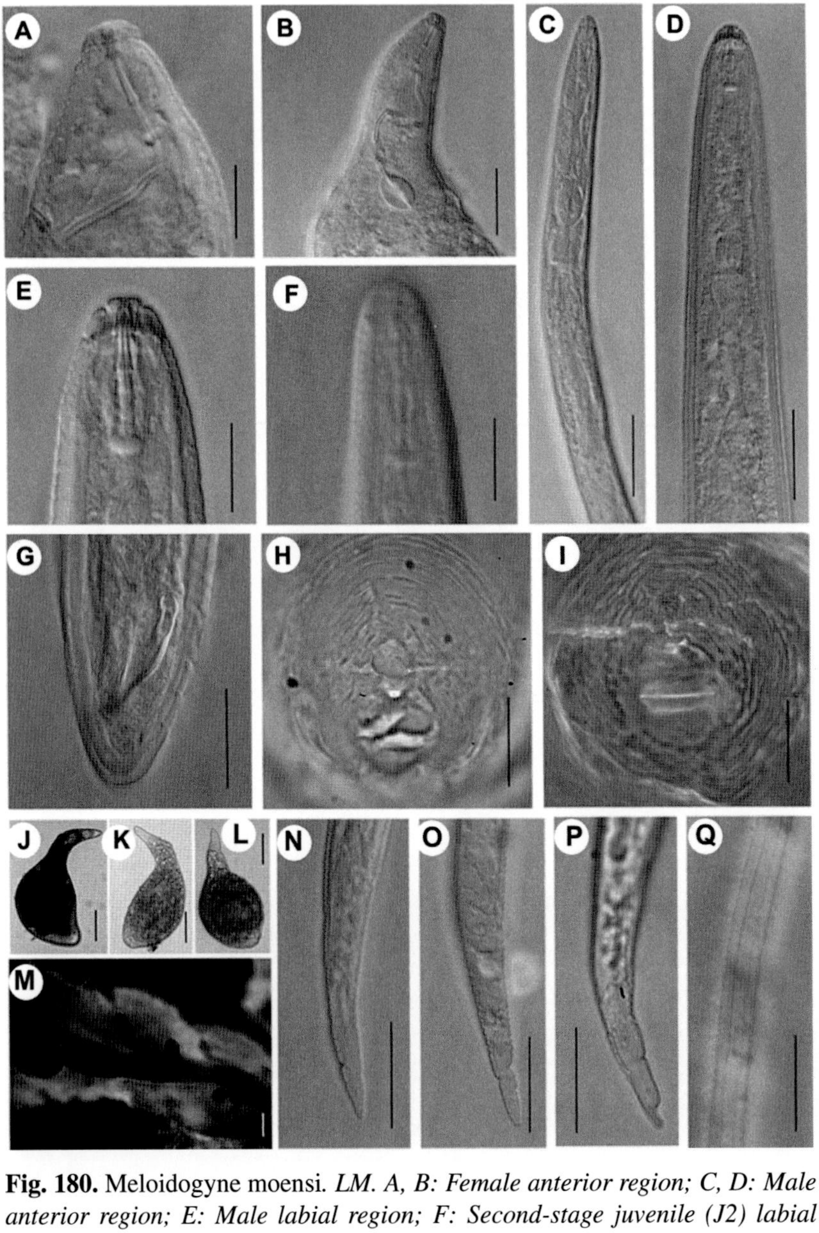

Fig. 180. Meloidogyne moensi. *LM. A, B: Female anterior region; C, D: Male anterior region; E: Male labial region; F: Second-stage juvenile (J2) labial region; G: Male tail; H, I: Perineal pattern; J-M: Entire female; N-P: J2 tail; Q: J2 lateral field. (Scale bars: F = 5 μm; A, E, G, N-Q = 10 μm; B, C, D, H, I = 20 μm.) (After Le* et al., *2019.)*

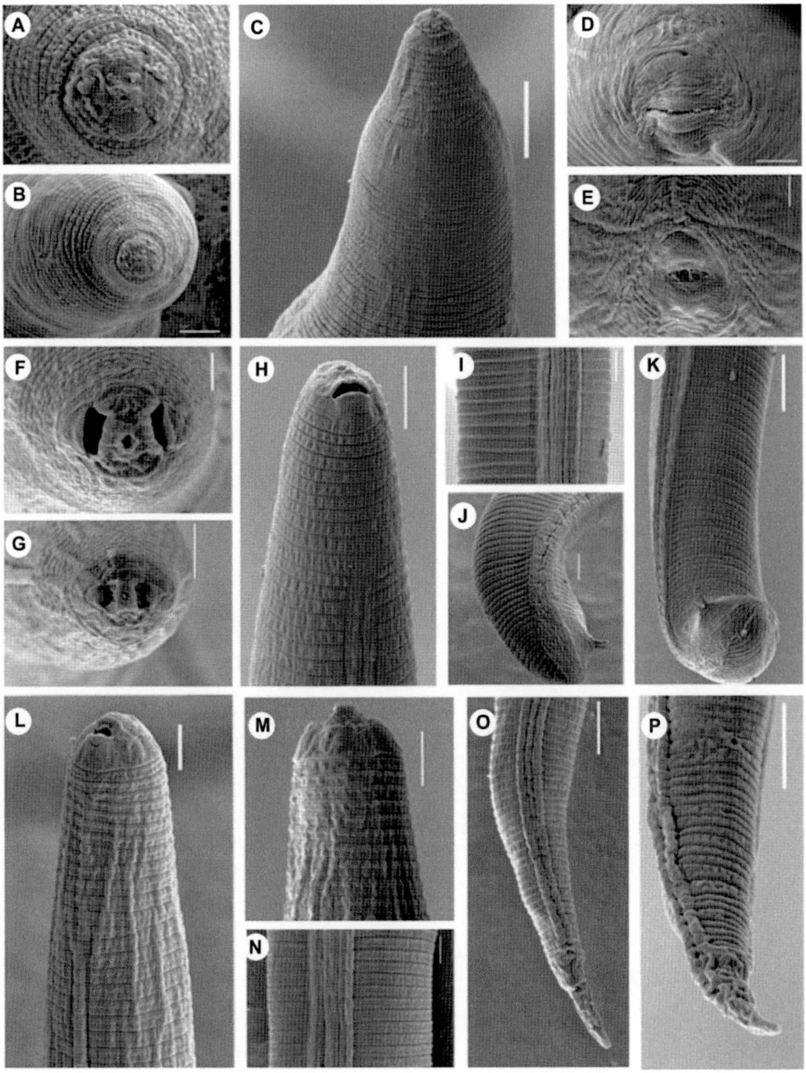

Fig. 181. Meloidogyne moensi. *SEM. A, B: Female* en face *view; C: Female anterior region; D, E: Perineal pattern; F, G: Male* en face *view; H: Male anterior region; I: Male lateral field; J, K: Male tail; L: Second-stage juvenile (J2) anterior region; M: J2 labial region; N: J2 lateral field; O, P: J2 tail. (Scale bars: A, F, L-N = 2 μm; B, H-J, O, P = 5 μm; C-E, G, K = 10 μm.) (After Le* et al., *2019.)*

- *Paratype females* (n = 10): L = 588 ± 133 (345-751) μm; W = 287 ± 73 (149-367) μm; stylet = 16 ± 1.4 (14.3-19.1) μm; DGO = 6.0 ± 1.0 (4.6-7.8) μm; median bulb width = 33 ± 6.8 (26-44) μm; median bulb to anterior end = 99.3 ± 11.7 (80.2-112.3) μm; excretory pore to anterior end = 37.9 ± 10.8 (27.9-61.5) μm; vulval slit = 18.8 ± 2.2 (16.0-24.4) μm; vulval slit to anus distance = 12.2 ± 2.6 (8.7-16.2) μm; interphasmid distance =15.7 ± 2.4 (12.0-19.0) μm; a = 1.7 ± 0.3 (1.4-2.6); EP/ST = 2.4.
- *Paratype males* (n = 10): L = 1198 ± 57 (1103-1266) μm; W = 34.7 ± 6.4 (27-43) μm; stylet = 15.4 ± 0.7 (14.0-16.0) μm; stylet knob width = 4.0 ± 0.4 (3.9-4.1) μm; stylet knob height = 4.0 ± 0.4 (3.4-4.7) μm; DGO = 3.9 ± 0.7 (3.0-4.8) μm; median bulb to anterior end = 72.8 ± 4.3 (67.0-80.3) μm; excretory pore to anterior end = 134 ± 11 (116-148) μm; spicules = 23.8 ± 2.6 (21.0-28.0) μm; tail = 11.8 ± 3.5 (5.9-16.6) μm; a = 35.7 ± 6.8 (26.8-44.8); b = 11.7 ± 0.8 (10.4-13.0); b′ = 5.3 ± 0.3 (4.7-5.7); c = 113.6 ± 46.5 (74.4-205); c′ = 0.5 ± 0.1 (0.3-0.6).
- *Paratype J2* (n = 20): L = 464 ± 21 (428-449) μm; W = 12.3 ± 0.6 (11.5-13.6) μm; stylet = 10.4 ± 0.7 (9.0-12.0) μm; DGO = 3.6 ± 0.4 (2.9-4.1) μm; median bulb to anterior end = 52.1 ± 1.6 (49.0-54.6) μm; excretory pore to anterior end = 73 ± 4 (68-85) μm; tail = 38.2 ± 5 (32.0-49.0) μm; hyaline region = 22.9 ± 3.9 (18.0-31.0) μm; a = 37.6 ± 1.8 (35.0-42.0); b = 6 ± 0.5 (5.6-7.6); b′ = 3.2 ± 0.2 (3.0-3.6); c = 12.3 ± 1.7 (9.4-15.1); c′ = 4.4 ± 0.5 (3.6-5.4).

DESCRIPTION

Female

Body swollen with a small posterior protuberance, pearly white varying in shape, elongated from ovoid to saccate. Neck prominent, bent at various angles to body. Labial region slightly offset from rest of body, stoma slit-like, located in prominent ovoid pre-stomatal cavity, surrounded by pit-like openings of six inner labial sensilla, labial disc round, slightly raised above median labials, labial cap and median labials slightly raised above lateral labials, median labials dumbbell-shaped (in SEM), lateral labials large, fused laterally with labial region, amphidial apertures oval-shaped, located between labial disc and lateral labials. Labial framework strong, hexaradiate. Stylet short, cone base triangular and wider than shaft, stylet tip normally straight, sometimes slightly curved dorsally, three oval stylet knobs sloping posteriorly. Secretory-excretory pore located at level of procorpus, posterior to stylet knobs, median bulb rounded or oval, with oval-shaped valve, pharyngeal glands with one large dorsal lobe, variable in shape, position and size. Perineal pattern round to oval with continuous, smooth, distinct striae, lateral field marked as a faint space, or linear depression junction of dorsal and ventral striate, dorsal arch low, rounded, covering distinct vulva and tail terminus, phasmids large, distinct, vulval slit centrally located at unstriated area, nearly as wide as vulva-anus distance, perivulval region free of striae, tail tip visible, wide, surrounded by concentric circles of striae, ventral striae concave, often free of striae.

Male

Body vermiform, anterior end tapering, posterior end bluntly rounded. Body annuli large, distinct. Lateral fields areolated with three incisures beginning near level of stylet knobs, two additional incisures starting near level of median bulb. Labial cap high and rounded, consisting of a large labial and two post-labial annuli, sometime with incomplete annuli (LM). Labial region continuous with body, stoma slit-like, located in ovoid to hexagonal cavity, surrounded by pit-like openings of six inner labial sensilla, subventral and subdorsal labials fusing to form median labials, each labial with two labial sensilla, lateral labials large, triangular, lower than labial disc and median labials, posterior edge of one or both lateral labials separated with labial region, crescentic, amphidial apertures elongated, located between labial disc and lateral labials. Stylet robust, cone pointed, smaller than shaft, slightly increasing in diam. posteriorly, knobs rounded, reniform, distinctly indented, posteriorly sloping, transversely elongated, merging gradually with shaft. Procorpus distinctly outlined, median bulb ovoid, with strong valve apparatus. Secretory-excretory pore distinct, located four to six annuli posterior to hemizonid. Testis one, occupying 58% body cavity, spicules slightly curved ventrally with bluntly rounded terminus, gubernaculum short, crescentic. Tail short, phasmids distinct, located at cloacal aperture level.

J2

Body slender, tapering to an elongated tail. Body annuli distinct, but fine. Lateral fields starting near level of procorpus as two lines, third line starting near median bulb and quickly dividing into four lines running entire length of body before gradually decreasing to two lines ending near hyaline region, lateral field areolated over entire body. Labial region narrower than body, weak and slightly offset. Under SEM, prestoma opening rounded, surrounded by small, pore-like openings of six inner labial sensilla, median labials and labial disc dumbbell-shaped, lateral labials large, triangular, lower than labial disc and median labials, labial sensilla not seen, amphidial apertures elongated, located between labial disc and lateral labials, labial region not annulated. Stylet slender, cone weakly expanding at junction with shaft, knobs small, oval-shaped and posteriorly sloping. Procorpus faintly outlined, median bulb broadly oval, valve large and heavily sclerotised, isthmus clearly defined, pharyngo-intestinal junction located posterior to level of secretory-excretory pore, gland lobe variable in length overlapping intestine ventrally. Secretory-excretory pore located posterior to hemizonid. Tail conoid with rounded unstriated tail tip, hyaline region clearly defined, rectum dilated, phasmids small, distinct.

TYPE PLANT HOST: Robusta coffee, *Coffea canephora* Pierre *ex* A.Froehner.

OTHER PLANTS: No other hosts were reported.

TYPE LOCALITY: Western Highland: Dak Lak Province, Vietnam.

DISTRIBUTION: *Asia*: Vietnam. It has been reported only in the type locality (Le *et al.*, 2019).

PATHOGENICITY: The coffee roots infected with *M. moensi* showed inconspicuous galls or slight swellings and had relatively small galls (1-2 mm diam.). The presence of the nematodes was detected on coffee by the distortion of the root system preventing absorption of nutrients that in turn lead to stunting and chlorosis. Each gall contained several females with egg masses remaining inside the gall.

CHROMOSOME NUMBER: No available data.

POLYTOMOUS KEY CODES: *Female*: A32, B213, C2, D3; *Male*: A34, B4, C32, D1, E1, F1; *J2*: A2, B32, C34, D1, E21, F32.

BIOCHEMICAL AND MOLECULAR CHARACTERISATION: No biochemical data are available for *M. moensi*. This species can be clearly distinguished by ribosomal RNA genes (ITS, 18S and D2-D3 segments of 28S rRNA) and mitochondrial (*COI*, *COII*-16S region) genes (Le *et al.*, 2019).

RELATIONSHIPS (DIAGNOSIS): *Meloidogyne moensi* is similar to *M. acronea*, *M. aberrans*, *M. africana*, *M. graminis*, *M. ichinohei*, *M. marylandi*, *M. megadora*, and *M. ottersoni*. However, it can be distinguished from each by certain features. *Meloidogyne moensi* differs from *M. acronea* in the females by perineal pattern, shorter body length (345-751 *vs* 527-1243 μm), and longer stylet length (14.3-19.1 *vs* 10-14 μm); from *M. aberrans* by females having striae in perineal pattern distinct and continuous *vs* extremely faint and broken, shorter body length (345-751 *vs* 806.2-1119.1 μm), longer stylet length (16 (14.3-19.1) *vs* 14.5 (13.6-15.5) μm) and in the J2 by shorter stylet length (10.4 (9.0-12.0) *vs* 16.3 (15.9-16.8) μm); from *M. africana* by longer female stylet length (14.3-19.1 *vs* 12.0-17.0 μm); from *M. graminis* by longer female stylet (14.3-19.1 *vs* 10.0-16.0 μm), excretory pore located posterior to stylet knobs *vs* about on a level with knobs and in the males by shorter average body length (1198 *vs* 1270-1512 μm); and from *M. ottersoni* by longer female stylet length (14.3-19.1 *vs* 9-13 μm).

65. *Meloidogyne morocciensis* Rammah & Hirschmann, 1990
(Figs 182-185)

COMMON NAME: Morrocan root-knot nematode.

Rammah & Hirschmann (1990b) described this species from roots of peach rootstock from Morocco.

MEASUREMENTS *(AFTER RAMMAH & HIRSCHMANN, 1990B)*

- *Holotype female*: L = 634 μm; W = 507 μm; stylet = 16.5 μm; DGO = 4.5 μm; excretory pore to anterior region = 37.5 μm; a = 1.3; EP/ST = 2.3.
- *Paratype females* (n = 35): L = 755 ± 90.3 (551-923) μm; W = 490 ± 72.8 (219-608) μm; stylet = 16.5 ± 0.8 (14.8-17.9) μm; DGO =

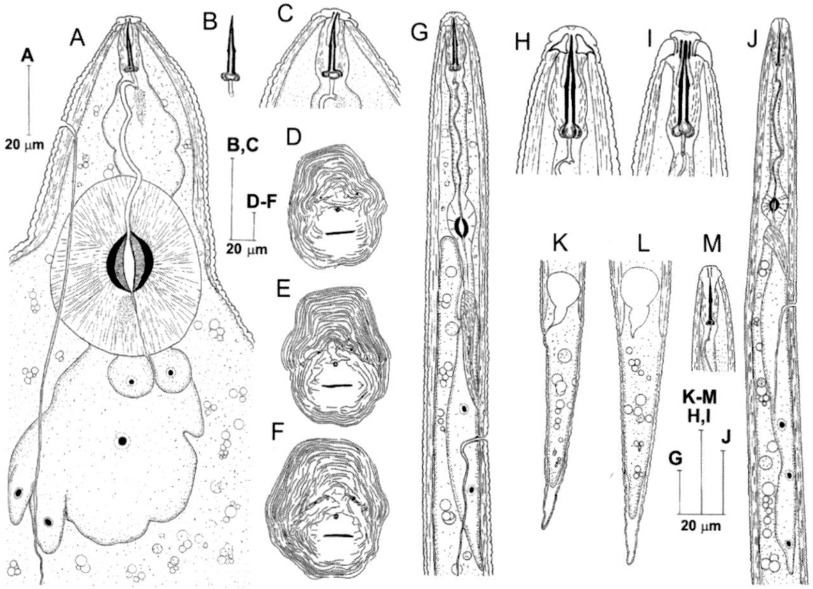

Fig. 182. Meloidogyne morocciensis. *A: Female anterior region; B: Female stylet; C: Female labial region; D-F: Perineal pattern; G: Male anterior region; H, I: Male labial region; J: Second-stage juvenile (J2) anterior region; K, L: J2 tail; M: J2 labial region. (After Rammah & Hirschmann, 1990.)*

3.9 ± 0.6 (2.5-5.2) μm; excretory pore to anterior end = 60.3 (23.7-125.8) μm; vulval slit = 25.2 ± 2.2 (22.2-29.6) μm; vulval slit to anus distance = 19.5 ± 2.7 (14.8-27.4) μm; interphasmid distance = 16.5 ± 2.0 (14.8-17.9) μm; a = 1.6 ± 0.4 (1.2-3.7); EP/ST = 3.6.

- *Allotype male*: L = 1754 μm; W = 38.5 μm; stylet = 24.0 μm; DGO = 5.0 μm; tail = 14.0 μm; a = 45.7; c = 119.5.
- *Paratype males* (n = 35): L = 1621 ± 155 (1296-1863) μm; W = 36.1 ± 4.5 (29.6-48.1) μm; stylet = 24.6 ± 0.7 (22.9-25.8) μm; DGO = 4.7 ± 0.7 (3.5-6.2) μm; median bulb to anterior end = 97.8 ± 3.9 (89.9-103.6) μm; excretory pore to anterior end = 177 ± 10.6 (150-195) μm; tail = 14.5 ± 1.5 (10.7-17.0) μm; spicules = 34.8 ± 2.1 (31.1-39.2) μm; gubernaculum = 9.4 ± 0.8 (8.1-10.7) μm; a = 45.2 ± 4.6 (36.5-57.5); c = 112.9 ± 14.9 (87.6-148.5).
- *Paratype J2* (n = 35): L = 401 ± 21.7 (374-454) μm; W = 15.2 (14.8-16.3) μm; stylet = 12.3 ± 0.5 (11.3-13.3) μm; DGO = 3.8 ± 0.4 (3.0-

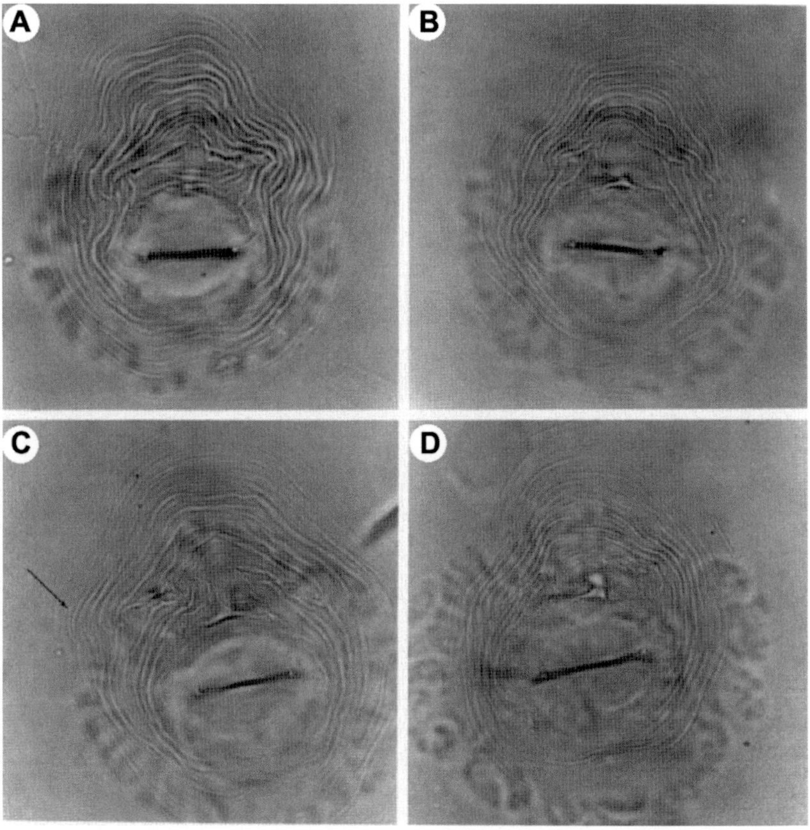

Fig. 183. Meloidogyne morocciensis. *A-D: Perineal patterns (arrow = 'shoulder'). (After Rammah & Hirschmann, 1990.)*

4.4) μm; median bulb to anterior end = 57.5 ± 5.7 (26.4-63.2) μm; excretory pore to anterior end = 88.1 ± 2.7 (84.4-94.7) μm; tail = 52.6 ± 2.7 (46.6-58.1) μm; a = 26.4 ± 1.4 (23.6-30.7); c = 7.6 ± 0.5 (6.4-8.9); c′ = 4.7 ± 0.3 (4.1-5.8).

- *Eggs* (n = 50): L = 93 ± 3.5 (84.4-101.3) μm; W = 42.6 ± 3.3 (38.5-50.3) μm; L/W = 2.2 ± 0.2 (1.7-2.6).

After Silva *et al.* (2020).

J2: L = 389.3 ± 3.8 (377.5-425.1) μm; stylet = 13.9 ± 0.2 (12.7-14.9) μm; DGO = 3.6 ± 0.1 (3.3-4.2) μm; tail = 47.5 ± 0.6 (45.3-48.9) μm; hyaline region = 14.1 ± 0.3 (12.5-17.0) μm; a = 25.6 ± 0.4 (23.8-28.8) μm; c = 8.6 ± 0.1 (8.0-9.3).

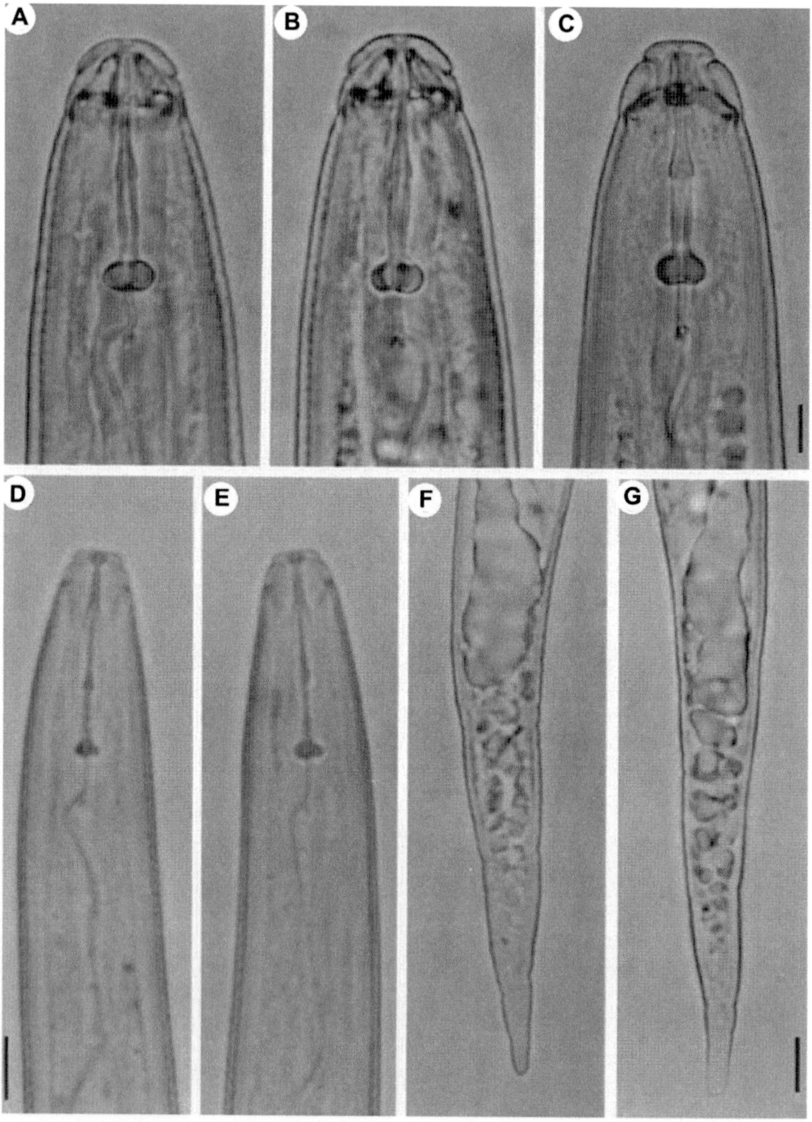

Fig. 184. Meloidogyne morocciensis. *LM. A-C: Male labial region; D, E: Second-stage juvenile (J2) labial region; F, G: J2 tail. (Scale bars: = 5 μm.) (After Rammah & Hirschmann, 1990.)*

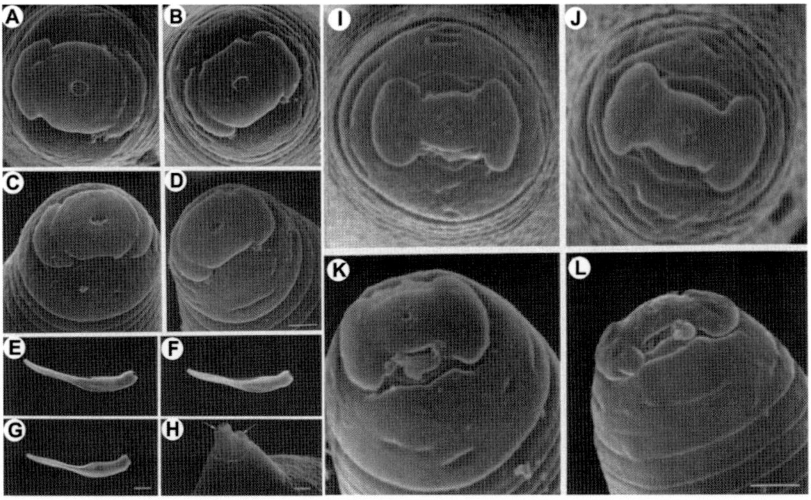

Fig. 185. Meloidogyne morocciensis. *SEM. A-D: Male* en face *view; E-H: Detail of spicules showing pores (arrows); I-L: Second-stage juvenile labial region. (Scale bars: A-D = 2 μm; E-G = 5 μm; H = 2 μm.) (After Rammah & Hirschmann, 1990.)*

DESCRIPTION

Female

Body globular, pearly white, variable in size, neck prominent, posterior end rounded, without distinct protuberance. Body cuticle distinctly annulated, annuli smaller in anterior neck region. Labial region offset, usually marked by incomplete annulation. Labial cap distinct, labial disc slightly elevated. Labial framework weakly sclerotised, vestibule and vestibule extension distinct. Stylet cone dorsally curved, shaft cylindrical. Stylet knobs distinctly separate, offset from shaft, transversely ovoid, with or without slight anterior indentation. Pharyngeal gland lobe large, three nuclei present, indicating one dorsal gland and two subventral glands. Two pharyngo-intestinal cells located near junction of median bulb and intestine. Excretory pore located between dorsal pharyngeal gland orifice and median bulb. Perineal patterns oval to squarish. Striae coarse, widely separated, usually continuous, sometimes broken. Tail tip distinct, with or without very fine broken striations. Fold over anus present. Vulval slit-like, usually without striae near lateral edges. Phasmids small, distinct. Dorsal arch moderately high to high, rounded to squarish, sometimes forming 'shoulders'. District lateral lines absent, indicated by slight interruption of striae. Sometimes lateral lines with short, vertical striae near phasmid area. Ventral pattern region rounded, striae smooth.

Male

Body vermiform, tapering anteriorly, bluntly rounded posteriorly. Heat-killed males assuming C-shape. Cuticle with distinct annulations. Lateral field with four incisures, areolated. Labial region offset, with incomplete, distinct annulations, usually at median sides. Labial cap with distinct labial disc. Amphidial openings elongated slits. Labial framework moderately sclerotised. In SEM (face view), labial disc elevated, almost circular, distinctly separated from median labials. Median labials crescentic, with distinct lateral indentations at junction with labial disc. Diam. of median labials smaller than that of labial disc. Labial sensilla obscure. Lateral labials absent. Stoma opening slit-like, situated below large, hexagonal prestoma. Inner labial sensilla indistinct, opening into prestomatal cavity. Stylet robust, large. Cone straight, slightly longer than shaft, tip pointed. Stylet opening situated at about 25% of cone length from tip. Base of cone broadened near junction with shaft. Shaft cylindrical, with same diam. along its length. Knobs offset from shaft, large, rounded, rarely slightly pear-shaped and sloping posteriorly. Procorpus well defined. Median bulb oval-shaped, with large valve. Pharyngo-intestinal junction obscure, at level of nerve ring. Gland lobe variable in length, with two nuclei. Caecum extending to level of median bulb. Excretory pore position variable, terminal excretory duct long. Hemizonid 2-4 annuli anterior to excretory pore. One or two testes, directed anteriorly, sometimes reflexed posteriorly. Phasmids pore-like, located at level of cloacal aperture. Spicules identical. Labial region cylindrical, offset, circular cytoplasmic core opening on outward lateral side. Shaft limits indistinct. Blade arcuate, tapering towards tip. Vela clearly visible on inward side of spicule. Distance between dorsal and ventral vela wide at beginning of blade, narrowing suddenly at middle of spicule length. Blade tip slightly curved ventrally, with two pores to exterior.

J2

Body vermiform, tapering at both ends, but more so posteriorly. Body annulation distinct, becoming larger and irregular in posterior tail region. Lateral field with four incisures, non-areolated. Labial region slightly offset, with incomplete annulation. Labial cap low, narrower than labial region. In SEM, labial disc elongated, slightly elevated. Median labials crescentic, with rounded corners, with or without indentations at their junction with labial disc. Labial sensilla distinct on median labials. Lateral labial margins rounded to slightly triangular, positioned below labial disc and median labials. Amphidial openings slit-like, just posterior to lateral sides of labial disc. Stoma slit-like, located below circular prestoma, surrounded by six pit-like inner labial sensilla. Labial framework weakly sclerotised. Vestibule and vestibule extension distinct. Stylet cone straight, pointed, gradually increasing in width posteriorly. Shaft cylindrical, may widen slightly posteriorly. Knobs distinctly separate, rounded, sloping posteriorly. Procorpus faintly outlined. Median bulb oval, with prominent valve. Pharyngo-intestinal junction indistinct, at level of nerve ring. Gland lobe variable in length, with three nuclei, dorsal nucleus smaller than two subventrals. Hemizonid two or three annuli anterior to excretory pore. Tail conical, ending in bluntly rounded tip. Tail annulations become irregular, increasing in size toward tip. Hyaline region distinct. Rectal dilation large. Phasmids obscure, at one-third tail length posterior to anal opening.

TYPE PLANT HOST: Peach rootstock, *Prunus persica* 'Missouri'.

OTHER PLANTS: Tomato (*Solanum lycopersicum*) (Carneiro *et al.*, 2008), soybean (Mattos *et al.*, 2016), kiwi (*Actinidia deliciosa* (A.Chev.) C.F.Liang & A.R.Ferguson), grapes (*Vitis vinifera*) (Divers *et al.*, 2019; Silva *et al.*, 2020), pumpkin (*Cucurbita pepo* L.) (Barros *et al.*, 2018) and beetroot (*Beta vulgaris* L.) (Machaca-Calsin *et al.*, 2021).

TYPE LOCALITY: Ain Taoujdate, Morocco.

DISTRIBUTION: *Africa*: Morocco; *Europe*: France; *South America*: Brazil (Carneiro *et al.*, 2008; Silva *et al.*, 2020).

PATHOGENICITY: Silva *et al.* (2020) reported pathogenicity of *M. morocciensis* to peach in an experimental test. Machaca-Calsin *et al.* (2021) showed the pathogenicity of this species to beetroot.

CHROMOSOME NUMBER: *Meloidogyne morocciensis* reproduces by mitotic parthenogenesis and has a somatic chromosome number of 42-49 (Rammah & Hirschmann, 1990b; Carneiro *et al.*, 2008).

POLYTOMOUS KEY CODES: *Female*: A213, B21, C1, D6; *Male*: A213, B12, C12, D1, E1, F2; *J2*: A23, B23, C3, D3, E32, F32.

BIOCHEMICAL AND MOLECULAR CHARACTERISATION: Biochemically, *M. morocciensis* has an A3 esterase phenotype and the malate dehydrogenase pattern is N1 (Rammah & Hirschmann, 1990b). The Est phenotype A3 of *M. arenaria* was identified as *M. morocciensis* (Carneiro *et al.*, 2008; Silva *et al.*, 2013, 2020; Mattos *et al.*, 2016).

Tigano *et al.* (2005) and Monteiro *et al.* (2017) provided 18S RNA, coxII-16S rRNA gene fragment, ITS rRNA and the D2-D3 of 28S rRNA gene sequences for *M. morocciensis*. Combined AFLP and RAPD markers allowed the separation of *M. morocciensis* from other species (Monteiro *et al.*, 2017). Specific SCAR primers developed to identify *M. arenaria* (Zijlstra *et al.*, 2000) also generated an amplicon with *M. morocciensis* (Carneiro *et al.*, 2008; Silva *et al.*, 2020).

RELATIONSHIPS (DIAGNOSIS): The species belongs to Molecular group I, the Incognita group, and resembles *M. incognita*, *M. floridensis*, *M. arenaria* and *M. paranaensis*. From *M. incognita* and *M. floridensis* it can be separated by EP/ST ratio, and from *M. arenaria* and *M. paranaensis* and other species by esterase pattern and molecular markers.

66. *Meloidogyne naasi* Franklin, 1965

(Figs 186, 187)

COMMON NAME: Barley root-knot nematode.

Franklin (1965a) described this species from the UK in roots of cereals, grasses and sugarbeet in west and south-west England and Wales. *Meloidogyne naasi* can cause serious cereal crop losses and is a problem on turfgrasses.

MEASUREMENTS

See Table 42.

- *Eggs*: L = 89 (71-99) μm; W = 41 (36-48) μm; L/W = 2.2.

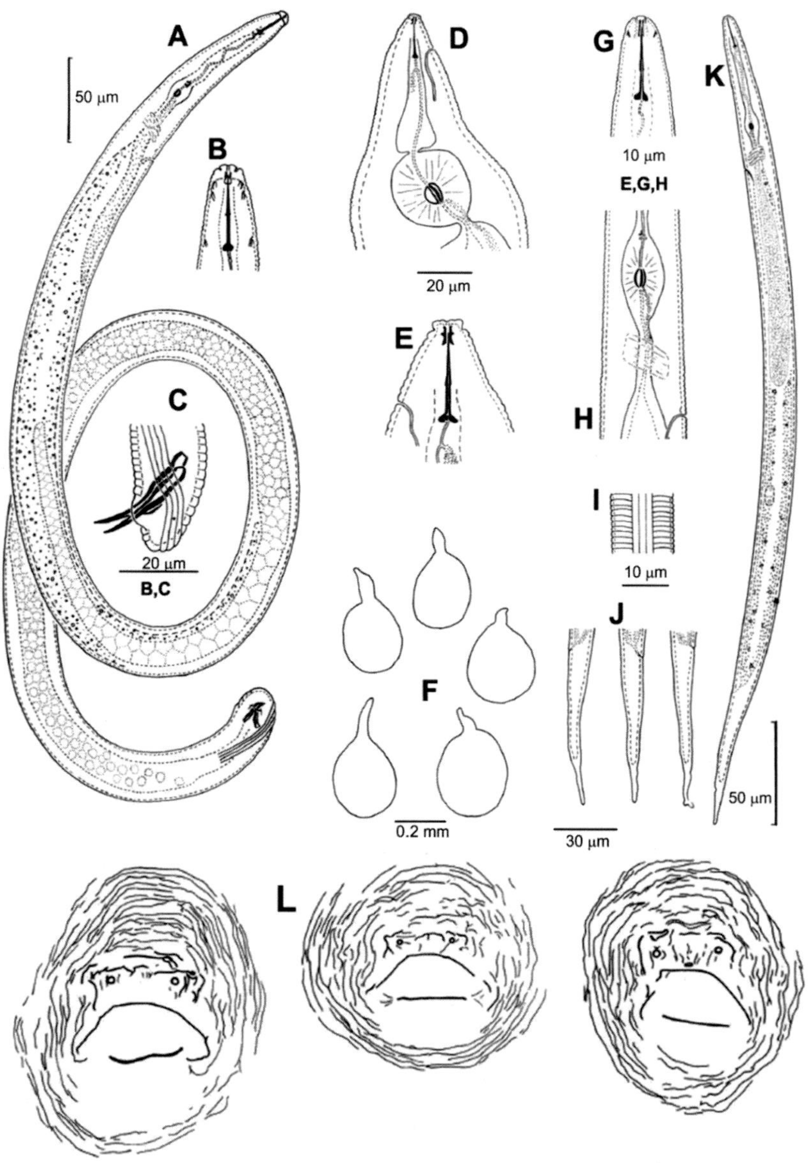

Fig. 186. Meloidogyne naasi. *A: Entire female; B: Male labial region; C: Male tail; D, E: Female anterior region; F: Entire female; G: Second-stage juvenile (J2) labial region; H: J2 median bulb region; I: J2 lateral field; J: J2 tail; K: Entire J2; L: Perineal pattern. (After Franklin, 1973.)*

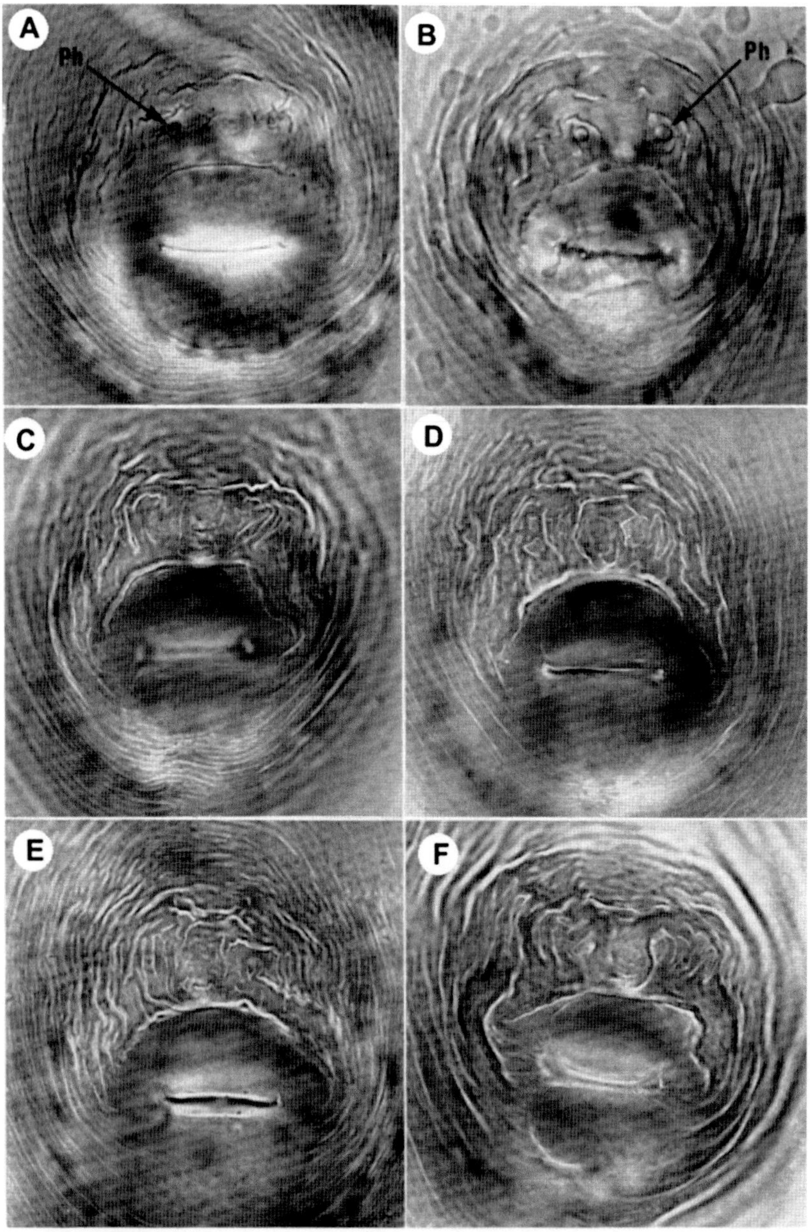

Fig. 187. Meloidogyne naasi. *LM. A-F: Perineal patterns. Ph = phasmid. (After Franklin, 1965.)*

Table 42. *Morphometrics of females, males and second-stage juveniles (J2) of* Meloidogyne naasi. *All measurements are in μm and in the form: mean ± s.d. (range).*

Character	Holotype, Allotype England & Wales, Franklin (1965a)	Paratypes, England & Wales, Franklin (1965a)	The Netherlands, Karssen (2002)	New Zealand, Zhao *et al.* (2017)	India, Suresh *et al.* (2017)	Portugal, Vieira dos Santos *et al.* (2020)
Female (n)	-	25	8	-	-	-
L	548	557 (455-705)	-	-	626 ± 105.2 (505-696)	-
W	305	330 (227-398)	-	-	331 ± 78.4 (243-394)	-
Stylet	15	13 (11.0-15.0)	13.5 ± 0.3 (13.3-13.9)	-	12.8 ± 1.3 (11.5-14.0)	-
DGO	3.0	3.0 (2.0-4.0)	3.7 ± 0.4 (3.2-4.4)	-	3.1 ± 0.4 (2.7-3.5)	-
Median bulb length	34	34 (28-46)	39.4 ± 3.3 (34.1-44.2)	-	31.7 ± 2.5 (30-43)	-
Median bulb width	27	28 (20-40)	37.3 ± 2.6 (31.6-39.2)	-	-	-
Anterior end to excretory pore	-	-	13.4 ± 0.9 (12.6-14.5)	-	-	-
Vulval slit	-	22.0 (17.0-25.0)	-	-	21.7 ± 2.5 (19-24)	20.7 (17.9-22.4)
Vulval slit to anus distance	-	-	-	-	25 ± 2.3 (23-27.5)	-
Interphasmid distance	-	22 (18-26)	-	-	-	18.1 (14.1-20.0)
EP/ST	-	-	1.0	-	-	-
Male (n)	1	25	5	-	-	-
L	1153	1148 (860-1316)	966 ± 167 (742-1088)	-	1252 ± 54 (1197-1305)	-
a	-	40 (32-48)	33.2 ± 2.5 (31.1-36.8)	-	32 ± 3.1	-
b	-	15 (11-17)	-	-	-	-
W	30	30.0 (23.0-34.0)	29.6 ± 7.0 (20.2-34.8)	-	32 ± 1.8 (30-33.5)	-
Labial region height	-	4	-	-	-	-
Labial region diam.	-	9	-	-	-	-
Stylet	18.0	18.0 (16.0-19.0)	17.8 ± 0.8 (16.4-18.3)	-	-	-
DGO	4.0	3.0 (2.0-4.0)	2.9 ± 0.4 (2.5-3.2)	-	1.7 ± 0.5	-
Labial end to excretory pore	60.0	-	-	-	-	-

Table 42. *(Continued.)*

Character	Holotype, Allotype England & Wales, Franklin (1965a)	Paratypes, England & Wales, Franklin (1965a)	The Netherlands, Karssen (2002)	New Zealand, Zhao *et al.* (2017)	India, Suresh *et al.* (2017)	Portugal, Vieira dos Santos *et al.* (2020)
Spicules	26.0	28.0 (25.0-30.0)	28.3 ± 1.4 (27.2-30.3)	-	27.8	-
Gubernaculum	-	6.0 (4.0-8.0)	7.5 ± 0.5 (7.0-8.2)	-	-	-
T	-	-	61 ± 4.1 (49-84)	-	-	-
J2 (n)	-	25	4	17	-	-
L	433	435 (418-465)	421 ± 8.1 (410-429)	429 ± 16.1 (397-467)	432 ± 13.2	457.3 (440.6-487.2)
a	-	28.0 (25.0-30.0)	30 ± 1.5 (28.3-31.8)	27.1 ± 1.9 (24.2-30.7)	28 ± 1.5 (26.5-29.5)	29.6 (27.8-32.6)
b	-	7.7 (7.0-8.0)	8.0 ± 0.1 (7.9-8.1)	7.2 ± 0.5 (6.5-8.0)	-	-
c	-	6.2	6.4 ± 0.3 (6.0-6.7)	6.4 ± 0.8 (5.6-8.0)	-	5.5 (5.2-5.8)
c′	-	6.4	6.8 ± 0.5 (6.1-7.4)	6.5 ± 0.6 (5.3-7.7)	-	-
W	15.0	15.0 (14.0-18.0)	14.1 ± 0.6 (13.3-14.5)	15.9 ± 0.7 (14.4-17.3)	15.3 ± 1.0 (14.5-16.5)	15.5 (14.5-16.7)
Stylet	14.0	14.0 (13.0-15.0)	13.3 ± 0.5 (12.6-13.9)	11.7 ± 0.6 (10.8-12.4)	12 ± 1.2	14.4 (13.6-15.2)
DGO	2.0	-	2.9 ± 0.4 (2.5-3.2)	-	-	-
Labial end to excretory pore	74	-	-	-	-	-
Tail	74	70.0 (52.0-78.0)	66.0 ± 3.9 (61.0-70.0)	68.2 ± 8.0 (55-78)	61.6 ± 4.8 (56.0-65.0)	82.8 (76.4-89.9)
Hyaline region	-	-	17.9 ± 1.8 (15.8-19.6)	24.7 ± 2.6 (19.6-29.8)	-	26.6 (23.2-32.0)

Description

Female

Body approximately spherical when mature, with slight prominence posteriorly and sharply offset neck region enclosing median pharyngeal bulb. Cuticle 7-18 μm thick, annulated on neck, two labial annuli, hard to distinguish, amphidial openings anterior, between small labial cap and first annulus. Excretory pore on annulus 7-11 posterior to labial region. Stylet slender, dorsally curved, with well-developed ovoid, posteriorly sloping knobs. Excretory pore near stylet knobs level. Procorpus separated by constriction from muscular bulb, conspicuous valve plates 7 μm across. Pharyngeal gland lobe overlapping intestine lateroventrally. Six large matrix glands surrounding rectum. Pattern characterised by large phasmids, usually a little closer together than width of vulva and coarse striae in dorsal region forming broken, irregular horizontal lines around and vertical lines over between phasmids. Striae over whole pattern coarse, widely spaced and ribbon-like only on outer part of pattern. Perivulval Zone I free from striae. Overall shape of pattern dorsoventrally ovoid. Lateral field indistinct and marked only by interrupted striae (Jepson, 1987).

Male

Body anteriorly tapering, in heat-relaxed specimens posterior tenth of body spirally twisted through about 180°. Cuticle annulated, four incisures on lateral field, which is about one third body diam. wide and with outer bands irregularly areolated. Labial region about 9 μm wide and 4 μm high, scarcely offset, with three annuli, amphidial openings anterior. Tail shorter than wide, hemispherical. Stylet with rounded knobs, anterior cephalids on 2nd annulus posterior to labial region, posterior ones about two annuli anterior to stylet knobs. Median pharyngeal bulb fusiform, with four or five small vesicle-like structures grouped irregularly round lumen anterior to valve plates. Pharyngeal gland lobe ventrolateral, overlapping intestine, three to four body diam. long. Only one testis in specimens examined, sometimes reflexed. Spicules paired, ventrally curved, tapering to a point, small saucer-shaped gubernaculum.

J2

Slender with relatively long, sharply-pointed tail. Cuticle annulated with four incisures in lateral field, which is not areolated. Labial region not offset, two faint annuli, amphidial openings anterior. Hemizonid immediately anterior to excretory pore and four or five annuli posterior to nerve ring, Stylet very slender with posteriorly sloping knobs, median bulb fusiform with vesicle-like structures similar to those in male, pharyngeal gland lobe lying ventrally over intestine and extending to a point about 40% of body length from terminus.

TYPE PLANT HOST: Spring barley, *Hordeum vulgare* L.

OTHER PLANTS: *Meloidogyne naasi* occurs mainly in temperate regions and is found principally on barley and other cereals (*Triticum durum* and *T. aestivum*) and grasses (*Poa* spp., *Agrostis* spp. and others). Gooris (1968) recorded more than 60 host plants for *M. naasi*, including some found in experiments, and Radewald *et al.* (1970), in host range studies, found 23 of 26 graminaceous species infected and hosts in six other families, *e.g.*, on onion, sugarbeet, lucerne, *Gossypium hirsutum* and *Solanum peruvianum* L. In England and Wales it has been found on cereals, sugarbeet, ryegrass and other common grasses and weeds (Franklin, 1965a; Lewis & Webley, 1966). It has also been reported in seedling onions (Gooris, 1968), *Phragmites communis* (Cav.) Trin. *ex* Steud. (Franklin, 1971), soybean (Taylor *et al.*, 1971), sorghum (Aytan & Dickerson, 1969), and cassava (Coyne *et al.*, 2005). A report from Italy of *Coronilla scorpioides* (L.) Koch, *Medicago hispida* L., *Melilotus sulcata* Desf., and *Vicia villosa* Roth. by Vovlas & Inserra (1979) was incorrect (Inserra, pers. comm.). Six species of grasses studied by Person-Dedryver & Fischer (1987) were good hosts of this nematode, including Italian ryegrass (*Lolium multiflorum*), perennial ryegrass (*L. perenne*), hybrid ryegrass (*Lolium* × *hybridum* Hauss. Kn), tall fescue (*Festuca arundinacea* Schreb.), meadow fescue (*F. pratensis* Huds). and cocksfoot (*Dactylis glomerata*). In India, this species was reported from orange jessamine (*Cestrum aurantiacum* L.) (Suresh *et al.*, 2017). In Portugal, this species was reported in a football field with perennial ryegrass, common meadow-grass (*Poa pratensis* L.) and annual meadow grass (*P. annua* L.) (Santos *et al.*, 2020). Resistance to *M. naasi* (Cook *et al.*, 1999) was reported in common bean and ryegrasses.

TYPE LOCALITY: Tytherington, Gloucestershire, UK.

DISTRIBUTION: *Europe*: UK (Franklin, 1965a; Lewis & Webley, 1996), Ireland, Belgium and France (Gooris, 1968; Caubel *et al.*, 1972; Vandenbossche *et al.*, 2011), Serbia, Czech Republic, former Yugoslavia (Karssen, 2002), The Netherlands (Franklin, 1971; Karssen, 2002), Hungary (Amin & Budal, 1994), Poland (Kornobis, 2001), Germany (Hallmann *et al.*, 2007), Portugal (Santos *et al.*, 2020); *North America*: USA (Golden & Taylor, 1967; Jensen *et al.*, 1968; Aytan & Dickerson, 1969;

Taylor *et al.*, 1971; McClure *et al.*, 2012), Canada (Bélair *et al.*, 2006; Simard *et al.*, 2008); *South America*: Argentina (Echevarría & Chaves, 1998), Chile (Kilpatrick *et al.*, 1976); *Africa*: Mozambique (Coyne *et al.*, 2006); *Oceania*: New Zealand (Sheridan & Grbavac, 1979; Yeates, 2010; Zhao *et al.*, 2017); *Asia*: Iran, Thailand, India (Suresh *et al.*, 2017).

SYMPTOMS: Galls induced by *M. naasi* on cereal roots are cylindrical, spindle-shaped, hooked, ring-shaped or, if terminal, club-shaped; on sugarbeet they are mainly on lateral roots and may be inconspicuous on mature beet. Several females are often found in one gall and the egg masses of the different individuals may become amalgamated (Franklin, 1973). Histopathological studies on different hosts showed the presence of J2 associated with necrosis, hyperplasia, and hypertrophy in the cortex.

PATHOGENICITY: Information on the economic importance of *M. naasi* on cereals is limited. In Europe and Chile it is an important pest of wheat (Kilpatrick *et al.*, 1976; Person-Dedryver, 1986). In barley, it causes up to 75% yield loss in California (Allen *et al.*, 1970). Yield of spring barley was reduced up to 50% with soil population density of 150 J2 (g soil)$^{-1}$ and yield loss in barley could be expected at small preplant densities with an economically significant loss at initial densities greater than 20 J2 (g soil)$^{-1}$ (York, 1980). Bélair *et al.* (2006) reported noticeable damage to turfgrass by *M. naasi* in Quebec, Canada.

BIOLOGY: Studies by Siddiqui & Taylor (1970) determined that at 26°C day and 20°C night temperatures, the life cycle of *M. naasi* on wheat 'Pawnee' was completed in 39-51 days. Under field conditions in England, hatching reaches a peak in April with a smaller peak in late summer; there is only one generation on barley (Franklin *et al.*, 1971). Eggs of *M. naasi* exhibit temperature dormancy, which is necessary to stimulate hatching in spring (Franklin *et al.*, 1971).

HOST RACES: Michell *et al.* (1973) studied the reactions of the five isolates of *M. naasi* (from different geographical origin, including England, California, Illinois, Kentucky and Kansas) on certain plant species and confirmed that populations from diverse geographical origins differ in their host range. These findings demonstrate the existence of at least five races within *M. naasi*.

CHROMOSOME NUMBER: 2n = 36, and reproduction was regularly by meiotic parthenogenesis (Triantaphyllou, 1969).

POLYTOMOUS KEY CODES: *Female*: A32, B324, C3, D3; *Male*: A34, B34, C23, D1, E1, F2; *J2*: A2, B21, C123, D1, E3, F12.

BIOCHEMICAL AND MOLECULAR CHARACTERISATION: Biochemically, *M. naasi* has a VF1 esterase phenotype and the malate dehydrogenase pattern is N1 (Esbenshade & Triantaphyllou, 1985a, b; Karssen & van Hoenselaar, 1998; Santos *et al.*, 2020).

Ziljstra *et al.* (2004) developed species-specific primers for accurate identification of *M. naasi* based on the ITS rRNA gene fragment. In addition, McClure *et al.* (2012) and Powers *et al.* (2018) studied the D2-D3 of 28S rRNA and mtDNA from *M. naasi* and both regions appeared to be distinctive enough to characterise this species. Ye *et al.* (2015) also developed a species-specific primer for this species based on the D2-D3 of 28S rRNA gene sequences (Table 10).

RELATIONSHIPS (DIAGNOSIS): The species belongs to Molecular group III and is clearly molecularly differentiated from all other *Meloidogyne* species. This species resembles *M. acronea* by perineal pattern but differs in EP/ST ratio. It can be also distinguished from other related species affecting grasses by biochemical markers.

67. *Meloidogyne nataliei* Golden, Rose & Bird, 1981
(Figs 188, 189)

COMMON NAME: Michigan grape root-knot nematode.

Meloidogyne nataliei was detected in 1977 from root samples of grape (*V. labrusca* 'Concord') from a declining vineyard in Mattawan, Van Buren County, Michigan. In 1980, the Michigan grape RKN became a state-mandated regulatory species for eradication. During 1983-1984 an attempt to eradicate this nematode species in the type locality was made, but it was found again on the original farm in 2012 and 2017. *Meloidogyne nataliei* has a unique morphology and does not induce visible plant hyperplastic symptoms, females generally being surrounded by a massive egg sac containing many eggs. *Meloidogyne*

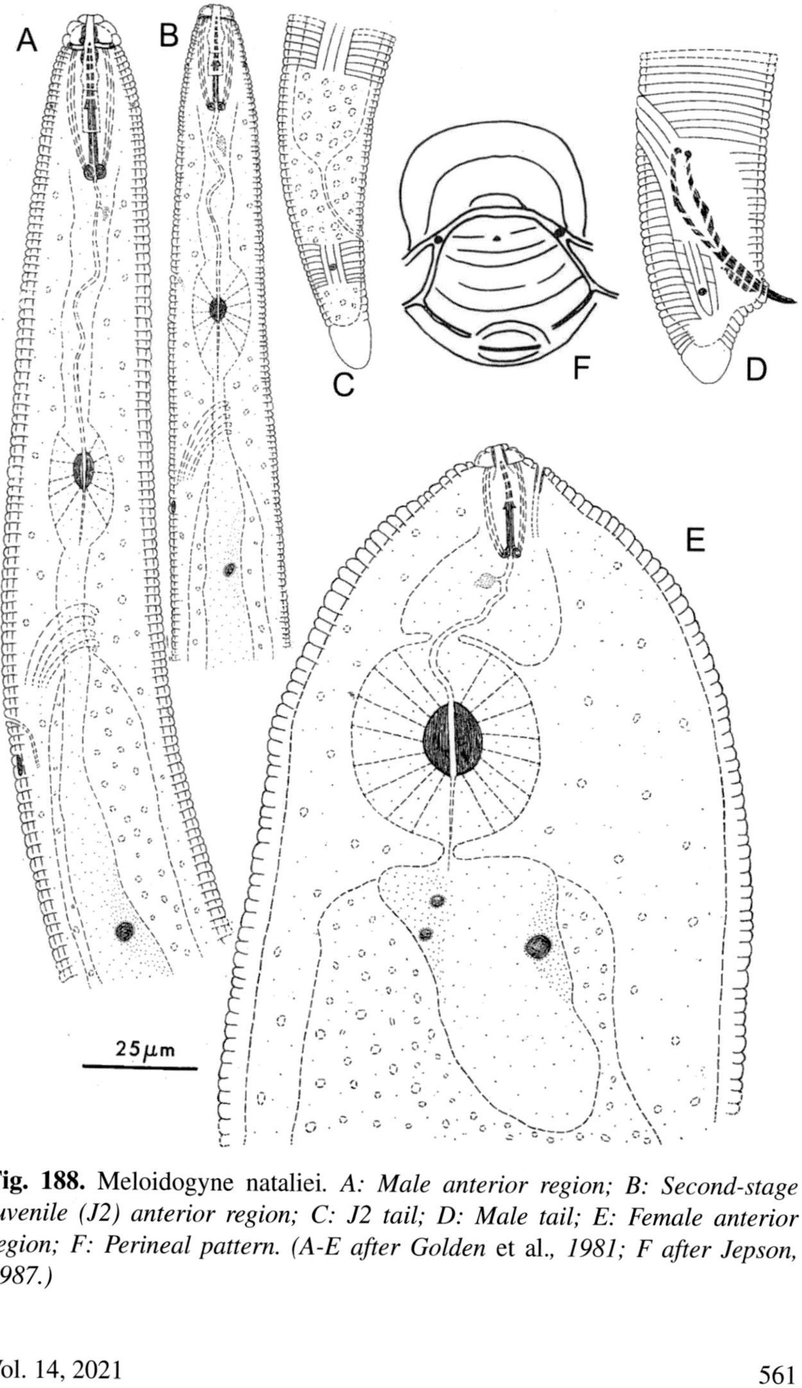

Fig. 188. Meloidogyne nataliei. *A: Male anterior region; B: Second-stage juvenile (J2) anterior region; C: J2 tail; D: Male tail; E: Female anterior region; F: Perineal pattern. (A-E after Golden* et al.*, 1981; F after Jepson, 1987.)*

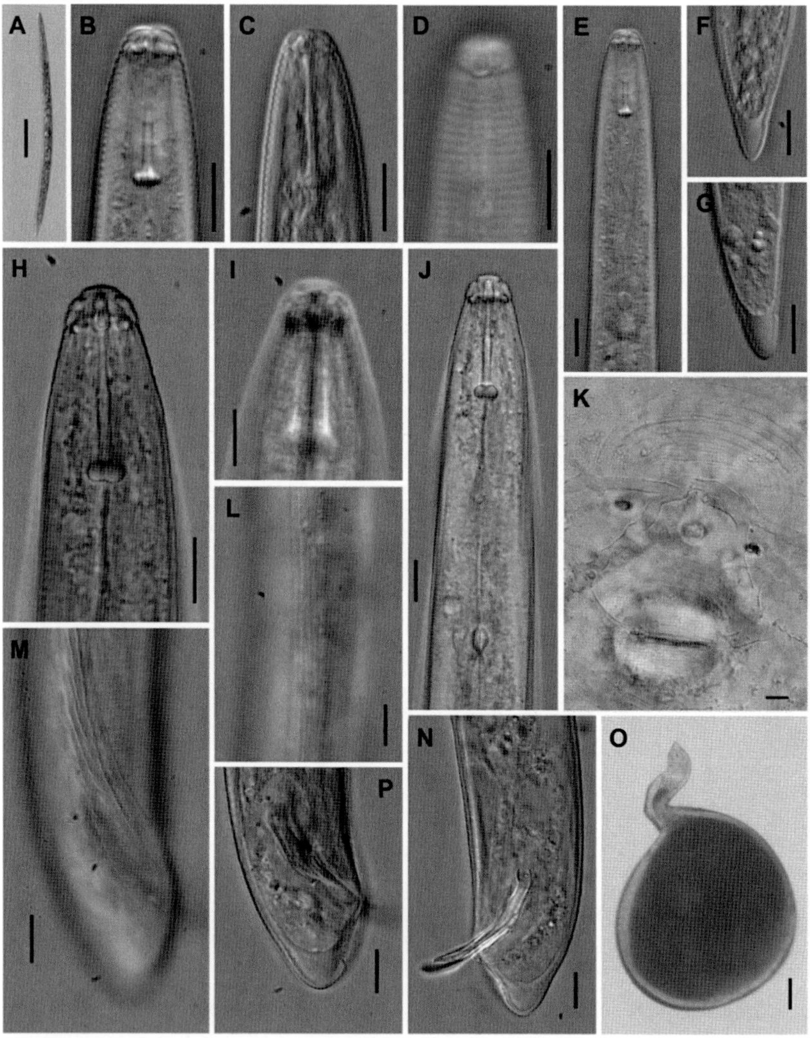

Fig. 189. Meloidogyne nataliei. *(LM). A-G: Second-stage juveniles; H-J, L-N, P: Male; K, O: Female; A, O: Entire body; B, C, H: Anterior region in median lateral view; D, I: Anterior region in surface lateral view; E, J: Pharyngeal region; K: Perineal pattern; L: Mid-body lateral field; F, G, M, N, P: Caudal region. (Scale bars: A, O = 100 μm; B-J, K-N, P = 10 μm.) (After Álvarez-Ortega* et al., *2019.)*

nataliei, together with *M. indica*, may represent ancestral species among other RKN.

MEASUREMENTS *(AFTER GOLDEN ET AL., 1981)*

- *Holotype female*: L = 835 μm; W = 465 μm; stylet = 21.9 μm; stylet knob width = 6.0 μm; DGO = 4.7 μm; vulval slit length = 37 μm; distance from vulva centre to anus = 49 μm; distance between phasmids = 83 μm; excretory pore to anterior end = 9.6 μm; a = 1.8; b = 6.0; EP/ST = 0.4.
- *Allotype male*: L = 1474 μm; stylet = 28.4 μm; DGO = 4.3 μm; spicules 43 μm; gubernaculum = 9 μm; tail = 14 μm; a = 35.7; b = 3.7; c = 107.
- *Paratype females* (n = 30): L = 933 $\pm$ 101 (731-1247) μm; W = 568 $\pm$ 103 (383-834) μm; excretory pore to anterior end = 9 $\pm$ 2 (6-11) μm; stylet = 21.8 $\pm$ 0.4 (21.0-22.5) μm; stylet knob width = 6.1 $\pm$ 0.5 (5.0-7.0) μm; DGO = 4.1 $\pm$ 0.9 (3.0-6.0) μm; anterior end to median bulb = 56 $\pm$ 6 (43-68) μm; vulva width = 38 $\pm$ 4 (31-45) μm; distance from vulva centre to anus = 60 $\pm$ 7 (49-77) μm; interphasmid distance = 78 (62-107) μm; a = 1.7 $\pm$ 0.1 (1.3-2.5); b = 6.7 $\pm$ 0.9 (5.0-8.3).
- *Paratype males*: L = 1489 $\pm$ 127 (1191-1757) μm; W = 38 $\pm$ 3.2 (33-43) μm; stylet = 28.9 $\pm$ 0.21 (28.4-29.2) μm; DGO = 5.0 $\pm$ 0.8 (4.0-6.5) μm; anterior end to median bulb = 85 $\pm$ 5.6 (71-95) μm; spicules = 43 $\pm$ 0.8 (41.3-44.3) μm; gubernaculum = 9.7 $\pm$ 1.6 (6.0-11.0) μm; tail = 15.0 $\pm$ 2.2 (12.0-18.0) μm; a = 39 $\pm$ 3.3 (33-45); b = 5.0 $\pm$ 0.5 (4.2-5.9); c = 101 $\pm$ 13.6 (86-125).
- *Paratype J2*: L = 599 $\pm$ 32 (539-641) μm; stylet = 22.4 $\pm$ 0.2 (21.9-22.8) μm; DGO = 3.8 $\pm$ 0.4 (3.0-4.3) μm; anterior end to median bulb = 68 $\pm$ 4.4 (60-83) μm; excretory pore to anterior end = 62 (60-65) μm; tail = 27.0 $\pm$ 1.6 (22.9-30.0) μm; hyaline region = 8.4 $\pm$ 0.9 (7.0-11.0) μm; a = 23.9 $\pm$ 2.1 (20.4-28.7); b = 3.6 $\pm$ 0.3 (3.2-4.1); c = 22 $\pm$ 1.6 (19-26).
- *Eggs* (n = 20): L = 168 $\pm$ 12 (147-186) μm; W = 56 $\pm$ 3 (50-62) μm.

DESCRIPTION

Female

Body pearly white, globular to pear-shaped, usually with exceptionally long neck. Vulva terminal and generally on a small protuberance located in a median plane with neck. Labial region offset from neck, bearing a labial cap and a prominent labial annulus, sometimes subdivided by a single striae. Labial framework distinct, stylet exceptionally strong with knobs sloping posteriorly. Massive egg mass extruded posteriorly, often three or more times size of female and containing 50-400 eggs. Cuticle thick, measuring 16 (10-21) μm. Perineal pattern large, highly distinctive, with heavy striae forming a rounded arch, and usually with two separated rope-like striae extending laterally from vulval and anal areas and forming a lateral field decreasing in size and prominence with increasing length. Phasmids prominent, commonly located at a level opposite or just anterior to distinct anus.

Male

Body long, vermiform, tapering slightly at both extremities. Body annuli distinct, measuring about 2 μm. Labial region markedly offset, with massive labial disc, prominent labials, and post-labial annulus often with a discontinuous striae. Labial sclerotisation heavy. Lateral field forming *ca* 25% body diam. at mid-body, with five lines, not areolated. Excretory pore anterior to hemizonid, ranging about 6-20 μm anteriorly. Testis one. Spicules arcuate, tips sharply pointed. Gubernaculum boat-shaped, distally with small projection extending dorsally. Distinct phasmids generally at level of cloacal aperture, tail short, reduced immediately past cloacal aperture, with narrowly rounded terminus.

J2

Body vermiform, tapering at both ends but more so posteriorly. Labial region offset, bluntly rounded, more than twice as wide as high, with heavy labial sclerotisation, and one post-labial annulus without striae. Body annuli fine but distinct. Lateral field with four lines, without areolation, and forming about 25% body diam., latter being 26 (22-29) μm at its widest part. Stylet strong, with stylet guide as shown in Figure 185. Excretory pore rather faint, located anterior to hemizonid at a level from centre to anterior portion of median bulb. Hemizonid distinct, situated 32 (22-36) μm posterior to centre of median bulb. Phasmids large, located in anterior 25% of tail. Rectum not dilated. Tail short, with exceptionally short hyaline region and a narrowly rounded terminus.

TYPE PLANT HOST: Fox grape, *Vitis labrusca* L., ‘Concord’ (Vitaceae).

OTHER PLANTS: Uncultivated grapes (Golden *et al.*, 1981; Bird *et al.*, 1994). *Vitis* rootstocks (5BB, St George, and Glorie), *Parthenocissus quinquefolia* (L.) Planch., and *P. tricuspidata* (Siebold & Zucc.) Planch. (Diamond & Bird, 1994).

TYPE LOCALITY: A vineyard at Mattawan, Michigan, USA.

DISTRIBUTION: *North America*: USA (Michigan).

PATHOGENICITY: *Meloidogyne nataliei* does not form galls or knots; the female protrudes from the root surface and become surrounded by a massive egg mass containing many eggs. Soil and root samples of grape (*V. labrusca* 'Concord') were obtained from a badly declining vineyard, although a strict pathogenicity test has not been studied for this species. *Meloidogyne nataliei* is tolerant of near freezing soil temperatures (Bird *et al.*, 1994).

CHROMOSOME NUMBER: *M. nataliei* reproduces exclusively by cross-fertilisation and has a haploid complement of four chromosomes, which are considerably larger than those of other *Meloidogyne* species (Triantaphyllou, 1985b).

POLYTOMOUS KEY CODES: *Female*: A12, B1, C4, D2; *Male*: A32, B1, C1, D1, E2, F2; *J2*: A1, B1, C4, D43, E1, F4.

BIOCHEMICAL AND MOLECULAR CHARACTERISATION: One strong band (EST = S1; Rm: 44.4) of EST activity was clearly detected when using homogenates of ten egg-laying females. This band is similar for that reported for *M. chitwoodi* (MDH = N1a) and *M. platani* (MDH = N1a), *M. nataliei* clearly differed from these RKN species because no MDH activity could be detected (Álvarez-Ortega *et al.*, 2019).

This species was characterised by partial 18S rRNA, the D2-D3 of 28S rRNA, ITS rRNA and partial *COI* mtDNA gene sequences (Álvarez-Ortega *et al.*, 2019). Phylogenetic relationships showed that *M. nataliei*, together with *M. indica*, occupied a basal clade within the RKN.

RELATIONSHIPS (DIAGNOSIS): The species belongs to Molecular group XI and is clearly molecularly differentiated from all other *Meloidogyne* species. It is morphologically similar to *M. baetica* and *M. caraganae* in female and juvenile morphology, but can be separated molecularly from the first species, and by the different perineal pattern and longer J2 (539-641 *vs* 417-482 μm) from *M. caraganae*.

68. *Meloidogyne oleae* Archidona-Yuste, Cantalapiedra-Navarrete, Liébanas, Rapoport, Castillo & Palomares-Rius, 2018

(Figs 190-193)

COMMON NAME: Spanish olive root-knot nematode.

Archidona-Yuste *et al.* (2018) described this species from roots of wild and cultivated olive in Tolox, Málaga province, Spain.

MEASUREMENTS

See Table 43.

Eggs (n = 30): L = 97.6 ± 3.3 (90-102) μm; W = 40.9 ± 1.5 (37-43) μm; L/W = 2.4 ± 0.09 (2.2-2.5).

DESCRIPTION

Female

Body usually completely embedded in galled tissue, pearly white, globose or pear-shaped, with long neck but no posterior protuberance. Labial region continuous with body contour. Labial cap variable in shape, with labial disc and postlabial annulus not elevated. In SEM view, labial disc appearing round-squared, slightly raised on median and lateral sectors, which are all fused together. Labial framework weakly sclerotised. Stylet short, with an almost straight, rarely curved, cone, cylindrical shaft, and knobs rounded and sloping posteriorly in most specimens. Excretory pore located at level of stylet knobs or a few body annuli anterior. Pharyngeal gland with a large mononucleate dorsal lobe and two subventral gland lobes, usually difficult to see. Perineal pattern mostly rounded-oval, dorsal arch generally low, with fine striae becoming coarser near perivulval region, lateral field and punctations not observed. Phasmids distinct, located just above level of anus. Vulval slit in middle of unstriated area, 18.0-22.0 μm long, slightly shorter than the vulva-anus distance, anal fold clearly visible, but not always present. Large egg sac commonly occurring outside root gall, containing up to 248 eggs.

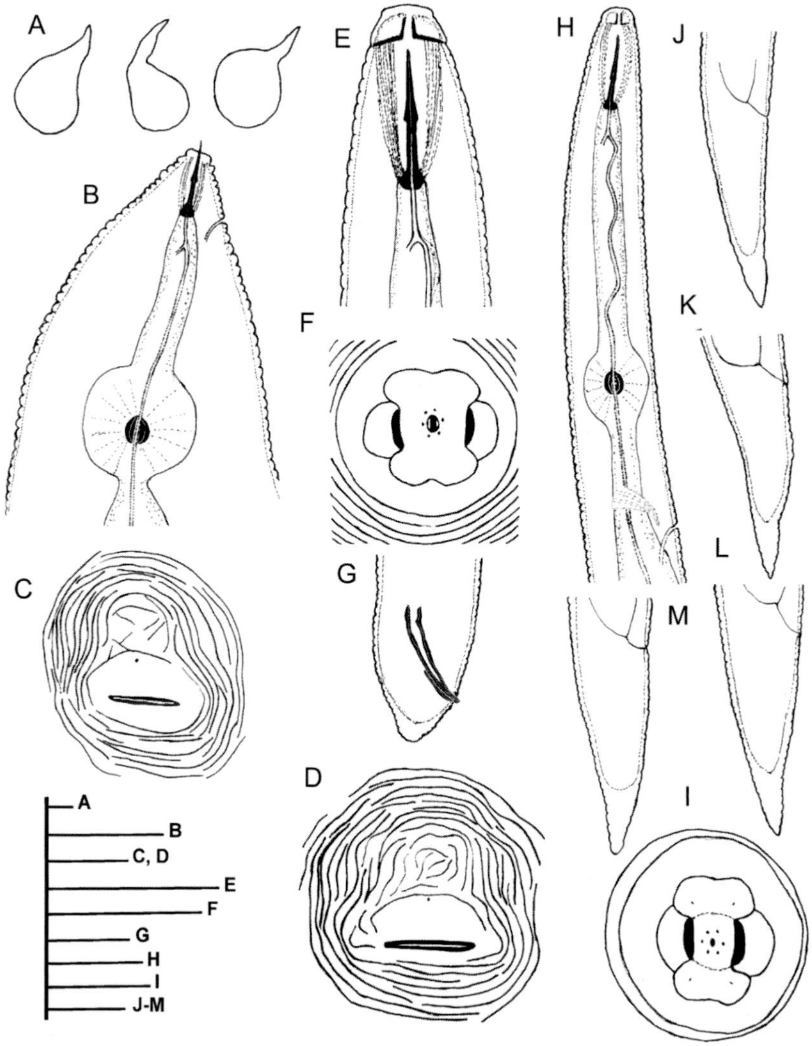

Fig. 190. Meloidogyne oleae. *A: Entire female; B: Female anterior region; C, D: Perineal pattern; E: Male anterior region; F: Male* en face *view; G: Male tail; H: Second-stage juvenile (J2) anterior region; I: J2* en face *view; J-M: J2 tail. (Scale bars: A = 200 μm; B-E, G, H = 20 μm; F = 5 μm; I = 2 μm; J-M = 10 μm.) (After Archidona-Yuste* et al., *2018.)*

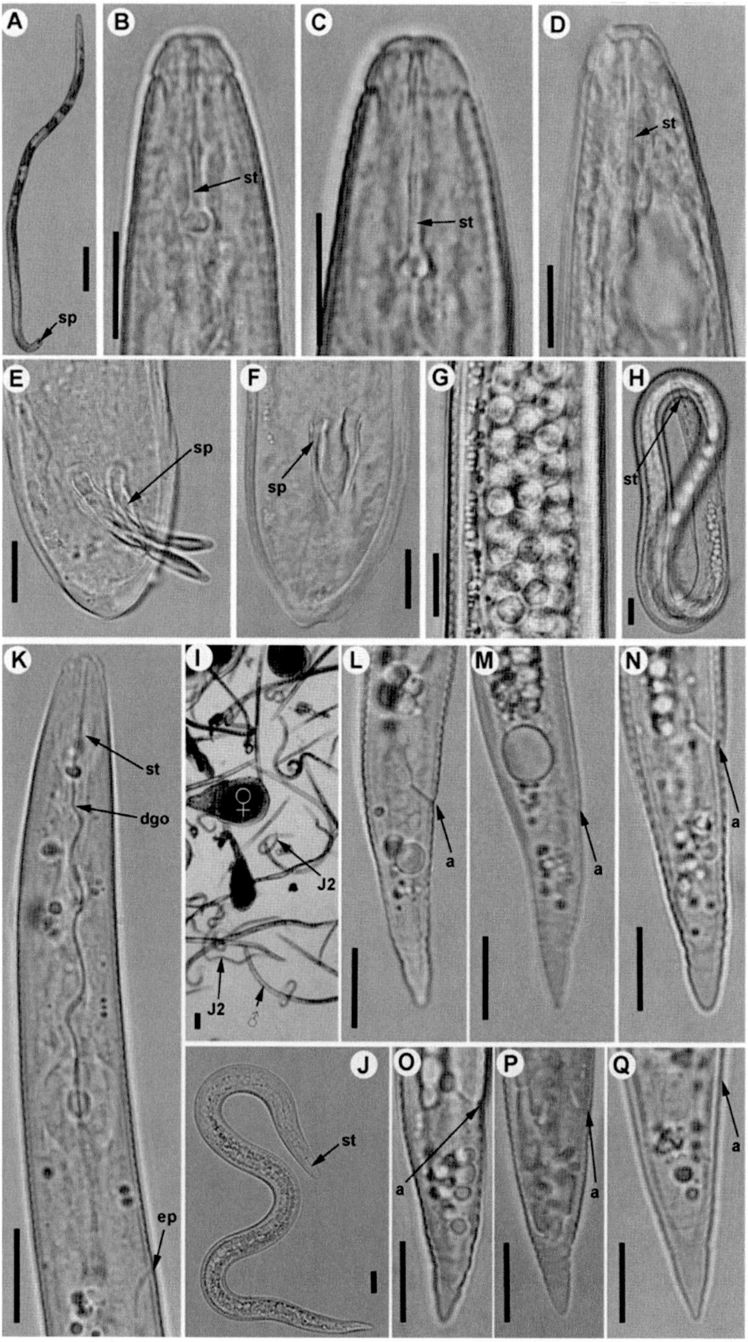
A
sp
B
st
C
st
D
st
E
sp
F
sp
G
H
st
K
st
dgo
ep
I
♀
J2
J2
♂
L
a
M
a
N
a
J
st
O
a
P
a
Q
a

Fig. 191. Meloidogyne oleae. *LM. A: Entire male; B-D: Male labial region; E, F: Male tail; G: Detail of sperm cells; H: Embryonated egg showing stylet of second-stage juvenile (J2); I: All life stages; J: Entire J2; K: J2 anterior region; L-Q: J2 tail. Abbreviations: a = anus; dgo = dorsal gland orifice; ep = excretory pore; sp = spicules; st = stylet. (Scale bars: A, I = 100 μm; B-H, J-Q = 10 μm; R = 100 μm; S-V = 20 μm.) (After Archidona-Yuste* et al., *2018.)*

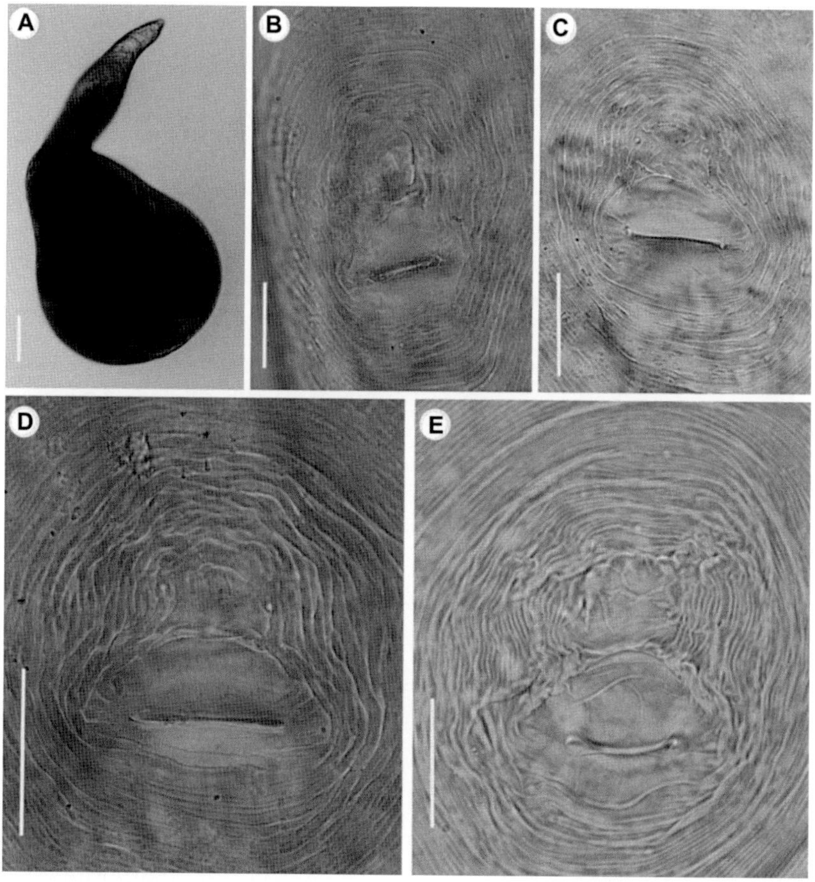

Fig. 192. Meloidogyne oleae. *A: Entire female; B-E: Perineal patterns. (Scale bars: A = 100 μm; B-E = 20 μm.) (After Archidona-Yuste* et al., *2018.)*

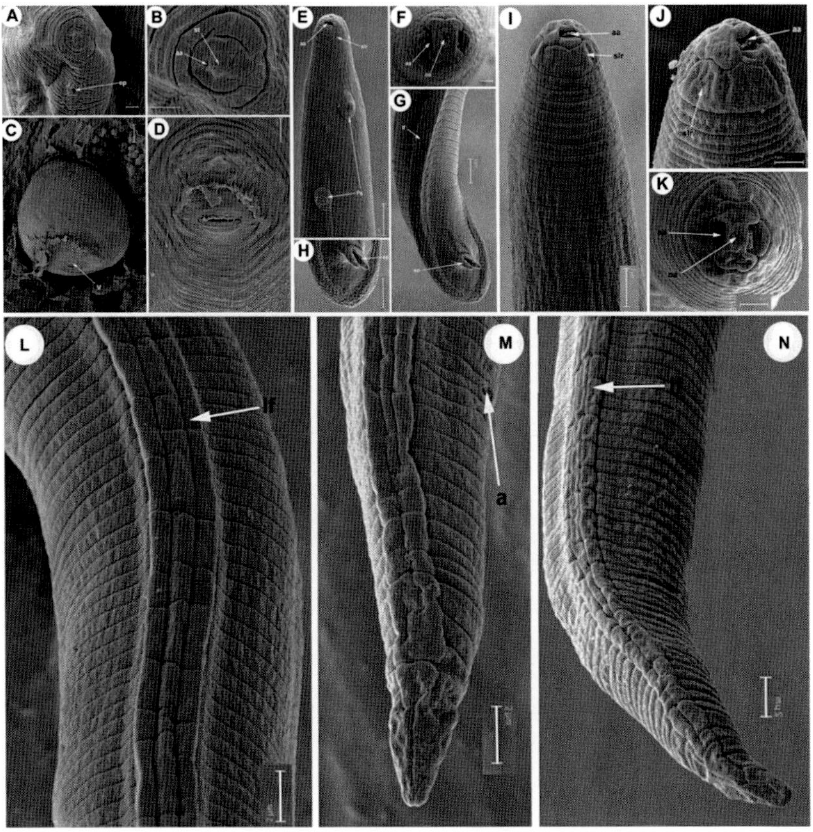

Fig. 193. Meloidogyne oleae. *SEM. A, B: Female labial region; C: Entire female included in wild olive root; D: Perineal pattern; E: Male anterior region showing endospore bacterial of* Pasteuria *sp.; F: Male* en face *view; G, H: Male tail; I, J: Second-stage juvenile (J2) labial region; K: J2* en face *view; L: J2 lateral field; M, N: J2 tail. Abbreviations: aa = amphidial aperture; ep = excretory pore; lf = lateral field; oa = oral aperture; Ps = endospore of* Pasteuria *sp.; slr = smooth labial region; st = stylet; V = vulva. (Scale bars: A, B = 2 μm; C = 500 μm; D = 10 μm; E, G, H = 5 μm; F = 1 μm; I, L-N = 2 μm; J, K = 1 μm.) (After Archidona-Yuste* et al.*, 2018.)*

Male

Body vermiform, tapering anteriorly, tail rounded, with twisted posterior body portion. Labial region slightly offset from body, labial cap relatively small, labial disc not elevated. Labial framework strong and sclerotised, vestibule extension distinct. Prominent slit-like amphidial

Table 43. *Morphometrics of females, males and second-stage juveniles (J2) of* Meloidogyne oleae. *All measurements are in μm and in the form: mean ± s.d. (range).*

Character	Holotype, Tolox, Spain, Archidona-Yuste *et al.* (2018)	Paratypes, Tolox, Spain, Archidona-Yuste *et al.* (2018)	JAO38, Antequera, Spain, Archidona-Yuste *et al.* (2018)	JAO31, Antequera, Spain, Archidona-Yuste *et al.* (2018)
Female (n)	-	19	-	-
L	701	591 ± 85.9 (445-790)	-	-
W	418	356 ± 54.1 (271-421)	-	-
Stylet	13	13.3 ± 0.5 (12.5-14)	-	-
DGO	4.5	4.4 ± 0.7 (3.0-5.5)	-	-
Median bulb length	32.0	31 ± 5.5 (23.0-39.0)	-	-
Median bulb width	28.0	27.1 ± 6.0 (20.0-38.0)	-	-
Ant. end to excretory pore	13.0	12.7 ± 0.5 (11.5-13.5)	-	-
Vulval slit	22.0	19.5 ± 1.7 (18.0-22.0)	-	-
Vulval slit to anus distance	24.0	23.3 ± 1.0 (22.0-24.0)	-	-
EP/ST	1.0	1.0 ± 0.05 (0.9-1.0)	-	-
Male (n)	-	20	5	8
L	-	1237 ± 238 (838-1762)	1056 ± 88 (995-1210)	1179 ± 162 (1024-1522)
a	-	48.6 ± 9.6 (27.3-64.1)	52.7 ± 3.4 (47.4-52.1)	51.4 ± 6.9 (43.8-63.2)
c	-	134.4 ± 24.2 (107-178.1)	142.9 ± 23.3 (118.8-175.5)	131.5 ± 19.2 (116.2-156.6)
W	-	25.8 ± 5.4 (17.5-38.5)	20.1 ± 1.7 (18.0-22.5)	23.1 ± 2.9 (17.5-26.0)

Table 43. *(Continued.)*

Character	Holotype, Tolox, Spain, Archidona-Yuste *et al.* (2018)	Paratypes, Tolox, Spain, Archidona-Yuste *et al.* (2018)	JAO38, Antequera, Spain, Archidona-Yuste *et al.* (2018)	JAO31, Antequera, Spain, Archidona-Yuste *et al.* (2018)
Labial region diam.	-	7.6 ± 0.6 (6.5-8.5)	7.8 ± 0.3 (7.5-8.5)	7.9 ± 0.3 (7.5-8.5)
Stylet	-	15.7 ± 1.1 (13.5-18.0)	14.7 ± 1.0 (13.5-16.0)	14.3 ± 1.2 (12.5-15.5)
Stylet knob width	-	2.5 ± 0.3 (2.0-3.0)	2.1 ± 0.3 (2.0-2.5)	2.7 ± 0.3 (2.5-3.0)
DGO	-	5.1 ± 0.7 (4.0-6.0)	4.3 ± 0.3 (4.0-4.5)	4.6 ± 0.3 (4.5-5.0)
Ant. end to excretory pore	-	118.3 ± 9.2 (79.0-140.5)	97.4 ± 20.9 (74.0-115.0)	116.7 ± 8.5 (108-127.5)
Spicules	-	26.7 ± 3.0 (21.0-32.0)	26.4 ± 1.9 (24.5-28.5)	26.6 ± 3.8 (21.0-29.0)
Gubernaculum	-	8.3 ± 1.4 (6.5-11.5)	7.6 ± 0.6 (7.0-8.5)	7.0 ± 0.5 (6.5-7.5)
Tail	-	9.5 ± 1.7 (8.0-12.5)	7.3 ± 1.0 (6.0-8.5)	9.5 ± 2.8 (7.0-13.0)
J2 (n)	-	16	6	6
L	-	371 ± 12.1 (351-385)	369 ± 11 (355-382)	391 ± 28.8 (360-437)
a	-	26.1 ± 1.5 (24.0-29.0)	28.7 ± 1.9 (27.3-31.5)	27.6 ± 1.9 (25.7-31.2)
b	-	5.3 ± 0.4 (4.8-6.1)	5.8 ± 0.4 (5.5-6.3)	5.0 ± 0.2 (4.8-5.2)
c	-	14.8 ± 0.8 (13.5-16.0)	16.1 ± 1.2 (14.2-17.6)	18 ± 1.2 (16-19.4)
c′	-	2.7 ± 0.2 (2.4-3.2)	2.6 ± 0.2 (2.4-2.9)	2.7 ± 0.2 (2.5-3.0)
W	-	14.3 ± 0.8 (13.0-16.0)	13 ± 0.9 (12.0-14.0)	14.2 ± 0.5 (13.5-15.0)
Stylet	-	12.3 ± 0.6 (11.0-13.0)	11.4 ± 0.6 (10.5-12.0)	11.3 ± 0.3 (11.0-11.5)
DGO	-	3.1 ± 0.5 (2.5-3.5)	2.9 ± 0.4 (2.5-3.5)	4.3 ± 0.4 (4.0-4.5)

Table 43. *(Continued.)*

Character	Holotype, Tolox, Spain, Archidona-Yuste *et al.* (2018)	Paratypes, Tolox, Spain, Archidona-Yuste *et al.* (2018)	JAO38, Antequera, Spain, Archidona-Yuste *et al.* (2018)	JAO31, Antequera, Spain, Archidona-Yuste *et al.* (2018)
Labial end to excretory pore	-	68 ± 4.2 (57.0-75.5)	74.6 ± 4.4 (70.5-80.0)	71.6 ± 2.5 (69.0-75.5)
Tail	-	25.1 ± 1.6 (23.0-28.5)	23 ± 2.0 (21.0-26.0)	21.8 ± 1.4 (19.5-23.5)
Hyaline region	-	7.4 ± 1.2 (6.5-11.0)	6.5 ± 0.7 (5.5-7.5)	6.0 ± 0.9 (5.0-7.0)

openings between labial disc and lateral labials. In SEM view, labial disc slightly narrower and raised above merged subventral and subdorsal median labial sectors, with a centred oval prestoma into which opens a slit-like dorsoventrally oriented stoma, lateral labial margins rounded. Labial region moderately high and lacking annulation. Stylet delicate and straight, with cone and shaft broadening slightly in distal part. Stylet knobs mostly rounded, laterally or obliquely directed, merging gradually with base of shaft. Lateral field consisting of four incisures with areolations along body but only few actually crossing central field. Procorpus distinct, 4.1-5.2 times length of median bulb. Median bulb ovoid, with strong valve apparatus. Excretory duct curved. Excretory pore distinct, usually located 3-4 annuli posterior to hemizonid. Testis single, long, monorchic, occupying 40-57% of body length. Tail usually curved ventrally, short, with bluntly rounded tip and finely annulated. Spicules of variable length, arcuate and with two pores clearly visible at tip. Gubernaculum distinct. Phasmids small and located at level of cloacal aperture.

J2

Body vermiform, tapering slightly towards posterior end. Labial region narrower than body and slightly offset. Labial cap slightly elevated. Labial framework weakly developed. Labial disc and median labials fused. In labial disc, a stoma-like slit located in an ovoid prestoma and surrounded by six inner labial sensilla. In SEM view, labial disc appearing oval to rectangular in shape, raised above median labials, to which it merges in a dumbbell-shaped structure. Labial region smooth and lacking annulation. Amphidial apertures elongated and located between labial disc and lateral labials. Body annulated from anterior end to terminus. Lateral field consisting of four incisures, with areolations along body. Stylet delicate, with cone straight, narrow, sharply pointed, shaft almost cylindrical, knobs small, rounded, laterally directed. Pharynx with a long, cylindrical procorpus (4.0-5.0 times length of median bulb), round-oval median bulb, short isthmus and rather long gland lobe, with three equally-sized nuclei and overlapping intestine ventrally. Hemizonid located anterior to excretory pore, extending for *ca* two body annuli. Excretory pore located posterior to nerve ring. Excretory duct curved and discernible when reaching intestine. Rectum slightly dilated. Tail short, conoid, tail terminus broad and rounded at tip, with several constrictions in hyaline region. Tail annulation fine, regular in proximal two-thirds, becoming slightly coarser and irregular in distal part. Hyaline region clearly defined and short, phasmids small, difficult to observe.

TYPE PLANT HOST: Wild olive, *Olea europaea* L. subsp. *europaea* var. *sylvestris*.

OTHER PLANTS: *Meloidogyne oleae* parasitises and establishes permanent, fully developed feeding sites on wild and cultivated olives, as well as on other host plants such as carob tree, rosebush, and lesser periwinkle (Archidona-Yuste *et al.*, 2018).

TYPE LOCALITY: Tolox, Málaga province, Spain.

DISTRIBUTION: *Europe*: Spain. It has been reported only in the type locality (Archidona-Yuste *et al.*, 2018).

PATHOGENICITY: Olive roots infected with *M. oleae* showed small galls of variable size but relatively small (almost two times the root diam.), and were commonly located along the root axis but also on the root tip. Both wild and cultivated olives and the other host plants showed a similar disease reaction. This species potentially constitutes an additional threat for olive in southern Spain, but further studies are necessary in order to confirm this assumption, as the observed levels of infection were not high.

CHROMOSOME NUMBER: No available data.

POLYTOMOUS KEY CODES: *Female*: A32, B3, C34, D3; *Male*: A324, B43, C23, D1, E1, F2; *J2*: A3, B23, C4, D43, E1, F4.

BIOCHEMICAL AND MOLECULAR CHARACTERISATION: The isozyme electrophoretic analysis of *M. oleae* revealed one very slow weak A1 Est band after repeated and prolonged staining and a N1c Mdh phenotype with a very weak-staining band that did not occur in the Est and Mdh phenotypes of *M. javanica* (Archidona-Yuste *et al.*, 2018).

The ITS, 18S, D2-D3 28S rRNA, and *COII*-16S rRNA and *COI* gene sequences are provided and analysed by Archidona-Yuste *et al.* (2018).

RELATIONSHIPS (DIAGNOSIS): The species belongs to Molecular group IX and is clearly molecularly differentiated from all other *Meloidogyne* species. In J2 tail length it is morphometrically similar to some members of group 1 (Table 12): *M. brevicauda*, *M. nataliei* and *M. indica*, from which it clearly differs by perineal pattern morphology. It can be separated from other *Meloidogyne* spp. parasitising olive, such as *M. lusitanica* (Abrantes & Santos, 1991), by clear differences in perineal pattern, excretory pore/stylet length ratio, J2 and male stylet length, J2 tail length, and isoenzyme phenotype (Abrantes & Santos, 1991).

69. *Meloidogyne oryzae* Maas, Sanders & Dede, 1978
(Figs 194-196)

COMMON NAME: Surinam rice root-knot nematode.

In 1971, a RKN infestation in irrigated rice was observed at the Wageningen rice scheme in Surinam. Although infestation always occurred at the higher spots within a field, the nematode appeared to be very well adapted to irrigated rice (Van Halteren, 1972).

MEASUREMENTS

See Table 44.

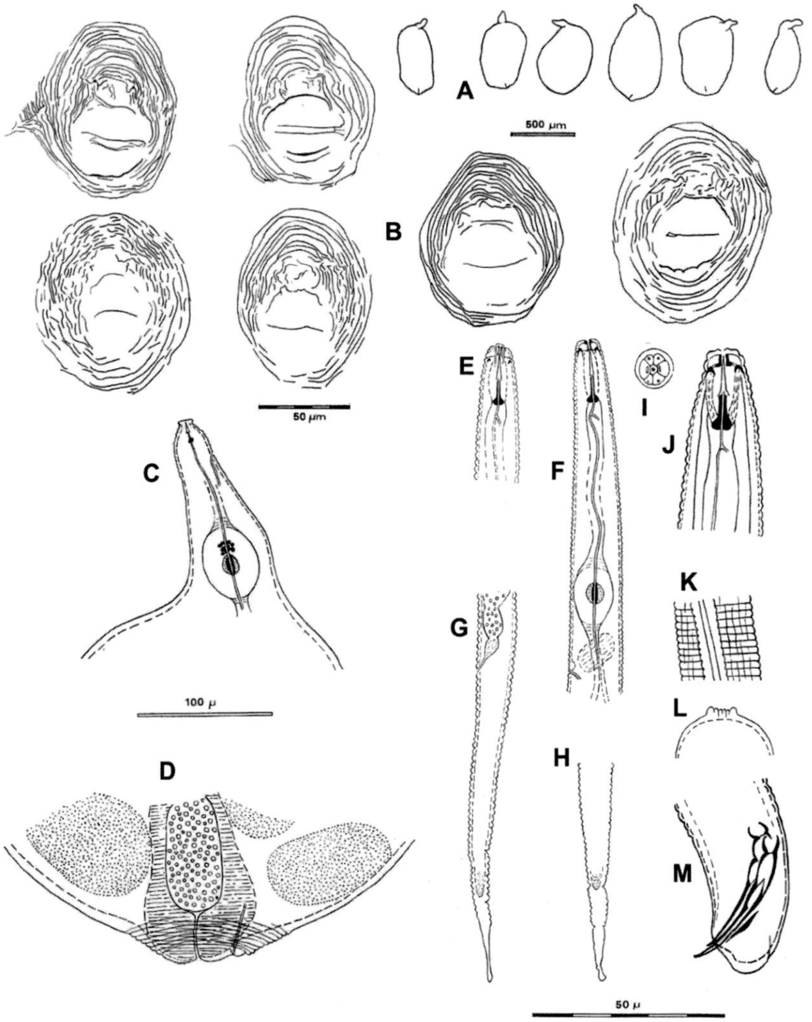

Fig. 194. Meloidogyne oryzae. *A: Entire female; B, D: Perineal pattern; C: Female anterior region; E: Second-stage juvenile (J2) labial region; F: J2 anterior region; G, H: J2 tail; I: Male* en face *view; J: Male labial region; K, L: Male lateral field; M: Male tail. (After Mas* et al.*, 1978.)*

DESCRIPTION (*AFTER MAAS* ET AL., *1978*)

Female

Mature females swollen and globular to pear-shaped with offset neck. Coarse annuli present in neck and tail region. No or slight posterior

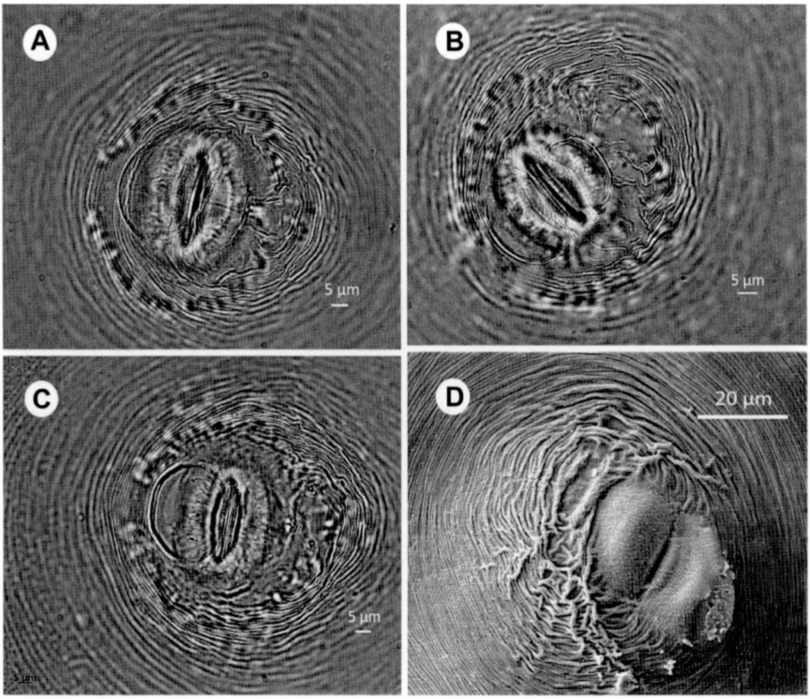

Fig. 195. Meloidogyne oryzae. *LM. A-D: Perineal pattern. (After Mattos* et al., *2018.)*

protuberance, if so, not formed by tail tip but by vulval lips. In some females there is a tendency to a conoid shape in posterior fifth of body. Females completely embedded in root tissue, where eggs are laid. Gelatinous matrix soft. Labial region low, with a distinct constriction posterior to oral disc. Stylet very delicate, knobs minute and sloping posteriorly. Procorpus separated by a constriction from a large muscular bulb, which has conspicuous valve plates. Just anterior to central valve plates, and surrounding lumen, are several small vesicle-like structures. Ovaries coiled, with oocytes in several rows. Glands associated with rectum. Perineal pattern oval in shape, arch moderate in height. No lateral incisures or gap observed. Phasmids small (not always visible). Ridges on cuticle in dorsal region forming wavy broken or unbroken irregular lines around and between phasmids. Striae rather coarse. A prominent fold covering anus dorsally, while in most specimens ventral side of zone 1 (according to Esser *et al.*, 1976) marked by a rather thick fold or at least an obvious regular line.

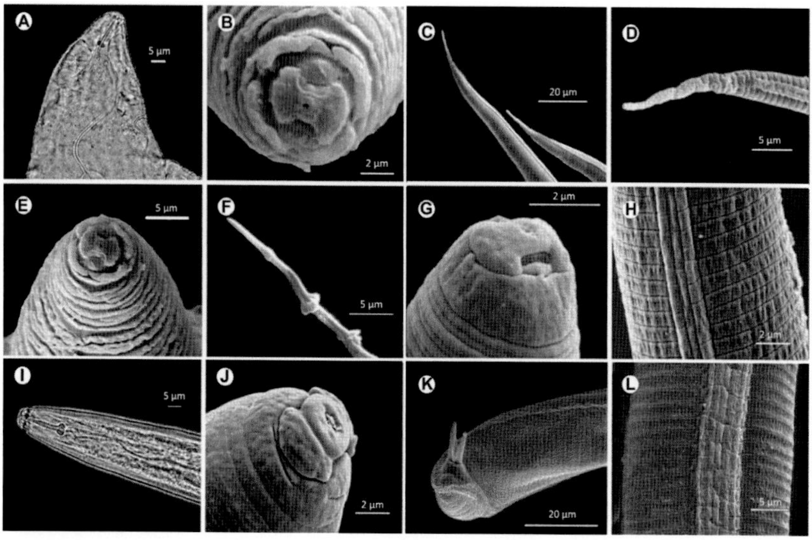

Fig. 196. Meloidogyne oryzae. *LM and SEM. A, B, E: SEM of anterior region of female; F: Excised female stylet; C, D: Second-stage juvenile (J2) tail; I, J: Anterior region of male; G: J2 head; H: Lateral field of J2; K: Male tail; L: Lateral field of male. (After Mattos* et al., *2018.)*

Male

Labial region hardly offset, in lateral view 1-3 annuli posterior to labial cap in dorsal and ventral profiles. Face view showing a hexaradiate framework, six labials, slit-like amphids, and perhaps four papillae on submedian labials. Stylet well developed with posteriorly-sloping rounded knobs, anterior part being a little shorter than posterior part. No vesicle-like structures associated with lumen observed at anterior part of median bulb. Fusiform median bulb with distinct valve plates. Gland lobe extending ventrolaterally along intestine. Lateral field occupying about one-third of body diam. and consisting of 3-7 bands (4-8 incisures) at mid-body, narrowing anteriorly, and continuing posteriorly to very end of tail. Cubic pattern may appear on cuticle, this caused by subcuticular structures. Tail bluntly rounded with an unstriated terminus. Hemizonid immediately (one annulus) anterior to excretory pore or at same level and lying 115-160 (135) μm from anterior end.

J2

Body slender and clearly annulated. Labial region rounded and not offset with a small cap and 3-4 annuli. Skeleton with a delicate guiding

Table 44. *Morphometrics of females, males, second-stage juveniles (J2) and eggs of* Meloidogyne oryzae. *All measurements are in* μm *and in the form: mean* ± *s.d. (range).*

Character	Holotype, Surinam, Maas *et al.* (1978)	Paratypes, Surinam, Maas *et al.* (1978)	Brazil, Mattos *et al.* (2018)
Female (n)	-	15	30
L	-	625 (475-750)	490 ± 11.7 (420-630)
a	-	1.7	1.6 ± 0.1 (1.0-2.0)
W	-	361 (250-432)	317 ± 12 (220-422)
Stylet	-	15 (14-18)	15 ± 0.2 (14-18)
Stylet knob width	-	-	3.5 ± 0.1 (3-4)
Stylet knob height	-	-	1.9 ± 0.1 (1-2)
Labial region diam.	-	4-5	-
DGO	-	7	4.4 ± 0.1 (4-6)
Median bulb length	-	43 (38-52)	-
Median bulb width	-	37 (28-45)	-
Central valve plates length	-	16	-
Central valve plates width	-	10	-
Ant. end to excretory pore	-	-	29 ± 0.6 (21-35)
Vulval slit	-	33 (29-42)	27 ± 0.3 (25-30)
Anus to vulva	-	20 (13-28)	19 ± 0.3 (15-21)
Interphasmid distance	-	21 (15-29)	18 ± 0.3 (16-23)
EP/ST	-	2.0-3.0	-
Male (n)	-	22	20

Table 44. *(Continued.)*

Character	Holotype, Surinam, Maas *et al.* (1978)	Paratypes, Surinam, Maas *et al.* (1978)	Brazil, Mattos *et al.* (2018)
L	-	1667 (1195-1915)	1226 ± 24 (1100-1520)
a	-	56 (44-68)	41.5 ± 1.0 (33-63)
b_1**	-	19 (14-27)	-
c	-	-	108 ± 3.0 (84.6-125)
O*	-	23 (20-26)	-
W	-	30 (24-43)	31 ± 1.9 (27-40)
Labial region height	-	4-5	-
Labial region diam.	-	10	-
Stylet	-	19 (19-20)	18 ± 0.4 (18-20)
Stylet knob width	-	-	3.9 ± 0.1 (3-4)
Stylet knob height	-	-	3.1 ± 0.1 (3-4)
DGO	-	4-5	4.5 ± 0.1 (4-6)
Ant. end to excretory pore	-	-	132 ± 0.1 (128-140)
Spicules	-	31 (25-34)	30 ± 0.2 (29-32)
Gubernaculum	-	8-10	8.5 ± 0.2 (6-9)
Tail	-	-	11.9 ± 0.3 (10-13)
J2 (n)	1	60	30
L	564	545 (500-615)	444 ± 4.0 (420-500)
a	38	37 (30-45)	27.6 ± 0.6 (23-41)

Table 44. *(Continued.)*

Character	Holotype, Surinam, Maas *et al.* (1978)	Paratypes, Surinam, Maas *et al.* (1978)	Brazil, Mattos *et al.* (2018)
b_1**	8.1	8.2 (7.2-9.8)	-
c	7.5	7.0 (6.8-8.6)	5.9 ± 0.1 (5.0-6.5)
c′	-	7.8 (6.8-9.0)	-
W	15	15 (12-17)	16 ± 0.4 (11-20)
Stylet	-	14.2 (14-15)	12.5 ± 0.2 (11-14)
DGO	-	-	3.7 ± 0.1 (3-5)
Stylet knob width	-	-	2.6 ± 0.1 (2-3)
Stylet knob height	-	-	1.6 ± 0.1 (1-2)
Ant. end to excretory pore	-	-	70.5 ± 1.1 (66-76)
Body diam. at anus	-	10 (10-11)	-
Tail	80	79 (70-90)	75.8 ± 0.8 (66-85)
Hyaline region	-	21 (14-26)	22 ± 0.3 (19-24)
Egg			
L	-	113 (97-124)	-
W	-	44 (42-49)	-

*Distance from stylet knobs to dorsal gland outlet as a percentage of stylet length.

**Body length/length from anterior end to middle of median bulb.

tube. Stylet very slender with minute posteriorly-sloping basal knobs. Dorsal gland duct opening hardly visible. Median pharyngeal bulb with distinct valve plates (4 μm long). No vesicle like structures associated with lumen observed at anterior part of median bulb. Pharyngeal gland lobe overlapping intestine ventrally. Hemizonid situated at same level as excretory pore or immediately anterior. Lateral field consisting of three bands (four incisures) and occupying one-third of body diam. Tail elongated, tapering to a narrow rounded terminus, often appearing slightly clavate, with a clear unstriated terminal region. Phasmids not seen. Rectum dilated. Genital primordium lying at 50-60% (58%) of body length.

TYPE PLANT HOST: Rice, *Oryza sativa* L.

OTHER PLANTS: Under field conditions it could also infest fimbristylis-grass (*Fimbristylis miliacea* (L.) Vahl.). Reproduction of this nematode has been observed in glasshouse host tests on plantain (*Musa* sp.), potato (*Solanum tuberosum*), tomato (*S. lycopersicum*), sorghum (*Sorghum bicolor*), and wheat (*Triticum* sp.) (Maas *et al.*, 1978). Other hosts include grasses (*Echinochloa colona* and *E. crus-pavonis* (Kunth) J.A. Schultes, *Eleocharis* sp. and *Hymenachne amplexicaulis* (Rudge) Nees) (Maas *et al.*, 1978; Segeren & Sanchit, 1984). No reproduction of this rice RKN occurred on corn, cotton, peanut, sweet pepper, sweet potato, or watermelon (Maas *et al.*, 1978).

TYPE LOCALITY: SML Wageningen Rice Scheme, Nickerie, Surinam.

DISTRIBUTION: *South America*: Surinam (Maas *et al.*, 1978; Carneiro *et al.*, 2000); French Guiana (Carneiro *et al.*, 2000); Brazil (Mattos *et al.*, 2018).

PATHOGENICITY: Infested plants were stunted and showed root-knots 2-3 mm in size. Tillering was reduced and the leaves were attacked by *Cochliobolus miyabeanus* (formerly known as *Helminthosporium oryzae*), a weakly parasitic fungus.

CHROMOSOME NUMBER: *Meloidogyne oryzae* has 3n = 50-56 chromosomes and a mitotic parthenogenesis mode of reproduction, in contrast with *M. graminicola*, which reproduces by facultative meiotic parthenogenesis and presents a haploid number of chromosomes (n = 18) (Eisenback & Triantaphyllou, 1991; Mattos *et al.*, 2018).

POLYTOMOUS KEY CODES: *Female*: A23, B213, C2, D3; *Male*: A213, B3, C23, D2, E2, F2; *J2*: A12, B12, C1, D12, E32, F1.

BIOCHEMICAL AND MOLECULAR CHARACTERISATION: Esbenshade & Triantaphyllou (1985a) found esterase and malate dehydrogenase (Mdh) isozyme phenotypes of VS1 and N1a, respectively. However, a recent integrative taxonomical analysis of *M. oryzae* from Brazil demonstrated that this species presents a specific esterase phenotype profile (Est O1), in contrast with *M. graminicola* (VS1) (Mattos *et al.*, 2018). The results obtained by Negretti *et al.* (2017) and Mattos *et al.* (2018) showed that the sample used for 18S rRNA gene sequencing was mistakenly identified and actually belonged to *M. graminicola*. The ITS rRNA gene sequence deposited by Mattos *et al.* (2018) showed that *M. oryzae* was placed in clade I with the Incognita group. Mattos *et al.* (2019) also developed a specific SCAR marker of 120 bp for *M. oryzae* (primers ORYA12 F/R), distinguishing this species from other RKN attacking rice (*viz.*, *M. graminicola* and *M. salasi*).

RELATIONSHIPS (DIAGNOSIS): The species belongs to Molecular group I, the Incognita group. *Meloidogyne oryzae* resembles *M. graminicola* and clearly differs from this species by molecular markers and esterase phenotype. In mature females of *M. oryzae* the labial region is offset more clearly and they have a longer stylet (14-18 *vs* 8.1-14.8 μm).

70. *Meloidogyne ottersoni* (Thorne, 1969) Franklin, 1971

(Figs 197-199)

COMMON NAME: Otterson's grass root-knot nematode.

Galls were observed on canary grass in the Wind Lake area near Kenosha, Wisconsin, USA, in 1958 by Dr Henry Otterson, Agricultural Chemist and Consultant, and later were recognised as belonging to the genus *Hypsoperine* Sledge & Golden, 1964. Arthur Weber undertook the study of this nematode in graduate work under the direction of Dr K.R. Barker, University of Wisconsin, Madison. This species was also found parasitising and causing damage to rice in several states in Brazil (Leite *et al.*, 2020).

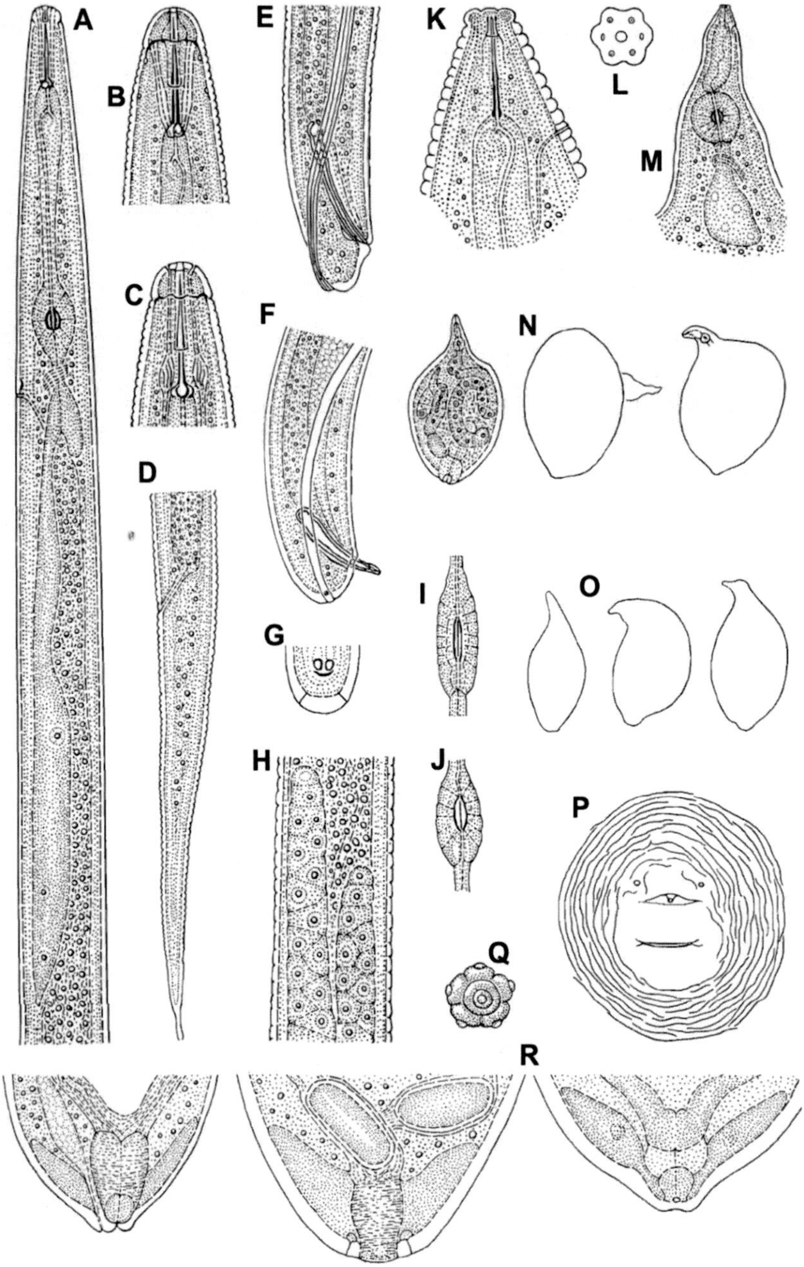
A
B
C
D
E
F
G
H
I
J
K
L
M
N
O
P
Q
R

Fig. 197. Meloidogyne ottersoni. *A: Male anterior region; B, C: Male labial region; D: Second-stage juvenile (J2) Tail; E-G: Male tail; H: Testes; I, J: J2 median bulb; K, M: Female anterior region; L: Female* en face *view; N, O: Entire female; P, R: Perineal pattern; Q: Female* en face *view. (After Thorne, 1969.)*

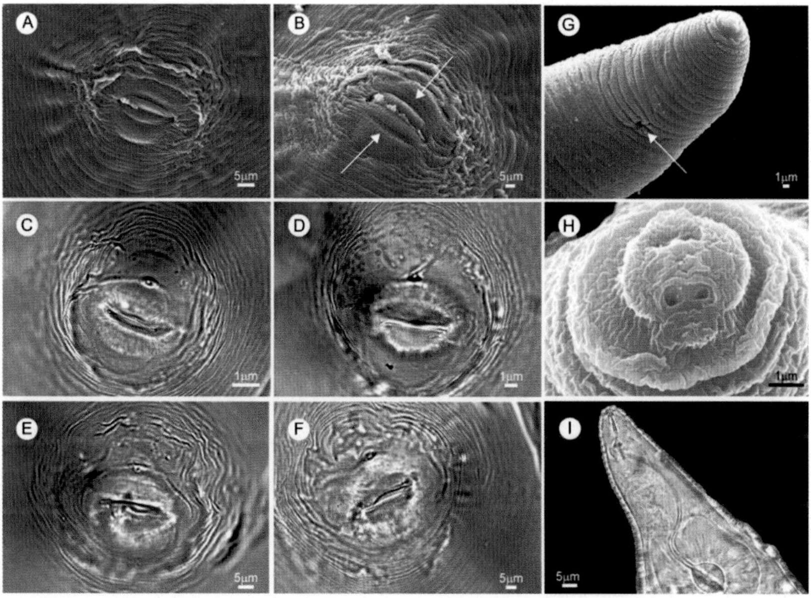

Fig. 198. Meloidogyne ottersoni. *SEM and LM. A-F: Perineal patterns; G-I: Anterior region of female. Arrow showing excretory pore. (After Leite* et al.*, 2020.)*

MEASUREMENTS

See Table 45.

DESCRIPTION *(AFTER THORNE, 1969)*

Female

Spheroid to elongate pyriform with neck projecting at different angles. Vulva and anus usually on a slight elevation. Cuticle 3 μm thick on main part of body, leathery with extremely fine striae except on labial region and neck where easily visible and forming annuli variable in width.

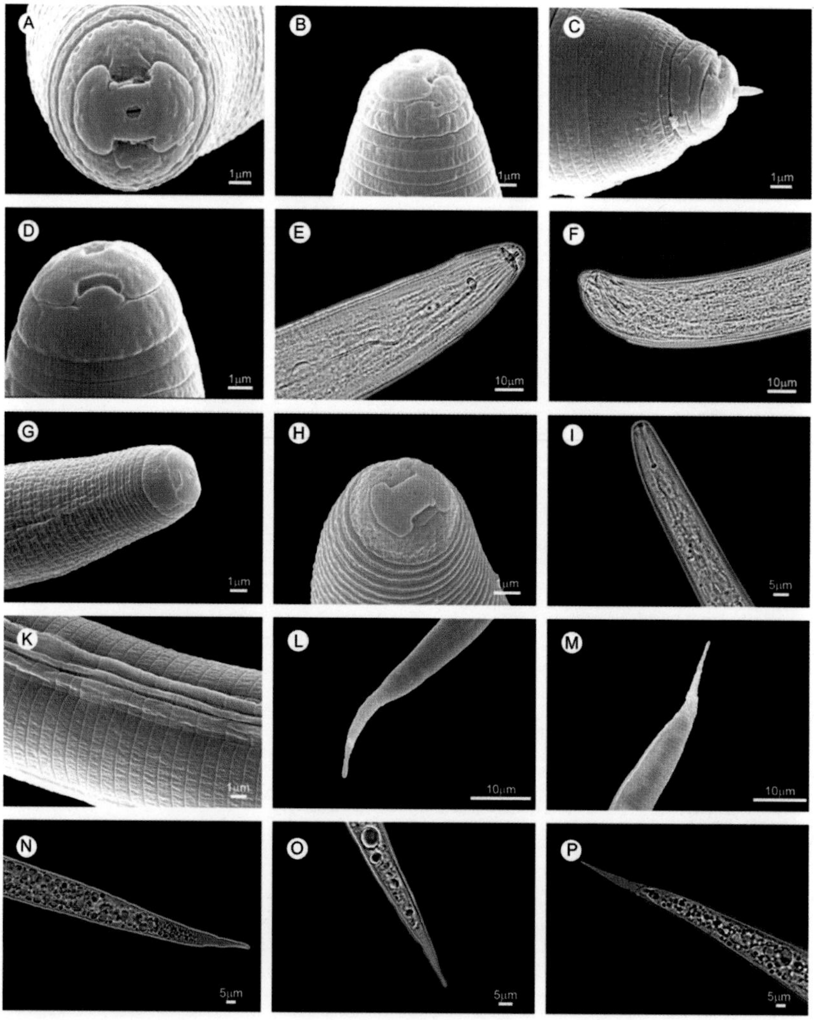

Fig. 199. Meloidogyne ottersoni. *SEM and LM. A-E: Anterior end of male; F: Posterior end of male; G-I: Anterior end of second-stage juvenile (J2); K: Lateral field of J2; L-P: Posterior end of J2. (After Leite* et al., *2020.)*

Labial region unstriated, offset by constriction. Face view showing six well-developed labials, laterals being somewhat larger than submedians. Amphidial apertures slightly oval, located at apices of lateral labials. Vestibule slightly sclerotised, forming a stylet-guiding apparatus. Stylet with small rounded knobs. DGO almost adjacent to stylet base. Anterior

Table 45. *Morphometrics of females, males, second-stage juveniles (J2) and eggs of* Meloidogyne ottersoni. *All measurements are in μm and in the form: mean ± s.d. (range).*

Character	Paratypes, Wisconsin, USA, Thorne (1969)	Brazil, Leite *et al.* (2020)
Female (n)	10	30
L	450 (390-520)	669 ± 134.8 (410-960)
a	-	1.4 ± 0.2 (0.95-1.78)
W	270 (180-320)	503 ± 95.2 (230-700)
Stylet	10-12	11.4 ± 1.4 (9-13)
Stylet knob width	-	3.4 ± 0.5 (3-4)
Stylet knob height	-	2.1 ± 0.3 (2-3)
DGO	-	4.4 ± 0.8 (3-6)
Ant. end to excretory pore	-	26.9 ± 3.2 (21-32)
Vulval slit	-	24.1 ± 1.6 (21-28)
Vulva to anus	-	17 ± 1.8 (15-20)
Interphasmid distance		17.1 ± 3.3 (12-24)
EP/ST	1.0-1.5	-
Male (n)	7	30
L	900-1000	1189 ± 195.9 (760-1430)
a	34	27.8 ± 6.2 (15.3-42.3)
c	-	129.9 ± 31.8 (66.7-197.1)
T	53-65	-
W	-	43.8 ± 7.2 (30-60)
Stylet	14-16	15.9 ± 1.4 (13-19)
Stylet knob width	-	4.2 ± 0.7 (3-5)
Stylet knob height	-	2.4 ± 0.5 (2-3)
DGO	-	4.4 ± 0.8 (3-6)
Ant. end to excretory pore	-	92.5 ± 13.9 (51-114)
Spicules	19-23	25.4 ± 3.94 (16-30)
Gubernaculum	3-4	7.1 ± 1.0 (5-9)
Tail	-	9.4 ± 1.8 (7-14)
J2 (n)	10	30
L	430-500	474 ± 45.7 (400-550)
a	23-30	23.8 ± 2.3 (20.5-27.9)
c	-	6.9 ± 0.7 (5.71-8.33)
W	-	19.9 ± 0.4 (19-21)
Stylet	13-14	10.7 ± 0.9 (9-13)
Stylet knob width	-	2.2 ± 0.4 (2-3)
Stylet knob height	-	1.8 ± 0.4 (1-2)

Table 45. *(Continued.)*

Character	Paratypes, Wisconsin, USA, Thorne (1969)	Brazil, Leite *et al.* (2020)
DGO	4	3.9 ± 0.9 (2-5)
Ant. end to excretory pore	-	73.2 ± 5.2 (63-80)
Tail	-	69 ± 3.3 (62-76)
Hyaline region	-	20.5 ± 2.4 (15-26)
Egg	15	-
L	105 (95-115)	-
W	45 (40-50)	-

portion of pharynx offset from median bulb by constriction. Bulb spheroid with refractive valve and strong radial muscles. Basal bulb a flattened, irregular-shaped lobe extending back over intestine, containing usual three nuclei that generally are difficult to observe. Isthmus very short. Nerve ring obscure. Excretory pore almost opposite to stylet base. Intestine and body contents densely granular, making observations very difficult. Ovaries two, elaborately convoluted, at maturity completely filling body cavity. Vulva transverse. Vagina cylindroid, muscular. Perineal pattern with simple rounded striae, usually on a slight terminal elevation. Anal opening visible from a lateral view and rarely from ventral. Neither spermatheca nor spermatazoa observed.

Male

Body twisted until a true lateral view of labial region and tail can be seen only by cutting specimens after fixation. Striae about 2 μm apart at mid-body, narrower near labial region and sometimes absent on tail. Lateral field about one-fifth body diam., marked by four incisures, two median ones sometimes very obscure. Labial region unstriated, labial disc visible only from a dorsal or ventral view when slit-like amphidial apertures set it off from labial contour. Cephalids obscure, visible only from a lateral view, located well anterior to stylet base. Stylet with strong, posteriorly-sloping knobs. Nerve ring one body diam., and excretory pore twice that distance posterior to median bulb. Lumen of pharynx joining intestine about one body diam. posterior to bulb. Hemizonid about two annuli anterior to excretory pore. Basal pharyngeal lobes average about three times as long as body diam. Testes 53-65% of body length, usually single, outstretched, rarely reflexed. Phasmids almost terminal. Tail variable in form and length. Sperm ducts packed with spheroid spermatozoa indicating that males may be functional although sperm not observed in female uteri.

J2

Lateral field marked by four fine incisures visible only on favourable specimens. Framework of rounded labial region practically undeveloped. Vestibule forming a minute sclerotised stylet guide. Stylet with well-developed knobs. Nerve ring adjacent to bulb, practically enveloping unusually short isthmus. Pharyngeal lumen joining intestine about one body diam. posterior to bulb. Dorsally, pharynx base short and rounded while ventrally extending in a long lobe 5-7 times body diam. Excretory pore slightly posterior to nerve ring. Anal opening very obscure, often unidentifiable. Tail ending in an irregularly clavate or knobbed terminus.

TYPE PLANT HOST: Canary grass, *Phalaris arundianacea* (L.) Reed.

OTHER PLANTS: *Echinochloa crus-galli* (barnyard grass) (Costilla & de Ojeda, 1986; Doucet & Pinochet, 1992), *Oryza sativa*, *E. colona* and *Phalaris canariensis* (Leite *et al.*, 2020).

TYPE LOCALITY: Horner Ranch, Wind Lake area, Wisconsin, USA.

DISTRIBUTION: *North America*: USA (Racine, Milwaukee, and Waukesha counties and near Waupun in Green Lake County); *South America*: Argentina (province of Tucumán) (Costilla & de Ojeda, 1986), Brazil (Leite *et al.*, 2020).

PATHOGENICITY: Infection always takes place just posterior to the apical cells and does not stop growth of the root, which generally forms an arcuate, elongated gall and then proceeds to develop a series of similar galls, each containing from 1-6 females and in rare instances 15 or 20 roots in turn are infected until a complicated, chainlike mass of roots results. Occasionally a progressive multiple infection occurs in a terminal gall with mature egg-producing females at the gall base, while elongated young females are located near the terminus, resulting in a series of various-aged individuals.

CHROMOSOME NUMBER: *Meloidogyne ottersoni* reproduced by meiotic parthenogenesis when males were absent but in the presence of males reproduction was by amphimixis (Triantaphyllou, 1973). This species has n = 18 chromosomes (Triantaphyllou, 1973).

POLYTOMOUS KEY CODES: *Female*: A3, B43, C3, D3; *Male*: A4, B4, C3, D1, E2, F1; *J2*: A2, B2, C2, D1, E3, F1.

BIOCHEMICAL AND MOLECULAR CHARACTERISATION: This species does not show any esterase profiles. The D2-D3 of 28S rRNA, ITS rRNA, *COII*-16S mtDNA genes of *M. ottersoni* by Leite *et al.* (2020). It can be differentiated from all other species of the Graminicola group by *COII*-16S mtDNA gene sequence. The species-specific SCAR primer markers for *M. graminicola*, developed by Belafiore *et al.* (2015), Htay *et al.* (2016) and Mattos *et al.* (2018), can amplify DNA of some populations of *M. ottersoni* (Leite *et al.*, 2020).

RELATIONSHIPS (DIAGNOSIS): *Meloidogyne ottersoni* belongs to Molecular group III, the Graminicola group. It is distinguished from *M. graminis* by perineal pattern, smaller males (0.9-1.0 mm), and EP/ST ratio. From *M. spartinae* (Rau & Fassuliotis, 1965) Whitehead, 1968, it is recognised by the absence of labial annuli (two in *M. spartinae*), and J2 body and tail lengths, and presence of the male hemizonid anterior to the excretory pore *vs* posterior in *M. spartinae*. The perineal pattern of this species also resembles *M. californiensis*, but differs in J2 stylet length, EP/ST ratio, and several morphometrics in the male and J2.

71. *Meloidogyne ovalis* Riffle, 1963

(Fig. 200)

COMMON NAME: Maple root-knot nematode.

This species was collected in August 1961 from swollen tips of rootlets of sugar maple *Acer saccharum* Marshall, growing in a woodlot 32 km northeast of Wausau, Wisconsin, USA (Riffle, 1963).

MEASUREMENTS

See Table 46.

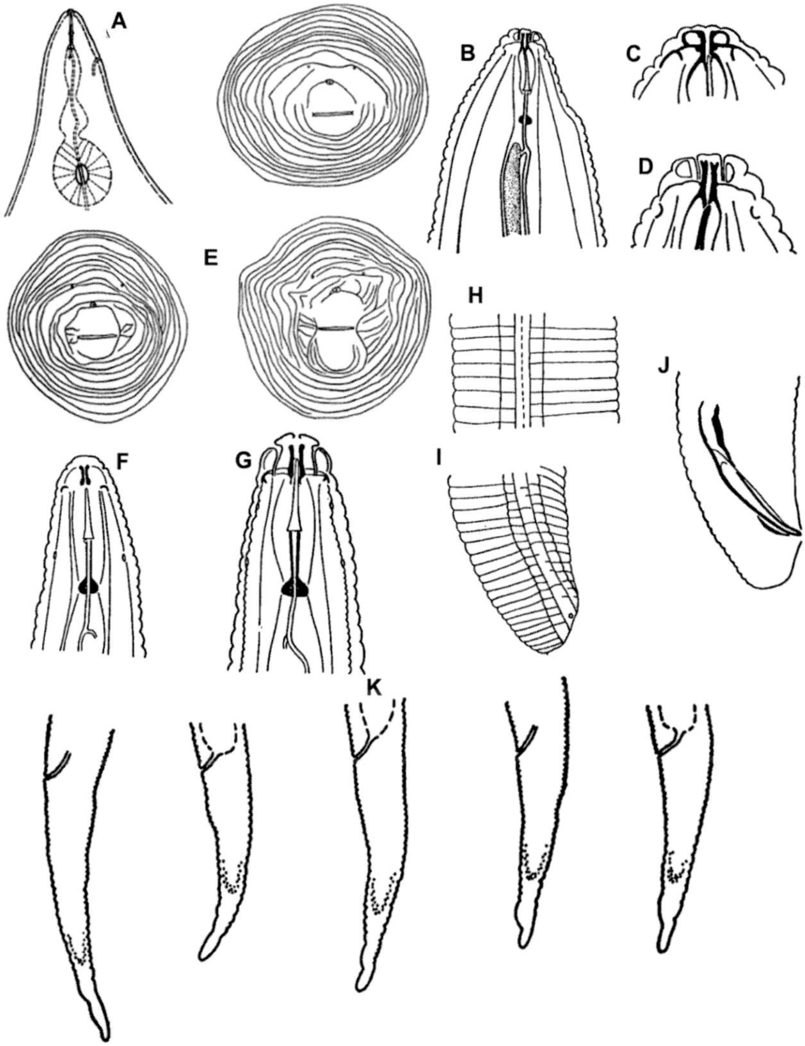

Fig. 200. Meloidogyne ovalis. *A, B: Female anterior region; C, D: Female labial region; E: Perineal pattern; F, G: Male labial region; H: Male lateral field; I, J: Male tail; K: Second-stage juvenile tail. (After Whitehead, 1968.)*

DESCRIPTION

Female

Body flask-shaped with neck tapering to a narrow labial region. Body cuticle thickening somewhat posteriorly, coarsely annulated. Labial

Table 46. *Morphometrics of females, males, second-stage juveniles (J2) and eggs of* Meloidogyne ovalis. *All measurements are in μm and in the form: mean ± s.d. (range).*

Character	Paratypes, Wisconsin, USA, Riffle (1963)	Revised paratypes, Wisconsin, USA, Whitehead (1968)
Female (n)	30	2
L	620 (460-820)	13, 14
W	410 (310-570)	-
Stylet	17-24	-
Stylet knob width	-	2.4
DGO	5	5
Vulval slit	19-23	-
Interphasmid distance	26-34	-
EP/ST	1.3	-
Male (n)	20	6
L	1500 (1300-1800)	1389 ± 109 (1230-1554)
a	41 (35-39)	38.4 ± 2.5 (35.1-42.4)
b	7 (5-8)	-
b_1^*	-	16.2 ± 1.1 (14.5-17.7)
c	102 (93-120)	156 ± 23.2 (134-196)
Labial region height	-	6.2 ± 0.6 (5.4-7.2)
Stylet	18-23	22.6 ± 2.1 (20.1-25.2)
Stylet knob width	4-5	4.0 ± 0.4 (3.6-4.7)
Stylet knob height	2-3	-
DGO	3-5	4.1 ± 0.7 (3.2-5.0)
Length median bulb	-	19.4 ± 2.6 (16.5-23.0)
Width median bulb	-	12.1 ± 1.4 (10.4-14.4)
Length median bulb valves	-	5.7 ± 1.0 (4.7-7.6)
Spicules	31-38	31.2 ± 2.0 (27.3-33.1)
Gubernaculum	7-10	8.6 ± 1.3 (7.2-10.1)
J2 (n)	10	22
L	370 (350-430)	342 ± 20 (302-377)
a	22 (21-24)	21.6 ± 1.1 (19.3-23.3)
b	-	2.39 ± 0.4 (2.0-2.7)
b_1^*	-	8.2 ± 0.8 (6.9-10.2)
c	8 (8-9)	8.2 ± 0.5 (7.4-9.2)
d	-	3.9 ± 0.3 (3.2-4.5)
Stylet	-	10.2 ± 0.8 (8.6-11.7)
DGO	-	2.6 ± 0.6 (2.0-4.6)

Table 46. *(Continued.)*

Character	Paratypes, Wisconsin, USA, Riffle (1963)	Revised paratypes, Wisconsin, USA, Whitehead (1968)
Length median bulb	-	10.7 ± 1.1 (8.6-12.2)
Width median bulb	-	6.7 ± 0.6 (5.8-7.9)
Length median bulb valves	-	3.7 ± 0.5 (3.2-5.0)
Tail	-	42 ± 4.0 (35-49)
Egg	-	-
L	85 (80-91)	-
W	46 (39-50)	-

* Body length/length from anterior end to middle of median bulb.

region low, not offset, with two annuli posterior to labial cap, labial region skeleton strong. Stylet with rounded knobs with anterior margins tending to be posteriorly-sloping. Excretory pore located at about 12 (9-14) annuli posterior to labial region. Perineal pattern oval-shaped, generally with low arch. Anus near centre of pattern. Phasmids in dorsal section. No punctations or lateral lines in pattern. A few short lateral striae present, some of them pointing toward vulva. Distance from anus to vulva *ca* three times that from anus to a line connecting phasmids.

Male

Body annuli averaging 3 μm in width. Lateral field marked by four incisures extending from pharyngeal region almost to terminus. Labial region not offset, hemispherical to truncate-cone shape; in lateral view of labial region, labial cap only faintly marked off from 'first annulus', 'first annulus' deeply marked off from wider basal annulus, which may be subdivided into two annuli of equal length, which are also marked by longitudinal striae, in dorsoventral view of labial region one or two annuli posterior to labial cap. Stylet knobs rounded with well-swept back anterior margins. Anterior cephalids located immediately posterior to labial framework, while posterior cephalids slightly anterior to stylet knobs. Nerve ring 1-2 bulb widths posterior to median bulb. Pharyngeal glands three body diam. in length, ventrally overlapping intestine. Hemizonid two body diam. posterior to median bulb. Excretory pore 2-3 annuli posterior to hemizonid. Excretory tube extending into body for a distance of about 3-4 body diam. Testis single or rarely double, extending three-fifths body length in specimens examined, occasionally reflexed. Lateral field with four main incisures for greater part of its length, also in mid-body region usually a central subsidiary fifth incisure, outer bands of lateral field areolated, inner band rarely cross-striated except near tail, one or two testes, phasmids at level of cloacal aperture, tail bluntly rounded, terminus not striated, spicules with smallish dilated heads and fairly thin-walled, tapering to subacute termini, gubernaculum crescentic. Anal body diam. about twice tail length. Phasmids small, located at level of anal opening.

J2

Labial region truncated cone-shaped not offset, labial cap just marked off from first labial region annulus, which is deeply separated from two basal annuli, labial region also separated from body by deep stria, a lateral 'cheek' and some longitudinal striation of labial region cuticle observed in lateral view of labial region by incident light, distal sclerotisation of buccal tube thicker than proximal, cephalids not seen, stylet knobs small with swept back anterior margins, posterior pharyngeal region overlapping intestine ventrally, lateral field with four incisures for greater part of its length, outer bands marked at irregular intervals by cross striae, hemizonid two annuli long, 0-2 annuli anterior to excretory pore, excretory pore 1.0-2.5 median bulb lengths posterior to level of posterior margin of median bulb, tail tapering to subacute terminus, rectum not dilated.

TYPE PLANT HOST: Sugar maple, *Acer saccharum* Marshall.

OTHER PLANTS: Under field conditions it could also infect American elm (*Ulmus americana* L.), and white ash (*Fraxinus americana* L.). Under glasshouse host tests the nematode infected and reproduced on box elder (*Acer negundo* L.), Norway maple (*A. platanoides* L.), red maple (*A. rubrum* L.), yellow birch (*Betula alleghaniensis* Britt.), paper birch (*B. papyrifera* Marshall) (Riffle, 1963; Riffle & Kuntz, 1967). This species also infected tomato, carrot, onion and geranium, but reproduced on none of them (Riffle & Kuntz, 1967).

TYPE LOCALITY: 32 km northeast of Wausau, Wisconsin, USA.

DISTRIBUTION: *North America*: USA (Wisconsin).

PATHOGENICITY: Many trees exhibited slow growth, sparse foliage, and light green leaves.

CHROMOSOME NUMBER: No available data.

POLYTOMOUS KEY CODES: *Female*: A23, B12, C3, D3; *Male*: A32, B213, C2, D1, E2, F1; *J2*: A3, B32, C34, D2, E32, F34.

BIOCHEMICAL AND MOLECULAR CHARACTERISATION: No available data.

RELATIONSHIPS (DIAGNOSIS): The perineal pattern of *M. ovalis* differs from *M. hapla* by absence of lateral lines and punctations. Males of *M. ovalis* were numerous and differ from those of *M. hapla* in having a higher labial region, longer spicules (27.3-38.0 *vs* 20.0-28.0 μm) and a longer body length (1230-1800 *vs* 791-1432 μm). It is also similar to *M. daklakensis* from which it differs in males in having longer body length (1300-1800 *vs* 1085-1365 μm), spicule length (27.3-38.0 *vs* 18.0-29.0 μm) and J2 body length (302-430 *vs* 280-373 μm).

72. *Meloidogyne panyuensis* Liao, Yang, Feng, & Karssen, 2005

(Figs 201, 202)

COMMON NAME: Chinese peanut root-knot nematode.

Liao *et al.* (2005) described this species from roots of peanut in Guangdong province, Panyu county, China.

MEASUREMENTS *(AFTER LIAO ET AL., 2005)*

- *Paratype females* (n = 20): L = 619 ± 77 (480-750) μm; W = 458 ± 75 (320-580) μm; a = 1.3 ± 0.2 (1.1-1.7); stylet = 13.0 ± 0.9 (12.0-15.0) μm; stylet knob width = 4.8 ± 0.3 (4.5-5.0) μm; DGO = 10.0 ± 1.4 (8.8-12.5) μm; excretory pore to anterior end = 35.6 ± 2.1 (32.5-37.5) μm; vulval slit length = 21.3 ± 1.1 (20.0-22.5) μm; anus to vulva distance = 15.9 ± 1.7 (15.0-17.5) μm; EP/ST = 2.7.
- *Paratype males* (n = 20): L = 1899 ± 89 (1710-2050) μm; W = 47.4 ± 2.2 (43.8-52.5) μm; a = 40.2 ± 2.4 (34.2-43.6); stylet =

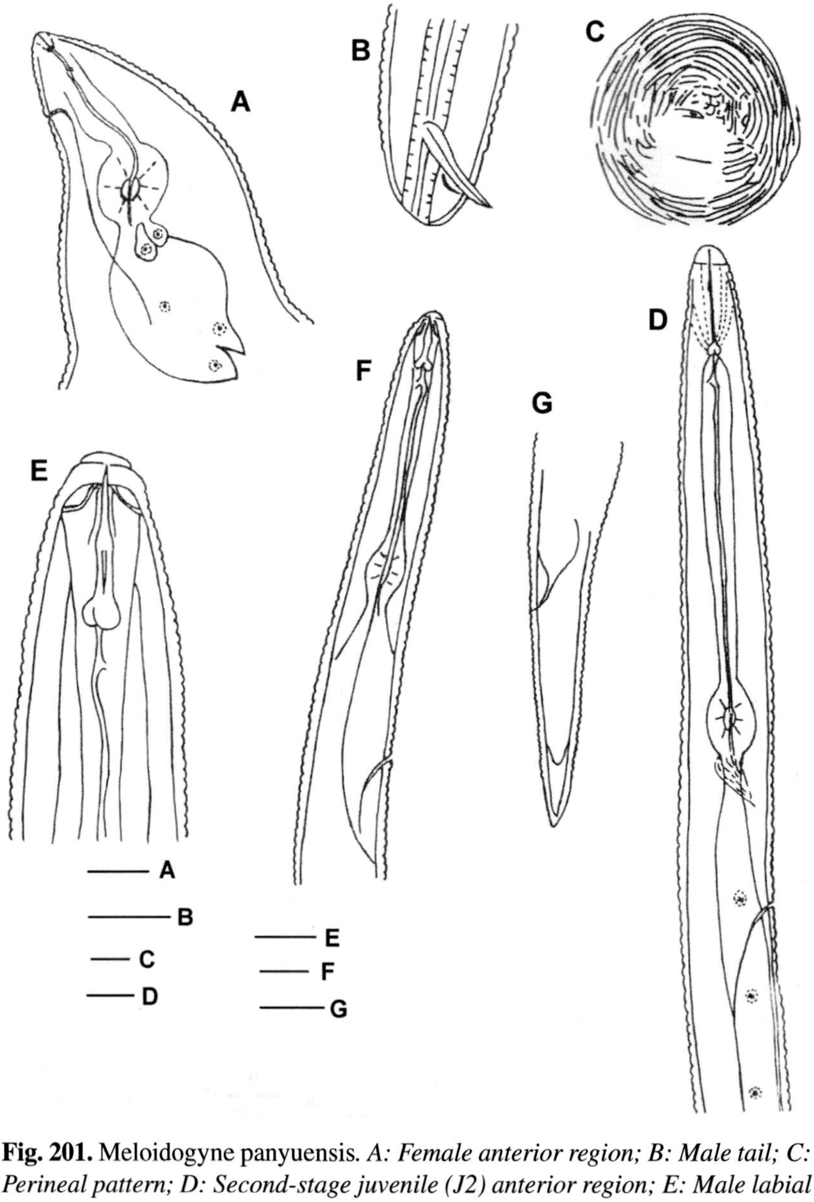

Fig. 201. Meloidogyne panyuensis. *A: Female anterior region; B: Male tail; C: Perineal pattern; D: Second-stage juvenile (J2) anterior region; E: Male labial region; F: Male anterior region; G: J2 tail. (Scale bars: A-C, F, G = 20 μm, D, E = 10 μm.) (After Liao* et al.*, 2005.)*

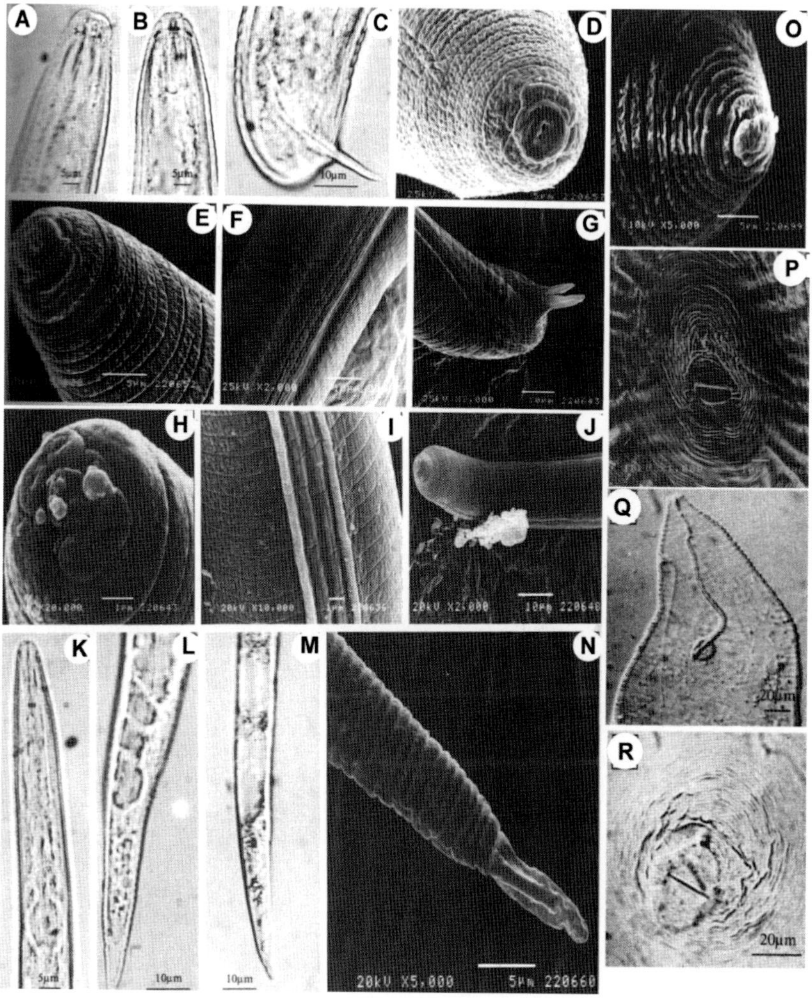

Fig. 202. Meloidogyne panyuensis. *LM and SEM. A, B: Male labial region; C, G: Male tail; D, E: Male* en face *view; H: Second-stage juvenile (J2)* en face *view; I: J2 lateral field; J, K: J2 anterior region; L-N: J2 tail; O: Female labial region; P, R: Perineal pattern; Q: Female anterior region. (After Liao* et al., *2005.)*

23.7 ± 1.1 (22.5-26.3) μm; stylet knob width = 5.1 ± 0.3 (4.5-5.8) μm; DGO = 5.8 ± 0.7 (5.0-7.0) μm; anterior end to median bulb valve = 98 ± 6 (86-105) μm; anterior end to excretory pore = 161 ± 19 (135-200) μm; spicules = 31.7 ± 2.9 (25.0-35.0) μm.

- *Paratype J2* (n = 20): L = 409 ± 28 (353-455) μm; W = 17 ± 1.2 (15.0-18.8) μm; a = 24.3 ± 2.4 (20.6-29.3); c = 7.5 ± 0.4 (6.7-8.1); stylet = 14.5 ± 0.6 (13.8-15.0) μm; stylet knob width = 2.0 ± 0.2 (1.5-2.5) μm; DGO = 4.1 ± 0.5 (3.0-4.5) μm; anterior end to median bulb valve = 61 ± 3 (55-68) μm; excretory pore to anterior end = 78 ± 6 (68-83) μm; tail = 55 ± 5 (48-63) μm; hyaline region = 9.1 ± 1.8 (7.5-10.0) μm.
- *Eggs* (n = 20): L = 100.4 ± 4.9 (90.0-105.0) μm; W = 45.8 ± 2.2 (40.0-47.5) μm.

Description

Female

Body annulated, pearly white, globular to pear-shaped, with slight posterior protuberance and distinct neck region. Labial framework weakly sclerotised. In face view in SEM, labial cap irregular and variable, labial disc fused with median labials. Labial region offset from body. Fine annuli in neck region, posterior body annuli unclear. Excretory pore located between labial region and median bulb level. Stylet well developed, cone slightly curved dorsally, with large rounded knobs, offset from shaft. Perineal pattern ovoid to oval-shaped, striae smooth to moderately coarse, dorsal arch relatively low, lateral lines indistinct (with SEM, lateral lines appearing as a weak indention), tail remnant area distinct with irregular striae, without punctations, phasmids very small, difficult to observe.

Male

Body vermiform, tapering, rounded at both extremities. Cuticle with distinct annulation. Lateral field with four incisures, areolated, incomplete incisure in middle of lateral field often present near mid-body. Labial region slightly offset, with a single postlabial annulus. Labial framework moderately sclerotised, labial disc large and oval, elevated and fused with crescentic median labials. Amphidial openings appearing as elongated slits between labial disc and small lateral labials. Vestibule extension distinct. Stylet strong, straight and pointed at anterior end. Shaft cylindrical, knobs large and rounded, offset from shaft. Pharynx with slender procorpus, median bulb large and oval, ventrally overlapping pharyngeal gland lobe. Testis usually long with reflexed or outstretched germinal zone. Tail short, twisted ventrally with rounded terminus. Spicules arcuate and strong, curved ventrally.

J2

Body vermiform, tapering slightly anteriorly and more so posteriorly. Body annuli small and distinct. Lateral field with four incisures, areolated. Labial region slightly offset from body. Labial framework weakly sclerotised, labial disc rounded, small, and fused with relatively large median labials. Median labials with four distinct labial sensilla, lateral labials small. Vestibule extension distinct. Stylet slender and moderately long, cone straight, knobs distinct and rounded, offset from shaft. Pharynx with slender procorpus and oval median bulb. Pharyngeal gland lobe overlapping intestine ventrally. Hemizonid anterior, adjacent to excretory pore, 2 μm in length. Rectum slightly dilated. Tail tapering, terminus slightly pointed, sometimes irregular shaped and marked with cuticular constrictions.

TYPE PLANT HOST: Peanut, *Arachis hypogaea* L.

OTHER PLANTS: No other hosts were reported.

TYPE LOCALITY: Panyu county, Guangzhou, Guangdong province, China.

DISTRIBUTION: *Asia*: China. It has been only reported in the type locality. However, the ITS rRNA gene sequence (MN383194) of unidentified *Meloidogyne* sp. from Chengmai county, Hainan Province, China, deposited by Long, H. (unpubl.), was very similar to that of *M. panyuensis*.

PATHOGENICITY: *Meloidogyne panyuensis* causes symptoms on peanut such as stunting, yellowed and smaller leaves, and slightly swollen roots (Liao *et al.*, 2005).

CHROMOSOME NUMBER: No available data.

POLYTOMOUS KEY CODES: *Female*: A23, B32, C2, D3; *Male*: A12, B21, C213, D1, E1, F2; *J2*: A23, B12, C32, D34, E3, F3.

BIOCHEMICAL AND MOLECULAR CHARACTERISATION: The isozyme phenotype of *M. panyuensis* is characterised by a unique S1-F1 esterase pattern and N1b malate dehydrogenase pattern (Liao *et al.*, 2005). The ITS rRNA gene sequence of this species (AY394719) was deposited in NCBI by Liao *et al.* (2005).

RELATIONSHIPS (DIAGNOSIS): The species belongs to Molecular group X and is clearly molecularly differentiated from all other *Meloidogyne* species. *Meloidogyne panyuensis* is morphologically very different from other *Meloidogyne* species parasitising peanut (*M. arenaria*, *M. hapla*). The most similar species is *M. fallax*, from which it differs by perineal pattern, a longer DGO in females, males and J2, a longer distance between excretory pore to anterior end in females, males and J2 and longer male stylet length. The female is also close to *M. ethiopica*, but clearly differs from it in several morphometrics of the male and J2.

73. *Meloidogyne paranaensis* Carneiro, Carneiro, Abrantes, Santos & Almeida, 1996a

(Figs 203-206)

COMMON NAME: Parana coffee root-knot nematode.

Meloidogyne paranaensis was described in 1996 in the state of Paraná, Brazil (Carneiro *et al.*, 1996a; Campos & Villain, 2005). This species is one of the most destructive RKN species parasitising coffee in Brazil and in the Americas. This species has probably been confused with *M. incognita*, Carneiro (1993) reporting a new pathotype of *M. incognita*, which was named “biotype IAPAR”. This population had been found attacking coffee in Paraná State and accounted for approximately 52% of all RKN infestations in Paraná. This nematode had been present on coffee in Brazil for many years and had been reported as “unidentified populations of *Meloidogyne* from coffee” (Janati *et al.*, 1982; Esbenshade & Triantaphyllou, 1985a; Carneiro *et al.*, 1996a).

MEASUREMENTS *(AFTER CARNEIRO* ET AL., *1996A)*

- *Holotype female*: L = 684 μm; W = 470 μm; stylet = 16.2 μm; stylet knob height = 2.4 μm; stylet knob width = 4.8 μm; DGO = 4.2 μm; excretory pore to anterior region = 32.5 μm EP/ST = 3.5.

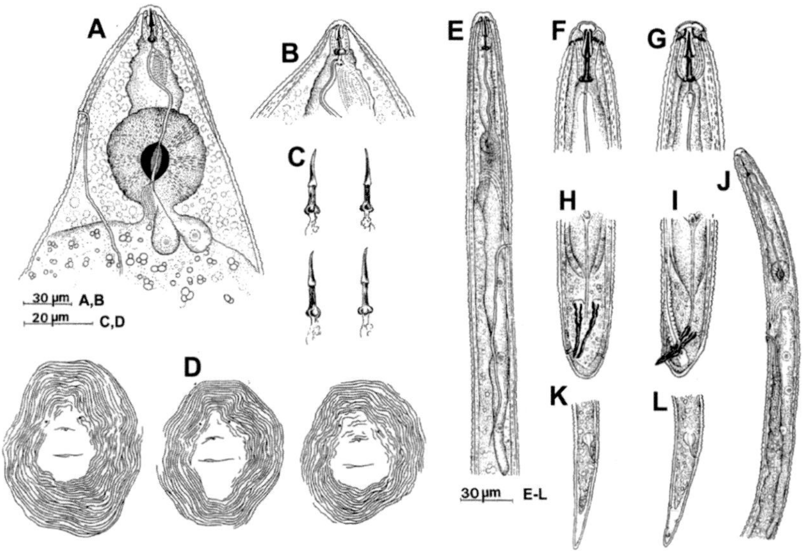

Fig. 203. Meloidogyne paranaensis. *A: Female anterior region; B: Female labial region; C: Female stylet; D: Perineal pattern; E: Male anterior region; F, G: Male labial region; H, I: Male tail; J: Second-stage juvenile (J2) anterior region; K, L: J2 tail. (After Carneiro* et al., *1996a.)*

- *Allotype male*: L = 1708 μm; W = 39 μm; body diam. at excretory pore = 27.6 μm; stylet = 22.2 μm; stylet knob width = 4.8 μm; stylet knob height = 2.4 μm; DGO = 4.2 μm; anterior region to median bulb valve = 86 μm; anterior region end to excretory pore = 157 μm; testis length = 897 μm; spicules = 26 μm.
- *Eggs* (n = 20): L = 90.5 ± 5.32 (82-106) μm; W = 43.3 ± 4.6 (37-51) μm.

See Table 47.

Description

Female

Body translucent white, variable in size, elongate, ovoid to pear-shaped. Neck sometimes prominent, cuticular annulation on body finer than on neck. Body posteriorly rounded, without tail protuberance. Labial region not offset from body, not annulated. In SEM, stoma slit-like, located in ovoid prestomatal cavity, central on labial disc. Pore-like openings of six inner labial sensilla surrounding prestoma. Labial

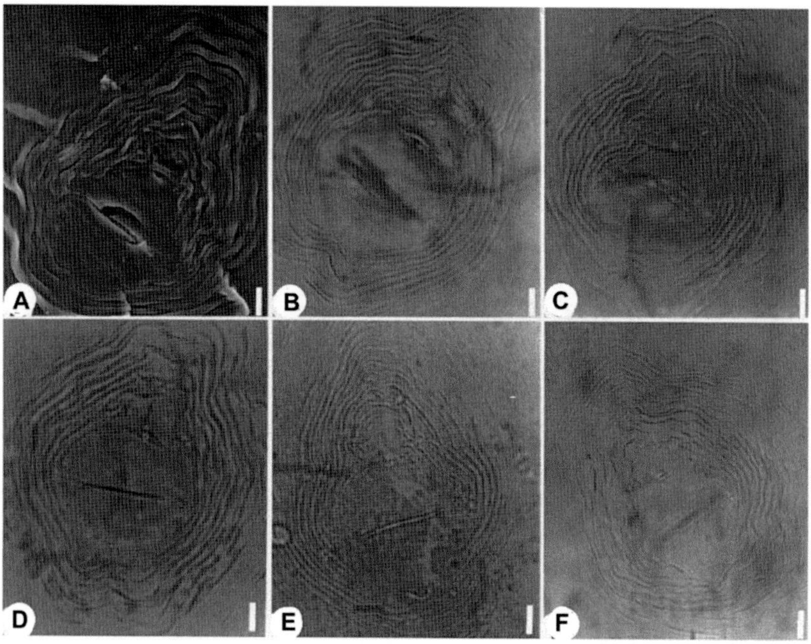

Fig. 204. Meloidogyne paranaensis. *LM. A-F: Perineal pattern. (Scale bars = 10 μm.) (After Carneiro* et al., *1996a.)*

disc and median labials fused, asymmetric and rectangular, forming two straight lateral edges in face view. Lateral labials small, triangular, fused laterally with labial region. Amphidial openings elongated slits between labial disc and lateral labials. In LM, labial framework weakly sclerotised, lateral sectors slightly enlarged, vestibule extension distinct. Anterior half of stylet cone pointed and slightly curved dorsally, posterior half conical. Shaft cylindrical, widening slightly near junction with knobs. Three large knobs tapering onto shaft. Pharynx with large, rounded median bulb, valve plates large. Pharyngeal gland with one large dorsal lobe with one nucleus, two small nucleated subventral gland lobes, variable in shape, position, and size, usually located posterior to dorsal gland lobe. Two large pharyngo-intestinal cells near junction of median bulb and intestine. Excretory pore at level of anterior median bulb. Perineal pattern variable, typically rectangular to oval-shaped, dorsal arch generally high, squarish, dorsal striae varying from fine to coarse, smooth to wavy. Lateral lines mostly discontinuous, without distinct incisures, sometimes

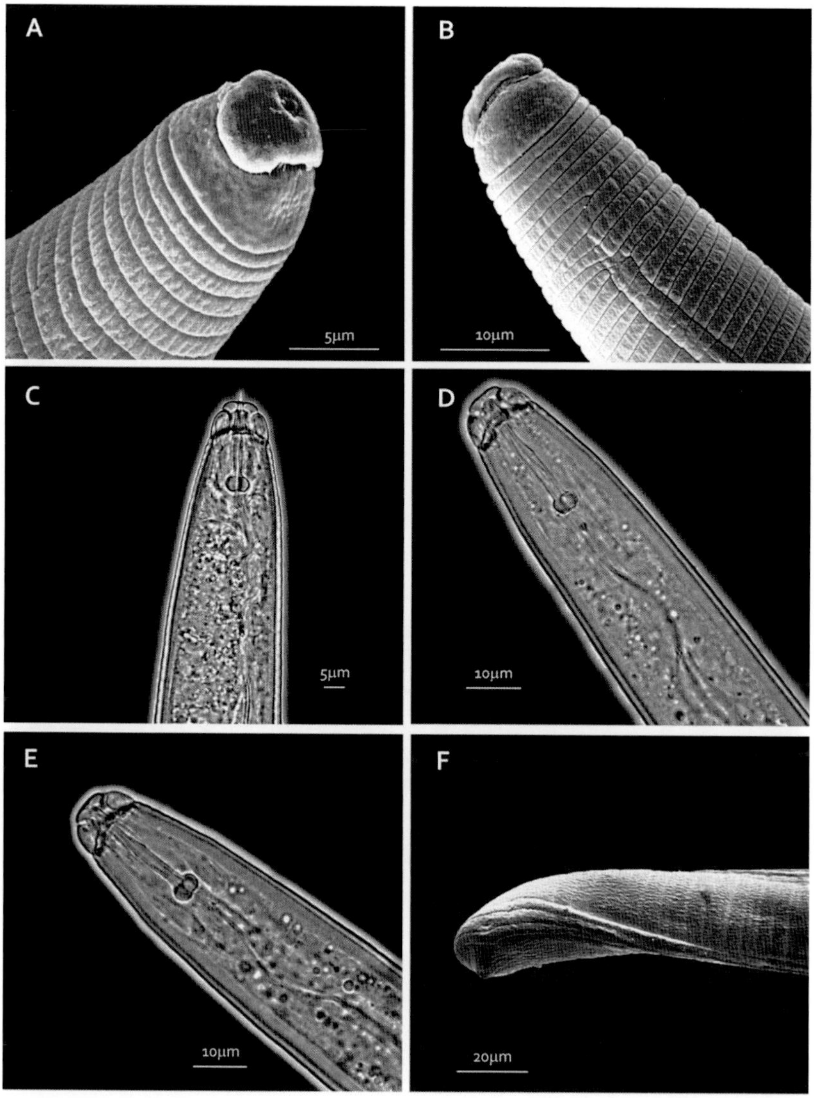

Fig. 205. Meloidogyne paranaensis. *LM and SEM. A-E: Male anterior region; F: Male tail. (After Santos* et al., *2020.)*

appearing as a discontinuous linear depression faintly marked by breaks and forks. All variants with a triangular postanal whorl. Phasmids distinct.

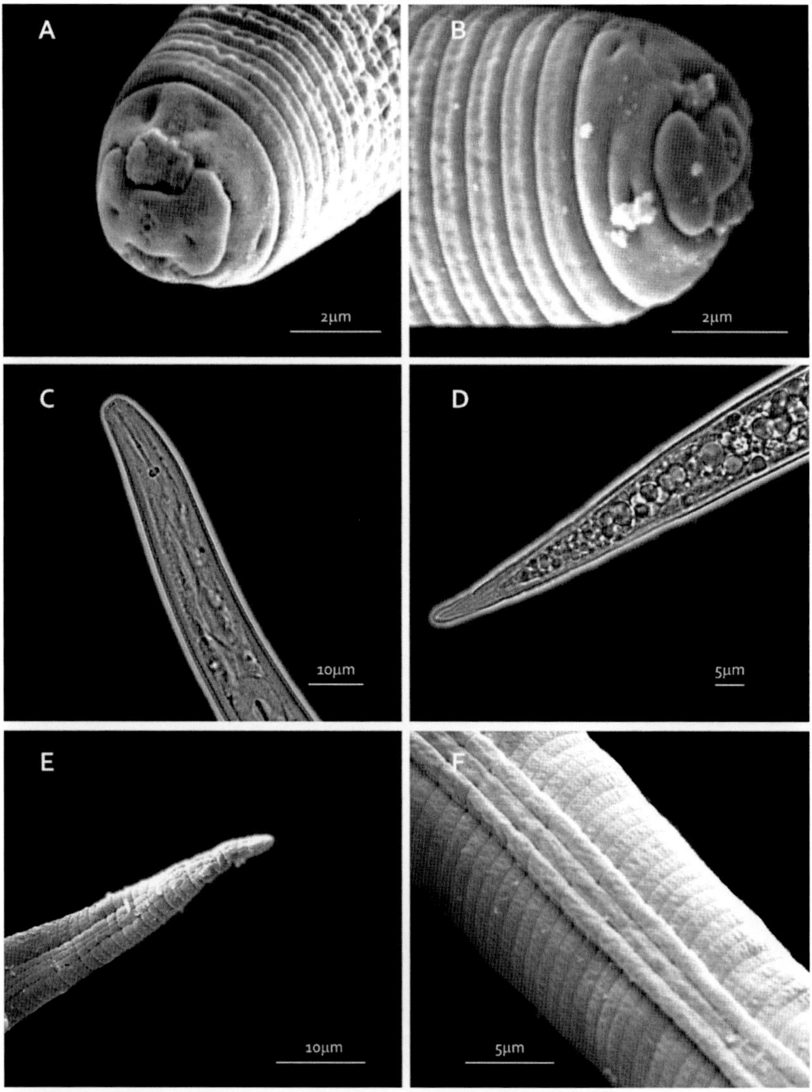

Fig. 206. Meloidogyne paranaensis. *LM and SEM. A, B: Labial region of second-stage juvenile (J2); C: J2 anterior region; D, E: J2 posterior region; F: J2 lateral field. (After Santos* et al., *2020.)*

Male

Body vermiform, length variable, body tapering anteriorly, bluntly rounded posteriorly, tail arcuate, twisting through 90°. Labial cap

Table 47. *Morphometrics of females, males and second-stage juveniles (J2) of* Meloidogyne paranaensis. *All measurements are in μm and in the form: mean ± s.d. (range).*

Character	Brazil, Paratypes, Carneiro *et al.* (1996a)	Brazil, Est P1, Santos *et al.* (2020)	Brazil, Est P2, Santos *et al.* (2020)	Guatemala, Est P2a, Santos *et al.* (2020)
Female (n)	30	30	30	30
L	681 ± 66.5 (512-780)	721 ± 137 (470-980)	791 ± 89.6 (550-1010)	703 ± 74.5 (420-1120)
a	1.6 ± 0.2 (1.1-2.1)	1.4 ± 0.2 (0.8-2.2)	1.4 ± 0.1 (0.9-2.4)	1.6 ± 0.2 (1.0-2.5)
W	428 ± 61.8 (320-532)	524.6 ± 84 (350-850)	547 ± 67 (300-780)	447 ± 36.9 (320-560)
Stylet	16.1 ± 0.6 (15-17.5)	15.4 ± 1.3 (15-17)	17.3 ± 0.8 (15-19)	16.1 ± 1.1 (13-20)
Stylet knob width	5.5 ± 0.5 (4.5-6.5)	4.0 ± 0.7 (2.0-6.0)	4.4 ± 0.5 (3.0-5.0)	5.0 ± 0.6 (4.0-7.0)
Stylet knob height	-	1.9 ± 0.3 (1.5-3.0)	3.4 ± 0.5 (2.0-4.0)	3.0 ± 0.6 (2.0-5.0)
DGO	5.0 ± 0.3 (4.2-5.5)	4.2 ± 0.7 (3.0-5.0)	4.1 ± 0.2 (4.0-5.0)	4.8 ± 0.1 (4.0-6.0)
Ant. end to excretory pore	56.2 ± 0.9 (30-73)	46.0 ± 13.3 (25-150)	41.0 ± 6.6 (25-68)	67.0 ± 19.9 (38-110)
Vulval slit	25.9 ± 3.4 (20-37)	27.4 ± 2.0 (21-32)	27.4 ± 4.1 (12-35)	31.0 ± 2.4 (26-40)
Vulval slit to anus	20.8 ± 2.6 (15-25)	22.0 ± 1.9 (16-26)	21.1 ± 2.9 (10-30)	28.0 ± 2.2 (25-35)
Interphasmid distance	23.2 ± 2.5 (18.1-29.6)	29.0 ± 2.4 (20-35)	25.6 ± 4.6 (18-35)	31.3 ± 2.4 (27-37)
Male (n)	30	19	20	16
L	1868 ± 284.7 (983-2284)	2074 ± 124 (1800-2400)	2405 ± 174 (1600-2690)	1537 ± 205 (1140-1950)
a	46.4 ± 6.4 (23.4-53.5)	47.4 ± 5.2 (39.2-60.0)	44.1 ± 4.2 (31.4-53.8)	47.7 ± 9.1 (28.5-97.5)
c	116 ± 23.9 (58-154)	-	-	-
W	40.3 ± 3.6 (31-46)	44.2 ± 4.9 (40-50)	55.0 ± 5.5 (50-70)	34.0 ± 6.2 (20-40)

Table 47. *(Continued.)*

Character	Brazil, Paratypes, Carneiro *et al.* (1996a)	Brazil, Est P1, Santos *et al.* (2020)	Brazil, Est P2, Santos *et al.* (2020)	Guatemala, Est P2a, Santos *et al.* (2020)
Stylet	24.7 ± 1.25 (20-27)	24.0 ± 1.6 (20-28)	28.3 ± 3.3 (25-30)	26 ± 0.3 (23-28)
Stylet knob width	5.8 ± 0.32 (4.5-7.0)	4.8 ± 0.5 (4.0-6.0)	6.0 ± 0.7 (4.0-8.0)	5.5 ± 0.4 (5.0-6.0)
Stylet knob height	-	3.3 ± 0.5 (3.0-5.0)	4.0 ± 0.2 (3.0-5.0)	4.0 ± 0.4 (3.0-5.0)
DGO	4.6 ± 0.5 (3.5-5.0)	4.0 ± 0.4 (3.0-5.0)	4.1 ± 0.5 (3.0-5.0)	4.1 ± 0.1 (3.0-5.5)
Ant. end to median bulb valve	98 ± 5.8 (82-107)	-	-	-
Ant. end to excretory pore	176 ± 16.5 (130-205)	171 ± 18 (135-217)	199 ± 17.2 (172-250)	154 ± 25.4 (101-185)
Spicules	26.0 ± 2.9 (22-35)	30.0 ± 1.7 (25-33)	34.1 ± 3.1 (28-42)	32.2 ± 2.9 (20-40)
Gubernaculum	-	8.4 ± 0.7 (7.0-10.0)	10.0 ± 0.6 (8.0-12.0)	8.2 ± 0.6 (7.0-9.0)
Tail	16.6 ± 3.0 (12-23)	40.3 ± 2.3 (35-44)	45.4 ± 3.7 (35-52)	43.8 ± 2.3 (40-50)
J2 (n)	30	30	29	30
L	458 ± 27.9 (389-513)	361 ± 14.8 (340-400)	436 ± 22.6 (380-506)	430 ± 33.6 (370-480)
a	28.7 ± 1.9 (24.6-31.7)	22.4 ± 3.9 (17.0-38.0)	21.2 ± 1.3 (17.9-25.0)	22 ± 1.0 (19.0-24.0)
b	7.8 ± 0.5 (6.6-8.9)	-	-	-
c	9.3 ± 0.6 (7.6-10.7)	-	-	-
c′	5.3 ± 0.4 (4.6-6.0)	-	-	-
W	15.9 ± 1.1 (15-20)	17.0 ± 2.8 (10-20)	21.0 ± 0.8 (19-23)	20.0 ± 0 (20-20)
Stylet	13.5 ± 0.9 (13-14)	14.3 ± 1.0 (11-18)	13.2 ± 1.2 (11-16)	15.0 ± 0.5 (13-16)
DGO	4.2 ± 0.3 (4.0-4.5)	2.3 ± 1.2 (2.0-4.0)	3.0 ± 0.2 (2.1-3.5)	3.1 ± 0.5 (2.0-4.0)

Table 47. *(Continued.)*

Character	Brazil, Paratypes, Carneiro *et al.* (1996a)	Brazil, Est P1, Santos *et al.* (2020)	Brazil, Est P2, Santos *et al.* (2020)	Guatemala, Est P2a, Santos *et al.* (2020)
Ant. end to median bulb valve	58.7 ± 3.1 (53-67)	-	-	-
Ant. end to excretory pore	92.1 ± 3.2 (85-98)	71 ± 6.3 (50-85)	83.2 ± 13.0 (62-110)	79 ± 5.7 (50-95)
Tail	49.0 ± 0.8 (48-51)	44.5 ± 5.4 (30-60)	36.5 ± 4.2 (30-50)	42 ± 5.6 (30-55)
Anal body diam.	10.0 ± 0.7 (8.5-11.0)	-	-	-
Hyaline region	10.1 ± 0.7 (9.0-10.0)	11.4 ± 1.1 (8.0-14.0)	14.1 ± 1.5 (11.0-17.0)	15 ± 0.7 (12.0-17.0)

high, rounded, continuous with body contour. In LM, labial framework strongly developed, vestibule and extension distinct. Stylet robust, large, cone straight, pointed, gradually increasing in diam. posteriorly, stylet opening marked by slight protuberance several micrometers from stylet tip, shaft cylindrical, sometimes with one or two large projections, knobs large, rounded, offset from shaft. Procorpus distinct, median bulb ovoid, sometimes covered by intestinal caecum extending anteriorly. Pharyngo-intestinal junction at level of nerve ring, indistinct. In SEM, labial cap flat, labial disc fused with median labials forming elongate, rectangular labial cap. Lateral labials absent. Labial region usually marked by a short, incomplete annulation in lateral view. Stoma opening slit-like, located in ovoid prestomatal cavity, surrounded by pit-like openings of six inner labial sensilla. Four labial sensilla marked by distinct cuticular depressions on median labials. Amphidial apertures elongate slits between labial disc and lateral sectors of labial region. Hemizonid distinct, three or four annuli anterior to excretory pore. Body annuli large, distinct. Areolated lateral field beginning near level of stylet base, usually with four incisures. Most males sex reversed with two testes, some normal with one testis. Testis(es) outstretched or distally reflexed. Spicules arcuate, gubernaculum distinct. Tail short, phasmids at level of cloacal aperture.

J2

Body vermiform, tapering more posteriorly than anteriorly, tail region distinctly narrowing. Body annuli distinct, increasing in size and becoming irregular in posterior tail region. Lateral field with four incisures. In LM, labial framework weak, hexaradiate. Vestibule and vestibule extension distinct. In SEM, stoma slit-like, located in oval prestomatal cavity, surrounded by pit-like openings of six inner labial sensilla. Labial disc and median labials fused, forming a dumbbell-shaped structure. Labial disc rounded, slightly elevated above median labials. Lateral labial sectors distinct, sometimes fused with labial region and labial disc at right angle. Labial region smooth, frequently with short broken annulations. Amphid openings slit-like, located between labial disc and lateral labials, often covered by exudate. Stylet cone increasing in width gradually, shaft cylindrical, knobs rounded and offset from shaft. Median bulb oval. Pharyngo-intestinal junction obscure. Gland lobe overlapping intestine ventrally, with three nuclei, hemizonid 1-2 annuli anterior to excretory pore. Tail usually conoid with rounded terminus. Hyaline region distinct. Rectal dilation large. Phasmids small, posterior to anus.

TYPE PLANT HOST: Isolated from roots of tomato (*Solanum lycopersicum* 'Santa Cruz'), glasshouse cultures. The isolate originated from a coffee plantation.

OTHER PLANTS: Coffee (*Coffea arabica* L.) is the primary host. Other hosts included robusta coffee (*C. canephora* Pierre *ex* A.Froehner), tobacco (*Nicotiana tabacum* L.), and watermelon (*Citrullus lanatus* (Thunb.) Matsum. & Nakai) (Carneiro *et al.*, 1996a). In Brazil, soybean (*Glycine max* (L.) Merr.), yerba mate (*Ilex paraguariensis* A.St.-Hil.), billygoat-weed (*Ageratum conizoides* L.) and lilac tasselflower (*Emilia sonchifolia* (L.) DC. *ex* Wight) were also hosts of *M. paranaensis* (Santiago *et al.*, 2000; Roese *et al.*, 2004; Campos & Villain, 2005; Moritz *et al.*, 2008). In Guatemala, garden balsam (*Impatiens balsamina* L.), a common weed in coffee plantations, was a good host of *M. paranaensis* and has been used successfully for rearing populations of this nematode in pots (Campos & Villain, 2005). Monaco *et al.* (2008) confirmed the status of *Raphanus raphanistrum* L. and *Eleusine indica* (L.) Gaertn. as hosts of *M. paranaensis*. They also found that the following weeds were susceptible to this species under experimental inoculation: billygoatweed (*Ageratum conyzoides* L.), low amaranth (*Amaranthus deflexus* L.), green amaranth (*A. hybridus* L.), slender amaranth (*A. viridis* L.), beggartick (*Bidens subalternans* DC.), lamb's quarters (*Chenopodium album* L.), clammy goosefoot (*C. carinatum* R.Br.), spiderwisp (*Cleome affinis* L.), Jamaican crabgrass (*Digitaria horizontalis* Willd.), catirinha (*Hyptis lophanta* Mart. *ex* Benth.), littlebell (*I. triloba* L.), mentruz (*Lepidium pseudodidymum* Thell. Ex Druce), balsam pear (*Momordica charantia* L.), cutleaf groundcherry (*Physalis angulata* L.), spotted lady's thumb (*Polygonum persicaria* L.), little hogweed (*Portulaca oleraceae* L.), marsh bristlegrass (*Setaria geniculata* (Poir.) Kerguélen), jewels of Opar (*Talinum paniculatum* (Jacq.) Gaertn.), and seashore vervain (*Verbena litoralis* Kunth). Sweet basil (*Ocimum basilicum* L.), oregano (*Origanum vulgare* L.), boldo (*Plectranthus barbatus* Andrews) and pennyroyal (*Mentha pulegium* L.) were also susceptible to *M. paranaensis* (Mônaco *et al.*, 2011). *Pfaffia glomerata* (Spreng.) Pedersen, *Hypericum perforatum* L., *Pogostemon cablin* Benth., and *Melissa officinalis* L. were considered as hosts (Mendonça *et al.*, 2014).

TYPE LOCALITY: Paranávai, Paraná State, Brazil.

DISTRIBUTION: *South America*: Brazil, Colombia; *Central America and Caribbean*: Costa Rica (Carneiro *et al.*, 2004a), Guatemala (Carneiro *et al.*, 2004c), Martinique (Quénéhervé *et al.*, 2011); *North America*: Mexico (López-Lima *et al.*, 2015); *Oceania*: Hawaii (USA) (Carneiro *et al.*, 2004a).

PATHOGENICITY: This nematode species is highly aggressive to *C. arabica* genotypes, which results in limited growth and reduced yield of plants cultivated in infested fields (Ferraz, 2008). *Meloidogyne paranaensis* produces symptoms such as splitting and cracking of the cortical root tissue, especially on the taproot, but does not produce typical RKN galls on coffee, in a syndrome called 'coffee corky-root disease' (Carneiro *et al.*, 1996a; López-Lima *et al.*, 2015). Necrotic spots occur along the roots where the female are located, and nematode feeding probably causes the death of tissues around the giant cells. Coffee plants could present different levels of general decline, reduced growth, nutritional deficiency symptoms, chlorosis, defoliation, and dieback (López-Lima *et al.*, 2015).

CHROMOSOME NUMBER: Chromosome numbers are 50-56. Freire *et al.* (2002) reported n = 28 chromosomes, suggesting reproduction by meiotic parthenogenesis associated to aneuploidy.

POLYTOMOUS KEY CODES: *Female*: A23, B2, C1, D6; *Male*: A1234, B123, C312, D1, E1, F1; *J2*: A213, B2, C3, D3, E23, F23.

BIOCHEMICAL AND MOLECULAR CHARACTERISATION: The perineal pattern and differential host tests could misidentify this species as *M. incognita*, as has happened for some years. For this reason, several additional identification approaches have been created. Several patterns have been described for esterase (P1, P2, and P2a) (Carneiro *et al.*, 1996b, 2000, 2004b; Santos *et al.*, 2018b, 2020) and a unique pattern (N1) for malate dehydrogenase (Carneiro *et al.*, 2000).

Species-specific primers have been developed by Randig *et al.* (2002) and validated by Carneiro *et al.* (2004c, 2005b) and Santos *et al.* (2018b, 2020). A specific length for the *coxII*-16S region (Powers & Harris, 1993) of 1.25 kb has been observed in comparison to other species of *Meloidogyne* (Fig. 13). Santos *et al.* (2018b) detected low molecular variability using neutral markers (AFLP and RAPD) among *M. paranaensis* populations in coffee from Brazil and Guatemala. Santos *et al.* (2020) showed that three populations of *M. paranaensis* with different esterase phenotypes presented similar morphological, morphometric and molecular characteristics that clustered them into a single species with small variation.

RELATIONSHIPS (DIAGNOSIS): The species belongs to Molecular group I, the Incognita group. *Meloidogyne paranaensis* is most similar to *M. konaensis* (Eisenback *et al.*, 1994) but differs from it in several morphological features, including female perineal pattern. Males of *M. paranaensis* differ from males of *M. konaensis* in body length (983-2690 *vs* 1149-2050 μm), and DGO (3.5-5.0 *vs* 5.0-9.0 μm). The J2 of *M. paranaensis* differ from *M. konaensis* in body length (340-513 *vs* 395-550 μm), DGO (2.0-4.5 *vs* 4.0-6.0 μm), and tail length (30-60 *vs* 35-76 μm). This species is also characterised by a high EP/ST ratio (3.5) and a perineal pattern with high squared dorsal arch, which is similar to *M. moroccienisis*, but can be separated from this species by esterase pattern and molecular markers.

74. *Meloidogyne partityla* Kleynhans, 1986
(Fig. 207)

COMMON NAME: Pecan root-knot nematode.

This species was found associated with pecan orchards sampled at Mataffin near Nelspruit, South Africa, in the Plant Protection Research Institute at Mataffin, and on the premises of the Citrus and Subtropical Fruit Research Institute, in 1984. It has been suggested that the nematode probably entered South Africa together with pecan seedlings imported from the USA in 1912, 1939 and 1940 (Kleynhans, 1986). Among RKN, *M. partityla* is the dominant species parasitising pecan with a greater incidence in the southern USA.

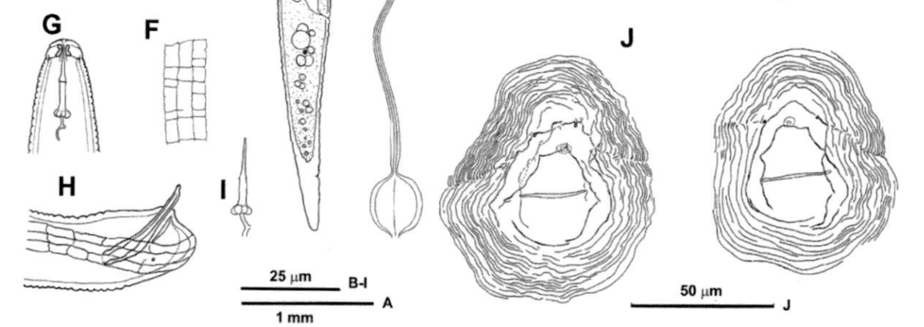

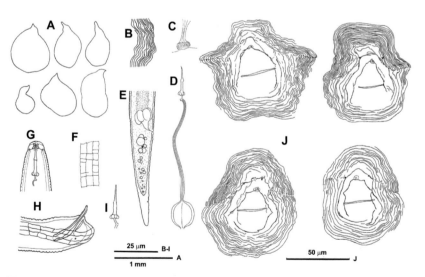

Fig. 207. Meloidogyne partityla. *A: Entire female; B: Female body annulation; C: Anus and rectum; D: Female stylet and metacorpus; E: Second-stage juvenile tail; F: Male lateral; G: Male labial region; H: Male tail; I: Male stylet; J: Perineal pattern. (After Kleynhans, 1986.)*

MEASUREMENTS

See Table 48.

DESCRIPTION

Female

Body white, pear-shaped, neck wide at base, short, posterior end without protuberance, excretory pore 15-27 annuli from labial region. Labial region with one postlabial transverse stria, stylet cone attenuates rapidly towards tip, often curved dorsally, with most of curve in anterior portion, junction between cone and shaft prominently thickened, stylet knobs large, transversely ovoid, each with a deep median longitudinal groove, anterior faces posteriorly sloping, pharyngeal lumen lining expanding gradually towards rear, usually constricted just anterior to rounded triradiate median bulb lining, subcuticular punctations around excretory duct not seen. Perineal pattern hexagonal, dorsal arch low to medium high, sometimes broadly rounded, pattern striae relatively coarse, annulation outside pattern fine, wavy to zigzag, lateral field areas often bulging, lateral field recessed, dorsal and ventral striae meeting at an angle, perineum free from striae, bordered laterally and dorsally by a wide, irregular ridge often forming irregular folds over indistinct tail terminus and adjacent areas, subcuticular rectal punctations positioned close to rectum.

Table 48. *Morphometrics of females, males, second-stage juveniles (J2) and eggs of* Meloidogyne partityla. *All measurements are in μm and in the form: mean ± s.d. (range).*

Character	Holotype, allotype, Mataffin, South Africa, Kleynhans (1986)	Paratypes, Mataffin, South Africa, Kleynhans (1986)	Arkansas, USA, Khanal *et al.* (2016b)
Female (n)	-	26	10
L	993	812 ± 93.6 (616-993)	819 ± 150 (552.1-995)
W	586	462 ± 67.6 (331-596)	505.3 ± 93.5 (352.4-642.7)
Stylet	16.4	16.6 ± 1.2 (14.6-20.0)	17.2 ± 2.7 (14.2-19.3)

Table 48. *(Continued.)*

Character	Holotype, allotype, Mataffin, South Africa, Kleynhans (1986)	Paratypes, Mataffin, South Africa, Kleynhans (1986)	Arkansas, USA, Khanal *et al.* (2016b)
Stylet cone length	9.5	9.1 ± 0.9 (7.7-11.2)	10.32 ± 1.69 (7.1-11.2)
DGO	4.9	3.9 ± 1.4 (2.1-8.0)	-
Anterior region end to excretory pore	32.3	27.9 ± 7.8 (17.3-41.0)	-
Dorsal gland opening to median bulb valve	-	66.5 ± 10.9 (46.0-90.8)	-
Vulval slit	24.1	23.0 ± 2.0 (19.3-27.5)	23.6 ± 3.48 (18.3-28.4)
Vulva to phasmids	-	21.6 ± 2.6 (18.0-30.0)	-
Vulva to anus	21.3	18.2 ± 1.8 (16.0-23.7)	17.57 ± 4.12 (12.2-24.4)
Interphasmid distance	23.4	18.6 ± 3.2 (13.4-25.3)	-
EP/ST	-	1.7*	-
Male (n)	1	21	
L	1030	987 ± 340 (723-1954)	-
W	25.7	26.3 ± 3.3 (22.3-32.0)	-
Stylet	18.2	18.6 ± 1.3 (16.9-21.4)	-
DGO	3.7	3.5 ± 1.0 (2.3-5.3)	-
Ant. end to median bulb valve	-	65.5 ± 11.9 (44.7-90.7)	-
Ant. end to excretory pore	98.0	120.8 ± 21.2 (98-186)	-
Tail	7.3	6.6 ± 1.7 (3.7-10.0)	-

Table 48. *(Continued.)*

Character	Holotype, allotype, Mataffin, South Africa, Kleynhans (1986)	Paratypes, Mataffin, South Africa, Kleynhans (1986)	Arkansas, USA, Khanal *et al.* (2016b)
Phasmids to tail end	7.7	8.9 ± 2.2 (4.8-13.0)	-
Spicules	-	29.2 ± 2.5 (26.6-34.5)	-
J2 (n)	-	30	25
L	-	437 ± 25.3 (383-494)	447.5 ± 32.2 (387.3-486.2)
W	-	17.1 ± 1.3 (14.7-19.6)	15.2 ± 1.4 (10.2-16.7)
Stylet	-	11.4 ± 0.5 (10.3-12.4)	13.2 ± 1 (10.7-15.2)
Stylet base to ant. end	-	15.8 ± 0.7 (14.1-17.5)	14.1 ± 0.7 (13.1-15.7)
DGO	-	2.9 ± 0.4 (1.7-3.7)	3.8 ± 0.7 (2.5-5.6)
Ant. end to median bulb valve	-	60.5 ± 2.7 (55.4-67.1)	62.3 ± 1.9 (58.4-66.9)
Ant. end to excretory pore	-	87.4 ± 4.8 (75.3-95.4)	86.8 ± 4.71 (79.2-97.4)
Pharynx	-	-	132 ± 7.3 (121.8-150.2)
Tail	-	50.5 ± 4.9 (45.0-63.5)	51.9 ± 4.1 (44.7-58.9)
Hyaline region	-	14 ± 1.7 (11.4-18.8)	15.1 ± 2.7 (11.7-18.3)
Body diam. at anus	-	11.7 ± 0.7 (10.7-13.2)	11.1 ± 1 (10.2-13.2)
Egg	-	30	25
L	-	99.8 ± 8.0 (82.0-116.0)	100.2 ± 8.5 (85.3-114.7)
W	-	43.7 ± 3.3 (36.7-49.2)	44.1 ± 3.8 (38.6-49.7)

* From means in the descriptions.

Male

Body vermiform, posterior portion twisted, lateral field width 16-31% of body diam., lateral lines four, in places a faint fifth median line, outer bands irregularly areolated throughout, inner band sparsely areolated, excretory pore 57-73 annuli from labial region, 0-3 annuli posterior to hemizonid, which extends over 1-2 annuli, phasmids opposite or slightly anterior to cloacal aperture. Labial cap sloping smoothly towards rear, narrower than postlabial region, which lacks transverse striae, vestibule expanding towards basal plate, stylet robust, stylet cone cylindrical for some distance, stylet opening at 24-34% of cone length posterior to a usually blunt tip, stylet shaft sometimes widened at middle, shaft and knobs as long as cone, knobs transversely ovoid, anterior faces inclined towards rear, each knob deeply grooved longitudinally, triradiate median bulb lining roundish, testes double, usually reflexed, spicule tips bluntly rounded.

J2

Annulation fine on body, coarse and irregular on tail, lateral field with four lines, lines fading opposite hyaline region, crenations in outer lines in places larger than body annuli, areolations not seen, excretory pore distinct, directly posterior to hemizonid, which extends over 3-4 annuli, phasmids in anterior one-third of tail, stylet cone longer than shaft and knobs, anterior faces of knobs sloping posteriorly, rectum greatly swollen, deeply grooved.

TYPE PLANT HOST: Pecan nut, *Carya illinoensis* (Wangenh.) K. Koch.

OTHER PLANTS: California black walnut (*Juglans hindsii* (Jeps.) Jeps. *ex* R.E. Sm.), English walnut (*J. regia* L.), and shagbark hickory (*Carya ovata* (Mill.) K.Koch) have been reported as hosts of this nematode (Starr *et al.*, 1996). Water oak (*Quercus nigra* L.) in Alachua County, Florida (USA) (Brito *et al.*, 2016), laurel oak (*Q. laurifolia* Michx.) (Brito *et al.*, 2013), *Carya illinoensis* (Wangenh.) K.Koch, (Brito *et al.*, 2008), and *Q. stellata* Wangenh. (Khanal *et al.*, 2016b) are also hosts. In a host status study conducted under glasshouse conditions, Marais & Heyns (1990) reported that beans (*Phaseolus vulgaris*), peas (*Pisum sativum*), cowpea (*Vigna* sp.), tomato (*Solanum lycopersicum*), weeping love grass (*Eragrostis curvula* (Schrad.) Nees), and peach (*Prunus persica*) were non-hosts of *M. partityla*.

TYPE LOCALITY: Transvaal Lowveld, South Africa.

DISTRIBUTION: *North America*: USA (Florida (Crow *et al.*, 2005), Texas (Starr *et al.*, 1996), South Carolina (Brito *et al.*, 2013; Eisenback *et al.*, 2015), New Mexico (Thomas *et al.*, 2001), Georgia (Nyczepir *et al.*, 2002), Arizona and Oklahoma (Brito *et al.*, 2006), Arkansas (Khanal *et al.*, 2016a)); *Africa*: South Africa (Kleynhans, 1986).

PATHOGENICITY: The damage seems more associated with seedlings, showing patches of stunted growth with leaf yellowing and dead branches in the upper canopy in the field (Nyczepir *et al.*, 2003, 2004; Brito *et al.*, 2006). This type of canopy damage is commonly referred to as 'mouse ear disorder' and has been associated with infection by *M. partityla* (Brito *et al.*, 2006). Infected plants exhibit root swellings and prominent galls on major roots as well as young roots (Brito *et al.*, 2006). *Meloidogyne partityla* can also cause decline in yields from mature pecan trees (Kleynhans, 1986; Thomas *et al.*, 2001) or dieback in laurel oak (Brito *et al.*, 2013; Eisenback *et al.*, 2015).

CHROMOSOME NUMBER: 2n = 40-42, obligatory mitotic parthenogenesis (Marais & Kruger, 1991).

POLYTOMOUS KEY CODES: *Female*: A21, B21, C3, D1; *Male*: A4123, B324, C2, D1, E1, F1; *J2*: A23, B32, C32, D213, E2, F3.

BIOCHEMICAL AND MOLECULAR CHARACTERISATION: EST phenotype (Mp3) from *M. partityla*, which is species-specific, can be easily used to distinguish this nematode from *M. arenaria*, *M. floridensis*, *M. graminis*, *M. graminicola*, *M. incognita*, *M. javanica* and *M. enterolobii*. Malate dehydrogenase (MDH) (N1) from *M. partityla* is identical to *M. arenaria*, *M. javanica*, *M. floridensis* and *M. incognita*. However, other esterase (three patterns) and malate dehydrogenase (two patterns) have been found for this species (Starr *et al.*, 1996). *Meloidogyne partityla* can be differentiated from all other species by *COII*-16S mtDNA (Powers *et al.*, 2005), ITS rRNA (Brito *et al.*, 2016) and *COI* (Powers *et al.*, 2018) gene sequences.

RELATIONSHIPS (DIAGNOSIS): The species belongs to Molecular group II and is clearly molecularly differentiated from all other *Meloidogyne* species. This species has a unique perineal pattern and biochemical markers, which differentiate this species from others.

75. *Meloidogyne petuniae* Charchar, Eisenback & Hirschmann, 1999
(Figs 208-210)

COMMON NAME: Petunia root-knot nematode.

This species was described by Charchar *et al.* (1999) associated with petunia plants (*Petunia hybrida* L.) in Brasilia, Brazil.

MEASUREMENTS *(AFTER CHARCHAR* ET AL., *1999)*

- *Holotype female* (in glycerin): L = 849 μm; W = 571 μm; stylet = 14.5 μm; stylet knob height = 1.9 μm; stylet knob width = 4.0 μm; DGO = 3.3 μm; anterior end to excretory pore = 35.3 μm; EP/ST = 2.4.
- *Allotype male* (in glycerin): L = 1988 μm; W = 42 μm; a = 47.2; c = 108; stylet = 21.1 μm; stylet knob height = 2.4 μm; stylet knob width = 4.1 μm; DGO = 2.1 μm; anterior end to excretory pore = 178.2 μm; tail = 18.4 μm; spicules = 28.9 μm; gubernaculum = 6.6 μm; testis length = 1092 μm; T = 54.9%.
- *Paratype females* (n = 30): L = 804 ± 86.3 (698-1002) μm; W = 565 ± 67.9 (454-721) μm; a = 1.4 ± 0.05 (1.3-1.5); b = 10 ± 0.68 (9.3-11.7); stylet = 14.3 ± 1.6 (12.9-16.5) μm; stylet knob height = 2.1 ± 0.4 (1.8-2.5) μm; stylet knob width = 3.9 ± 0.6 (3.5-4.8) μm; DGO = 3.3 ± 0.5 (2.4-4.2) μm; anterior end to excretory pore = 33.7 ± 7.3 (15.4-53.6) μm; vulva length = 26.5 ± 3.72 (20.6-34.8) μm; vulva to anus = 20.5 ± 2.89 (15.5-28.4) μm; distance between phasmids = 30.5 ± 4.26 (25.8-45.1) μm; number of body annuli from anterior end to excretory pore = 22.8 ± 6.24 (14-42).
- *Paratype males* (n = 30): L = 1904 ± 312.2 (849-2202) μm; W = 55.2 ± 5.1 (46.0-62.4) μm; a = 37.4 ± 2.72 (29.8-43.2); c = 92.2 ± 7.9 (70.6-102.4); stylet = 23.2 ± 1.9 (21.1-26.0) μm; stylet knob height = 3.4 ± 0.5 (2.7-3.8); stylet knob width = 4.7 ± 0.8 (3.5-5.3); DGO = 2.3 ± 0.7 (1.3-3.4) μm; anterior end to excretory pore = 162 ± 16.1 (111-203) μm; tail = 14.0 ± 2.1 (12.6-16.5) μm; spicules = 33.7 ± 3.1 (28.4-36.7) μm; gubernaculum = 7.0 ±

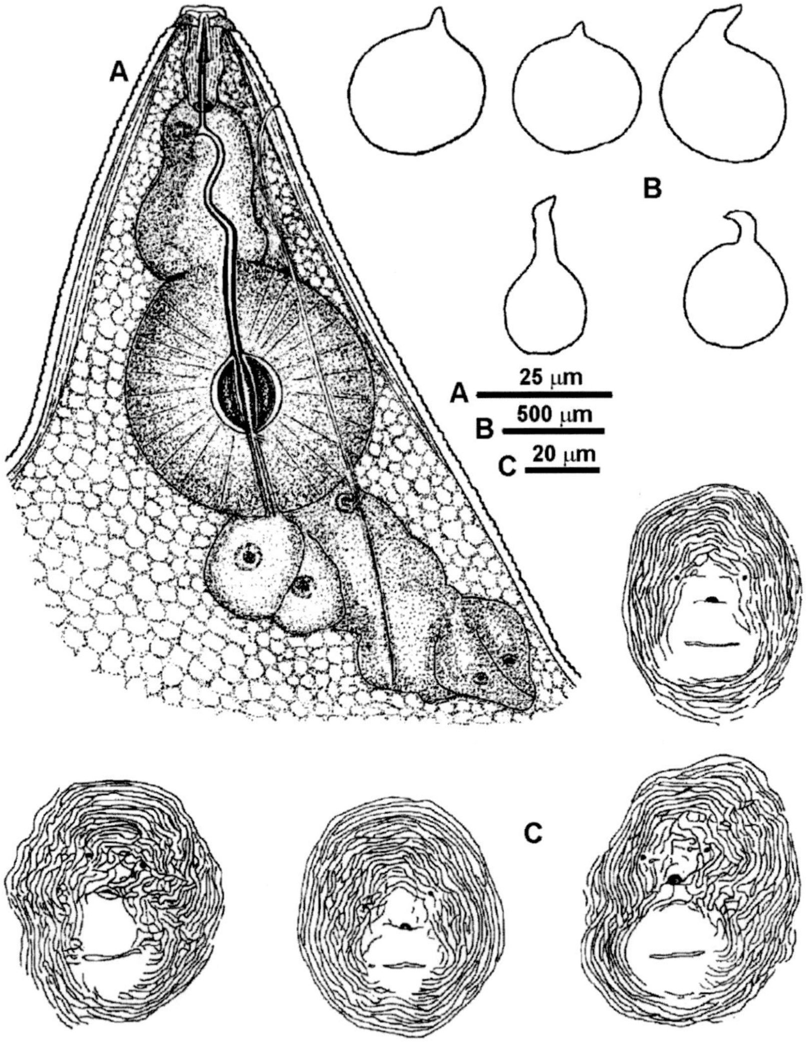

Fig. 208. Meloidogyne petuniae. *A: Female anterior region; B: Entire female; C: Perineal pattern. (After Charchar* et al., *1999.)*

1.0 (5.6-8.1); testis length = 994 ± 73.3 (654-1544) μm; body length/anterior end to posterior end of median bulb = 17.9 ± 1.0 (14.5-20.1); T = 55.8 ± 3.34 (47.8-76.6)%.

- *Paratype J2* (n = 30): L = 392 ± 27.3 (353-464) μm; W = 16.4 ± 0.9 (15.5-18.2) μm; stylet = 10 ± 0.6 (9.2-10.8) μm; stylet knob height = 0.9 ± 0.1 (0.8-1.2) μm; stylet knob width = 1.7 ± 0.2 (1.5-

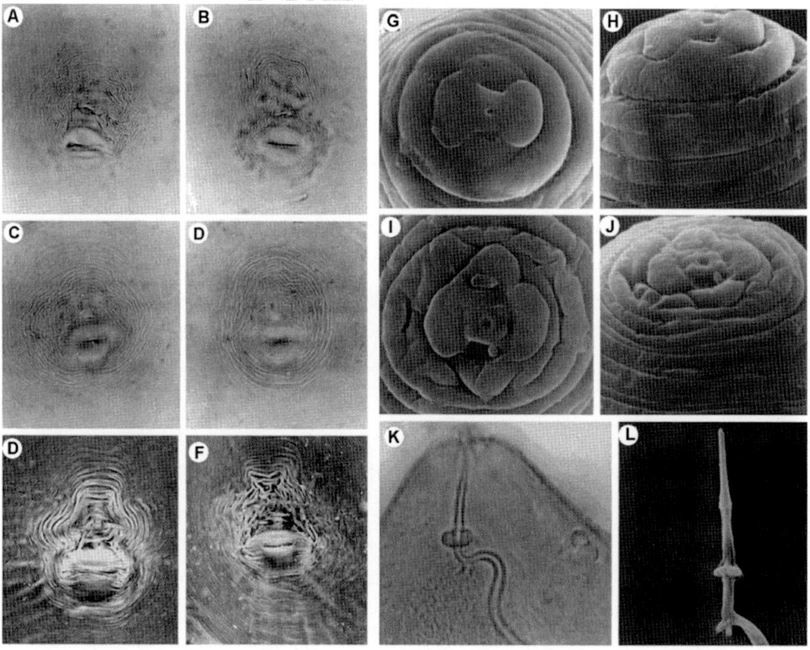

Fig. 209. Meloidogyne petuniae. *LM and SEM. A-F: Perineal pattern; G-K: Female anterior region; L: Female stylet. (After Charchar* et al., *1999.)*

2.1) μm; DGO = 3.4 ± 0.4 (2.8-4.0) μm; anterior end to excretory pore = 80.1 ± 3.7 (72.7-89.6) μm; tail = 47 ± 3.7 (46.4-57.2) μm; hyaline region = 12.4 ± 1.4 (10.3-13.5) μm; a = 24.1 ± 1.1 (21.8-26.1); c = 8.0 ± 0.2 (7.6-8.4); body length/anterior end to posterior end of median bulb = 6.8 ± 0.1 (6.6-7.1).

- *Eggs* (n = 30): L = 87 ± 4.9 (75.4-98.8) μm; W = 41 ± 2.3 (36.4-46.8) μm.

Description

Female

Body translucent white, variable in size, pear-shaped to ovoid with short neck, posteriorly rounded, without posterior protuberance. In SEM, stoma slit-like, located in ovoid to hexagonal prestoma, surrounded by pit-like openings of six inner labial sensilla. Labial disc in most specimens small, rounded, slightly raised above median labials, labial disc and median labials fused, dumbbell-shaped to form a narrow labial

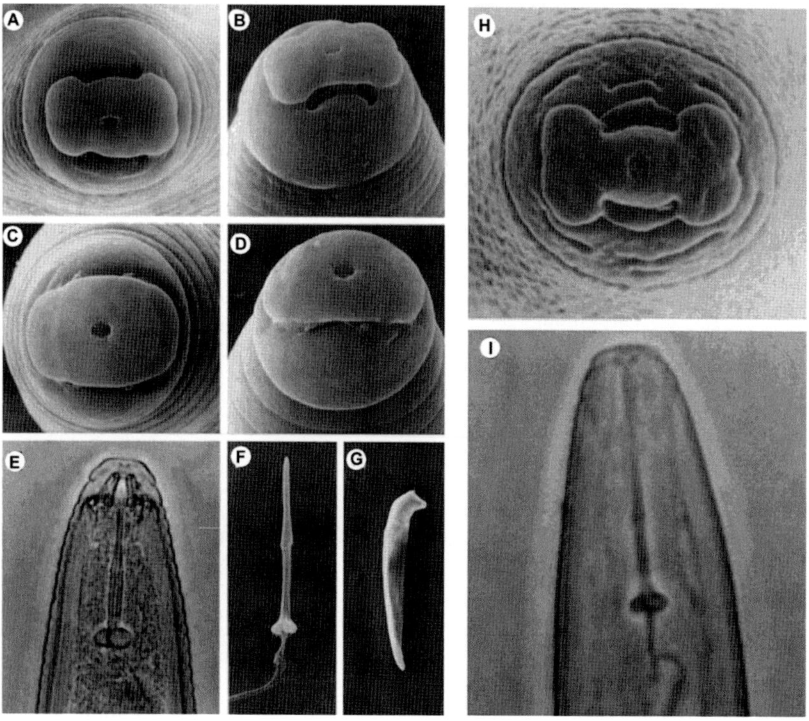

Fig. 210. Meloidogyne petuniae. *LM and SEM. A-E: Male labial region; F: Male stylet; G: Spicule; H: Second-stage juvenile (J2)* en face *view; I: J2 labial region. (After Charchar* et al., *1999.)*

structure in face view. Median labials crescentic. Lateral labials usually triangular, sometimes crescentic, partially fused with median labials. Some females with labial disc distinctly demarcated by four small bumps. Labial region slightly offset from regular body annuli, rarely marked with one incomplete annulation. In LM, labial framework distinct, hexaradiate, lateral sectors enlarged. Vestibule and vestibule extension moderately sclerotised. Cephalids and hemizonids not observed. Excretory pore located at or posterior to level of stylet base. Stylet robust, widening gradually posteriorly, cone tip straight to slightly curved dorsally. Shaft cylindrical, robust, narrower near junction with cone, knobs wide and narrow, distinctly offset from shaft with deep longitudinal indentations in the middle of each knob both posteriorly and anteriorly. Indentations dividing shaft base into six distinct ridges. Orifice branching into three ducts, dorsal gland ampulla small. Pharyngeal glands large, with one large uninucleate dorsal lobe and two smaller nucleated subventral lobes usually posterior to dorsal lobe, variable in size and shape. Perineal patterns elongated to ovoid. Dorsal arch flattened to very high, sometimes squarish, striae widely spaced, coarse. Rarely present, lateral field with broken striae on both sides, ventral striae varying from wavy to coarse striations. Tail tip well defined, with few striae. Perivulval region not striated, few striae near lateral edges of vulva. Vulva slightly sunken, surrounded by wide cuticular striae, often branched into two or more striae. Phasmids small, ducts distinct within cuticle, surface structure not apparent. Anus distinct, sometimes within a cuticular fold, surrounded by thick cuticular layer.

Male

Body translucent white, vermiform, robust, tapering anteriorly, bluntly rounded posteriorly, tail twisting through 90° in heat-killed specimens. Labial cap high in face and lateral view, tapering anteriorly, extending posteriorly onto distinctly offset labial region. Hexaradiate labial framework sclerotised, vestibule and vestibule extension distinct. Stoma slit-like, prestoma hexagonal, surrounded by pore-like openings of six inner labial sensilla. In most specimens, labial disc small, rounded, slightly raised above or fused with median labials; median labials narrow, fused partially with labial disc. Median labials crescentic to reniform-shaped, with outer margins sometimes slightly indented medianly. Lateral labials sometimes slightly demarcated by short, irregular grooves, usually almost completely fused with first labial region annulus. Amphidial apertures wide, elongated, slit-like, located between labial disc and first labial region annulus. Some specimens with large, rounded labial disc, slightly raised above median labials. Labial region smooth or marked by an incomplete annulation, distinctly offset from regular body annuli. Body annuli distinct. Lateral field marked with four incisures beginning near level of stylet knobs, areolated, encircling tail. Stylet slender, long, cone straight, pointed, gradually increasing in diam. posteriorly, orifice located several micrometers from stylet tip. Shaft cylindrical, slender. Knobs large, wide, distinctly offset from shaft, separated from each other, deep longitudinal indentations in middle of each knob posteriorly and anteriorly dividing shaft base into three distinct ridges. Dorsal gland duct branched into three, ampulla poorly defined. Procorpus distinct, median bulb elongated, oval-shaped with enlarged, triradiate cuticular lumen lining, subventral pharyngeal gland orifices in median bulb.

Pharyngo-intestinal junction indistinct, at level of nerve ring. Distance of excretory pore to anterior end variable. Usually one testis, rarely two, generally outstretched or rarely reflexed anteriorly. Spicules robust, arcuate, with extremely thick blade, labial region rectangular, sharply curved anteriorly, velum absent. Gubernaculum distinct, crescentic, tail short and rounded, phasmids pore-like, at level of cloacal aperture.

J2

Body translucent white, vermiform, slender, tapering posteriorly to a sharp point. Stoma slit-like, located in oval-shaped prestoma, surrounded by pore-like openings of six inner labial sensilla. Labial disc rounded, slightly raised above median labials, fused to form dumbbell-shaped labial structure, median labials reniform. Lateral labials long, crescentic, partially fused with labial disc and median labials. Amphidial apertures elongated, located between labial disc and lateral labials. Labial region high, slightly offset from body, marked by incomplete annuli. Body annulation distinct, increasing in width and becoming irregularly spaced in posterior end. Lateral field beginning as a ridge near level of stylet base increasing to four lines, areolated. Labial framework weak. Vestibule and extension distinct. Stylet moderately sized, but delicate, stylet cone very narrow at tip and slightly pointed, gradually increasing in width posteriorly, knobs distinctly sclerotised, sloping posteriorly. Orifice branching into three ducts, ampulla poorly defined. Procorpus distinct, median bulb ovoid with large lumen lining, subventral pharyngeal glands opening posterior to triradiate lumen lining, ampulla distinct. Pharyngo-intestinal junction indistinct, at level of nerve ring. Gland lobes variable in length. Excretory pore distinct, variable in position, terminal duct long. Hemizonid distinct, anterior to excretory pore. Tail slender, ending in slightly rounded tip, annuli increasing in width posteriorly, Hyaline region short. Phasmids small, distinct, located below level of anus, rectum dilated.

TYPE PLANT HOST: Petunia, *Petunia hybrida* L.

OTHER PLANTS: Tomato, tobacco, pea and bean are good hosts; pepper, watermelon and sweet corn are poor hosts; peanut, cotton and soybean are non-hosts. Galls produced by this species are smaller on petunia than on tomato. Potato (*Solanum tuberosum* L.), sweet potato (*Ipomoea batatas* (L.) Poir.) and carrot (*Daucus carota* L.) are good hosts (Charchar *et al.*, 1999).

TYPE LOCALITY: Flower garden of the EMBRAPA/CNPH, National Research Centre for Vegetable Crops, Brasilia, Brazil.

DISTRIBUTION: *South America*: Brazil (Charchar *et al.*, 1999).

PATHOGENICITY: Symptoms included chlorotic and stunted plants with small galls less than 4 mm in diam. In addition, the size and shape of galls produced by this species on petunia compared to tomato are useful diagnostic characters for this species; galls produced on tomato roots (*S. lycopersicum* 'Rutgers') were often more than twice as large.

CHROMOSOME NUMBER: Reproduction of *M. petuniae* occurs by obligatory mitotic parthenogenesis, and the somatic chromosome number is 2n = 41 (Charchar *et al.*, 1999) and 2n = 47 (Chitwood & Perry, 2009).

POLYTOMOUS KEY CODES: *Female*: A21, B32, C2, D3; *Male*: A1234, B21, C21, D1, E1, F2; *J2*: A32, B3, C3, D23, E3, F3.

BIOCHEMICAL AND MOLECULAR CHARACTERISATION: The esterase phenotype of *M. petuniae* is VS1-S1, with S1 being a weak band and the malate dehydrogenase phenotype is N1 (Charchar *et al.*, 1999). The fragment of *COII*-16S gene sequence is characterised by Pagan *et al.* (2015).

RELATIONSHIPS (DIAGNOSIS): *Meloidogyne petuniae* can be distinguished from the most common species of RKN (*M. arenaria*, *M. hapla*, *M. incognita* and *M. javanica*) and other species in the genus by the perineal pattern, the shape and morphology of the labial region and stylet of the female, the shape and morphology of the male labial region, stylet and spicules, and the J2 labial region morphology. The species is also similar to *M. sinensis* from which it differs in female stylet length (12.9-16.5 *vs* 16.5-24.5 μm), and EP/ST ratio (2.4 *vs* 2.1), and several morphometrics in the male and J2.

76. *Meloidogyne phaseoli* Charchar, Eisenback, Charchar & Boiteux, 2008

(Figs 211-213)

COMMON NAME: Bean root-knot nematode.

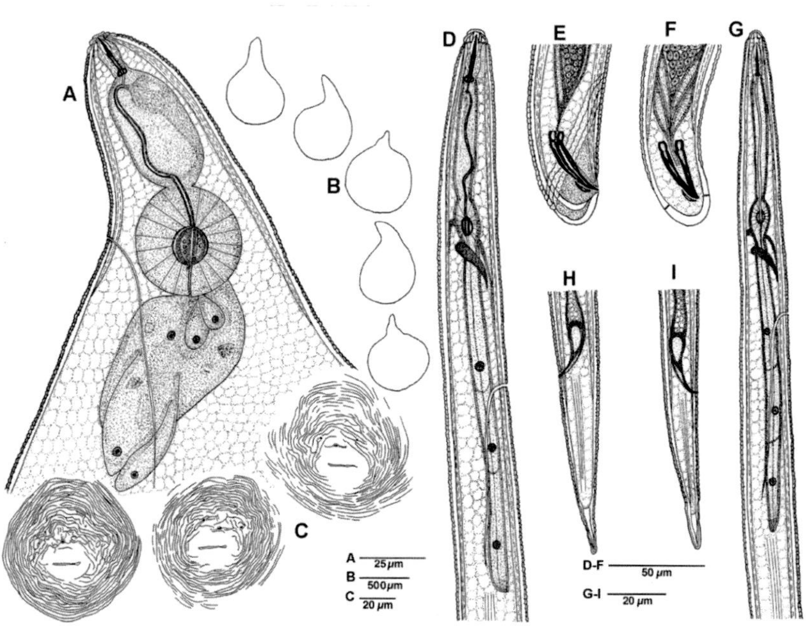

Fig. 211. Meloidogyne phaseoli. *A: Female anterior region; B: Entire female; C: Perineal pattern; D: Male anterior region; E, F: Male labial region; G: Second-stage juvenile (J2) anterior region; H, I: J2 tail. (After Charchar* et al., *2008.)*

Charchar *et al.* (2008b) described this species from bean plants collected from Brasilia, Brazil.

Measurements (after Charchar et al., 2008b)

- *Holotype female*: L = 787 μm; a = 1.4; stylet = 13.5 μm; stylet knob height = 2.1 μm; stylet knob width = 3.9 μm; DGO = 2.5 μm; anterior end to excretory pore = 71 μm; EP/ST = 3.6.
- *Paratype females* (n = 30): L = 657 $\pm$ 71.3 (546-848) μm; stylet = 16.5 $\pm$ 1.5 (13.9-19.3) μm; stylet knob height = 2.5 $\pm$ 0.4 (2.1-3.4) μm; stylet knob width = 4.5 $\pm$ 0.33 (4.2-5.5) μm; DGO = 3.4 $\pm$ 0.5 (2.5-4.2) μm; anterior end to excretory pore = 59 $\pm$ 11.1 (34-82) μm; vulva length = 22.4 $\pm$ 3.3 (16.1-32.2) μm; vulva to anus = 17.4 $\pm$ 2.6 (10.3-23.2) μm; distance between phasmids = 30.6 $\pm$ 3.3 (24.5-37.4) μm; a = 1.4 $\pm$ 0.04 (1.3-1.5).

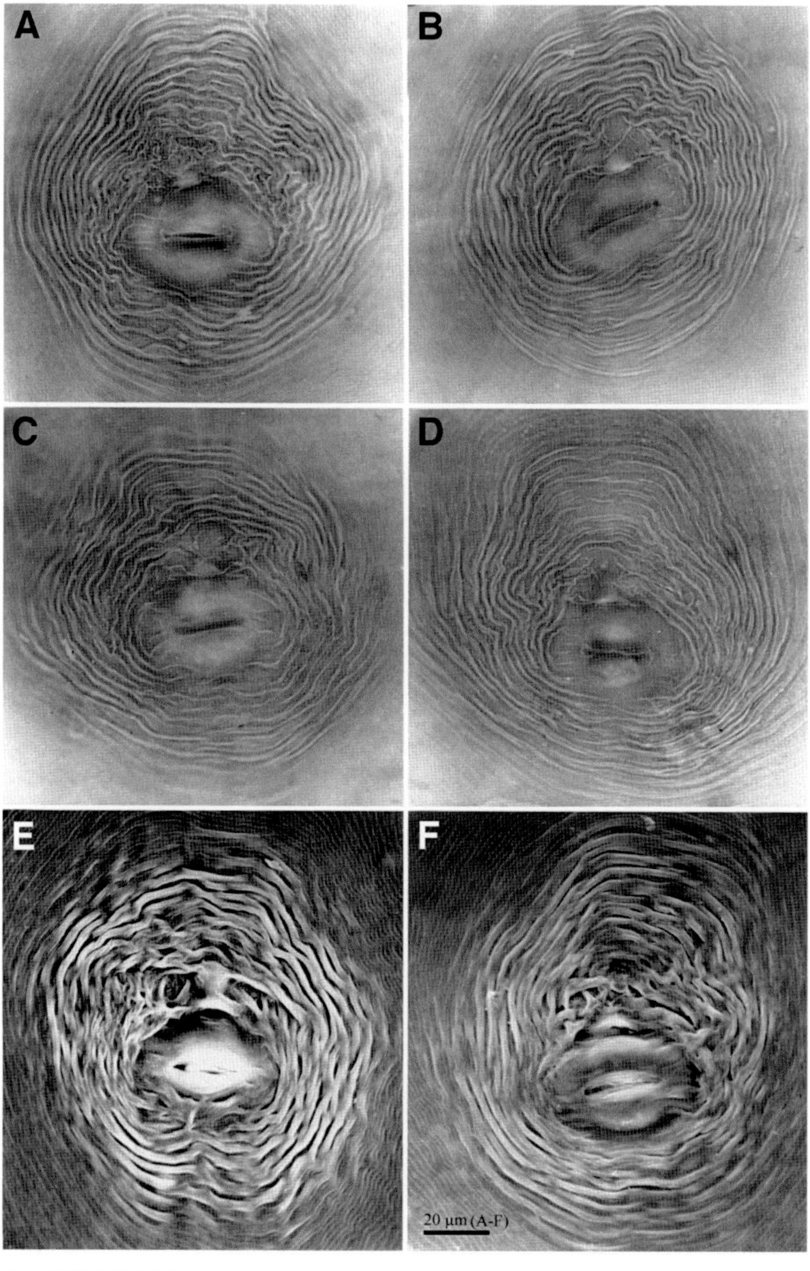

Fig. 212. Meloidogyne phaseoli. *LM and SEM. A-F: Perineal patterns. (After Charchar* et al.*, 2008.)*

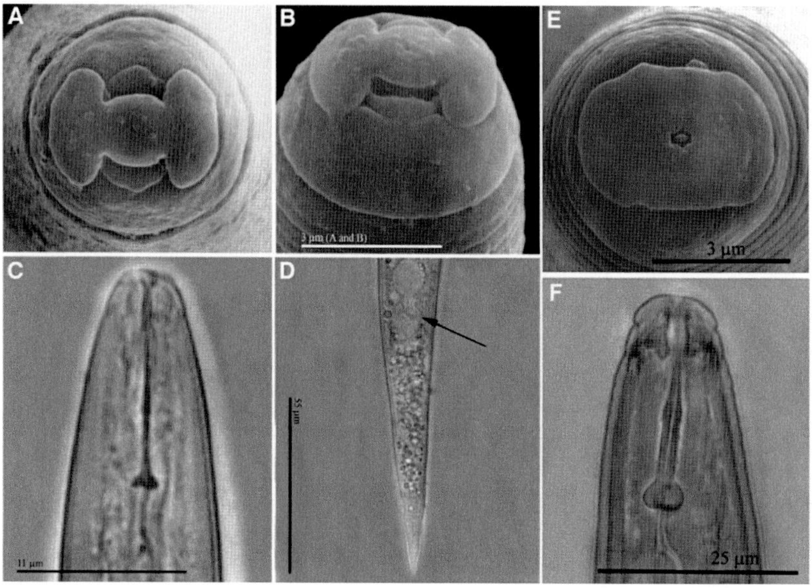

Fig. 213. Meloidogyne phaseoli. *LM and SEM. A, B: Second-stage juvenile (J2) labial region, face and lateral views; C: J2 labial region; D: J2 tail, inflated rectum arrowed; E: Male* en face *view; F: Male labial region. (After Charchar* et al., *2008.)*

- *Paratype males* (n = 30): L = 1779 ± 260.1 (999-2105) μm; stylet = 24.3 ± 1.3 (20.2-26.0) μm; stylet knob height = 3.3 ± 0.3 (2.5-3.8) μm; stylet knob width = 5.0 ± 0.5 (4.2-5.9) μm; DGO = 4.9 ± 0.6 (3.8-5.9) μm; anterior end to excretory pore = 170 ± 18.9 (121-197) μm; tail = 19.7 ± 3.0 (14.3-25.2) μm; spicules = 33.8 ± 3.0 (27.7-39.5) μm; gubernaculum = 8.2 ± 1.0 (5.9-10.1) μm; testis length = 786 ± 128.6 (261-1041) μm; T = 45.2 ± 6.2 (28.9-51.3)%; a = 48.3 ± 3.1 (42.6-52.6); c = 90.6 ± 7.1 (70-100.1).
- *Paratype J2* (n = 30): L = 464 ± 29.3 (400-546) μm; stylet = 11.4 ± 0.5 (10.5-12.2) μm; stylet knob height = 1.2 ± 0.2 (0.8-1.7) μm; stylet knob width = 2.0 ± 0.3 (1.7-2.5) μm; DGO = 3.4 ± 0.3 (2.9-4.2) μm; anterior end to excretory pore = 94 ± 6.1 (84-108) μm; tail = 54.4 ± 4.9 (46.2-63.8) μm; hyaline region = 14.4 ± 2.0 (10.5-19.3) μm; a = 25.6 ± 2.1 (22.7-28.7); c = 8.5 ± 2.0 (7.9-9.4).
- *Eggs* (n = 30): L = 103.4 ± 4.5 (93.6-114.4) μm; W = 44.1 ± 3.0 (36.4-49.4) μm.

DESCRIPTION

Female

Body translucent white, variable in size, pear-shaped with short neck, posteriorly rounded, without tail protuberance. In SEM, stoma slit-like, located in ovoid prestoma, surrounded by pit-like openings of six inner labial sensilla. Labial disc squarish, often marked by two bumps. Labial disc and median labial dumbbell-shaped in face view. Median labials may be indented medianly, crescentic. Lateral labials large, triangular, separated from median labials and labial region. Labial region not offset from regular body annuli, often marked with two distinct incomplete annulations, with several transverse striae. In LM, labial framework distinct, hexaradiate, lateral sectors enlarged. Vestibule and extensions prominent. Cephalids and hemizonid not observed. Distance of excretory pore from labial region variable, located in most specimens near level of median bulb, terminal excretory duct very long. Stylet long and robust, cone slightly longer than shaft, tip straight or slightly curved dorsally, widening gradually posteriorly, junction of cone and shaft uneven. Shaft cylindrical and same diam. throughout, or widening slightly near junction with knobs, with small rounded projections on shaft, knobs broad laterally, offset from shaft, distinctly separated from each other, knobs very slightly indented, dorsal knob slightly sloping posteriorly. Deep, wide indentation on shaft base corresponding to middle of each knob. Distance between stylet base and dorsal pharyngeal gland orifice short to moderate, gland orifice branching into three ducts, dorsal gland ampulla large, subventral gland orifices branched, located posteriorly to enlarged triradiate lumen lining of median bulb. Pharyngeal lumen lining with small rounded thickenings anterior to DGO and median bulb. Median bulb rounded with round granules of unknown constitution between lumen lining and median bulb muscles. Pharyngeal glands large, trilobed, dorsal lobe largest, uninucleate, two subventral nucleated lobes, variable in size, shape and position, located posterior to dorsal gland lobe. Pharyngo-intestinal cells two, small, rounded, nucleated, located between median bulb and intestine. Two gonads and six rectal glands. Perineal pattern rounded to oval-shaped. Dorsal arch flattened to moderately high and with moderately spaced striae. Lateral field often with rounded shoulder and distinctly forked striae. Tail tip area well defined, generally free of striae. Perivulval region not striated, rarely striae near lateral edges of vulva. Vulva located in depression, surrounded by wide cuticular ridges. Phasmidial ducts distinct, no phasmid surface structure apparent in SEM. Anus distinct, surrounded by a thick cuticular layer.

Male

Body translucent white, vermiform, body tapering anteriorly, bluntly rounded posteriorly, tail twisting through 90° in heat-killed specimens. Labial cap high in lateral view, extending posteriorly on to distinctly offset labial region. Labial region high in lateral view, tapering posteriorly, distinctly offset from body. Hexaradiate labial framework well sclerotised, vestibule and extension distinct. Stoma slit-like, located in large, hexagonal prestomatal cavity, surrounded by pore-like openings of six inner labial sensilla. In SEM, labial disc rounded, very large. Median labials very wide, outer margins crescentic, sloping posteriorly. Labial disc and median labials partially fused to form elongate and wide labial structures extending posteriorly on to labial region. Four labial sensilla present on median labials as shallow, elongated ovoid depressions. Amphidial apertures large, elongated, slit-like, located between labial disc and lateral sectors of labial region. Lateral labials fused with smooth labial region. Body annuli large, distinct. Lateral field with four incisures, two beginning near level of stylet knobs and two near level of median bulb, lateral field areolated, encircling tail. Stylet robust, large, cone straight, pointed, gradually increasing in diam. posteriorly, opening located several micrometres from stylet tip, cone same length as shaft. Shaft cylindrical, posterior end wider than anterior end. Small, rounded projections on middle of shaft. Knobs large, wide laterally, offset from shaft, rarely sloping posteriorly. Dorsal pharyngeal gland orifice to stylet base variable in distance, dorsal gland duct branched into three ducts, gland ampulla distinct. Procorpus distinctly outlined, median bulb elongated, oval-shaped with valve enlarged, triradiate cuticular lumen lining, subventral pharyngeal gland orifices branched, located posteriorly to median bulb. Pharyngo-intestinal junction indistinct. Gland lobe variable in length, with two or three nuclei. Excretory pore distinct, variable in position, terminal duct long, curved, ending in excretory cell with large nucleus. Hemizonid located near excretory pore. Intestinal caecum extending anteriorly on dorsal side to level of median bulb. Usually one testis, rarely two testes, outstretched, or reflexed anteriorly. Spicules long, slender, slightly arcuate with single tip, short labial region, wide velum and indistinct shaft. Gubernaculum distinct, crescentic. Tail short, rounded. Phasmids slit-like, opening at level of cloacal aperture.

J2

Body translucent white, long, slender, tapering anteriorly but more so posteriorly, tail region distinctly narrowing. Body annuli distinct, increasing in size and becoming irregular in posterior tail region. Lateral field starting approximately at mid-procorpus, extending to near phasmids, with four incisures, areolated in some specimens. Stoma slit-like, located in oval-shaped prestomatal depression, surrounded by pore-like openings of six inner labial sensilla. Labial cap high, narrower than labial region. In SEM, labial disc elongated, round-shaped, completely fused with median labials. Median labials with outer margins crescentic, smooth. Median labials and labial disc dumbbell-shaped. Lateral labials distinct, lower than median labials, margins crescentic, may slightly fuse with labial region. Labial region smooth, without annulation. Amphidial apertures elongate, located between labial disc and lateral labials. Labial region high, distinctly offset from body. Hexaradiate framework weakly sclerotised in LM, vestibule and vestibule extension distinct. Stylet moderately long, but delicate, stylet cone sharply pointed, gradually increasing in width posteriorly, shaft cylindrical, may widen slightly posteriorly, knobs rounded, distinctly separated, dorsal knob slightly sloping posteriorly. Distance of dorsal pharyngeal gland orifice to stylet base moderately long, orifice branching into three ducts, ampulla poorly defined. Procorpus faintly outlined, median bulb ovoid with distinct valve, subventral pharyngeal gland orifices posterior to valve, ampulla distinct. Pharyngo-intestinal junction indistinct, at level of nerve ring. Pharyngeal gland lobe variable in length with three small nuclei of similar size. Excretory pore distinct, variable in position, terminal duct very long. Hemizonid distinct located anterior to excretory pore. Tail slender, ending in slightly rounded tip, tail annuli larger and irregular posteriorly. Hyaline region long, variable in length. Rectal dilation large. Phasmids small, indistinct, located posterior to anus.

TYPE PLANT HOST: Bean plant, *Phaseolus vulgaris* L., 'Carioca'.

OTHER PLANTS: Tomato, tobacco, bean and pea are good hosts, corn is a very poor host, whilst pepper, watermelon, peanut, cotton and soybean are non-hosts (Charchar *et al.*, 2008b).

TYPE LOCALITY: Collected from a commercial field called 'Fazenda Irmïos Maldane'-PADEF, Brasilia, Brazil.

DISTRIBUTION: *South America*: Brazil (Charchar *et al.*, 2008b).

PATHOGENICITY: Symptoms expressed included chlorosis, stunting, root rot and the total absence of *Rhizobium* nodulation (Charchar *et al.*, 2008b).

CHROMOSOME NUMBER: No available data.

POLYTOMOUS KEY CODES: *Female*: A23, B213, C1, D6; *Male*: A2134, B123, C21, D1, E1, F2; *J2*: A21, B32, C32, D213, E23, F3.

BIOCHEMICAL AND MOLECULAR CHARACTERISATION: The esterase phenotype of *M. phaseoli* is E3 compared to *M. javanica*. However, all three bands of *M. javanica* are strong, whereas *M. phaseoli* has one weak band and two strong bands and slight differences in rate of migration (*M. javanica* has a rate of migration value (Rm) of 46.0, 48.0 and 63.0, whereas the esterase bands of *M. phaseoli* are 48.0, 53.0 and 58.0, respectively) (Charchar *et al.*, 2008b).

Monteiro *et al.* (2018) sequenced the partial 28S and ITS rRNA genes of a population from Brasilia, Brazil, but these sequences showed a high similarity with other *Meloidogyne* spp. from the Incognita group. Consequently, additional molecular markers need to be provided to specifically characterise this species.

RELATIONSHIPS (DIAGNOSIS): The species belongs to Molecular group I, the Incognita group, and is characterised by a large EP/ST ratio. It is similar to *M. morocciensis* and *M. paranaensis*, from which it can be separated by biochemical and molecular markers.

77. *Meloidogyne pini* Eisenback, Yang & Hartman, 1985
(Figs 214, 215)

COMMON NAME: Sand pine root-knot nematode.

This species was described by Eisenback *et al.* (1985) in severely stunted commercial stands of sand pine (*Pinus clausa* (Chapm.) Vasey)

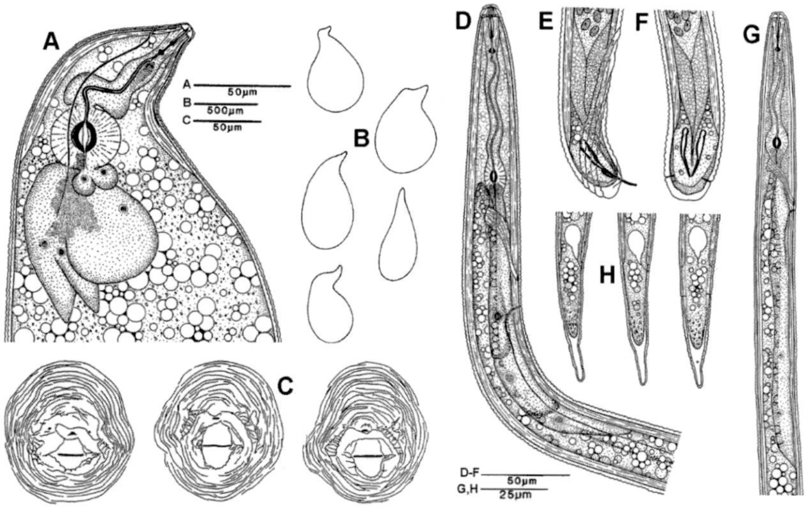

Fig. 214. Meloidogyne pini. *A: Female anterior region; B: Entire female; C: Perineal pattern; D: Male anterior region; E, F: Male tail; G: Second-stage juvenile (J2) anterior region; H: J2 tail. (After Eisenback* et al.*, 1985.)*

in south-eastern Georgia, USA. This species may have been introduced into Georgia from Florida nursery transplants.

MEASUREMENTS *(AFTER EISENBACK* ET AL., *1985)*

- *Holotype female* (in glycerin): L = 871.1 μm; W = 439 μm; stylet = 16.1 μm; stylet knob height = 2.4 μm; stylet knob width = 4.3 μm; DGO = 4.6 μm; anterior end to excretory pore = 8.5 μm; EP/ST = 1.0.
- *Allotype male* (in glycerin): L = 1501.4 μm; a = 38.8; c = 118.1; W = 38.7 μm; stylet = 21.0 μm; stylet knob height = 2.7 μm; stylet knob width = 4.7 μm; DGO = 4.9 μm; anterior end to excretory pore = 141.8 μm; tail = 12.7 μm; spicules = 31.2 μm; testis length = 681.7 μm; T = 45%.
- *Paratype females* (n = 30): L = 696.9 ± 109.2 (463.4-925.1) μm; W = 424.4 ± 67.9 (304.9-636.7) μm; stylet = 14.6 ± 1.3 (12.8-18.4) μm; stylet knob height = 2.8 ± 0.3 (2.3-3.7) μm; stylet knob width = 4.7 ± 0.4 (4.1-5.6) μm; DGO = 4.6 ± 0.7 (3.2-6.3) μm; anterior end to excretory pore = 15.1 ± 7.2 (3.0-32.6) μm; vulva length = 21.1 ± 2.8 (16.2-26.1) μm; vulva to anus = 19.2 ± 1.9 (14.5-22.9) μm; distance between phasmids = 31.3 ± 2.3 (22.6-48.6) μm;

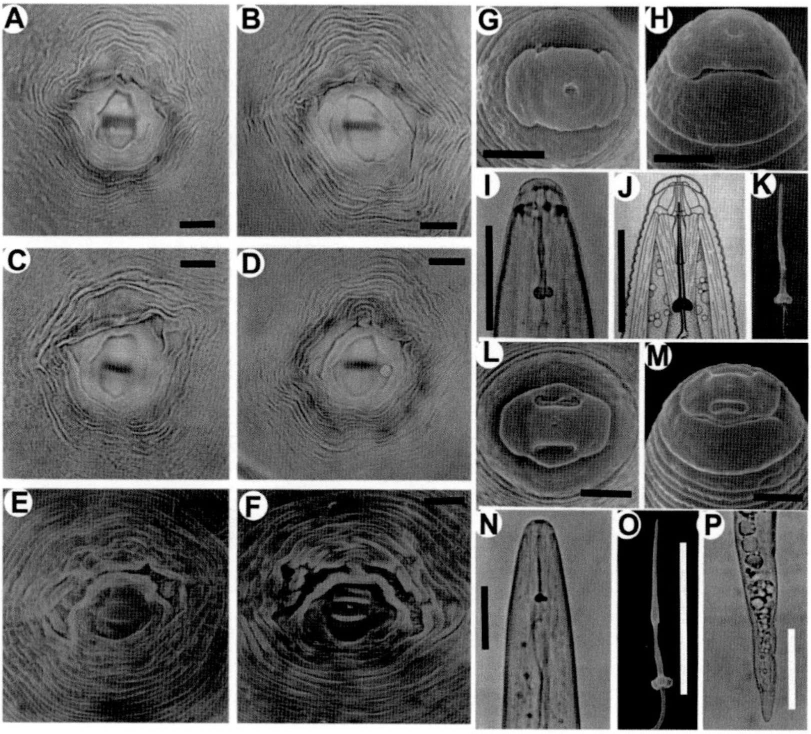

Fig. 215. Meloidogyne pini. *LM and SEM. A-F: Perineal pattern; G-J: Male labial region; K: Male stylet; L-N: Second-stage juvenile (J2) labial region; O: J2 stylet; P: J2 tail. (Scale bars A-F = 20 μm; G, H = 5 μm; I-K = 20 μm; L, M = 5 μm; N, O = 10 μm; P = 20 μm). (After Eisenback* et al.*, 1985.)*

body length/anterior end to posterior end of median bulb = 8.6 ± 1.9 (5.2-11.6); number of body annuli from anterior end to excretory pore = 7.0 ± 0.5 (2.0-13.0); a = 1.7 ± 0.3 (0.8-2.2).

- *Paratype males* (n = 30): L = 1374.5 ± 158.4 (989.6-1637.6) μm; W = 35.3 ± 3.4 (29.0-43.2); stylet = 20.8 ± 1.3 (18.1-22.9) μm; stylet knob height = 2.7 ± 0.2 (2.3-3.1); stylet knob width = 4.5 ± 0.3 (3.8-5.0); DGO = 4.2 ± 0.9 (2.3-7.0) μm; anterior end to excretory pore = 150.0 ± 15.1 (119.1-173.6) μm; tail = 11.5 ± 2.3 (7.6-19.0) μm; spicules = 28.9 ± 3.3 (21.3-35.4) μm; gubernaculum = 7.8 ± 1.0 (6.5-10.2) μm; testis length = 623.4 ± 167.5 (263.5-952) μm; body length/anterior end to posterior end of median bulb = 14.7 ± 1.8 (9.8-19.1); T = 45.5 ± 0.1 (17.4-63.8)%; a = 39 ± 3.0 (31.5-44.4); c = 123.3 ± 21 (66.0-165.9).

- *Paratype J2* (n = 30): L = 434.2 ± 32.9 (376.3-493.0) μm; W = 16.9 ± 1.0 (14.7-18.5); stylet = 12.8 ± 0.7 (11.4-14.1) μm; stylet knob height = 1.7 ± 0.1 (1.5-2.1); stylet knob width = 3.0 ± 0.2 (2.6-3.4); DGO = 3.7 ± 0.4 (3.1-4.4) μm; anterior end to excretory pore = 80 ± 5.9 (64.8-91.2) μm; tail = 44.4 ± 3.6 (37.0-53.4) μm; hyaline region = not measured; a = 25.7 ± 1.9 (21.8-29.1); c = 9.8 ± 0.9 (7.5-11.8); body length/anterior end to posterior end of median bulb = 6.8 ± 0.5 (5.4-7.9).
- *Eggs* (n = 30): L = 102.9 ± 1.41 (87.8-117.2) μm; W = 47.4 ± 0.52 (41.3-52.7) μm.

DESCRIPTION

Female

Body translucent white, variable in size, pear-shaped to ovoid with prominent neck sometimes twice as long as body, without tail protuberance. In SEM, stoma slit-like, located in ovoid prestoma, surrounded by pit-like openings of six inner labial sensilla. Labial disc slightly raised above labials, rectangular, indented medianly on one or both sides, often marked by two or four bumps. Labial disc and median labial dumbbell-shaped in face view. Median labials reniform, lateral labials large, triangular, separated from median labials and labial region. Labial region distinctly offset from body annuli, often marked with 1-2 incomplete annulations. In LM, labial framework weak, hexaradiate, lateral sectors slightly enlarged. Vestibule and vestibule extension prominent. Cephalids and hemizonid not observed. Excretory pore near level of stylet base. Stylet delicate, cone twice as long as shaft, tip straight or slightly curved dorsally. Shaft cylindrical, knobs rounded, offset from shaft, distinctly separated from each other. Dorsal pharyngeal gland orifice 4.6 μm from stylet base, orifice branching into three ducts, dorsal gland ampulla large. Subventral gland orifices branched, located immediately posterior to enlarged triradiate lumen lining of median bulb. Pharyngeal glands with one large uninucleate dorsal lobe, two smaller nucleated subventral lobes usually posterior to dorsal lobe but extremely variable in size, shape and position. Two small, rounded, nucleated pharyngo-intestinal cells located between median bulb and intestine. Two gonads and six rectal glands as characteristic of genus. Perineal pattern rounded to ovoid, sometimes hexagonal. Striae coarse to fine.

Dorsal arch flattened to high. Vulva located in depression, surrounded by wide cuticular, circular ridge, ridge often subdivided by 3-8 fine striae. Deep trough located between cuticular ridge and dorsal arch, marked by transverse striae. Perivulval region free of striae. Phasmids small, surface structure not apparent. Anus covered by cuticular flap.

Male

Body translucent white, vermiform, tapering anteriorly, bluntly rounded posteriorly, tail twisting through 90° in heat-killed specimens. Labial cap low in lateral view, tapering posteriorly, labial region low, slightly offset from body. Hexaradiate labial framework moderately developed, vestibule and extension distinct. Stoma slit-like, prestoma hexagonal, surrounded by pore-like openings of six inner labial sensilla. Labial disc rounded in face view, fused with median labials. Median labials small, crescentic, sloping posteriorly. Four labial sensilla marked on median labials by shallow, elongate ovoid, depressions. Amphidial apertures elongate, slit-like, between labial disc and lateral sectors of labial region. Lateral labials absent. Labial region generally smooth, occasionally with one or two small annuli. Body annuli distinct. Lateral field incisures four, two beginning near level of stylet knobs and two near level of median bulb, lateral field areolated, encircling tail. Stylet moderate in size, cone twice length of shaft, straight, pointed, opening located several micrometers from tip, cone base wider than shaft, junction of cone and shaft uneven. Shaft cylindrical, posterior end wider than anterior end. Knobs large, offset from shaft, sometimes indented anteriorly. Dorsal pharyngeal gland orifice to stylet base variable in distance, orifice branching into three ducts, ampulla poorly defined. Procorpus distinct, median bulb elongate, oval-shaped with enlarged, triradiate cuticular lumen lining, subventral pharyngeal gland orifices posterior to enlarged lining of median bulb, branched into several ducts, ampullae distinct. Pharyngo-intestinal junction indistinct, at level of nerve ring. Gland lobe variable in length, with two or three nuclei. Excretory pore duct delicate. Hemizonid 2-5 annuli anterior to excretory pore. Intestinal caecum extending anteriorly beyond level of nerve ring. Usually one testis, sometimes two, generally outstretched. Spicules arcuate with rounded base, single tip. Gubernaculum short and smooth. Tail short and rounded. Phasmids pore-like, at level of cloacal aperture.

J2

Body translucent white, vermiform, tapering posteriorly to a sharp point. Stoma slit-like, located in oval-shaped prestomatal depression, surrounded by pore-like openings of six inner labial sensilla. Labial disc rounded or rectangular, raised slightly above median and lateral labials. Labial disc, median labials and elongate lateral labials fused. Amphidial apertures elongate oval, located between labial disc and lateral labials. Labial region smooth, only slightly offset from body. Body annulation distinct. Lateral field beginning as a ridge near level of stylet base. Two additional incisures appearing near level of median bulb, disappearing near anal opening, outer two merging near tail terminus. Lateral field areolated. Hexaradiate framework weak (LM), vestibule and extension distinct. Stylet cone pointed, increasing in width gradually, straight or slightly curved dorsally, shaft cylindrical, increasing in width posteriorly, knobs rounded and offset from shaft. Procorpus faintly outlined, median bulb ovoid valve enlarged, subventral pharyngeal gland orifices posterior to valve, ampullae distinct. Pharyngo-intestinal junction indistinct, at level of nerve ring. Gland lobe variable in length, with three nuclei, overlapping intestine ventrally. Excretory pore distinct, hemizonid 1-2 annuli anterior to excretory pore. Tail annuli larger and irregular posteriorly. Rectum rarely dilated. Hyaline region clearly defined, tail tip broad, bluntly rounded, 1-2 fat droplets may occur in hyaline region. Phasmids small, indistinct, located posterior to anus level.

TYPE PLANT HOST: Sand pine tree, *Pinus clausa* (Chapm.) Vasey.

OTHER PLANTS: Loblolly pine (*Pinus taeda*) and slash pine (*P. elliottii* Engelm.) were parasitised and a few barren females developed on some of the seedlings (Eisenback *et al.*, 1985). *Meloidogyne pini* from eastern Georgia parasitised both the Ocala and Choctawhatchie races of sand pine in the glasshouse. The North Carolina differential host plants: tobacco, cotton, pepper, watermelon and tomato, as well as azalea and blueberry, were non-hosts.

TYPE LOCALITY: Tatnall County, Georgia, USA, in a commercial plantation bordered by the Ohoopee River, Thomas Creek, and Georgia Route 121 south of Reidsville, USA.

DISTRIBUTION: *North America*: USA (Georgia) (Eisenback *et al.*, 1985).

PATHOGENICITY: *Meloidogyne pini* is an economically important species. Diseased trees have been observed in commercial plantings at several locations in south-eastern Georgia, USA. Affected plants were chlorotic, dwarfed, with tufted foliage and stunting. Root segments had large compound galls up to 15 cm long and 3 cm in diam. (Eisenback *et al.*, 1985).

CHROMOSOME NUMBER: Reproduction probably occurs by amphimixis, and the approximate somatic chromosome number is 2n = 18 (Triantaphyllou, 1979).

POLYTOMOUS KEY CODES: *Female*: A213, B213, C3, D2; *Male*: A324, B23, C213, D1, E1, F2; *J2*: A23, B213, C3, D2, E23, F3.

BIOCHEMICAL AND MOLECULAR CHARACTERISATION: Biochemically, according to the original description, this species has a unique phenotype of esterase activity (Eisenback *et al.*, 1985), but no other data were provided in any publications.

RELATIONSHIPS (DIAGNOSIS): The perineal pattern of this species resembles *M. indica* and *M. luci*, but differs in a smaller EP/ST ratio (1.7, 2.0, respectively, *vs* 1.0), and several morphometrics in the male and J2.

78. *Meloidogyne piperi* Sahoo, Ganguly & Eapen, 2000
(Fig. 216)

COMMON NAME: Pepper root-knot nematode.

Shahoo *et al.* (2000) described this species from black pepper plants (*Piper nigrum* L.) collected from the Calicut, Kerala, India.

MEASUREMENTS (*AFTER SAHOO* ET AL., *2000*)

- *Holotype female*: L = 780 μm; W = 515 μm; stylet = 13 μm; length median bulb = 32 μm; width of median bulb = 28.5 μm; length of

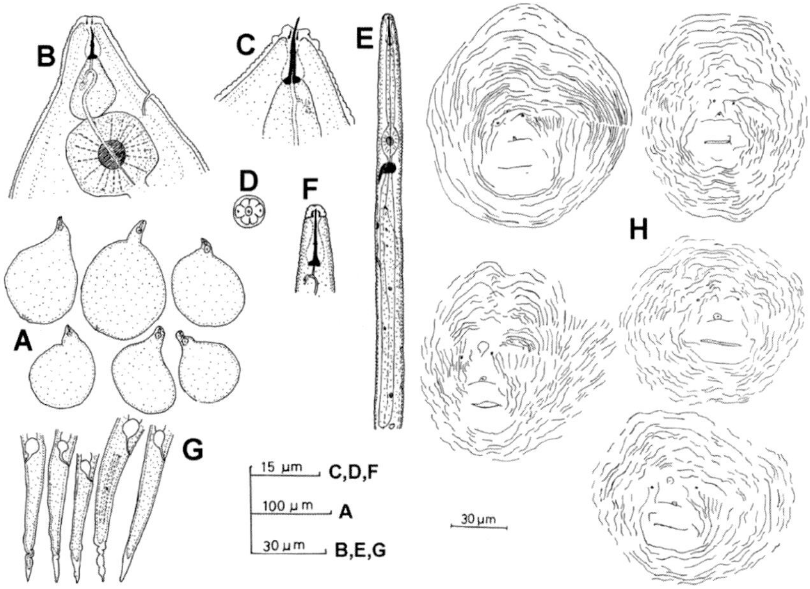

Fig. 216. Meloidogyne piperi. *A: Entire females; B, C: Female anterior region; D: Second-stage juvenile (J2)* en face *view; E: J2 anterior region; F: J2 labial region; G: J2 tail; H: Perineal pattern. (After Sahoo* et al.*, 2000.)*

median valve = 12 μm; width of median valve = 9.5 μm; DGO = 4 μm; anterior end to excretory pore = 57 μm; a = 1.5.

- *Paratype females* (n = 20): L = 720 ± 42 (555-874) μm; W = 533 ± 43 (400-685) μm; stylet = 14.3 ± 0.5 (13-16) μm; median bulb length = 38 ± 2.1 (30-45) μm; median bulb width = 36 ± 2.8 (30-50) μm; anterior end to median bulb = 50 (30-70) μm; DGO = 4.5 ± 0.3 (4.0-6.0) μm; anterior end to excretory pore = (33-64) μm; vulva length = 19.9 ± 1.3 (18-22) μm; vulva to anus = 13.3 ± 0.8 (12.0-14.5) μm; distance between phasmids = 19 ± 1.9 (15-22) μm; a = 1.4 ± 0.1 (1.1-1.6); b′ = 15.4 ± 2.3 (7.9-26.8); EP/ST = 2.0-2.5.
- *Paratype J2* (n = 20): L = 366 ± 15 (310-400) μm; W = 12.9 ± 0.4 (12-14) μm; stylet = 12.2 ± 0.3 (11-13) μm; DGO = 3.6 ± 0.2 (3.0-4.0) μm; length of median bulb = 12.6 (11-14) μm; width of median bulb = 7.2 (6.0-8.0) μm; anterior end to median bulb = 45 ± 1.7 (40-50) μm; anterior end to median valve = 51 ± 1.8 (46-56) μm; anterior end to excretory pore = 74 ± 2.8 (62-81) μm; tail = 47 ± 1.9 (40-53) μm; hyaline region = 8.4 ± 1.5 (5.0-16) μm; anal body diam. = 8.6 ± 0.4 (7.5-10) μm; a = 28.4 ± 1.1 (26-33); b = 4.9 ± 0.1 (4.3-

5.4); $b' = 2.5 \pm 0.2$ (1.9-3.1); $c = 7.8 \pm 0.2$ (7.5-8.5), $c' = 5.5 \pm 0.1$ (5.1-6.0).
- *Eggs* (n = 20): L = 85 ± 2.4 (77-92) μm; W = 37 ± 1.1 (33-42) μm.

DESCRIPTION

Female

Body opaque, pyriform with short neck. In some specimens, dorsal curvature of body more than ventral. Posterior protuberance absent. Labial cap present. Labial region with three annuli, first annulus being anteriorly directed, second one low and regressed while third annulus wide and outwardly directed. Stylet slender with dorsally curved conus. Stylet knobs short. Excretory pore located about 2.0-2.5 stylet lengths from anterior end. Perineal pattern oval with high dorsal arch in most specimens, with broken wavy striae in both dorsal and ventral arches. Lateral field and lateral lines absent. Lateral forking of a few striations of ventral arch seen near lateral region on one side. Broken striae in ventral arch directed toward right lateral region on one side only and sometimes spreading up to middle of dorsal arch. Also, broken striae spreading toward perineum making this region very narrow. Tail whorl absent, phasmids closely spaced, located near tail tip.

Male

Not found.

J2

Body slender, tapering at both ends. Labial region hemispherical slightly offset from body with labial cap followed by two labial annuli. Basal plate thin. Dorsal gland orifice located more than one-fourth of stylet length posterior to stylet base, cephalids absent, hemizonid three annuli long. Excretory pore located 8-9 annuli posterior to hemizonid. Pharyngeal glands overlapping intestine ventrally, lateral field with four incisures, phasmids located at mid-tail region, rectum dilated, tail tapering with notch just posterior to hyaline region and ending with subacute terminus. In an aberrant form, tail narrowing abruptly posterior to hyaline region forming a distinct notch and then ending in a digitate tail terminus.

TYPE PLANT HOST: Black pepper, *Piper nigrum* L.

OTHER PLANTS: Eggplant, *Solanum melongena* L.

TYPE LOCALITY: Calicut, Kerala, India.

DISTRIBUTION: *Asia*: India.

CHROMOSOME NUMBER: No available data.

POLYTOMOUS KEY CODES: *Female*: A213, B32, C2, D3; *Male*: (not found); *J2*: A32, B23, C3, D4123, E32, F2.

BIOCHEMICAL AND MOLECULAR CHARACTERISATION: No available data.

RELATIONSHIPS (DIAGNOSIS): *Meloidogyne piperi* resembles *M. megadora*, but differs from this species by perineal pattern and shorter J2 body length.

79. *Meloidogyne pisi* Charchar, Eisenback, Charchar & Boiteux, 2008

(Figs 217-220)

COMMON NAME: Pea root-knot nematode.

Charchar *et al.* (2008a) described this species from pea plants (*Pisum sativum* L.) collected from Brasilia, Brazil.

MEASUREMENTS *(AFTER CHARCHAR ET AL., 2008A)*

- *Holotype female*: L = 712 μm; W = 532 μm; stylet = 13.5 μm; stylet knob height = 2.3 μm; stylet knob width = 3.4 μm; DGO = 3.2 μm; anterior end to excretory pore = 22 μm; EP/ST = 2.6.
- *Paratype females* (n = 30): L = 755 ± 75.7 (655-900) μm; W = 518 ± 50.8 (406-624); stylet = 15.3 ± 1.0 (13-17) μm; stylet knob height = 2.2 ± 0.4 (1.7-2.9) μm; stylet knob width = 4.1 ± 0.5 (3.4-5.0) μm; DGO = 4.2 ± 0.8 (3.4-6.7) μm; anterior end to excretory

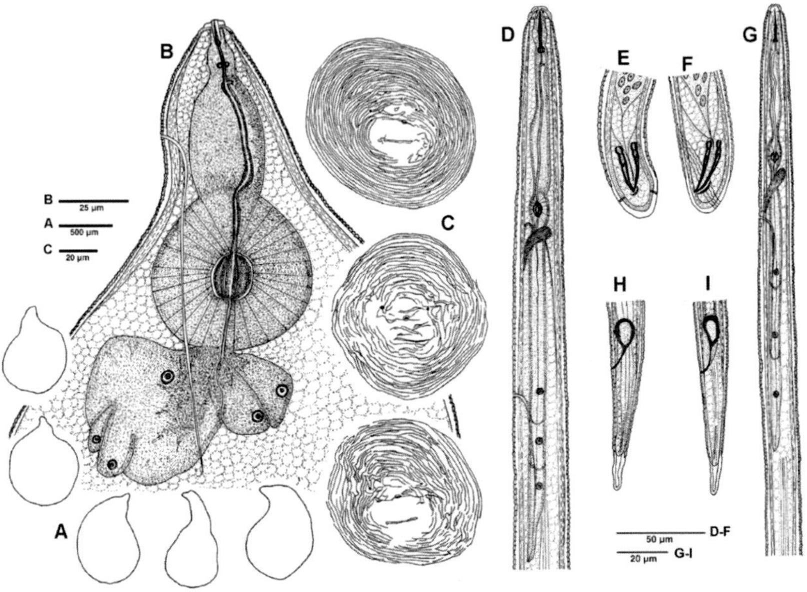

Fig. 217. Meloidogyne pisi. *A: Entire female; B: Female anterior region; C: Perineal pattern; D: Male anterior region; E, F: Male tail; G: Second-stage juvenile (J2) anterior region; H, I: J2 tail. (After Charchar* et al., *2008.)*

pore = 40 ± 9.7 (27-59) μm; vulva length = 22 ± 2.1 (17-26) μm; vulva to anus = 19 ± 2.1 (15-23) μm; distance between phasmids = 33.1 ± 3.3 (28-41) μm; a = 1.4 ± 0.1 (1.4-1.6).

- *Paratype males* (n = 30): L = 1835 ± 383 (893-2510) μm; W = 39 ± 7.0 (26-57); stylet = 23.3 ± 1.6 (19-26) μm; stylet knob height = 3.1 ± 1.6 (2.5-4.2) μm; stylet knob width = 4.3 ± 0.3 (4.2-5.5) μm; DGO = 5.9 ± 1.2 (4.2-8.4) μm; anterior end to excretory pore = 192 ± 24.4 (124-234) μm; tail = 16 ± 1.5 (13-19) μm; spicules = 31.6 ± 3.0 (26.9-37.0) μm; gubernaculum = 7.7 ± 0.6 (6.7-8.4) μm; testis length = 702 ± 128.6 (425-1096) μm; T = 40.5 ± 3.3 (36.6-50.6)%; a = 46.5 ± 3.4 (34.3-51.1); c = 112.2 ± 15.4 (70.9-138.0).
- *Paratype J2* (n = 30): L = 417 ± 20.5 (374-463) μm; W = 17 ± 1.5 (15.6-20.8) μm; stylet = 10.7 ± 0.4 (10-11) μm; stylet knob height = 1.0 ± 0.2 (0.8-1.3) μm; stylet knob width = 1.7 ± 0.2 (1.3-2.1); DGO = 3.5 ± 0.5 (2.5-4.2) μm; anterior end to excretory pore = 85 ± 6.6 (75-95) μm; tail = 49 ± 5.0 (42-66) μm; hyaline region = 13.1 ± 1.8 (9.2-16.8) μm; a = 24.8 ± 1.4 (22.2-26.5); c = 8.6 ± 0.4 (7.1-9.1).

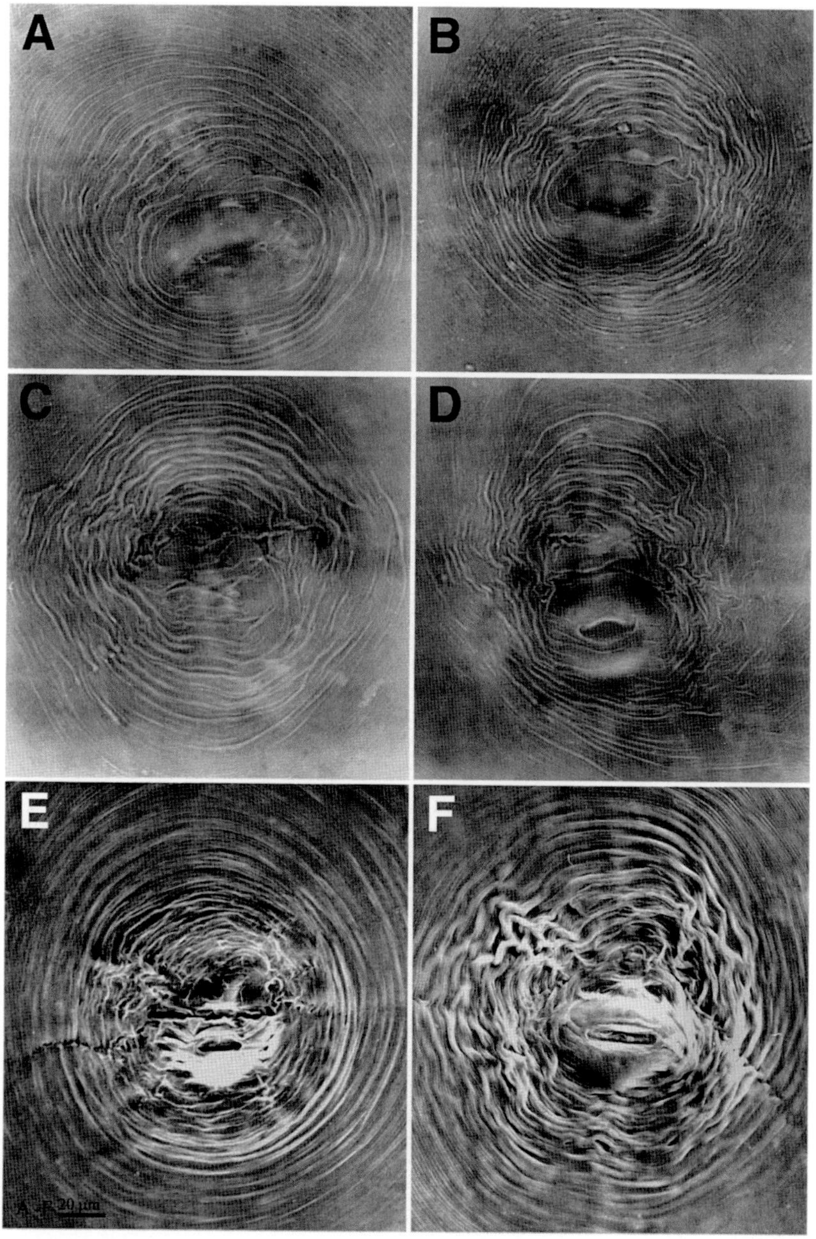

Fig. 218. Meloidogyne pisi. *A-F: LM and SEM of perineal pattern. (After Charchar* et al., *2008.)*

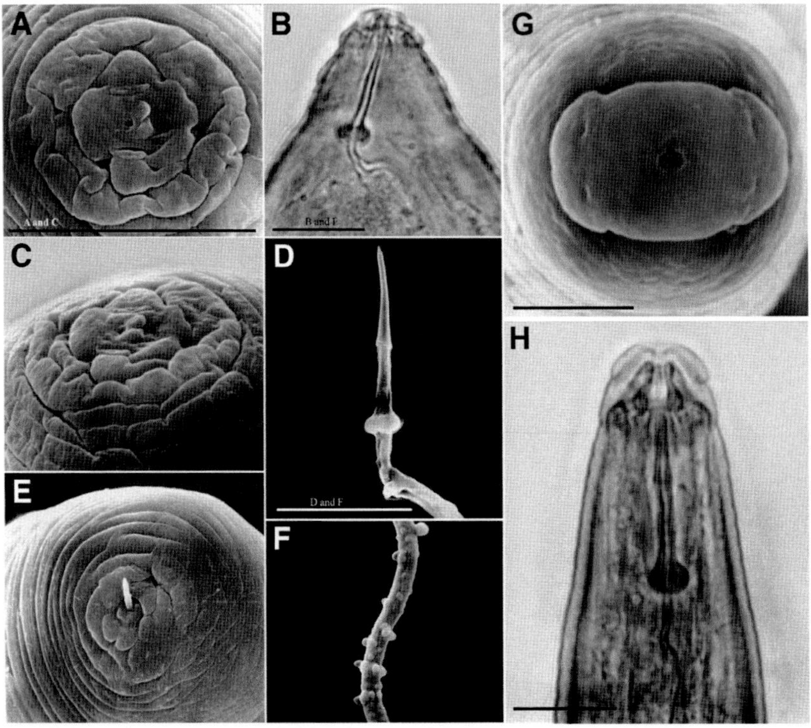

Fig. 219. Meloidogyne pisi. *LM and SEM. A-C, E: Female labial region; D: Female stylet; F: Pharynx lumen lining; G, H: Male labial region. (Scale bars A-F, H = 10 μm; G = 5 μm.) (After Charchar* et al.*, 2008.)*

- *Eggs* (n = 30): L = 95.4 ± 4.4 (85.8-104.0) μm; W = 42.7 ± 2.9 (39.0-46.8) μm.

DESCRIPTION

Female

Body translucent white, variable in size, pear-shaped to ovoid with short, thick neck, without tail protuberance. In SEM, stoma slit-like, located in ovoid prestoma, surrounded by pit-like openings of six inner labial sensilla. Labial disc squarish, slightly raised above labials, indented medianly on one or both sides, partially fused with median labials in some specimens. Median labials long, wide, with indented edges on one or both sides, some specimens with two indentations on one edge of median labials. Lateral labials large, triangular, separated from

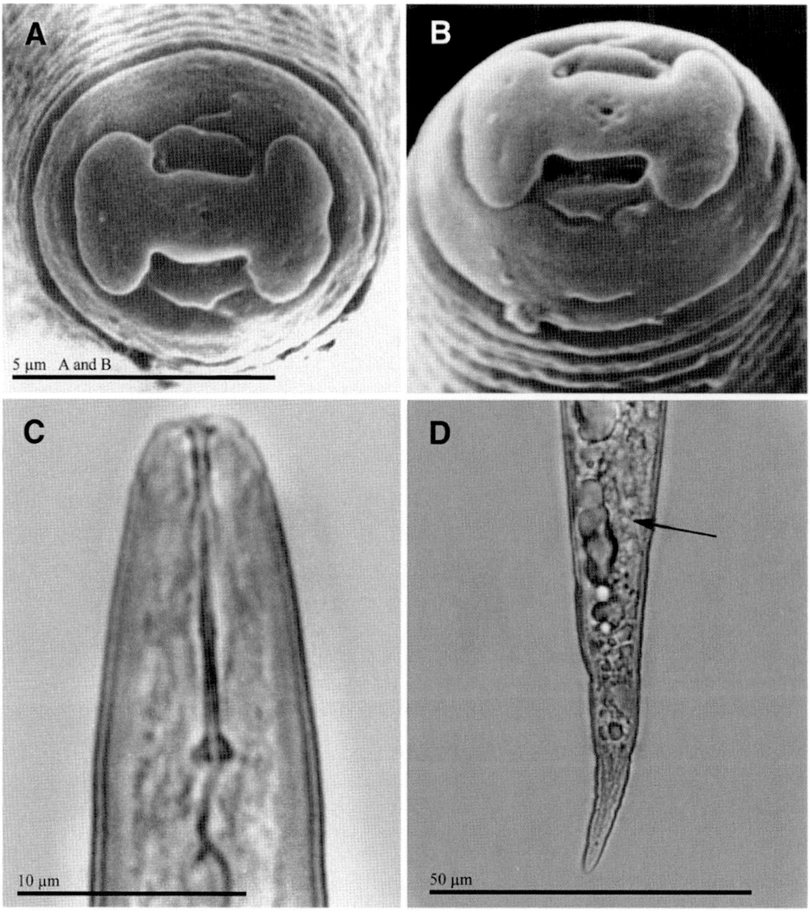

Fig. 220. Meloidogyne pisi. *LM and SEM. A, B: Second-stage juvenile (J2)* en face *view; C: J2 labial region; D: J2 tail (rectum arrowed). (Scale bars A, B = 5 µm; C = 10 µm; D = 50 µm.) (After Charchar* et al., *2008.)*

median labials and labial region. Labial region distinctly offset from body annuli, often marked by two distinct, incomplete annuli interrupted by transverse striae. In LM, labial framework strong, hexaradiate, lateral sectors slightly enlarged. Vestibule and extension prominent. Cephalids and hemizonid not observed. Excretory pore located posterior to stylet base in most specimens, variable in distance from anterior end, terminal excretory duct very long. Stylet robust, cone base wider than shaft, tip slightly curved dorsally. Shaft cylindrical, same width throughout, knobs wide, slightly divided by a central indentation. Lumen lining with large, rounded thickenings between stylet base and median bulb. Dorsal pharyngeal gland orifice branching into three ducts, dorsal gland ampulla large. Subventral gland orifices branched, located immediately posterior to enlarged triradiate lumen lining of median bulb. Median bulb rounded to ovoid. Pharyngeal glands large, trilobed, with one large uninucleate dorsal lobe, two smaller nucleated subventral lobes, usually posterior to dorsal lobe, two pharyngo-intestinal cells. Two gonads and six rectal glands characteristic of genus. Perineal pattern circular to ovoid. Dorsal arch rounded, sometimes flattened to high and squarish. Cuticle with rounded ridges, external ridges widely spaced, coarse with nearly circular striae in many specimens, whereas internal ridges near perivulval area are wavy. Lateral field occasionally marked by a few transverse striae. Tail tip well defined, with striations in most specimens. Vulva located in depression, surrounded by wide, circular striae in most specimens, ridges often subdivided by several fine striae. Perivulval region marked by transverse striae. Phasmids small, surface structure not apparent. Anus covered by elevated and thickened cuticle.

Male

Body translucent white, vermiform, tapering anteriorly, bluntly rounded posteriorly, tail twisting through 90° in heat-killed specimens. Labial cap high, tapering posteriorly, labial region elevated, distinctly offset from body. Hexaradiate labial framework well developed, vestibule and extension distinct. Stoma slit-like, located in large hexagonal prestoma, surrounded by pore-like openings of six inner labial sensilla. Labial disc rounded and large in face view, partially fused with median labials. Median labials large, crescentic, sloping posteriorly. Four large labial sensilla marked on median labials by deep and elongate depressions. Amphidial apertures elongate, slit-like, located between labial disc and lateral sectors of labial region. Lateral labials fused with labial region. Labial region without annulation. Body annuli distinct. Four incisures in areolated lateral field. Stylet long, delicate, straight, pointed, gradually increasing in diam. posteriorly, opening located several micrometers from tip, cone base wider than shaft, junction of cone and shaft uneven. Shaft cylindrical, marked by small projections throughout. Knobs rounded, large, offset from shaft. Dorsal pharyngeal gland orifice to stylet base variable in distance, orifice branching into three ducts, ampulla distinct. Procorpus distinct, median bulb elongated, oval-shaped with enlarged, triradiate cuticular lumen lining, subventral pharyngeal gland orifices posterior to enlarged lining of median bulb, branching into several ducts, ampulla distinct. Pharyngo-intestinal junction indistinct, at level of nerve ring. Gland lobe variable in length, with two or three nuclei. Excretory pore 124-234 μm from anterior end, terminal duct distinct, long. Hemizonid not observed. Intestinal caecum extending anteriorly beyond level of nerve ring. Usually one testis, but some specimens with two testes, generally outstretched. Spicules arcuate, very long, slender with narrow head region, long and narrow shaft, narrow blade and without velum, single tip. Gubernaculum short, smooth, crescentic. Tail short, rounded. Phasmids pore-like, at level of cloacal aperture.

J2

Body translucent white, vermiform, tapering posteriorly to a slightly round tip. Stoma slit-like, located in oval-shaped prestomatal depression, surrounded by pore-like openings of six inner labial sensilla. In SEM, labial disc elongated, rectangular-shaped, raised slightly above median labials. Labial disc and median labials fused forming a dumbbell-shaped structure. Median labials with outer margins indented and reniform. Lateral labials distinct, lower than median labials, crescentic, margins indented or marked by a short groove. Amphidial apertures elongate with round corners, located between labial disc and lateral labials. Anterior region elevated, distinctly offset from body, with one incomplete annulation. Body annulation distinct. Lateral field marked by four incisures and areolation. Hexaradiate framework weak in LM, vestibule and extension distinct. Stylet cone sharply pointed, shaft cylindrical, increasing in width posteriorly, knobs elongated, offset from shaft, dorsal knobs sloping slightly. Distance of dorsal pharyngeal gland orifice variable, moderately long, orifice branching into three ducts, ampulla indistinct. Procorpus distinctly outlined, median bulb ovoid, valve enlarged, subventral pharyngeal gland orifices posterior to pump lining, ampullae distinct. Pharyngo-intestinal junction indistinct, at nerve ring level. Gland lobe variable in length, with three nuclei, overlapping intestine ventrally. Excretory pore distinct, variable in position, terminal duct very long. Hemizonid not observed. Tail annuli large and irregular posteriorly. Rectum dilated. Hyaline region clearly defined, variable in length, tail ending in slightly rounded tip. Phasmids small, distinct, located posterior to anal level.

TYPE PLANT HOST: Pea, *Pisum sativum* L., 'Mikado'.

OTHER PLANTS: The differential host test showed that *M. pisi* can be easily differentiated from the four most common RKN by its inability to reproduce on peanut and cotton and by poor reproduction on pepper. The reproduction of *M. pisi* on watermelon was similar to that of *M. arenaria* race 2, *M. incognita* race 4, and *M. javanica*, but *M. pisi* can be differentiated from *M. incognita* by its inability to infect cotton and its only slight ability to infect pepper.

TYPE LOCALITY: It was collected from a commercial field called 'Nucleo Rural do Jardim', in Brasilia, Brazil.

DISTRIBUTION: *South America*: Brazil (Charchar *et al.*, 2008a).

PATHOGENICITY: Symptoms included chlorosis, stunting, root rot, and a complete lack of nodulation by *Rhizobium* spp. (Charchar *et al.*, 2008a).

CHROMOSOME NUMBER: No available data.

POLYTOMOUS KEY CODES: *Female*: A21, B23, C2, D5; *Male*: A1234, B213, C21, D1, E1, F2; *J2*: A23, B3, C32, D213, E23, F3.

BIOCHEMICAL AND MOLECULAR CHARACTERISATION: *Meloidogyne pisi* has an E5 phenotype of esterase activity, characterised by three strong bands (SB) and two weak bands (WB). Esterase bands of *M. javanica* have a rate of migration value (Rm) of 43.0, 55.0 and 62.5, whereas esterase bands of *M. pisi* have Rm values of 37, 43, 47, 57 and 61. No molecular data are available.

RELATIONSHIPS (DIAGNOSIS): This species is similar in female and J2 morphology to *M. kikuyensis*, *M. duytsi* and *M. cruciani*. They have a similar perineal pattern (group 5, Jepson, 1987), but differ from *M. kikuyensis* by longer J2 body (374-463 *vs* 290-360 μm), and from *M. duytsi* by shorter J2 tail (42-66 *vs* 65.1-76.5 μm), and from *M. cruciani* and other species by the esterase patterns.

80. *Meloidogyne platani* Hirschmann, 1982
(Figs 221, 222)

COMMON NAME: Sycamore root-knot nematode.

Hirschmann (1982) described this species from roots of American sycamore, *Platanus occidentalis* L., in Virginia, USA.

MEASUREMENTS *(AFTER HIRSCHMANN, 1982)*

- *Holotype female*: L = 710 μm; W = 450 μm; neck length = 290 μm; neck width = 155 μm; stylet = 15.2 μm; stylet knob height = 2.2 μm; stylet knob width = 4.9 μm; DGO = 3.7 μm; anterior end to excretory pore = 24.8 μm; a = 1.6; EP/ST = 1.6.
- *Allotype male*: L = 1740 μm; W = 35.4 μm; stylet = 21.8 μm; stylet knob height = 3.0 μm; stylet knob width = 5.2 μm; DGO = 3.6 μm; anterior region to excretory pore = 153.7 μm; anterior end to median bulb valve: 93.2 μm; tail = 15.8 μm; testis length = 742 μm; a = 49.2; and T = 42.6%.
- *Paratype females* (n = 30): L = 678.4 ± 80.2 (540-860) μm; W = 453 ± 71.0 (320-585) μm; neck length = 190.7 ± 52.4 (95-320) μm; neck width = 136.8 ± 36.4 (50-198) μm; stylet = 16.5 ± 0.5 (15.8-17.3) μm; stylet knob height = 2.5 ± 0.2 (2.2-3.1) μm; stylet knob width = 4.8 ± 0.2 (4.3-5.2) μm; DGO = 3.7 ± 0.5 (2.8-4.6) μm; anterior end to excretory pore = 26.8 ± 6.7 (15.8-39.5) μm; vulval slit length = 26.7 ± 2.9 (21.6-33.2) μm; vulva to anus = 16.4 ± 2.6 (10.3-23.1) μm; interphasmid distance = 27.6 ± 3.8 (19.8-35.3) μm; a = 1.5 ± 0.2 (1.1-2.0).
- *Paratype males* (n = 30): L = 1354.8 ± 275.6 (960-2010) μm; W = 34.1 ± 4.2 (27.7-42.7) μm; stylet = 22.0 ± 1.1 (19.4-24.3) μm; stylet knob height = 3.3 ± 0.3 (2.8-3.8) μm; stylet knob width = 5.4 ± 0.3 (4.8-6.0) μm; DGO = 3.7 ± 0.6 (2.3-4.5) μm; anterior end to median bulb = 90.9 ± 7.9 (75.1-107.0) μm; anterior end to excretory pore = 136.9 ± 13.5 (111.9-161.5) μm; tail = 12.9 ± 1.5 (9.5-17.1) μm; phasmids to tail end = 13.4 ± 2.8 (9.5-22.2) μm; spicules = 27.5 ± 1.4 (25.1-31.6) μm; gubernaculum = 8.5 ± 0.6 (7.9-9.7) μm; testis length = 756.1 ± 165.8 (435-1146) μm; a = 39.6 ± 5.4 (27.3-51.9); c = 104.8 ± 17.1 (78.5-160.3); T = 56.8 ± 12.2 (37.2-81.8)%.

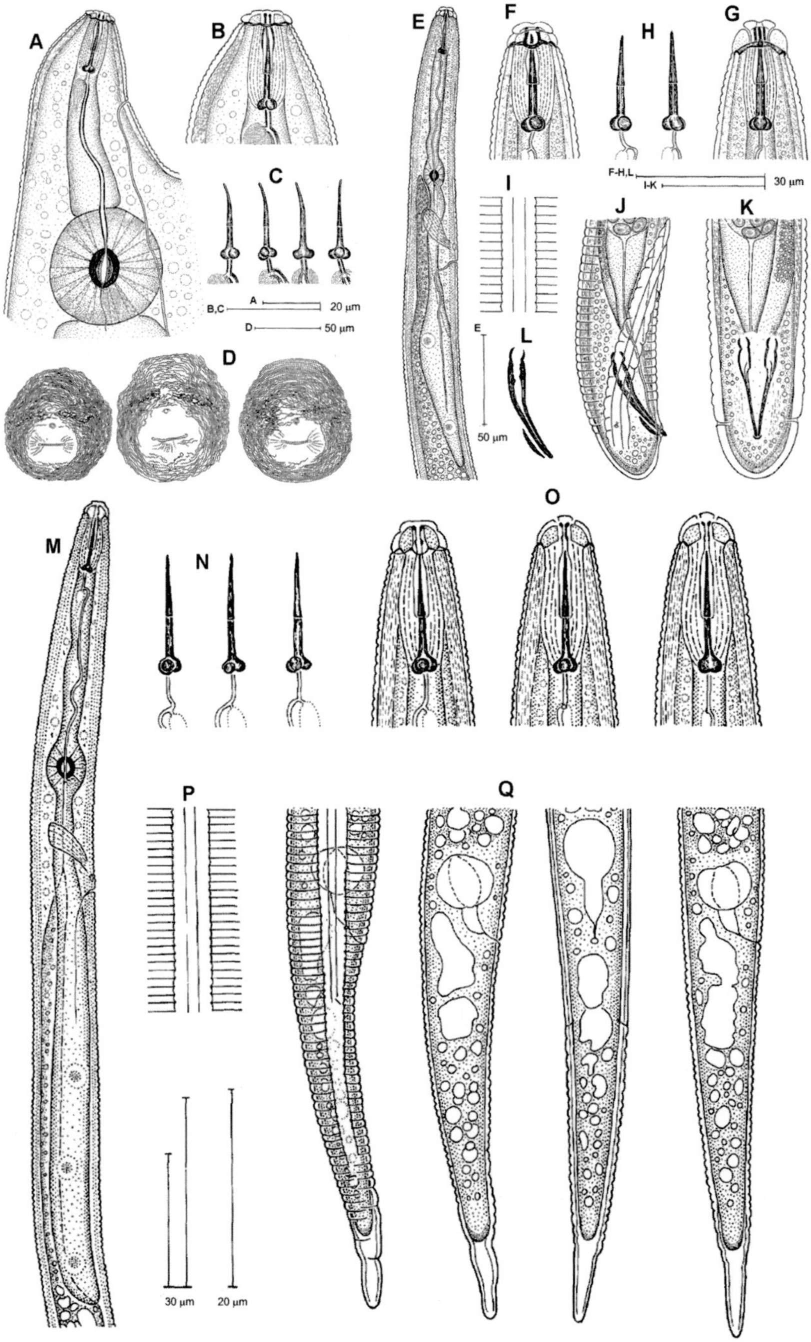
A
B
C
A 20 µm
B,C
D 50 µm
D
E
F
G
H
F-H,L
I-K
30 µm
I
J
K
E
L
50 µm
M
N
O
P
Q
30 µm
20 µm

Fig. 221. Meloidogyne platani. *A: Female anterior region; B: Female labial region; C: Female stylet; D: Perineal pattern; E: Male anterior region; F, G: Male labial region; H: Male stylet; I: Male lateral field; J, K: Male tail; L: Spicules; M: Second-stage juvenile (J2) anterior region; N: J2 stylet; O: J2 labial region; P: J2 lateral field; Q: J2 tail. (After Hirschmann, 1982.)*

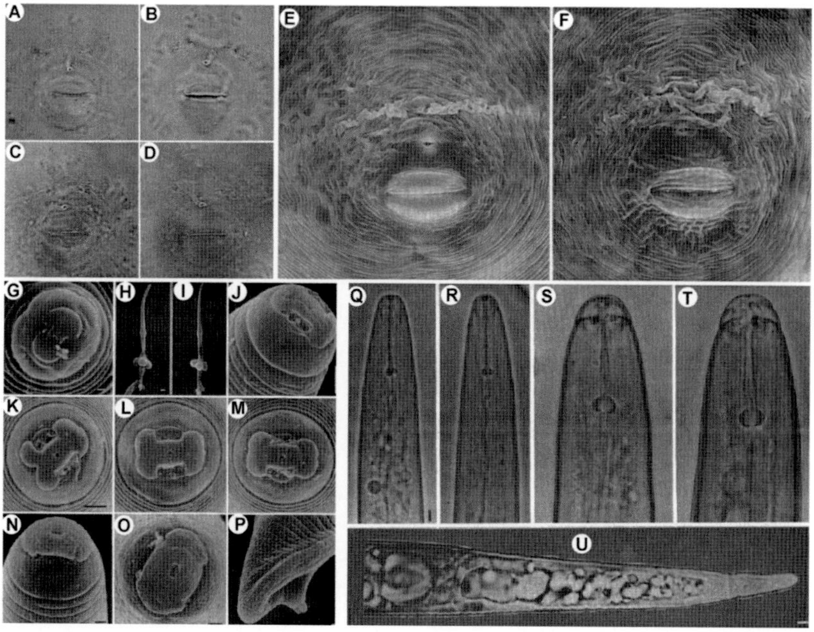

Fig. 222. Meloidogyne platani. *LM and SEM. A-F: Perineal pattern; G: Female* en face *view; H, I: Female stylet; J-M: Second-stage juvenile (J2)* en face *view; N, O: Male labial region; P: Male tail; Q, R: J2 labial region; S, T: Male labial region; U: J2 tail. (After Hirschmann, 1982.)*

- *Paratype J2* (n = 50): L = 443.0 ± 22.5 (395.3-496.8) μm; W = 17 ± 1.2 (14.9-19.8) μm; stylet = 12.2 ± 0.3 (11.6-12.6) μm; stylet knob height = 1.6 ± 0.1 (1.4-1.7) μm; stylet knob width = 2.7 ± 0.1 (2.6-2.9) μm; DGO = 3.5 ± 0.3 (2.7-4.0) μm; anterior end to median bulb = 63.9 ± 2.8 (57.6-69.5) μm; anterior end to excretory pore = 88.4 ± 3.6 (80.5-98.1) μm; tail = 57.3 ± 3.8 (49.6-64.8) μm; hyaline region = 12.4 ± 1.4 (9.3-15.3) μm; a = 26.2 ± 1.5 (22.6-29.7); b = 6.9 ± 0.4 (5.9-7.7); c = 7.7 ± 0.4 (7.1-8.5).

- *Eggs* (n = 50): L = 98.8 ± 4.7 (92-107) μm; W = 44.3 ± 2.9 (40-53) μm.

DESCRIPTION

Female

Body white, variable in size, globular to pear-shaped, with prominent neck. Body posteriorly rounded, without tail protuberance. Cuticle with maximum thickness about 4 μm finely striated, cephalids not observed. Labial region distinctly offset from body. In SEM labial disc slightly elevated. Median labials and labial disc forming dumbbell-shaped structure in face view. Median labials usually rounded, in some specimens slightly indented medianly. No lateral labials distinguishable. Amphid openings small. Prestoma small, oval-shaped, stoma obscure, slit-like. Labial sensilla not observed. Labial region without annulation. Labial framework approximately hexaradiate with slightly enlarged lateral sectors. Vestibule and vestibule extension distinct. Stylet delicate, conical part distinctly curved dorsally, tapering towards tip, stylet shaft slightly enlarged posteriorly, stylet knobs offset from shaft, large, evenly rounded, distinctly separate. DGO closely posterior to stylet knobs, branched into three ducts, dorsal gland ampulla large, subventral gland orifices immediately posterior to median bulb valve and also branched. Pharyngeal gland lobe distinct with one large dorsal nucleus and two smaller subventral nuclei. Two pharyngo-intestinal cells present near junction of median bulb and intestine. Position of excretory pore variable, closely posterior to dorsal pharyngeal gland orifice in many females. Perineal pattern with closely spaced, mostly fine, continuous striae. Dorsal arch low, entire pattern appearing much rounded. Slightly forked striae at lateral lines present in only few specimens; usually no lateral lines distinguishable and striae in dorso- and ventrolateral areas slightly wavy to pronounced zigzag and continuous from dorsal to ventral region. Zigzag striae, characteristic for this species, often more pronounced on one side than other. Inner lateral line regions with raised, irregular, closely looped and folded striae. Ventral pattern area composed of fine curved, continuous striae, usually interrupted midway. Tail tip not visible, central tail area mostly free of striation. Vulval edges distinctly crenate, very fine vulval striations radiating outwards from edges. Phasmidial ducts very distinct, no phasmid surface structure distinguishable in SEM.

Male

Body slender, vermiform, tapering to rounded, posterior end more rounded than anterior. Heat-killed males curving ventrally into C-shape, tail twisting through 90°. Labial cap in lateral view low, rounded, narrower than slightly offset labial region. In SEM, large rounded labial disc, distinctly elevated. Median labials and labial disc forming elongate labial cap in face view. In some specimens lateral indentation at junction of labial disc and median labials. Diam. of labial disc not larger than that of median labials. Median labials crescentic to irregularly pointed in face view, extending some distance onto labial region. No indication of lateral labials. Labial region without annulation. In most specimens, inner labial sensilla obscure, opening into prestoma. Labial sensilla not distinct. Prestoma oval-shaped to hexagonal. Body annuli distinct, about 8.1-8.6 μm wide at mid-body. Lateral field not clearly visible, basically with four incisures, in some specimens up to eight incisures for some distance, additional incisures short, broken, and fainter, inner incisures may fork, outer incisures straight or very slightly crenate, starting approximately at level of stylet knobs, no areolation except for few short lines in tail end, no areolation or crenation seen in tail by SEM. Labial framework moderately developed. Stylet robust with rounded, large knobs, distinctly offset from shaft and slightly sloping posteriorly. Stylet shaft of same diam. throughout. Dorsal pharyngeal gland orifice distance variable, mostly opening a short distance posterior to stylet knob base, dorsal gland duct branched, gland ampulla poorly defined. Pharynx lumen lining narrow between stylet knobs and median bulb valve, procorpus distinctly outlined, median bulb elongate, oval-shaped with large valve plates, pharyngo-intestinal junction distinct, gland lobe variable in length with two *ca* equal-sized nuclei: anterior nucleus near beginning or middle of gland lobe and posterior nucleus in end of lobe, position of nuclei variable. Intestinal caecum extending on dorsal side to beginning or mid-level of median bulb. Excretory pore distinct, terminal duct long, typically curved, large sinus nucleus in right lateral chord. Hemizonid 1-4 annuli anterior to excretory pore. One testis or two testes, outstretched or reflexed anteriorly. Sperm rounded and granular. Spicules arcuate, long and slender, with small rounded base and blunt single tip, typically tylenchoid. Gubernaculum distinct, simple. Tail elongate, sometimes slightly digitate. Phasmids frequently at level of cloacal aperture, appearing slit-like in SEM.

J2

Rather large and long. Body vermiform, tapering slightly anteriorly but more so posteriorly, tail region distinctly narrowing. Heat-killed specimens slightly curved ventrally. Labial region truncate, slightly offset from body. In SEM, labial disc distinctly elevated. Median labials and labial disc dumbbell-shaped. Median labials with squared off margins in one-third of specimens examined, median labials slightly indented to indicate two labials in 50% of specimens and deeply indented forming a labial pair in 20% of specimens, only two specimens with two labial pairs each. Lateral labials fused at right angle with median labials and lower than median labials, occasionally one lateral labial fused with labial region. Labial region not annulated. Prestoma large, oval-shaped. Inner labial sensilla distinct, opening on labial disc, arranged symmetrically around prestoma, prestoma and inner labial sensilla in slight depression. Labial sensilla faint. Body annuli small but distinct, increasing in size and becoming irregular in posterior tail region. Lateral field 4.9-5.3 μm wide, with four incisures, starting approximately at middle of procorpus and extending near beginning of hyaline region, not areolated, outer lines slightly crenate. Labial framework very weakly sclerotised, lateral sectors slightly enlarged. Stylet delicate, but stylet knobs large, rounded, well separated from each other and slightly sloping posteriorly, one subventral knob frequently sloping more posteriad than other. Cephalids not observed. DGO distance long, gland ampulla poorly defined. Procorpus faintly outlined, median bulb oval-shaped with large valve plates, isthmus not clearly outlined. Pharyngo-intestinal junction poorly defined, between nerve ring and excretory pore. Gland lobe variable in length with three distinct nuclei of about equal size. Hemizonid 1-3 annuli anterior to excretory pore. Tail slender, in some specimens narrowing abruptly posterior to anal opening. Tail annulations irregular and increasing in size near hyaline region. Hyaline region distinct, slightly offset without annulations, or not offset and with few, large annulations, tail tip broad, bluntly rounded. Rectal dilation very large, filled with matrix material. Phasmids small, obscure, located a short distance posterior to anal opening.

TYPE PLANT HOST: American sycamore, *Platanus occidentalis* L.

OTHER PLANTS: Tobacco ('NC 95') is a good host with galls and a moderate egg mass production; tomato and watermelon were moderately galled with very few egg masses, and pepper was lightly galled but with no reproduction. No infection was found on peanut, strawberry, cotton, sweet potato, or corn. White ash shows moderate to high galling and moderate egg mass production, while other hardwood species (red maple, dogwood, sweet gum, yellow poplar) were not hosts (Al-Hazma & Sasser, 1982; Hirschmann, 1982). The London plane tree, *Platanus* × *acerifolia*, in Washington, D.C., seems to be a good host (Clemens & Krusberg, 1971).

TYPE LOCALITY: Nursery plots owned by Union Camp Corporation, Franklin, Virginia, USA.

DISTRIBUTION: *North America*: USA (Virginia) (Hirschmann, 1982).

PATHOGENICITY: This nematode causes severe galling. In pathogenicity studies, a significant negative correlation was shown to exist between fresh shoot and root weights and inoculum density (Al-Hazma & Sasser, 1982).

CHROMOSOME NUMBER: Reproduction is by mitotic parthenogenesis, and the somatic chromosome number is 2n = 42-44 (Hirschmann, 1982).

POLYTOMOUS KEY CODES: *Female*: A213, B2, C3, D4; *Male*: A3124, B213, C23, D1, E1, F2; *J2*: A23, B2, C32, D23, E32, F3.

BIOCHEMICAL AND MOLECULAR CHARACTERISATION: Esbenshade & Triantaphyllou (1985a) provided S1 esterase phenotype and N1a malate dehydrogenase phenotypes No molecular data are available for this species.

RELATIONSHIPS (DIAGNOSIS): *Meloidogyne platani* belongs to perineal pattern group 4 (Jepson, 1987). This species is similar to *M. arenaria*, *M. enterolobii*, *M. mali*, *M. mersa* and *M. tadshikistanica*, but differs by a larger EP/ST ratio, and several morphometrics in male and J2.

81. *Meloidogyne propora* Spaull, 1977
(Fig. 223)

COMMON NAME: Sedge root-knot nematode.

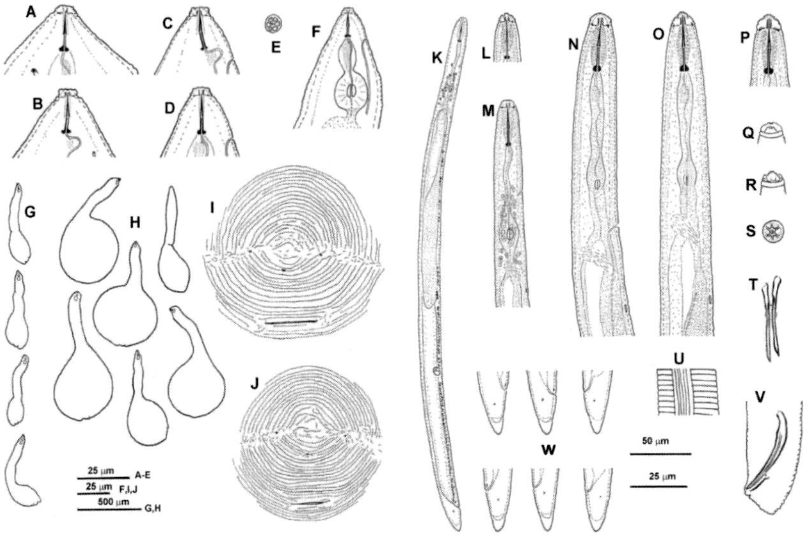

Fig. 223. Meloidogyne propora. *A-D: Female labial region; E: Female* en face *view; F: Female anterior region; G: Young female; H: Entire female; I, J: Perineal pattern; K: Entire second-stage juvenile (J2); L, M: J2 anterior region; N, O: Male anterior region; P-R: Male labial region; S: Male* en face *view; T: Spicules; U: Male lateral field; V: Male tail; W: J2 tail. (After Spaull, 1977.)*

This species was found during a general survey of the soil and litter invertebrates on Aldabra, Seychelles, where two species of *Meloidogyne* were found parasitising the roots of *Cyperus obtusiflorus* Vahl.

MEASUREMENTS (*AFTER SPAULL, 1977*)

- *Holotype J2*: L = 410.8 μm; W = 18.7 μm; pharynx = 75.9 μm; anterior end to centre of median bulb = 58.6 μm; stylet = 16.5 μm; anterior end to excretory pore = 35.5 μm; anterior end to hemizonid = 82 μm; anterior end to mid genital primordium = 281 μm; tail = 17.7 μm.
- *Paratype females* (n = 9): L = 871 ± 75 (735-965) μm; W = 420 ± 92 (225-511) μm; stylet = 17.4 ± 1.5 (13.8-19.6) μm; stylet knob height = 1-2 μm; stylet knob width = 4-5 μm; DGO = 2.8 (2.2-3.6) μm; median bulb length = 30.4 ± 2.9 (27-34) μm; median bulb width = 27.8 ± 2.0 (26-30) μm; median bulb valve length = 9.6 ± 0.5 (9-10) μm; median bulb valve width = 7.1 ± 0.7 (6-8) μm; anterior

end to base of median bulb = 66.5 ± 7.3 (59.9-78.0) μm; anterior end to excretory pore = 25.2 (19-30) μm; vulva length = 29.2 ± 5.9 (20-38) μm; vulva to anus = 40.4 ± 7.7 (32-53) μm; distance between phasmids = 47.8 ± 5.9 (37-54) μm; distance between vulva and an imaginary line between phasmids = 54.2 ± 8.1 (44-68) μm; EP/ST = 1.5.

- *Paratype males* (n = 29): L = 1168 ± 133 (810-1370) μm; W = 27.9 ± 3.3 (22-35) μm; stylet = 21.0 ± 1.6 (18.1-23.2) μm; DGO = 4.5 (2.4-6.3) μm; anterior end to the centre of median bulb = 67.6 ± 4.5 (58-77) μm; pharynx length = 94.2 ± 7.1 (81-107) μm; anterior end to excretory pore = 117.7 ± 18.8 (47-147) μm; median bulb length = 14.2 ± 1.6 (11-17) μm; median bulb width = 8.9 ± 0.9 (6.0-10.0) μm; median bulb valve length = 4.1 (3.0-5.0) μm; tail = 11.6 ± 1.9 (6.0-15) μm; spicules = 33.4 ± 1.5 (32-35) μm; gubernaculum = 8.7 (7.0-10.0) μm; a = 42.0 ± 4.0 (36-49); b = 12.4 ± 1.1 (10.0-14.2); c = 103 ± 17.8 (80-149); c′ = 0.6 (0.5-0.8).
- *Paratype J2* (n = 27): L = 393.1 ± 21 (335-430) μm; W = 18.1 ± 1.1 (15-20) μm; pharynx length (to pharyngo-intestinal junction) = 73.9 ± 5.3 (63-83) μm; anterior end to centre of median bulb = 57.1 ± 3.7 (50-63) μm; median bulb length = 14.5 ± 1.7 (12-18) μm; median bulb width = 8.3 ± 0.8 (7.0-10.0) μm; median bulb valve length: 4.0 ± 0.3 (3.0-5.0) μm; median bulb valve width = 2.6 ± 0.6 (1.0-3.0) μm; stylet = 17.3 ± 0.6 (16.5-18.5) μm; stylet knob height = (1.0-2.0) μm; stylet knob width = (2.0-3.0) μm; DGO = (1.0-2.0) μm; anterior end to excretory pore = 35.8 ± 2.4 (33-41) μm; tail = 18.2 ± 1.2 (15.9-21.7) μm; a = 21.8 + 1.9 (17-25); b (anterior end to pharyngo-intestinal junction) = 5.3 ± 0.2 (4.9-5.9); c = 21.6 ± 1.6 (17.3-24.1); c′ = 1.5 ± 0.1 (1.3-1.8).
- *Eggs* (n = 24): L = 104.1 ± 4.0 (98-113) μm; W = 42.3 ± 1.8 (41-48) μm.

DESCRIPTION

Female

Body globular to saccate usually with a short tail protuberance. Cuticle 9-14 μm thick on swollen part of body. Neck relatively long, offset from body. Neck cuticle narrowing just anterior to median bulb. Annuli anterior to median bulb sometimes indistinct. Labial region offset. In lateral view, three annuli posterior to labial cap, in ventrolateral view, two or three annuli posterior to labial cap. Basal annulus of labial region usually smaller than first body annulus. Basal plate hexaradiate in end-on view, lateral labial region sectors slightly larger than sublaterals. Cephalids not seen. Conical part of stylet *ca* twice length of shaft. Procorpus swollen. Median bulb almost spherical, its boundary indistinct in three specimens. Excretory pore *ca* 18-22 annuli posterior to labial region. Pharyngeal gland lobe globular in young females, indistinct in mature females, overlapping intestine ventrally. Pharyngo-intestinal junction *ca* 6 μm posterior to base of median bulb. Nerve ring and rectal glands not seen. Posterior cuticular pattern circular or nearly so, with faint, generally smooth striae forming a tail whorl. Just dorsal to tail, striae tend to be wavy. Striae not compressed into bundles. Lateral field faint with a single broken, weak incisure, or merely represented by a band of irregular striae. Lateral field not seen at mid-body but visible on neck to within a few annuli of labial region. Vulva slightly below general contour of body. Phasmids wider apart than width of vulva. Anal opening small.

Male

Body twisted up to 360°. Annuli 1.1-2.7 μm wide near median bulb 1.6-2.5 μm at mid-body and 1.4-1.9 μm near spicules. Lateral field usually with six or seven, rarely four or five, incisures at mid-body. Outer incisures sometimes crenate and outer bands may be areolated, particularly near tail, inner incisures sometimes broken and disordered. Labial region not offset, cone-shaped 9.7 (7.9-11.4) μm broad and 5.2 (4.3-6.1) μm high. Labial cap dumbbell-shaped, overlapping first or first and second labial annuli dorsally and ventrally. Dorsal and ventral portions of labial cap apparently trilobed. Due to overlapping labial cap only one annulus visible in optical section posterior to cap in lateral view, two or three annuli visible in optical section posterior to labial region labial cap in dorsoventral view. Basal plate hexaradiate in end-on view with lateral labial sectors slightly larger than sublaterals. Cephalids not seen. Anterior part of stylet *ca* half total length of stylet. Knobs with anterior margins concave or slightly sloping posteriorly. Median bulb and median bulb valves weakly developed, virtually same size as in preparasitic juveniles. Valves frequently situated in posterior half of bulb. Nerve ring close to pharyngo-intestinal junction, junction usually distinct. Pharyngeal gland lobe overlapping intestine ventrolaterally. Position of excretory pore variable, level with or 4-11 μm posterior to the hemizonid. In one specimen, pore anterior to median bulb. Hemizonid 1.5-2.5 annuli long, 30 (19-45) μm posterior to pharyngo-intestinal junction. Hemizonion rarely seen, *ca* 20 annuli posterior to hemizonid. Tail short, phasmids situated 0.2-1.0 tail lengths anterior to cloacal aperture. Cloacal lips raised with unusual transverse crescentic slit-like opening on posterior lip. Striae of lateral field passing round tail terminus. Single outstretched testis. Spicules arcuate, thin-walled. Gubernaculum slightly arcuate in lateral view, almost circular in ventral view, proximal end usually with a short dorsal process.

J2

Body rather stout. Cuticle annulated except for terminal hyaline region of tail. Annuli 0.9-1.4 μm wide. Lateral field with four incisures for most of body length, narrowing to two incisures in region of median bulb and anus and fading away anterior to stylet knobs and *ca* mid-way along tail. Labial region not offset, truncated cone shape. Labial cap dumbbell-shaped, overlapping first labial annulus dorsally and ventrally. In lateral view one annuli posterior to labial cap, in dorsoventral view two annuli posterior to labial cap. Sclerotisation of basal plate and vestibule lining relatively well developed. Basal plate hexaradiate in end-on view. Cephalids not seen. Stylet relatively long, shaft 5 μm in length or just over half length of anterior conical part. Stylet knobs with anterior margins flat, concave or, occasionally, slightly posteriorly sloping. Pharyngo-intestinal junction usually distinct. Pharyngeal gland lobe overlapping intestine ventrolaterally. Pharyngeal region with numerous small spherical to oval granular bodies in pseudocoel and within median bulb, these bodies, which may be nuclei, also occurring along ventral wall of pseudocoel for greater part of body. Excretory pore distinct in freshly killed individuals, less so in fixed specimens, situated at a level about halfway along procorpus 31 (28-39) annuli posterior to labial region and 16 (8-19) annuli anterior to centre of median bulb. Hemizonid two annuli long, 19 (17-24) annuli or 24 (21-29) μm posterior to hemizonid. Rectum not dilated. Tail short conoid with blunt, unstriated hyaline region. Phasmids situated within anterior third of tail. Caudalid visible in some specimens.

TYPE PLANT HOST: Yellow-flowered sedge, *Cyperus obtusiflorus* Vahl.

OTHER PLANTS: Also found on the feeder roots of *Casuarina equisetilolia* L. naturally infected and cultured experimentally on the roots of *Cyperus conglomeratus* Rottb. and *Solanum nigrum* L. (Spaull, 1977).

TYPE LOCALITY: Amongst mixed scrub growing on a brown loam soil inland of the Royal Society's research station, approximately 200 m northeast of the flagpole on the southern tip of Île Picard; Aldabra grid reference: 060E, 094N (Westoll & Stoddart, 1971).

DISTRIBUTION: *Africa*: Seychelle Islands (Spaull, 1977).

PATHOGENICITY: *Meloidogyne propora* causes slight galling in *Solanum* but no galls were found on the roots of *Cyperus* or *Casuarina*.

CHROMOSOME NUMBER: No available data.

POLYTOMOUS KEY CODES: *Female*: A12, B213, C3, D2; *Male*: A34, B23, C21, D2, E1, F2; *J2*: A32, B1, C4, D4, E1, F4.

BIOCHEMICAL AND MOLECULAR CHARACTERISATION: No available data.

RELATIONSHIPS (DIAGNOSIS): *Meloidogyne propora* is distinguished from *M. brevicauda* by having a J2 with a shorter body and a slightly longer stylet and from *M. indica* by having a J2 with a longer stylet, females with a relatively long neck, two or three annuli posterior to the labial cap, and males with a different labial region and labial cap shape, six or seven incisures in the lateral field, and a short dorsal process on the gubernaculum.

82. *Meloidogyne querciana* Golden, 1979
(Figs 224-226)

COMMON NAME: Oak root-knot nematode.

This species was found in 1965 on a pin oak (*Quercus palustris* Muenchh.) seedling from a nursery in Virginia, USA.

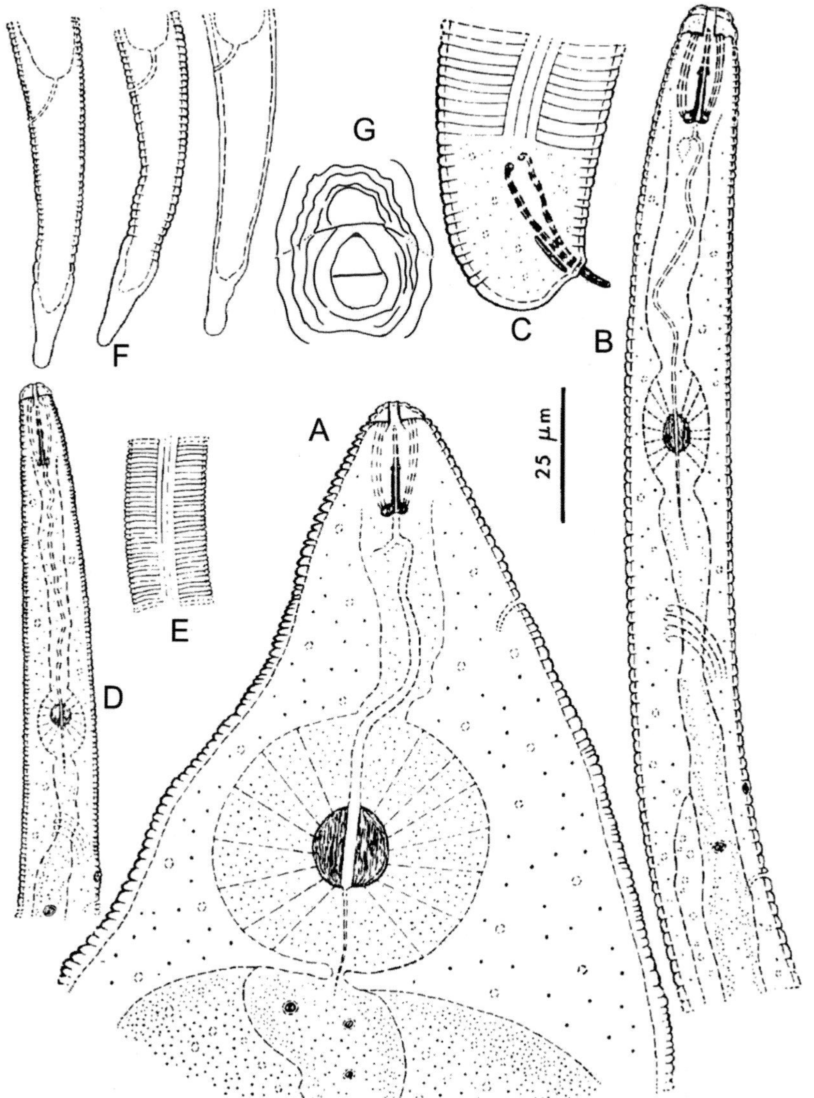

Fig. 224. Meloidogyne querciana. *A: Female anterior region; B: Male anterior region; C: Male tail; D: Second-stage juvenile (J2) anterior region; E: J2 lateral field; F: J2 tail. (After Golden, 1979.) G: Perineal pattern. (After Jepson, 1987.)*

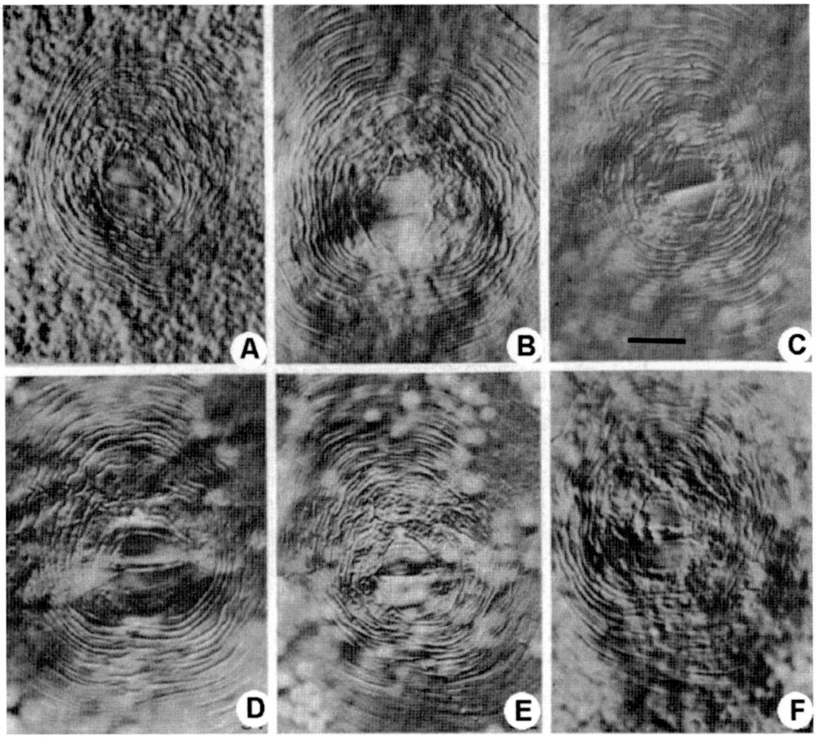

Fig. 225. Meloidogyne querciana. *A-F: Perineal pattern. (Scale bar = 20 μm). (After Golden, 1979.)*

MEASUREMENTS (AFTER GOLDEN, 1979)

- *Holotype female*: L = 955 μm; W = 731 μm; stylet = 17.2 μm; width of stylet knobs = 4.7 μm; DGO = 4.7 μm; anterior end to excretory pore = 43 μm; vulval slit length = 33 μm; distance from vulval slit to anus = 21 μm; centre of median bulb from anterior end = 89 μm; a = 1.3; b = 4.8.
- *Allotype male* (in glycerin): L = 1836 μm; stylet = 19 μm; DGO = 2.8 μm; spicules = 32 μm; gubernaculum = 10 μm; tail = 10 μm; a = 49; b = 9; c = 182.
- *Paratype females* (n = 30): L = 891 ± 163 (533-1170) μm; W = 639 ± 106 (387-877) μm; stylet = 17.5 ± 0.5 (17.0-18.7) μm; stylet knob width = 4.3 ± 0.2 (3.9-4.7) μm; DGO = 4.5 ± 0.5 (4.3-6.0) μm; centre of median bulb to anterior end = 84 ± 7.1 (73-95); anterior end to excretory pore = 36 ± 6.4 (26-51) μm; vulva length = 29 ± 3.3

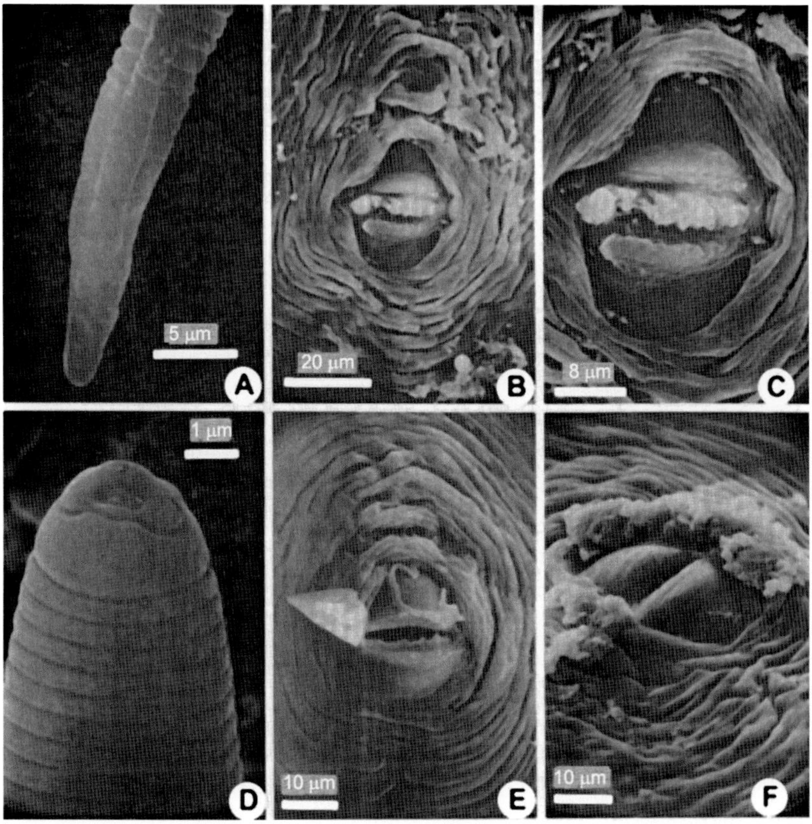

Fig. 226. Meloidogyne querciana. *SEM. A: Second-stage juvenile (J2) tail; B, C, E, F: Perineal patterns; D: J2 labial region. (After Golden, 1979.)*

(20-33) μm; vulva to anus = 20 ± 2.3 (17-23) μm; a = 1.4 ± 0.2 (1.1-1.9); b = 5 ± 1.2 (3.7-7.1); EP/ST = 2.0.

- *Paratype males* (n = 4): L = 1635 ± 248 (1392-1862) μm; stylet = 19.3 ± 0.4 (19-19.6) μm; stylet knob width = 4.1 ± 0.3 (3.9-4.5) μm; DGO = 2.6 ± 0.3 (2.2-2.8) μm; centre of median bulb from anterior end = 83 ± 3.3 (80-87) μm; tail = 11 ± 1.0 (10-12) μm; spicules = 32.3 ± 0.4 (32.0-32.5) μm; gubernaculum = 8.4 ± 1.2 (7.3-10.1); a = 53 ± 4.5 (49-58); b = 8 ± 1.3 (6-9); c = 148 ± 25 (137-182).
- *Paratype J2* (n = 70): L = 467 ± 25.5 (411-541); anterior end to centre of median bulb = 56 ± 2.5 (51-63) μm; stylet = 11.1 ± 0.3 (10.2-11.6) μm; DGO = 3.5 ± 0.4 (2.6-4.3) μm; labial region height = 5.2 ± 0.2 (4.3-5.6) μm; tail = 46 ± 3 (39-52) μm; hyaline region =

11 ± 1.4 (8-14) μm, caudal ratio A = 2.1 ± 0.3 (1.4-2.8); caudal ratio B = 3.3 ± 0.8 (2.4-5.6); a = 30 ± 3.0 (23-39); b = 2.6 ± 0.2 (2.2-3.4); c = 10 ± 0.3 (7-13).

- *Eggs* (n = 30): L = 101 ± 3.5 (94-110) μm; W = 51 ± 2.6 (47-62) μm.

DESCRIPTION

Female

Body pearly white, globular to pear-shaped, without posterior protuberance, and with well defined but small neck situated anteriorly on a median plane with terminal vulva. Pharyngeal and anterior region commonly appearing as illustrated. Labial region more or less continuous with neck, bearing a prominent labial cap and two annuli. Labial framework weak. Stylet strong with rather prominent knobs sloping posteriorly. Excretory pore distinct, about two stylet lengths from anterior end. Vulva sunken, surrounded by a prominent, squarish to egg-shaped area devoid of striae. Perineal pattern often oval, but more elongate in dorsoventral plane, sometimes squarish with indication of a low arch just above anal area, striae generally well spaced, fairly coarse and often continuous, commonly surrounding vulval area, which is located almost in centre of pattern.

Male

Body quite slender, vermiform, tapering slightly at both extremities. Labial region slightly offset, with massive labial annuli (labial cap), and without postlabial annuli. Stylet, knobs, cephalids, hemizonid, excretory pore, and anterior portion commonly appearing as illustrated. Lateral field with four lines, generally not areolated, forming about 20% body diam. at mid-body, latter being 25-38 μm (31 ± 5.1 μm). Testes two. Spicules arcuate, with rounded tips. Phasmids prominent, at level of cloacal aperture. Tail short, rounded. Males are extremely rare, with only four found on type host (pin oak) and two, in poor condition, on red oak.

J2

Body vermiform, tapering at both extremities but much more so posteriorly. Labial region not offset, with weak labial framework, and without annuli. Cuticular annulation on most of body very fine, measuring about 1 μm, but becoming much coarser in vicinity of anus and then disappearing about one-third of tail length from terminus. Lateral field with four lines, not areolated, forming 25% body diam. at mid-body. Stylet, knobs, hemizonid, excretory pore, and anterior portion usually appearing as illustrated. Cephalids indistinct and not shown. Phasmids small, difficult to see, located about one anal body diam. posterior to anus level. Rectum not dilated. Tail rather variable in shape but with a characteristic slight swelling near beginning of hyaline region and with a slightly swollen, rounded terminus.

TYPE PLANT HOST: Pin oak, *Quercus palustris* Muenchh.

OTHER PLANTS: American chestnut (*Castanea dentata* (Marsh.) Borkh.), and red oak (*Quercus rubra* L.) were hosts, while barley (*Hordeum vulgare*), corn (*Zea mays* 'Golden Bantam'), Kentucky bluegrass (*Poa pratensis*), oats (*Avena sativa*), pansy (*Viola tricolor* L.), peanut (*Arachis hypogaea*), rye (*Secale cereale* 'Gates'), strawberry (*Fragaria* × *ananassa* Duchesne 'Fairfax no. 2') and wheat (*Triticum aestivum* 'Red coat') were not hosts.

TYPE LOCALITY: Nursery in Augusta County, Virginia, USA.

DISTRIBUTION: *North America*: USA (Virginia) (Golden, 1979).

PATHOGENICITY: On pin oak, the numerous galls on young roots were rather small and discrete but on older roots the galls coalesced and formed massive knots. Females of *M. querciana* were generally embedded in the galls, although on young roots egg masses could sometimes be seen protruding from the gall surface. Host plants infected by *M. querciana* were smaller than non-infected controls, as indicated by red oak.

CHROMOSOME NUMBER: Reproduction is by obligatory mitotic parthenogenesis, and the somatic chromosome number is 2n = 30-32 (Triantaphyllou, 1985b).

POLYTOMOUS KEY CODES: *Female*: A123, B21, C2, D6; *Male*: A213, B3, C2, D1, E1, F2; *J2*: A21, B32, C3, D324, E213, F423.

BIOCHEMICAL AND MOLECULAR CHARACTERISATION: This species has unique F1 esterase phenotype and N3a malate dehydrogenase phenotypes (Esbenshade & Triantaphyllou, 1985a). No molecular data of this species are available.

RELATIONSHIPS (DIAGNOSIS): *Meloidogyne querciana* has a characteristic perineal pattern (group 6, Jepson, 1987) and sunken vulva with the surrounding area devoid of striae. Female morphology and morphometrics are close to *M. izalcoensis* and *M. floridensis*, but there are several differences in the morphometrics of the male and J2.

83. *Meloidogyne salasi* López, 1984
(Figs 227, 228)

COMMON NAME: Costa Rican root-knot nematode.

This species was reported in Central and South America, and causes severe losses in rice fields in Costa Rica, Panama, and Venezuela (López, 1984; Sancho & Salazar, 1985; Medina *et al.*, 2009, 2011).

MEASUREMENTS

See Table 49.

DESCRIPTION

Female

Body pearly white, with body length (excluding neck)/max. body diam. (ratio a) with an average value of 1.4 (1-2). Distinct posterior protuberance present. Cuticle distinctly annulated, often with incomplete annulations in labial region and neck region. Labial region offset from body. In SEM, labial disc appearing slightly elevated with a rounded and relatively large prestoma located in middle. Labial region appearing as a single annulus, often marked by longitudinal lines. Amphidial openings clearly distinct, rectangular. Lateral labials arched, slightly larger than ventral or dorsal sectors. Vestibule and vestibule extension distinct when observed with LM. Stylet delicate, cone usually straight, with triangular

Fig. 227. Meloidogyne salasi. *A: Perineal pattern; B: Entire female; C, D: Female anterior region; E, F: Male anterior region; G, H: Male tail; I, J: Second-stage juvenile (J2) anterior region; K, L: J2 tail. (After López, 1984.)*

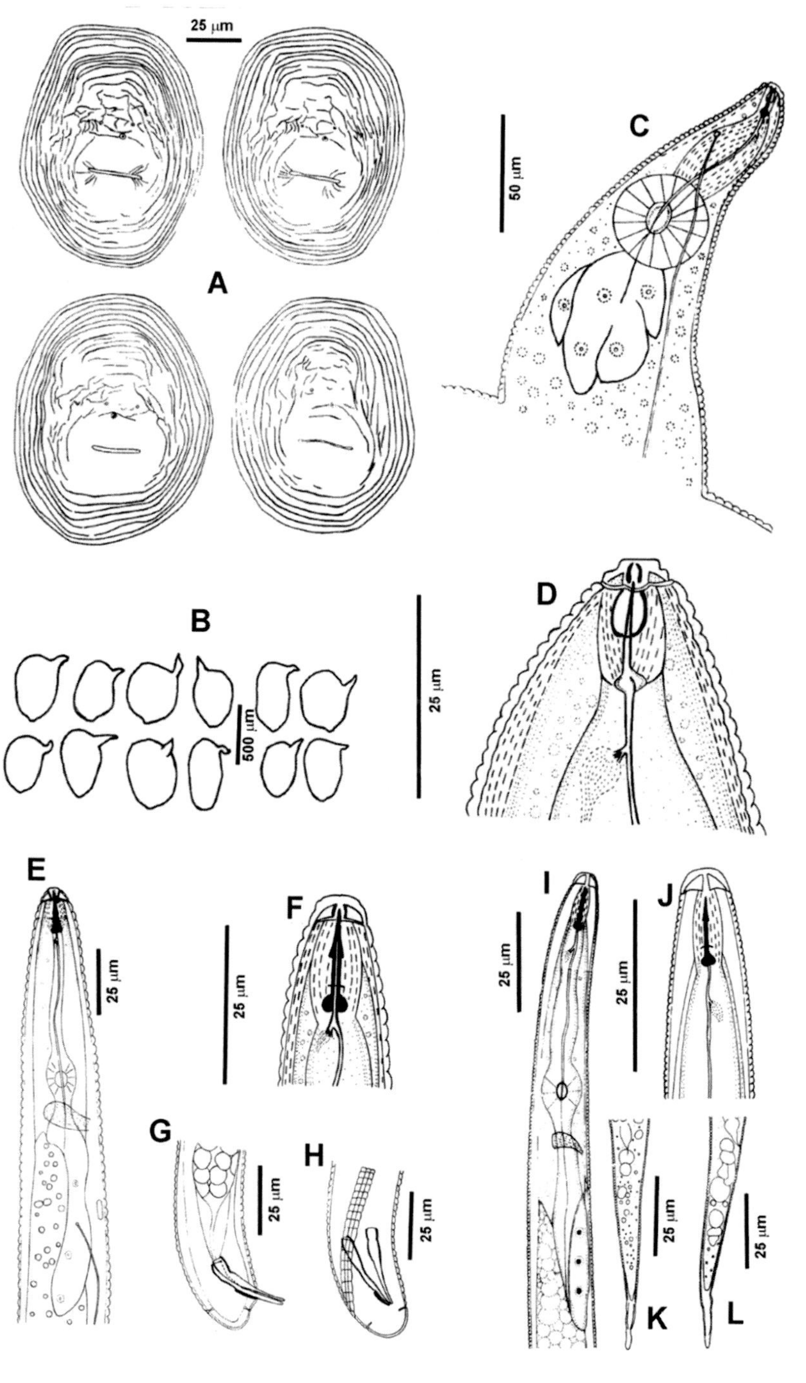
25 μm
A
50 μm
C
B
500 μm
25 μm
D
E
25 μm
F
25 μm
G
25 μm
H
25 μm
I
25 μm
J
25 μm
25 μm
K
25 μm
L

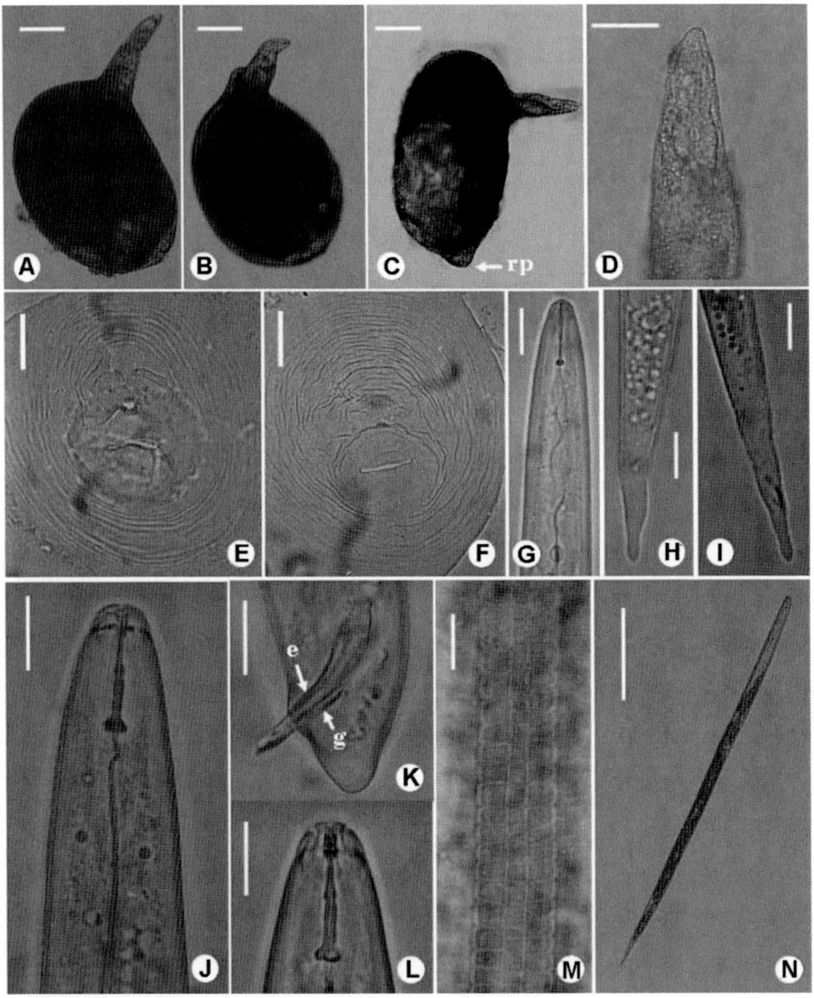

Fig. 228. Meloidogyne salasi. *LM. A-C: Entire female; D: Female anterior region; E, F: Perineal pattern; G: Second-stage juvenile (J2) anterior region; H, I: J2 tail; J, L: Male anterior region; K: Male tail; M: Male lateral field; N: Entire J2. Abbreviations: e = spicules, g = gubernaculum; rp = perineal region. (Scale bars: A-C, N = 100 μm; D = 50 μm; E, F = 20 μm; G-M = 10 μm.) (After Medina* et al., *2011.)*

base about 25% of its length, tapering to a fine, pointed tip. Opening of stylet near tip, in anterior 25% of cone. Outlet of dorsal pharyngeal gland branched, with dorsal ampulla relatively large. Excretory pore position

Table 49. *Morphometrics of females, males, second-stage juveniles (J2) and eggs of* Meloidogyne salasi. *All measurements are in μm and in the form: mean ± s.d. (range).*

Character	Holotype, La Cuesta, Costa Rica, López (1984)	Paratypes, La Cuesta, Costa Rica, López (1984)	Calabozo, Venezuela, Medina *et al.* (2011)
Female (n)	-	50	15
L	422	486.3 ± 63.1 (372-625)	588 ± 72.5 (464-735)
a	1.37	1.4 ± 0.2 (1.0-2.0)	1.9 ± 0.4 (1.4-3.1)
W	306	338.1 ± 46.8 (209-425)	-
Stylet	10.9	10 ± 0.9 (8.1-12.5)	13 ± 0.9 (11-14)
Stylet knob width	3.2	3.4 ± 0.4 (2.5-4.5)	-
Stylet knob height	2.1	2.1 ± 0.4 (1.5-3.4)	-
DGO	4.5	4.9 ± 1.0 (3.4-6.8)	3.1 ± 0.6 (2.4-4.8)
Ant. end to median bulb valve	71.9	78.2 ± 10.2 (60.9-99.9)	-
Anterior end to excretory pore	-	32.1 ± 10.3 (18.7-62.5)	29 ± 6.9 (19-40)
Vulval slit	-	21.9 ± 2.4 (15.9-26.5)	-
Vulva to anus	-	16.4 ± 2.9 (9.0-24)	17 ± 1.9 (13-19)
Interphasmid distance	-	15.2 ± 2.4 (10.6-21.8)	-
EP/ST	-	1.0-1.5	2.1 ± 0.4 (1.5-2.6)
Male (n)	1	50	15
L	1711	1619 ± 289 (992-2093)	1069 ± 99.4 (858-1200)
a	48.4	47.5 ± 6.5 (31.8-58.1)	44 ± 3.5 (36.5-48.6)

Table 49. *(Continued.)*

Character	Holotype, La Cuesta, Costa Rica, López (1984)	Paratypes, La Cuesta, Costa Rica, López (1984)	Calabozo, Venezuela, Medina *et al.* (2011)
b	7.9	-	7.2 ± 1.0 (5.8-9.4)
c	123.9	132.8 ± 38.6 (46.6-254.7)	100 ± 13.3 (77-124)
c′	-	-	0.7 ± 0.1 (0.6-0.8)
T	60.4	55.0 ± 7.9 (32.0-71.6)	-
W	35.3	33.9 ± 3.5 (25.4-41.8)	-
Stylet	19	18.2 ± 2.1 (12.1-21.8)	15 ± 0.9 (13-17)
Stylet knob width	3.3	4.6 ± 0.7 (3.5-7.5)	-
Stylet knob height	2.3	3.1 ± 0.5 (2.1-4.2)	-
DGO	4.4	4.1 ± 0.7 (2.8-5.9)	4.0 ± 1.0 (3.5-6.0)
Ant. end to median bulb valve	94.5	101.7 ± 18.5 (64-134)	-
Ant. end to excretory pore	136.7	156.9 ± 32.3 (88-227)	96.4 ± 10 (78-116)
Spicules	28.8	25.8 ± 4.5 (17.5-34.5)	22 ± 2.6 (18-26)
Gubernaculum	9	7.8 ± 1.3 (5.6-11.8)	7.0 ± 1.0 (6.0-9.0)
Tail	13.8	13.0 ± 4.7 (6.5-39)	-
J2 (n)	-	50	15
L	-	464.4 ± 18.4 (422-503)	418 ± 29.3 (367-461)
a	-	28.6 ± 1.7 (23.9-32.2)	27 ± 1.8 (24.6-31.0)
b	-	3.8 ± 0.3 (3.0-4.4)	4.1 ± 0.4 (3.2-5.1)

Table 49. *(Continued.)*

Character	Holotype, La Cuesta, Costa Rica, López (1984)	Paratypes, La Cuesta, Costa Rica, López (1984)	Calabozo, Venezuela, Medina *et al.* (2011)
c	-	6.8 ± 0.4 (5.9-7.7)	7.0 ± 0.5 (6.0-8.0)
c′	-	5.7 ± 0.5 (4.2-6.8)	6.3 ± 0.9 (4.8-7.8)
W	-	16.2 ± 9.3 (15.3-19.3)	-
Stylet	-	11.4 ± 1.1 (9.2-13.3)	11 ± 1.4 (9.0-13.0)
Stylet knob width	-	2.3 ± 0.3 (1.5-2.8)	-
Stylet knob height	-	1.5 ± 0.2 (1.0-2.1)	-
DGO	-	3.7 ± 0.6 (2.1-5.3)	3 ± 0.5 (2.5-4.0)
Ant. end to median bulb valve	-	56.7 ± 2.6 (50.6-62.1)	-
Ant. end to excretory pore	-	80.3 ± 3.3 (71.5-89.6)	66 ± 4.4 (57-78)
Pharyngeal lobe base to ant. end	-	121.8 ± 9.3 (103-153)	-
Tail	-	67.8 ± 5.2 (56.5-80.2)	-
Hyaline region	-	19.7 ± 3.3 (11.8-26.2)	-
Egg	-	50	-
L	-	94.5 ± 5.2 (82.8-113.2)	-
W	-	41.1 ± 1.5 (38.2-44.5)	-

variable, about 1.0-1.5 times stylet length posterior to stylet knobs in 66% of specimens observed. In a few females (4%) excretory pore about 50% stylet length posterior to stylet knobs, while in others (6%) about three times stylet length posterior to stylet knobs. Median bulb relatively large and rounded, with a strong, oval central valve. Pharyngeal glands appearing as a massive, globose structure with five nucleated lobes, often difficult to observe with bright field illumination but distinct with Nomarski differential interference contrast optics. Perineal pattern oval-shaped, with fine outer striae and somewhat coarse striae in inner portion. Striae mostly unbroken, smooth, relatively few in numbers and far apart. Perineum with no or only one striae, and only a few in roughly circular central area of pattern. Vulva a transverse, smooth slit, with no or few lateral striae. Phasmids small, closely spaced. Dorsal arch high and wide, usually rectangular in shape, but somewhat square in some specimens. No evidence of lateral lines or interrupted striae. Tail tip prominent in freshly mounted perineal patterns.

Male

Body length variable, tapering at anterior end and relatively rounded at posterior end. Labial region slightly offset from body, bearing a variable number of incomplete annuli, with distinct labial cap. In SEM, large, rounded labial disc slightly elevated above median labials, with lateral edges slightly arcuate. Oval prestoma in centre of labial disc, encircled by six inner labial sensilla with pit-like openings. Stoma with a slit-like opening. Median labials wider than labial disc, the two forming a continuous labial cap with no discernible indentations at lateral junctions. Four labial sensilla appearing as slight, small cuticular depressions on median labials, two on each. Amphidial openings relatively long slits posterior to lateral edges of labial disc. Lateral labials almost inconspicuous, marked by short grooves starting near lateral junction of median labials and labial disc, and extending into labial region. Cuticle with distinct annuli, about 1.9 μm wide near labial region, 2 μm wide around mid-body and 1.6 μm wide near tail. Lateral field about 6, 7.5 and 5 μm wide near anterior, middle portion and tail areas of the body, respectively. Four lateral lines in lateral field, one at each edge on ridge and two in inner portion, but in some specimens five or up to six lines are visible for some distance in middle of body, the additional lines being fainter. Lateral field starting as two lines with crenate edges near base of stylet, some 4-10 body annuli posterior to labial region, where inner two lines appearing, and continuing to posterior end, where they twist through 90°. Lateral field areolated, usually corresponding with body annulations, but in some areas, especially middle portion, there is no correspondence. Labial framework sclerotised, with lateral sectors slightly larger than labial cap. Stylet robust, with a pointed cone, slightly longer than shaft, with opening near the tip and a triangular base in basal 25% of its length. Stylet knobs rounded, offset from shaft, with an ascending slope toward its base. Lumen of stylet almost as wide as that of procorpus, but narrowing at cone. Distinct excretory pore, with long, curved excretory duct disappearing in intestine. Basal lobe of pharynx overlapping intestine ventrally, with three nuclei, anterior nucleus near beginning of lobe; posterior nucleus near end of lobe. Hemizonid 1-2 annuli anterior to excretory pore, 1-2 annuli long. Intestinal caecum extending on dorsal side of body to about same level or posterior to nerve ring. Sperm globular, granular. Spicules long, arcuate, typical of genus. In SEM each spicular tip showing one transverse opening. Gubernaculum simple. Phasmids typically posterior to cloacal aperture, with a pore-like opening. Body twisting about 90° near cloacal region.

J2

Body tapering at both ends but much more so posteriorly. Labial region slightly offset from body, lateral sectors slightly narrower than body, and elevated labial cap. In SEM, elongated labial disc slightly elevated above median labials, with lateral edges straight or almost so. Oval prestoma in centre of labial disc encircled by six inner labial sensilla with pit-like openings. Stoma with a small slit-like opening. Median labials crescentic in most specimens, wider than labial disc, with no discernible indentations at lateral junctions with it, forming a dumbbell-shaped cap. Amphidial opening slit-like, located posterior to lateral edges of labial disc. Labial framework weakly developed. Body distinctly annulated, the annuli discernible with LM up to beginning of tail terminus. Lateral field areolated, with four lines, two external lines slightly crenate. Beginning as two lines at about middle of procorpus, then three and finally four lines that continue as far as beyond the anus where the two central lines disappear and the two lateral ones continue for a short distance up to beginning of hyaline region. Stylet weakly developed, with small rounded knobs, one slightly larger and lower than other two. Knobs posteriorly sloping. Ring-like structure encircling shaft near base. Dorsal ampulla weakly developed. Procorpus *ca* 2.0-2.5 times as long as muscular oval median bulb, this with a sclerotised central valve. Nerve ring encircling narrow isthmus. Hemizonid 1-2 annuli anterior to excretory pore, *ca* one annulus long. Excretory pore located at about same level as or posterior to nerve ring. Basal lobe of pharynx rather short, with three nuclei, anterior one located near its beginning and posterior one near its end. Basal pharyngeal lobe overlapping intestine ventrally. Anal opening a small pore on cuticle. Rectum weakly dilated. Tail relatively long, tapering to a fine, rounded, slightly clavate terminus.

TYPE PLANT HOST: Rice, *Oryza sativa* L. 'C.R.1113'.

OTHER PLANTS: *Homolepis aturensis* (Kunth) Chase and *Echinochloa colona* are also hosts for *M. salasi* (Figueroa, 1973; López, 1984). *Cynodon plectostachyus* (K.Schum.) Pilg., *C. dactylon*, *Ischaemum ciliare* Retz., *Digitaria eriantha* Steud., *Tripsacum laxum* Nash, *E. polystachya* (Kunth) A.S. Hitchc., *Leucaena leucocephala* (Lam.) de Wit, *Kazungula* sp., *Brachiaria ruziziensis* Germ. & C.M.Evrard, *B. decumbens* Stapf, *B. rugulosa* Stapf, *Panicum maximum* Jacq. and *Saccharum sinensis* Roxb. are poor hosts (Tarte, 1981). *Meloidogyne salasi* did not infect any of the plant species used in the North Carolina differential host test (Sasser & Carter, 1982).

TYPE LOCALITY: La Cuesta, Puntarenas, Costa Rica.

DISTRIBUTION: *Central America*: Costa Rica and Panama (López, 1984); *South America*: Venezuela (Medina *et al.*, 2011).

PATHOGENICITY: This nematode produces significant damage on rice. Sancho *et al.* (1987) demonstrated that increasing inoculum densities from 333 to 999 eggs (100 cm^3 soil)$^{-1}$ showed a linear decrease in plant growth.

POLYTOMOUS KEY CODES: *Female*: A32, B43, C3, D3; *Male*: A2134, B324, C32, D1, E1, F2; *J2*: A21, B32, C213, D123, E3, F213.

BIOCHEMICAL AND MOLECULAR CHARACTERISATION: *Meloidogyne salasi* from Costa Rica showed a VS1-2 esterase profile (Rm: 0.64 extending from 0.60 to 0.70), slightly different from the *M. graminicola* profile, and three-bands of malate-dehydrogenase S3 phenotype (Rm 1.4, 1.6, 1.8) (Mattos *et al.*, 2019). Mattos *et al.* (2018) provided and analysed ITS and D2-D3 of 28S rRNA gene of this species. *Meloidogyne salasi* formed a clade with *M. graminicola* and *M. trifoliophila*. Mattos *et al.* (2019) also developed a specific SCAR marker for *M. salasi* (primers SALR12-1 F/R), distinguishing this species from other RKN attacking rice (*viz.*, *M. graminicola* and *M. oryzae*).

CHROMOSOME NUMBER: Populations of this nematode from both Costa Rica and Panama have been studied cytologically (Triantaphyllou, 1982) and found to reproduce by obligatory mitotic parthenogenesis and have a diploid chromosome number of 2n = 36.

RELATIONSHIPS (DIAGNOSIS): The species belongs to Molecular group III, the Graminicola group. *Meloidogyne salasi* can be distinguished from *M. kralli* by longer female (L = 486 (372-735) *vs* 463 μm; max. body diam. of 338 *vs* 306 μm), the straight and shorter stylet (10 *vs* 13.1 μm), longer excretory pore of the female (32 *vs* 15.8 μm), by the higher dorsal arch of the perineal pattern, and absence of a posterolaterally directed irregular double incisure on either side of the tail region of the perineal pattern, longer males (1619 *vs* 1076 μm), and greater a and c ratios in the males (47.5 and 132.8 *vs* 31.7 and 117, respectively). *Meloidogyne salasi* can be distinguished from *M. acronea* by the female body length (486 *vs* 980-1040 μm), max. body diam. (338 *vs* 530-750 μm), shorter female stylet (10 *vs* 12 μm) and shorter spicules (25.8 *vs* 33-35 μm). *Meloidogyne salasi* can be differentiated from *M. graminis* by the body length of the female, max. body diam. and DGO (486, 338 and 4.9 μm *vs* 726, 472 and 3.7 μm, respectively), the absence of lateral lines in the perineal pattern, and the fine striae in the perineal pattern.

84. *Meloidogyne sasseri* Handoo, Huettel & Golden, 1993
(Figs 229-231)

COMMON NAME: Beachgrass root-knot nematode.

In August 1990, a RKN was found parasitising the roots of beachgrasses, *Ammophila breviligulata* Fern. and *Panicum amarulum* Hitchcock & Chase, at Henlopen State Park and Fenwick Island near the Maryland state line in Delaware, USA (Seliskar & Huettel, 1992). The

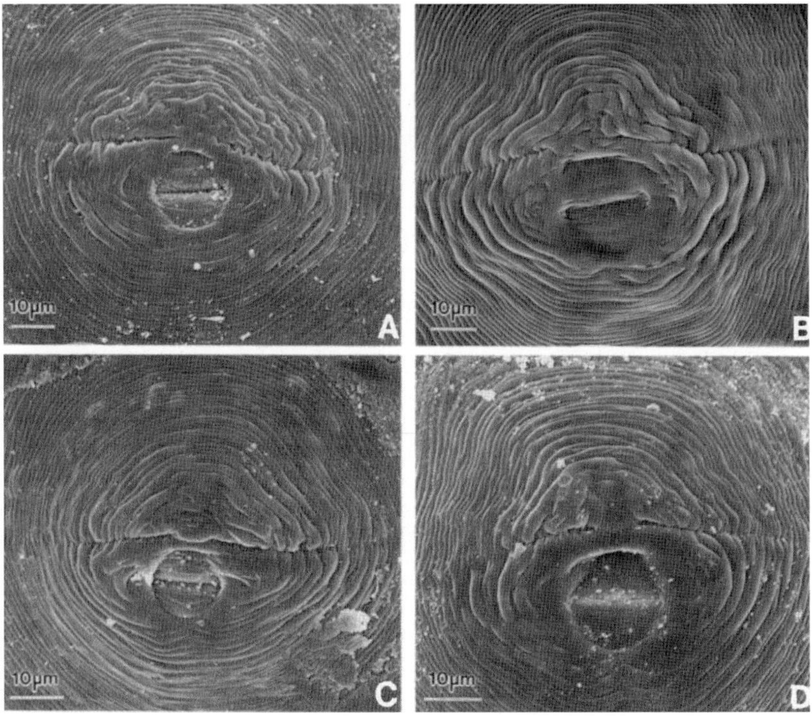

Fig. 229. Meloidogyne sasseri. *SEM. A-D: Perineal patterns. (After Handoo* et al., *1994.)*

nematode did not produce galls, and females were generally surrounded by a massive egg sac.

MEASUREMENTS *(AFTER HANDOO* ET AL., *1993)*

See Table 50.

- *Holotype female*: L = 822 μm; W = 493 μm; anterior end to posterior end of median bulb = 172 μm; stylet = 14 μm; stylet knob height = 2 μm; stylet knob width = 5 μm; DGO = 4 μm; excretory pore from anterior end 13.5 μm; vulval slit length = 30 μm; distance from vulval slit to anus = 18 μm; EP/ST ratio = 1.0.
- *Allotype male*: L = 1700 μm; excretory pore to anterior end = 140 μm; centre of median bulb from anterior end = 94 μm; stylet = 20 μm; stylet knob height = 3 μm; stylet knob width = 5 μm;

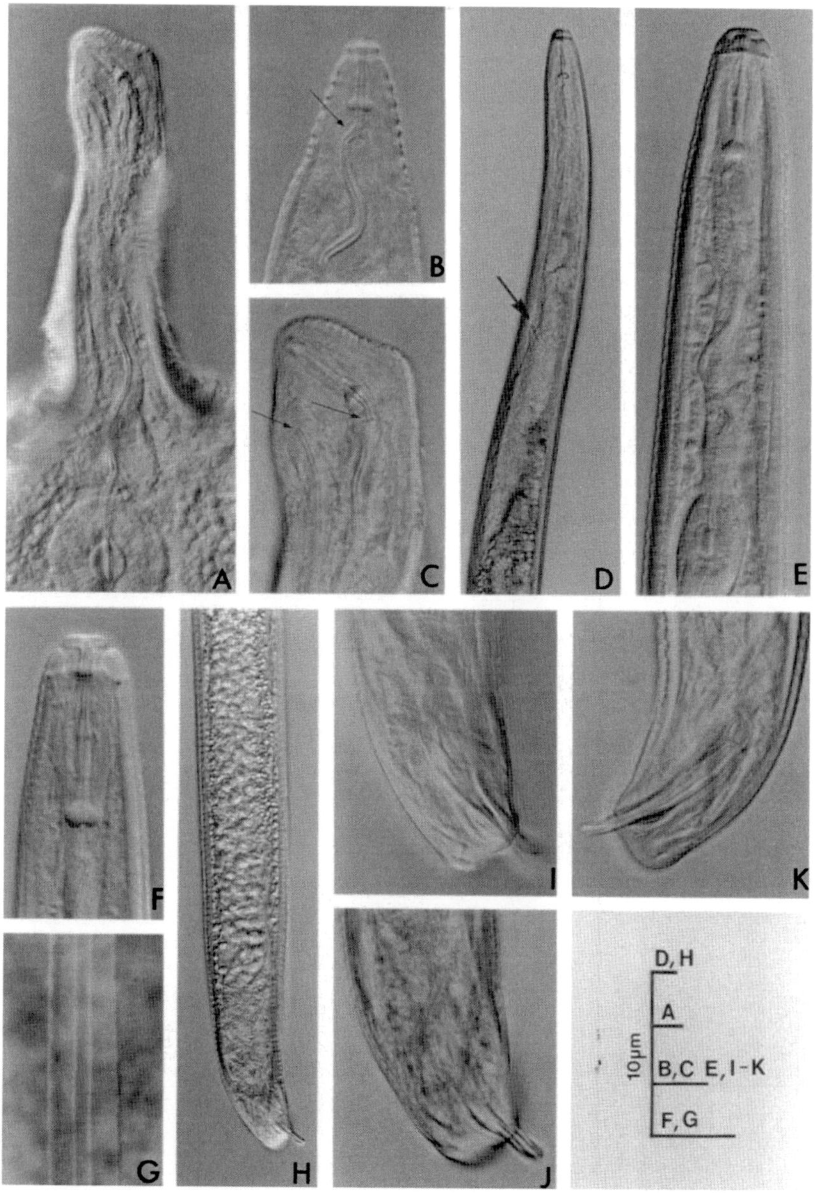

Fig. 230. Meloidogyne sasseri. *LM. A-C: Female anterior region (excretory pore and dorsal gland orifice arrowed); D, E: Male anterior region (excretory pore arrowed); F: Male labial region; G: Male lateral field; H-K: Male tail. (After Handoo* et al., *1994.)*

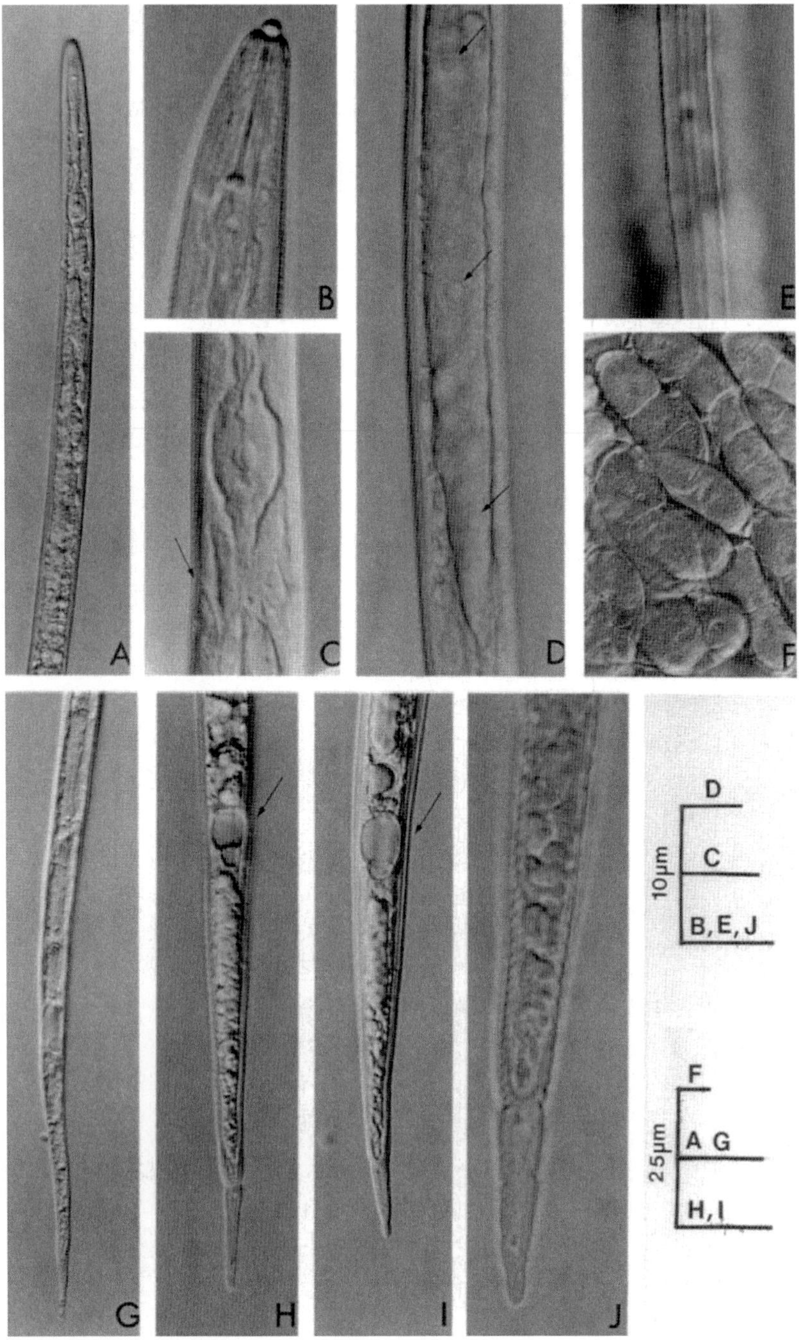
A
B
C
D
E
F
G
H
I
J
D
C
10µm
B, E, J
F
A G
25µm
H, I

Fig. 231. Meloidogyne sasseri. *LM. A, B: Second-stage juvenile (J2) anterior region; C: J2 part of anterior region showing excretory pore (arrow); D: Basal pharyngeal bulb showing three nuclei (arrows) in J2; E: J2 lateral field; F: Eggs; G-J: J2 posterior region showing inflated rectum (arrows). (After Handoo* et al., *1994.)*

spicules = 32 μm; gubernaculum = 8.5 μm; tail = 8 μm; a = 37; b = 9.1; c = 213.

- *Paratype females* (n = 60): L = 860 ± 120.7 (644-1150) μm; W = 524.5 ± 72.6 (395-685) μm; stylet = 14.0 ± 0.6 (13-15) μm; stylet knob height = 2.0 ± 0.2 (1.5-2.5) μm; stylet knob width = 4.8 ± 0.4 (4.0-5.5) μm; DGO = 4.2 ± 0.6 (3.0-5.5) μm; anterior end to excretory pore = 15.2 ± 2.9 (7.5-31.0) μm; vulva length = 30.2 ± 2.3 (25-35) μm; vulva to anus = 16.9 ± 2.4 (13-22) μm; a = 1.6 ± 0.2 (1.3-2.4); EP/ST = 1.3 ± 0.3 (0.5-2.2).
- *Paratype males* (n = 45): L = 1746 ± 192.6 (1290-2175) μm; W = 43.8 ± 4.1 (30-50); stylet = 20.0 ± 0.5 (19-21.5) μm; stylet knobs height = 3.0 ± 0.2 (2.5-3.5) μm, stylet knob width = 4.8 ± 0.2 (4.5-5.0) μm; excretory pore to anterior end = 148.1 ± 10.6 (121-173) μm; centre of median bulb from anterior end = 94.3 ± 7.3 (71-108) μm; tail = 8.7 ± 1.0 (8-11) μm; spicules = 33 ± 1.8 (30-36) μm; gubernaculum = 8.6 ± 0.9 (8-11) μm; a = 39.9 ± 3.6 (32.2-51.7); b = 8.6 ± 1.3 (6.1-12.9); c = 203.2 ± 27 (161-257).
- *Eggs* (n = 40): L = 113.1 ± 9.9 (82.8-113.2) μm; W = 41.1 ± 3.8 (38.2-44.5) μm.

DESCRIPTION

Female

Body white, round to pear-shaped, with slight posterior protuberance, neck long, distinct, tapering anteriorly. Labial framework weak, slightly offset from neck, bearing one large smooth annulus. Labial region variable in shape with circumoral elevation. Labial disc slightly raised above median labials, disc and median labials dumbbell-shaped. Lateral labials smaller than and adjacent to median labials. Amphidial openings oval, located between labial disc and lateral labials. Stylet strong, basal knobs robust, rounded, sloping posteriorly. Excretory pore distinct, generally

Table 50. *Morphometrics of second-stage juveniles (J2) of* Meloidogyne sasseri. *All measurements are in μm and in the form: mean ± s.d. (range).*

Character	American beachgrass	Bitter panicum	Rice	Wheat	Oat	Bermudagrass	Zoysiagrass
J2 (n)	50	50	50	50	50	50	50
L	541.9 ± 19.4	543.0 ± 36.2	536.9 ± 23.5	549.7 ± 30.2	512.3 ± 11.3	554.3 ± 27.5	619.6 ± 19.2
	(490-575)	(470-640)	(485-610)	(485-620)	(490-540)	(490-605)	(575-650)
a	29.9 ± 1.6	31.1 ± 1.7	30.7 ± 1.5	31.4 ± 1.7	31.1 ± 1.3	31.1 ± 1.5	33.2 ± 1.8
	(25-32.8)	(27.2-35.9)	(28.3-35.8)	(26.9-34.6)	(27.7-33.7)	(28.8-35.9)	(28.7-36.8)
b	2.4 ± 0.2	2.4 ± 0.2	2.2 ± 0.1	2.3 ± 0.1	2.2 ± 0.1	2.3 ± 0.1	2.4 ± 0.1
	(2.0-3.0)	(2.2-3.3)	(2.0-2.6)	(2.0-2.6)	(2.0-2.6)	(2.1-2.6)	(2.1-2.8)
c	5.6 ± 0.2	5.5 ± 0.3	5.6 ± 0.2	5.5 ± 0.2	5.8 ± 0.2	5.8 ± 0.2	5.8 ± 0.2
	(5.2-6.3)	(5.0-6.5)	(5.2-6.3)	(5.2-6.2)	(5.5-6.3)	(5.4-6.5)	(5.4-6.6)
Stylet	13.7 ± 0.4	13.9 ± 0.2	13.4 ± 0.4	13.5 ± 0.4	13.5 ± 0.3	13.7 ± 0.3	13.7 ± 0.3
	(13-14.5)	(13-14.5)	(13-14)	(13-14)	(13-14)	(13-14)	(13-14.5)
DGO	2.4 ± 0.4	2.7 ± 0.4	2.6 ± 0.2	2.8 ± 0.3	2.5 ± 0.1	2.8 ± 0.4	2.8 ± 0.5
	(1.5-3.0)	(2.5-4.0)	(2.5-3.5)	(2.5-3.5)	(2.5-3.0)	(2.5-4.0)	(2.5-4.5)

Table 50. *(Continued.)*

Character	American beachgrass	Bitter panicum	Rice	Wheat	Oat	Bermudagrass	Zoysiagrass
Ant. end to median bulb valve	63.7 ± 2.4 (60-68)	62.3 ± 5.0 (55-80)	65.8 ± 2.6 (60-76)	65.4 ± 2.9 (56-70)	64.1 ± 3.0 (54-70)	66.2 ± 2.6 (60-72)	71.0 ± 2.5 (66-76)
Pharyngeal lobe base to ant. end	228.6 ± 17.5 (175-257)	216.6 ± 16.8 (170-265)	229.4 ± 15.5 (204-250)	231.7 ± 12.2 (192-264)	224.2 ± 14.5 (194-250)	231.8 ± 11.1 (204-248)	249.3 ± 16.4 (210-280)
W	18.1 ± 0.9 (16-20)	17.2 ± 0.8 (16-20)	17.5 ± 0.7 (16-20)	17.5 ± 0.8 (16-18)	16.4 ± 0.7 (16-18)	17.4 ± 0.6 (16-18)	18.6 ± 1.0 (16-20)
Tail	96.6 ± 5.3 (85-107)	97.4 ± 5.7 (85-106)	94.1 ± 5.2 (85-104)	96.5 ± 5.3 (85-107)	87.4 ± 2.9 (83-95)	93.4 ± 4.7 (83-102)	105.3 ± 4.3 (96-115)
Hyaline region	19.9 ± 1.6 (16-23)	20.9 ± 2.7 (16.5-27)	19.5 ± 2.2 (15-24)	19.6 ± 1.6 (16-23)	16.7 ± 1.5 (15-21)	18.8 ± 2.1 (15-22.5)	21.5 ± 1.6 (18-26)

located posterior to stylet, sometimes at level of or above retracted stylet base. Pharynx well developed, procorpus long, cylindrical and median bulb spherical with heavily sclerotised valve. Cuticle thick at neck and mid-body. Perineal pattern usually with high to rounded arch with shoulders and widely spaced lateral line interrupting transverse striations, vulva and anus sunken with coarse broken striae around and near vulval area. Phasmids small, nor prominent, seen only in some specimens.

Male

Body slender, vermiform, tapering gradually towards either end. Labial region slightly offset, labial disc in face view slightly rounded, raised above crescentic median labials extending some distance into labial region. Each median labial with one pair of faint labial sensilla, marked by small cuticular depressions. Amphidial openings appearing as long slits. Cuticular annulation distinct. Annuli 2.5-3.5 μm wide at mid-body, becoming smaller towards ends of body. Mid-body diam. averaging 44 μm. Lateral field with four incisures, areolated. Stylet heavy, knobs massive, rounded, sloping posteriorly. Hemizonid prominent, about two annuli long, located one annulus posterior to excretory pore. Excretory pore usually near middle of basal pharyngeal bulb, in some specimens more posteriorly. Phasmids about 6 μm from tail terminus. Spicules arcuate, tips rounded, gubernaculum short, simple. Tail rounded.

J2

Body cylindrical, vermiform, tapering at both extremities, more so posteriorly. Labial region truncate, slightly offset, without annulation, labial framework weak. Labial disc slightly raised above median labials, lateral labials in same contour as labial region. Labial disc and median labials fused, dumbbell-shaped in face view. Stylet large, heavy, knobs rounded, usually sloping posteriorly. Body annulation fine, distinct. Lateral field with four incisures, areolated. Excretory pore posterior to nerve ring in isthmus region. Hemizonid located 1-2 annuli posterior to excretory pore. Rectum dilated. Vesicles present in median bulb but difficult to observe in specimens mounted in glycerin. Vesicles occurring in same position in median bulb as in *M. naasi* although their number and size are slightly smaller (Karssen, 1996b). Tail very long, tapering gradually to finely rounded or variable shaped terminus. Morphometrics of J2 from *Panicum amarulum* and *Ammophila breviligulata* from

Delaware, and from the populations obtained from cultures originating from these plants at Beltsville showed both short- and long-tail forms (Table 50).

TYPE PLANT HOST: American beachgrass, *Ammophila breviligulata* Fern.

OTHER PLANTS: Reproduced on wheat (*Triticum aestivum* 'Coldwell' and 'F1 302'), rice (*Oryza sativa* 'Lemont'), oat (*Avena sativa* 'Fl 501'), bitter panicum (*Panicum amarulum*), bermudagrass (*Cynodon dactylon*), zoysiagrass (*Zoysia japonica*), and St Augustinegrass (*Stenotaphrum secundatum*). Corn (*Zea mays*) was not a host (Handoo *et al.*, 1993).

TYPE LOCALITY: Henlopen State Park in Delaware, USA.

DISTRIBUTION: *North America*: USA (Delaware, Maryland) (Handoo *et al.*, 1993).

PATHOGENICITY: *Meloidogyne sasseri* did not cause galling on the hosts in tests, the females protruding from the roots with a large egg mass attached at the posterior end.

CHROMOSOME NUMBER: No available data.

POLYTOMOUS KEY CODES: *Female*: A12, B32, C324, D5; *Male*: A213, B32, C21, D1, E1, F2; *J2*: A12, B21, C1, D1, E3, F1.

BIOCHEMICAL AND MOLECULAR CHARACTERISATION: Sequence of a partial heat shock protein 90 gene (AF457581) was deposited for this species and is different from those of all other *Meloidogyne* spp. (Skantar & Carta, 2004).

RELATIONSHIPS (DIAGNOSIS): *Meloidogyne sasseri* is morphologically similar to *M. graminis*, *M. spartinae* and *M. californiensis*. From *M. graminis* it differs by longer J2 body length (470-650 *vs* 365-510 μm), longer stylet (13.0-14.5 *vs* 9.0-13.5 μm), and longer tail (83-115 *vs* 52-88 μm). From *M. spartinae* it differs by shorter J2 body length (470-650 *vs* 612-912 μm), location of hemizonid in relation to excretory pore at 4 *vs* 20 μm posterior to excretory pore, tail tapering to a finely rounded terminus *vs* asymmetrical, spiked bulbous tail; in females, the perineal pattern has a high to rounded arch with occasional shoulders, widely spaced lateral line interrupting transverse striations, sunken vulva and anus and coarse broken striae around and near the anal area *vs* perineal pattern often indistinct with a more truncate pattern above anal area without any shoulders and no lateral line and (or) coarse lines. *Meloidogyne sasseri* differs from *M. californiensis* primarily in having females with longer body, no posterior cuticular protuberance and perineal pattern with a high to rounded arch with shoulders, widely spaced lateral line interrupting the transverse striae, sunken vulva and anus, and coarse broken striae around and near the anal area.

85. *Meloidogyne sewelli* Mulvey & Anderson, 1980
(Figs 232, 233)

COMMON NAME: Spike-rush root-knot nematode.

Specimens were collected in Canada from root galls on spike-rush (*Eleocharis acicularis* (L.) R & S) growing in shallow soil on a rocky substrate about 3 m from the Ottawa River on the Quebec side.

MEASUREMENTS (*AFTER MULVEY & ANDERSON, 1980*)

- *Holotype female*: L = 490 μm; W = 350 μm; neck length = 115 μm; stylet = 15 μm; DGO = 7 μm.
- *Allotype male*: L = 1500 μm; stylet = 18 μm; DGO = 5 μm; spicules = 30 μm; gubernaculum = 8 μm; tail = 8 μm; a = 42; b = 8.0; c = 90.
- *Paratype females* (n = 20): L = 442 (410-510) μm; W = 266 (200-350) μm; neck length = 93 (70-115) μm; stylet = 14.5 (14.0-15.0) μm; stylet knob width = 3.5-4.0 μm; DGO = 4.2 (7-8) μm; vulva length = 23-29 μm; vulva to anus = 15-17 μm; EP/ST = 1.0-1.3.
- *Paratype males* (n = 20): L = 1400 (1200-1650) μm; stylet = 19 (18-20) μm; stylet knob width = 4.0; DGO = 5-6 μm; excretory pore to anterior end = (40-50) μm; centre of median bulb from anterior end = 83-92 μm; tail = 15-18 μm; spicules = 29 (28-30) μm; gubernaculum = 8-9 μm; a = 40 (36-44); b = 7.9 (7.5-8.3); c = 80-92.

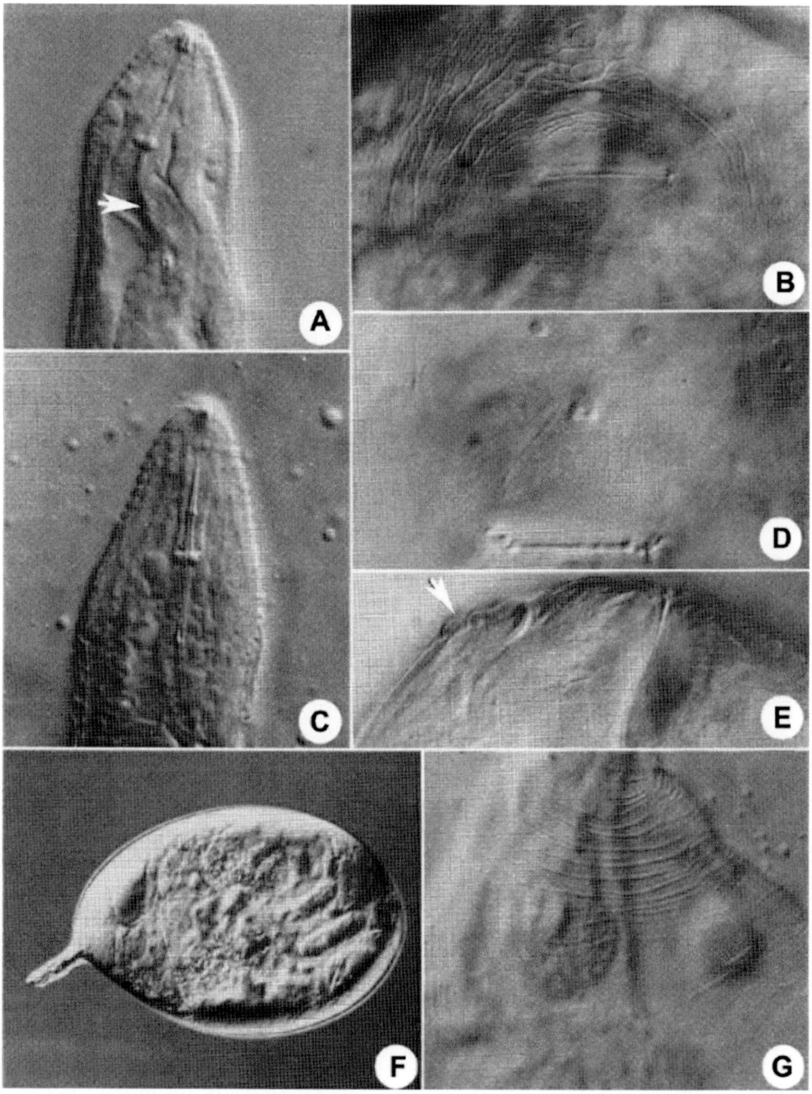

Fig. 232. Meloidogyne sewelli. *LM. A, C: Female anterior region (dorsal gland orifice arrowed); B, D, E, G: Perineal pattern (anus arrowed); F: Entire female. (After Mulvey & Anderson, 1980.)*

- *Paratype J2* (n = 40): L = 505 (460-540) μm; W = 13-15 μm; pharyngeal lobe base to anterior end = 177 (168-190) μm; stylet = 12 (11-13) μm; DGO = 7-8 μm; excretory pore to anterior end = 15-

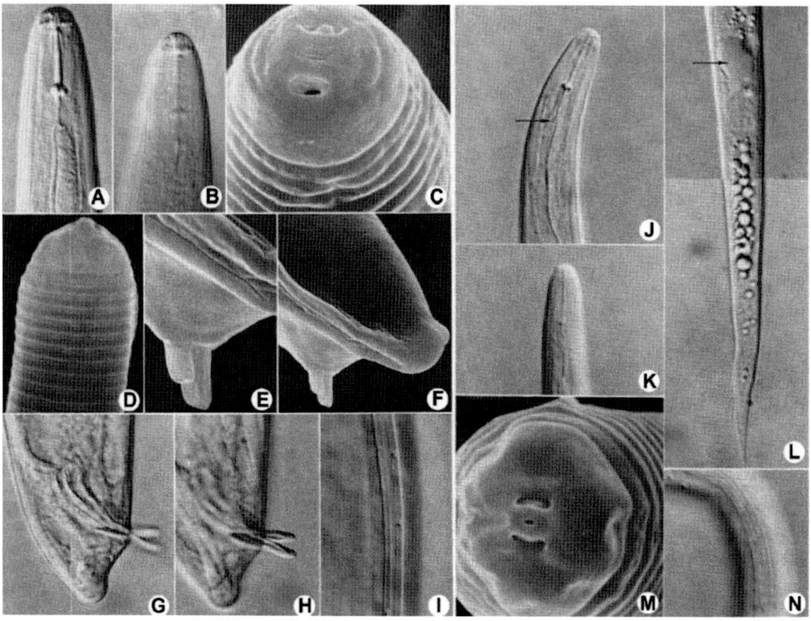

Fig. 233. Meloidogyne sewelli. *LM and SEM. A-D: Male labial region; E-H: Male tail; I: Male lateral field; J, K, M: Second-stage juvenile (J2) labial region (dorsal gland orifice arrowed); L: J2 tail (inflated rectum arrowed); N: J2 lateral field. (After Mulvey & Anderson, 1980.)*

24 μm; centre of median bulb from anterior end = 70 (68-72) μm; tail = 74 (68-78) μm; a = 39 (37-41); b = 2.9 (2.8-3.2); c = 7.3 (6.6-7.9).

- *Eggs* (n = 20): L = 93 (85-100) μm; W = 42 (40-45) μm.

Description

Female

Body pearly white, globular to pear-shaped, without a posterior protuberance. Cuticle thin, 2-3 μm thick at mid-body. Stylet slender, knobs small, rounded. Excretory pore usually slightly posterior to base of stylet. Perineal pattern often indistinct with rounded dorsal arch and many discontinuous, closely spaced striae. Phasmids large, closely spaced. Tail terminus marked by striae. Egg sac attached to mature female, 2-3 times size of female and containing several hundred eggs. Eggshell hyaline, without markings.

Male

Labial region slightly offset, labial cap prominent in dorsoventral view, labial annuli two or three, weakly defined. Lateral field with four incisures, not areolated. Stylet knobs rounded. Cephalids not observed. Testis outstretched. Spicules arcuate, tips bluntly rounded. Gubernaculum short, linear. Tail short, distinctively subdigitate.

J2

Labial region slightly offset from body, bearing two or three faint annuli, oral disc poorly delineated, amphids cylindroid, large, about four times length of oral opening. Stylet slender with medium-sized knobs. Hemizonid not observed. Lateral field with four incisures, not areolated. Rectum always dilated. Phasmid small, 10-15 μm posterior to anus. Tail long, tapering to narrowly rounded terminus.

TYPE PLANT HOST: Spike-rush, *Eleocharis acicularis* (L.) R & S.

OTHER PLANTS: No other hosts were reported.

TYPE LOCALITY: Shore of the Ottawa River on the outskirts of Deschênes, Quebec, Canada.

DISTRIBUTION: *North America*: Canada (Quebec).

PATHOGENICITY: Numerous galls on roots of length of 1.5-4.0 mm, containing from 2-10 females and about an equal number of males, and with third- and fourth-stage juveniles.

CHROMOSOME NUMBER: No available data.

POLYTOMOUS KEY CODES: *Female*: A3, B23, C3, D1; *Male*: A32, B3, C2, D1, E2, F2; *J2*: A12, B23, C12, D23, E3, F1.

BIOCHEMICAL AND MOLECULAR CHARACTERISATION: No available data.

RELATIONSHIPS (DIAGNOSIS): *Meloidogyne sewelli* differs from other described species by its perineal pattern (female group 1, Jepson, 1987).

The body length of *M. sewelli* J2 is similar to those of *M. graminis*, *M. graminicola*, *M. oryzae*, *M. naasi*, and *M. spartinae*. It differs from these species by the more posterior position of DGO (7-8 μm posterior to the stylet *vs* less than 7 μm) and by a different perineal pattern. It is further differentiated from *M. graminis*, *M. graminicola*, and *M. naasi* in having a dilated rectum *vs* not dilated, and from *M. graminis* and *M. oryzae* by the EP/ST ratio.

86. *Meloidogyne silvestris* Castillo, Vovlas, Troccoli, Liébanas, Palomares-Rius & Landa, 2009a

(Figs 234-236)

COMMON NAME: Holly root-knot nematode.

In August 2006 and 2007, a RKN was found parasitising the roots of European holly, *Ilex aquifolium*, at Arévalo de la Sierra, Soria province, northern Spain.

MEASUREMENTS *(AFTER CASTILLO ET AL., 2009A)*

- *Holotype female*: L = 583 μm; W = 366 μm; stylet = 20 μm; DGO = 6 μm; excretory pore from anterior end = 20 μm; vulval slit length = 17 μm; distance from vulval slit to anus = 19 μm; EP/ST = 1.0.
- *Paratype females* (n = 20): L = 531 ± 77 (401-557) μm; W = 319 ± 75 (189-424) μm; stylet = 19.5 ± 0.5 (19-20) μm; anterior end to excretory pore = 15.5 ± 3.5 (13-22) μm; vulval slit = 15.8 ± 1.6 (14-19) μm; vulva to anus = 19.1 ± 1.6 (15-24) μm; a = 1.6 ± 0.3 (1.3-2.2); EP/ST ratio = 0.8 ± 0.2 (0.6-1.1).
- *Paratype males* (n = 12): L = 1864 ± 184 (1565-2141) μm; W = 38 ± 3.2 (30-42) μm; stylet = 25.6 ± 1.1 (24.0-27.3) μm; stylet knob width = 5.0 ± 0.4 (4.5-5.5) μm; DGO = 6.0 ± 0.5 (5.3-6.7) μm; excretory pore to anterior end = 158 ± 8.8 (145-176) μm; centre of median bulb from anterior end = 109 ± 6.1 (97-117) μm; tail = 10.5 ± 1.2 (8.5-12.5) μm; spicules = 33.3 ± 2.5 (28.7-38.0) μm; gubernaculum = 9.7 ± 1.3 (6.7-11.3) μm; testis length = 668 ± 125.1 (505-932) μm; a = 49.2 ± 4.5 (41.9-56.4); b = 12.1 ± 1.0 (10.6-13.5); b′ = 7.4 ± 1.1 (5.8-8.9); c = 178.4 ± 28.4 (146.7-234.1); c′ = 0.4 ± 0.1 (0.3-0.5).

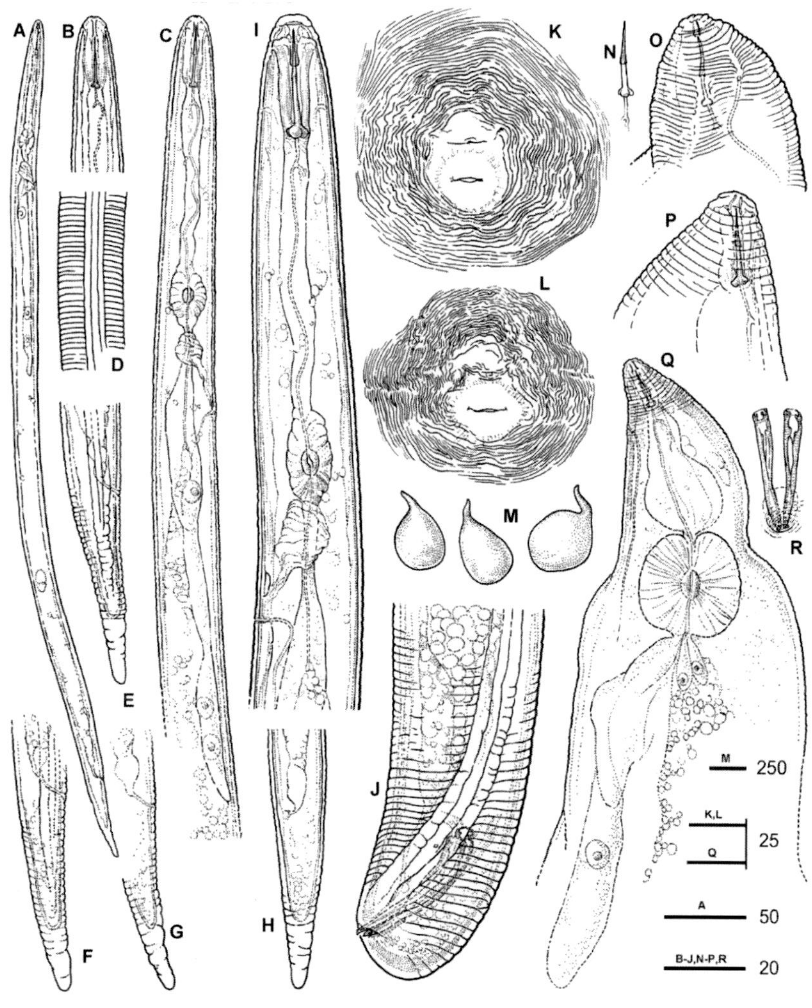

Fig. 234. Meloidogyne silvestris. *A: Entire second-stage juvenile (J2); B: J2 labial region; C: J2 anterior region; D: J2 lateral field; E-H: J2 tail; I: Male anterior region; J: Male tail; K, L: Perineal pattern; M: Entire female; N: Female stylet; O-Q: Female anterior region; R: Spicules. (Scale bars in microns.) (After Castillo* et al.*, 2009a.)*

- *Paratype J2* (n = 15): L = 506 ± 21.3 (465-539) μm; W = 16.5 ± 0.8 (15.3-18.0) μm; pharyngeal lobe base to anterior end = 190 ± 15 (150-215) μm; middle of median bulb to anterior end = 66 ± 5.2

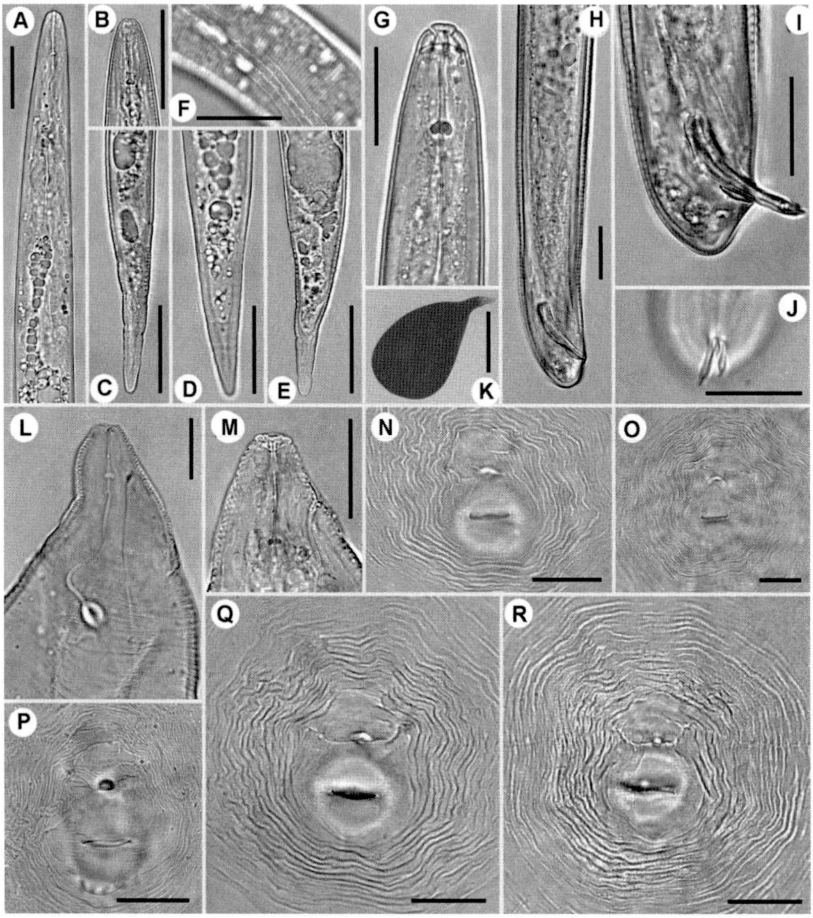

Fig. 235. Meloidogyne silvestris. *LM. A: Second-stage juvenile (J2) anterior region; B: J2 labial region; C-E: J2 tail; F: J2 lateral field; G: Male labial region; H-J: Male tail; K: Entire female; L, M: Female anterior region; N-R: Perineal pattern. (Scale bars: A-J, L-R = 25 μm; K = 250 μm.) (After Castillo* et al., *2009a.)*

(53-72) μm; stylet = 12.8 ± 0.5 (12.3-14.3) μm; stylet knob width = 2.4 ± 0.3 (2.0-2.7) μm; DGO = 3.0 ± 0.6 (2.0-4.0) μm; excretory pore to anterior end = 94 ± 4.9 (84-104) μm; tail = 44 ± 2.8 (37.3-48.7) μm; hyaline region = 15.3 ± 1.2 (14.0-18.7) μm; a = 30.8 ± 1.5 (27.9-32.9); b = 5.2 ± 0.2 (4.7-5.5); b′ = 2.7 ± 0.3 (2.4-3.4); c = 11.5 ± 0.7 (10.4-13.7); c′ = 4.0 ± 0.4 (3.3-4.7).

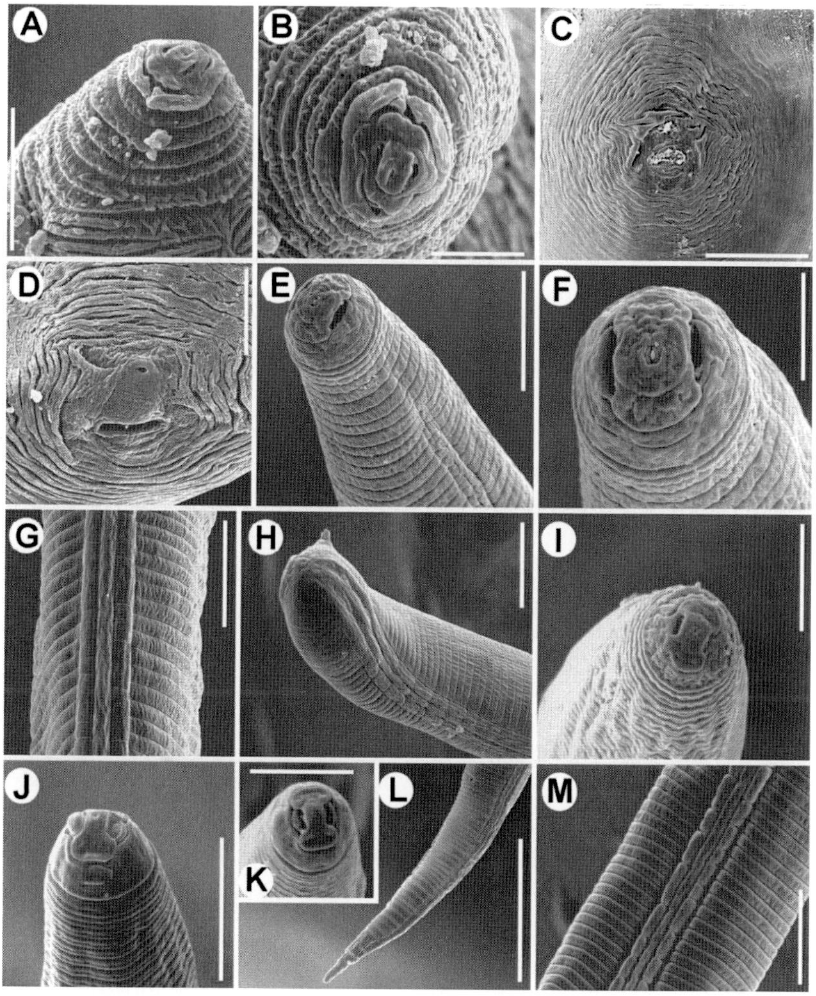

Fig. 236. Meloidogyne silvestris. *SEM. A, B: Female labial region; C, D: Perineal pattern; E, F: Male labial region; G: Male lateral field; H: Male tail; I-K: Second-stage juvenile (J2) labial region; L: J2 tail; M: J2 lateral field. (Scale bars: A, E, G = 10 μm; B, F, I-K, M = 5 μm; C = 50 μm; D, H, L = 20 μm.) (After Castillo* et al.*, 2009a.)*

Description

Female

Body usually completely embedded in galled tissue, pearly white, globose or, rarely, pear-shaped, with long neck but no posterior protuberance. Labial region slightly offset from body. Labial cap variable in shape, with labial disc and postlabial annulus elevated. In SEM view, labial disc appearing rounded-squared, slightly raised on median and lateral sectors, which are all fused together. Labial framework weakly sclerotised. Stylet fairly long, with an almost straight, rarely curved, cone, cylindrical shaft and knobs transversely ovoid, sometimes with concave anterior surfaces, posteriorly sloping in several specimens. Excretory pore usually at level of stylet knobs, or a few body annuli anterior (rarely posterior). Pharyngeal gland with a large mononucleate dorsal lobe and two subventral gland lobes, usually difficult to see. Perineal pattern mostly rounded, dorsal arch generally low, with fine, sinuous cuticle striae becoming coarser in vicinity of perivulval region, lateral field not clearly visible. In a few specimens, striae forming two wings or shoulders ending near lateral field, which, in this case, are made more visible by fine and small zigzag striae. Phasmids distinct, located just anterior to anus level. Vulval slit in middle of unstriated area, slightly shorter than vulva-anus distance, anal fold clearly visible when present. Large egg sac commonly occurring outside root gall, containing up to 400 eggs.

Male

Body vermiform, tapering anteriorly; tail rounded, with twisted posterior body portion. Labial region slightly offset, 6.1 (5.3-6.7) μm high, 12.2 (11.3-13.3) μm diam., with large labial annulus. Prominent slit-like amphidial openings between labial disc and lateral labials. In SEM view, labial disc slightly narrower and raised above merged subventral and subdorsal median labial sectors, with a central oval prestoma into which opens a slit-like dorsoventrally oriented stoma, lateral labials reduced to a very narrow strip, largely fused, in middle part, with postlabial annulus. Amphidial apertures large, dorsoventrally elongated, just posterior to lateral edge of labial disc. Distinct slit-like labial sensilla visible in median labials. Postlabial annulus marked by usually short, incomplete incisures, sometimes obliquely oriented. Labial framework strongly sclerotised, vestibule extension distinct. Stylet straight, with cone and shaft broadening slightly in distal part. Stylet knobs mostly rounded, sometimes with angular edges, laterally or obliquely directed, merging gradually with base of shaft. Body annulation distinct, 1.5 (1.3-1.7) μm wide. Lateral field with four incisures, outer bands areolated, forming blocks extending for 2-3 body annuli. Testis single, monorchic, rather long and outstretched. Spicules strong, slightly curved ventrally, with their proximal end often showing a minute, lateral apophysis. In one specimen, a small terminal process was observed at tip of each spicule. Gubernaculum simple, almost straight. Tail slightly curved ventrally, short, with bluntly rounded tip and surrounded by lateral fields joining at bottom of tail terminus. Phasmids small, located a few body annuli anterior to cloacal aperture.

J2

Body vermiform, tapering posteriorly with a rather short and stout tail. Anterior end elevated, subhemispherical, labial region continuous with body contour, 6.0 (5.3-6.7) μm high, 12.8 (12.3-14.3) μm wide. In SEM view, labial disc appearing oval to rectangular in shape, raised above median labials with which it merges in a dumbbell-shaped structure. Slit-like stoma centrally located, surrounded by six pore-like openings of inner labial sensilla. Lateral labials small and narrow, lower than labial disc, delimiting wide and elongate amphidial apertures. Postlabial annulus smooth, sometimes with short, horizontal incisures, body annuli distinct but fine. Lateral field beginning at level of procorpus as two areolated lines, near median bulb third line beginning and shortly after splitting to form three bands delimited by four lines, running entire length of body until ending, as an irregularly areolated field, near hyaline region. Stylet delicate, with cone straight, narrow, sharply pointed, shaft almost cylindrical, knobs small, rounded, separate, laterally directed. Pharynx with a long, cylindrical procorpus, rounded-oval median bulb, short isthmus and rather long gland lobe, with three equally-sized nuclei and overlapping intestine ventrally. Pharyngo-intestinal junction just anterior to excretory pore level. Excretory pore distinct, at level with distal end of isthmus, hemizonid located just anterior, extending for two additional body annuli. Tail conoid, relatively short, subdigitate, with no constrictions in hyaline region, annulation fine, regular in proximal two-thirds, becoming coarser and irregular in distal part. Hyaline region clearly defined and fairly long, tail tip broadly rounded and smooth. Rectum dilated. Phasmids small, difficult to observe.

TYPE PLANT HOST: Holotype female and additional paratypes from a population extracted from soil samples and infected roots of European holly (*Ilex aquifolium*) collected from a woodland at Arévalo de la Sierra (Soria province), northern Spain.

OTHER PLANTS: Not detected. Tomato is not a host for this species.

TYPE LOCALITY: Woodland at Arévalo de la Sierra (Soria province), northern Spain.

DISTRIBUTION: *Europe*: Spain (Castillo *et al.*, 2009a).

PATHOGENICITY: No disease symptoms could be observed on the stems or leaves of nematode-infected European holly trees as compared to non-infected ones. However, holly trees infected by the nematode showed some decline and low growth. Infected feeder roots were distorted and showed numerous root galls of large (8-10 mm) to moderate (2-3 mm) size. Galls occurred either singly or in clusters on the root. Generally, large, irregular galls were present on root tips, but were also present along the root axis. In many cases, up to five mature globose females were found associated with the largest galls.

CHROMOSOME NUMBER: No available data.

POLYTOMOUS KEY CODES: *Female*: A3, B1, C43, D3; *Male*: A12, B12, C21, D1, E1, F2; *J2*: A12, B21, C3, D21, E21, F34.

BIOCHEMICAL AND MOLECULAR CHARACTERISATION: The isozyme electrophoretic analysis of five-specimen groups of young egg-laying females of *M. silvestris* revealed one very slow weak A1 Est band after prolonged staining and a N1c Mdh phenotype with strong and very weak-staining bands (Castillo *et al.*, 2009a). The partial 18S rRNA, ITS1-5·8S-ITS2 and the D2-D3 of 28S rRNA genes were sequenced for this species by Castillo *et al.* (2009a).

RELATIONSHIPS (DIAGNOSIS): The species belongs to Molecular group VII and is clearly molecularly differentiated from all other *Meloidogyne* species. The female perineal pattern morphology of *M. silvestris* places it in Jepson's group 3 (Jepson, 1987) and the EP/SP is rather short. It is morphometrically closer to *M. ardenensis*, from which it differs by perineal pattern shape (roundish with fine striae *vs* oval with coarse striae and high dorsal arch in *M. ardenensis*), longer J2 body (465-539 *vs* 365-453 μm), hemizonid anterior *vs* posterior to excretory pore. The two species also show different isozyme patterns. *Meloidogyne silvestris* is also similar to *M. lusitanica*, from which it differs mainly in the morphology of the perineal patterns (strikingly trapezoid with medium-high dorsal arch in *M. lusitanica*), length of J2 stylet (12.3-14.3 *vs* 13-16 μm in *M. lusitanica*), J2 hyaline region (14.0-18.7 *vs* 10-14 μm), and spicule length (28.7-38.0 *vs* 32.0-44.5 μm).

87. *Meloidogyne sinensis* Zhang, 1983
(Fig. 237)

COMMON NAME: Chinese potato root-knot nematode.

This RKN was collected from roots of potato. Infected potatoes were planted in an experimental field that was previously used to test Chinese medicinal herbs.

MEASUREMENTS *(AFTER ZHANG, 1983)*

- *Paratype females* (n = 30): L = 790 (629-1496) μm; W = 598 (416-768) μm; stylet = 19.2 (16.5-24.5) μm; stylet knob height = 3.2-4.5 μm; DGO = 5.3 (4.3-7.0) μm; anterior end to excretory pore = about 40 μm; interphasmid distance = 19.2-24.0 μm; EP/ST = 2.1.

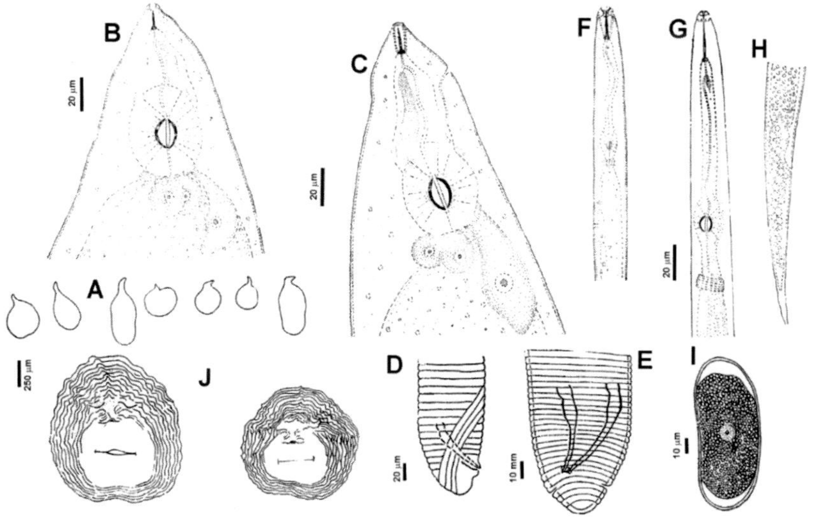

Fig. 237. Meloidogyne sinensis. *A: Entire females; B, C: Female anterior region; D, E: Male tail; F: Male anterior region; G: Second-stage juvenile (J2) anterior region; H: J2 tail; I: Egg; J: Perineal pattern. (After Zhang, 1983.)*

- *Paratype males* (n = 12): L = 1730 (800-2460) μm; W = 40 (24-60) μm; stylet = 28.0 (24-32) μm; stylet knob height = 2.4-5.2 μm; stylet knob width = 4.9-7.5 μm; labial region height = 1.7-2.0 μm, DGO = 5.4 (4.0-8.0) μm; spicules = 31.8 (27.2-40.0) μm; gubernaculum = 9.1 (7.5-12.8) μm; a = 57.2 (50.3-64.4); b = 16.1 (15.9-16.5).
- *Paratype J2* (n = 26): L = 536.1 (481.0-590.4) μm; W = 18.5 (16.0-22.8) μm; stylet = 18.8 (17.6-19.2) μm; DGO = 4.7 (3.7-5.8) μm; tail = 65.2 (52.0-76.2) μm; a = 28.5 (24.1-35.7), b = 7.1 (6.3-7.7), c = 8.2 (7.0-11.1), c′ = 5.2 (4.6-6.3).
- *Eggs*: L = 108 (80-123) μm; W = 43 (35-49) μm.

DESCRIPTION

Female

Body white, pear-shaped to globular. Excretory pore distinct, variable in position, excretory pore to labial region distance usually about 40 μm. Perineal pattern more or less circular, dorsal arch low and rounded. Striae in dorsal sector smooth and continued. Often there are sawtooth-like striae from near centre and both sides of dorsal sectors. Lateral lines not distinct, some striae short and irregular near lateral lines, sometimes dorsal and ventral striae meeting at a small angle at lateral lines, numerous short striae over anus, cuticle near anus forming a distinct fold. Striae in ventral sector smooth, continuous, and wavy, striae near edge of perineal pattern are distinctly wavy, sometimes striae in ventral sector extending and meeting with striae in dorsal sector to form a 'wing'.

Male

Body vermiform, tapering at anterior end, tail end short and rounded. Lateral cheek length 6.4 μm (6.9), labial region annulated, usually 3-4 annuli. Hemizonid length 5.9 μm and 7 μm anterior to excretory pore or about 3-4 annuli. Four incisures in lateral field. Two spicules of same shape and size, arcuate, with rounded base and tapering tip. Gubernaculum short, arched in lateral view. Phasmids distinct and symmetric.

J2

Hemizonid length 2.4-7.2 μm, very near and posterior to excretory pore. Rectum dilated, posterior of rectum narrow tube-shaped. Tail tip broad.

TYPE PLANT HOST: Potato, *Solanum tuberosum* L.

OTHER PLANTS: No other hosts were reported.

TYPE LOCALITY: Jinan city, Shandong province, China.

DISTRIBUTION: *Asia*: China (Zhang, 1983).

PATHOGENICITY: The galls are generally small and usually have secondary roots below or above the galls instead of coming directly from the galls. Egg masses are brown and coarse with many grains of sand adhering. Distinct galls on highly infected roots can totally change the shape of the roots. All developmental stages of *M. sinensis* are found with the galls of highly infected roots. No trace of gall formation may show on slightly infected roots, but all developmental stages of *M. sinensis* were also present.

CHROMOSOME NUMBER: No available data.

POLYTOMOUS KEY CODES: *Female*: A21, B12, C2, D3; *Male*: A2134, B12, C21, D1, E2, F1; *J2*: A12, B1, C213, D2, E32, F213.

BIOCHEMICAL AND MOLECULAR CHARACTERISATION: No available data.

RELATIONSHIPS (DIAGNOSIS): This species is close to *M. acronea* and *M. petuniae*, from which it differs by the female stylet length at 16.5-24.5 *vs* 9.9-14.0 and 12.9-16.5 μm, respectively, and several morphometrics in the male and J2. It also resembles *M. vandervegtei*, but differs in the J2 tail length.

88. *Meloidogyne spartelensis* Ali, Tavoillot, Mateille, Chapuis, Besnard, El-Bakkali, Cantalapiedra-Navarrete, Liebanas, Castillo & Palomares-Rius, 2015

(Figs 238-240)

COMMON NAME: Spartel root-knot nematode.

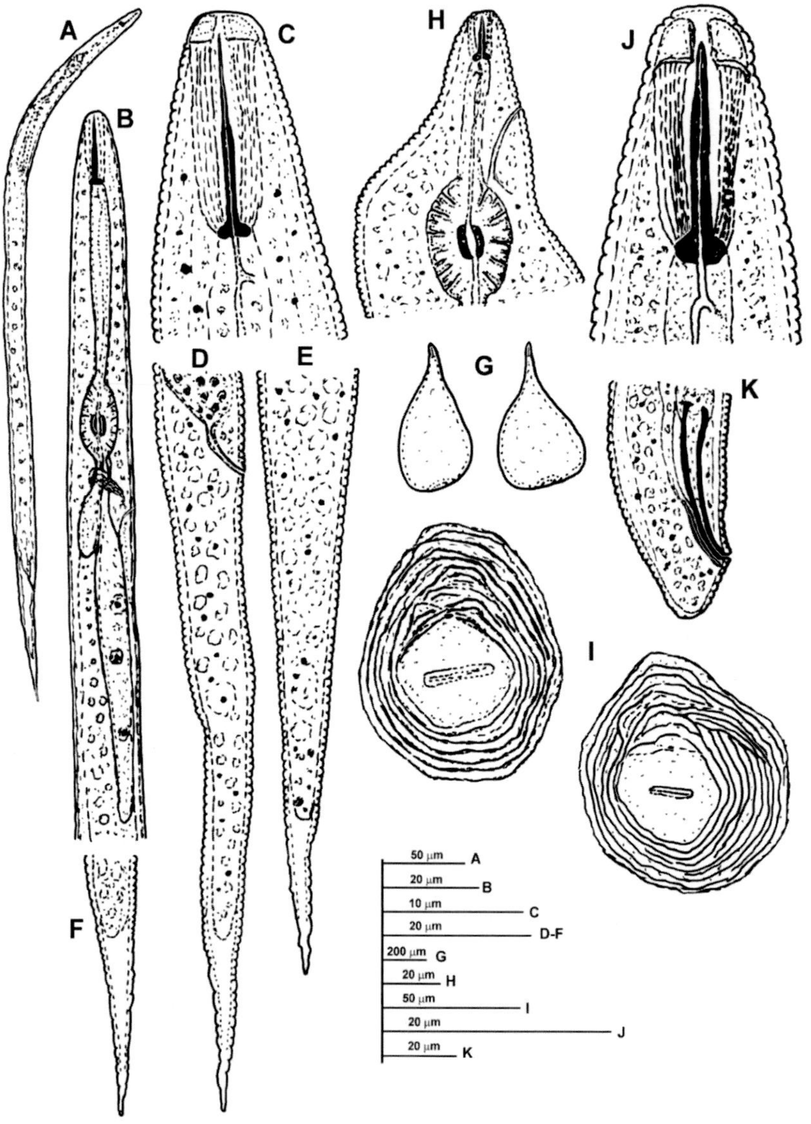

Fig. 238. Meloidogyne spartelensis. *A: Entire second-stage juvenile (J2); B: J2 anterior region; C: J2 labial region; D-F: J2 tail; G: Entire female; H: Female anterior region; I: Perineal pattern; J: Male labial region; K: Male tail. (After Ali* et al., *2015.)*

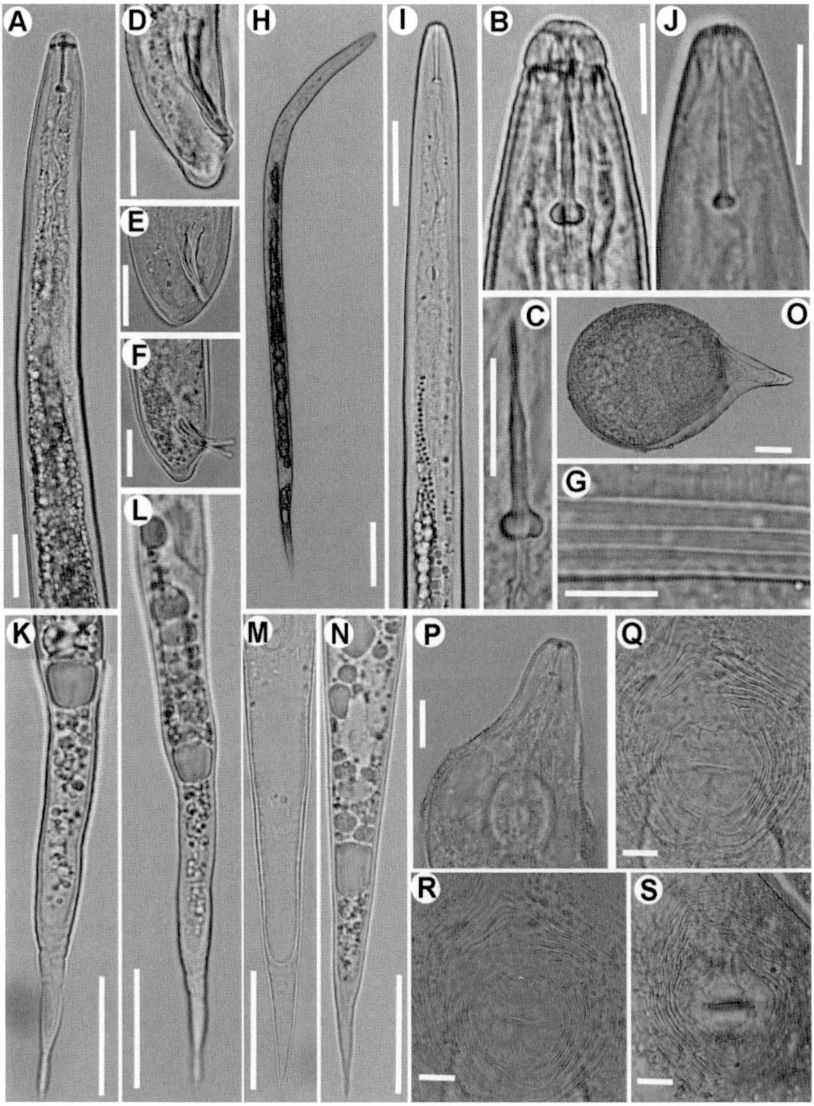

Fig. 239. Meloidogyne spartelensis. *LM. A: Male anterior region; B: Male labial region; C: Male stylet; D-F: Male tail; G: Male lateral field; H: Entire second-stage juvenile (J2); I: J2 anterior region; J: J2 labial region; K-N: J2 tail; O: Entire female; P: Female anterior region; Q-S: Perineal pattern. (Scale bars: A, D-F, I, K-N, P-S = 20 μm; B, C, G, J = 10 μm; H = 50 μm; O = 100 μm.) (After Ali* et al., *2015.)*

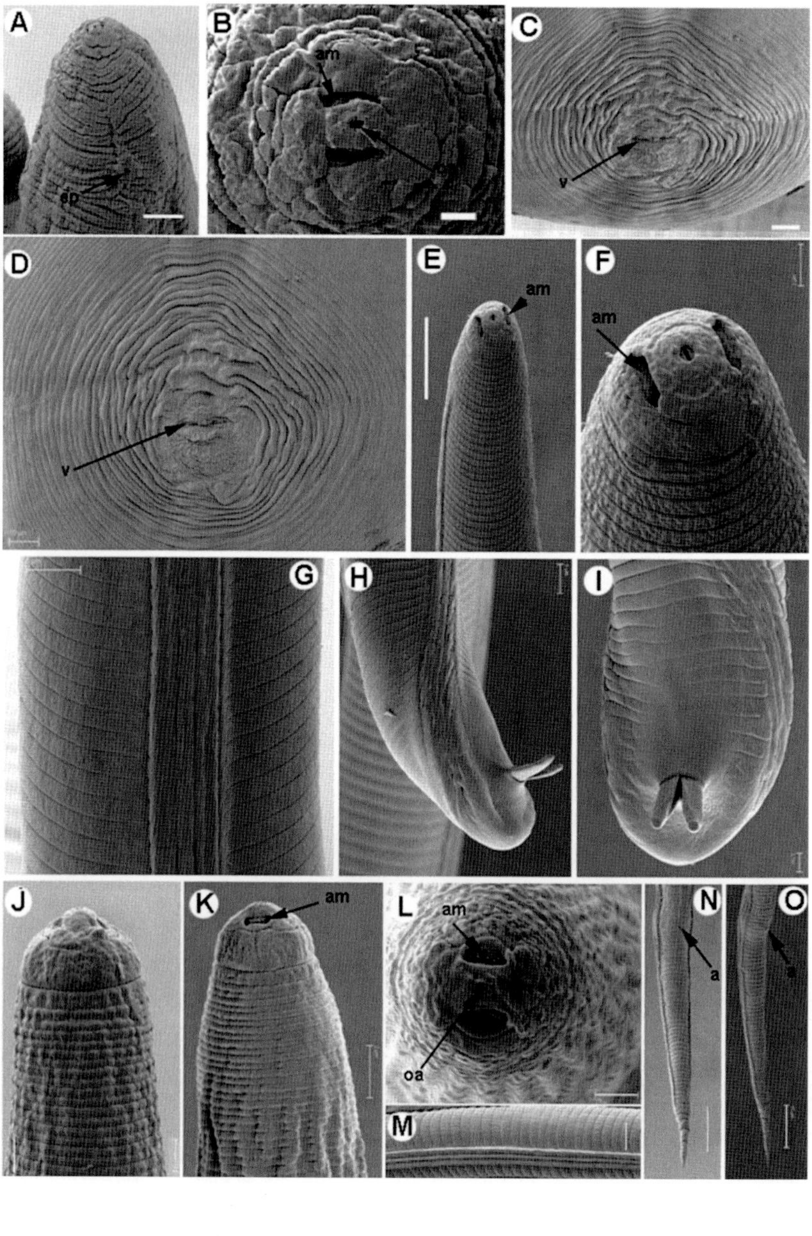
A
ep
B
am
C
v
D
v
E
am
F
am
G
H
I
J
K
am
L
am
oa
M
N
a
O

Fig. 240. Meloidogyne spartelensis. *SEM. A, B: Female anterior region; C, D: Perineal pattern; E, F: Male labial region; G: Male lateral field; H, I: Male tail; J-L: Second-stage juvenile (J2) labial region; M: J2 lateral field; N, O: J2 tail. Abbreviations: a = anus; am = amphid; oa = oral aperture; v = vulva. (Scale bars: A, G, H, M = 5 μm; B, J, L = 1 μm; C-E, N, O = 10 μm; F, I, K = 2 μm.) (After Ali* et al.*, 2015.)*

Specimens were collected from soil and roots associated with wild olive in northern Morocco.

MEASUREMENTS *(AFTER ALI* ET AL.*, 2015)*

- *Holotype female*: L = 712 μm; W = 409 μm; stylet = 14.5 μm; DGO = 3.5 μm; excretory pore from anterior end = 21.5 μm; vulval slit length = 22 μm; vulval slit to anus = 26.5 μm; a = 1.5; EP/ST = 1.5.
- *Paratype females* (n = 16): L = 621 ± 89 (489-772) μm; W = 386 ± 45 (267-455) μm; stylet = 15.1 ± 0.3 (14.5-15.5) μm; anterior end to excretory pore = 23.1 ± 2.2 (21.0-30.0) μm; vulval slit = 22.0 ± 1.6 (18.0-25.0) μm; vulva to anus = 21.3 ± 2.2 (16.5-35.0) μm; a = 1.6 ± 0.1 (1.4-1.8); EP/ST ratio = 1.5 ± 0.1 (1.4-2.0).
- *Paratype males* (n = 9): L = 1497 ± 154 (1144-1633) μm; W = 33 ± 3.2 (28-37) μm; stylet = 18.6 ± 0.7 (17.5-19.5) μm; stylet knob width = 4.5 ± 0.6 (4.0-5.5) μm; DGO = 3.3 ± 0.6 (2.5-4.0); excretory pore to anterior end = 145 ± 9.7 (133-159); gland lobe base from anterior end = 194 ± 20.7 (167-226) μm; tail = 13.2 ± 1.8 (10.0-16.0) μm; spicules = 27.5 ± 1.8 (25.0-30.5) μm; gubernaculum = 6.9 ± 0.3 (6.5-7.5); a = 42.5 ± 2.1 (40.1-46.3); b′ = 7.3 ± 1.0 (6.4-9.1); c = 107.8 ± 20.9 (75.0-141.1); c′ = 0.5 ± 0.1 (0.4-0.6).
- *Paratype J2* (n = 15): L = 451 ± 25.1 (398-489); W = 15.3 ± 1.1 (14.0-17.5) μm; pharyngeal lobe base to anterior end = 141 ± 17.1 (109-165) μm; stylet = 14.0 ± 0.5 (13.0-14.5) μm; DGO = 2.1 ± 0.6 (1.5-3.0) μm; excretory pore to anterior end = 84 ± 12.7 (60-109) μm; tail = 78.8 ± 7.1 (69-93) μm; hyaline region = 28.1 ± 10.8 (19.5-46.0) μm; a = 28.9 ± 3.1 (22.7-34.9); b = 3.7 ± 0.9 (2.8-5.2); b′ = 3.3 ± 0.4 (2.7-3.8); c = 5.8 ± 0.5 (5.0-6.4); c′ = 7.3 ± 0.8 (6.3-8.8).

DESCRIPTION

Female

Body usually completely embedded in galled tissue, pearly white, varying in shape from ovoid to saccate and with a variable neck diam. and length. Labial region continuous with body contour. Labial cap variable in shape, with labial disc and postlabial annulus not elevated. In SEM view, labial disc appearing rounded-squared, slightly raised on median and lateral sectors, which are all fused together. Labial framework weakly sclerotised. Amphidial apertures elongated, located between labial disc and lateral labials. Stylet fairly long, with an almost straight, rarely curved, cone, cylindrical shaft and knobs oval and sloping posteriad, sometimes with concave anterior surfaces. Excretory pore usually at level of anterior end of procorpus. Pharyngeal gland with a large mononucleate dorsal lobe and two subventral gland lobes, usually difficult to see. Perineal pattern mostly rounded to oval, with moderately high dorsal arch that is mostly rounded and sometimes squarish and generally low, with fine, sinuous cuticle striae, which become coarser in vicinity of perivulval region; lateral field not clearly visible. However, in some specimens lateral field slightly marked. In a few specimens, striae forming two wings or shoulders ending near lateral field, which, in this case, are made more visible by fine and small zigzag striae. Phasmids distinct, located just anterior to anal level. Vulval slit in middle of unstriated area, almost as long as vulva-anus distance, anal fold clearly visible, but not always present. Punctations and striae absent in perineum. Large egg sac commonly occurring outside root gall, containing up to 350-450 eggs.

Male

Body vermiform, tapering anteriorly, tail rounded, with twisted posterior body portion. Labial region slightly offset from body and with a high labial cap. Labial framework strong and sclerotised. Prominent slit-like amphidial openings between labial disc and lateral labials. In SEM view, labial disc slightly narrower and raised above merged subventral and subdorsal median labial sectors, with a centred oval prestoma into which opens a slit-like dorsoventrally oriented stoma, lateral labials reduced to a very narrow strip, largely fused, in middle part, with postlabial annulus. Labial region high and lacking annulation. Stylet robust and straight, with cone and shaft broadening slightly in distal part. Stylet knobs mostly rounded, laterally or obliquely directed, merging gradually with the base of the shaft. Lateral field consisting of four incisures with areolations along body but only few actually crossing central field. Procorpus distinctly outlined, three times larger than median bulb. Median bulb ovoid, with strong valve apparatus. Excretory duct curved. Excretory pore distinct, usually located four to six annuli posterior to hemizonid. Normally only one testis extending anteriorly. Spicules of variable length, arcuate and with two pores clearly visible at tip. Gubernaculum distinct. Phasmids at level of cloacal aperture and located in central lateral field, showing slit-like openings.

J2

Body vermiform, tapering more towards posterior end than anterior. Labial region narrower than body and slightly offset. Labial cap slightly elevated. Labial framework weakly developed. Labial disc and median labials fused. In labial disc, stoma-like slit located in an ovoid prestoma and surrounded by six inner labial sensilla. In SEM view, labial disc appearing oval to rectangular in shape, raised above median labials, to which it merges in a dumbbell-shaped structure. Labial region smooth and lacking annulation. Amphidial apertures elongated and located between labial disc and lateral labials. Body annulated from anterior end to terminus. Lateral field consisting of four incisures, with areolations along body but only few extending across field. Stylet delicate, with cone straight, narrow, sharply pointed, shaft almost cylindrical, knobs small, rounded, separate from each other, laterally directed. Pharynx with a long, cylindrical procorpus (3.0-4.0 times length of median bulb), rounded-oval median bulb, short isthmus and rather long gland lobe, with three equally sized nuclei and overlapping intestine ventrally. Hemizonid located anterior to excretory pore, extending for *ca* two body annuli. Excretory pore located posterior to nerve ring. Excretory duct curved and discernible when reaching intestine. Rectum slightly dilated. Tail long, conoid, terminus usually pointed, with several constrictions in hyaline region. Tail annulation fine, regular in proximal two-thirds, becoming slightly coarser and irregular in distal part. Hyaline region clearly defined and long, phasmids small, difficult to observe.

TYPE PLANT HOST: *Meloidogyne spartelensis* was found in a loamy-clay soil associated with wild olive (*Olea europaea* L. subsp. *europaea* var. *sylvestris*) without evidence of root parasitism. However the RKN parasitised and reproduced on tomato (*Solanum lycopersicum* L.), the only known host, which is designated as type host.

OTHER PLANTS: No other hosts were reported.

TYPE LOCALITY: Cape Spartel, near Tangier, northern Morocco (35.790583°N; 5.924983°W; altitude 16 m a.s.l.).

DISTRIBUTION: *Africa*: Morocco (Ali *et al.*, 2015).

PATHOGENICITY: Pathogenicity on olive trees not demonstrated, but numerous galls were found on tomato roots.

CHROMOSOME NUMBER: No available data.

POLYTOMOUS KEY CODES: *Female*: A23, B2, C32, D3; *Male*: A32, B3, C23, D1, E1, F2; *J2*: A23, B21, C12, D1, E3, F1.

BIOCHEMICAL AND MOLECULAR CHARACTERISATION: The isozyme electrophoretic analysis of five-specimen groups of young egg-laying females of *M. spartelensis* revealed one strong Est band (Rm = 42.65) after prolonged staining and one N1b Mdh band. There are partial 18S, ITS1-5.8S-ITS2, D2-D3 region of 28S rDNA, and *COII*-16S rRNA region and partial *COI* mtDNA sequences characterising this species (Ali *et al.*, 2015).

RELATIONSHIPS (DIAGNOSIS): The species belongs to Molecular group II and is clearly molecularly differentiated from all other *Meloidogyne* species. The female perineal pattern morphology of *M. spartelensis* is roundish without marked lateral lines, which places it in Jepson's group 3. This species is similar to *M. kralli* and differs by EP/ST (1.4-2.0 *vs* 0.8 in *M. kralli*), longer J2 stylet (13.0-14.5 *vs* 10.5-11.5 μm), and shorter DGO for J2 (2.1 *vs* 4.4 μm), longer male (1144-1633 *vs* 947-1143 μm) and a longer distance from the anterior end to the excretory pore (133-159 *vs* 120-133.5 μm) in the male. *Meloidogyne spartelensis* is also similar to *M. dunensis*, from which it differs mainly by the perineal pattern (lateral field clearly visible on *M. dunensis*). The J2 stylet is longer (13.0-14.5 *vs* 11.0-12.5 μm) as is its hyaline region (19.5-

46.0 *vs* 9.5-16.5 μm). In males, c ratio (107.8 *vs* 280.6) and spicules (25.0-30.5 *vs* 29.0-38.0 μm) were smaller. The new species is also morphologically similar to *M. sewelli*, from which it differs mainly in higher female EP/ST ratio (1.4-2.0 *vs* 1.2), longer length of J2 stylet (13.0-14.5 *vs* 11.0-12.0 μm), shorter length of male gubernaculum (6.5-7.5 *vs* 8.0-9.0 μm) and shorter DGO (2.5-4.0 *vs* 5.0-6.0 μm).

89. *Meloidogyne spartinae* (Rau & Fassuliotis, 1965) Whitehead, 1968

(Figs 241-244)

COMMON NAME: Cordgrass root-knot nematode.

This species was first observed on roots of smooth cordgrass (*Spartina alterniflora* Loiselius) in Florida, USA, in 1958 and described as *Hypsoperine spartinae* from South Carolina, USA, by Rau & Fassuliotis (1965). Additional notes on the morphology of this species were given by Eisenback & Hirschmann (2001).

MEASUREMENTS

See Table 51.

DESCRIPTION

Female

Females oval to lemon-shaped with a well-defined, protruding neck usually situated toward one side and with a protruding perineal area. Cuticle thin and annulated. Labial region offset with an indistinct labial framework and a distinct vestibule and extension. Stylet with small, rounded knobs, which usually slope posteriorly but may vary in shape to slightly angular, often with one or more anterior surfaces slightly indented. Excretory pore always close to anterior end, usually at level of stylet knobs. Pharyngeal structure typical for *Meloidogyne*. Perineal pattern with a dorsal arch low and rounded with striae that fork slightly near lateral lines. Lateral lines distinct, starting from area of tail tip. Tail remnant rounded and without striae, may protrude from perineal pattern. Striae encircling tail tip. Vulva often characteristically shaped at a slightly lower level, always wide open in prepared slides. At a

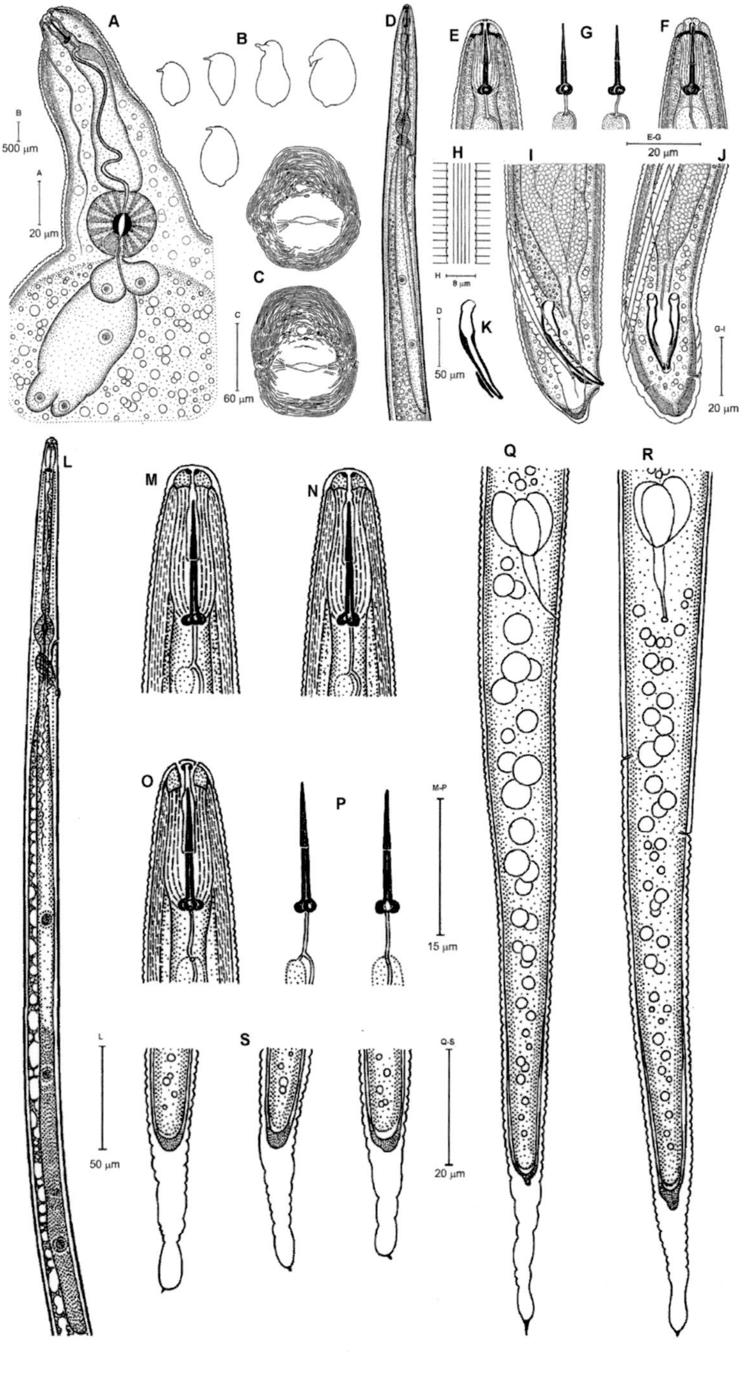
A
B
C
D
E
F
G
H
I
J
K
L
M
N
O
P
Q
R
S
500 µm
20 µm
60 µm
50 µm
8 µm
20 µm
15 µm
50 µm
20 µm

Fig. 241. Meloidogyne spartinae. *A: Female anterior region; B: Entire female; C: Perineal pattern; D: Male anterior region; E, F: Male labial region; G: Male stylet; H: Male lateral field; I, J: Male tail; K: Spicules; L: Second-stage juvenile (J2) anterior region; M-O: J2 labial region; P: J2 stylet; Q-S: J2 tail. (After Eisenback & Hirschmann, 2001.)*

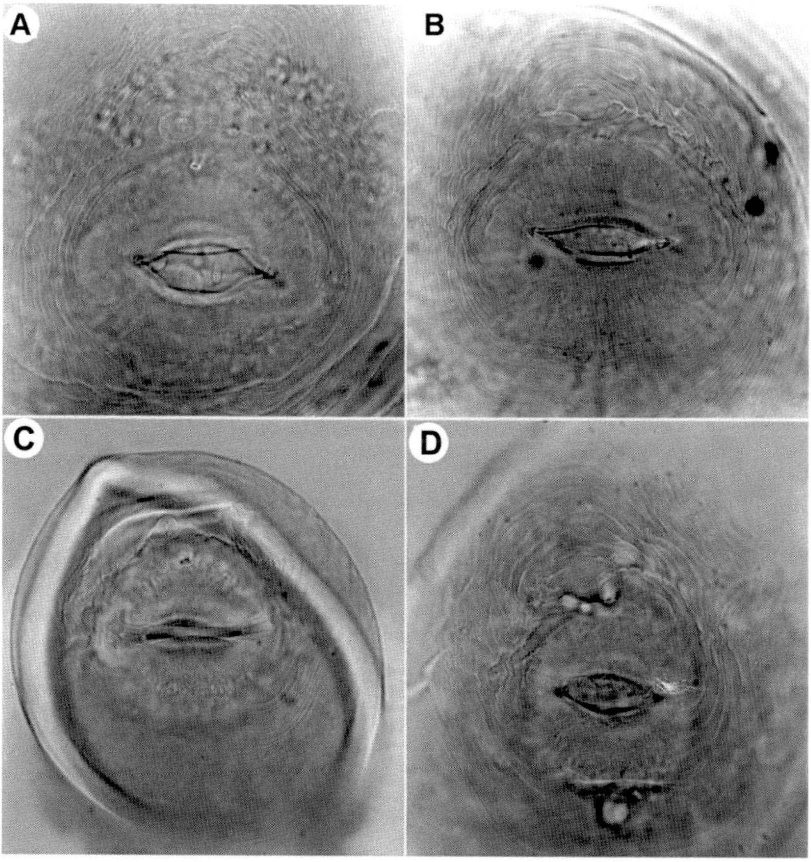

Fig. 242. Meloidogyne spartinae. *LM. A-C: Perineal pattern. (After Eisenback & Hirschmann, 2001.)*

higher focal level, vulva appearing more rectangular where perivulval striae extend to form vulval edges, otherwise perivulval area free of striae. Anus often covered by a cuticular fold. Striae fine and very faint, mostly continuous but with some breaks. A well-defined tail

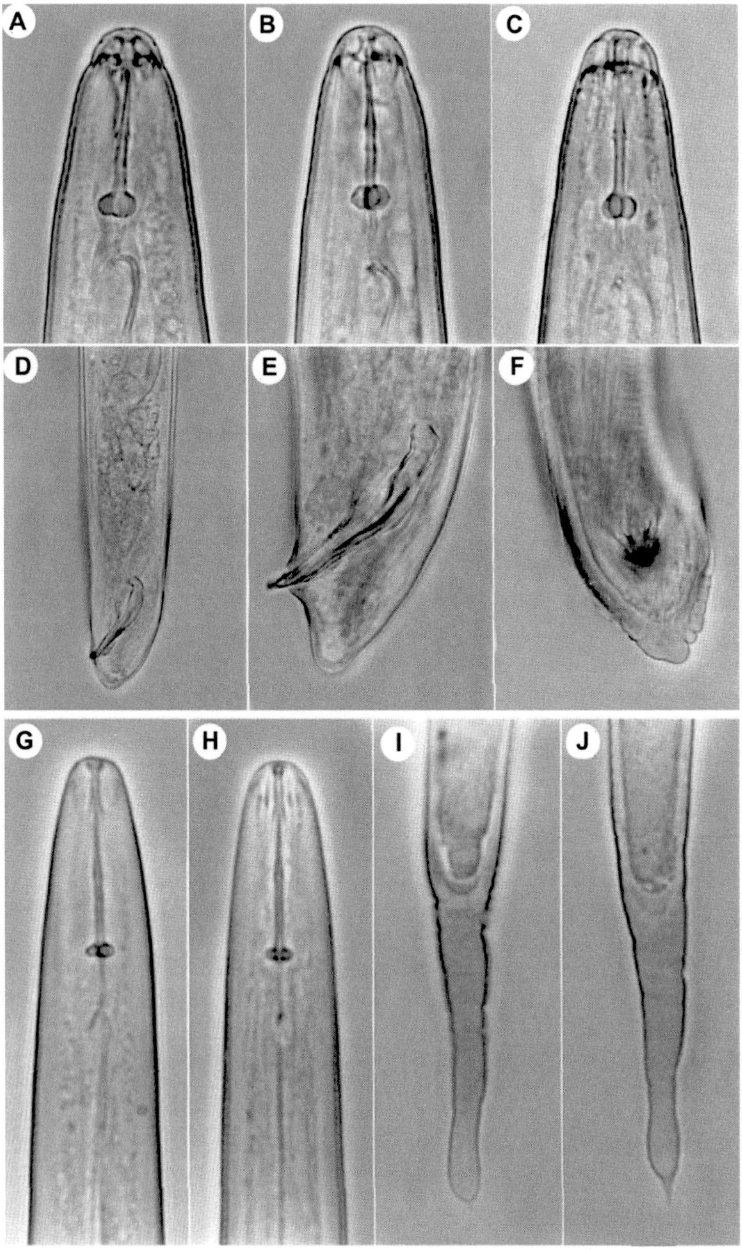

Fig. 243. Meloidogyne spartinae. *LM. A-C: Male labial region; D-F: Male tail; G, H: Second-stage juvenile (J2) labial region; I, J: J2 tail. (After Eisenback & Hirschmann, 2001.)*

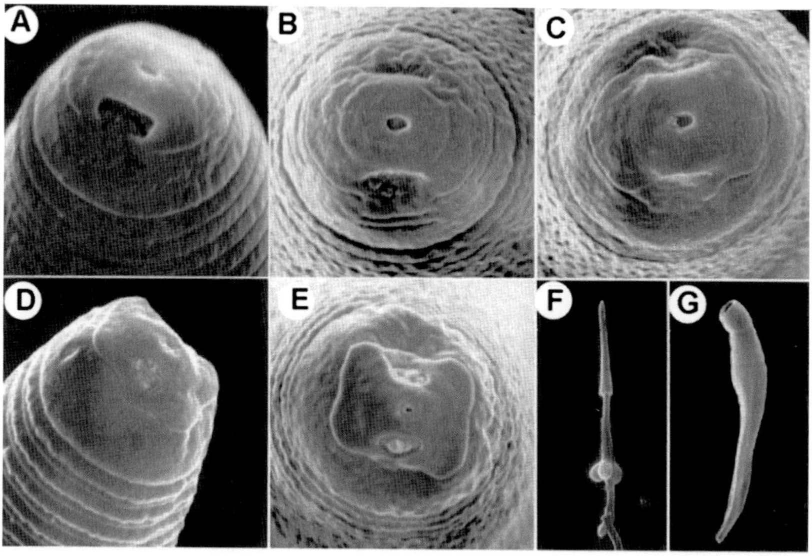

Fig. 244. Meloidogyne spartinae. *SEM. A-C: Male labial region; D, E: Second-stage juvenile (J2) labial region; F: Male stylet; G: Spicules. (After Eisenback & Hirschmann, 2001.)*

remnant often present. Ovary followed by oviduct-spermatheca region, which is composed of numerous cells. Oviduct made up of 6-8 cells and spermatheca formed by 28-34 large, rounded cells with undulating junctions.

Male

Males very long, tapering sharply in region of the pharynx and especially near stylet. Body nearly straight except for a slight twist in tail region. Small, dome-shaped labial region not offset from body and having a slightly sclerotised labial framework. Stylet small and delicate. Knobs varying in shape, rounded to angular, often with slightly irregular outlines. Malformed knobs present in live nematodes observed in water mounts. Attachment of vestibule extension to stylet cone always very distinct. Cone distinctly offset from slender, cylindrical shaft. DGO large and ampulla often filled with secretory granules all the way through median bulb to gland cell. Amphidial ducts distinct and broad, their openings very large, especially as compared to small labial region. Pharyngeal glands variable in shape and length and usually with three relatively small, equally-sized nuclei. Nuclei difficult to see in fixed specimens

Table 51. *Morphometrics of females, males, second-stage juveniles (J2) and eggs of* Meloidogyne spartinae. *All measurements are in µm and in the form: mean ± s.d. (range).*

Character	Paratypes, Charleston, South Carolina, USA, Rau & Fassuliotis (1965)	North Carolina, USA, Eisenback & Hirschman (2001)	Cape Cod, USA, LaMondia & Elmer (2007)
Female (n)	158	100	-
L	851 ± 104 (600-1140)		-
a	1.7 (1.2-2.7)	-	-
W	504 ± 95 (270-870)	-	-
Stylet	13.0 ± 0.8 (11.0-17.0)	-	-
DGO	4.2 ± 0.3 (2.8-5.6)	3.0-5.2	-
Ant. end to excretory pore	18.8 ± 3.7 (11.2-28.0)	-	-
Median bulb length	31 ± 4.6 (21.0-44.8)	-	-
Median bulb width	30 ± 3.5 (23.8-35.0)	-	-
Vulva to anus	21.7 ± 2.58 (15.4-29.4)	-	-
Interphasmid distance	-	24-32	-
EP/ST	1.5	1.0	-
Male (n)	110	100	40
L	2250 ± 212 (1676-2592)	2210-2780	2203
a	52.9 (33.8-76.4)	-	-
c	178.6 (111.0-240.6)	-	-
W	43 ± 7.1 (25-59)	-	-
Stylet	19.7 ± 1.0 (16.8-21.0)	17.7-20.0	-
DGO	4.8 ± 0.64 (3.5-7.0)	4.3-6.7	-
Ant. end to median bulb valve	97.9 ± 6.81 (81.2-116.2)	-	-
Ant. end to excretory pore	131.3 ± 13.6 (84-172)	-	-
Spicules	31.4 ± 2.81 (25.2-40.0)	32-35	-
Gubernaculum	8.3 ± 0.9 (7.0-9.8)	-	-
Tail	12.6 ± 1.2 (9.8-15.4)	-	-
Testis	1086 ± 170 (466-1356)	-	-

Table 51. *(Continued.)*

Character	Paratypes, Charleston, South Carolina, USA, Rau & Fassuliotis (1965)	North Carolina, USA, Eisenback & Hirschman (2001)	Cape Cod, USA, LaMondia & Elmer (2007)
J2 (n)	105	100	60
L	773 ± 153 (612-912)	767-857	703.3
a	53.3 (43.2-65.0)	-	-
c	7.6 (6.7-9.0)	-	-
W	15.0 ± 0.9 (14.0-16.8)	-	-
Stylet	15.4 ± 0.3 (14.0-6.8)	-	-
DGO	4.0	4.2-5.4	-
Ant. end to median bulb valve	75.8 ± 5.0 (63-105)	-	-
Ant. end to excretory pore	101 ± 5.1 (88.2-117.6)	-	-
Tail	99.7 ± 7.44 (77.0-113.4)	-	-
Hyaline region	22.3 ± 2.5 (16.8-28.0)	-	-
Egg	10	-	-
L	121 (100.8-132.3)	-	-
W	54 (47.3-63.0)	-	-

where sometimes only two are visible, but in live specimens they are easily seen. Procorpus bulging and swollen at junction with metacorpus. Lumen lining of junction with intestine distinct. An intestinal caecum observed in some live specimens, reaching anteriorly to median bulb. Hemizonid always posterior to excretory pore. Excretory pore narrow, duct slightly sclerotised through somatic muscle, then lining becoming rather thick for a short distance. In some specimens, beginning of lateral field bulging out and lateral chord appearing to indicate a cephalid-like structure. Lateral field containing 4-11 very fine short and broken incisures, depending on location on body. Tail twisting only slightly toward very tail end and outline frequently irregular in shape where it fuses with tail tip. Phasmids very inconspicuous in lateral view, but more distinct in ventral view. Spicules long and slender, head region characteristically shaped, pointing forward. Likewise, gubernaculum characteristically shaped and in some aberrant specimens central portion appearing less sclerotised. Posterior end of gonad always very distinct, sometimes lumen widened distinctly, cloaca rather short. Terminal portion of *vas deferens*, with or without sperm, apparently consisting of several large cells filled with globules. Sperm and spermatocytes very large. Always only one testis present, which is rarely reflexed. Lateral labials usually lacking and median labials sloping and frequently fused with labial region. Labial region may be smooth or marked with short, incomplete transverse incisures. Amphidial openings large. Characteristic features of stylet and spicules more clearly seen when dissected from body.

J2

Elongate, especially on tail, which narrows more than labial region. Labial region is not offset and apparently not annulated, labial framework very faint and only slightly indicated, although vestibule and extension are very pronounced. Stylet knobs variable in shape. In most specimens they are discrete and rounded, quite separate and sticking out, but in others they are slanted. Shaft slender, same diam. throughout and junction with cone emphasised. Ampulla of dorsal pharyngeal gland quite distinct with some granulation. Hemizonid always located posteriorly to excretory pore, which is at level of or slightly posterior to median bulb. Excretory duct very long and wavy, ending in an excretory cell. Lateral field very faint with four incisures and lateral chord apparently very broad. Characteristic long bulbous tail with a distinct hyaline terminus, usually spicate in most specimens. Rectal dilation large and chambered. Phasmids are distinct in ventral view, but difficult to see in lateral view. In SEM, labial region of J2 typical. Labial region structures rectangular and lateral labials apparently at same level as median labials. Labial region smooth, not marked by additional annulation.

TYPE PLANT HOST: Smooth cordgrass, *Spartina alterniflora* Loisel.

OTHER PLANTS: No other hosts were reported (LaMondia & Elmer, 2008).

TYPE LOCALITY: Banks of Long Branch Creek, east of the US Vegetable Breeding Laboratory farm 3 miles southwest of the city limits of Charleston, South Carolina, USA.

DISTRIBUTION: *North America*: USA (New York, New Jersey, Georgia, Florida, North Carolina, South Carolina, Connecticut, Maine and Massachusetts) (Rau & Fassuliotis, 1965; Eisenback & Hirschmann, 2001; LaMondia & Elmer, 2007, 2008).

PATHOGENICITY: *Meloidogyne spartinae* galls were reported to differ in number and size on different geographical populations of *S. alterniflora* (Seneca, 1974). *Spartina alterniflora* and *M. spartinae* survive harsh conditions in the low marsh. LaMondia & Elmer (2008) measured salt concentrations as high as 35 000 ppm in the water in the sediments, similar to seawater, and the peat and tidal sediments are often flooded, waterlogged and anaerobic. An additional study (Elmer & LaMondia, 2014) determined that *M. spartinae* could survive even at 1.0 M NaCl for at least 12 days.

CHROMOSOME NUMBER: Fourteen rod-shaped chromosomes were present in oogonia and spermatogonia, whereas seven bivalents were observed in oocytes and spermatocytes (Triantaphyllou, 1987a).

POLYTOMOUS KEY CODES: *Female*: A12, B324, C3, D3; *Male*: A12, B324, C213, D1, E1, F1; *J2*: A1, B12, C1, D1, E32, F1.

BIOCHEMICAL AND MOLECULAR CHARACTERISATION: No biochemical data of this species are available. Plantard *et al.* (2007) sequenced and analysed 18S rRNA gene of this species. *Meloidogyne spartinae* is closely related to *M. maritima* (Plantard *et al.*, 2007) and *M. graminis* and *M. marylandi* (Álvarez-Ortega *et al.*, 2019).

RELATIONSHIPS (DIAGNOSIS): The species belongs to Molecular group IV and is clearly molecularly differentiated from all other *Meloidogyne* species. The most distinctive character of this species is the extremely long J2, with its asymmetrical, spiked, bulbous tail. *Meloidogyne spartinae* can be distinguished from *M. acronea*, *M. graminis* and *M. ottersoni* by J2 body length. Additionally, *M. graminis* and *M. mersa* have a different perineal pattern and EP/ST ratio. From *M. mersa*, it can be separated by perineal pattern and EP/ST ratio.

90. *Meloidogyne subarctica* Bernard, 1981
(Figs 245, 246)

COMMON NAME: Subarctic grass root-knot nematode.

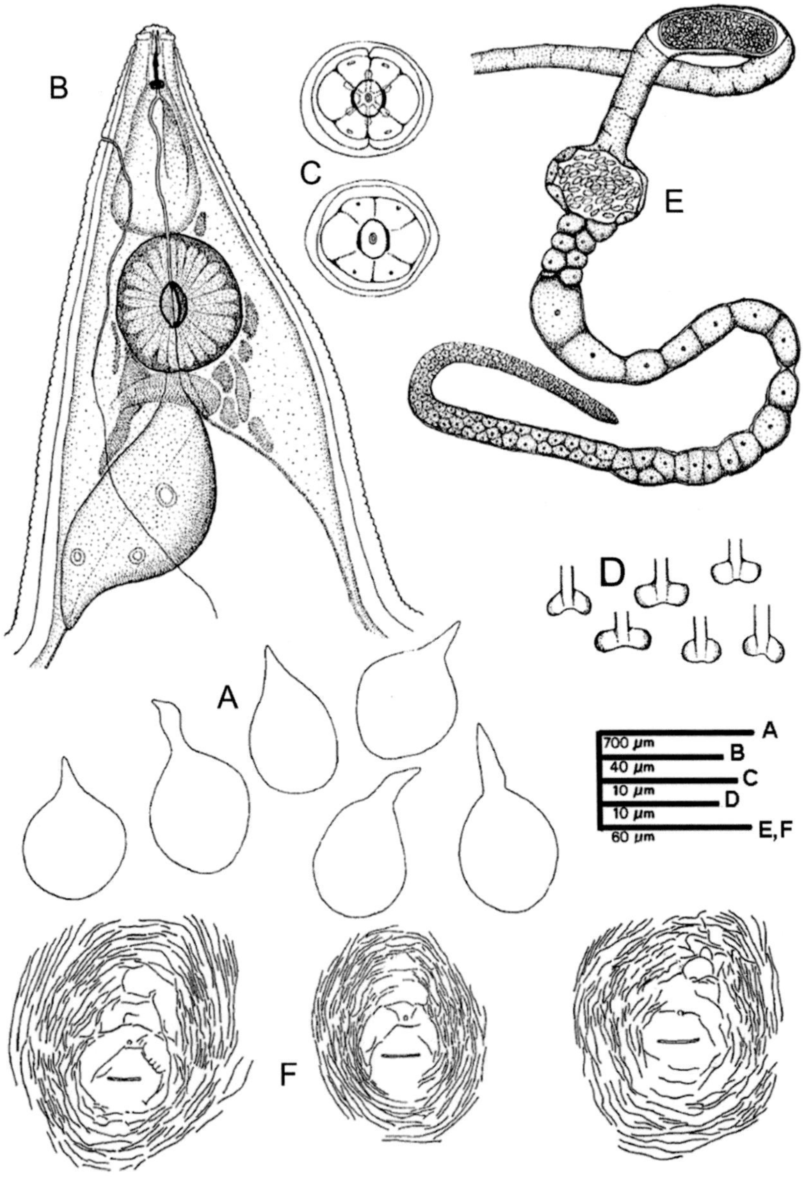

Fig. 245. Meloidogyne subarctica. *A: Entire female; B: Female anterior region; C: Female* en face *view; D: Female stylet; E: Female gonad; F: Perineal pattern. (After Bernard, 1981.)*

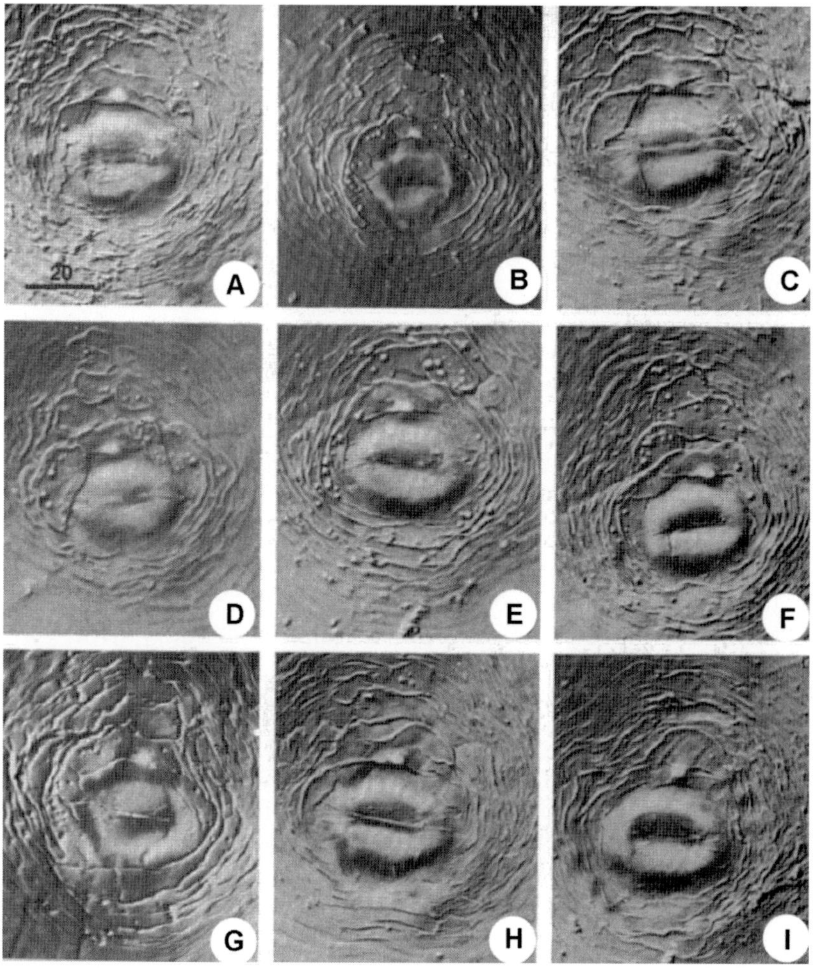

Fig. 246. Meloidogyne subarctica. *LM. A-I: Perineal patterns. (After Bernard, 1981.)*

Collected in the Aleutian Islands from galled roots of a dunegrass, *Leymus mollis* (Trin.) Pilg during the summers of 1977-1979.

MEASUREMENTS *(AFTER BERNARD, 1981)*

- *Holotype female*: L = 629 μm; W = 393 μm; neck length = 226 μm; stylet = 13.6 μm; stylet knobs width = 3.5 μm; stylet knob height =

1.8 μm; DGO = 3.8 μm; excretory pore from anterior end = 27.2 μm; a = 1.6; b = 6.8; b′ = 4.1.

- *Allotype male*: L = 1870 μm; W = 35 μm; stylet = 19.6 μm; stylet knob width = 4.2 μm; stylet knob height = 2.7 μm; DGO = 4.9 μm; excretory pore from anterior end = 150 μm; spicules = 35.5 μm; gubernaculum = 8.4 μm; tail = 8 μm; a = 53.4; b = 14.4; b′ = 9.0.
- *Paratype females* (n = 12): L = 709 ± 82.5 (608-836) μm; W = 466 ± 69.8 (356-622) μm; stylet = 14.2 ± 0.73 (12.7-15.7) μm; stylet knob width = 4.0 ± 1.2 (3.7-4.4) μm; stylet knob height = 2.0 ± 0.2 (1.7-2.3) μm; DGO = 5.2 ± 1.04 (3.7-7.1) μm; anterior end to excretory pore = 31.3 ± 3.97 (22.4-38.5) μm; a = 1.5 ± 0.14 (1.3-1.8); b′ = 4.9 ± 0.5 (4.2-5.7); c = 9.2 ± 0.6 (8.2-11.0); c′ = 4.5 ± 0.4 (3.9-5.2); EP/ST = 1.5-2.5; vulval width = 20.9 ± 1.9 (18-25.3) μm; interphasmid distance = 22.1 ± 4.9 (11.4-28.8) μm; vulva to anus = 19.2 ± 2.5 (12.8-23.4) μm.
- *Paratype males* (n = 6): L = 1648 (1403-1755) μm; W = 31.4 (27.2-36.9) μm; stylet = 18.9 (17.2-19.7) μm; stylet knob width = 3.4 (3.1-4.8) μm; stylet knob height = 2.5 (2.0-2.8) μm; DGO = 4.7 (4.1-5.8) μm; excretory pore to anterior end = 138 (125-141) μm; spicules = 34.9 (32.9-36.9) μm; gubernaculum = 8.0 (7.7-8.4); a = 53 (45.3-61); b = 13.6 (12.7-14.5); b′ = 8.2 (7.6-9.0).
- *Paratype J2* (n = 20): L = 439 ± 30.1 (349-507) μm; W = 15.0 ± 1.42 (12.2-17.0) μm; stylet = 14.4 ± 0.44 (13.5-15.4) μm; stylet knob width = 2.1 ± 0.17 (1.9-2.5) μm; stylet knob height = 1.2 ± 0.16 (1.0-1.4) μm; DGO = 3.9 ± 0.62 (2.9-4.8) μm; excretory pore to anterior end = 79.8 ± 4.87 (64.1-86.0) μm; phasmid to tail end = 34.6 ± 2.43 (27.2-38.8) μm; a = 29.4 ± 2.5 (24.7-34.8); b = 6.6 ± 0.16 (6.5-6.9); b′ = 2.1 ± 0.14 (2.0-2.5); c = 9.2 ± 0.6 (8.2-11); tail length/body diam. at anus = 4.5 ± 0.4 (3.9-5.2) μm.

Description

Female

Females milky-white, spheroid or ovoid, necks short to long, not directed ventrally. Posterior protuberance absent. Face view showing small labial disc surrounded by six labials, labial region wider laterally. Labial region with one wide annulus, somewhat angular in profile. Stylet knobs variable in shape, occasionally asymmetric. Dorsal gland orifice slightly posterior to stylet knobs, ampulla often very large.

Median bulb ovate to round, valve large. Pharyngeal glands distinct and overlapping intestine ventrally. Each gonad highly differentiated, spermatheca enlarged and filled with sperm. Perineal pattern generally oval to rounded, arch with few unbroken striae, resulting in large, oval or angular regions devoid of striae, tail region without tail whorl. Lateral field marked by short, broken striae, or devoid of striae. Vulval width and perivulval height less variable than other morphometric perineal characters.

Male

Males vermiform, nearly straight to strongly curved when heat-relaxed. Face view showing fused labial disc and median labials. Laterally, amphidial apertures large, labial margins present. Four labial sensilla present. Labial region slightly offset with one annulus dorsally and ventrally, one or two laterally. Stylet with rounded knobs. Dorsal gland orifice about one-fourth stylet length posterior to knobs. Median bulb reduced, valve large. Junction of pharynx and intestine not clearly seen, terminating in vicinity of hemizonid. Excretory pore 5-8 annuli posterior to hemizonid. Pharyngeal glands indistinct distally. Lateral field with four incisures and areolated, short, shallow incisures sporadically present to give effect of six incisures. Spicules and gubernaculum similar to those of other *Meloidogyne* spp., cloacal aperture occurring a little anterior to phasmids. Tail broadly conoid-rounded.

J2

Juveniles nearly straight when heat-relaxed. In face view, labial disc thinner than in male and fused with median labials, margins of lateral labials not seen. Four labial sensilla present. Labial region slightly offset, with one annulus. Stylet slender, with rounded knobs. Dorsal gland orifice about one-fourth of stylet length posterior to knobs. Valve of median bulb in a central position. Excretory pore just posterior to hemizonid. Pharyngeal gland lobe very long, extending to mid-body, gland nuclei evenly distributed. Rectum not dilated. Lateral field similar to that of male. Tail tapering evenly to a finely rounded tip. Phasmids offset, near ventral edge of lateral field.

TYPE PLANT HOST: American dune grass, *Leymus mollis* (Trin.) Pilg.

OTHER PLANTS: No galling was observed on oat, fescue, or tomato, but these cultures were kept in the glasshouse and thus may have been grown at unsuitable temperatures.

TYPE LOCALITY: Kuluk Bay, Adak Island, Alaska, USA.

DISTRIBUTION: *North America*: USA (Bernard, 1981).

PATHOGENICITY: *Meloidogyne subarctica* forms large terminal and intercalary galls on the roots of *L. mollis*. Intercalary galls usually contain only one female, but terminal galls may have a dozen or more females and numerous swollen juveniles each. Males are found most often in small, terminal galls. Egg masses produced by mature females are very large. No females were observed to protrude from root tissue. Normal J2 hatched from eggs maintained in water at room temperature, but cultures of *M. subarctica* on *L. mollis* died out in the glasshouse (24-26°C). However, cultures were increased in a growth chamber at 15°C.

CHROMOSOME NUMBER: Chromosomes of oocytes very small and rather difficult to distinguish, $n = 18$, reproduction by amphimixis.

POLYTOMOUS KEY CODES: *Female*: A2, B32, C23, D3; *Male*: A23, B34, C12, D1, E1, F2; *J2*: A213, B12, C3, D4, E3, F2.

BIOCHEMICAL AND MOLECULAR CHARACTERISATION: No available data.

RELATIONSHIPS (DIAGNOSIS): The perineal pattern of *M. subarctica* resembles that of *M. sewelli*, but the peripheral striae are less distinct and the pattern is a more flattened oval in *M. sewelli*. In addition, the J2 of *M. subarctica* is stockier, has a longer stylet, and has DGO closer to the stylet knobs. Females of *M. subarctica* are much larger and have longer necks, while males are more slender and possess longer spicules.

91. *Meloidogyne suginamiensis* Toida & Yaegashi, 1984
(Figs 247, 248)

COMMON NAME: Suginami root-knot nematode.

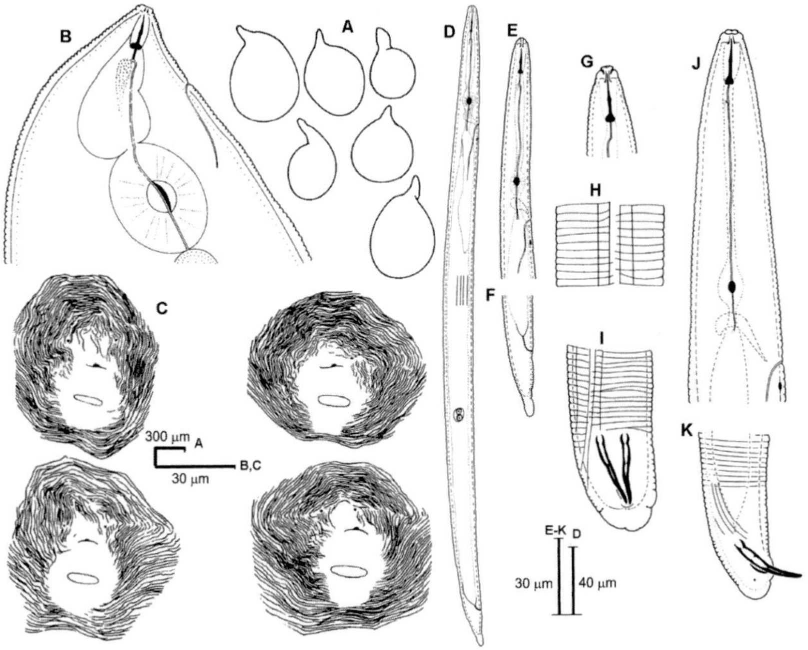

Fig. 247. Meloidogyne suginamiensis. *A: Entire female; B: Female anterior region; C: Perineal pattern; D: Entire second-stage juvenile (J2); E: J2 anterior region; F: J2 tail; G: J2 labial region; H: J2 lateral field; I, K: Male tail; J: Male anterior region. (After Toida & Yaegashi, 1984.)*

This species was described from galled roots of mulberry at the Sericultural Experiment Station in Wada, Suginami-ku, Tokyo, Japan, by Toida & Yaegashi (1984).

MEASUREMENTS *(AFTER TOIDA & YAEGASHI, 1984)*

- *Holotype female*: L = 804 μm; W = 603 μm; neck length = 201 μm; stylet = 14 μm; DGO = 3.5 μm; excretory pore from anterior end = 34 μm; a = 1.3; EP/ST = 2.7.
- *Allotype male*: L = 1426 μm; W = 34 μm; neck length = 201 μm; stylet = 20 μm; DGO = 6.3 μm; excretory pore from anterior end = 156 μm; spicules = 33 μm; gubernaculum = 8.9 μm; a = 42; c = 96.
- *Paratype females* (n = 15): L = 950 (670-1220) μm; W = 740 (600-1010) μm; stylet = 14.1 (12.0-17.0) μm; DGO = 4.7 (3.0-6.0) μm; vulval slit = 20 (15-24) μm; a = 1.3 (1.0-1.5).

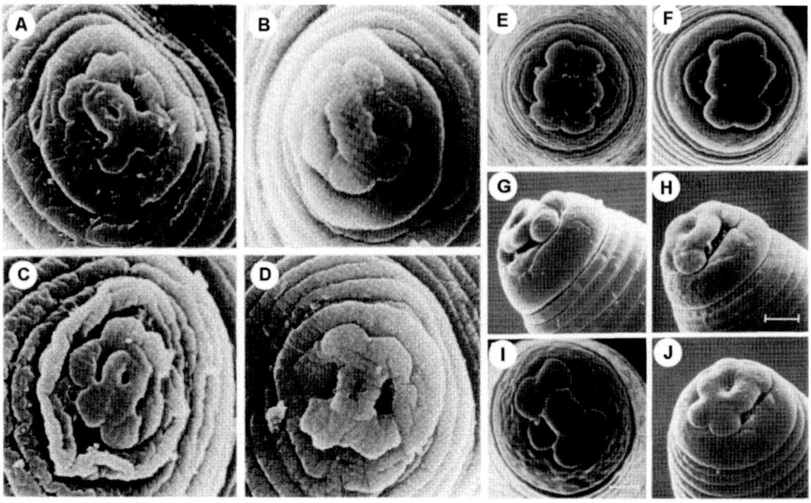

Fig. 248. Meloidogyne suginamiensis. *SEM. A-D: Female* en face *view; E-H, J: Male labial region; I: Second-stage juvenile labial region. (After Toida & Yaegashi, 1984.)*

- *Paratype males* (n = 20): L = 1510 (1040-1770) μm; W = 33.0 (26.0-37.0) μm; stylet = 20.0 (17.0-21.0) μm; DGO = 5.4 (4.0-8.0) μm; excretory pore to anterior end = 145 (110-170) μm; tail = 17.0 (13-19) μm; spicules = 32 (25-37) μm; gubernaculum = 8.0 (6.0-10) μm; a = 45 (40-50); c = 97 (60-115).
- *Paratype J2* (n = 50): L = 420 (370-490) μm; W = 16.0 (14.0-19.0) μm; stylet = 14.0 (12.0-15.0) μm; DGO = 4.1 (3.0-5.0) μm; excretory pore to anterior region = 76.0 (69.0-82.0) μm; tail = 28.0 (24.0-33.0) μm; hyaline region = 3-5 μm; a = 27 (23-29), c = 15.8 (14.5-18.2); c′ = 2.9 (2.6-3.2); diam. at anus = 8.9 (8-10) μm.
- *Eggs* (n = 50): L = 92 (83-100) μm; W = 40 (37-43) μm; L/W = 2.3 (1.9-2.6).

Description

Female

Body pearly white, sometimes yellow brown, usually rather elongate ovoid to pear-shaped, with projecting neck and slightly flattened to rounded posteriorly. Neck tapering to labial region tip, being less than one-fourth of body length. Labial region small, slightly offset from body. Labial cap low, in SEM observation of face view, dumbbell-shaped, with labial disc fused to median labials in same contour and lateral labials fused to upper surface of labial region. Cuticle comparatively thick. Excretory pore generally located posterior to DGO, being on 22-28th annulus or at 3-7% of total body length from anterior end. Stylet slender, short, often curved dorsally, with small rounded knobs usually sloping posteriorly. Perineal pattern slightly squarish in outline especially in dorsal arch, with wavy, very fine, thin striae. Lateral field not clear. Phasmids rather large, being 22 (15-28) μm apart, slightly wider than length of vulval slit which nearly equals distance to anus.

Male

Body long, slender, tapering at both labial and tail ends, narrower anteriorly. Body almost straight when killed by gentle heat with twist in posterior portion. Labial region squarish, offset from body, with one or two annuli, narrower than first body annulus. Labial cap, slightly protruding, smaller than labial region. In SEM observation of face view, labial disc large, lozenge-shaped and lateral labials clear. Lateral field less than one-third of body diam. with four incisures areolated on tail. Stylet slender with small knobs (4 μm wide and 3.5 μm high) directed posteriorly. Excretory pore located 18-27 annuli posterior to valve of median bulb or at 9-12% of body length from anterior end. Hemizonid located 1-3 annuli posterior to excretory pore, 1.0-1.5 annuli long. Labial region with six labials. Testis single, 550-840 μm long extending to near middle portion of body. Spicules rather arcuate ventrally. Gubernaculum less than one-third of spicule length long. Tail shorter than cloacal body diam., with rounded broad tip. Phasmids distinct.

J2

Body slender, tapering at both labial and tail regions, but being narrower posteriorly. Cuticular annulation fine but fairly distinct. Lateral field more than one-third of body diam., with four incisures, of which outer two slightly crenate and not areolated. Labial region slightly offset from body, with thin labial cap. In SEM observation of face view, labial disc round, fused to median labials in same contour forming a dumbbell-shape, lateral labials never fused to the median labials in same contour. Stylet delicate, with small knobs sloping posteriorly. Excretory pore located at 17-20% (mean 17.9) of body length from anterior end. Hemizonid not distinct, 4-7 annuli posterior to excretory pore. Tail conoid, short, usually deeply constricted at one-third of tail length from terminus, with rounded, blunt tip. Hyaline region short (3.5 μm).

TYPE PLANT HOST: Mulberry, *Morus alba* L.

OTHER PLANTS: Woody plants: Mulberries (*Morus bombycis*, *M. latifolia* Poir., *M. australis* Poir., *M. nigra* L.), elm (*Ulmus davidiana* var. *japonica* Rehder), fig (*Ficus carica*), paper mulberry (*Broussonetia kazinoki* Sieb.), *B. papyrifera* (L.) Vent., *Maclura tricuspidata* (Carrière) Bureau, raspberry (*Rubus* sp.) and cherry (*Prunus yedoensis*). Weed plants: *Achyranthes japonica* (Miq.) Nakai, sorrel vine (*Cayratia japonica* (Thunb.) Gagnep.), ink-bush (*Phylolacca americana* L.), goosefoot (*Chenopodium ficifolium* Sm.), yellow cress (*Rorippa indica* Hiern.) and skunkvine (*Paederia foetida* L.). Vegetable plants: tomato (*Solanum lycopersicum*), eggplant (*S. melongena*), pepper (*Capsicum. annuum*), cucumber (*Cucumis sativus*), pumpkin (*Cucurbita* sp.), cabbage (*Brassica oleracea* var. *capitata*), great burdock (*Arcutium lappa* L.) and carrot (*Daucus carota* var. *sativa*) (Toida & Yaegashi, 1984). Gu *et al.* (2020) reported this species from maple trees (*Acer palmatum* Thunb.) imported from Japan.

TYPE LOCALITY: Wada, Suginami-ku, Tokyo, Japan.

DISTRIBUTION: *Asia*: Japan (Toida & Yaegashi, 1984; Orui, 1998).

PATHOGENICITY: Numerous galls on roots of mulberry and other plants.

CHROMOSOME NUMBER: No available data.

POLYTOMOUS KEY CODES: *Female*: A12, B32, C2, D3; *Male*: A324, B324, C213, D1, E2, F2; *J2*: A23, B21, C4, D4, E1, F4.

BIOCHEMICAL AND MOLECULAR CHARACTERISATION: No biochemical data of this species are available. Orui (1998) studied RFLP of *COII*-16S mtDNA fragments from several *Meloidogyne* species from Japan and distinguished a specific RFLP for this species. Gu *et al.* (2020) sequenced a *COII*-16S mtDNA fragment, 18S rRNA, ITS rRNA and partial 28S rRNA genes.

RELATIONSHIPS (DIAGNOSIS): *Meloidogyne suginamiensis* is characterised by a short J2 conoid deeply constricted, usually at one-third of tail length from a broad terminus. It can be differentiated from the short tail length groups 1 and 2 (Jepson, 1987) by perineal pattern and EP/ST ratio.

92. *Meloidogyne tadshikistanica* Kirjanova & Ivanova, 1965
(Fig. 249)

COMMON NAME: Tadzhik root-knot nematode.

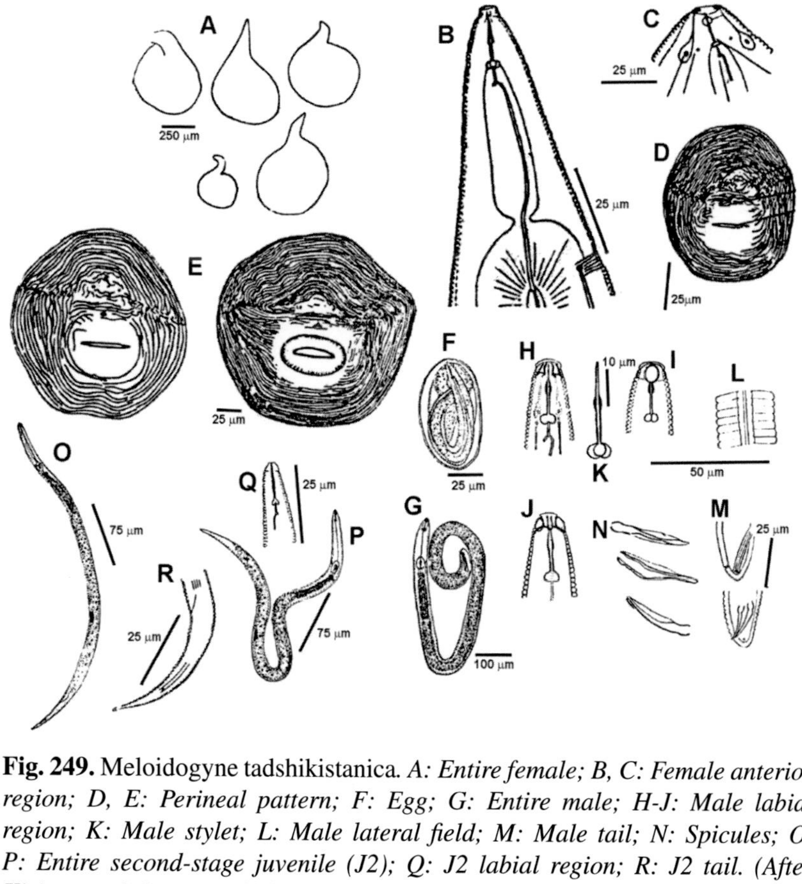

Fig. 249. Meloidogyne tadshikistanica. *A: Entire female; B, C: Female anterior region; D, E: Perineal pattern; F: Egg; G: Entire male; H-J: Male labial region; K: Male stylet; L: Male lateral field; M: Male tail; N: Spicules; O, P: Entire second-stage juvenile (J2); Q: J2 labial region; R: J2 tail. (After Kirjanova & Ivanova, 1965.)*

This species was described from *Pelargonium roseum* L. in Tajikistan (Kirjanova & Ivanova, 1965).

MEASUREMENTS (AFTER KIRJANOVA & IVANOVA, 1965)

- *Paratype females* (n = 10): L = 630.3 (410-830) μm; W = 434.9 (320-540) μm; stylet = (14.5-15.0) μm; stylet knob width = *ca* 4 μm; DGO = 4.4 (3.0-5.3) μm; vulval slit = 24.1 (22-26) μm; distance from anus to vulva = 18 (15.2-22.8) μm; interphasmid distance = 14.4 (11.4-19.0) μm; EP/ST = 5.
- *Paratype males* (n = 8): L = 1247.5 (934-1405) μm; W = 31.7 (28-35) μm; stylet = 23.5 (22.4-25.0) μm; stylet knob height = 2.8-3.0 μm; stylet knob width = 4.0-4.5 μm; labial region height = 5.6 (5.3-7.0) μm; DGO = 5.0-5.4 μm; spicules = 32.2 (27.0-36.6) μm; gubernaculum = 7-8 μm; tail = 13.2 (11.4-16.0) μm; width at the anus level = 21.6 (16-26) μm; a = 39.3; c = 94.6.
- *Paratypes J2*: L = 398.5 (350-435); stylet = 12.0-15.0 μm; DGO = 3.6 μm; labial region = 3.0-3.8 μm; tail = 45 (42-48) μm.
- *Eggs*: L = 98.5 (87.5-107.8) μm; W = 40.4 (35-42) μm.

DESCRIPTION

Female

Onion-shaped with neck of variable length. Labial region about 3 μm high with four annuli, one labial and three postlabial. Stylet robust with clearly defined segments of base (telorhabdions) about 4 μm wide. Amphid diam. 3.5 (3.0-3.8) μm. Cuticle thick, 3.0-3.8 μm on labial region and adjacent areas, 5 μm in middle part of body and 6.5-7.0 μm on posterior end. Excretory pore opening opposite upper part of bulb, slightly anterior to valve. Anal-vulval plate oval, but can be almost circular and even wider (in older specimens), its size varying from 80-114 (88.7) μm long and 80-99 (88.8) μm wide. In older specimens, anal-vulval plate noticeably larger than in younger ones. Presence of more or less straight lines anterior to anus and between latter and the tail rudiment is characteristic. Area of cuticle containing tail rudiment always isolated from other parts of anal-vulval plate. Side field marked by short branching lines.

Male

Labial region with four annuli. Labial annulus wide and flattened, postlabial annuli slightly narrower and rather shallow. Cuticle strongly annulated, annuli in middle part of body 2 μm wide. Stylet robust with well-developed telorhabdions. Cylindrical part of stylet in certain views seems sharply narrowed to its base. Amphidial aperture oval, 6.0-7.0 $\times$ 3.8-5.5 μm, and occupying half of labial region diam. Lateral field consisting of four lines, forming three bands. Width of two outer bands = 2.6 μm, width middle band = 1.5 μm. Lateral field occupying about 25% of body diam. Phasmids located level with cloacal aperture or slightly posterior. Spicules weakly bent in middle part. Proximal ends of spicules widened, distal ends strongly narrowed. Velum extending from head region of spicules.

J2

Labial region consisting of four minute annuli. Tail bluntly rounded. Lateral field with four incisures, occupying about one-third of body diam.

TYPE PLANT HOST: Geranium, *Pelargonium roseum* L.

OTHER PLANTS: Also found on the roots of *Tradescantia* sp., accompanied by the simultaneous infection by *M. incognita*.

TYPE LOCALITY: Experimental station, Gissar valley (former Pakhtaabad district), Tajikistan.

DISTRIBUTION: *Asia*: Tajikistan.

PATHOGENICITY: Galls were reported to be rather small.

CHROMOSOME NUMBER: No available data.

POLYTOMOUS KEY CODES: *Female*: A23, B2, C1, D4; *Male*: A34, B21, C21, D1, E2, F1; *J2*: A32, B21, C3, D3, E2, F4.

BIOCHEMICAL AND MOLECULAR CHARACTERISATION: No available data.

RELATIONSHIPS (DIAGNOSIS): This species is characterised by a large EP/ST ratio (5.0), one of the longest in the genus and a perineal pattern characteristic of group 4 (Jepson, 1987). The perineal pattern of this species also resembles *M. enterolobii* and *M. platani*, but differs from *M. platani* by EP/ST ratio and several male and J2 morphometrics. From *M. enterolobii* it differs in c′ ratio and J2 hyaline region.

93. *Meloidogyne thailandica* Handoo, Skantar, Carta & Erbe, 2005 (Figs 250-252)

COMMON NAME: Ginger root-knot nematode.

In October 2002, a RKN was discovered on roots of ginger (*Zingiber* spp.) from Thailand, the shipment being intercepted by the Animal and Plant Health Inspection Service (APHIS) at the port of San Francisco, USA. The plants came from Bangkok local nursery growers and were a variegated variety of *Zingiber* sp.

MEASUREMENTS (*AFTER HANDOO ET AL., 2005B*)

- *Holotype female*: L = 700 μm; W = 405 μm; stylet = 13.5 μm; stylet knob width = 5.0 μm; stylet knob height = 2.5 μm; DGO = 5.0 μm; excretory pore from anterior end = 38.0 μm; vulval slit length = 35.0 μm; distance from vulval slit to anus = 20.0 μm; EP/ST = 2.8.
- *Allotype male*: L = 1380 μm; stylet = 18.0 μm; stylet knob width = 4.5 μm; stylet knob height = 3.0 μm; excretory pore from anterior end = 138 μm; centre of median bulb from anterior end = 77.5 μm; spicules = 31 μm; gubernaculum = 10.0 μm; tail = 10.0 μm; a = 53; b = 7.2; c = 110.
- *Paratype females* (n = 25): L = 762 ± 115.8 (570-955) μm; W = 459 ± 109.1 (272-690) μm; stylet = 13.8 ± 1.2 (12.0-15.5) μm; stylet knob width = 4.9 ± 0.3 (4.0-5.0) μm; stylet knob height = 2.4 ± 0.5 (1.5-3.5) μm; DGO = 4.0 ± 0.9 (3.0-5.0) μm; anterior end to excretory pore = 30.6 ± 8.4 (15.0-47.0) μm; vulval slit = 29.1 ± 2.6 (25.0-35.1) μm; vulva to anus = 18.0 ± 2.4 (15-22) μm; a = 1.7 ± 0.2 (1.4-2.3); EP/ST ratio = 2.3 ± 0.5 (1.7-3.6).
- *Paratype males* (n = 26): L = 1240 ± 205.2 (950-1510) μm; W = 31.3 ± 6.0 (21-48) μm; stylet = 18.7 ± 1.1 (17-20) μm; stylet knob width = 4.7 ± 0.3 (4.0-5.0) μm; stylet knob height = 2.6 ± 0.2

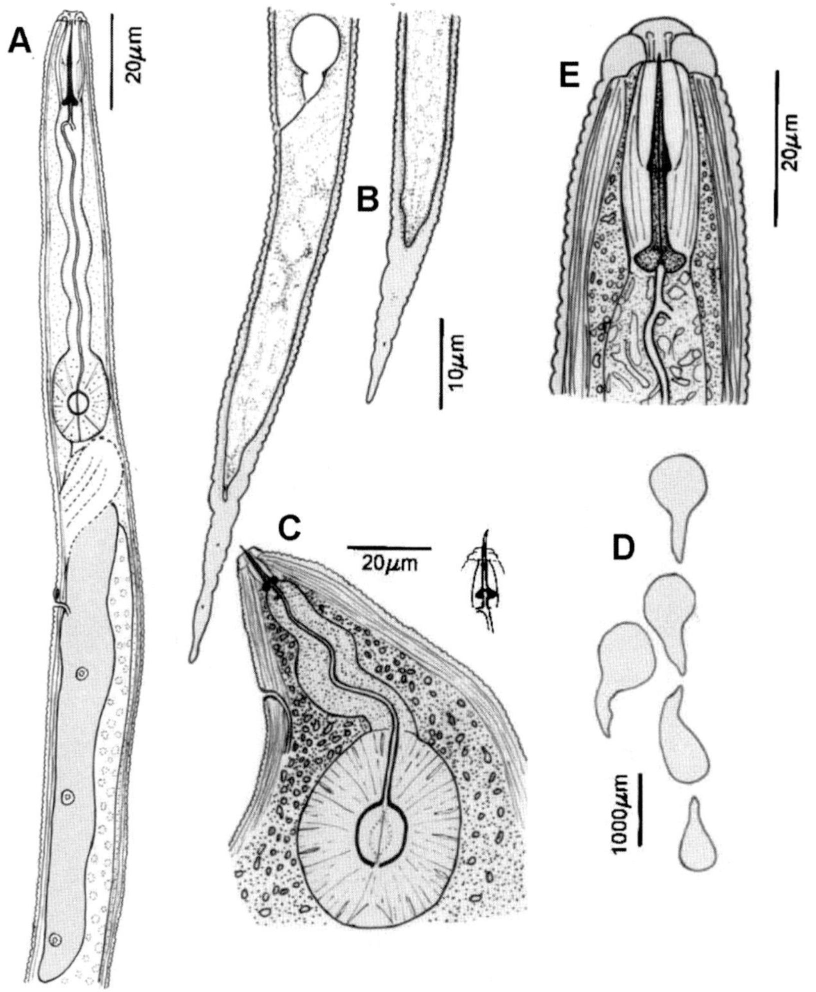

Fig. 250. Meloidogyne thailandica. *A: Second-stage juvenile (J2) anterior region; B: J2 tail; C: Female anterior region; D: Entire female; E: Male labial region. (After Handoo* et al.*, 2005b.)*

(2.5-3.0) μm; DGO = 3.7 ± 0.9 (2.5-5.0) μm; excretory pore to anterior end = 154.6 ± 11.7 (138-170) μm; centre of median bulb from anterior end = 87.1 ± 9.4 (75-100) μm; tail = 9.7 ± 1.2 (7.5-12.5) μm; spicules = 31.2 ± 3.8 (25-38) μm; gubernaculum = 8.9 ± 1.2 (7.5-11.0) μm; a = 40.4 ± 6.6 (28.4-53.0); b = 5.8 ± 0.9 (10.6-13.5); c = 131.3 ± 26 (95.0-179.7).

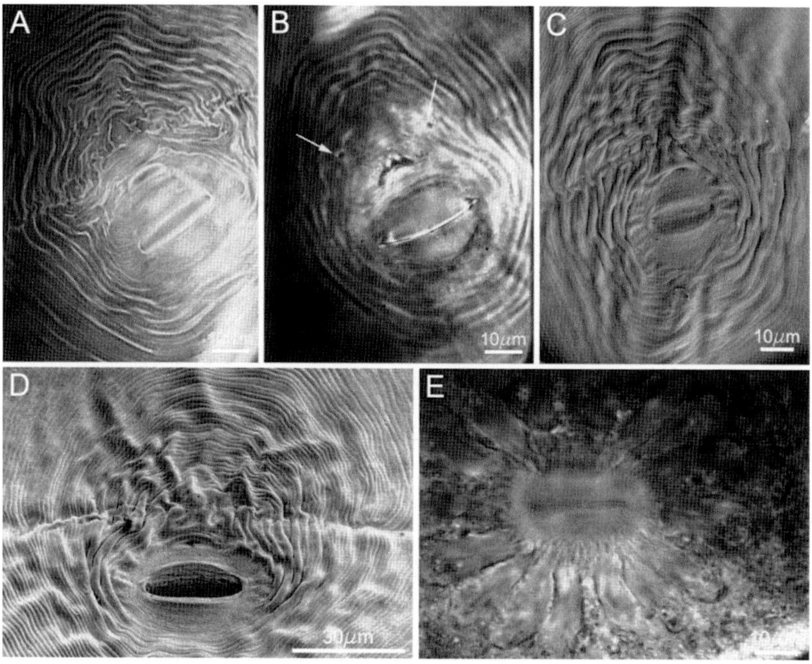

Fig. 251. Meloidogyne thailandica. *LM and SEM. A-E: Perineal pattern (phasmid arrowed). (After Handoo* et al., *2005.)*

- *Paratype J2* (n = 25): L = 484 ± 25.5 (450-540) μm; W = 14.3 ± 0.6 (13-15) μm; stylet = 10.2 ± 0.3 (10-11) μm; DGO = 2.9 ± 0.3 (2.5-3.5) μm; pharyngeal lobe base to anterior end = 141.5 ± 14.8 (117-170) μm; middle of median bulb to anterior end = 60.6 ± 1.8 (55.0-62.5) μm; excretory pore from anterior end = 89.2 ± 6.7 (80-110) μm; tail = 61.2 ± 3.0 (55-65) μm; hyaline region = 18.3 ± 1.9 (15-20) μm; a = 33.9 ± 1.9 (30.3-37.5); b = 3.4 ± 0.3 (2.8-4.0); c = 7.9 ± 0.4 (7.2-8.6).

DESCRIPTION

Female

Body pearly white, variable in size, round to pear-shaped with relatively distinct and variable-size neck, sometimes bent at various angles to body. Labial framework weak, hexaradiate, lateral sectors slightly enlarged, vestibule and extension prominent. Cephalids not observed. Labial region not offset, with labial disc, labial region with two an-

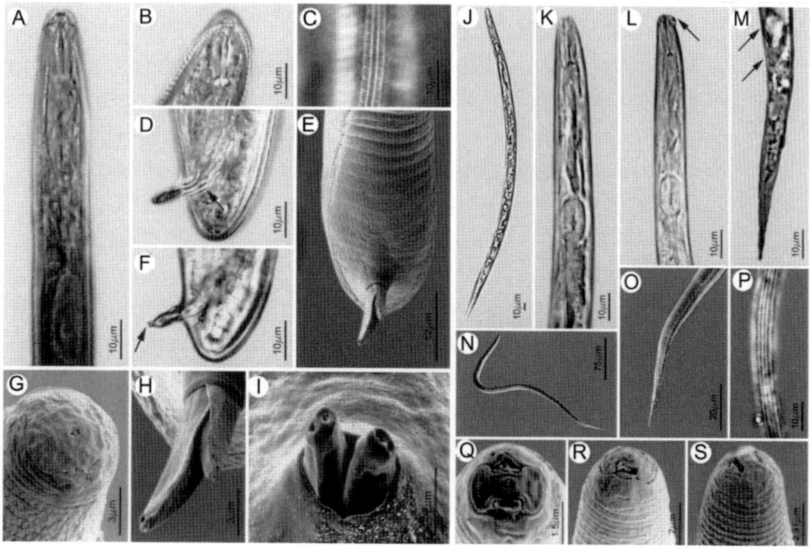

Fig. 252. Meloidogyne thailandica. *LM and SEM. A: Male anterior region; B: Male labial region; C: Male lateral field; D-F: Male tail (spicules arrowed); G: Male* en face *view; H, I: Spicules; J, N: Entire second-stage juvenile (J2); K, L: J2 anterior region (*Pasteuria *sp. spore arrowed); M, O: J2 tail (anus and rectal glands arrowed); P: J2 lateral field; Q-S: J2 labial region. (After Handoo* et al.*, 2005.)*

nuli, first annulus being slightly larger and more expanded than second. SEM observations revealing labial disc fused with median labials and dumbbell-shaped, lateral labials indistinct and fused with median labials, and amphidial openings elongate, located outside indistinct fused lateral labials. Stylet strong, with rounded, posteriorly sloping knobs, cone and shaft usually straight. However, occasionally cone slightly bent. Excretory pore distinct, generally located 1-3 stylet lengths posterior to stylet base. Pharynx well developed with elongate cylindrical procorpus and large, rounded median bulb with heavily sclerotised valve. Body cuticle thick at mid-body, thinner near anterior end of neck. Perineal pattern oval to rectangular with smooth to moderately wavy and coarse striae and characteristic radial structures present beneath pattern area, dorsal arch high, sometimes round to rectangular, striae in and around anal area forming a thick network-like pattern interrupted by lateral lines from each lateral side, vulva and anus slightly sunken. Phasmids large and distinct.

Male

Body cylindrical, vermiform, length variable, with both long and short forms, tapering anteriorly; bluntly rounded posteriorly. Labial region continuous, rounded with three or four annuli. In SEM (face view) labial disc high and narrower than labial region, continuous with median labials, median labials extending some distance into labial region, lateral labials absent; prestoma hexagonal, stomatal opening slit-like, located in large hexagonal prestoma and amphidial openings appearing as long slits. Body cuticle with transverse annulation. Lateral field with four areolated incisures. Stylet robust, cone straight, pointed, knobs large, rounded. Hemizonid indistinct. Excretory pore variable in position, usually near anterior half of basal pharyngeal bulb, in some specimens more posterior. Median pharyngeal bulb large, oval-shaped, measuring about 30-35 μm. Spicules arcuate, commonly with an acutely angled shaft with a bidentate tip, gubernaculum distinct, short, simple. Tail short, rounded to conoid.

J2

Body small, vermiform, tapering at both extremities, but more so posteriorly. Labial region truncate, slightly offset with labial disc; labial framework weak. SEM observations confirming presence of two or three incomplete striations on labial region and on large postlabial annulus. In SEM, stoma slit-like, located in round-shaped prestoma, surrounded by six pore-like openings of inner labial sensilla, median labials and labial disc bow-tie-shaped, labial disc slightly rounded, raised above crescentic median labials, lateral labials large and triangular, lower than labial disc and median labials, amphidial openings appearing as long slits located between labial disc and lateral labials. Stylet delicate, with small, posteriorly sloping rounded knobs. Cuticular annulations fine, distinct. Lateral field prominent, with four incisures. Excretory pore usually near beginning of basal pharyngeal bulb. Hemizonid prominent, about two annuli long, 1-2 annuli anterior to excretory pore. Phasmids indistinct. Rectal glands dilated. Tail long and slender with a long, gradually tapering hyaline region ending in a rounded terminus.

TYPE PLANT HOST: Ginger, *Zingiber* sp.

OTHER PLANTS: No other hosts were reported.

TYPE LOCALITY: Thailand. Infected plants were intercepted by Animal and Plant Health Inspection Service at the port of San Francisco, California, USA. The plants had been purchased at a Bangkok market supplied by local growers.

DISTRIBUTION: *Asia*: Thailand (Handoo *et al.*, 2005b).

PATHOGENICITY: The roots exhibited galling typical of RKN. Heavily infected roots were dark brown to black and from each infected root area clusters of 1-4 RKN females with attached egg masses were recovered.

CHROMOSOME NUMBER: No available data.

POLYTOMOUS KEY CODES: *Female*: A213, B32, C213, D6; *Male*: A34, B34, C213, D1, E1, F1; *J2*: A21, B3, C23, D12, E32, F2.

BIOCHEMICAL AND MOLECULAR CHARACTERISATION: No biochemical data are available for this species. The IGS, ITS1 and D2-D3 of 28S rRNA gene sequences and *COII*-16S rRNA region separate this species from *M. arenaria*, *M. incognita* and *M. javanica*, but differences occur in only a few nucleotide positions (Handoo *et al.*, 2005b; Skantar *et al.*, 2008).

RELATIONSHIPS (DIAGNOSIS): The species belongs to Molecular group I, the Incognita group. *Meloidogyne thailandica* is morphologically similar to *M. incognita*, *M. arenaria*, *M. microcephala*, *M. enterolobii* and *M. lopezi*. From *M. arenaria*, *M. microcephala* and *M. enterolobii* it can be separated by perineal pattern and other morphological characters. From *M. incognita* it differs by having, in the J2, a longer body and longer tail with a long, gradually tapering, broadly rounded terminus *vs* shorter tail with a shorter, subacute terminus. From *M. lopezi* it differs in female stylet length, J2 hyaline region and lateral field areolation in males.

94. ***Meloidogyne trifoliophila*** **Bernard & Eisenback, 1997**

(Figs 253, 254)

COMMON NAME: Clover root-knot nematode.

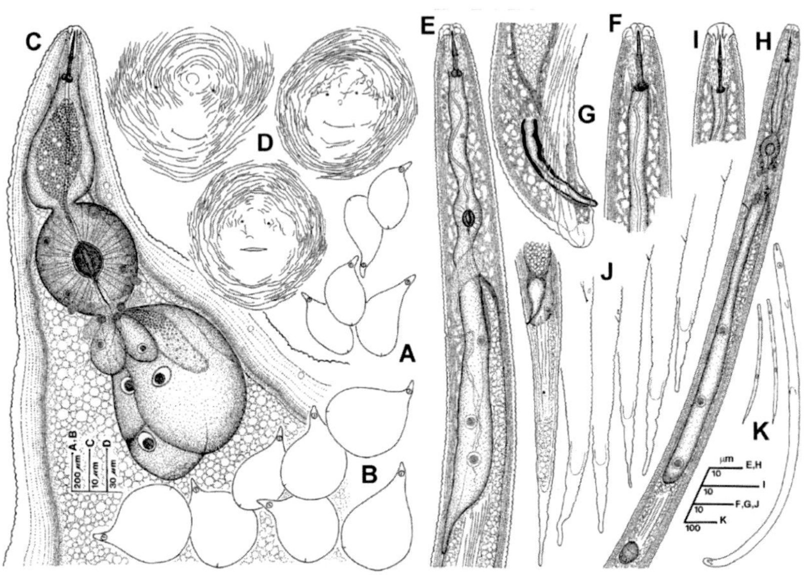

Fig. 253. Meloidogyne trifoliophila. *A, B: Entire female; C: Female anterior region; D: Perineal pattern; E: Male anterior region; F: Male labial region; G: Male tail; H: Second-stage juvenile (J2) anterior region; I: J2 labial region; J: J2 tail; K: Entire J2 and male. (After Bernard & Eisenback, 1997.)*

Meloidogyne trifoliophila was originally collected from a mixed tall fescue-white clover pasture at Ames Plantation, Fayette County, Tennessee, USA. This species produced spongy, spherical galls on white clover but not on fescue. Clover RKN is an important parasite in *Trifolium* spp. and soybean (Bernard & Jennings, 1997). This species may have been misidentified in the past as *M. graminicola*, which has been reported to parasitise many dicotyledonous hosts.

MEASUREMENTS *(AFTER BERNARD & EISENBACK, 1997)*

See Table 52.

DESCRIPTION

Female

Body milky white, oval to pear-shaped, neck usually short, not bent, vulval region not protuberant. Labial framework weakly developed, labial region low, rounded. In SEM, labial disc obscure, fused with median labials. Median labials divided into distinct labial pairs, each labial

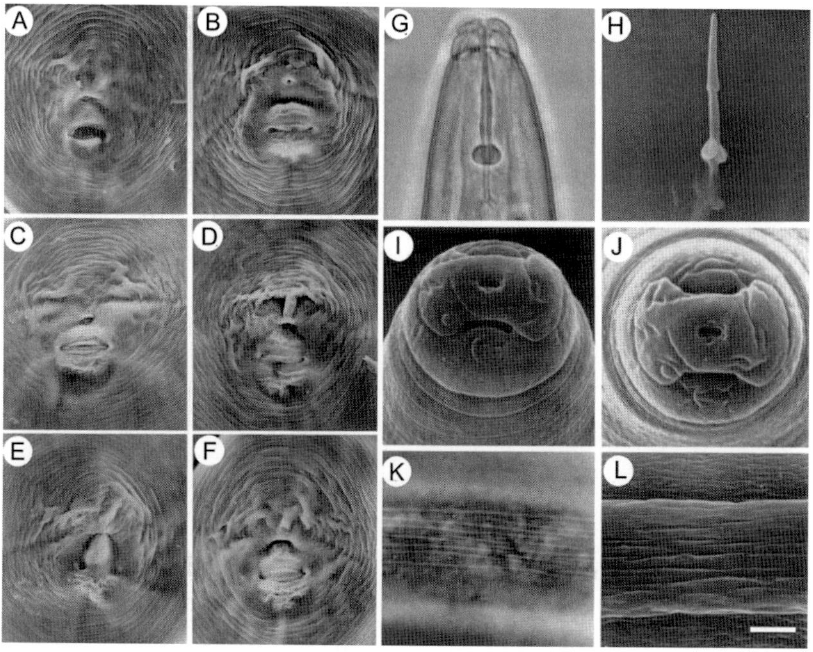

Fig. 254. Meloidogyne trifoliophila. *LM. A-F: Perineal pattern; G, I, J: Male labial region; H: Male stylet; K, L: Male lateral field. (Scale bar = 15 μm.) (After Bernard & Eisenback, 1997.)*

bilobed. Lateral labials distinct, triangular. Labial annulus not clearly demarcated, anterior annuli sometimes with bead-like extensions. Excretory pore located level with stylet knobs or up to one stylet length posterior to knobs, excretory canal visible well past pharyngeal glands. Stylet cone curved dorsally, knobs bilobed, with only slight tapering onto stylet shaft. DGO to stylet knobs less than one shaft length. Dorsal gland ampulla large. Vesicles not observed along lumen in anterior median bulb. Dorsal pharyngeal gland large, *ca* twice volume of smaller subventral glands. Two pharyngo-intestinal cells attached dorsally to median bulb. Vulva of mature females offset from body axis. Perineal patterns typically rounded, striae fine, smooth to slightly wavy, infrequently forked. Dorsal arch smoothly rounded, in SEM with thickened striae posterior to anus in phasmid region. Lateral field obscure or absent in LM, sometimes indicated by a few short, forked striae in SEM, lateral field indicated by weak depression and forking of striae. Perivulval region usually free of striae. Phasmids posterior to anus or rarely at anus level.

Table 52. *Morphometrics of females, males, second-stage juveniles (J2) and eggs of* Meloidogyne trifoliophila. *All measurements are in μm and in the form: mean ± s.d. (range).*

Character	Holotype, Tennessee, USA, Bernard & Eisenback. (1997)	Paratypes, USA, Bernard & Eisenback. (1997)	Australia, Zahid *et al.* (2001)	New Zealand, Zhao *et al.* (2017)
Female (n)	1	20	25	
L	491	537 ± 50.1 (475-626)	534.6 (479-626)	-
a	1.4	1.5 ± 0.2 (1.3-2.1)	-	-
W	371	365 ± 52.4 (237-429)	368.5 (240-430)	-
Stylet	13.3	14.1 ± 0.9 (12.6-15.5)	14.3 (12.5-15.7)	-
Stylet knob width	2.4	2.4 ± 0.3 (1.9-2.9)	-	-
Stylet knob height	1.8	1.9 ± 0.3 (1.5-2.3)	-	-
DGO	4.2	3.8 ± 0.5 (3.0-4.4)	3.8 (3.0-4.6)	-
Ant. end to median bulb valve	64.6	64.8 ± 2.4 (61.6-68.9)	-	-
Ant. end to excretory pore	23.7	22.6 ± 2.6 (17.0-26.2)	-	-
Vulval slit	-	26.3 ± 3.2 (20-34.7)	-	-
Vulva to anus	-	19.2 ± 3.7 (12.4-29.1)		-
Interphasmid distance	-	17.9 ± 2.31 (14.1-23.8)	-	-
EP/ST		1.0-2.0*	-	-
Male (n)		20	25	
L	-	1077 ± 80.5 (915-1207)	1082.6 (916-1205)	-
a	-	32.3 ± 2.4 (26.5-35.8)	-	-

Table 52. *(Continued.)*

Character	Holotype, Tennessee, USA, Bernard & Eisenback. (1997)	Paratypes, USA, Bernard & Eisenback. (1997)	Australia, Zahid *et al.* (2001)	New Zealand, Zhao *et al.* (2017)
c	-	115 ± 17.8 (89-151)	-	-
W	-	33.5 ± 3.6 (26.8-39.0)	32.3 (28-39)	-
Stylet	-	18 ± 0.6 (17.0-18.9)	18.0 (17.0-19.0)	-
Stylet knob width	-	2.5 ± 0.3 (1.9-3.0)	-	-
Stylet knob height		2.6 ± 0.4 (2.0-3.1)	-	-
DGO	-	4.5 ± 0.9 (3.4-6.5)	4.6 (3.5-6.7)	-
Ant. end to median bulb valve		68.6 ± 5.8 (58.2-77.6)	-	-
Ant. end to excretory pore	-	112 ± 7.4 (97-123)	-	-
Tail	-	9.7 ± 1.3 (7.3-11.6)	-	-
Spicules	-	30 ± 2.1 (27.2-33.5)		-
Gubernaculum	-	8.7 ± 0.9 (7.3-9.7)		-
J2 (n)		20	25	14
L	-	379 ± 15.2 (357-400)	384.9 (357-408)	413 ± 20.9 (378-461)
a	-	29.4 ± 2.3 (26.9-33.6)	-	28.1 ± 3.0 (23.2-33.9)
b				7.1 ± 0.6 (6.1-8.2)
c	-	5.5 ± 0.7 (4.2-6.2)	-	6.6 ± 0.4 (5.9-7.2)
c′	-	6.9 ± 0.8 (5.7-8.6)	-	6.1 ± 1.1 (4.0-7.1)

Table 52. *(Continued.)*

Character	Holotype, Tennessee, USA, Bernard & Eisenback. (1997)	Paratypes, USA, Bernard & Eisenback. (1997)	Australia, Zahid *et al.* (2001)	New Zealand, Zhao *et al.* (2017)
W	-	12.9 ± 1.0 (11.0-14.3)	13.9 (11.1-15.0)	14.8 ± 1.2 (12.6-16.3)
Stylet	-	12.7 ± 0.5 (11.9-13.6)	13 (11.1-13.8)	10.9 ± 0.6 (10.1-11.9)
DGO	-	3.6 ± 0.6 (2.9-4.6)	3.5 (3.0-4.5)	-
Ant. end to median bulb valve		49.8 ± 3.4 (44.6-53.8)		-
Ant. end to excretory pore	-	70.1 ± 3.8 (65.5-76.1)	-	70.1 ± 4.6 (62-73)
Pharynx length	-	-	-	58.4 ± 5.6 (51-71)
Tail	-	69.9 ± 8.7 (60.6-87.3)	-	63.2 ± 5.2 (59-79)
Hyaline region	-	16.4 ± 0.6 (15.1-19.3)	-	20.8 ± 2.0 (18.6-26.2)
Body diam. at anus	-	10.1 ± 1.1 (8.7-12.9)		10.7 ± 1.9 (9.2-11.7)
Egg				
L	-	-	83.9 (72.0-91.0)	-
W	-	-	39.1 (34.0-40.0)	-

* From description.

Male

Body vermiform, tapering anteriorly, tail conoid, bluntly rounded, twisting 90°. Labial cap high, rounded-truncated. Labial region distinct from first body annulus, without annulation or with a few short, irregular lines. In SEM, stoma slit-like, prestoma narrowly oval and surrounded by pits of inner labial sensilla. Labial disc shape irregular, partially demarcated by short grooves, median labials fused on each side, rounded, with several straight and curved grooves possibly associated with labial sensilla. Lateral labials weakly demarcated, not protuberant. Amphid apertures elongate slits. Vestibule and vestibule extension distinct. Stylet robust, cone bluntly pointed at tip, widening at junction with shaft, shaft cylindrical, smooth, slightly narrowing anteriorly. Knobs oval, indented longitudinally, tapering onto shaft, rounded posteriorly. Dorsal gland ampulla indistinct. Procorpus thick, almost as wide as median bulb. Pharyngo-intestinal junction at level of nerve ring. Pharyngeal gland lobe long, with three nuclei, one nucleus level with excretory pore, other two usually in posterior half of lobe, occasionally, one of posterior nuclei in anterior half of lobe just posterior to anteriormost nucleus. Anterior intestinal caecum absent. Excretory canal prominent, twisted irregularly, visible to end of gland lobe. Excretory pore 1-2 annuli posterior to hemizonid, hemizonion about 13 annuli posterior to hemizonid. Lateral field not areolated, composed of numerous frequently broken or forked incisures, appearing as about eight lines in LM, appearing more numerous in SEM. In posterior region, incisures more broken, with posteriormost incisures merging with lateral field terminus. Spicules weakly cephalated, distal two-thirds nearly straight, gubernaculum weakly curved in lateral view. Phasmids not seen.

J2

Body slender, tapering to an elongate tail. Labial framework weak, vestibule and vestibule extension distinct but not strongly sclerotised, except at anterior terminus. Stylet slender, cone weakly expanding at junction with shaft, knobs rounded and tapering. Distance from stylet knobs to dorsal gland orifice less than shaft length. Ampulla obscure. Median bulb broadly oval, valve large and heavily sclerotised. Gland lobe very long, nearly reaching genital primordium, nuclei equally spaced in posterior half. Pharyngo-intestinal junction at level of nerve ring. Excretory pore slightly posterior to nerve ring, opening through hemizonid. Rectum dilated. Lateral field with four incisures, not areolated. On tail, lateral field with two incisures not fusing terminally. Phasmids small, distinct, slightly offset ventrally within lateral field, about one-third tail length from anus. Caudalid sometimes visible, three annuli anterior to anus. Tail elongate-conoid, tapering to a slender, terminal, digitiform process.

TYPE PLANT HOST: White clover, *Trifolium repens*, growing in a mixed clover-tall fescue pasture.

OTHER PLANTS: An extensive study on additional hosts was conducted by Bernard & Jennings (1997). All species of clover (*Trifolium* spp.) were severely galled, regardless of species or cultivar (Bernard & Eisenback, 1997; Bernard & Jennings, 1997). Among other legumes, broad bean, garden pea, Korean lespedeza, sweet clover and common vetch were good hosts, but alfalfa, bird's-foot trefoil, peanut and pole bean were poor or non-hosts. Only galls without further development of swollen J2 were found on *Poa annua*, *Paspalum dilatatum* and *Eleusine indica* (Mercer *et al.*, 1997), while in rice, *Alopecurus pratensis* L. and *Echinochloa crus-galli*, only females without egg masses were found in a few galls (Mercer *et al.*, 1997). *Meloidogyne trifoliophila* rarely parasitised most grasses and did not establish itself on rice, wheat, or barley (Bernard & Eisenback, 1997). *Meloidogyne trifoliophila* failed to reproduce on any of the standard North Carolina hosts: cotton, tobacco, pepper, watermelon, peanut and tomato, but reproduced on two of the eight dicotyledon weeds, spear thistle (*Cirsium vulgare*) and pigweed (*Portulaca oleraceae*) (Mercer *et al.*, 1997; Zahid *et al.*, 2001).

TYPE LOCALITY: Ames Plantation, Fayette County, Tennessee, USA.

DISTRIBUTION: *North America*: USA (Tennessee); *Oceania*: New Zealand (Mercer *et al.*, 1997), Australia (north coast of New South Wales and southern Queensland) (Zahid *et al.*, 2000).

PATHOGENICITY: Galls on white clover roots caused by *M. trifoliophila* were elongate (mean length = 4 mm, mean width = 1.5 mm). The *M. trifoliophila* galls were a cream colour and lighter than the rest of the root, and less than 1% had a lateral root. Each gall contained 1-9 females and egg masses. Egg masses of *M. trifoliophila* were typically embedded within the galls rather than protruding from the gall surface (Mercer *et al.*, 1997; Zahid *et al.*, 2011). The glasshouse study indicated that a moderate to high density (>100 nematode/500 cm^3 of soil) of nematodes would be required to markedly reduce white clover yield. The ability of *M. trifoliophila* to suppress white clover growth under controlled conditions emphasises the potential importance of this nematode (Zahid *et al.*, 2011).

POLYTOMOUS KEY CODES: *Female*: A32, B32, C32, D3; *Male*: A43, B34, C2, D2, E2, F2; *J2*: A32, B2, C12, D12, E3, F12.

BIOCHEMICAL AND MOLECULAR CHARACTERISATION: Esterase pattern of a single band and a malate dehydrogenase isozyme phenotype of two bands (Mercer *et al.*, 1997). Nucleotide sequences of ITS and D2-D3 of 28S rRNA genes are deposited in GenBank (Hugall *et al.*, 1994; Mercer *et al.*, unpubl.).

RELATIONSHIPS (DIAGNOSIS): The species belongs to Molecular group III, the Graminicola group, and is morphologically similar to *M. graminicola*. The presently available molecular data do not allow the reliable separation of these species from each other. The host ranges of *M. trifoliophila* and *M. graminicola* also overlap. These species can be differentiated from each other mainly by female characters. Females of *M. trifoliophila* differ from *M. graminicola* by the perineal pattern: round, striae smooth, dorsal arch without prominent interruptions or ridges *vs* dorsoventrally elongated with prominent ridges and angled striae in the dorsal arch of *M. graminicola*, and by the position of the female excretory pore at one stylet length or less posterior to stylet knobs in *M. trifoliophila vs* more than one stylet length in *M. graminicola*.

95. *Meloidogyne triticoryzae* Gaur, Saha & Khan, 1993
(Figs 255, 256)

COMMON NAME: Indian rice root-knot nematode.

Meloidogyne triticoryzae was first described by Gaur *et al.* (1993) and its hosts include rice, wheat, barley, oats, sorghum, and a number of other graminaceous and some dicotyledonous crops and weeds in India. This nematode is adapted to the cropping system in which a rice crop in the rainy season is followed by wheat or another graminaceous host in the winter (Gaur *et al.*, 1993; Gaur & Sharma, 1998).

MEASUREMENTS *(AFTER GAUR ET AL., 1993)*

- *Holotype female*: L = 395 μm; W = 280 μm; stylet = 13 μm; DGO = 3.5 μm; excretory pore from anterior end = 19 μm; median bulb length = 27 μm, median bulb width = 24 μm; anterior end to median bulb valve = 55 μm; a = 1.4; b′ = 2.8; EP/ST = 1.5.
- *Paratype females* (n = 20): L = 425 $\pm$ 46.1 (330-480) μm; W = 242 $\pm$ 33.4 (200-320) μm; stylet = 13.2 $\pm$ 0.7 (12-14) μm; median bulb length = 26.1 $\pm$ 1.5 (24-28) μm; median bulb width = 21.7 $\pm$ 0.8 (21-

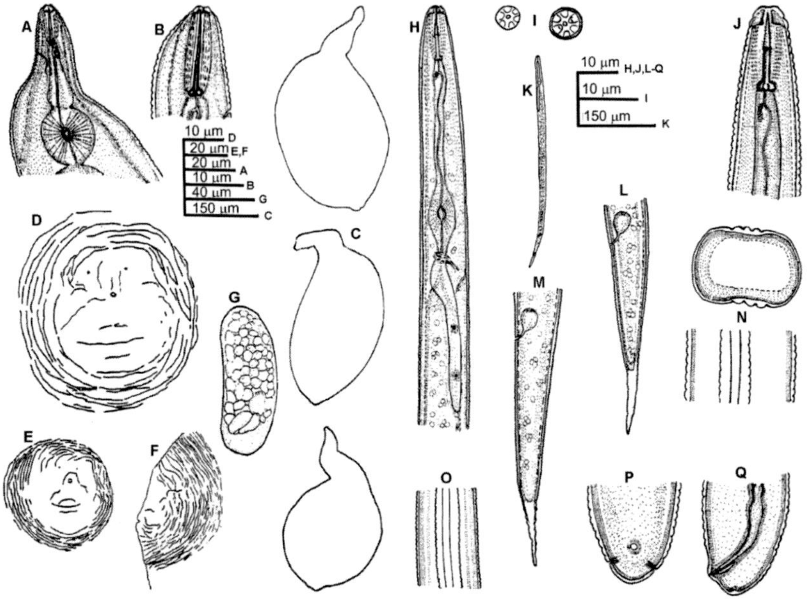

Fig. 255. Meloidogyne triticoryzae. *A: Female anterior region; B: Female labial region; C: Entire female; D-F: Perineal pattern; G: Egg; H: Male anterior region; I: Male* en face *view; J: Male labial region; K: Entire second-stage juvenile (J2); L, M: J2 tail; N: J2 lateral field; O: Male lateral field; P, Q: Male tail. (After Gaur* et al.*, 1993.)*

23) μm; vulval slit = 18 ± 1.6 (16-21) μm; interphasmid distance = 9.4 ± 0.8 (8-11) μm; vulva to anus = 19.1 ± 1.6 (15-24) μm; a = 1.8 ± 0.3 (1.4-2.2); EP/ST = 1.5 ± 0.1 (1.4-1.6).

- *Paratype males* (n = 15): L = 1305 ± 180 (1100-1590) μm; stylet = 17.5 ± 0.5 (17-19) μm; excretory pore to anterior end = 90-128 μm; centre of median bulb from anterior end = 62.5 ± 7.9 (56-76) μm; median bulb length = 18.3 ± 1.1 (17-20) μm; median bulb width = 9.0 ± 1.6 (7-11) μm; tail = 4.6 ± 1.0 (3.5-6.5) μm; spicules = 29 ± 2.2 (26-32) μm; gubernaculum = 9.8 ± 1.1 (8-11) μm; a = 63.5 ± 10.5 (55-74); b′ = 7.9 ± 2.2 (5.4-10.5); c = 293 ± 69.6 (216-397); c′ = 0.3 ± 0.1 (0.2-0.4).
- *Paratype J2* (n = 20): L = 395.7 ± 12.8 (380-425) μm; stylet = 12.1 ± 0.5 (11.5-13.0) μm; DGO = 2.4 ± 0.5 (2.0-3.0) μm; middle of median bulb to anterior end = 45.6 ± 1.2 (44-48) μm; median bulb length = 11.3 ± 0.5 (11-12) μm; median bulb width = 5.7 ±

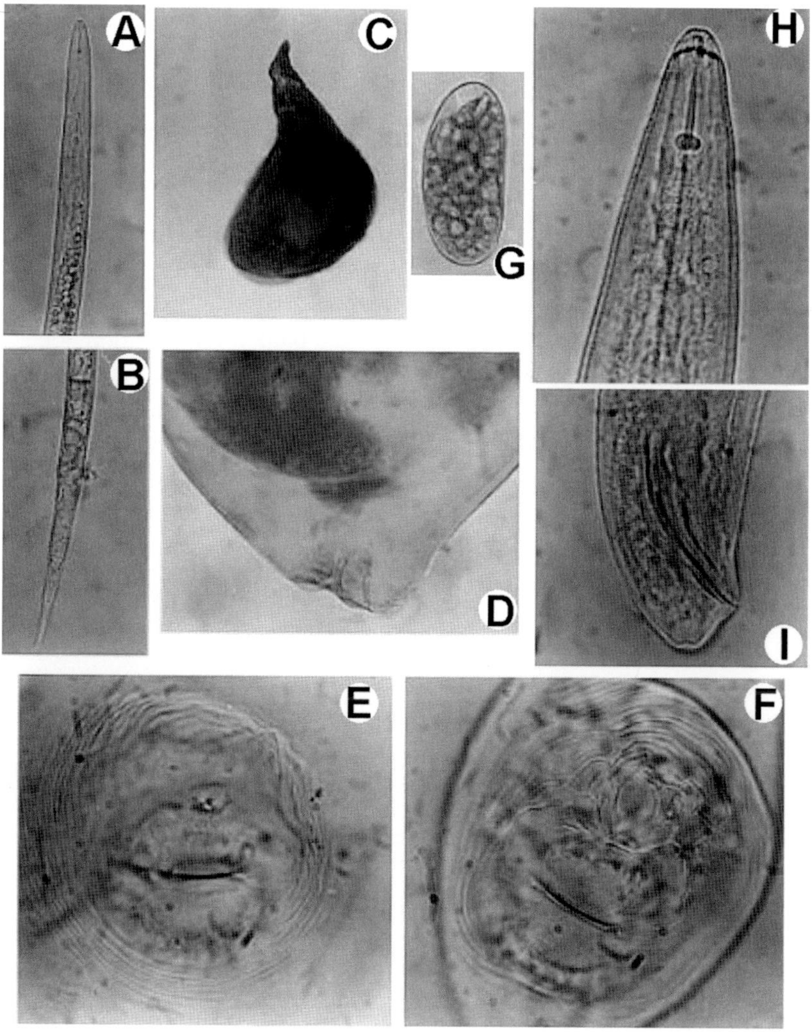

Fig. 256. Meloidogyne triticoryzae. *LM. A: Second-stage juvenile (J2) anterior region; B: J2 tail; C: Entire female; D-F: Perineal pattern; G: Egg; H: Male labial region; I: Male tail. (After Gaur* et al., *1993.)*

0.5 (5.0-6.0) μm; excretory pore to anterior end = 65.6 ± 1.8 (63-68) μm; tail = 61.7 ± 3.6 (50-68) μm; hyaline region = 17.6 ± 1.0 (16-19) μm; a = 36.4 ± 2.5 (29.7-39.7); b = 8.7 ± 0.2 (8.3-9.1); b′ = 3.7 ± 0.5 (3.4-4.0); c = 6.4 ± 0.5 (5.7-7.4); c′ = 6.2 ± 0.4 (5.5-6.8).

- *Eggs*: L = 76-93 μm; W = 28-35 μm.

DESCRIPTION

Female

Females elongate lemon-shaped to nearly rounded with elevated perineum, neck asymmetrically located and bent to ventral side of longitudinal body axis, cuticular striation more clear in labial and perineal regions. Labial region continuous (appearing slightly offset in some specimens), weakly sclerotised, stylet with posteriorly sloping rounded knobs, procorpus irregularly swollen, offset from well-developed rounded to slightly ovoid median bulb. Excretory pore located near or slightly posterior to level of DGO. Ovaries two, prodelphic, highly coiled, perineal pattern on top of protuberance rounded to slightly ovoid dorsoventrally, dorsal arch hemispherical with widely spaced striae. Striae smooth and mostly continuous except near tail tip and phasmids where they are broken and wavy forming angular shapes. No distinct tail whorl and punctations, anus with a slight dorsal cuticular fold. Lateral field not marked by lateral lines, but some broken striae without forking seen, ventral arch hemispherical and marked by smooth but relatively closer striae.

Male

Highly variable in length. Labial region high with 2-3 annuli, continuous with body contour, labial cap narrower in dorsoventral view, hexaradiate labial framework moderately sclerotised, six labials with large slit-like amphidial openings on lateral lines. Buccal vestibule and its extension distinct. Stylet moderately sclerotised with elongate posteriorly-sloping rounded knobs. Pharynx well developed with ovoid median bulb. Excretory pore well posterior to nerve ring level, hemizonid preceding it by one annulus. Tail rounded to asymmetrically conoid with a smooth terminus, slightly bulging cloacal aperture, spicules ventrally arcuate, phasmids on mid-tail. Upon killing by gentle heat, posterior portion of body twisting by about 90°. Lateral field with four incisures, no areolation seen but outermost incisures crenate. Males rare and reproduction may be mostly parthenogenetic.

J2

Slightly ventrally arcuate upon killing by gentle heat. Labial region weakly sclerotised, hemispherical, continuous with body contour and smooth. Stylet weakly sclerotised with rounded, posteriorly-sloping knobs. Median bulb ovoid with weakly sclerotised valve, posterior pharyngeal gland lobe ventral to intestine. Rectum dilated, tail long, tapering to a finely rounded terminus, constricted at one or two places. Lateral field marked by four incisures, no areolation seen, *en face* view showing six labials.

TYPE PLANT HOST: Wheat, *Triticum aestivum* L.

OTHER PLANTS: Rice (*Oryza sativa* L.), common weeds including *Phalaris minor* Retz., *Cynondon dactylon*, *Cyperus rotundus*, *Echinochloa* spp., *Poa annua*, *Spergula arvensis* L., *Anagallis arvensis* L., *Melilotus indicus* (L.) All., and *Rumex dentatus* (Gaur *et al.*, 1993; Rich *et al.*, 2008). *Echinochloa colona*, *E. crus-galli* and *Leptochloa coloniculus* P. Beauv. were reported as hosts (Gaur & Sharma, 1998). This species apparently prefers graminaceous hosts but can also reproduce on some dicotyledons including legumes, soybean (Gaur *et al.*, 1993). Radish (*Raphanus sativus*), fennel (*Foniculum vulgare* Mill.), cowpea (*Vigna unguiculata*), green-gram (*V. radiata* (L.) R.Wilczek) and sunflower (*Helianthus annuus*) were good hosts for *M. triticoryzae* (Sabir & Gaur, 2005).

TYPE LOCALITY: Indian Agricultural Research Institute Farm, New Delhi, India.

DISTRIBUTION: *Asia*: India (Gaur *et al.*, 1993).

PATHOGENICITY: Unthrifty growth of rice and wheat crops occurred under wheat-rice cropping systems for 4-5 years. Both crops exhibited stunted growth, chlorosis, poor tillering and declining yield, the symptoms being more apparent in young plants and generally more pronounced on irrigated upland rice than on wheat. The roots of both wheat and rice plants had numerous small, rounded elongated galls and club-shaped root tips that contained juvenile and adult stages and egg sacs embedded in root tissue. Fine lateral roots in large numbers were often produced near the gall. The club-root symptom was more common in rice than on wheat, while the cortical hypertrophy occurred more in wheat, sometimes producing irregularly-shaped galls (Gaur *et al.*, 1993).

BIOLOGY: Gaur & Chandel (1997) found that *M. triticoryzae* survived in moist and wet soil for over 18 months in the absence of a host.

CHROMOSOME NUMBER: No available data.

POLYTOMOUS KEY CODES: *Female*: A3, B3, C3, D3; *Male*: A324, B34, C23, D1, E2, F1; *J2*: A32, B2, C23, D1, E3, F21.

BIOCHEMICAL AND MOLECULAR CHARACTERISATION: No available data.

RELATIONSHIPS (DIAGNOSIS): *Meloidogyne triticoryzae* is similar to *M. graminis* but has a different perineal pattern, shorter females (330-480 *vs* 541-841 μm) and males (1100-1590 *vs* 900-1743 μm). *Meloidogyne triticoryzae* is also morphologically quite distinct from *M. graminicola* from which it differs by shorter J2 (380-425 *vs* 403-535 μm), shorter females (330-480 *vs* 445-857 μm) and EP/ST ratio (1.4-1.6 *vs* 2.9).

96. *Meloidogyne turkestanica* Shagalina, Ivanova & Krall, 1985
(Figs 257, 258)

COMMON NAME: Turkmenistan root-knot nematode.

This species was described after a survey of plants in the Karakum Desert, in the sands of Kyrdzhalakum, and on the slopes of Vaksh Mountain, Turkmenistan (Shagalina *et al.*, 1985).

MEASUREMENTS

See Table 53.

DESCRIPTION

Female

Body shape varying from spherical to pear-shaped. Neck short, slightly bent to ventral side. No projection on posterior end. Anterior end offset from body, basal annulus 5.5-9.0 (6.7) μm wide. Six labials, labial disc and hexaradiate skeleton. Amphidial apertures slit-like, located on

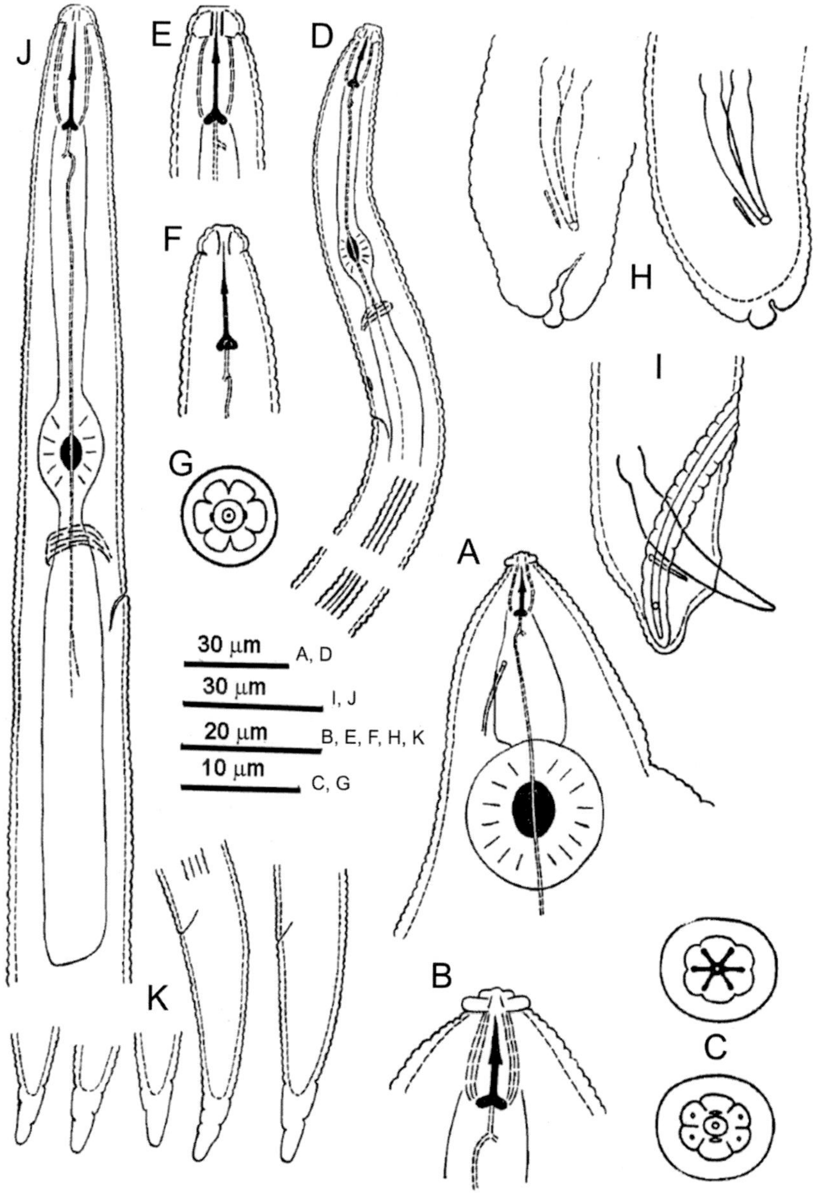

Fig. 257. Meloidogyne turkestanica. *A: Female anterior region; B: Female labial region; C: Female* en face *view; D: Male anterior region; E, F: Male labial region; G: Male* en face *view; H, I: Male tail; J: Second-stage juvenile (J2) anterior region; K: J2 tail. (After Shagalina* et al., *1985.)*

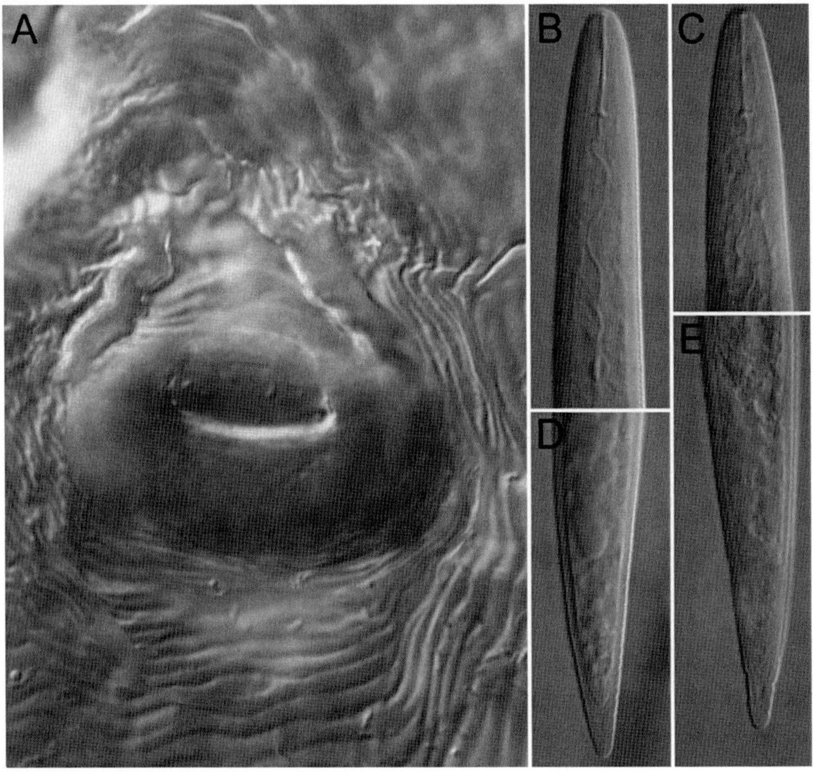

Fig. 258. Meloidogyne turkestanica. *A: Perineal pattern; B, C: Second-stage juvenile (J2) anterior region; D, E: J2 posterior region. (Original photos.)*

inside edge of lateral labials. Small papillae on sublateral labials. Excretory pore located 15-40 μm (1.0-2.5 stylet lengths) from anterior end. Thickness of cuticle at basal knob level = 1.8 (1.5-2.5) μm, near base of neck = 3.7 (2.5-5.5) μm, in middle part of body = 7.4 (5.0-9.0) μm. Stylet slightly bent, metenchium and telenchium of same length. Basal knobs rounded, projecting posteriorly. Median bulb rounded with developed valve apparatus. Cuticle in anal-vulval area often elongated and resembling symbol 8, but sometimes rounded and flattened. Lines thin and wave-like; only around tail rudiment and anus are they slightly broken, from ventral arc often directed to side, forming 'wings'. External lines very thin, at level of dorsal arch directed straight up or fan-like to either side. From tail rudiment to vulval opening and below, and from both sides, body covered by smooth cuticle without folds. Length of 'smooth

Table 53. *Morphometrics of females, males, second-stage juveniles (J2) and eggs of* Meloidogyne turkestanica. *All measurements are in μm and in the form: mean ± s.d. (range).*

Character	Holotype, allotype, Kulbakan, Turkmenistan, Shagalina *et al.* (1985)	Paratypes, Kulbakan, Turkmenistan, Shagalina *et al.* (1985)	Kabadien, Tajikistan, Shagalina *et al.* (1985)
Female (n)	-	30	8
L	728.5	720.3 (398.7-904)	679.2 (636.4-758.2)
W	573.2	515.1 (348.0-745.8)	379.1 (311.4-433.3)
Stylet	13.8	14.1 (13.8-15.4)	14.5 (13.8-15.2)
Stylet knob width	-	4.0	-
Stylet knob height	-	2.0	-
DGO	-	3.7 (3.5-4.1)	-
Median bulb length	-	34.2 (29.0-41.4)	-
Median bulb width	-	30.3 (22.1-37.5)	-
Ant. end to excretory pore	-	15-40	-
Vulval slit	-	25.2 (20.0-30.1)	-
Vulva to anus	-	20.7 (15.0-26.3)	-
Interphasmid distance	-	15.2 (11.1-18.8)	-
EP/ST	-	1.0-1.3	-
Male (n)	1	10	10
L	811.2	1012.9 (886.4-1288)	1207.8 (954.9-1427.6)
a	43.6	40.5 (32.5-51.6)	41.5 (29.7-55.6)
b	10.2	12.6 (9.1-16.0)	15.1 (13.1-17.1)
c	80.0	88.2 (80-100)	91.9 (70.6-127.1)
T	44.8	53.3 (42.0-60.9)	49.4 (47.0-52.2)
W	18.6	25.2 (22.2-30.5)	29.9 (24.6-33.8)
Stylet	17.1	18.1 (16.3-18.8)	17.6 (17.1-17.9)
Stylet knob width	-	4.3 (3.8-5.0)	-
Stylet knob height	-	2.1 (1.9-2.5)	-
DGO	-	3.3 (3.0-3.7)	-
Ant. end to median bulb valve		80.3 (69.3-92.4)	-

Table 53. *(Continued.)*

Character	Holotype, allotype, Kulbakan, Turkmenistan, Shagalina *et al.* (1985)	Paratypes, Kulbakan, Turkmenistan, Shagalina *et al.* (1985)	Kabadien, Tajikistan, Shagalina *et al.* (1985)
Ant. end to excretory pore	-	106.1 (95.4-130.3)	-
Spicules	25.0	29.8 (25.0-33.9)	26.7 (25.0-27.6)
Gubernaculum	5.5	6.5 (6.1-7.5)	6.4 (5.2-7.6)
J2 (n)	-	14	-
L	-	401.5 (334.2-441.9)	-
a	-	24.2 (20.5-28.7)	-
b′		8.7 (7.1-10.3)	-
c	-	12.6 (11.5-15.4)	-
c′	-	3.0 (2.8-3.2)	-
W	-	16.8 (13.9-19.4)	-
Stylet	-	14.5 (13.4-15.6)	-
DGO	-	3.0 (2.8-3.4)	-
Ant. end to median bulb valve		47.1 (36.9-58.2)	-
Ant. end to excretory pore	-	81.1 (71.1-91.8)	-
Tail	-	32.9 (21.7-35.2)	-
Hyaline region	-	7.0 (6.2-8.3)	-
Body diam. at anus	-	10.9 (10.3-12.5)	-
Egg	-	-	-
L	-	106.0 (91.4-121.6)	-
W	-	58.1 (46.2-68.5)	-

field' is 26.4 (20.0-34.5) μm, widest part is 6.8 (5.0-10.4) μm. Phasmids clearly visible. Lateral fields usually absent; if present, then usually only on one side, difficult to see and very short.

Male

Vermiform, cuticle with narrow annuli 1.5 μm wide. Lateral field occupying 20-25% of body diam., consisting of four incisures that meet on terminus of tail with jagged edges. In middle part of body, a fifth incisure occurs between two internal incisures. Labial region 3.9

(3.8-5.0) μm in height and 8.4 (7.5-10.0) μm in diam., slightly offset from body, with two annuli of which lower one is wider, sclerotisation absent. Six labials, on lateral labials slit-like amphidial apertures are visible. Labial disc rounded. Stylet thin, basal knobs rounded and projecting posteriorly, 2.1 (1.9-2.5) μm high and 4.3 (3.8-5.0) μm diam., metenchium (8.3 μm) shorter than telenchium (9.4 μm). Median bulb oblong-oval with developed valve apparatus. Hemizonid located 4-5 annuli anterior to excretory pore. Nerve ring located just posterior to base of median bulb. Tail very short and blunt; in lateral view near base it is finger-like, on ventral side, close to terminus it is forked and has a small cuticular projection. Cuticle anterior and posterior to cloacal aperture smooth. Phasmids small, located in intermediate part of tail. Spicules slightly bent, gubernaculum small and simple. Anterior end of testis sometimes curved.

J2

Body slightly curved in dorsoventral plane, cuticle with fine annuli. Anterior end slightly offset from body, lacking annuli and not sclerotised, 2.3 (2.1-2.5) μm high and 5.7 (4.8-6.2) μm diam. with six labials. Lateral field occupying 20-25% of body diam., consisting of four incisures. Stylet thin with rounded and posteriorly-sloping basal knobs (1.5 × 2.2 μm), metenchium (7.4-10.0 μm) longer than telenchium (6.0-6.7 μm). Hemizonid not visible. Tail short and conical, 19-23 annuli on ventral side. Phasmids located 7.5-10.5 μm posterior to anus level.

TYPE PLANT HOST: Not indicated.

OTHER PLANTS: Wild trees and bushes: *Sasola richeri* (Moq.) Kar., *Haloxylon perscium* Bge., and *H. aphyllum* (Minkw). Iljin. from Chenopodiaceae, *Calligonum rubescens* Mattei, *C. setosetosum* Litv. from Polygonaccae, and *Tamarix* sp. from Tamaricacae (Shagalina *et al.*, 1985).

TYPE LOCALITY: Kulbakan, Turkmenistan.

DISTRIBUTION: *Asia*: Turkmenistan, Tajikistan (Shagalina *et al.*, 1985).

PATHOGENICITY: The thin roots had small, thickened galls that were darker in colour than the normal root, and only the anterior end of the female was inside the root. The rest is on the surface of the root and is surrounded by an enormous egg sac.

CHROMOSOME NUMBER: No available data.

POLYTOMOUS KEY CODES: *Female*: A213, B32, C3, D3; *Male*: A43, B34, C23, D1, E2, F2; *J2*: A23, B12, C43, D4, E12, F4.

BIOCHEMICAL AND MOLECULAR CHARACTERISATION: No available data.

RELATIONSHIPS (DIAGNOSIS): It differs from *M. ardenensis* and *M. querciana* by the shorter J2 tail and different EP/ST ratio. Additionally, it can be separated from *M. africana* by the shorter J2 tail and c′ ratio, and from *M. vandervegtei* by the c′ ratio.

97. *Meloidogyne vandervegtei* Kleynhans, 1988
(Fig. 259)

COMMON NAME: van der Vegte's root-knot nematode.

This species was described from material collected by Mr F.A. van der Vegte from the Vegetable and Ornamental Plant Research Institute. Galled roots of an unidentified woody plant were found in the coastal subtropical forest in Trafalgar, Natal, South Africa (Kleynhans, 1988).

MEASUREMENTS *(AFTER KLEYNHANS, 1988)*

- *Holotype female*: L = 981 μm; W = 865 μm; stylet = 17.5 μm; DGO = 6.1 μm; excretory pore from anterior end = 14.3 μm; vulval slit = 22.4 μm; vulva to anus = 19.9 μm; vulva to phasmids = 28.8 μm; interphasmid distance = 11.2 μm; EP/ST = 2.6.

Fig. 259. Meloidogyne vandervegtei. *A: Male labial region; B: Male stylet; C, D: Male tail; E: Spicules; F: Entire second-stage juvenile (J2); G: J2 labial region; H: J2 lateral field; I: J2 tail; J: Entire female; K: Female labial region; L: Female tubular pharyngeal lumen; M: Female stylet; N: Perineal pattern. (After Kleynhans, 1988.)*

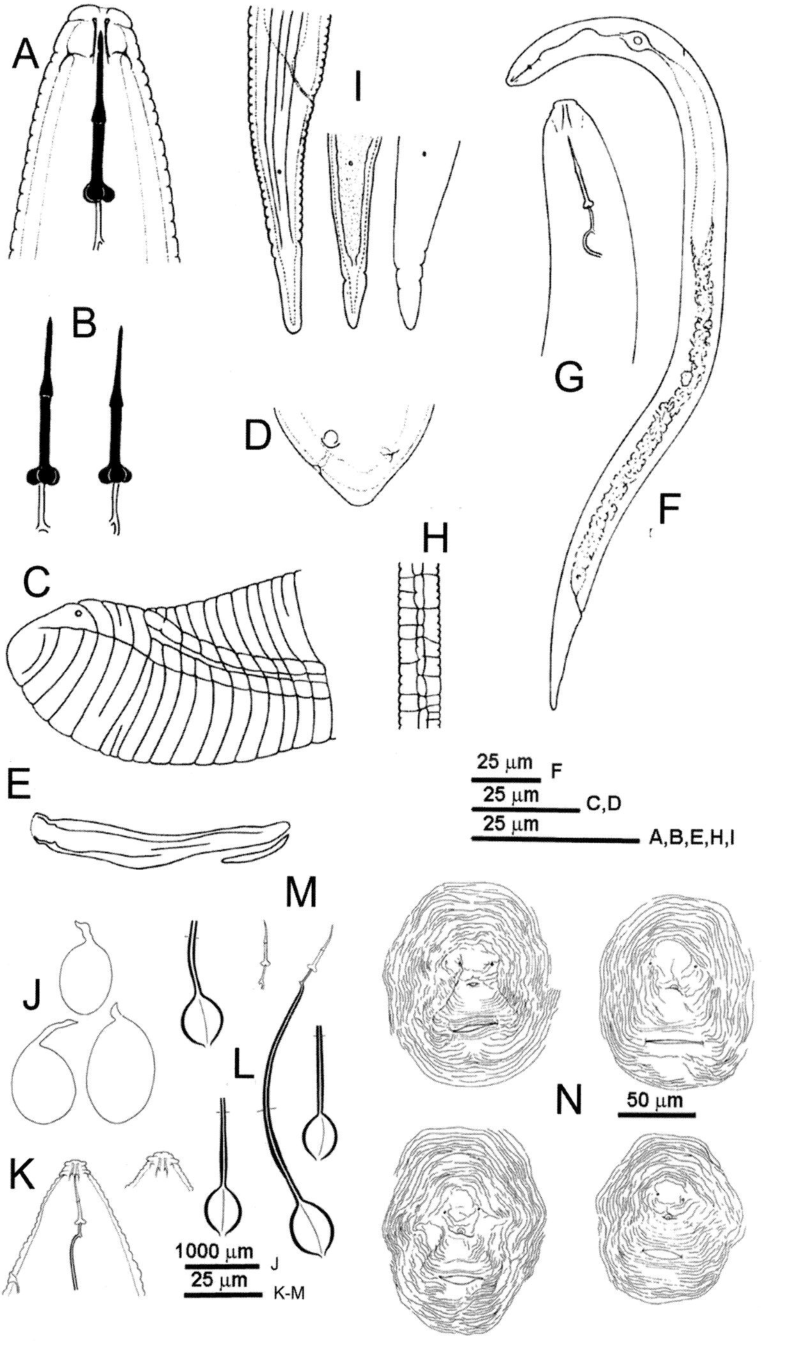
A
B
C
D
E
F
G
H
I
J
K
L
M
N
25 μm F
25 μm C,D
25 μm A,B,E,H,I
1000 μm J
25 μm K-M
50 μm

- *Allotype male*: L = 2051 μm; stylet = 24.8 μm; stylet cone length = 11.6 μm; DGO = 5.8 μm; excretory pore to anterior end = 161 μm; tail = 14.2 μm; phasmids to tail end = 6.4 μm; spicules = 44.1 μm.
- *Paratype females* (n = 16): L = 1249 ± 348 (824-1730) μm; W = 649 ± 168 (391-886) μm; stylet = 17.9 ± 1.3 (16.4-20.1) μm; stylet cone length = 9.6 ± 1.2 (7.2-11.8) μm; DGO = 6.2 ± 1.4 (4.1-9.1) μm; excretory pore from anterior end = 22.7 ± 7.9 (11.4-44.3) μm; vulval slit = 23.7 ± 2.9 (17.4-28.0) μm; interphasmid distance = 16.6 ± 3.8 (11.1-28.9) μm; vulva to anus = 19.2 ± 2.8 (13.7-24.1) μm; vulva to phasmids = 28 ± 3.6 (22.0-34.2) μm.
- *Paratype males* (n = 16): L = 1933 ± 281 (1473-2420) μm; stylet = 24.6 ± 1.3 (21.7-26.9) μm; stylet cone length = 12.1 ± 0.6 (11.2-13.2) μm; DGO = 5.7 ± 1.2 (3.3-7.6) μm; excretory pore to anterior end = 183 ± 24.2 (149-223) μm; tail = 16.6 ± 2.4 (11.8-21.5) μm; phasmids to tail end = 8.7 ± 2.3 (5.7-14.5) μm; spicules = 38.6 ± 2.5 (34.9-44.1) μm.
- *Paratype J2* (n = 11): L = 349 ± 27.8 (313-388) μm; W = 14.2 ± 0.5 (13.9-15.1) μm; stylet = 11.5 ± 0.6 (10.5-12.2) μm; stylet knobs to anterior end = 15.9 ± 0.7 (14.7-17.0) μm; DGO = 5.0 ± 0.4 (4.2-5.3) μm; excretory pore to anterior end = 66.8 ± 1.8 (64.3-68.5) μm; body diam. at anus = 8.4 ± 0.3 (8.0-8.8) μm; tail = 32.6 ± 1.8 (30.7-35.6) μm.

DESCRIPTION

Female

Body ovoid, posterior region smoothly rounded, neck on same axis as tail, neck length variable, body annulation fine, regular, excretory pore 6-14 annuli from base of labial region. Labial disc slightly raised above labials, postlabial labial region with two annuli, stylet very slender, cone generally slightly longer than half stylet length, moderately curved dorsally, shaft cylindrical or slightly narrowed anteriorly, knobs slightly sloping posteriorly, vestibule lining prominent, pharyngeal lumen lining expanding posteriorly, usually narrowing before transforming into ovoid lining of triradiate median bulb lumen. Perineal pattern striae coarse, interspersed with fine striae, or all striae fine, dorsal arch medium high, squarish to broadly rounded, with or without very shallow dorsolateral indentations and slight to pronounced bulges above lateral field, tail end large, with few, faint markings, lateral field indistinct, sometimes indicated by interruptions between dorsal and ventral striae, or by forking of striae, phasmids immediately outside tail end area, rectal lining thin, dilating slightly towards small anal opening, anal fold inconspicuous, subsurface rectal punctations obscure, edges of vulva finely serrated, striae of ventral arch turned inwards, traverse perineum appearing as very fine lines, in some specimens, areas bordering perineum laterally folding inwards as a pair of oblique ridges between phasmids and corners of vulval slit.

Male

Body twisted posteriorly through about 90°, labial region continuous with body contour, labial cap shallowly rounded, as wide as, and about half as high as postlabial region of labial region, latter usually without, sometimes with a transverse stria dorsally, excretory pore 45-61 annuli posterior to basal plate, 0-3 annuli posterior to hemizonid which extends for about one annulus, tail end conical, phasmids posterior to level of cloacal aperture, at about mid-tail, lateral field areolated, each with four lines, sometimes seven in mid-body region, outer lines crenate, ending anteriorly about seven annuli posterior to basal plate, inner lines merging about 30 annuli posterior to basal plate, all lines ending posteriorly before phasmids, or one or both outer lines extending to tail terminus, labial framework weak, vestibule lining narrowing slightly expanded at basal plate, stylet cone about half of stylet length, stylet lumen opening at 26-39% of cone length posterior to pointed tip, shaft cylindrical, knobs moderately posteriorly sloping, rounded, testis single, spicule tips blunt, posterior surface of gubernaculum appearing uneven.

J2

Body annulation fine, prominent, lateral field irregularly areolated, with four lines, outer lines with irregular crenation, ending anteriorly opposite middle of stylet, posteriorly near hyaline region, inner lines irregular, only one line extending past excretory pore level anteriorly, and past phasmid posteriorly, inner band narrower than outer bands, postlabial region of labial region without transverse striae, stylet knobs slightly posteriorly sloping, excretory pore immediately posterior to indistinct hemizonid, 57-64 annuli posterior to labial region, tail with 15-21 annuli, phasmids inconspicuous, located about eight annuli posterior to anus level, hyaline region indistinctly delimited anteriorly, rectum not dilated.

TYPE PLANT HOST: Unidentified woody plant.

OTHER PLANTS: No other hosts were reported.

TYPE LOCALITY: Natal at Trafalgar, South Africa.

DISTRIBUTION: *Africa*: South Africa (Kleynhans, 1988).

CHROMOSOME NUMBER: No available data.

POLYTOMOUS KEY CODES: *Female*: A12, B12, C2, D3; *Male*: A123, B12, C1, D1, E1, F1; *J2*: A3, B23, C43, D3, E2, F3.

BIOCHEMICAL AND MOLECULAR CHARACTERISATION: No available data.

RELATIONSHIPS (DIAGNOSIS): It differs from *M. ardenensis* and *M. querciana* by the perineal pattern and other morphometrics in the female and J2. Additionally, it can be separated from *M. turkestanica* by EP/ST ratio and c′ ratio in the J2, and from *M. africana* by the longer female body and stylet.

98. *Meloidogyne vitis* Yang, Hu, Liu, Chen, Peng, Wang & Zhang, 2021

(Figs 260-262)

COMMON NAME: Chinese grape root-knot nematode.

This nematode was found in vineyards in Luliang County, Yunnan Province, China. More than 90% of the grape roots collected from vineyards were severely damaged by root-knot nematodes (Yang *et al.*, 2021).

MEASUREMENTS *(AFTER YANG ET AL., 2021)*

- *Holotype female*: L = 823 μm, W = 532 μm; stylet = 15.3 μm; stylet knob width = 4.3 μm; stylet knob height = 1.9 μm; DGO = 5.3 μm; anterior end to centre of median bulb = 77.6 μm; anterior end to excretory pore = 38.8 μm; a = 1.55; EP/ST = 2.5.

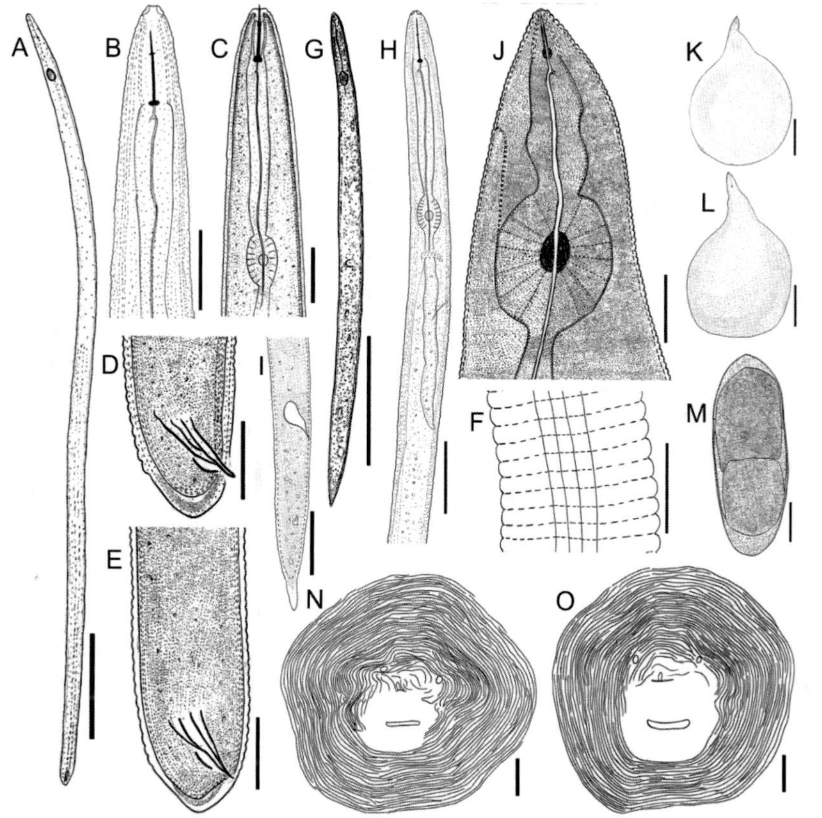

Fig. 260. Meloidogyne vitis. *Male (A-F). A: Entire body; B, C: Anterior region; D, E: Tail region; F: Lateral field. Second-stage juvenile (G-I). G: Entire body; H: Anterior region; I: Tail region. Female (J-O). J: Anterior region; K, L: Entire body; M: Egg; N, O: Perineal pattern. (Scale bars: A, K, L = 200 μm; B-E, G-I, M-O = 20 μm; F = 10 μm; J = 100 μm). (After Yang* et al., *2021.)*

- *Paratype females* (n = 25): L = 959 ± 132 (823-1245.2) μm; W = 609 ± 43.6 (532-688) μm; a = 1.58 ± 0.2 (1.3-1.95); stylet = 15.7 ± 3.7 (8.1-26.6) μm; stylet knob width = 4.4 ± 1.0 (2.7-6.0) μm; stylet knob height = 2.1 ± 0.5 (1.3-3.3) μm; DGO = 4.1 ± 0.8 (2.6-5.3) μm; median bulb to anterior end = 73 ± 12.7 (44-86) μm; excretory pore to anterior end = 39 ± 4.2 (34-45) μm; vulval slit to anus = 19.9 ± 1.6 (17.4-23.5) μm; vulva to phasmid distance = 28.0 ± 2.3 (23.3-33.0) μm; anus to phasmid distance = 6.8 ± 1.59 (4.7-9.8) μm.

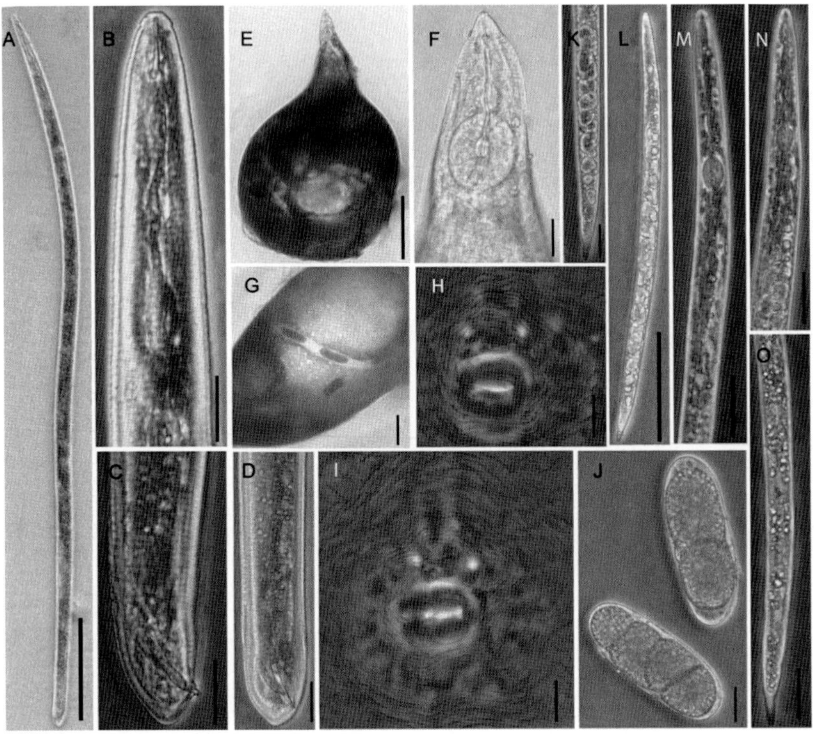

Fig. 261. Meloidogyne vitis. *Male (A-D). A: Entire body; B: Anterior region; C, D: Tail region. Female (E-J). E: Entire body; F: Anterior region; G: Partial region of body; H, I: Perineal pattern; J: Eggs. Second-stage juvenile (K-O). K, O: Tail region; L: Entire body; M, N: Anterior region. (Scale bars: A, E = 200 μm; B-D, F, H-K, M-O = 20 μm; G, L = 100 μm.) (After Yang* et al., *2021.)*

- *Paratype males* (n = 10): L = 1330 ± 179.2 (1032-1593) μm; W = 36.8 ± 6.2 (25.7-43.9) μm; a = 36.8 ± 6.0 (30.7-50.2); c =103.6 ± 13.8 (81.7-127.7); stylet = 19.3 ± 1.7 (17.0-21.4) μm; stylet knob width = 3.5 ± 0.6 (2.7-4.7) μm; DGO = 3.0 ± 0.5 (2.4-3.9) μm; anterior end to median bulb valve = 99 ± 5.9 (91-109) μm; anterior end to excretory pore = 133 ± 4.9 (136-141) μm; tail = 12.9 ± 0.8 (11.8-14.2); spicules = 30.9 ± 26.9 (27.9-35.8) μm; gubernaculum = 10.2 ± 1.9 (8.2-14.9) μm.
- *Paratype J2* (n = 26): L = 397 ± 18.3 (353-426) μm; W = 16.2 ± 1.9 (12.8-22.4) μm; a = 24.8 ± 2.5 (19.0-28.4); c = 7.0 ± 0.7 (6.2-8.8); c′ = 4.6 ± 0.6 (3.4-5.3); stylet = 13.3 ± 0.3 (12.7-14.1) μm; stylet

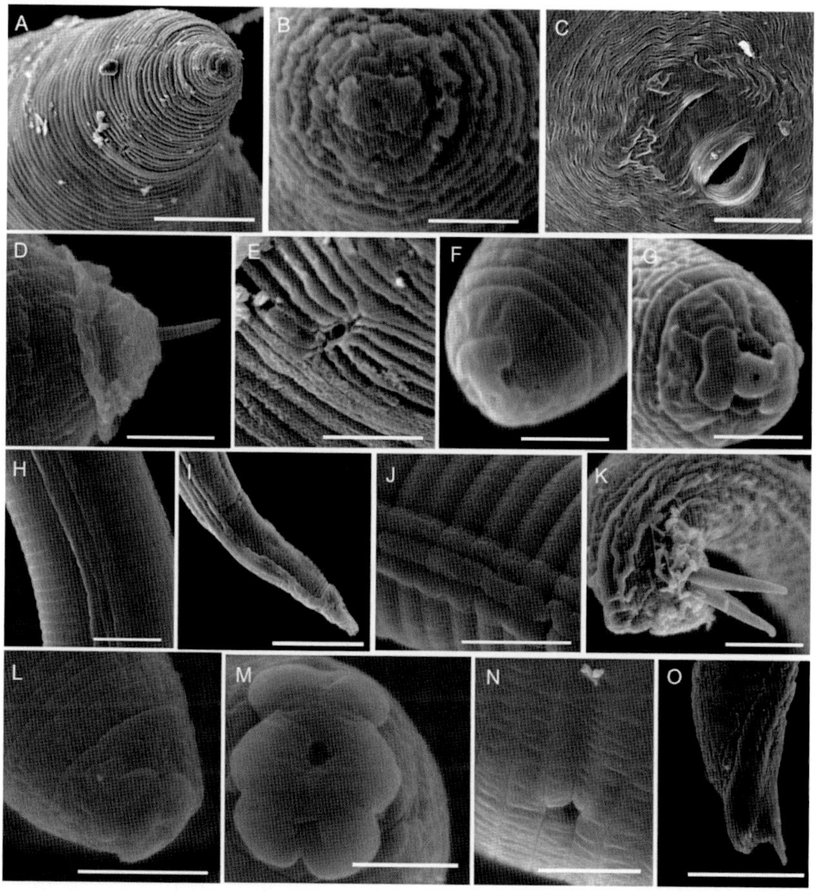

Fig. 262. *SEM of* Meloidogyne vitis. *Female (A-E). A: Anterior end in lateral view; B: Anterior end in face view; C: Perineal pattern; D: Anterior end in lateral view showing protruded stylet; E: Excretory pore. Second-stage juvenile (F-I). F: Anterior end in lateral view; G: Anterior end in face view; H: Lateral field; I: Tail region. Male (J-O). J: Lateral field; K: Tail region; L: Anterior end in lateral view; M: Anterior end in face view; N: Excretory pore; O: Tail region. (Scale bars: A, C, O* = *20* μ*m; B, D, M* = *3* μ*m; F, G* = *2* μ*m; E, H, L, N* = *5* μ*m; I-K* = *10* μ*m.) (After Yang* et al., *2021.)*

knob width = 1.6 ± 0.3 (1.2-2.2) μm; DGO = 1.4 ± 0.3 (1.0-2.0) μm; anterior end to median bulb valve = 55 ± 2.0 (51-59) μm; tail = 57 ± 3.9 (47-64) μm; hyaline region = 12.2 ± 1.7 (9.7-15.7) μm.

- *Eggs* (n = 20): L = 99 ± 6.4 (86-111) μm; W = 35 ± 2.4 (31-39) μm.

DESCRIPTION

Female

Body pear-shaped and milky white, with a prominent and variably sized neck. Posterior part of body round, anal region without protuberances. Under SEM, labial disc ovoid-squared, slightly raised on medial lips, fused with medial lips forming a dumbbell-shaped structure. No obvious lateral lips, oral aperture slit-like, located in middle of labial disc, surrounded by six inner labial sensilla. Excretory pore located between 23rd and 25th annulus posterior to lips. Perineal pattern round to ovoid with a moderately high dorsal arch and smooth and fine striae that are extremely dense and faint. Lateral fields not clearly visible, and without lateral lines, although a few specimens have faint striae on two shoulders or wings in lateral field. Two large phasmids, prominent and round, with diam. accounting for 2-5 annular striae. Straight lines of two phasmids parallel or nearly parallel to vulva, vulval slit wide. Anal fold clearly visible. Area of vulva and anus smooth, lacking striae.

Male

Body vermiform, anterior end of body tapering. Labial cap obvious and slightly separated from body, stylet developed and with clear boundary with stylet shaft. Stylet knobs oblate-spheroid. Tail mostly straight and short with a humped end, spicules well developed, arch-shaped, slightly curved. Gubernaculum crescentic. Under SEM, lip region lacking annuli. Labial disc horizontally ovoid-squared, slightly raised on medial lips, slightly wider than medial lips and fused with medial lips to form a dumbbell-shaped structure. Prominent slit-like opening between labial disc and medial lips. Distinct depression apparent in middle of medial lips, oral aperture slit-like, located in middle of labial disc, surrounded by six inner labial sensilla. Lateral field consisting of four incisures forming three lateral bands filled with reticular striae. Spicules resembling a figure eight, tip slightly curving, hook-like.

J2

Body tapering at both ends, but more towards tail than anterior end. Stylet straight, slender, sharply pointed, with clear boundary with stylet shaft, stylet knobs well developed and spherical. Tail variable, exhibiting a range of variation in tail lateral fields, conical and constricted at tip. Hyaline region short. Anus difficult to distinguish. Rectum and caudal sensory organ difficult to observe. Under SEM, lip region not smooth and slightly folded. Labial disc appearing round, slightly raised on medial lips and fused with medial lips to form a dumbbell-shaped structure. Lateral field forming three lateral bands delimited by four incisures. Anal opening elliptical, located in cuticular depression on tail.

TYPE PLANT HOST: Grape, *Vitis vinifera* L.

OTHER PLANTS: No other hosts were reported.

TYPE LOCALITY: Luliang County, Yunnan Province, China.

DISTRIBUTION: *Asia*: China.

SYMPTOMS: Severely nematode-infected plants showed symptoms of dwarfing, leaf yellowing and shedding, small fruit, and declining and low growth. Roots were atrophied and distorted with severe root knots and other symptoms. Both the axial and side roots were damaged. Aged roots were rotten and necrotic.

CHROMOSOME NUMBER: No available data.

POLYTOMOUS KEY CODES: *Female*: A12, B1234, C2, D4; *Male*: A324, B324, C21, D1, E1, F2; *J2*: A32, B21, C32, D213, E32, F324.

BIOCHEMICAL AND MOLECULAR CHARACTERISATION: *Meloidogyne vitis* is characterised by one EST band (VF1) and three MDH bands (N3d). The ITS, D2-D3 of 28S rRNA, intergenic *COII*-16S region and *COI* of mtDNA gene sequences clearly distinguished this species from all others. PCR with specific primers were also developed for this species (Yang *et al.*, 2021).

RELATIONSHIPS (DIAGNOSIS): The species belongs to Molecular group VIII and is clearly molecularly differentiated from all other *Meloidogyne* species. *Meloidogyne vitis* resembles *M. mali* and differs in the perineal pattern, which has a moderately high dorsal arch rather than a low and flat dorsal arch and there are no lateral lines rather than clear single or double lateral lines in the lateral fields, and longer J2 tail (47-64 *vs* 24-39 μm).

References

ABAD, P., GOUZY, J., AURY, J.M., CASTAGNONE-SERENO, P., DANCHIN, E.G.J., DELEURY, E., PERFUS-BARBEOCH, L., ANTHOUARD, V., ARTIGUENAVE, F., BLOK, V.C. ET AL. (2008). Genome sequence of the metazoan plant-parasitic nematode *Meloidogyne incognita*. *Nature Biotechnology* 26, 909-915. DOI: 10.1038/nbt.1482

ABDEL-RAHMAN, F. & MAGGENTI, A.R. (1987). *Meloidogyne californiensis* n. sp. (Nemata: Meloidogyninae), parasitic on bulrush, *Scirpus robustus* Pursh. *Journal of Nematology* 19, 207-217.

ABRANTES, I.M. DE O. & SANTOS, M.S.N. DE A. (1991). *Meloidogyne lusitanica* n. sp. (Nematoda: Meloidogynidae), a root-knot nematode parasitizing olive tree (*Olea europaea* L.). *Journal of Nematology* 23, 210-224.

ABRANTES, I.M. DE O., SANTOS, M.S.N. DE A. & VOVLAS, N. (1995). Studies on *Meloidogyne megadora* found in coffee plantations in Républica de S. Tomé e Príncipe. *Nematologica* 41, 278. [Abstr.] DOI: 10.1163/003925995X00242

ABRANTES, I.M. DE O., SANTOS, M.C.V. DOS, CONCEIÇÃO, I.L.P.M. DA, CUNHA, M.J.M. DA & SANTOS, M.S.N. DE A. (2004). Biochemical and molecular characterization of plant-parasitic nematodes. *Phytopathologia Mediterranea* 43, 232-258.

ABRANTES, I.M. DE O., SANTOS, M.C.V. DOS, CONCEIÇÃO, I.L.P.M. DA, SANTOS, M.S.N. DE A. & VOVLAS, N. (2008). Root-knot and other plant-parasitic nematodes associated with fig trees in Portugal. *Nematologia Mediterranea* 36, 131-136.

ADAM, M.A.M., PHILLIPS, M.S. & BLOK, V.C. (2005). Identification of *Meloidogyne* spp. from North East Libya and comparison of their inter- and intra-specific genetic variation using RAPDs. *Nematology* 7, 599-609. DOI: 10.1163/156854105774384840

ADAM, M.A.M., PHILLIPS, M.S. & BLOK, V.C. (2007). Molecular diagnostic key for identification of single juveniles of seven common and economically important species of root-knot nematode (*Meloidogyne* spp.). *Plant Pathology* 56, 190-197. DOI: 10.1111/j.1365-3059.2006.01455.x

ADAMS, B.J., DILLMAN, A.R. & FINLINSON, C. (2009). Molecular taxonomy and phylogeny. In: Perry, R.N., Moens, M. & Starr, J.L. (Eds). *Root-knot nematodes*. Wallingford, UK, CAB International, pp. 119-138.

AGUDELO, P., LEWIS, S.A. & FORTNUM, B.A. (2011). Validation of a real-time polymerase chain reaction assay for the identification of *Meloidogyne arenaria*. *Plant Disease* 95, 835-838. DOI: 10.1094/PDIS-09-10-0668

AGUIRRE, Y., CROZZOLI, R. & GRECO, N. (2003). Effect of the root-knot nematode *Meloidogyne incognita* on parsley. *Russian Journal of Nematology* 11, 27-31.

AHMED, M., VAN DE VOSSENBERG, B.T.L.H., CORNELISSE, C. & KARSSEN, G. (2013). On the species status of the root-knot nematode *Meloidogyne ulmi* Palmisano & Ambrogioni, 2000 (Nematoda, Meloidogynidae). *ZooKeys* 362, 1-27. DOI: 10.3897/zookeys.362.6352

AIHARA, T., YUHARA, I. & YAMAZAKI, K. (1981). [*Meloidogyne camelliae* found on three species of Theaceae: the first record from the native land of the nematode.] *Japanese Journal of Nematology* 10, 8-15.

AIHARA, T., YUHARA, I. & YAMAZAKI, K. (1983). [Seasonal changes in population density and some host plants of the camellia root-knot nematode, *Meloidogyne camelliae*.] *Japanese Journal of Nematology* 12, 33-40.

AÏT HAMZA, M., ALI, N., TAVOILLOT, J., FOSSATI-GASCHIGNARD, O., BOUBAKER, H., EL MOUSADIK, A. & MATEILLE, T. (2017). Diversity of root-knot nematodes in Moroccan olive nurseries and orchards: does *Meloidogyne javanica* disperse according to invasion processes? *BMC Ecology* 17, 41. DOI: 10.1186/s12898-017-0153-9

AKYAZI, F., JOSEPH, S., FELEK, A.F. & MEKETE, T. (2017). Mitochondrial haplotype-based identification of root-knot nematodes, *Meloidogyne arenaria* and *Meloidogyne hapla*, infecting kiwifruit in Turkey. *Nematropica* 47, 34-48.

AL-HAZMA, A.S. & SASSER, J.N. (1982). Biology of *Meloidogyne platani* Hirschmann parasitic on sycamore, *Platanus occidentalis*. *Journal of Nematology* 14, 154-161.

ALI, N., TAVOILLOT, J., MATEILLE, T., CHAPUIS, E., BESNARD, G., EL BAKKALI, A., CANTALAPIEDRA-NAVARRETE, C., LIÉBANAS, G., CASTILLO, P. & PALOMARES-RIUS, J.E. (2015). A new root-knot nematode *Meloidogyne spartelensis* n. sp. (Nematoda: Meloidogynidae) in Northern Morocco. *European Journal of Plant Pathology* 143, 25-42. DOI: 10.1007/s10658-015-0662-3

ALI, N., TAVOILLOT, J., CHAPUIS, E. & MATEILLE, T. (2016). Trend to explain the distribution of root-knot nematodes *Meloidogyne* spp. associated with olive trees in Morocco. *Agriculture, Ecosystems & Environment* 225, 22-32. DOI: 10.1016/j.agee.2016.03.042

ALLEN, M.W., HART, W.H. & BAGHOTT, K.V. (1970). Crop rotation controls the barley root-knot nematode at Tukelake. *California Agriculture* 24, 4-5.

ALMEIDA, A.M.S.F. DE & SANTOS, M.S.N. DE A. (2002). Resistance and host-response of selected plants to *Meloidogyne megadora*. *Journal of Nematology* 34, 140-142.

ALMEIDA, A.M.S.F. DE, SANTOS, M.S.N. DE A. & RYAN, M.F. (1997). Host status of selected plant species for *Meloidogyne megadora*. *Nematropica* 27, 1-6.

ÁLVAREZ-ORTEGA, S., BRITO, J.A. & SUBBOTIN, S.A. (2019). Multi-gene phylogeny of root-knot nematodes and molecular characterization of *Meloidogyne nataliei* Golden, Rose & Bird, 1981 (Nematoda: Tylenchida). *Scientific Reports* 9, 11788. DOI: 10.1038/s41598-019-48195-0

AMIN, A.W. (1993). A new local race of the root-knot nematode *Meloidogyne thamesi* Chitwood *in* Chitwood, Specht & Havis, 1952 in Hungary. *Opuscula Zoologica (Budapest)* 26, 3-8.

AMIN, A.W. & BUDAL, C. (1994). Some weed host plants of the root-knot nematode *Meloidogyne* species in south-eastern Hungary. *Pakistan Journal of Nematology* 12, 59-65.

ANON. (2006). PM 3/69 (1) *Meloidogyne chitwoodi* and *M. fallax*: sampling potato tubers for detection. *Bulletin OEPP/EPPO Bulletin* 36, 421-422.

ANON. (2008). An emerging root-knot nematode, *Meloidogyne enterolobii*: addition to the EPPO Alert List. EPPO Reporting Service no. 5, 2008/105. (https://gd.eppo.int/reporting/article-690) Accessed on 5 June 2018.

ANON. (2009). *Meloidogyne chitwoodi* and *Meloidogyne fallax*. *OEPP/EPPO Bulletin* 39, 5-17.

ANON. (2010a). *Meloidogyne acronea*. *Distribution maps of plant diseases*, Map 1087. Wallingford, UK, CAB International.

ANON. (2010b). *Meloidogyne coffeicola*. *Distribution maps of plant diseases*, Map 1088. Wallingford, UK, CAB International.

ANON. (2011). *Meloidogyne enterolobii*. *Bulletin OEPP/EPPO Bulletin* 41, 329-339.

ANON. (2012a). *Meloidogyne chitwoodi*. *Distribution maps of plant diseases*, Map 803. Wallingford, UK, CAB International.

ANON. (2012b). *Meloidogyne brevicauda*. *Distribution maps of plant diseases*, Map 1122. Wallingford, UK, CAB International.

ANON. (2013a). *Meloidogyne artiellia*. *Distribution maps of plant diseases*, Map 1141. Wallingford, UK, CAB International.

ANON. (2013b). PM 9/17 (1) *Meloidogyne chitwoodi* and *Meloidogyne fallax*. *OEPP/EPPO Bulletin* 43, 527-533.

ANON. (2014). *Meloidogyne enterolobii*. *Bulletin OEPP/EPPO Bulletin* 44, 159-163.

ANON. (2015). *Meloidogyne exigua*. *Distribution maps of plant diseases*, Map 825. Wallingford, UK, CAB International.

ANON. (2016a). PM 7/41 (3) *Meloidogyne chitwoodi* and *Meloidogyne fallax*. *OEPP/EPPO Bulletin* 46, 171-189.

ANON. (2016b). *Meloidogyne graminicola* (rice root knot nematode). Wallingford, UK, CAB International.

ANON. (2016c). PM 7/103 (2) *Meloidogyne enterolobii*. *Bulletin OEPP/EPPO Bulletin* 46, 190-201. DOI: 10.1111/epp.12293

ANON. (2018a). *Meloidogyne graminicola*. *Distribution maps of plant diseases*, Map 826. Wallingford, UK, CAB International.

ANON. (2018b). *Meloidogyne mali*. *Bulletin OEPP/EPPO Bulletin* 48, 438-445.

ANWAR, S.A. & VAN GUNDY, S.D. (1989). Influence of four nematodes on root and shoot growth parameters in grape. *Journal of Nematology* 21, 276-283.

ANWAR, S.A., ZIA, A. & JAVED, N. (2009). *Meloidogyne incognita* infection of five weed genotypes. *Pakistan Journal of Zoology* 41, 95-100.

ARAGON, B.C.A., GOMES, M.B. & CAICEDO, J.E. (1978). Plantas de la zona cafeteria colombiana hospedantes de espécies de *Meloidogyne* Goeldi, 1887. *Cenicafé* 29, 35-45.

ARAKI, M. (1992a). Description of *Meloidogyne ichinohei* n. sp. (Nematoda: Meloidogynidae) from *Iris laevigata* in Japan. *Japanese Journal of Nematology* 22, 11-20.

ARAKI, M. (1992b). The first record of *Meloidogyne marylandi* Jepson & Golden, 1987 from *Zoysia* sp. in Japan. *Japanese Journal of Nematology* 22, 49-52.

ARCHIDONA-YUSTE, A., CANTALAPIEDRA-NAVARRETE, C., LIÉBANAS, G., RAPOPORT, H.F., CASTILLO, P. & PALOMARES-RIUS, J.E. (2018). Diversity of root-knot nematodes of the genus *Meloidogyne* Göeldi, 1892 (Nematoda: Meloidogynidae) associated with olive plants and environmental cues regarding their distribution in southern Spain. *PLoS ONE* 13, e0198236. DOI: 10.1371/journal.pone.0198236

ARTAVIA-CARMONA, R. & PERAZA-PADILLA, W. (2020). Identificación morfológica, morfométrica y molecular de *Meloidogyne exigua* (Göeldi, 1887) en café (*Coffea arabica*). *Agronomía Mesoamericana* 31, 531-545. DOI: 10.15517/am.v31i3.38733

ASUAJE, L., JIMÉNEZ, M.A., JIMÉNEZ-PÉREZ, N. & CROZZOLI, R. (2004). Efecto del nematodo agallador, *Meloidogyne incognita*, sobre el crecimiento de tres cultivares de lechuga. *Fitopatología Venezolana* 17, 2-5.

ATKINSON, G.F. (1889). A preliminary report upon the life history and metamorphoses of a root-gall nematode, *Heterodera marioni* (Greeff) Müll., and the injuries caused by it upon the roots of various plants. *Scientific Contributions of the Agricultural Experiment Station, Agricultural and Mechanical College of Alabama* 1, 177-266.

AVISE, J.C., SHAPIRO, J.F., DANIEL, S.W., AQUADRO, C.F. & LANSMAN, R.A. (1983). Mitochondrial DNA differentiation during the speciation process in *Peromyscus*. *Molecular Biology and Evolution* 1, 38-56. DOI: 10.1093/oxfordjournals.molbev.a040301

AYDINLI, G. (2018). Detection of the root-knot nematode *Meloidogyne luci* Carneiro *et al.*, 2014 (Tylenchida: Meloidogynidae) in vegetable fields of Samsun Province, Turkey. *Türkiye Entomoloji Dergisi* 42, 229-237. DOI: 10.16970/entoted.409941

AYDINLI, G., MENNAN, S., DEVRAN, Z., ŠIRCA, S. & UREK, G. (2013). First report of the root-knot nematode *Meloidogyne ethiopica* on tomato and cucumber in Turkey. *Plant Disease* 97, 1262. DOI: 10.1094/PDIS-01-13-0019-PDN

AYTAN, S. & DICKERSON, O.J. (1969). *Meloidogyne naasi* on sorghum in Kansas. *Plant Disease Reporter* 53, 737.

BAČIĆ, J., GERIČ STARE, B., STRAJNAR, P., ŠIRCA, S. & UREK, G. (2016). First report of a highly damaged potato crop from Serbia caused by *Meloidogyne incognita*. *Plant Disease* 100, 1021. DOI: 10.1094/PDIS-09-15-1072-PDN

BACK, M.A., HAYDOCK, P.P.J. & JENKINSON, P. (2002). Disease complexes involving plant parasitic nematodes and soilborne pathogens. *Plant Pathology* 51, 683-697. DOI: 10.1046/j.1365-3059.2002.00785.x

BALDWIN, J.G. & SASSER, J.N. (1979). *Meloidogyne megatyla* n. sp., a root-knot nematode from loblolly pine. *Journal of Nematology* 11, 47-56.

BALI, S., HU, S., VINING, K., BROWN, C., MOJTAHEDI, H., ZHANG, L., GLEASON, C. & SATHUVALLI, V. (2021). Nematode genome announcement: Draft genome of *Meloidogyne chitwoodi*, an economically important pest of potato in the Pacific Northwest. *Molecular Plant-Microbe Interactions*. Published Online: 29 Mar 2021. DOI: 10.1094/MPMI-12-20-0337-A

BAOJUN, Y., HU, K., CHEN, H. & ZHU, W. (1990). A new species of root-knot nematode *Meloidogyne jianyangensis* n. sp. parasitizing mandarin orange. *Acta Phytopathologica Sinica* 20, 259-264.

BARBOSA, D.H.S.G., VIEIRA, H.D., SOUZA, R.M., VIANA, A.P. & SILVA, C.P. (2004). Field estimates of coffee yield losses and damage threshold by *Meloidogyne exigua*. *Nematologia Brasileira* 28, 49-54.

BARKER, K.R., CARTER, C.C. & SASSER, J.N. (Eds) (1985). *An advanced treatise on* Meloidogyne. *Volume 2. Methodology*. Raleigh, NC, USA, North Carolina State University Graphics.

BARROS, A.F., CAMPOS, V.P., SOUZA, L.N., COSTA, S.S., TERRA, W.C. & LESSA, J.H. (2018). Morphological, enzymatic and molecular characterization of root-knot nematodes parasitizing vegetable crops. *Horticultura Brasileira* 36, 473-479. DOI: 10.1590/s0102-053620180408

BARTLEM, D.G., JONES, M.G.K. & HAMMES, U.Z. (2014). Vascularization and nutrient delivery at root-knot nematode feeding sites in host roots. *Journal of Experimental Botany* 65, 1789e1798. DOI: 10.1093/jxb/ert415

BASTIDAS, H. & MONTEALEGRE, S.F.A. (1994). Aspectos generales de la nueva enfermedad del arroz llamada entorchamiento. *Arroz* 43, 30-35.

Been, T.H. & Schomaker, C.H. (2006). Distribution patterns and sampling. In: Perry, R.N. & Moens, M. (Eds). *Plant nematology*, 1st edition. Wallingford, UK, CAB International, pp. 302-326.

Bélair, G. (1996). Using crop rotation to control *Meloidogyne hapla* Chitwood and improve marketable carrot yield. *HortScience* 31, 106-108. DOI: 10.21273/HORTSCI.31.1.106

Bélair, G., Simard, L. & Eisenback, J.D. (2006). First report of the barley root-knot nematode *Meloidogyne naasi* infecting annual bluegrass on a golf course in Quebec, Canada. *Plant Disease* 90, 1109. DOI: 10.1094/PD-90-1109A

Bellafiore, S., Jougla, C., Chapuis, E., Besnard, G., Suong, M., Nguyen Vu, P., De Waele, D., Gantet, P. & Thi, X. (2015). Intraspecific variability of the facultative meiotic parthenogenetic root-knot nematode (*Meloidogyne graminicola*) from rice fields in Vietnam. *Comptes Rendus Biologies* 338, 471-483. DOI: 10.1016/j.crvi.2015.04.002

Bellé, C., Kaspary, T.E., Schmitt, J. & Kuhn, P.R. (2016a). *Meloidogyne ethiopica* and *Meloidogyne arenaria* parasitizing *Oxalis corniculata* in Brazil. *Australasian Plant Disease Notes* 11, 24. DOI: 10.1007/s13314-016-0212-7

Bellé, C., Brum, D., Groth, M.Z., Barros, D.R., Kaspary, T.E., Schafer, J.T. & Gomes, C.B. (2016b). First report of *Meloidogyne luci* parasitizing *Glycine max* in Brazil. *Plant Disease* 100, 2174. DOI: 10.1094/PDIS-05-16-0624-PDN

Bellé, C., Kulczynski, S.M., Kuhn, P.R., Garneiro, M.D.G., Lima-Medina, I. & Gomes, C.B. (2017a). First report of *Meloidogyne ethiopica* parasitizing sugarcane in Brazil. *Plant Disease* 101, 635. DOI: 10.1094/PDIS-09-16-1303-PDN

Bellé, C., Kaspary, T.E., Croth, M.Z. & Cocco, K.L.T. (2017b). *Meloidogyne ethiopica* parasitizing melon fields in Rio Grande do Sul State, Brazil. *Journal of Plant Disease and Protection* 124, 393-397. DOI: 10.1007/s41348-017-0087-7

Bellé, C., Ramos, R.F., Balardin, R.R., Kaspary, T.E. & Antoniolli, Z.I. (2019a). Reproduction of *Meloidogyne enterolobii* on weeds found in Brazil. *Tropical Plant Pathology* 44, 380-384. DOI: 10.1007/s40858-019-00278-z

Bellé, C., Ramos, R.F., Balardin, R.R., Nora, D.D. & Kaspary, T.E. (2019b). Host weed species range of *Meloidogyne ethiopica* Whitehead (Tylenchida: Meloidogynidae) found in Brazil. *European Journal of Plant Pathology* 156, 979-985. DOI: 10.1007/s10658-019-01900-1

Bellé, C., Balardin, R.R., Nora, D.D., Schmitt, J., Gabriel, M., Ramos, R.F. & Antoniolli, Z.I. (2019c). First report of *Meloidogyne graminicola* (Nematoda: Meloidogynidae) on barley (*Hordeum vulgare*) in Brazil. *Plant Disease* 103, 1045. DOI: 10.1094/PDIS-11-18-2010-PDN

BELLÉ, C., KASPARY, T.E., BALARDIN, R.R. & ANTONIOLLI, Z.I. (2019d). Detection of *Meloidogyne graminicola* parasitizing *Cyperi rotundus* in Rio Grande do Sul, Brazil. *Australasian Plant Disease Notes* 14, 2. DOI: 10.1007/s13314-018-0333-2

BENDEZU, I., MORGAN, T. & STARR, J. (2004). Hosts for *Meloidogyne haplanaria*. *Nematropica* 34, 205-209.

BERG, R.H., FESTER, T. & TAYLOR, C.G. (2008). Development of the root-knot nematode feeding cell. In: Berg, R.H. & Taylor, C.G. (Eds). *Plant cell monographs. Cell biology of plant nematode parasitism.* Heidelberg, Berlin, Springer, pp. 115-152.

BERGÉ, J.B. & DALMASSO, A. (1975). Caractéristiques biochimiques de quelques populations de *Meloidogyne hapla* et *Meloidogyne* spp. *Cahiers – ORSTOM. Série Biologie* 10, 263-271.

BERGÉ, J.B. & DALMASSO, A. (1976). Variations génétiques associées à un double mode de reproduction parthénogénétique et amphimictique chez le nématode *Meloidogyne hapla*. *Comptes Rendus Hebdomadaires des Séances de l'Académie des Sciences, Paris* 282, D, 2087-2090.

BERGESON, G.B. (1966). Mobilization of minerals to the infection site of root knot nematodes. *Phytopathology* 56, 1287-1289.

BERKELEY, M.J. (1855). Vibrio forming cysts on the roots of cucumbers. *Gardener's Chronicle and Agricultural Gazette* 14, 220.

BERNARD, E. (1981). Heteroderoidea (Nematoda) from the Aleutian Islands. *Journal of Nematology* 13, 499-513.

BERNARD, E. & EISENBACK, J. (1997). *Meloidogyne trifoliophila* n. sp. (Nemata: Meloidogynidae), a parasite of clover from Tennessee. *Journal of Nematology* 29, 43-54.

BERNARD, E. & JENNINGS, P. (1997). Host range and distribution of the clover root-knot nematode, *Meloidogyne trifoliophila*. *Supplement to the Journal of Nematology* 29(4S), 662-672.

BERNARDO, E.R.A., SANTOS, J.M., SILVA, R.A., CASSETARI NETO, D., SANTOS, S.S., DELMADI, L. & ROCHA, V.F. (2003). Levantamento de *Meloidogyne exigua* na cultura da seringueira em São José do Rio Claro, MT, Brasil. *Ciência Rural* 33, 157-159.

BERRY, S.D., FARGETTE, M., SPAULL, V.W., MORAND, S. & CADET, P. (2008). Detection and quantification of root-knot nematode (*Meloidogyne javanica*), lesion nematode (*Pratylenchus zeae*) and dagger nematode (*Xiphinema elongatum*) parasites of sugarcane using real-time PCR. *Molecular and Cellular Probes* 22, 168-176. DOI: 10.1016/j.mcp.2008.01.003

BERT, W., KARSSEN, G., VAN DRIESSCHE, R. & GERAERT, E. (2002). The cellular structure of the female reproductive system within the Heteroderinae and Meloidogyninae (Nematoda). *Nematology* 4, 953-963. DOI: 10.1163/156854102321122575

BERTRAND, B., AGUILAR, G., BOMPARD, E., RAFINON, A. & ANTHONY, F. (1997). Comportement agronomique et résistance aux principaux déprédateurs des lignées de Sarchimors et Catimors au Costa Rica. *Plantations, Recherche, Développement* 4, 312-321.

BERTRAND, B., NUNEZ, C. & SARAH, J.-L. (2000). Disease complex in coffee involving *Meloidogyne arabicida* and *Fusarium oxysporum*. *Plant Pathology* 49, 383-388. DOI: 10.1046/j.1365-3059.2000.00456.x

BESNARD, G., JÜHLING, F., CHAPUIS, E., ZEDANE, L., LHUILLIER, E., MATEILLE, T. & BELLAFIORE, S. (2014). Fast assembly of the mitochondrial genome of a plant parasitic nematode (*Meloidogyne graminicola*) using next generation sequencing. *Comptes Rendus Biologies* 337, 295-301. DOI: 10.1016/j.crvi.2014.03.003

BESSEY, E.A. (1911). Root-knot and its control. *Bulletin 217, U.S. Department of Agriculture. Bureau of Plant Industry*. Washington DC, USA, Government Printing Office.

BIRCHFIELD, W. (1964). Histopathology of nematode-induced galls of *Echinochloa colonum*. *Phytopathology* 54, 888.

BIRCHFIELD, W. (1965). Host-parasite relations and host range studies of a new *Meloidogyne* species in southern USA. *Phytopathology* 55, 1359-1361.

BIRD, G., DIAMOND, C., WARNER, F. & DAVENPORT, J. (1994). Distribution and regulation of *Meloidogyne nataliei*. *Supplement to Journal of Nematology* 26, 727-730.

BIRD, G.W. & WARNER, F. (2018). Nematodes and nematologists of Michigan. In: Subbotin, S.A. & Chitambar, J.J. (Eds). *Plant parasitic nematodes in sustainable agriculture of North America. Vol. 2 – Northeastern, Midwestern and Southern USA*. Springer, pp. 57-86.

BLANC-MATHIEU, R., PERFUS-BARBEOCH, L., AURY, J.-M., DA ROCHA, M., GOUZY, J., SALLET, E., MARTIN-JIMENEZ, C., BAILLY-BECHET, M., CASTAGNONE-SERENO, P., FLOT, J.-F. ET AL. (2017). Hybridization and polyploidy enable genomic plasticity without sex in the most devastating plant-parasitic nematodes. *PLoS Genetics* 13, e1006777. DOI: 10.1371/journal.pgen.1006777

BLOK, V.C. & POWERS, T.O. (2009). Biochemical and molecular identification. In: Perry, R.N., Moens, M. & Starr, J.L. (Eds). *Root-knot nematodes*. Wallingford, UK, CABI Publishing, pp. 98-103.

BLOK, V.C., PHILLIPS, M.S., MCNICOL, J.W. & FARGETTE, M. (1997a). Genetic variation in tropical *Meloidogyne* spp. as shown by RAPDs. *Fundamental and Applied Nematology* 20, 127-133.

BLOK, V.C., PHILLIPS, M.S. & FARGETTE, M. (1997b). Comparison of sequence differences in the intergenic region of the ribosomal cistron of

Meloidogyne mayaguensis and the major tropical root-knot nematodes. *Journal of Nematology* 29, 16-22.

Blok, V.C., Wishart, J., Fargette, M., Berthier, K. & Phillips, M.S. (2002). Mitochondrial DNA differences distinguishing *Meloidogyne mayaguensis* from the major species of tropical root-knot nematodes. *Nematology* 4, 773-781. DOI: 10.1163/156854102760402559

Bojang, S., Ghani, I.A., Kadir, J., Paiko, A.S., Iftikhar, Y. & Kamran, M. (2019). Ultrastructural characterization of *Meloidogyne graminis* from golf course turf grasses in peninsular Malaysia. *Pakistan Journal of Zoology* 51, 1591-1594. DOI: 10.17582/journal.pjz/2019.51.4.sc5

Boydston, R.A., Mojtahedi, H., Crosslin, J.M., Brown, C.R., Anderson, T. & Sanyal, D. (2008). Effect of hairy nightshade (*Solanum sarrachoides*) presence on potato nematodes, diseases, and insect pests. *Weed Science* 56, 151-154. DOI: 10.1614/WS-07-035.1

Braun-Kiewnick, A. & Kiewnick, S. (2018). Real-time PCR, a great tool for fast identification, sensitive detection and quantification of important plant-parasitic nematodes. *European Journal of Plant Pathology* 152, 271-283. DOI: 10.1007/s10658-018-1487-7

Braun-Kiewnick, A., Viaene, N., Folcher, L., Ollivier, F., Anthoine, G., Niere, B., Sapp, M., van de Vossenberg, B., Toktay, H. & Kiewnick, S. (2016). Assessment of a new qPCR tool for the detection and identification of the root-knot nematode *Meloidogyne enterolobii* by an international test performance study. *European Journal of Plant Pathology* 144, 97-108. DOI: 10.1007/s10658-015-0754-0

Bridge, J. (1984). Coffee nematode survey of Tanzania. *Report on a visit to examine plant parasitic nematodes of coffee in Tanzania*, February-March 1984. St Albans, UK, Commonwealth Institute of Parasitology.

Bridge, J. & Page, S. (1982). The rice root-knot nematode, *Meloidogyne graminicola*, on deep water rice (*Oryza sativa* subsp. *indica*). *Revue de Nématologie* 5, 225-232.

Bridge, J., Jones, E. & Page, L.J. (1976). *Meloidogyne acronea* associated with reduced growth of cotton in Malawi. *Plant Disease Reporter* 60, 5-7.

Bridge, J., Plowright, R.A. & Peng, D. (2005). Nematode parasites of rice. In: Luc, M., Sikora, R.A. & Bridge, J. (Eds). *Plant parasitic nematodes in subtropical and tropical agriculture*, 2nd edition. Wallingford, UK, CABI Publishing, pp. 87-130.

Brinkman, H. (1985). Planteparasitaire aaltjes bij helmgras (*Ammophila arenaria*). *Verslagen en Mededelingen Plantenziektenkundige Dienst Wageningen* 164, 173.

Brinkman, H., Goossens, J.J.M. & van Riel, H.R. (1996). Comparative host suitability of selected crop plants to *Meloidogyne chitwoodi* and *M.*

fallax. Anzeiger für Schädlingskunde, Planzenschutz, Umweltschutz 69, 127-129.

BRITO, J.A., POWERS, T.O., MULLIN, P.G., INSERRA, R.N. & DICKSON, D.W. (2004a). Morphological and molecular characterization of *Meloidogyne mayaguensis* isolates from Florida. *Journal of Nematology* 36, 232-240.

BRITO, J.A., STANLEY, J., CETINTAS, R., POWERS, T., INSERRA, R., MCAVOY, G., MENDES, M.L., CROW, B. & DICKSON, D. (2004b). Identification and host preference of *Meloidogyne mayaguensis* and other root-knot nematodes from Florida, and their susceptibility to *Pasteuria penetrans*. *Journal of Nematology* 36, 308-309.

BRITO, J.A., KAUR, R., DICKSON, D.W., RICH, J.R. & HALSEY, L.A. (2006). The pecan root-knot nematode, *Meloidogyne partityla* Kleynhans, 1986. *Nematology Circular* 222, Gainesville, FL, USA, Florida Department of Agriculture & Consumer Services Division of Plant Industry.

BRITO, J.A., KAUR, R., CETINTAS, R., STANLEY, J.D., MENDES, M.L., MCAVOY, E.J., POWERS, T.O. & DICKSON, D.W. (2008). Identification and isozyme characterisation of *Meloidogyne* spp. infecting horticultural and agronomic crops, and weed plants in Florida. *Nematology* 10, 757-766. DOI: 10.1163/156854108785787253

BRITO, J.A., KAUR, R., CETINTAS, R., STANLEY, J.D., MENDES, M.L., POWERS, T.O. & DICKSON, D.W. (2010). *Meloidogyne* spp. infecting ornamental plants in Florida. *Nematropica* 40, 87-103.

BRITO, J.A., SUBBOTIN, S.A., HAN, H., STANLEY, J.D. & DICKSON, D.W. (2015a). Molecular characterization of *Meloidogyne christiei* Golden & Kaplan, 1986 (Nematoda, Meloidogynidae) topotype population infecting Turkey oak (*Quercus laevis*) in Florida. *Journal of Nematology* 47, 169-175.

BRITO, J.A., DICKSON, D.W., KAUR, R., VAU, S. & STANLEY, J.D. (2015b). The peach root-knot nematode: *Meloidogyne floridensis*, and its potential impact for the peach industry in Florida. *Nematology Circular* 224, Gainesville, FL, USA, Florida Department of Agriculture & Consumer Services Division of Plant Industry.

BRITO, J.A., SMITH, T., ACHINELLY, M., MONTEIRO, T. & DICKSON, D.W. (2016). First report of *Meloidogyne partityla* infecting water oak (*Quercus nigra*) in Florida. *Plant Disease* 100, 1246. DOI: 10.1094/PDIS-11-15-1314-PDN

BROWN JR, W.M.G. & WILSON, A.C. (1979). Rapid evolution of animal mitochondrial DNA. *Proceedings of the National Academy of Sciences of the United Staes of America* 76, 1967-1971.

BRUESKE, C.H. & BERGESON, G.B. (1972). Investigation of growth hormones in xylem exudate and root tissue of tomato infected with *M. incognita*. *Journal of Experimental Botany* 23, 14-22.

BRZESKI, M.W. (1998). *Nematodes of Tylenchina in Poland and temperate Europe*. Warsaw, Poland, Muzeum I Instytut Zoologii PAN.

BUANGSUWON, D., TONBOON-EKE, P., RUJIRACHOON, G., BRAUN, A.J. & TAYLOR, A.L. (1971). Nematodes. In: *Rice diseases and pests of Thailand*, English edition. Rice Protection Research Centre, Rice Department, Ministry of Agriculture, Thailand, pp. 61-67.

BUISSON, A., CHABERT, A., RUCK, L. & FOURNET, S. (2014). Nematodes associated with damage in oilseed rape: new data on the biology and geographical distribution of *Meloidogyne artiellia. Nematology* 16, 201-206. DOI: 10.1163/15685411-00002758

BUSTILLO, Y.Y., CROZZOLI, R., GRECO, N. & LAMBERTI, F. (2000). Efecto del nematodo agallador *Meloidogyne incognita* sobre el crecimiento de la lechosa (*Carica papaya*) en vivero. *Nematologia Mediterranea* 28, 163-170.

CABASAN, M.T.N., FERNANDEZ, L. & DE WAELE, D. (2015). Host response of *Oryza glaberrima* and *O. sativa* rice genotypes to the rice root-knot nematode *Meloidogyne graminicola* in a hydroponic system under growth chamber. *Archives of Phytopathology and Plant Protection* 48, 740-750. DOI: 10.1080/03235408.2016.1140608

CABRERA, J., BARCALA, M., FENOLL, C. & ESCOBAR, C. (2014a). Transcriptomic signatures of transfer cells in early developing nematode feeding cells of *Arabidopsis* focused on auxin and ethylene signaling. *Frontiers in Plant Science* 5, 107. DOI: 10.3389/fpls.2014.00107

CABRERA, J., BUSTOS, R., FAVERY, B., FENOLL, C. & ESCOBAR, C. (2014b). NEMATIC: a simple and versatile tool for the *in silico* analysis of plant-nematode interactions. *Molecular Plant Pathology* 15, 627-636. DOI: 10.1111/mpp.12114

CABRERA, J., DÍAZ-MANZANO, F.E., BARCALA, M., ARGANDA-CARRERAS, I., DE ALMEIDA ENGLER, J., ENGLER, G., FENOLL, C. & ESCOBAR, C. (2015a). Phenotyping nematode feeding sites: three dimensional reconstruction and volumetric measurements of giant cells induced by root-knot nematodes in Arabidopsis. *New Phytologist* 206, 868-880. DOI: 10.1111/nph.13249

CABRERA, J., DÍAZ-MANZANO, F.E., FENOLL, C. & ESCOBAR, C. (2015b). Developmental pathways mediated by hormones in nematode feeding sites. *Advances in Botanical Research* 73, 167-188. DOI: 10.1016/bs.abr.2014.12.005

CAI, Y., ZHOU, Q.J., GU, J.F., CHEN, X.F. & CHEN, J. (2016). Rapid and sensitive detection of *Meloidogyne camelliae* by LAMP-LFD. *Journal of Agricultural Biotechnology* 24, 770-780.

CAIN, S.C. (1974). *Meloidogyne exigua. CIH Descriptions of plant-parasitic nematodes*, Set 4, No. 49. St Albans, UK, Commonwealth Institute of Helminthology.

CAMARA, G.R., CARVALHO, A.H.O., TEIXEIRA, A.G., FERREIRA, M.L.S.M., DE OLIVEIRA, F.L., MORAES, W.B. & ALVES, F.R. (2019). First report of *Meloidogyne inornata* on *Smallanthus sonchifolius* in Brazil. *Plant Disease* 104, 595. DOI: 10.1094/PDIS-08-19-1739-PDN

CAMPOS, V.P. & VILLAIN, L. (2005). Nematode parasites of coffee and cocoa. In: Luc, M., Sikora, R.A. & Bridge, J. (Eds). *Plant parasitic nematodes in subtropical and tropical agriculture*, 2nd edition. Wallingford, UK, CABI Publishing, pp. 529-579.

CARNEIRO, R.G. (1993). Fitonematoides na cafeicultura paranaense: Situaçao atual. *Resumos do XVII Congresso Brasileiro de Nematologia, Fevereiro. Jaboticabal, SP, Brazil.*

CARNEIRO, R.M.D.G. & ALMEIDA, M.R.A. (2000). Caracterização isoenzimática e variabilidade intraespecífica dos nematóides de galhas do cafeeiro no Brasil. *I Simpósio de Pesquisa dos Cafés do Brasil, 26-29 setembro 2000, Poços de Caldas, MG, Brasil*, pp. 280-282.

CARNEIRO, R.M.D.G. & COFCEWICZ, E.T. (2008). Taxonomy of coffee-parasitic root-knot nematodes, *Meloidogyne* spp. In: Souza, R.M. (Ed.). *Plant-parasitic nematodes of coffee.* Springer Science and Business Media B.V., pp. 87-121.

CARNEIRO, R.M.D.G., CARNEIRO, R.G., ABRANTES, I.M.O., SANTOS, M.S.N.A. & ALMEIDA, M.R.A. (1996a). *Meloidogyne paranaensis* n. sp. (Nematoda: Meloidogynidae) a root-knot nematode parasitizing coffee in Brazil. *Journal of Nematology* 28, 177-189.

CARNEIRO, R.M.D.G., ALMEIDA, M.R.A. & CARNEIRO, R.G. (1996b). Enzyme phenotypes of Brazilian populations of *Meloidogyne* spp. *Fundamental and Applied Nematology* 19, 555-560.

CARNEIRO, R.M.D.G., CASTAGNONE-SERENO, P. & DICKSON, D.W. (1998). Variability among four populations of *Meloidogyne javanica* from Brazil. *Fundamental and Applied Nematology* 21, 319-326.

CARNEIRO, R.M.D.G., ALMEIDA, M.R.A. & QUÉNÉHERVÉ, P. (2000). Enzyme phenotypes of *Meloidogyne* spp. populations. *Nematology* 2, 645-654. DOI: 10.1163/156854100509510

CARNEIRO, R.M.D.G., MOREIRA, W.A., ALMEIDO, M.R.A. & GOMES, A.C.M.M. (2001). Primeiro registro de *Meloidogyne mayaguensis* em goiabeira no Brasil. *Nematologia Brasileira* 25, 223-228.

CARNEIRO, R.M.D.G., MAZZAFERA, P., FERRAZ, L.C.C.B., MURAOKA, T. & TRIVELIN, P.C.O. (2002). Uptake and translocation of nitrogen, phosphorus and calcium in soybean infected with *Meloidogyne incognita* and *M. javanica*. *Fitopatologia Brasileira* 27, 141-150.

CARNEIRO, R.M.D.G., CAMEIRO, R.G., DAS NEVES, D.I. & ALMEIDA, M.R.A. (2003a). A new race of *Meloidogyne javanica* on *Arachis pintoi* in the state of Parana. *Nematologia Brasileira* 27, 219-221.

CARNEIRO, R.M.D.G., GOMES, C.B., ALMEIDA, M.R.A., GOMES, A.C.M.M. & MARTINS, I. (2003b). First report of *Meloidogyne ethiopica* Whitehead, 1968 on kiwi in Brazil and reaction on different plant species. *Nematologia Brasileira* 27, 151-158.

CARNEIRO, R.M.D.G., RANDIG, O., ALMEIDA, M.R.A. & GOMES, A.C.M.M. (2004a). Additional information on *Meloidogyne ethiopica* Whitehead, 1968 (Tylenchida: Meloidogynidae), a root-knot nematode parasitising kiwi fruit and grape-vine from Brazil and Chile. *Nematology* 6, 109-123. DOI: 10.1163/156854104323072982

CARNEIRO, R.M.D.G., ALMEIDA, M.R.A. & GOMES, A.C.M.M. (2004b). First record of *Meloidogyne hispanica* Hirschmann, 1986 on squash in State of Bahia, Brazil. *Nematologia Brasileira* 28, 215-218.

CARNEIRO, R.M.D.G., TIGANO, M.S., RANDIG, O., ALMEIDA, M.R.A. & SARAH, J.L. (2004c). Identification and genetic diversity of *Meloidogyne* spp. (Tylenchida: Meloidogynidae) on coffee from Brazil, Central America and Hawaii. *Nematology* 6, 287-298. DOI: 10.1163/1568541041217942

CARNEIRO, R.M.D.G., ALMEIDA, M.R.A., GOMES, A.C.M.M. & HERNANDEZ, A. (2005a). *Meloidogyne izalcoensis* n. sp. (Nematoda: Meloidogynidae), a root-knot nematode parasitising coffee in El Salvador. *Nematology* 7, 819-832. DOI: 10.1163/156854105776186361

CARNEIRO, R.M.D.G., RANDIG, O., ALMEIDA, M.R.A. & GONÇALVES, W. (2005b). Identificação e caracterização de espécies de *Meloidogyne* em cafeeiros nos Estados de São Paulo e Minas Gerais através dos fenótipos de esterase e SCAR-multiplex PCR. *Nematologia Brasileira* 29, 233-241.

CARNEIRO, R.M.D.G., ALMEIDA, M.R.A., COFCEWICZ, E.T., MAGUNACELAYA, J.C. & ABALLAY, E. (2007). *Meloidogyne ethiopica*, a major root-knot nematode parasitising *Vitis vinifera* and other crops in Chile. *Nematology* 9, 635-641. DOI: 10.1163/156854107782024794

CARNEIRO, R.M.D.G., SANTOS, M.F.A. DOS, ALMEIDA, M.R.A., MOTA, F.C., GOMES, A.C.M.M. & TIGANO, M.S. (2008). Diversity of *Meloidogyne arenaria* using morphological, cytological and molecular approaches. *Nematology* 10, 819-834. DOI: 10.1163/156854108786161526

CARNEIRO, R.M.D.G., CORREA, V., ALMEIDA, M.R.A., GOMES, A.C.M.M., MOHAMMAD DEIMI, A., CASTAGNONE-SERENO, P. & KARSSEN, G. (2014). *Meloidogyne luci* n. sp. (Nematoda: Meloidogynidae), a root-knot nematode parasitising different crops in Brazil, Chile and Iran. *Nematology* 16, 289-301. DOI: 10.1163/15685411-00002765

CARRILLO-FASIO, J.A., MARTINEZ-GALLARDO, J.A., ALLENDE-MOLAR, R., VELARDE-FELIX, S., ROMERO-HIGAREDA, C.E. & RETES-MANJARREZ, J.E. (2019). Distribution of *Meloidogyne* species (Tylenchida: Meloidogynidae) in tomato crop in Sinaloa, Mexico. *Nematropica* 49, 71-82.

CASASSA, A.M., CROZZOLI, R., MATHEUS, J., BRAVO, V. & MARIN, M. (1998). Efecto del nematodo agallador *Meloidogyne incognita* sobre el crecimiento del guayabo (*Psidium* spp.) en vivero. *Nematologia Mediterranea* 26, 237-242.

CASTAGNONE-SERENO, P. (2006). Genetic variability and adaptive evolution in parthenogenetic root-knot nematodes. *Heredity* 96, 282-289. DOI: 10.1038/sj.hdy.6800794

CASTAGNONE-SERENO, P., BONGIOVANNI, M. & DALMASSO, A. (1993). Stable virulence against the tomato resistance *Mi* gene in the parthenogenetic root-knot nematode *Meloidogyne incognita*. *Phytopathology* 83, 803-805. DOI: 10.1094/Phyto-83-803

CASTAGNONE-SERENO, P., LEROY, F., BONGIOVANNI, M., ZIJLSTRA, C. & ABAD, P. (1999). Specific diagnosis of two root-knot nematodes, *Meloidogyne chitwoodi* and *M. fallax*, with satellite DNA probes. *Phytopathology* 89, 380-384. DOI: 10.1094/PHYTO.1999.89.5.380

CASTAGNONE-SERENO, P., DANCHIN, E.G.J., PERFUS-BARBEOCH, L. & ABAD, P. (2013). Diversity and evolution of root-knot nematodes, genus *Meloidogyne*: new insights from the genomic era. *Annual Review of Phytopathology* 51, 203-220. DOI: 10.1146/annurev-phyto-082712-102300

CASTILLO, P. & VOVLAS, N. (2005). *Bionomics and identification of the genus* Rotylenchus *(Nematoda: Hoplolaimidae). Nematology Monographs and Perspectives 3* (Series editors: Hunt, D.J. & Perry, R.N.). Leiden, The Netherlands, Brill.

CASTILLO, P. & VOVLAS, N. (2007). Pratylenchus *(Nematoda: Pratylenchidae): diagnosis, biology, pathogenicity and management. Nematology Monographs and Perspectives 6*. (Series editors: Hunt, D.J. & Perry, R.N.). Leiden, The Netherlands, Brill.

CASTILLO, P., DI VITO, M., VOVLAS, N. & JIMÉNEZ-DÍAZ, R.M. (2001). Host-parasite relationships in root-knot disease of white mulberry. *Plant Disease* 85, 277-281. DOI: 10.1094/PDIS.2001.85.3.277

CASTILLO, P., VOVLAS, N., SUBBOTIN, S. & TROCCOLI, A. (2003a). A new root-knot nematode: *Meloidogyne baetica* n. sp. (Nematoda: Heteroderidae) parasitizing wild olive in Southern Spain. *Phytopathology* 93, 1093-1102. DOI: 10.1094/PHYTO.2003.93.9.1093

CASTILLO, P., NAVAS-CORTÉS, J.A., GOMAR TINOCO, D., DI VITO, M. & JIMÉNEZ-DÍAZ, R.M. (2003b). Interactions between *Meloidogyne artiellia*, the cereal and legume root-knot nematode, and *Fusarium oxysporum* f. sp. *ciceris* race 5 in chickpea. *Phytopathology* 93, 1513-1523. DOI: 10.1094/PHYTO.2003.93.12.1513

CASTILLO, P., RAPOPORT, H., PALOMARES-RIUS, J.E. & JIMÉNEZ DÍAZ, R.M. (2008). Suitability of weed species prevailing in Spanish vineyards as hosts for root-knot nematodes. *European Journal of Plant Pathology* 120, 43-51. DOI: 10.1007/s10658-007-9195-8

CASTILLO, P., VOVLAS, N., TROCCOLI, N., LIÉBANAS, G., PALOMARES-RIUS, J.E. & LANDA, B.B. (2009a). A new root-knot nematode, *Meloidogyne silvestris* n. sp. (Nematoda: Meloidogynidae), parasitizing European holly in northern Spain. *Plant Pathology* 58, 606-619. DOI: 10.1111/j.1365-3059.2008.01991.x

CASTILLO, P., GUTIÉRREZ-GUTIÉRREZ, C., PALOMARES-RIUS, J.E., CANTALAPIEDRA NAVARRETE, C. & LANDA, B.B. (2009b). First report of root-knot nematode *Meloidogyne hispanica* infecting grapevines in southern Spain. *Plant Disease* 93, 1353. DOI: 10.1094/PDIS-93-12-1353B

CASTRO, J.M.C., LIMA, R.D. DE & CARNEIRO, R.M.D.G. (2003). Isoenzymatic variability in Brazilian populations of *Meloidogyne* spp. from soybean. *Nematologia Brasileira* 27, 1-12.

CASTRO, J.M.C., CAMPOS, V.P. & DUTRA, M.R. (2004). Ocorrência de *Meloidogyne coffeicola* em cafeeiros do município de Coromandel, Região do Alto Paranaíba em Minas Gerais. *Fitopatologia Brasileira* 29, 227. DOI: 10.1590/S0100-41582004000200022

CAUBEL, G., RITTER, M. & RIVOAL, R. (1972). Observations relatives à des attaques du nématode *Meloidogyne naasi* Franklin sur céréales et graminées fourragères dans l'Ouest de la France en 1970. *Comptes Rendus des Séances de l'Académie d'Agriculture de France* 58, 351-356.

CHAPUIS, E., BESNARD, G., ANDRIANASETRA, S., RAKOTOMALALA, M., NGUYEN, H.T. & BELLAFIORE, S. (2016). First report of the root-knot nematode (*Meloidogyne graminicola*) in Madagascar rice fields. *Australasian Plant Disease Notes* 11, 32. DOI: 10.1007/s13314-016-0222-5

CHARCHAR, J.M. & EISENBACK, J.D. (2002). *Meloidogyne brasilensis* n. sp. (Nematoda: Meloidogynidae), a root-knot nematode parasitising tomato cv. Rossol in Brazil. *Nematology* 4, 629-643. DOI: 10.1163/15685410260438926

CHARCHAR, J.M., EISENBACK, J.D. & HIRSCHMANN, H. (1999). *Meloidogyne petuniae* n. sp. (Nemata: Meloidogynidae), a root-knot nematode parasitic on petunia in Brazil. *Journal of Nematology* 31, 81-91.

CHARCHAR, J.M., EISENBACK, J.D., CHARCHAR, M.J. & BOITEUX, M.E.N.F. (2008a). *Meloidogyne pisi* n. sp. (Nematoda: Meloidogynidae), a root-knot nematode parasitising pea in Brazil. *Nematology* 10, 479-493. DOI: 10.1163/156854108784513905

CHARCHAR, J.M., EISENBACK, J.D., CHARCHAR, M.J. & BOITEUX, M.E.N.F. (2008b). *Meloidogyne phaseoli* n. sp. (Nematoda: Meloidogynidae), a root-knot nematode parasitising bean in Brazil. *Nematology* 10, 525-538. DOI: 10.1163/156854108784513842

CHARCHAR, J.M., EISENBACK, J.D., VIEIRA, J.V., FONSECA-BOITEUX, M.E. & BOITEUX, L.S. (2009). *Meloidogyne polycephannulata* n. sp. (Nematoda: Meloidogynidae), a root-knot nematode parasitizing carrot in Brazil. *Journal of Nematology* 41, 174-186.

CHARCHAR, J.M., FONSECA, M.E.N., PINHEIRO, J.B., BOITEUX, L.S. & EISENBACK, J.D. (2010). Epidemics of *Meloidogyne brasilensis* in Central Brazil on processing tomato hybrids that have the root-knot nematode *Mi* resistance gene. *Plant Disease* 94, 781. DOI: 10.1094/PDIS-94-6-0781B

CHAVES, A., MELO, L.J.O.T., SIMÕES NETO, D.E., COSTA, I.G. & PEDROSA, E.M.R. (2007). Declínio severo do desenvolvimento da cana-de-açúcar em tabuleiros costeiros do Estado de Pernambuco. *Nematologia Brasileira* 31, 93-95.

CHAVES, R.L. (1994). Differentiation with SEM of 6 species of *Meloidogyne* (Nemata: Heteroderidae) found in Costa Rica. *Revista De Biologia Tropical* 42, 113-120.

CHEN, J.-M., COOPER, D.N., CHUZHANOVA, N., FÉREC, C. & PATRINOS, G.P. (2007). Gene conversion: mechanisms, evolution and human disease. *Nature Reviews Genetics* 8, 762-775. DOI: 10.1038/nrg2193

CHEN, P., PENG, D.L. & ZHENG, J.W. (1990). [Description of a new root-knot nematode *Meloidogyne fanzhiensis* sp. n.] *Journal of Shanxi Agricultural University* 10, 55-60.

CHI, Y.-K., ZHAO, W., YE, M.-D., ALI, F., WANG, T. & QI, R.D. (2020). Evaluation of recombinase polymerase amplification assay for detecting *Meloidogyne javanica. Plant Disease* 104, 801-807. DOI: 10.1094/PDIS-07-19-1473-RE

CHIANG, H.C. (1979). A general model of the economic threshold level of pest populations. *FAO Plant Protection Bulletin* 27, 71-73.

CHITAMBO, O., HAUKELAND, S., FIABOE, K.K.M., KARIUKI, G.M. & GRUNDLER, F.M.W. (2016). First report of the root-knot nematode *Meloidogyne enterolobii* parasitizing African nightshades in Kenya. *Plant Disease* 100, 1954. DOI: 10.1094/PDIS-11-15-1300-PDN

CHITWOOD, B.G. (1949). Root-knot nematodes, part I. A revision of the genus *Meloidogyne* Göldi, 1887. *Proceedings of the Helminthological Society of Washington* 16, 90-104.

CHITWOOD, B.G. & OTEIFA, B.A. (1952). Nematodes parasitic on plants. *Annual Review of Microbiology* 6, 151-184. DOI: 10.1146/annurev.mi.06.100152.001055

CHITWOOD, B.G. & TOUNG, M. (1960). *Meloidogyne* from Taiwan and New Delhi. *Phytopathology* 50, 631-632.

CHITWOOD, B.G., SPECHT, A.W. & HAVIS, L. (1952). Root-knot nematodes. III. Effect of *Meloidogyne incognita* and *M. javanica* on some peach rootstocks. *Plant and Soil* 4, 77-95.

CHIZHOV, V.N. (1981). [Some peculiarities of the structure of the female reproductive system in some species of Meloidogynidae and Heteroderidae.] *Byulleten Vsesoyuznogo Instituta Gelmintologii K.I. Skrjabina* 31, 66-73.

CHIZHOV, V.N. & TURKINA, A. YU. (1986). [*Meloidogyne ardenensis* – a parasite of *Betula pendula* in Moscow region.] *Byulleten Vsesoyuznogo Instituta Gelmintologii K.I. Skrjabina* 45, 100-103.

CHRISTIE, I.J. & ALBIN, F.E. (1944). Host-parasite relationships of the root-knot nematode, *Heterodera marioni*. I. The question of races. *Proceedings of the Helminthological Society of Washington* 11, 31-37.

CHRISTIE, J.R. (1936). The development of root-knot nematode galls. *Phytopathology* 26, 1-22.

CHRISTIE, J.R. (1946). Host-parasite relationships of the root-knot nematode, *Heterodera marioni*; some effects of the host on the parasite. *Phytopathology* 36, 340-352.

CHRISTIE, J.R. (1959). *Plant nematodes, their bionomics and control.* Gainesville, FL, USA, Agricultural Experiment Station, University of Florida.

CHURCH, G.T. (2005). First report of the root-knot nematode *Meloidogyne floridensis* on tomato (*Lycopersicon esculentum*) in Florida. *Plant Disease* 89, 527. DOI: 10.1094/PD-89-0527B

CLEMENS, G.P. & KRUSBERG, L.R. (1971). Root-knot nematode found on London plane trees. *Plant Disease Reporter* 55, 280.

CLIFF, G.M. & HIRSCHMANN, H. (1984). *Meloidogyne microcephala* n. sp. (Meloidogynidae), a root-knot nematode from Thailand. *Journal of Nematology* 16, 183-193.

COBB, N.A. (1924). The amphids of *Caconema* (nom. nov.) and other nemas. *Journal of Parasitology* 11, 118-120.

COETZEE, V. (1956). *Meloidogyne acronea*, a new species of root-knot-nematode. *Nature* 177, 899-900. DOI: 10.1038/177899a0

COETZEE, V. & BOTHA, H.J. (1966). A redescription of *Hypsoperine acronea* (Coetzee, 1956) Sledge & Golden, 1964 (Nematoda: Heteroderidae), with a note on its biology and host specificity. *Nematologica* 11, 480-484. DOI: 10.1163/187529265X00654

COFCEWICZ, E.T., CARNEIRO, R.M.D.G., RANDIG, O., CHABRIER, C. & QUÉNÉHERVÉ, P. (2005). Diversity of *Meloidogyne* spp. on *Musa* in Martinique, Guadeloupe, and French Guiana. *Journal of Nematology* 37, 312-322.

COLAGIERO, M. & CIANCIO, A. (2011). Climate changes and nematodes: Expected effects and perspectives for plant protection. *Redia* 94, 113-118.

CONCEIÇÃO, I.L.P.M. DA, CUNHA, M.J.M., FEIO, G., CORREIA, M., SANTOS, M.C.V. DOS, ABRANTES, I.M. DE O. & SANTOS, M.S.N. DE A. (2009). Root-knot nematodes, *Meloidogyne* spp., on potato in Portugal. *Nematology* 11, 311-313. DOI: 10.1163/156854109X415515

CONCEIÇÃO, I.L.P.M. DA, TZORTZAKAKIS, E.A., GOMES, P., ABRANTES, I. & CUNHA, M.J. (2012). Detection of the root-knot nematode *Meloidogyne ethiopica* in Greece. *European Journal of Plant Pathology* 134, 451-457. DOI: 10.1007/s10658-012-0027-0

CONN, H.J. (1942). Validity of the genus *Alcaligenes*. *Journal of Bacteriology* 44, 353-360.

COOK, R., MIZEN, K.A. & PERSON-DEDRYVER, F. (1999). Resistance in ryegrasses, *Lolium* spp., to three European populations of the root-knot nematode, *Meloidogyne naasi*. *Nematology* 1, 661-671. DOI: 10.1163/156854199508621

CORNU, M. (1878). *Études sur le* Phylloxéra vastatrix. *Mémoires présentés par divers savants à l'Académie des Sciences de l'Institut de France*. Paris, France, Imprimerie Nationale.

CORREA, V.R., SANTOS, M.F.A. DOS, ALMEIDA, M.R.A., PEIXOTO, J.R., CASTAGNONE-SERENO, P. & CARNEIRO, R.M.D.G. (2013). Species-specific DNA markers for identification of two root-knot nematodes of coffee: *Meloidogyne arabicida* and *M. izalcoensis*. *European Journal of Plant Pathology* 137, 305-313. DOI: 10.1007/s10658-013-0242-3

CORREA, V.R., MATTOS, V.S., ALMEIDA, M.R.A., SANTOS, M.F.A., TIGANO, M.S., CASTAGNONE-SERENO, P. & CARNEIRO, R.M.D.G. (2014). Genetic diversity of the root-knot nematode *Meloidogyne ethiopica* and development of a species-specific SCAR marker for its diagnosis. *Plant Pathology* 63, 476-483. DOI: 10.1111/ppa.12108

CORREIA, E.C.S.S., SILVA, M.F.A., AIRES, B.C., SORATTO, R.P. & WILCKEN, S.R.S. (2016). Report of *Meloidogyne inornata* in common bean in São Paulo State, Brazil. *Summa Phytopathologica* 42, 273-273.

COSTILLA, M.A. & DE OJEDA, S.G. (1986). Primera cita para Tucumán (República Argentina) de dos especies de nematodos del nudo: *Meloidogyne ottersoni* (Thorne, 1969) Franklin, 1971 y *M. decalineata* Whitehead, 1968 (Nematoda-Meloidogynidae). *Revista Industrial y Agricola de Tucumán* 63, 175-182.

COYNE, D.L., TOKO, M., ANDRADE, M., HANNA, R., SITOLE, A., KAGODA, F., AL BANNA, L. & MARAIS, M. (2005). *Meloidogyne* spp. and associated galling and damage on cassava in Kenya and Mozambique. *African Plant Protection* 12, 35-36.

COYNE, D.L., TCHABI, A., BAIMEY, H., LABUSCHAGNE, N. & ROTIFA, I. (2006). Distribution and prevalence of nematodes (*Scutellonema bradys* and *Meloidogyne* spp.) on marketed yam (*Dioscorea* spp.) in West Africa. *Field Crops Research* 96, 142-150. DOI: 10.1016/j.fcr.2005.06.004

CROW, W.T., LEVIN, R., HALSEY, L.A. & RICH, J.R. (2005). First report of *Meloidogyne partityla* on pecan in Florida. *Plant Disease* 89, 1128. DOI: 10.1094/PD-89-1128C

CROZZOLI, R. & PARRA, N. (1999). Patogenicidad del nematodo agallador, *Meloidogyne incognita*, en yuca (*Manihot esculentum*). *Nematologia Mediterranea* 27, 95-100.

CROZZOLI, R., GRECO, N., SUÁREZ, A. & RIVAS, D. (1997). Patogenicidad del nematodo agallador, *Meloidogyne incognita*, en cultivares de *Phaseolus vulgaris* y *Vigna unguiculata*. *Nematropica* 27, 61-67.

CROZZOLI, R., GRECO, N., SUÁREZ, A.C. & RIVAS, D. (1998). Pathogenicity of the root-knot nematode *Meloidogyne incognita* to *Vigna unguiculata*. *Nematropica* 29, 99-103.

CUEVAS, O.J. (1995). *Distribución, rango de hospedantes y determinación de las razas de* Meloidogyne chitwoodi *Golden, O'Bannon & Finley, 1980 (Nematoda: Meloidogyninae) en el Valle de Huamantla, Tlaxcala.* M.S. Thesis, Chapingo, Mexico, Universidad Autónoma Chapingo.

CURRAN, J., BAILLIE, D.L. & WEBSTER, J.M. (1985). Use of genomic DNA restriction fragment length differences to identify nematode species. *Parasitology* 90, 137-144. DOI: 10.1017/S0031182000049088

CURRAN, J., DRIVER, F., BALLARD, J.W.O. & MILNER, R.J. (1994). Phylogeny of *Metarhizium*: analysis of ribosomal DNA sequence data. *Mycological Research* 98, 547-555. DOI: 10.1016/S0953-7562(09)80478-4

CURTO, G., EVLICE, E., GUITIAN CASTRILLON, J.M., KARSSEN, G., DEB NIJS, L., MAGNUSSON, C., PRIOR, T.J., WESEMAEL, W. & GROUSSET, F. (2017). Pest Risk Analysis for *Meloidogyne mali* (Tylenchida: Meloidogynidae), apple root-knot nematode. EPPO, Paris. http://www.eppo.int/QUARANTINE/Pest_Risk_Analysis/PRA_intro.htm.

DA PONTE, J.J. (1969). *Meloidogyne lordelloi* n. sp., a nematode parasite of *Cereus macrogonus* Salm-Dick. *Boletin Cearense da Agronomia* 10, 59-63.

DA PONTE, J.J. (1977). Nematóides das galhas: espécies ocorrentes no Brasil e seus hospedeiros. *Coleção Mossoroense (Brasil)* 54, 7-99.

DABUR, K.R., TAYA, A.S. & BAJAJ, H.K. (2004). Life cycle of *Meloidogyne graminicola* on paddy and its host range studies. *Indian Journal of Nematology* 34, 80-84.

DADAZIO, T.S., DA SILVA, S.A., DORIGO, O.F., WILCKEN, S.R.S. & MACHADO, A.C.Z. (2016). Host-parasite relationships in root-knot disease caused by *Meloidogyne inornata* in common bean (*Phaseolus vulgaris*). *Journal of Phytopathology* 164, 735-744. DOI: 10.1111/jph.12494

DAHER, R.K., STEWART, G., BOISSINOT, M. & BERGERON, M.G. (2016). Recombinase polymerase amplification for diagnostic applications. *Clinical Chemistry* 62, 947-958. DOI: 10.1373/clinchem.2015.245829

DALMASSO, A. & BERGÉ, J.B. (1975). Variabilité génétique chez les *Meloidogyne* et plus particulièrement chez *M. hapla*. *Cahiers ORSTROM. Série biologie* 10, 233-238.

DALMASSO, A. & BERGÉ, J.B. (1977). Variabilité liée aux phénomènes de reproduction chez les *Meloidogyne. Annales de Zoologie Ecologie Animale* 9, 568-569.

DALMASSO, A. & BERGÉ, J.B. (1978). Molecular polymorphism and phylogenetic relationship in some *Meloidogyne* spp.: application to the taxonomy of *Meloidogyne. Journal of Nematology* 10, 323-332.

DALMASSO, A. & BERGÉ, J.B. (1979). Genetic approach to the taxonomy of *Meloidogyne* species. In: Lamberti, F. & Taylor, C.E. (Eds). *Root-knot nematodes* (Meloidogyne species). *Systematics, biology and control.* New York & London, Academic Press, pp. 111-113.

DALMASSO, A. & BERGÉ, J.B. (1983). Enzyme polymorphism and the concept of parthenogenetic species, exemplified by *Meloidogyne*. In: Stone, A.R., Platt, H.M. & Khalil, L.F. (Eds). *Concepts in nematode systematics.* Systematics Association, Special Volume No. 22, London & New York, Academic Press, pp. 187-196.

DAMME, N., WAEYENBERGE, L., VIAENE, N., VAN HOENSELAAR, T. & KARSSEN, G. (2013). First report of the root-knot nematode *Meloidogyne artiellia* in Belgium. *Plant Disease* 97, 152. DOI: 10.1094/PDIS-08-12-0720-PDN

DAMMINI PREMACHANDRA, W.T.S. & GOWEN, S.R. (2015). Influence of preplant densities of *Meloidogyne incognita* on growth and root infestation of spinach (*Spinacia oleracea* L.) (Amaranthaceae) – an important dimension towards enhancing crop production. *Future of Food: Journal on Food, Agriculture & Society* 3, 18-26.

DANG-NGOC, K., HUONG, N.M. & UT, N.V. (1982). Root-knot disease of rice in the Mekong Delta, Vietnam. *International Rice Research Newsletter* 7, 15.

DAULTON, R.A.C. (1963). Controlling *M. javanica* in Southern Rhodesia. *Rhodesian Agricultural Journal* 60, 150-152.

DAULTON, R.A.C. & CURTIS, R.F. (1964). The effects of *Tagetes* spp. on *Meloidogyne javanica* in Southern Rhodesia. *Nematologica* 10, 61-62. DOI: 10.1163/187529263X00890

DAVIS, R.F., EARL, H.J. & TIMPER, P. (2014). Effect of simultaneous water deficit stress and *Meloidogyne incognita* infection on cotton yield and fiber quality. *Journal of Nematology* 46, 108-118.

DE GIORGI, C., VERONICO, P., DE LUCA, F., NATILLA, A., LANAVE, C. & PESOLE, G. (2002). Structural and evolutionary analysis of the ribosomal genes of the parasitic nematode *Meloidogyne artiellia* suggests its ancient origin. *Molecular and Biochemical Parasitology* 124, 91-94. DOI: 10.1016/S0166-6851(02)00161-5

DE GRISSE, A. (1961). *Meloidogyne kikuyensis* n. sp., a parasite of kikuyu grass (*Pennisetum clandestinum*) in Kenya. *Nematologica* 5, 303-308. DOI: 10.1163/187529260X00118

De Guiran, G. & Netscher, C. (1970). Les nématodes du genre *Meloidogyne*, parasites de cultures tropicales. *Cahiers ORSTOM, Série Biologie* 11, 151-185.

De Haan, E.G., Dekker, C.C.E.M., Tameling, W.I.L., Den Nijs, L.J.M.F., Van Den Bovenkamp, G.W. & Kooman-Gersmann, M. (2014). The MeloTuber test: a real-time TaqMan® PCR-based assay to detect the root-knot nematodes *Meloidogyne chitwoodi* and *M. fallax* directly in potato tubers. *Bulletin OEPP/EPPO* 44, 166-175. DOI: 10.1111/epp.12128

De Ley, I.T., Karssen, G., De Ley, P., Vierstraete, A., Waeyenberge, L., Moens, M. & Vanfleteren, J. (1999). Phylogenetic analyses of internal transcribed spacer region sequences within *Meloidogyne. Journal of Nematology* 31, 530-531.

De Ley, I.T., De Ley, P., Vierstraete, A., Karssen, G., Moens, M. & Vanfleteren, J. (2002). Phylogenetic analyses of *Meloidogyne* small subunit rDNA. *Journal of Nematology* 34, 319-327.

De Luca, F., Di Vito, M., Fanelli, E., Reyes, A., Greco, N. & De Giorgi, C. (2009). Characterization of the heat shock protein 90 gene in the plant parasitic nematode *Meloidogyne artiellia* and its expression as related to different developmental stages and temperature. *Gene* 440, 16-22. DOI: 10.1016/j.gene.2009.03.020

De Weerdt, M., Kox, L., Waeyenberge, L., Viaene, N. & Zijlstra, C. (2011). A real-time PCR assay to identify *Meloidogyne minor. Journal of Phytopathology* 159, 80-84. DOI: 10.1111/j.1439-0434.2010.01717.x

Decker, H. & Rodriguez Fuentes, M.E. (1989). Uber das Auftreten des Wurzelgallenematoden *Meloidogyne mayaguensis* an *Coffea arabica* in Kuba. *Wissenschaftliche Zeitschrift der Wilhelm-PieckUniversitat, Rostock, Naturwissenschaftliche Reihe* 38, 32-34.

Decker, H., Yassin, A.M. & El-Amin, E.T.M. (1980). Plant nematology in the Sudan – a review. *Beitrage zur Tropischen Landwirtschaft und Veterinarmedizin* 18, 271-290.

Den Nijs, L.J.M.F., Brinkman, H. & van der Sommen, A.T.C. (2004). A Dutch contribution to knowledge on phytosanitary risk and host status of various crops for *Meloidogyne chitwoodi* Golden *et al.*, 1980 and *M. fallax* Karssen, 1996: an overview. *Nematology* 6, 303-312. DOI: 10.1163/1568541042360492

D'Errico, G., Crescenzi, A. & Landi, S. (2014). First report of the southern root-knot nematode *Meloidogyne incognita* on the invasive weed *Araujia sericifera* in Italy. *Plant Disease* 98, 1593. DOI: 10.1094/PDIS-06-14-0584-PDN

Derycke, S., Remerie, T., Vierstraete, A., Backeljau, T., Vanfleteren, J., Vincx, M. & Moens, T. (2005). Mitochondrial DNA variation and cryptic speciation within the free-living marine nematode *Pelliodi-*

tis marina. Marine Ecology Progress Series 300, 91-103. DOI: 10.3354/meps300091

DESAEGER, J. & RAO, M.R. (2000). Infection and damage potential of *Meloidogyne javanica* on *Sesbania sesban* in different soil types. *Nematology* 2, 169-178. DOI: 10.1163/156854100509060

D'HERDE, J. (1965). Een nieuw wortelknoppelaaltje, parasiet van de bieteteelt. *Mededelingen van de Landbouwhogeschool, Gent* 30, 1429-1436.

DI VITO, M. (1986). Population densities of *Meloidogyne incognita* and growth of susceptible and resistant pepper plants. *Nematologia Mediterranea* 14, 217-221.

DI VITO, M. & EKANAYAKE, H.M.R.K. (1983). Relationship between population densities of *Meloidogyne incognita* and growth of resistant and susceptible tomato. *Nematologia Mediterranea* 11, 151-155.

DI VITO, M. & GRECO, N. (1988a). The relationship between initial population densities of *Meloidogyne artiellia* and yield of winter and spring chickpea. *Nematologia Mediterranea* 16, 163-166.

DI VITO, M. & GRECO, N. (1988b). Effect of population densities of *Meloidogyne artiellia* on yield of wheat. *Nematologia Mediterranea* 16, 167-169.

DI VITO, M. & GRECO, N. (1988c). Investigation on the biology of *Meloidogyne artiellia. Revue de Nématologie* 11, 223-227.

DI VITO, M. & ZACCHEO, G. (1991). Population density of root-knot nematode, *Meloidogyne incognita* and growth of artichoke (*Cynara scolimus*). *Advances in Horticultural Science* 5, 81-82.

DI VITO, M., VOVLAS, N. & INSERRA, R.N. (1980). Influence of *Meloidogyne incognita* on growth of corn in pots. *Plant Disease* 64, 1025-1026. DOI: 10.1094/PD-64-1025

DI VITO, M., GRECO, N. & CARELLA, A. (1981). Relationship between population densities of *Meloidogyne incognita* and yield of sugarbeet and tomato. *Nematologia Mediterranea* 9, 99-103.

DI VITO, M., GRECO, N. & CARELLA, A. (1983). The effect of population densities of *Meloidogyne incognita* on the yield of cantaloupe and tobacco. *Nematologia Mediterranea* 11, 169-174.

DI VITO, M., GRECO, N. & CARELLA, A. (1985a). Population densities of *Meloidogyne incognita* and yield of *Capsicum annuum. Journal of Nematology* 17, 45-49.

DI VITO, M., GRECO, N. & ZACCHEO, G. (1985b). On the host range of *Meloidogyne artiellia. Nematologia Mediterranea* 13, 207-212.

DI VITO, M., GRECO, N. & CARELLA, A. (1986). Effect of *Meloidogyne incognita* and importance of the inoculum on the yield of eggplant. *Journal of Nematology* 18, 487-490.

Di Vito, M., Vovlas, N. & Simeone, A.M. (1988). Effect of root-knot nematode *Meloidogyne incognita* on the growth of kiwi (*Actinidia deliciosa*) in pots. *Advances in Horticultural Sciences* 2, 109-112.

Di Vito, M., Cianciotta, V. & Zaccheo, G. (1991). The effect of population densities of *Meloidogyne incognita* on yield of susceptible and resistant tomato. *Nematologia Mediterranea* 19, 265-268.

Di Vito, M., Cianciotta, V. & Zaccheo, G. (1992). Yield of susceptible and resistant pepper in microplots infested with *Meloidogyne incognita*. *Nematropica* 22, 1-6.

Di Vito, M., Vovlas, N., Lamberti, F., Zaccheo, G. & Catalano, F. (1996a). Pathogenicity of *Meloidogyne javanica* on Asian and African rice. *Nematologia Mediterranea* 24, 95-99.

Di Vito, M., Zaccheo, G., Della Gatta, C. & Catalano, F. (1996b). Relationship between initial population densities of *Meloidogyne javanica* and yield of sunflower in microplots. *Nematologia Mediterranea* 24, 109-112.

Di Vito, M., Piscioneri, I. Pace, S. Zaccheo, G. & Catalano, F. (1997). Pathogenicity of *Meloidogyne incognita* on kenaf in microplots. *Nematologia Mediterranea* 25, 165-168.

Di Vito, M., Vovlas, N., Lamberti, F., Zaccheo, G. & Catalano, F. (1999). Difference in growth reduction potential of three Indian populations of *Meloidogyne javanica* on peanut. *International Journal of Nematology* 9, 168-173.

Di Vito, M., Crozzoli, R. & Vovlas, N. (2000). Pathogenicity of *Meloidogyne exigua* on coffee (*Coffea arabica* L.) in pots. *Nematropica* 30, 55-61.

Di Vito, M., Parisi, B. & Catalano, F. (2004a). Effect of population densities of *Meloidogyne incognita* on common bean. *Nematologia Mediterranea* 32, 81-85.

Di Vito, M., Vovlas, N. & Castillo, P. (2004b). Host-parasite relationships of *Meloidogyne incognita* on spinach. *Plant Pathology* 53, 508-514. DOI: 10.1111/j.1365-3059.2004.01053.x

Di Vito, M., Simeone, A.M. & Catalano, F. (2005). Effect of the root-knot nematode, *Meloidogyne javanica*, on the growth of a peach (*Prunus persica*) rootstock in pots. *Nematologia Mediterranea* 33, 87-90.

Di Vito, M., Parisi, B. & Catalano, F. (2007). Pathogenicity of *Meloidogyne javanica* on common bean (*Phaseolus vulgaris* L.) in pots. *Nematropica* 37, 339-344.

Diamond, C.J. & Bird, G.W. (1994). Observations on the host range of *Meloidogyne nataliei*. *Plant Disease* 78, 1050-1051.

Dickerson, O.J. (1966). Some observations on *Hypsoperine graminis* in Kansas. *Plant Disease Reporter* 50, 396-398.

DICKSON, D.W., HUISINGH, D. & SASSER, J.N. (1971). Dehydrogenases, acid and alkaline phosphatases, and esterases for chemotaxonomy of selected *Meloidogyne, Ditylenchus, Heterodera* and *Aphelenchus* spp. *Journal of Nematology* 3, 1-16.

DING, W.C., CHEN, J., SHI, Y.H., LU, X.J. & LI, M.Y. (2010). Rapid and sensitive detection of infectious spleen and kidney necrosis virus by loop-mediated isothermal amplification combined with a lateral flow dipstick. *Archives of Virology* 155, 385-389. DOI: 10.1007/s00705-010-0593-4

DIVERS, M., GOMES, C.B., MENEZES-NETTO, A.C., LIMA-MEDINA, I., NONDILLO, A., BELLÉ, C. & ARAÚJO-FILHO, J.V. (2019). Diversity of plant-parasitic nematodes parasitizing grapes in southern Brazil. *Tropical Plant Pathology* 44, 401-408. DOI: 10.1007/s40858-019-00301-3

DODGE, D.J. (2014). Cell wall dissolution of feeding cells in feeding sockets caused by *Meloidogyne kikuyensis*. *Nematology* 16, 1237-1239. DOI: 10.1163/15685411-00002842

DONG, K., DEAN, R.A., FORTNUM, B.A. & LEWIS, S.A. (2001). Development of PCR primers to identify species of root-knot nematodes: *Meloidogyne arenaria, M. hapla, M. incognita* and *M. javanica*. *Nematropica* 31, 271-280.

DOUCET, M.E. & PINOCHET, J. (1992). Occurrence of *Meloidogyne* spp. in Argentina. *Supplement to Journal of Nematology* 24(4S), 765-770.

DROPKIN, V.H. (1953). Studies on the variability of the anal plate patterns in pure lines of *Meloidogyne* spp., the root-knot nematode. *Proceedings of the Helminthological Society of Washington* 20, 32-39.

DUARTE, A., MALEITA, C., EGAS, C., ABRANTES, I. & CURTIS, R. (2017). Significant effects of RNAi silencing of the venom allergen-like protein (*Mhi-vap-1*) of the root-knot nematode *Meloidogyne hispanica* in the early events of infection. *Plant Pathology* 66, 1329-1337. DOI: 10.1111/ppa.12673

DUNCAN, L.W. & FERRIS, H. (1983). Validation of a model for prediction of host damage by two nematode species. *Journal of Nematology* 15, 227-234.

DUTTA, T., GANGULY, A.K. & GAUR, H.S. (2012). Global status of rice root-knot nematode, *Meloidogyne graminicola*. *African Journal of Microbiology Research* 6, 6016-6021.

EBSARY, B.A. (1986). Species and distribution of Heteroderidae and Meloidogynidae (Nematoda: Tylenchida) in Canada. *Canadian Journal of Plant Pathology* 8, 170-184. DOI: 10.1080/07060668609501823

EBSARY, B.A. & EVELEIGH, E.S. (1983). *Meloidogyne aquatilis* n. sp. (Nematoda: Meloidogynidae) from *Spartina pectinata* with a key to the Canadian species of *Meloidogyne*. *Journal of Nematology* 15, 349-353.

ECHEVERRÍA, M.M. & CHAVES, E.J. (1998). Identification of *Meloidogyne naasi* Franklin, 1965 from Argentina. *Nematologica* 44, 219-220. DOI: 10.1163/005325998X00090

EISENBACK, J.D. (1982). Description of the blueberry root-knot nematode, *Meloidogyne carolinensis* n. sp. *Journal of Nematology* 14, 303-317.

EISENBACK, J.D. (1985). Diagnostic characters useful in the identification of the four most common species of root-knot nematodes (*Meloidogyne* spp.). In: Sasser, J.N. & Carter, C.C. (Eds). *An advanced treatise on* Meloidogyne. *Vol. 1. Biology and control.* Raleigh, NC, USA, North Carolina State University Graphics, pp. 95-112.

EISENBACK, J.D. (1987). Reproduction of northern root-knot nematode (*Meloidogyne hapla*) on marigolds. *Plant Disease* 71, 281. DOI: 10.1094/PD-71-0281E

EISENBACK, J.D. (1993). Morphological comparisons of females, males, and second-stage juveniles of cytological races A and B of *Meloidogyne hapla* Chitwood, 1949. *Fundamental and Applied Nematology* 16, 259-271.

EISENBACK, J.D. (1997). *Root-knot nematode taxonomic database*. Wallingford, UK, CAB International.

EISENBACK, J.D. (2010). A new technique for photographing perineal patterns of root-knot nematodes. *Journal of Nematology* 42, 33-34.

EISENBACK, J.D. & DODGE, D.J. (2012). Description of a unique, complex feeding socket caused by the putative primitive root-knot nematode, *Meloidogyne kikuyensis*. *Journal of Nematology* 44, 148-152.

EISENBACK, J.D. & GNANAPRAGRASAM, N. (1992). Additional notes on the morphology of *Meloidogyne brevicauda*. *Fundamental and Applied Nematology* 15, 347-353.

EISENBACK, J.D. & HIRSCHMANN, H. (1979a). Morphological comparison of second-stage juveniles of six populations of *Meloidogyne hapla* by SEM. *Journal of Nematology* 11, 5-16.

EISENBACK, J.D. & HIRSCHMANN, H. (1979b). Morphological comparison of second-stage juveniles of several *Meloidogyne* species (root-knot nematodes) by scanning electron microscopy. *Scanning Electron Microscopy* 3, 223-229.

EISENBACK, J.D. & HIRSCHMANN, H. (1981). Identification of *Meloidogyne* species on the basis of head shape and stylet morphology of the male. *Journal of Nematology* 13, 513-521.

EISENBACK, J.D. & HIRSCHMANN, H. (2001). Additional notes on the morphology of *Meloidogyne spartinae* (Nematoda: Meloidogynidae). *Nematology* 3, 303-312. DOI: 10.1163/156854101317020222

EISENBACK, J.D. & HUNT, D.J. (2009). General morphology. In: Perry, R.N., Moens, M. & Starr, J.L. (Eds). *Root-knot nematodes*. Wallingford, UK, CAB International, pp. 18-54.

EISENBACK, J.D. & SPAULL, V.W. (1988). Additional notes on the morphology of males of *Meloidogyne kikuyensis*. *Journal of Nematology* 20, 633-634.

EISENBACK, J.D. & TRIANTAPHYLLOU, H.H. (1991). Root-knot nematodes: *Meloidogyne* species and races. In: Nickle, W.R. (Ed.). *Manual of agricultural nematology*. New York, NY, USA, Marcel Dekker, pp. 191-274.

EISENBACK, J.D. & VIEIRA, P. (2020). Additional notes on the morphology and molecular data of the Kikuyu root-knot nematode, *Meloidogyne kikuyensis* (Nematoda: Meloidogynidae). *Journal of Nematology* 52, e2020-67. DOI: 10.21307/jofnem-2020-067

EISENBACK, J.D., HIRSCHMANN, H. & TRIANTAPHYLLOU, A.C. (1980). Morphological comparison of *Meloidogyne* female head structures, perineal patterns, and stylets. *Journal of Nematology* 12, 300-313.

EISENBACK, J.D., HIRSCHMANN, H., SASSER, J.N. & TRIANTAPHYLLOU, A.C. (1981). *A guide to the four most common species of root-knot nematodes (*Meloidogyne *species) with a pictorial key*. A cooperative Publication of the Departments of Plant Pathology and Genetics, North Carolina State University and the United States Agency for International Development, North Carolina State University Graphics, NC, USA.

EISENBACK, J.D., YANG, B. & HARTMAN, K.M. (1985). Description of *Meloidogyne pini* n. sp., a root-knot nematode parasitic on sand pine (*Pinus clausa*), with additional notes on the morphology of *M. megatyla*. *Journal of Nematology* 17, 206-219.

EISENBACK, J.D., BERNARD, E.C. & SCHMITT, D.P. (1994). Description of the Kona coffee root-knot nematode, *Meloidogyne konaensis* n. sp. *Journal of Nematology* 26, 363-374.

EISENBACK, J.D., BERNARD, E.C., STARR, J.J., LEE, T.A. & TOMASZEWSKI, E.K. (2003). *Meloidogyne haplanaria* n. sp. (Nematoda: Meloidogynidae), a root-knot nematode parasitizing peanut in Texas. *Journal of Nematology* 35, 395-403.

EISENBACK, J.D., PAES-TAKAHASHI, PAES-TAKAHASHI, V. DOS S. & GRANEY, L.S. (2015). First report of the pecan root-knot nematode, *Meloidogyne partityla*, causing dieback to laurel oak in South Carolina. *Plant Disease* 99, 1041. DOI: 10.1094/PDIS-11-14-1122-PDN

EISENBACK, J.D., GRANEY, L.S. & VIEIRA, P. (2017). First report of the apple root-knot nematode (*Meloidogyne mali*) in North America, found parasitizing *Euonymus* in New York. *Plant Disease* 101, 510. DOI: 10.1094/PDIS-06-16-0894-PDN

EISENBACK, J.D., HOLLAND, L.A., SCHROEDER, J., THOMAS, S.H., BEACHAM, J.M., HANSON, S.F., PAES-TAKAHASHI, V.S. & VIEIRA, P. (2019). *Meloidogyne aegracyperi* n. sp. (Nematoda: Meloidogynidae), a root-knot nematode parasitizing yellow and purple nutsedge in New Mexico. *Journal of Nematology* 51, e2019-2071. DOI: 10.21307/jofnem-2019-071

EKANAYAKE, H.M.R.K. & DI VITO, M. (1984). Effect of population densities of *Meloidogyne incognita* on growth of susceptible and resistant tomato plants. *Nematologia Mediterranea* 12, 1-6.

EL-GHORE, A.A., HAROON, S., EL RHEEM, M.A. & ABDELLA, E. (2004). Development of specific SCAR-markers for *Meloidogyne incognita* and *Meloidogyne javanica*. *Arab Journal of Biotechnology* 7, 37-44.

EL-SHERIF, A.G., REFAEI, A.R. & GAD, S.B. (2009). The role of different inoculum levels of *Meloidogyne javanica* juveniles on nematode reproduction and host response of peanut plant. *Pakistan Journal of Biological Sciences* 12, 551-553.

ELHADY, A., ADSS, S., HALLMANN, J. & HEUER, H. (2018). Rhizosphere microbiomes modulated by pre-crops assisted plants in defense against plant-parasitic nematodes. *Frontiers in Microbiology* 9, 1133. DOI: 10.3389/fmicb.2018.01133

ELLING, A.A. (2013). Major emerging problems with minor *Meloidogyne* species. *Phytopathology* 103, 1092-1102. DOI: 10.1094/PHYTO-01-13-0019-RVW

ELMER, W.H. & LAMONDIA, J.A. (2014). Comparison of saline tolerance among genetically similar species of *Fusarium* and *Meloidogyne* recovered from marine and terrestrial habitats. *Estuarine, Coastal and Shelf Science* 149, 320-324.

ELMI, A.A., WEST, C.P., ROBBINS, R.T. & KIRKPATRICK, T.L. (2000). Endophyte effects on reproduction of a root-knot nematode (*Meloidogyne marylandi*) and osmotic adjustment in tall fescue. *Grass and Forage Science* 55, 166-172. DOI: 10.1046/j.1365-2494.2000.00210.x

ELMILIGY, I.A. (1968). Three new species of the genus *Meloidogyne* Goeldi, 1887 (Nematoda: Heteroderidae). *Nematologica* 14, 577-590. DOI: 10.1163/187529268X00282

EPPO (2017a). Pest risk analysis for *Meloidogyne mali*, apple root-knot nematode. https://gd.eppo.int/download/doc/1262_pra_exp_MELGMA.pdf.

EPPO (2017b). EPPO Alert List: addition of *Meloidogyne luci* together with *M. ethiopica*. EPPO Reporting Service, 2017/218, accessed from: https://gd.eppo.int/reporting/article-6186.

EROSHENKO, A.S. & LEBEDEVA, E.V. (1992). [Description of a new species of root-knot nematode, *Meloidogyne chosenia* sp. n. (Nematoda: Meloidogynidae), a parasite of willow in Kamchatka.] *Parazitologia* 4, 340-343.

ESBENSHADE, P.R. & TRIANTAPHYLLOU, A.C. (1985a). Use of enzyme phenotypes for identification of *Meloidogyne* species. *Journal of Nematology* 17, 6-20.

ESBENSHADE, P.R. & TRIANTAPHYLLOU, A.C. (1985b). Identification of major *Meloidogyne* species employing enzyme phenotypes as differentiating characters. In: Sasser, J.N. & Carter, C.C. (Eds). *An advanced treatise on*

Meloidogyne. *Vol. I. Biology and control.* Raleigh, NC, USA, North Carolina State University Graphics, pp. 135-140.

ESBENSHADE, P.R. & TRIANTAPHYLLOU, A.C. (1987). Enzymatic relationships and evolution in the genus *Meloidogyne* (Nematoda: Tylenchida). *Journal of Nematology* 19, 8-18.

ESBENSHADE, P.R. & TRIANTAPHYLLOU, A.C. (1990). Isozyme phenotypes for the identification of *Meloidogyne* species. *Journal of Nematology* 22, 10-15.

ESCOBAR, C., BARCALA, M., CABRERA, J. & FENOLL, C. (2015). Overview of root-knot nematodes and giant cells. *Advances in Botanical Research* 73, 1-32. DOI: 10.1016/bs.abr.2015.01.001

ESSER, R.P., PERRY, V.G. & TAYLOR, A.L. (1976). A diagnostic compendium of the genus *Meloidogyne* (Nematoda: Heteroderidae). *Proceedings of the Helminthological Society of Washington* 43, 138-150.

EVANS, J.R. (1989). Photosynthesis and nitrogen relationships in leaves of C_3 plants. *Oecologia* 78, 9-19. DOI: 10.1007/BF00377192

FANELLI, E., COTRONEO, A., CARISIO, L., TROCCOLI, A., GROSSO, S., BOERO, C., CAPRIGLIA, F. & DE LUCA, F. (2017). Detection and molecular characterization of the rice root-knot nematode *Meloidogyne graminicola* in Italy. *European Journal of Plant Pathology* 149, 467-476. DOI: 10.1007/s10658-017-1196-7

FARGETTE, M. (1984). *Utilisation de l'électrophorèse dans l'étude de la systématique de deux organismes d'intérêt agricole:* Trichogramma *supersp.* evanescens *(Hymenoptera, Chalcidoidea) et* Meloidogyne *spp. (Nematoda, Tylenchida).* Thèse Dr. Ing., École Nationale Supérieure d'Agronomomie de Montpellier, France.

FARGETTE, M. (1987a). Use of the esterase phenotype in the taxonomy of the genus *Meloidogyne*. 1. Stability of the esterase phenotype. *Revue de Nématologie* 10, 39-43.

FARGETTE, M. (1987b). Use of the esterase phenotype in the taxonomy of the genus *Meloidogyne*. 2. Esterase phenotypes observed in West African populations and their characterization. *Revue de Nématologie* 10, 45-55.

FARGETTE, M. & BRAAKSMA, R. (1990). Use of the esterase phenotype in the taxonomy of the genus *Meloidogyne*. 3. A study of some 'B' race lines and their taxonomic position. *Revue de Nématologie* 13, 375-386.

FARGETTE, M., PHILLIPS, M.S., BLOK, V.C., WAUGH, R. & TRUDGILL, D.L. (1996). An RFLP study of relationships between species, populations and resistance-breaking lines of tropical species of *Meloidogyne*. *Fundamental and Applied Nematology* 19, 193-200.

FASKE, T.R. & STARR, J.L. (2009). Reproduction of *Meloidogyne marylandi* and *M. incognita* on several Poaceae. *Journal of Nematology* 41, 2-4.

FENG, H., NIE, G., CHEN, X., ZHANG, J.F., ZHOU, D.M. & WEI, L.H. (2017). [Morphological and molecular identification of *Meloidogyne graminicola* isolated from Jiangsu province.] *Jiangsu Journal of Agricultural Sciences* 4, 794-801.

FERRAZ, L.C.C.B. (2008). Plant parasitic nematodes of coffee in Brazil. In: Souza, R.M. (Ed.). *Plant parasitic nematodes of coffee*. New York, NY, USA, APS Press & Springer, pp. 225-248.

FERRIS, H. (1981). Dynamic action thresholds for diseases induced by nematodes. *Annual Review of Phytopathology* 19, 427-436. DOI: 10.1146/annurev.py.19.090181.002235

FERRIS, H., CARLSON, H.L., VIGLIERCHIO, D.R., WESTERDAHL, B.B., WU, F.W., ANDERSON, C.E., JUURMA, A. & KIRBY, D.W. (1993). Host status of selected crops to *Meloidogyne chitwoodi. Journal of Nematology* 25 (Supplement), 849-857.

FERRIS, V.R., FERRIS, J.M. & FAGHIHI, J. (1993). Variation in spacer ribosomal DNA in some cyst-forming species of plant parasitic nematodes. *Fundamental and Applied Nematology* 16, 177-184.

FIGUEIREDO, M.B. (1958). Algumas observações sobre os nematóides que atacam o fumo no estado de São Paulo. *Revista de Agricultura* 33, 69-73.

FIGUEROA, A. (1973). *Estudio morfométrico y biológico sobre el nematodo cecidógeno del arroz* Hypsoperine *sp. (Nematoda: Heteroderidae) y pruebas de susceptibilidad al mismo de doce variedades y una línea de arroz (*Oryza sativa *L.).* Ing. Agr. Thesis San Pedro de Montes de Oca, Universidad de Costa Rica, Costa Rica.

FLORES, L. & LÓPEZ, R. (1989a). Caracterización morfológica del nematodo nodulador del cafeto *Meloidogyne exigua* (Nemata: Heteroderidae). I. Hembras y huevos. *Turrialba (Costa Rica)* 39, 352-360.

FLORES, L. & LÓPEZ, R. (1989b). Caracterización morfológica del nematodo nodulador del cafeto *Meloidogyne exigua* (Nemata: Heteroderidae). II. Machos. *Turrialba (Costa Rica)* 39, 361-368.

FLORES, L. & LÓPEZ, R. (1989c). Caracterización morfológica del nematodo nodulador del cafeto *Meloidogyne exigua* (Nemata: Heteroderidae). III. Segundos estados juvenils. *Turrialba (Costa Rica)* 39, 369-376.

FLORES-ROMERO, P. & NAVAS, A. (2005). Enhancing taxonomic resolution: distribution dependent genetic diversity in populations of *Meloidogyne. Nematology* 7, 571-530. DOI: 10.1163/156854105774384831

FONSECA, H.S., FERRAZ, L.C.C.B. & MACHADO, S.R. (2003). Ultraestrutura comparada de raízes de seringueira parasitadas por *Meloidogyne exigua* e *M. javanica. Nematologia Brasileira* 27, 199-206.

FOURIE, H., ZIJLSTRA, C. & MC DONALD, A.H. (2001). Identification of root-knot nematode species occurring in South Africa using SCAR-PCR technique. *Nematology* 3, 675-680.

FOX, J.A. (1967). *Biological studies of the blueberry root-knot nematode (*Meloidogyne carolinensis *n. sp.).* Ph.D. Thesis, North Carolina State University, Raleigh, NC, USA.

FRANCE, R.A. & ABAWI, G.S. (1994). Interaction between *Meloidogyne incognita* and *Fusarium oxysporum* f. sp. *phaseoli* on selected bean genotypes. *Journal of Nematology* 26, 467-74.

FRANKLIN, M.T. (1957). Review of the genus *Meloidogyne. Nematologica* 2 (Suppl.), 387-397.

FRANKLIN, M.T. (1961). A British root-knot nematode, *Meloidogyne artiellia* n. sp. *Journal of Helminthology, R.T. Leiper Supplement*, 85-92.

FRANKLIN, M.T. (1965a). A root-knot nematode, *Meloidogyne naasi* n. sp., on field crops in England and Wales. *Nematologica* 11, 79-86. DOI: 10.1163/187529265X00500

FRANKLIN, M.T. (1965b). *Meloidogyne*–root-knot eelworms. In: Southey, J.F. (Ed.). *Plant nematology*. London, UK, Her Majesty's Stationery Office, pp. 59-88.

FRANKLIN, M.T. (1971). Taxonomy of Heteroderidae. In: Zuckerman, B.M., Mai, W.F. & Rohde, R.A. (Eds). *Plant parasitic nematodes. Volume I. Morphology, anatomy, taxonomy and ecology*. New York, NY, USA, Academic Press, pp. 139-162.

FRANKLIN, M.T. (1972). The present position in the systematics of *Meloidogyne. OEPP/EPPO Bulletin* 6, 5-15.

FRANKLIN, M.T. (1973). *Meloidogyne naasi. CIH Descriptions of plant-parasitic nematodes.* Set 2, No. 19. Farnham Royal, UK, Commonwealth Agricultural Bureaux.

FRANKLIN, M.T. (1976). *Meloidogyne.* In: Southey, J.F. (Ed.). *Plant nematology.* London, UK, Her Majesty's Stationery Office, pp. 98-124.

FRANKLIN, M.T. (1979). Taxonomy of the genus *Meloidogyne*. In: Lamberti, F. & Taylor, C.E. (Eds). *Root-knot nematodes (*Meloidogyne *species); systematics, biology and control.* London & New York, Academic Press, pp. 37-54.

FRANKLIN, M.T., CLARK, S.A. & COURSE, J.A. (1971). Population changes and development of *Meloidogyne naasi* in the field. *Nematologica* 17, 575-590. DOI: 10.1163/187529271X00297

FREIRE, C., DAVIDE, L., CAMPOS, V., SANTOS, C.D. & FREIRE, P. (2002). Cromosomas de tres especies brasileiras de *Meloidogyne*. *Ciencia e Agrotecnologia* 26, 900-903.

FREITAS, V.M., SILVA, J.G.P., GOMES, C.B., CASTRO, J.M.C., CORREA, V.R. & CARNEIRO, R.M.D.G. (2017). Host status of selected cultivated fruit crops to *Meloidogyne enterolobii. European Journal of Plant Pathology* 148, 307-319.

Fu, M.Y., Chen, M.C., Xiao, T.B. & Wang, H.F. (2011). Characterisation of *Meloidogyne* species on Southern Herbs in Hainan island using perineal pattern and esterase phenotype and amplified mitochondrial DNA restriction fragment length polymorphism analysis. *Russian Journal of Nematology* 19, 173-180.

Gamel, S., Huchet, E., Le Roux-Nio, A.C. & Anthoine, G. (2014). Assessment of PCR-based tools for the specific identification of some temperate *Meloidogyne* species including *M. chitwoodi*, *M. fallax* and *M. minor*. *European Journal Plant Pathology* 138, 807-817. DOI: 10.1007/s10658-013-0355-8

Ganaie, M.A., Rather, A.A. & Siddiqui, M.A. (2011). Pathogenicity of root knot nematode *Meloidogyne incognita* on okra and its management through botanicals. *Archives of Phytopathology and Plant Protection* 44, 1683-1688. DOI: 10.1080/03235408.2010.496583

García, L.E. & Sánchez-Puerta, M.V. (2012). Characterization of a root-knot nematode population of *Meloidogyne arenaria* from Tupungato (Mendoza, Argentina). *Journal of Nematology* 44, 291-301.

García, L.E. & Sánchez-Puerta, M.V. (2015). Comparative and evolutionary analyses of *Meloidogyne* spp. based on mitochondrial genome sequences. *PLoS ONE* 10, e0121142. DOI: 10.1371/journal.pone.0121142

García-Martinez, R. (1982). Post-infection development and morphology of *Meloidogyne cruciani*. *Journal of Nematology* 14, 332-338.

García-Martinez, R., Taylor, A.L. & Smart Jr, G.C. (1982). *Meloidogyne cruciani* n. sp. a root-knot nematode from St. Croix (U.S. Virgin Islands) with observations on morphology of this and two other species of the genus. *Journal of Nematology* 14, 292-302.

Gaur, H.S. & Chandel, S.T. (1997). Effect of soil moisture content on the survival of the root-knot nematode, *Meloidogyne triticoryzae*, without host. *National Symposium on Integrated Pest Management in India – Constraints and Opportunities, 23-24 October, 1997*, p. 50.

Gaur, H.S. & Sharma, S.N. (1998). Studies on the host range of the root-knot nematode, *Meloidogyne triticoryzae*, among cultivated crops and weeds. *Annals of Plant Protection Science* 6, 41-47.

Gaur, H.S., Saha, M. & Khan, E. (1993). *Meloidogyne triticoryzae* sp. n. (Nematoda: Meloidogynidae) a root-knot nematode damaging wheat and rice in India. *Annals of Plant Protection Science* 1, 18-26.

Gaur, H.S., Beane, J. & Perry, R.N. (2000). The influence of root diffusate, host age and water regimes on hatching of the root-knot nematode, *Meloidogyne triticoryzae*. *Nematology* 2, 191-199. DOI: 10.1163/156854100509088

Gergon, E.B., Miller, S.A., Halbrendt, J.M. & Davide, R.G. (2002). Effect of rice root-knot nematode on growth and yield of Yellow Granex onion. *Plant Disease* 86, 1339-1344. DOI: 10.1094/PDIS.2002.86.12.1339

GERIČ STARE, B., STRAJNAR, P., SUSIČ, N., UREK, G. & ŠIRCA, S. (2017). Reported populations of *Meloidogyne ethiopica* in Europe identified as *Meloidogyne luci*. *Plant Disease* 101, 1627-1632. DOI: 10.1094/PDIS-02-17-0220-RE

GERIČ STARE, B., AYDINH, G., DEVRAN, Z., MENNAN, S., STRAJNAR, P., UREK, G. & ŠIRCA, S. (2019). Recognition of species belonging to *Meloidogyne ethiopica* group and development of a diagnostic method for its detection. *European Journal of Plant Pathology* 154, 621-633. DOI: 10.1007/s10658-019-01686-2

GHADERI, R. & KARSSEN, G. (2020). An updated checklist of *Meloidogyne* Göldi, 1887 species, with a diagnostic compendium for second-stage juveniles and males. *Journal of Crop Protection* 9, 183-193.

GHARABADIYAN, F., JAMALI, S., YAZDI, A., HADIZADEH, M.H. & ESKANDARI, A. (2012). Weed hosts of root-knot nematodes in tomato fields. *Journal of Plant Protection Research* 52, 230-234.

GILLARD, A. (1961). Onderzoekingen omtrent de biologie, de verspreiding en de bestrijding van wortelknobbelaaltjes (*Meloidogyne* spp.). *Mededelingen van de Landbouwhogeschool, Gent* 26, 515-646.

GINÉ, A., LÓPEZ-GÓMEZ, M., VELA, M.D., ORNAT, C., TALAVERA, M., VERDEJO-LUCAS, S. & SORRIBAS, F.J. (2014). Thermal requirements and population dynamics of root-knot nematodes on cucumber and yield losses under protected cultivation. *Plant Pathology* 63, 1446-1453. DOI: 10.1111/ppa.12217

GINÉ, A., GONZÁLEZ, C., SERRANO, L. & SORRIBAS, F.J. (2017). Population dynamics of *Meloidogyne incognita* on cucumber grafted onto the *Cucurbita* hybrid RS841 or ungrafted and yield losses under protected cultivation. *European Journal of Plant Pathology* 148, 795-805. DOI: 10.1007/s10658-016-1135-z

GODFREY, G.H. (1923). *Root-knot, its cause and control*. Farmers' bulletin, United States Department of Agriculture 1345.

GOLDEN, A.M. (1979). Descriptions of *Meloidogyne camelliae* n. sp. and *M. querciana* n. sp. (Nematoda: Meloidogynidae), with SEM and host-range observations. *Journal of Nematology* 11, 175-189.

GOLDEN, A.M. (1989). Further details and SEM observations on *Meloidogyne marylandi* (Nematoda: Meloidogynidae). *Journal of Nematology* 21, 453-461.

GOLDEN, A.M. (1992). Large phasmids in the female of *Meloidogyne ethiopica* Whitehead. *Fundamental and Applied Nematology* 15, 189-191.

GOLDEN, A.M. & BIRCHFIELD, W. (1965). *Meloidogyne graminicola* (Heteroderidae) a new species of root-knot nematode from grass. *Proceedings of the Helminthological Society of Washington* 32, 228-231.

GOLDEN, A.M. & KAPLAN, D.T. (1986). Description of *Meloidogyne christiei* n. sp. (Nematoda: Meloidogynidae) from oak with SEM and host-range observations. *Journal of Nematology* 18, 533-540.

GOLDEN, A.M. & SLANA, L.J. (1978). *Meloidogyne grahami* n. sp. (Meloidogynidae), a root-knot nematode on resistant tobacco in South Carolina. *Journal of Nematology* 10, 355-361.

GOLDEN, A.M. & TAYLOR, D.P. (1967). The barley root-knot nematode in Illinois. *Plant Disease Reporter* 51, 974-975.

GOLDEN, A.M., O'BANNON, J.H., SANTO, G.S. & FINLEY, A.M. (1980). Description and SEM observations of *Meloidogyne chitwoodi* n. sp. (Meloidogynidae), a root-knot nematode on potato in the Pacific Northwest. *Journal of Nematology* 12, 319-327.

GOLDEN, A.M., ROSE, L.M. & BIRD, G.W. (1981). Description of *Meloidogyne nataliei* n. sp. (Nematoda: Meloidogynidae) from grape (*Vitis labrusca*) in Michigan, with SEM observations. *Journal of Nematology* 13, 393-400.

GÖLDI, E.A. (1887). Relatório sôbre a moléstia do cafeeiro na provincia do Rio de Janeiro. Extrahido do VIII Vol. dos. *Archivos do Museu Nacional, Rio de Janeiro, Imprensa Naciona*, pp. 1-121.

GOLDSTEIN, P. & TRIANTAPHYLLOU, A.C. (1982). The synaptonemal complexes of *Meloidogyne*: relationship of structure and evolution of parthenogensis. *Chromosoma* 87, 117-124. DOI: 10.1007/BF00333513

GOLDSTEIN, P. & TRIANTAPHYLLOU, A.C. (1986). The synaptonemal complex of *Meloidogyne nataliei* and its relationship to that of other *Meloidogyne* species. *Chromosoma* 93, 261-266. DOI: 10.1007/BF00292747

GOMES, C.B., CARBONARI, J.J., MEDINA, I.L. & LIMA, D.L. (2005). Survey *Meloidogyne ethiopica* in kiwi nurseries in Rio Grande do Sul State and record the occurrence in tobacco (*Nicotiana tabacum*) and *Sida rhombifolia*. *Nematologia Brasileira* 25, 69.

GOODEY, J.B. (1963). *Soil and freshwater nematodes*, 2nd edition. London, UK, Methuen & Co.

GOODEY, J.B., FRANKLIN, M.T. & HOOPER, D.J. (1965). *T. Goodey's the nematode parasites of plants catalogued under their hosts*, 3rd edition. Farnham Royal, UK, Commonwealth Agricultural Bureaux.

GOORIS, J. (1968). Host plants and non-host plants of the gramineae root-knot nematode *Meloidogyne naasi* Franklin. *Mededelingen van de Rijksfaculteit Landbouwetenschappen Gent* 33, 85-100.

GOOSSENS, J.J.M. (1995). Host range test of *Meloidogyne* n. sp. In: Annual Report 1994. Diagnostic Centre, Plant Protection Service, Wageningen, The Netherlands, pp. 95-97.

GORNY, A.M., WANG, X.H., HAY, F.S. & PETHYBRIDGE, S.J. (2019). Development of a species-specific PCR for detection and quantification of

Meloidogyne hapla in soil using the *16D10* root-knot nematode effector gene. *Plant Disease* 103, 1902-1909. DOI: 10.1094/PDIS-09-18-1539-RE

GOVERSE, A. & BIRD, D. (2011). The role of plant hormones in nematode feeding cell formation. In: Jones, J., Gheysen, G. & Fenoll, C. (Eds). *Genomics and molecular genetics of plant-nematode interactions*. Dordrecht, The Netherlands, Springer, pp. 325-347.

GRECO, N. (1984). Presenza di *Meloidogyne artiellia* su cese in Italia. *Nematologia Mediterranea* 12, 235-238.

GRECO, N. & DI VITO, M. (2009). Population dynamics and damage levels. In: Perry, R.N., Moens, M. & Starr, J.L. (Eds). *Root-knot nematodes*. Wallingford, UK, CABI, pp. 246-274.

GRECO, N., DI VITO, M., REDDY, M.V. & SAXENA, M.C. (1984). A preliminary report of survey of plant parasitic nematodes of leguminous crops in Syria. *Nematologia Mediterranea* 12, 87-93.

GRECO, N., VOVLAS, N., DI VITO, M. & INSERRA, R.N. (1992). *Meloidogyne artiellia*: a root-knot nematode parasite of cereals and other field crops. *Nematology Circular* 201.

GREEFF, R. (1872). Über Nematoden in Wurzelanschwellungen (Gallen) verschiedener Pflanzen. *Sitzungsberichte der Gesellschaft zur Beförderung der Gesamten Naturwissenschaften in Marburg* 11, 172-174.

GRIFFIN, G.D., INSERRA, R.N. & DI VITO, M. (1982). Comparative relationship between *Meloidogyne chitwoodi* and *M. hapla* population densities and growth of sugarbeet seedlings. *Journal of Nematology* 14, 409-411.

GRIFFIN, G.D., INSERRA, R.N. & VOVLAS, N. (1984). Rangeland grasses as hosts of *Meloidogyne chitwoodi*. *Journal of Nematology* 16, 399-402.

GRISHAM, M.P., DALE, J.L. & RIGGS, R.D. (1974). *Meloidogyne graminis* and *Meloidogyne* spp. on zoysia; infection, reproduction, disease development, and control. *Phytopathology* 64, 1485-1489.

GU, J.F. & HE, J. (2015). First report of the apple root-knot nematode, *Meloidogyne mali*, infecting crape myrtle from Japan. *Plant Disease* 99, 893-894. DOI: 10.1094/PDIS-11-14-1145-PDN

GU, J.F., WANG, J.L., SHAO, F., GAO, F.F. & GE, J.J. (2013). [Identification of *Meloidogyne mali* detected in *Acer palmatum* from Japan.] *Plant Quarantine* 1, 43-49.

GU, J.F., FANG, Y. & LIU, L. (2020). Morphological and molecular analyses of a *Meloidogyne mali* population with high intragenomic rRNA polymorphism. *Journal of Nematology* 52, e2020-105. DOI: 10.21307/jofnem-2020-105

GUGINO, B.K., ABAWI, G.S. & LUDWIG, J.W. (2006). Damage and management of *Meloidogyne hapla* using oxamyl on carrot in New York. *Journal of Nematology* 38, 483-490.

GUO, H. & GE, F. (2016). Root nematode infection enhances leaf defense against whitefly in tomato. *Arthropod-Plant Interactions* 11, 23-33. DOI: 10.1007/s11829-016-9462-8

GUZMAN-PLAZOLA, R.A., DIOS JARABA NAVAS, J. DE, CASWELL-CHEN, E., ZAVALETA-MEJÍA, E. & CID DEL PRADO-VERA, I. (2006). Spatial distribution of *Meloidogyne* species and races in the tomato (*Lycopersicon esculentum* Mill.) producing region of Morelos, Mexico. *Nematropica* 36, 215-229.

HALLMANN, J., FRANKENBERG, A., PAFFRATH, A. & SCHMIDT, H. (2007). Occurrence and importance of plant-parasitic nematodes in organic farming in Germany. *Nematology* 9, 869-879. DOI: 10.1163/156854107782331261

HAN, H., BRITO, J.A. & DICKSON, D.W. (2012). First report of *Meloidogyne enterolobii* infecting *Euphorbia punicea* in Florida. *Plant Disease* 96, 1706. DOI: 10.1094/PDIS-05-12-0497-PDN

HANDOO, Z.A., HUETTEL, R.N. & GOLDEN, A.M. (1993). Description and SEM observations of *Meloidogyne sasseri* n. sp. (Nematoda: Meloidogynidae), parasitizing beach grasses. *Journal of Nematology* 25, 628-641.

HANDOO, Z.A., NYCZEPIR, A.P., ESMENJAUD, D., VAN DER BEEK, J.G., CASTAGNONE-SERENO, P., CARTA, L.K., SKANTAR, A.M. & HIGGINS, J.A. (2004). Morphological, molecular and differential-host characterization of *Meloidogyne floridensis* n. sp. (Nematoda: Meloidogynidae), a root-knot nematode parasitizing peach in Florida. *Journal of Nematology* 36, 20-35.

HANDOO, Z.A., SKANTAR, A.M., CARTA, L.K. & SCHMITT, D.P. (2005a). Morphological and molecular evaluation of a *Meloidogyne hapla* population damaging coffee (*Coffea arabica*) in Maui, Hawaii. *Journal of Nematology* 37, 136-145.

HANDOO, Z.A., SKANTAR, A.M., CARTA, L.K. & ERBE, E.F. (2005b). Morphological and molecular characterization of a new root-knot nematode, *Meloidogyne thailandica* n. sp. (Nematoda: Meloidogynidae), parasitizing ginger (*Zingiber* sp.). *Journal of Nematology* 37, 343-353.

HARPREET, K. & RAJNI, A. (2012). Morphological and morphometrical characterization of *Meloidogyne graminicola* (Golden & Birchfield) from rice host plant in the four districts of Punjab. *Trends in Biosciences* 5, 252-254.

HARTMAN, R.M. & SASSER, J.N. (1985). Identification of *Meloidogyne* species on the basis of differential host test and perineal pattern morphology. In: Barker, K.R., Carter, C.C. & Sasser, J.N. (Eds). *An advanced treatise on* Meloidogyne. *Vol. 2. Methodology*. Raleigh, NC, USA, North Carolina State University Graphics, pp. 69-77.

HASEEB, A., SRIVASTAVA, N.K. & PANDEY, R. (1990). The influence of *Meloidogyne incognita* on growth, physiology, nutrient concentration and alkaloid yield of *Hyoscyamus niger*. *Nematologia Mediterranea* 18, 127-129.

He, Q.C., Wang, D.W., Tang, B., Wang, J., Zhang, D., Liu, Y. & Cheng, F.X. (2021). Rapid and sensitive detection of *Meloidogyne graminicola* in soil using conventional PCR, loop-mediated isothermal amplification, and real-time PCR methods. *Plant Disease* 105, 456-463. DOI: 10.1094/PDIS-06-20-1291-RE

He, X.F., Peng, H., Ding, Z., He, W.T., Huang, W.K. & Peng, D.L. (2013). Loop-mediated isothermal amplification assay for rapid diagnosis of *Meloidogyne enterolobii* directly from infected plants. *Scientia Agricultura Sinica* 46, 534-544.

Heald, C.M. (1969). Pathogenicity and histopathology of *Meloidogyne graminis* infecting 'Tifdwarf' bermudagrass roots. *Journal of Nematology* 1, 31-34.

Helder, J., Mooijman, P.J.W., van den Elsen, S.J.J., van Megen, H.H.B., van Vervoort, M.T.W., Quist, C.W., Bert, W., Karegar, A., Karssen, G. & Decraemer, W. (2014). Biological and systematic implications of phylogenetic analysis of ~2,800 full length small subunit ribosomal DNA sequences. *Proceedings of the 6th International Congress of Nematology, Cape Town, South Africa*, p. 26.

Hernandez, A., Fargette, M. & Sarah, J.-L. (2004). Characterisation of *Meloidogyne* spp. (Tylenchida: Meloidogynidae) isolated from coffee plantations in Central America and Brazil. *Nematology* 6, 193-204. DOI: 10.1163/1568541041217933

Herrera, I., Bryngelsson, T., Monzón, A. & Geleta, M. (2011). Identification of coffee root-knot nematodes based on perineal patterns, SCAR markers and nuclear ribosomal DNA sequences. *Nematologia Mediterranea* 39, 101-110.

Heve, W.K., Been, T.H., Schomaker, C.H. & Teklu, M.G. (2015). Damage thresholds and population dynamics of *Meloidogyne chitwoodi* on carrot (*Daucus carota*) at different seed densities. *Nematology* 17, 501-514. DOI: 10.1163/15685411-00002884

Hewlett, T.E. & Tarjan, A.C. (1983). Synopsis of the genus *Meloidogyne* Goeldi, 1887. *Nematropica* 13, 79-102.

Hirschmann, H. (1982). *Meloidogyne platani* n. sp. (Meloidogynidae), a root-knot nematode parasitizing American sycamore. *Journal of Nematology* 14, 84-95.

Hirschmann, H. (1985). The genus *Meloidogyne* and morphological characters differentiating its species. In: Sasser, J.N. & Carter, C.C. (Eds). *An advanced treatise on* Meloidogyne. *Vol. I. Biology and control*. Raleigh, NC, USA, North Carolina State University Graphics, pp. 79-93.

Hirschmann, H. (1986). *Meloidogyne hispanica* n. sp. (Nematoda: Meloidogynidae), the 'Seville root-knot nematode'. *Journal of Nematology* 18, 520-532.

HISAMUDDIN, S.S. & AZAM, T. (2010). Pathogenicity of root-knot nematode, *Meloidogyne incognita* on *Lens culinaris* (Medik.). *Archives of Phytopathology and Plant Protection* 43, 1504-1511. DOI: 10.1080/03235400802583537

HOCKLAND, S., INSERRA, R.N., MILLAR, L. & LEHMAN, P.S. (2006). International plant health – putting legislation into practice. In: Perry, R.N. & Moens, M. (Eds). *Plant nematology*. Wallingford, UK, CAB International, pp. 327-345.

HODGETTS, J., OSTOJÁ-STARZEWSKI, J.C., PRIOR, T., LAWSON, R., HALL, J. & BOONHAM, N. (2016). DNA barcoding for biosecurity: case studies from the UK plant protection program. *Genome* 59, 1033-1048. DOI: 10.1139/gen-2016-0010

HOLGADO, R. & HAMMERAAS, B. (2001). First report of the root-knot nematode *Meloidogyne ardenensis* on Lady's Mantle in Norway. *Plant Disease* 85, 1289. DOI: 10.1094/PDIS.2001.85.12.1289D

HOLOVACHOV, O., CAMP, L. & NADLER, S.A. (2015). Sensitivity of ribosomal RNA character sampling in the phylogeny of Rhabditida. *Journal of Nematology* 47, 337-355.

HOLTERMAN, M., VAN DER WURFF, A., VAN DEN ELSEN, S., VAN MEGEN, H., BONGERS, T., HOLOVACHOV, O., BAKKER, J. & HELDER, J. (2006). Phylum-wide analysis of SSU rDNA reveals deep phylogenetic relationships among nematodes and accelerated evolution toward crown Clades. *Molecular Biology and Evolution* 23, 1792-1800. DOI: 10.1093/molbev/msl044

HOLTERMAN, M., KARSSEN, G., VAN DEN ELSEN, S., VAN MEGEN, H., BAKKER, J. & HELDER, J. (2009). Small subunit rDNA-based phylogeny of the Tylenchida sheds light on relationships among some high-impact plant-parasitic nematodes and the evolution of plant feeding. *Phytopathology* 99, 227-235. DOI: 10.1094/PHYTO-99-3-0227

HOLTERMAN, M.H.M., OGGENFUSS, M., FREY, J.E. & KIEWNICK, S. (2012). Evaluation of high-resolution melting curve analysis as a new tool for root-knot nematode diagnostics. *Journal of Phytopathology* 160, 59-66. DOI: 10.1111/j.1439-0434.2011.01859.x

HOOPER, D.J. (1977). Spicule and stylet protrusion induced by ammonia solution in some plant and soil nematodes. *Nematologica* 23, 126-127. DOI: 10.1163/187529277X00327

HTAY, C., PENG, H., HUANG, W., KONG, L., HE, W., HOLGADO, R. & PENG, D. (2016). The development and molecular characterization of a rapid detection method for rice root-knot nematode (*Meloidogyne graminicola*). *European Journal of Plant Pathology* 146, 281-291. DOI: 10.1007/s10658-016-0913-y

HU, M. & GASSER, R.B. (2006). Mitochondrial genomes of parasitic nematodes – progress and perspectives. *Trends in Parasitology* 22, 78-84. DOI: 10.1016/j.pt.2005.12.003

HUGALL, A., MORITZ, C., STANTON, J. & WOLSTENHOLME, D.R. (1994). Low, but strongly structured mitochondrial DNA diversity in root knot nematodes (*Meloidogyne*). *Genetics* 136, 903-912.

HUGALL, A., STANTON, J. & MORITZ, C. (1999). Reticulate evolution and the origins of ribosomal internal transcribed spacer diversity in apomictic *Meloidogyne*. *Molecular Biology and Evolution* 16, 157-164. DOI: 10.1093/oxfordjournals.molbev.a026098

HUMPHREYS-PEREIRA, D.A. & ELLING, A.A. (2013). Intraspecific variability and genetic structure in *Meloidogyne chitwoodi* from the USA. *Nematology* 15, 315-327. DOI: 10.1163/15685411-00002684

HUMPHREYS-PEREIRA, D.A. & ELLING, A.A. (2014a). Morphological variability in second-stage juveniles and males of *Meloidogyne chitwoodi*. *Nematology* 16, 149-162. DOI: 10.1163/15685411-00002753

HUMPHREYS-PEREIRA, D.A. & ELLING, A.A. (2014b). Mitochondrial genomes of *Meloidogyne chitwoodi* and *M. incognita* (Nematoda: Tylenchina): Comparative analysis, gene order and phylogenetic relationships with other nematodes. *Molecular and Biochemical Parasitology* 194, 20-32. DOI: 10.1016/j.molbiopara.2014.04.003

HUMPHREYS-PEREIRA, D.A., WILLIAMSON, V.M., SALAZAR, L., FLORES-CHAVES, L. & GÓMEZ-ALPIZAR, L. (2012). Presence of *Meloidogyne enterolobii* Yang & Eisenback (= *M. mayaguensis*) in guava and acerola from Costa Rica. *Nematology* 14, 199-207. DOI: 10.1163/138855411X584151

HUMPHREYS-PEREIRA, D.A., FLORES-CHAVES, L., GÓMEZ, M., SALAZAR, L., GÓMEZ-ALPIZAR, L. & ELLING, A.A. (2014). *Meloidogyne lopezi* n. sp. (Nematoda: Meloidogynidae), a new root-knot nematode associated with coffee (*Coffea arabica* L.) in Costa Rica, its diagnosis and phylogenetic relationship with other coffee-parasitising *Meloidogyne* species. *Nematology* 16, 643-661. DOI: 10.1163/15685411-00002794

HUNT, D.J. & HANDOO, Z.A. (2009). Taxonomy, identification and principal species. In: Perry, R.N., Moens, M. & Starr, J.L. (Eds). *Root-knot nematodes*. Wallingford, UK, CAB International, pp. 55-97.

HUNTER, A.H. (1958). Nutrient absorption and translocation of phosphorus as influenced by the root-knot nematode (*Meloidogyne incognita acrita*). *Soil Science* 86, 245-250.

IMREN, M., ÖZARSLANDAN, A., KASAPOĞLU, E.B., TOKTAY, H. & ELEKCIOĞLU, I.H. (2014). Morphological and molecular identification of a new species *Meloidogyne artiellia* (Franklin) on wheat fauna in Turkey. *Turkiye Entomoloji Dergisi* 38, 189-196.

INAGAKI, H. (1978). Apple root-knot nematode *Meloidogyne mali*, its taxonomy, ecology, damage and control. Second Asian Regional Conference on root-knot nematodes. *Thailand Kasetsart Journal* 12, 25-30.

INSERRA, R.N., PERROTTA, G., VOVLAS, N. & CATARA, A. (1978). Reaction of *Citrus* rootstocks to *Meloidogyne javanica*. *Journal of Nematology* 10, 181-184.

INSERRA, R.N., O'BANNON, J.H., DI VITO, M. & FERRIS, H. (1983). Response of two alfalfa cultivars to *Meloidogyne hapla*. *Journal of Nematology* 15, 644-646.

INSERRA, R.N., BRITO, J.A., DONG, K., HANDOO, Z.A., LEHMAN, P.S., POWERS, T. & MILLAR, L. (2003a). Exotic nematode plant pests of agricultural and environmental significance to the United States: *Meloidogyne citri*. Society of Nematologists. Available on-line at http://nematode.unl.edu/wgroup.htm.

INSERRA, R.N., BRITO, J.A., DONG, K., HANDOO, Z.A., LEHMAN, P.S., POWERS, T. & MILLAR, L. (2003b). Exotic nematode plant pests of agricultural and environmental significance to the United States: *Meloidogyne donghaiensis*. Society of Nematologists. Available on-line at http://nematode.unl.edu/wgroup.htm.

INSERRA, R.N., BRITO, J.A., DONG, K., HANDOO, Z.A., LEHMAN, P.S., POWERS, T. & MILLAR, L. (2003c). Exotic nematode plant pests of agricultural and environmental significance to the United States: *Meloidogyne fujianensis*. Society of Nematologists. Available on-line at http://nematode.unl.edu/wgroup.htm.

INSERRA, R.N., BRITO, J.A., DONG, K., HANDOO, Z.A., LEHMAN, P.S., POWERS, T. & MILLAR, L. (2003d). Exotic nematode plant pests of agricultural and environmental significance to the United States: *Meloidogyne indica*. Society of Nematologists. Available on-line at http://nematode.unl.edu/wgroup.htm.

INSERRA, R.N., BRITO, J.A., DONG, K., HANDOO, Z.A., LEHMAN, P.S., POWERS, T. & MILLAR, L. (2003e). Exotic nematode plant pests of agricultural and environmental significance to the United States: *Meloidogyne jiangyangensis*. Society of Nematologists. Available on-line at http://nematode.unl.edu/wgroup.htm.

INSERRA, R.N., BRITO, J.A., DONG, K., HANDOO, Z.A., LEHMAN, P.S., POWERS, T. & MILLAR, L. (2003f). Exotic nematode plant pests of agricultural and environmental significance to the United States: *Meloidogyne kongi*. Society of Nematologists. Available on-line at http://nematode.unl.edu/wgroup.htm.

INSERRA, R.N., BRITO, J.A., DONG, K., HANDOO, Z.A., LEHMAN, P.S., POWERS, T. & MILLAR, L. (2003g). Exotic nematode plant pests of agricultural and environmental significance to the United States: *Meloidogyne mingnan-*

ica. Society of Nematologists. Available on-line at http://nematode.unl.edu/wgroup.htm.

ITOH, Y., OHSHIMA, Y. & ICHINOHE, M. (1969). A root-knot nematode, *Meloidogyne mali* n. sp. on apple tree from Japan (Tylenchida: Heteroderidae). *Applied Entomology and Zoology* 4, 194-202. DOI: 10.1303/aez.4.194

IWAHORI, H., TRUC, N.T.N., BAN, D.V. & ICHINOSE, K. (2009). First report of root-knot nematode *Meloidogyne enterolobii* on guava in Vietnam. *Plant Disease* 93, 675. DOI: 10.1094/PDIS-93-6-0675C

JABBAR, A., JAVED, N., KHAN, S.A. & ALI, M.A. (2015). *Meloidogyne graminicola* an emerging threat to rice and wheat in Punjab province in Pakistan. *Pakistan Journal of Nematology* 33, 227-228.

JAEHN, A., REBEL, E.K. & LORDELLO, L.G.E. (1980). A origem do nematóide *Meloidogyne coffeicola*. *Nematologia Brasileira* 4, 159-161.

JAHANSHAHI AFSHAR, F., SASANELLI, N., HOSSEININEJAD, S.A. & TANHA MAAFI, Z. (2014). Effects of the root-knot nematodes *Meloidogyne incognita* and *M. javanica* on olive plants growth in glasshouse conditions. *Helminthologia* 51, 46-52. DOI: 10.2478/s11687-014-0207-x

JAMES, A. & MACDONALD, J. (2015). Recombinase polymerase amplification: Emergence as a critical molecular technology for rapid, low-resource diagnostics. *Expert Review of Molecular Diagnostics* 15, 1475-1489. DOI: 10.1586/14737159.2015.1090877

JANATI, A., BERGÉ, J.B., THANTAPHYLLOU, A.C. & DALMASSO, A. (1982). Nouvelles données sur l'utilisation des isoestérases pour l'identification des *Meloidogyne*. *Revue de Nématologie* 5, 147-154.

JANSSEN, T., KARSSEN, G., VERHAEVEN, M., COYNE, D. & BERT, W. (2016). Mitochondrial coding genome analysis of tropical root-knot nematodes (*Meloidogyne*) supports haplotype based diagnostics and reveals evidence of recent reticulate evolution. *Scientific Reports* 6, 22591. DOI: 10.1038/srep22591

JANSSEN, T., KARSSEN, G., TOPALOVIĆ, O., COYNE, D. & BERT, W. (2017). Integrative taxonomy of root-knot nematodes reveals multiple independent origins of mitotic parthenogenesis. *PLoS ONE* 12, e0172190. DOI: 10.1371/journal.pone.0172190

JENSEN, H.J., HOPPER, W.E.R. & LORING, L.B. (1968). Barley root-knot nematode discovered in western Oregon. *Plant Disease Reporter* 52, 169.

JEPSON, S.B. (1983a). The use of second-stage juvenile tails as an aid in the identification of *Meloidogyne* species. *Nematology* 29, 11-28. DOI: 10.1163/187529283X00140

JEPSON, S.B. (1983b). *Meloidogyne kralli* n. sp. (Nematoda: Meloidogynidae) a root-knot nematode parasitising sedge (*Carex acuta* L.). *Revue de Nématologie* 6, 239-245.

JEPSON, S.B. (1983c). Identification of *Meloidogyne*: a general assessment and a comparison of male morphology using light microscopy, with a key to 24 species. *Revue de Nématologie* 6, 291-309.

JEPSON, S.B. (1987). *Identification of root-knot nematodes (*Meloidogyne *species).* Wallingford, UK, CAB International.

JINDAPUNNAPAT, K., CHINNASRI, B. & KWANKUAE, S. (2013). Biological control of root-knot nematodes (*Meloidogyne enterolobii*) in guava by the fungus *Trichoderma harzianum. Journal of Developments in Sustainable Agriculture* 8, 110-118.

JOBERT, C. (1878). Sur une maladie du caféier observée au Brésil. *Comptes Rendus de l'Academie des Sciences Paris* 87, 941-943.

JOHNSON, A.W., BURTON, G.W., WILSON, J.P. & GOLDEN, A.M. (1995). Rotations with coastal bermudagrass and fallow for management of *Meloidogyne incognita* and soilborne fungi on vegetable crops. *Journal of Nematology* 27, 457-464.

JONES, J.T., HAEGEMAN, A., DANCHIN, E.G.J., GAUR, H.S., HELDER, J., JONES, M.G.K., KIKUCHI, T., MANZANILLA-LÓPEZ, R., PALOMARES-RIUS, J.E., WESEMAEL, W.M.L. ET AL. (2013). Top 10 plant-parasitic nematodes in molecular plant pathology. *Molecular Plant Pathology* 14, 946-961. DOI: 10.1111/mpp.12057

JONES, M.G.K. & GUNNING, B.E.S. (1976). Transfer cells and nematode induced giant cells in *Helianthemum*. *Protoplasma* 87, 273-279. DOI: 10.1007/BF01623973

JORGE JUNIOR, A.S., CARES, J.E., MATTOS, V.S., COYNE, D., SANTOS, M.F.A. DOS, CARNEIRO, R.M.D.G. (2016). First report of *Meloidogyne izalcoensis* (Nematoda: Meloidogynidae) on coffee, cabbage, and other crops in Africa. *Plant Disease* 100, 2173. DOI: 10.1094/PDIS-03-16-0294-PDN

JOSEPH, S., MEKETE, T., DANQUAH, W.B. & NOLING, J. (2016). First report of *Meloidogyne haplanaria* infecting *Mi*-resistant tomato plants in Florida and its molecular diagnosis based on mitochondrial haplotype. *Plant Disease* 100, 1438-1445. DOI: 10.1094/PDIS-09-15-1113-RE

JU, Y.L., LIN, Y., YANG, G.G., WU, H.P. & PAN, Y.M. (2019). Development of recombinase polymerase amplification assay for rapid detection of *Meloidogyne incognita*, *M. javanica*, *M. arenaria*, and *M. enterolobii*. *European Journal of Plant Pathology* 155, 1155-1163. DOI: 10.1007/s10658-019-01844-6

KAPLAN, D.T. & KOEVENIG, J.L. (1989). Description of the host-parasite relationship of *Meloidogyne christiei* with *Quercus laevis*. *Revue de Nématologie* 12, 57-61.

KARNKOWSKI, W., DOBOSZ, R., STADNICKA, M. & SALDAT, M. (2013). Occurrence of the northern root-knot nematode *Meloidogyne hapla* Chit-

wood, 1949 (Nematoda: Meloidogynidae) in seed potatoes on the territory of Poland. *Progress in Plant Protection* 53, 371-375.
KARSSEN, G. (1994). The use of isozyme phenotypes for the identification of root-knot nematodes (*Meloidogyne* species). In: Annual Report 1992. Diagnostic Centre, Plant Protection Service, Wageningen, The Netherlands, pp. 85-88.
KARSSEN, G. (1995). Morphological and biochemical differentiation in *Meloidogyne chitwoodi* populations in The Netherlands. *Nematologica* 41, 314-315. [Abstr.]
KARSSEN, G. (1996a). Description of *Meloidogyne fallax* n. sp. (Nematoda: Heteroderidae), a root-knot nematode from The Netherlands. *Fundamental and Applied Nematology* 19, 593-599.
KARSSEN, G. (1996b). On the morphology of *Meloidogyne sasseri* Handoo, Huettel & Morgan Golden, 1993. *Nematologica* 42, 262-264. DOI: 10.1163/004325996X00093
KARSSEN, G. (2002). *The plant-parasitic nematode genus* Meloidogyne *Göldi, 1892 (Tylenchida) in Europe*. Leiden, The Netherlands, Brill.
KARSSEN, G. (2004). Nieuwe wortelknobbelaaltjes en opvallende waarnemingen in Europa. *Gewasbescherming* 5, 245-246.
KARSSEN, G. & GRUNDER, J. (2002). First report of the root-knot nematode *Meloidogyne kralli* in Switzerland. *Plant Disease* 86, 919. DOI: 10.1094/PDIS.2002.86.8.919C
KARSSEN, G. & MOENS, M. (2006). Root-knot nematodes. In: Perry, R.N. & Moens, M. (Eds). *Plant nematology*, 1st edition. Wallingford, UK, CAB International, pp. 59-90.
KARSSEN, G. & VAN AELST, A.C. (2001). Root-knot nematode perineal pattern development: a reconsideration. *Nematology* 3, 95-111. DOI: 10.1163/156854101750236231
KARSSEN, G. & VAN HOENSELAAR, T. (1998). Revision of the genus *Meloidogyne* Göldi, 1892 (Nematoda: Heteroderidae) in Europe. *Nematologica* 44, 713-788. DOI: 10.1163/156854101750236231
KARSSEN, G., VAN AELST, A. & COOK, R. (1998a). Redescription of the root-knot nematode *Meloidogyne maritima* Jepson, 1987 (Nematoda: Heteroderidae), a parasite of *Ammophila arenaria* (L.) Link. *Nematologica* 44, 241-253. DOI: 10.1163/005425998X00026
KARSSEN, G., VAN AELST, A. & VAN DER PUTTEN, W.H. (1998b). *Meloidogyne duytsi* n. sp. (Nematoda: Heteroderidae), a root-knot nematode from Dutch coastal foredunes. *Fundamental and Applied Nematology* 21, 299-306.
KARSSEN, G., BOLK, R.J., VAN AELST, A.C., VAN DEN BELD, I., KOX, L.F.F., KORTHALS, G., MOLENDIJK, L., ZIJLSTRA, C., VAN HOOF, R. & COOK, R. (2004). Description of *Meloidogyne minor* n. sp. (Nematoda:

Meloidogynidae), a root-knot nematode associated with yellow patch disease in golf courses. *Nematology* 6, 59-72. DOI: 10.1163/156854104323072937

KARSSEN, G., KEULEN, I., HOENSELAAR, T. & HEESE, E. (2008). *Meloidogyne ulmi*: een nieuwe iepenparasiet in Nederland? *Boomzorg* 2, 62-63.

KARSSEN, G., LIAO, J.L., KAN, Z., VAN HEESE, E.Y.J. & DEN NIJS, LJ.M.F. (2012). On the species status of the root-knot nematode *Meloidogyne mayaguensis* Rammah & Hirschmann, 1988. *ZooKeys* 181, 67-77. DOI: 10. 3897/zookeys.181.2787

KARSSEN, G., WESEMAEL, W. & MOENS, M. (2013). Root-knot nematodes. In: Perry, R.N. & Moens, M. (Eds). *Plant nematology*, 2nd edition, Wallingford, UK, CAB International, pp. 73-108.

KARURI, H.W., OLAGO, D., NEILSON, R., MARARO, E. & VILLINGER, J. (2017). A survey of root knot nematodes and resistance to *Meloidogyne incognita* in sweet potato varieties from Kenyan fields. *Crop Protection* 92, 114-121. DOI: 10.1016/j.cropro.2016.10.020

KATSUTA, A., TOYOTA, K., MIN, Y.Y. & MAUNG, T.T. (2016). Development of real-time PCR primers for the quantification of *Meloidogyne graminicola*, *Hirschmanniella oryzae* and *Heterodera cajani*, pests of the major crops in Myanmar. *Nematology* 18, 257-263. DOI: 10.1163/15685411-00002957

KAUR, H. & ATTRI, R. (2013). Morphological and morphometrical characterization of *Meloidogyne incognita* from different host plants in four districts of Punjab, India. *Journal of Nematology* 45, 122-127.

KAUR, R., BRITO, J.A. & RICH, J.R. (2007). Host suitability of selected weed species to five *Meloidogyne* species. *Nematropica* 37, 107-120.

KAVITHA, P.G., UMADEVI, M., SURESH, S. & RAVI, V. (2016). The rice root-knot nematode (*Meloidogyne graminicola*) – life cycle and histopathology. *International Journal of Science and Nature* 7, 483-486.

KEPENEKCI, I. & DURA, O. (2017). *Anethum graveolens*, a new host of root-knot nematode (*Meloidogyne incognita*) in Turkey. *Pakistan Journal of Nematology* 35, 215-216.

KHAN, M.R., PAL, S., MANOHAR, G.T., BHATTACHARYYA, S., SINGH, A., SARKAR, P. & LALLIANSANGA, S. (2017). Detection, diagnosis and pathogenic potential of *Meloidogyne incognita* on passion fruit from Mizoram, India. *Pakistan Journal of Zoology* 49, 1207-1214.

KHAN, M.R., PAL, S., DHANANJAYBHAI PATEL, A., AMBARAM PATEL, B., MANOHAR GHULE, T., SINGH, A. & PHANI, V. (2018). Further observation on *Meloidogyne indica* Whitehead, 1968 from India. *Pakistan Journal of Zoology* 50, 2009-2017.

KHANAL, C., SZALANSKI, A.L. & ROBBINS, R.T. (2016a). First report of *Meloidogyne partityla* parasitizing pecan in Arkansas and confirmation of *Quercus stellate* as a host. *Nematropica* 46, 1-7.

KHANAL, C., ROBBINS, R.T., FASKE, T.R., SZALANSKI, A.L., MCGAWLEY, E.C. & OVERSTREET, C. (2016b). Identification and haplotype designation of *Meloidogyne* spp. of Arkansas using molecular diagnostics. *Nematropica* 46, 261-270.

KIATPATHOMCHAI, W., JAROENRAM, W., ARUNRUT, N., JITRAPAKDEE, S. & FLEGEL, T.W. (2008). Shrimp Taura syndrome virus detection by reverse transcription loop-mediated isothermal amplification combined with a lateral flow dipstick. *Journal of Virological Methods* 153, 214-217. DOI: 10.1016/j.jviromet.2008.06.025

KIEWNICK, S., KARSSEN, G., BRITO, J.A., OGGENFUSS, M. & FREY, J.E. (2008). First report of root-knot nematode *Meloidogyne enterolobii* on tomato and cucumber in Switzerland. *Plant Disease* 92, 1370. DOI: 10.1094/PDIS-92-9-1370A

KIEWNICK, S., WOLF, S., WILLARETH, M. & FREY, J.E. (2013). Identification of the tropical root-knot nematode species *Meloidogyne incognita*, *M. javanica* and *M. arenaria* using a multiplex PCR assay. *Nematology* 15, 891-894. DOI: 10.1163/15685411-00002751

KIEWNICK, S., HOLTERMAN, M.H.M., VAN DEN ELSEN, S.J.J., VAN MEGEN, H.H.B., FREY, J.E. & HELDER, J. (2014). Comparison of two short DNA barcoding loci (COI and COII) and two longer ribosomal DNA genes (SSU & LSU rRNA) for specimen identification among quarantine root-knot nematodes (*Meloidogyne* spp.) and their close relatives. *European Journal of Plant Pathology* 140, 97-110. DOI: 10.1007/s10658-014-0446-1

KIEWNICK, S., FREY, J.E. & BRAUN-KIEWNICK, A. (2015). Development and validation of LNA-based quantitative real-time PCR assays for detection and identification of the root-knot nematode *Meloidogyne enterolobii* in complex DNA backgrounds. *Phytopathology* 105, 1245-1249. DOI: 10.1094/PHYTO-12-14-0364-R

KILPATRICK, R.A., GILCHRIST, L. & GOLDEN, A.M. (1976). Root knot on wheat in Chile. *Plant Disease Reporter* 60, 135.

KIM, D.G. & FERRIS, H. (2002). Relationship between crop losses and initial population densities of *Meloidogyne arenaria* in winter-grown oriental melon in Korea. *Journal of Nematology* 34, 43-49.

KIRJANOVA, E.S. (1963). [Collection and taxonomy of root nematodes of the family Heteroderidae (Skarbilovich, 1947) Thorne, 1949.] In: [*Methods of Investigating Nematodes of Plants, Soil and Insects*.] Leningrad, USSR, USSR Academy of Sciences, Zoological Institute, pp. 6-32.

KIRJANOVA, E. & IVANOVA, T. (1965). [Eelworm fauna of *Pelargonium roseum* L. in the Tadzhik S.S.R.] *Izvestiya Akademii Nauk Tadzhikskoi SSR. Otdel Biologicheski Nauk* 1(18), 24-31.

KIRKPATRICK, T.L., OOSTERHUIS, D.M. & WULLSCHLEGER, S.D. (1991). Interaction of *Meloidogyne incognita* and water stress in two cotton cultivars. *Journal of Nematology* 23, 462-467.

KLEYNHANS, K.P.N. (1986). *Meloidogyne partityla* sp. nov. from pecan nut (*Carya illinoensis* (Wangenh.) C. Koch) in the Transvaal lowveld (Nematoda: Meloidogynidae). *Phytophylactica* 18, 103-106.

KLEYNHANS, K.P.N. (1988). *Meloidogyne vandervegtei* sp. nov. from subtropical coastal forest in Natal (Nemata: Heteroderidae). *Phytophylactica* 20, 263-267.

KLEYNHANS, K.P.N. (1991). *The root-knot nematodes of South Africa.* Technical Communication No. 231. Pretoria, South Africa, Department of Agricultural Development.

KLEYNHANS, K.P.N. (1993). *Meloidogyne hispanica* Hirschmann, 1986 and *M. ethiopica* Whitehead, 1968 in South Africa (Nemata: Heteroderidae). *Phytophylactica* 25, 283-288.

KOFOID, C.A. & WHITE, W.A. (1919). A new nematode infection of man. *Journal of the American Medical Association* 72, 567-569.

KOLOMBIA, Y.A., KUMAR, P.L., CLAUDIUS-COLE, A.O., KARSSEN, G., VIAENE, N., COYNE, D. & BERT, W. (2016). First report of *Meloidogyne enterolobii* causing tuber galling damage on white yam (*Dioscorea rotundata*) in Nigeria. *Plant Disease* 100, 2173. DOI: 10.1094/PDIS-03-16-0348-PDN

KOLOMBIA, Y.A., KARSSEN, G., VIAENE, N., AVA KUMAR, P.L., DE SUTTER, N., JOOS, L., COYNE, D.L. & BERT, W. (2017). Diversity of root-knot nematodes associated with tubers of yam (*Dioscorea* spp.) established using isozyme analysis and mitochondrial DNA-based identification. *Journal of Nematology* 49, 177-188.

KORNOBIS, S. (2001). [Root-knot nematodes, *Meloidogyne* spp. in Poland.] *Progress in Plant Protection* 41, 189-192.

KORUNES, K.L. & NOOR, M. (2017). Gene conversion and linkage: effects on genome evolution and speciation. *Molecular Ecology* 26, 351-364. DOI: 10.1111/mec.13736

KOUTSOVOULOS, G., MARQUES, E., ARGUEL, M.-J., DURET, L., MACHADO, A.C.Z., CARNEIRO, R.M.D.G., KOZLOWSKI, D.K., BAILLY-BECHET, M., CASTAGNONE-SERENO, P., ALBUQUERQUE, E.V.S. ET AL. (2018). Population genomics supports clonal reproduction and multiple gains and losses of parasitic abilities in the most devastating nematode pest. *bioRxiv* 362129. DOI: 10.1101/362129

KOUTSOVOULOS, G.D., POULLET, M., EL ASHRY, A., KOZLOWSKI, D.K., SALLET, E., DA ROCHA, M., MARTIN-JIMENEZ, C., PERFUS-BARBEOCH, L., FREY, J.-E., AHRENS, C. ET AL. (2020). Genome assembly and annotation of *Meloidogyne enterolobii*, an emerging parthenogenetic root-knot nematode. *Scientific Data* 7, 324. DOI: 10.1038/s41597-020-00666-0

KUMAR, A., SINGH, P. & JAIN, R.K. (2016). Comparative study on damage potential of *Meloidogyne* species and races on tomato hybrids. *Annals of Biology* 32, 201-208.

KUMAR, V., VERMA, K.K. & KUMAR, A. (2017). Studies on pathogenicity of *Meloidogyne graminicola* in different soil types on scented and non-scented rice. *International Journal of Pure & Applied Bioscience* 5, 687-693. DOI: 10.18782/2320-7051.2506

KUTYWAYO, V. & BEEN, T.H. (2006). Host status of six major weeds to *Meloidogyne chitwoodi* and *Pratylenchus penetrans*, including a preliminary field survey concerning other weeds. *Nematology* 8, 647-657. DOI: 10.1163/156854106778877839

KYROU, N.C. (1969). First record of occurrence of *Meloidogyne artiellia* on wheat in Greece. *Nematologica* 15, 432-433. DOI: 10.1163/187529269X00560

LAMBERTI, F. & TAYLOR, C.E. (Eds) (1979). *Root-knot nematodes (*Meloidogyne *species); systematics, biology and control.* London & New York, Academic Press.

LAMBERTI, F., EKANAYAKE, H.M.R.K. & DI VITO, N. (1987). The root-knot nematodes, *Meloidogyne* spp., found in Sri Lanka. *FAO Plant Protection Bulletin* 35, 27-31.

LAMONDIA, J.A. (2002). Seasonal populations of *Pratylenchus penetrans* and *Meloidogyne hapla* in strawberry roots. *Journal of Nematology* 34, 409-413.

LAMONDIA, J.A. & ELMER, W.H. (2007). Occurrence of *Meloidogyne spartinae* on *Spartina alterniflora* in Connecticut and Massachusetts. *Plant Disease* 91, 327. DOI: 10.1094/PDIS-91-3-0327C

LAMONDIA, J.A. & ELMER, W.H. (2008). Ecological relationships between *Meloidogyne spartinae* and salt marsh grasses in Connecticut. *Journal of Nematology* 40, 217-220.

LANDA, B.B., PALOMARES-RIUS, J.E., VOVLAS, N., CARNEIRO, R.M.D.G., MALEITA, C.M.N., ABRANTES, I.M.O. & CASTILLO, P. (2008). Molecular characterization of *Meloidogyne hispanica* (Nematoda, Meloidogynidae) by phylogenetic analysis of genes within the rDNA in *Meloidogyne* spp. *Plant Disease* 92, 1104-1110. DOI: 10.1094/PDIS-92-7-1104

LATHA, M.S., SHARMA, S.B., GOUR, T.B. & REDDY, D.D.R. (1998). Comparison of two populations of *Meloidogyne javanica* based on morphology, biology and pathogenicity to groundnut. *Nematologia Mediterranea* 26, 189-94.

LAUGHLIN, C.W. & WILLIAMS, A.S. (1971). Population behavior of *Meloidogyne graminis* in field-grown 'Tifgreen' bermudagrass. *Journal of Nematology* 3, 386-389.

LAVERGNE, G. (1901). L'anguillule du Chili (*Anguillula vialae*). *Revue de Viticulture* 16, 445-452.

LE, H.T.T., PADGHAM, J.L. & SIKORA, R.A. (2009). Biological control of the rice root-knot nematode *Meloidogyne graminicola* on rice, using endophytic and rhizosphere fungi. *International Journal of Pest Management* 55, 31-36. DOI: 10.1080/09670870802450235

LE, T.M.L., NGUYEN, T.D., NGUYEN, H.T., LIEBANAS, G., NGUYEN, T.A.D. & TRINH, Q.P. (2019). A new root-knot nematode, *Meloidogyne moensi* n. sp. (Nematoda: Meloidogynidae), parasitizing Robusta coffee from Western Highlands, Vietnam. *Helminthologia* 56, 229-246. DOI: 10.2478/helm-2019-0014

LEHMAN, P.S. & LORDELLO, L.G.E. (1982). *Meloidogyne exigua*, a root-knot nematode of coffee. *Nematology Circular* 88. Gainesville, FL, USA, Florida Department of Agriculture & Consumer Services Division of Plant Industry.

LEHMAN, P.S., SANTO, G.S. & O'BANNON, J.H. (1983). The Columbia root-knot nematode, *Meloidogyne chitwoodi*. *Nematology Circular* 96, Gainesville, FL, USA, Florida Department of Agriculture & Consumer Services Division of Plant Industry.

LEITE, R.R., MATTOS, V.S., GOMES, A.C.M.M., PY, L.G., SOUZA, D.A., CASTAGNONE-SERENO, P., CARES, J.E. & CARNEIRO, R.M.D.G. (2020). Integrative taxonomy of *Meloidogye ottersoni* (Thorne, 1969) Franklin, 1971 (Nematoda: Meloidogynidae) parasitizing flooded rice in Brazil. *European Journal of Plant Pathology* 157, 943-959. DOI: 10.1007/s10658-020-02049-y

LEWIS, S. & WEBLEY, D. (1966). Observations on two nematodes infesting grasses. *Plant Pathology* 15, 184-186. DOI: 10.1111/j.1365-3059.1966.tb00347.x

LEWIS, S.A. (1987). Nematode-plant compatibility. In: Veech, A. & Dickson, D.W. (Eds). *Vistas on nematology*. Hyattsville, MD, USA, Society of Nematologists, pp. 246-252.

LI, S.J. & YU, Z. (1991). [A new species of root-knot nematode (*Meloidogyne actinidiae*) on *Actinidia chinensis* in Henan Province.] *Acta Agriculturae Universitatis Henanensis* 25, 251-253.

LIAO, J.L. (2001). [*Study on the identification and polymorphism of root-knot nematodes (*Meloidogyne*).*] Ph.D. Thesis, South China Agricultural University, Guangzhou, P.R. China.

LIAO, J.L. & FENG, Z.X. (1995). [A root-knot nematode *Meloidogyne hainanensis* sp. nov. (Nematoda: Meloidogynidae) parasitizing rice in Hainan, China.] *Journal of South China Agricultural University* 16, 34-39.

LIAO, J.L., YANG, W.C., FENG, Z.X. & KARSSEN, G. (2005). Description of *Meloidogyne panyuensis* sp. n. (Nematoda: Meloidogynidae), parasitic on peanut (*Arachis hypogaea* L.) in China. *Russian Journal of Nematology* 13, 107-114.

LIMA, E.A., MATTOS, J.K., MOITA, A.W., CARNEIRO, R.G. & CARNEIRO, R.M.D.G. (2009). Host status of different crops for *Meloidogyne ethiopica* control. *Tropical Plant Pathology* 34, 152-157. DOI: 10.1590/S1982-56762009000300003

LIMA, R.D.A. DE & FERRAZ, S. (1985). Análise comparativa das variações morfométricas entre diferentes populações de *Meloidogyne exigua. Revista Ceres* 32, 362-373.

LIMA, R.D.A. DE, CAMPOS, V.P., HUANG, S.P. & MELLES, C.A.A. (1985). Reprodutividade e parasitismo de *Meloidogyne exigua* em ervas daninhas que ocorrem em cafezais. *Nematologia Brasileira* 9, 63-72.

LIN, L., DENG, Y., JIANG, N. & HU, X. (2004). Distribution and damage level of root-knot on medicinal plant in China. *Journal of Yunnan Agricultural University* 19, 666-669.

LIU, G.K. & ZHANG, S.S. (2001). [Description of a new species of root-knot nematode, *Meloidogyne dimocarpus* n. sp.] *Journal of Shenyang Agricultural University* 32, 167-172.

LIU, Q.L. & WILLIAMSON, V.M. (2006). Host-specific pathogenicity and genome differences between inbred strains of *Meloidogyne hapla. Journal of Nematology* 38, 158-164.

LOMBARDO, S., COLOMBO, A. & RAPISARDA, C. (2011). Severe damages caused by *Meloidogyne artiellia* to cereals and leguminous in Sicily. *Redia* 94, 149-151.

LONG, H., WANG, X. & XU, J. (2006). Molecular cloning and life-stage expression pattern of a new chorismate mutase gene from the root-knot nematode *Meloidogyne arenaria. Plant Pathology* 55, 559-563. DOI: 10.1111/j.1365-3059.2006.01362.x

LONG, H.B., FENG, T.Z., SUN, Y.F., PEI, Y.L. & PENG, D.L. (2017). First report of *Meloidogyne graminicola* on soybean (*Glycine max*) in China. *Plant Disease* 101, 1554. DOI: 10.1094/PDIS-03-17-0334-PDN

LOOF, P.A.A. & LUC, M. (1993). A revised polytomous key for the identification of species of the genus *Xiphinema* Cobb, 1913 (Nematoda: Longidoridae) with exclusion of the *X. americanum*-group: Supplement 1. *Systematic Parasitology* 24, 185-189. DOI: 10.1007/BF00010531

LOOS, C.A. (1953). *Meloidogyne brevicauda*, n. sp. a cause of root-knot of mature tea in Ceylon. *Proceedings of the Helminthological Society of Washington* 20, 83-91.

LÓPEZ, R. (1984). *Meloidogyne salasi* sp. n. (Nematoda: Meloidognidae), a new parasite of rice (*Oryza sativa* L.) from Costa Rica and Panama. *Turrialba* 34, 275-286.

LÓPEZ, R. (1985). Observaciones sobre la morfologia de *Meloidogyne exigua* con el microscópio electronico de rastreo. *Nematropica* 15, 27-36.

LÓPEZ, R. & SALAZAR, L. (1989). *Meloidogyne arabicida* sp. n. (Nemata: Heteroderidae) nativo de Costa Rica: un nuevo y severo patogeno del cafeto. *Turrialba* 39, 313-323.

LÓPEZ, R. & VILCHEZ, H. (1991). Two new hosts of the coffee root-knot nematode, *Meloidogyne exigua*, in Costa Rica. *Agronomia-Costarricense* 15, 163-166.

LÓPEZ-GÓMEZ, M., GINE, A., VELA, M.D., ORNAT, C., SORRIBAS, F.J., TALAVERA, M. & VERDEJO-LUCAS, S. (2014). Damage functions and thermal requirements of *Meloidogyne javanica* and *Meloidogyne incognita* on watermelon. *Annals of Applied Biology* 165, 466-473. DOI: 10.1111/aab.12154

LÓPEZ-GÓMEZ, M., FLOR-PEREGRÍN, E., TALAVERA, M., SORRIBAS, F.J. & VERDEJO-LUCAS, S. (2015). Population dynamics of *Meloidogyne javanica* and its relationship with the leaf chlorophyll content in zucchini. *Crop Protection* 70, 8-14. DOI: 10.1016/j.cropro.2014.12.015

LÓPEZ-LIMA, D., SÁNCHEZ-NAVA, P., CARRION, G., ESPINOSA DE LOS MONTEROS, A. & VILLAIN, L. (2015). Corky-root symptoms for coffee in central Veracruz are linked to the root-knot nematode *Meloidogyne paranaensis*, a new report for Mexico. *European Journal of Plant Pathology* 141, 623-629. DOI: 10.1007/s10658-014-0564-9

LÓPEZ-PÉREZ, J.A., M. ESCUER, M., DÍEZ-ROJO, A., ROBERTSON, L. PIEDRA BUENA, A., LÓPEZ-CEPERO, J. & BELLO, A. (2011). Host range of *Meloidogyne arenaria* (Neal, 1889) Chitwood, 1949 (Nematoda: Meloidogynidae) in Spain. *Nematropica* 41, 130-140.

LORDELLO, L.G.E. (1956). *Meloidogyne inornata* sp. n., a serious pest of soybean in the state of São Paulo, Brazil. (Nematoda, Heteroderidae). *Revista Brasileira de Biològia* 16, 65-70.

LORDELLO, L.G.E. & LORDELLO, R.R.A. (1972). Duas plantas hospedeiras novas do nematóide *Meloidogyne coffeicola*. *Anais da Escola Superior de Agicultura Luiz de Queriroz* 29, 61-62.

LORDELLO, L.G.E. & ZAMITH, A.P.L. (1958). On the morphology of the coffee root-knot nematode, *Meloidogyne exigua* Goeldi, 1887. *Proceedings of the Helminthological Society of Washington* 25, 133-137.

LORDELLO, L.G.E. & ZAMITH, A.P.L. (1960). "*Meloidogyne coffeicola*" sp. n. a pest of coffee tree in the state of Panara, Brazil (Nematoda, Heteroderidae). *Revista Brasileira Biologica* 20, 375-379.

LORDELLO, R.R.A. & FAZUOLI, L.C. (1980). [*Meloidogyne decalineata* in coffee roots from São Tomé Island (Africa).] *Revista de Agricultura* 55, 238.

LOVEYS, B.R. & BIRD, A.F. (1973). The influence of nematodes on photosynthesis in tomato plants. *Physiological Plant Pathology* 3, 525-529. DOI: 10.1016/0048-4059(73)90063-5

LU, P., DAVIS, R.F., KEMERAIT, R.C., VAN IERSE, M.W. & SCHERM, H. (2014). Physiological effects of *Meloidogyne incognita* infection on cotton genotypes with differing levels of resistance in the greenhouse. *Journal of Nematology* 46, 352-359.

LUC, M., MAGGENTI, A.R. & FORTUNER, R. (1988). A reappraisal of Tylenchina (Nemata). 9. The family Heteroderidae Filip'ev & Schuurmans Stekhoven, 1941. *Revue de Nématologie* 11, 159-176.

LUNT, D.H. (2008). Genetic tests of ancient asexuality in root knot nematodes reveal recent hybrid origins. *BMC Evolionary Biology* 8, 194. DOI: 10.1186/1471-2148-8-194

LUNT, D.H., KUMAR, S., KOUTSOVOULOS, G. & BLAXTER, M.L. (2014). The complex hybrid origins of the root knot nematodes revealed through comparative genomics. *PeerJ* 2, e356. DOI: 10.7717/peerj.356

LUO, M., LI, B.X. & WU, H.Y. (2020). Incidence of the rice root-knot nematode, *Meloidogyne graminicola*, in Guangxi, China. *Plant Pathology Journal* 36, 297-302. DOI: 10.5423/PPJ.NT.02.2020.0034

MAAS, P.W.TH., SANDERS, H. & DEDE, J. (1978). *Meloidogyne oryzae* n. sp. (Nematoda, Meloidogynidae) infesting irrigated rice in Surinam (South America). *Nematologica* 24, 305-311. DOI: 10.1163/187529278X00272

MACGOWAN, J.B. (1984). *Meloidogyne graminis*, a root-knot nematode of grass. *Nematology Circular* 107, Gainesville, FL, USA, Florida Department of Agriculture & Consumer Services Division of Plant Industry.

MACGOWAN, J.B. (1989). The rice root-knot nematode *Meloidogyne graminicola* Golden & Birchfield 1965. *Nematology Circular* 166, Gainesville, FL, USA, Florida Department of Agriculture & Consumer Services Division of Plant Industry.

MACGOWAN, J.B. & LANGDON, K.R. (1989). Host of the rice root-knot nematode *Meloidogyne graminicola*. *Nematology Circular* 172, Gainesville, FL, USA, Florida Department of Agriculture & Consumer Services Division of Plant Industry.

MACGUIDWIN, A.E., BIRD, G.W., HAYNES, D.L. & GAGE, S.H. (1987). Pathogenicity and population dynamics of *Meloidogyne hapla* associated with *Allium cepa*. *Plant Disease* 71, 446-449.

MACHACA-CALSIN, C.P., SILVA, W.R., GRINBERG, P.S., ARAÚJO FILHO, J.V. DE & GOMES, C.B. (2021). Occurrence of *Meloidogyne morocciensis* parasitizing beetroot in Brazil. *European Journal of Plant Pathology* 160, 239-242. DOI: 10.1007/s10658-021-02231-w

MACHADO, A.C.Z., DORIGO, O.F. & MATTEI, D. (2013). First report of the root knot nematode, *Meloidogyne inornata*, on common bean in Paraná State, Brazil. *Plant Disease* 97, 431. DOI: 10.1094/PDIS-09-12-0832-PDN

MACHADO, A.C.Z., DADAZIO, T., MARINI, P.M. & SILVA, S.A. (2014). Patogenicidade comparativa de *M. incognita* e *M. javanica* em feijao (online). Internet Resource: http://www.conafe2014.com.br/trabalhos-anais ID118.

Machado, A.C.Z., Dorigo, O.F., Carneiro, R.M.D.G. & De Araújo Filho, J.V. (2016). *Meloidogyne luci*, a new infecting nematode species on common bean fields at Paraná State, Brazil. *Helminthologia* 53, 207-210. DOI: 10.1515/helmin-2016-0014

Mai, W.F. & Abawi, G.S. (1987). Interactions among root-knot nematodes and Fusarium wilt fungi on host plants. *Annual Review of Phytopathology* 25, 317-338. DOI: 10.1146/annurev.py.25.090187.001533

Maleita, C., Curtis, R. & Abrantes, I. (2012a). Thermal requirements for the embryonic development and life cycle of *Meloidogyne hispanica. Plant Pathology* 61, 1002-1010. DOI: 10.1111/j.1365-3059.2011.02576.x

Maleita, C.M., Santos, M.C.V. dos & Abrantes, I.M. de O. (2005). Resistência de plantas cultivadas ao nemátodedas-galhas-radiculares *Meloidogyne hispanica*. In: *A Produção Integrada e a Qualidade e Segurança Alimentar-Actas do VII Encontro Nacional de Protecção Integrada, Sociedade Portuguesa de Fitopatologia, Coimbra, Portugal.*

Maleita, C.M., Santos, M.C.V. dos, Curtis, R.H.C., Powers, S.J. & Abrantes, I.M.D.O. (2011). Effect of the *Mi* gene on reproduction of *Meloidogyne hispanica* on tomato genotypes. *Nematology* 13, 939-949. DOI: 10.1163/138855411X566449

Maleita, C.M., Curtis, R.H.C., Powers, S.J. & Abrantes, I. (2012b). Host status of cultivated plants to *Meloidogyne hispanica. European Journal of Plant Pathology* 133, 449-460. DOI: 10.1007/s10658-011-9918-8

Maleita, C.M., Simões, M.J., Egas, C., Curtis, R.H.C. & Abrantes, I.M. de O. (2012c). Biometrical, biochemical, and molecular diagnosis of Portuguese *Meloidogyne hispanica* isolates. *Plant Disease* 96, 865-874. DOI: 10.1094/PDIS-09-11-0769-RE

Maleita, C.M., de Almeida, A., Vovlas, N. & Abrantes, I. (2016). Morphological, biometrical, biochemical, and molecular characterization of the coffee root-knot nematode *Meloidogyne megadora. Plant Disease* 100, 1725-1734. DOI: 10.1094/PDIS-01-16-0112-RE

Maleita, C.M., Esteves, I., Cardoso, J.M.S., Cunha, M.J., Carneiro, R.M.D.G. & Abrantes, I. (2018). *Meloidogyne luci*, a new root-knot nematode parasitizing potato in Portugal. *Plant Pathology* 67, 366-376. DOI: 10.1111/ppa.12755

Mandefro, W. & Dagne, K. (2000). Cytogenetic and esterase isozyme variation of root-knot nematode populations from Ethiopia. *African Journal of Plant Protection* 10, 39-47.

Manser, P.D. (1968). *Meloidogyne graminicola* a cause of root knot of rice. *FAO Plant Protection Bulletin* 16, 11.

Mantelin, S., Bellafiore, S. & Kyndt, T. (2017). *Meloidogyne graminicola*: a major threat to rice agriculture. *Molecular Plant Pathology* 18, 3-15. DOI: 10.1111/mpp.12394

MARAIS, M. & HEYNS, J. (1990). Host plant tests with *Meloidogyne partityla* Kleynhans, 1986. *Phytophylactica* 22, 261-162.

MARAIS, M. & KRUGER, J.C. DE W. 1991. The cytogenetics of some South African root-knot nematodes (Heteroderidae: Nematoda). *Phytophylactica* 23, 265-272.

MARCINOWSKI, K. (1909). Parasitisch und semiparasitisch an Pflanzen lebende Nematoden. *Arbeiten aus der Kaiserlichen Biologischen Anstalt für Land- und Forstwirtschaft* 7, 1-192.

MARINARI-PALMISANO, A. & AMBROGIONI, L. (2000). *Meloidogyne ulmi* sp. n., a root-knot nematode from elm. *Nematologia Mediterranea* 28, 279-293.

MARQUEZ, J., FORGHANI, F. & HAJIHASSANI, A. (2020). First report of the root-knot nematode, *Meloidogyne floridensis*, on tomato in Georgia, USA. *Plant Disease*, in press. DOI: 10.1094/PDIS-10-20-2286-PDN

MARSHALL, J.W., ZIJLSTRA, C. & KNIGHT, K.W.L. (2001). First record of *Meloidogyne fallax* in New Zealand. *Australian Plant Pathology* 30, 283-284. DOI: 10.1071/AP01033

MATTOS, V.S., FURLANETTO, C., SILVA, J.G.P., SANTOS, D.F. DOS, ALMEIDA, M.R., CORREA, V.R., MOITA, A.W., CASTAGNONE-SERENO, P. & CARNEIRO, R.M.D.G. (2016). *Meloidogyne* spp. populations from native Cerrado and soybean cultivated areas: genetic variability and aggressiveness. *Nematology* 18, 505-515. DOI: 10.1163/15685411-00002973

MATTOS, V. DA S., CARES, J.E., GOMES, C.B., GOMES, A.C.M.M., MONTEIRO, J. DA M. DOS S., GOMEZ, G.M., CASTAGNONE-SERENO, P. & CARNEIRO, R.M.D.G. (2018). Integrative taxonomy of *Meloidogyne oryzae* (Nematoda: Meloidogyninae) parasitizing rice crops in Southern Brazil. *European Journal of Plant Pathology* 151, 649-662. DOI: 10.1007/s10658-017-1400-9

MATTOS, V. DA S., MULET, K., CARES, J.H., GOMES, C.B., FERNANDEZ, D., SÁ, M.F.G. DE, CARNEIRO, R.M.D.G. & CASTAGNONE-SERENO, P. (2019). Development of diagnostic SCAR markers for *M. graminicola*, *M. oryzae* and *M. salasi* associated with irrigated rice fields in Americas. *Plant Disease* 103, 83-88. DOI: 10.1094/PDIS-12-17-2015-RE

MAUNG, M.O. & JENKINS, W.R. (1959). Effect of the root knot nematode *Meloidogyne incognita acrita* Chitwood, 1949 and a stubby root nematode *Trichodorus christiei* Allen, 1957 on the nutrient status of tomato *Lycopersicon esculentum* hort. var. Chesapeak. *Plant Disease Reporter* 43, 27-33.

MCCLURE, M.A., NISCHWITZ, C., SKANTAR, A.M., SCHMITT, M.E. & SUBBOTIN, S.A. (2012). Root-knot nematodes in golf course greens of the western United States. *Plant Disease* 96, 635-647. DOI: 10.1094/PDIS-09-11-0808

MCSORLEY, R. & PHILLIPS, M.S. (1993). Modelling population dynamics and yield losses and their use in nematode management. In: Evans,

K., Trudgill, D.L. & Webster, J.M. (Eds). *Plant-parasitic nematodes in temperate agriculture*. Wallingford, UK, CAB International, pp. 63-85.

McSorley, R., Dickson, D.W., Candanedo-Lay, E.M., Hewlett, T.E. & Frederick, J.J. (1992). Damage functions for *Meloidogyne arenaria* on peanut. *Journal of Nematology* 24, 193-198.

Medina, A., Crozzoli, R. & Perichi, G. (2009). Nematodos fitoparásitos asociados a los arrozales en Venezuela. *Nematologia Mediterranea* 37, 59-66.

Medina, A., Crozzoli, R., Perichi, G. & Jáuregui, D. (2011). *Meloidogyne salasi* (Nematoda: Meloidogynidae) on rice in Venezuela. *Fitopatología Venezolana* 24, 46-51.

Medina, I.L., Coila, V.H.C., Gomes, C.B., Pereira, A.S. & Nazareno, N.R.X. (2014). [*Meloidogyne ethiopica* report in Paraná state, Brazil, and reaction of potato cultivars to root-knot nematode.] *Horticultura Brasileira* 32, 482-485. DOI: 10.1590/S0102-053620140000400018

Medina, I.L., Gomes, C.B., Correa, V.R., Mattos, V.S., Castagnone-Sereno, P. & Carneiro, R.M.D.G. (2017). Genetic diversity of *Meloidogyne* spp. parasitising potato in Brazil and aggressiveness of *M. javanica* populations on susceptible cultivars. *Nematology* 19, 69-80. DOI: 10.1163/15685411-00003032

Medina-Canales, M.G., Ramírez-San Juan, E., Torres-Coronel, R. & Tovar-Soto, A. (2012). Pathogenicity of *Meloidogyne arenaria* against two varieties of carrot (*Daucus carota* L.) in Mexico. *Nematropica* 42, 337-342.

Mekete, T., Mandefro, W. & Greco, N. (2003). Relationship between initial population densities of *Meloidogyne javanica* and damage to pepper and tomato in Ethiopia. *Nematologia Mediterranea* 31, 169-171.

Melakeberhan, H. & Webster, J.M. (1993). The phenology of plant nematode interaction and yield loss. In: Khan, M.W. (Ed.). *Nematode interactions*. London, UK, Chapman & Hall, pp. 26-41.

Melakeberhan, H., Webster, J.M., Brooke, R.C., D'Auria, J.M. & Cackette, M. (1987). Effect of *Meloidogyne incognita* on plant nutrient concentration and its influence on the physiology of beans. *Journal of Nematology* 19, 324-330.

Mendonça, C.I. de, De Abreu Mattos, J.K. de A. & Carneiro, R.M.D.G. (2017). Hospedabilidade de plantas medicinais a *Meloidogyne paranaensis*. *Nematropica* 47, 49-54.

Meng, Q.P., Long, H. & Xu, J.H. (2004). PCR assays for rapid and sensitive identification of three major root-knot nematodes, *Meloidogyne incognita*, *M. javanica* and *M. arenaria*. *Acta Phytopathologica Sinica* 34, 204-210.

MERCER, C.F., STARR, J.L. & MILLER, K.J. (1997). Host-parasite relationships of *Meloidogyne trifoliophila* isolates from New Zealand. *Journal of Nematology* 29, 55-64.

MERESSA, B.H., DEHNE, H.W. & HALLMANN, J. (2014). Host suitability of cut-flowers to *Meloidogyne* spp. and population dynamics of *M. hapla* on the rootstock *Rosa corymbifera* 'Laxa'. *American Journal of Experimental Agriculture* 4, 1397-1409. DOI: 10.9734/AJEA/2014/10884

MERESSA, B.H., DEHNE, H.W. & HALLMANN, J. (2015). Molecular and morphological characterisation of *Meloidogyne hapla* populations from Ethiopia. *Russian Journal of Nematology* 23, 1-20.

MERESSA, B.H., DEHNE, H.W. & HALLMANN, J. (2016). Population dynamics and damage potential of *Meloidogyne hapla* to rose rootstock species. *Journal of Phytopathology* 164, 711-721. DOI: 10.1111/jph.12492

MEZA, P., SOTO, B., ROJAS, L. & ESMENJAUD, D. (2016). Identification of *Meloidogyne* Species from the Central Valley of Chile and interaction with stone fruit rootstocks. *Plant Disease* 100, 1358-1363. DOI: 10.1094/PDIS-11-15-1331-RE

MICHELL, R.E., MALEK, R.B., TAYLOR, D.P. & EDWARDS, D.I. (1973). Races of the barley root-knot nematode, *Meloidogyne naasi*. I. Characterisation by host preference. *Journal of Nematology* 5, 41-44.

MOENS, M., PERRY, R.N. & STARR, J.L. (2009). *Meloidogyne* species – a diverse group of novel and important plant parasites. In: Perry, R.N., Moens, M. & Starr, J.L. (Eds). *Root-knot nematodes*. Wallingford, UK, CAB International, pp. 1-17.

MOJTAHEDI, H., SANTO, G.S. & WILSON, J.H. (1988). Host tests to differentiate *Meloidogyne chitwoodi* races 1 and 2 and *M. hapla. Journal of Nematology* 20, 468-473.

MOJTAHEDI, H., SANTO, G.S., BROWN, C.R., FERRIS, H. & WILLIAMSON, V. (1994). A new host race of *Meloidogyne chitwoodi* from California. *Plant Disease* 78, 1010. DOI: 10.1094/PD-78-1010E

MOLINARI, S., LAMBERTI, F., CROZZOLI, R., SHARMA, S.B. & SÁNCHEZ PORTALES, L. (2005). Isozyme patterns of exotic *Meloidogyne* spp. populations. *Nematolologia Mediterranea* 33, 61-65.

MÔNACO, A.P. DO A., CARNEIRO, R.G., SCHERER, A. & SANTIAGO, D.C. (2011). Host suitability of medicinal plants to *Meloidogyne paranaensis. Nematologia Brasileira* 35, 46-49.

MONTEIRO, J.M.S., CARES, J.E., GOMES, A.C.M.M., CORREA, V.R., MATTOS, V.S., SANTOS, M.F.A., ALMEIDA, M.R.A., SANTOS, C.D.G., CASTAGNONE-SERENO, P. & CARNEIRO, R.M.D.G. (2016). First report of, and additional information on, *Meloidogyne konaensis* (Nematoda: Meloidogyninae) parasitizing various crops in Brazil. *Nematology* 18, 831-844. DOI: 10.1163/15685411-00002997

MONTEIRO, J.M.S., CARES, J.E., CORREA, V.R., PINHEIRO, J.B., MATTOS, V.S., SILVA, J.G.P., GOMES, A.C.M.M., SANTOS, M.F.A., CASTAGNONE-SERENO, P. & CARNEIRO, R.M.D.G. (2017). *Meloidogyne brasiliensis* Charchar & Eisenback, 2002 is a junior synonym of *M. ethiopica* Whitehead, 1968. *Nematology* 19, 655-669. DOI: 10.1163/15685411-00003078

MONTEIRO, J.M.S., MATTOS, V.S., SANTOS, M.F.A., GOMES, A.C.M.M., CORREA, V.R., SOUSA, D.A., CARES, J.E., PINHEIRO, J.B. & CARNEIRO, R.M.D.G. (2019). Additional information on *Meloidogyne polycephannulata* and its proposal as a junior synonym of *M. incognita*. *Nematology* 21, 129-146. DOI: 10.1163/15685411-00003202

MOORE, M.R., BRITO, J.A., QIU, S., ROBERTS, C.G. & COMBEE, L.A. (2020). First report of *Meloidogyne enterolobii* infecting Japanese blue berry tree (*Elaeocarpus decipiens*) in Florida, USA. *Journal of Nematology* 52, 1-3.

MOOSAVI, M.R. (2015). Damage of the root-knot nematode *Meloidogyne javanica* to bell pepper, *Capsicum annuum*. *Journal of Plant Diseases and Protection* 122, 244-249. DOI: 10.1007/BF03356559

MORAES DE, M.V., LORDELLO, L.G.E., PICCININ, O.A. & LORDELLO, R.R.A. (1972). Host plants for the coffee root-knot nematode *Meloidogyne exigua*. *Ciência-e-Cultura* 24, 658-660.

MORITZ, C., DOWLING, T.E. & BROWN, W.M. (1987). Evolution of animal mitochondrial DNA: Relevance for population biology and systematics. *Annual Review* of *Ecology and Systematics* 18, 269-292. DOI: 10.1146/annurev.es.18.110187.001413

MORITZ, M.P., CARNEIRO, R.G., SANTIAGO, D.C., NAKAMURA, K.C., PIGNONI, E. & GOMES, J.C. (2008). Estudo comparative da penetracao e reproducao de *Meloidogyne paranaensis* em raizes de cultivares de soja resistente e suscetivel. *Nematologia Brasileira* 32, 33-40.

MORRIS, K., HORGAN, F.G. & GRIFFIN, C.T. (2013). Spatial and temporal dynamics of *Meloidogyne minor* on creeping bentgrass in golf greens. *Plant Pathology* 62, 1166-1172. DOI: 10.1111/ppa.12025

MORRIS, K.S., HORGAN, F.G., DOWNES, M.J. & GRIFFIN, C.T. (2011). The effect of temperature on hatch and activity of second-stage juveniles of the root-knot nematode, *Meloidogyne minor*, an emerging pest in north-west Europe. *Nematology* 13, 985-993. DOI: 10.1163/138855411X571902

MUKHTAR, T., AROOJ, M., ASHFAQ, M. & GULZAR, A. (2017). Resistance evaluation and host status of selected green gram germplasm against *Meloidogyne incognita*. *Crop Protection* 92, 198-202. DOI: 10.1016/j.cropro.2016.10.004

MULK, M. (1976). *Meloidogyne graminicola*. *CIH Descriptions of plant-parasitic nematodes*, Set 6, No. 87. Farnham Royal, UK, Commonwealth Agricultural Bureaux.

MÜLLER, C. (1884). Mittheilungen über die unseren Kulturpflanzen schädlichen, das Geschlecht *Heterodera* bildenden Würmer. *Landwirtschaftliche Jarhbücher, Berlin* 13, 1-42.

MULVEY, R.H. & ANDERSON, R.V. (1980). Description and relationships of a new root-knot nematode, *Meloidogyne sewelli* n. sp. (Nematoda: Meloidogynidae) from Canada and a new host record for the genus. *Canadian Journal of Zoology* 58, 1551-1556. DOI: 10.1139/z80-214

MULVEY, R.H., TOWNSHEND, J.L. & POTTER, J.W. (1975). *Meloidogyne microtyla* sp. nov. from southwestern Ontario, Canada. *Canadian Journal of Zoology* 53, 1528-1536. DOI: 10.1139/z75-188

MUNIZ, M.F.S., CAMPOS, V.P., CASTAGNONE-SERENO, P., CASTRO, J.M.C.C., ALMEIDA, M.R.A. & CARNEIRO, R.M.D.G. (2008). Diversity of *Meloidogyne exigua* (Tylenchida: Meloidogynidae) populations from coffee and rubber tree. *Nematology* 10, 897-910. DOI: 10.1163/156854108786161418

MUNIZ, M. DE F.S., CAMPOS, V.P., ALMEIDA, M.R., GOMES, A.C.M.M., SANTOS, M.F. DOS, CASTRO MOTA, F. DE & CARNEIRO, R.M.D.G. (2009). Additional information on an atypical population of *Meloidogyne exigua* Göldi, 1887 (Tylenchida: Meloidogynidae) parasitising rubber tree in Brazil. *Nematology* 11, 95-106. DOI: 10.1163/156854108X398444

MURGA-GUTIERREZ, S.N., COLAGIERO, M., ROSSO, L.C., FINETTI-SIALER, M.M. & CIANCIO, A. (2012). Root-knot nematodes from asparagus and associated biological antagonists in Peru. *Nematropica* 42, 57-62. DOI: 10.1163/15685411-00003154

MWAMULA, A.O., LEE, G., KIM, Y.H., KIM, Y.H., LEE, K.-S. & LEE, D.W. (2021). Molecular phylogeny of several species of Hoplolaimina (Nematoda: Tylenchida) associated with turfgrass in Korea, with comments on their morphology. *Nematology* 23, 559-576. DOI: 10.1163/15685411-bja10061

NAALDEN, D., VERBEEK, R. & GHEYSEN, G. (2018). *Nicotiana benthamiana* as model plant for *Meloidogyne graminicola* infection. *Nematology* 20, 491-499. DOI: 10.1163/15685411-00003154

NEAL, J.C. (1889). *The root-knot disease of the peach, orange and other plants in Florida, due to the work of* Anguillula. Bulletin 20, Division of Entomology, US Department of Agriculture.

NEGRETTI, R.R.D., MANICA-BERTO, R., AGOSTINETTO, D., THURMER, L. & GOMES, C.B. (2014). Host suitability of weeds and forage species to root-knot nematode *Meloidogyne graminicola* as a function of irrigation management. *Planta Daninha* 32, 555-561. DOI: 10.1590/S0100-83582014000300011

NEGRETTI, R.R.D., GOMES, C.B., MATTOS, V.S., SOMAVILLA, L., MANICA-BERTO, R., AGOSTINETTO, D., CASTAGNONE-SERENO, P. & CARNEIRO, R.M.D.G. (2017). Characterisation of a *Meloidogyne* species complex parasitising rice in southern Brazil. *Nematology* 19, 403-412. DOI: 10.1163/15685411-00003056

NIBLACK, T., HUSSEY, R.S. & BOERMA, H.R. (1986). Effects of environments, *Meloidogyne incognita* inoculum levels, and *Glycine max* cultivar on root-knot nematode-soybean interactions in field microplots. *Journal of Nematology* 18, 338-346.

NICOL, J.M., TURNER, S.J., COYNE, D.L., DEN NIJS, L., HOCKLAND, S. & TANHA MAAFI, Z. (2011). Current nematode threats to world agriculture. In: Jones, J., Gheysen, G. & Fenoll, C. (Eds). *Genomics and molecular genetics of plant-nematode interactions*. Springer, Dordrecht, The Netherlands, pp. 21-43. DOI: 10.1007/978-94-007-0434-3_2

NISCHWITZ, C., SKANTAR, A., HANDOO, Z.A., HULT, M.N., SCHMITT, M.E. & MCCLURE, M.A. (2013). Occurrence of *Meloidogyne fallax* in North America, and molecular characterization of *M. fallax* and *M. minor* from U.S. golf course greens. *Plant Disease* 97, 1424-1430. DOI: 10.1094/PDIS-03-13-0263-RE

NIU, J.-H., GUO, Q.-X., JIAN, H., CHEN, C.L., YANG, D., LIU, Q. & GUO, Y.-D. (2011). Rapid detection of *Meloidogyne* spp. by LAMP assay in soil and roots. *Crop Protection* 30, 1063-1069. DOI: 10.1016/j.cropro.2011.03.028

NIU, J.-H., JIAN, H., GUO, Q.-X., CHEN, C.L., WANG, X.Y., LIU, Q. & GUO, Y.-D. (2012). Evaluation of loop-mediated isothermal amplification (LAMP) assays based on 5S rDNA-IGS2 regions for detecting *Meloidogyne enterolobii*. *Plant Pathology* 61, 809-819. DOI: 10.1111/j.1365-3059.2011.02562.x

NOBBS, J.M., LIU, Q., HARTLEY, D., HANDOO, Z., WILLIAMSON, V.M., TAYLOR, S., WALKER, G. & CURRAN, J. (2001). First record of *Meloidogyne fallax* in Australia. *Australasian Plant Pathology* 30, 373. DOI: 10.1071/AP01060

NOE, J.P. & BARKER, K.R. (1985). Relation of within-field spatial variation of plant-parasitic nematode population densities and edaphic factors. *Phytopathology* 75, 247-252. DOI: 10.1094/Phyto-75-247

NOTOMI, T., OKAYAMA, H., MASUBUCHI, H., YONEKAWA, T., WATANABE, K., AMINO, N. & HASE, T. (2000). Loop-mediated isothermal amplification of DNA. *Nucleic Acids Research* 28, E63. DOI: 10.1093/nar/28.12.e63

NUGALIYADDE, L., DISSANAYAKE, D.M.N., HERATH, H.M.D.N., DHARMASENA, C.M.D., JAYASUNDERA, D.M., PREMACHANDRA, M.M., DASSANAYAKE, D.M.M., EMITIYAGODA, G.A.M.S.S., AMARASINGHE, A.A.L. & EKANAYAKE, H.M.R.K. (2001). Outbreak of rice root knot nematode, *Meloidogyne graminicola* (Golden and Birchfield) in Nikewaratiya, Kurunegala in *maha* 2000/2001. *Annals of the Sri Lanka Department of Agriculture* 3, 373-374.

NUNN, G.B. (1992). *Nematode molecular evolution*. Ph.D. Thesis, University of Nottingham, Nottingham, UK.

NYCZEPIR, A.P. & BECKER, J.O. (1998). Fruit and citrus trees. In: Barker, K.R., Pederson, G.A. & Windham, G.L. (Eds). *Plant nematode interactions, Agronomy Monograph 36*. Madison, WI, USA, American Society of Agronomy-Crop Science Society of America-Soil Science Society of America, pp. 637-684.

NYCZEPIR, A.P. & HALBRENDT, J.M. (1993). Nematode pests of deciduous fruit and nut trees. In: Evans, K., Trudgill, D.L. & Webster, J.M. (Eds). *Plant parasitic nematodes in temperate agriculture*. Wallingford, UK, CAB International, pp. 381-425.

NYCZEPIR, A.P., INSERRA, R.N., O'BANNON, J.H. & SANTO, G.S. (1984). Influence of *Meloidogyne chitwoodi* and *M. hapla* on wheat growth. *Journal of Nematology* 16, 162-165.

NYOIKE, T.W., MEKETE, T., MCSORLEY, R., WEIBELZAHL-KARIGI, E. & LIBURD, O.E. (2012). Identification of the root-knot nematode, *Meloidogyne hapla*, on strawberry in Florida using morphological and molecular methods. *Nematropica* 42, 253-259.

O'BANNON, J.H., SANTO, G.S. & NYCZEPIR, A.P. (1982). Host range of the Columbia root-knot nematode. *Plant Disease* 66, 1045-1048. DOI: 10.1094/PD-66-1045

OKA, Y., KARSSEN, G. & MOR, M. (2003). Identification, host range and infection process of *Meloidogyne marylandi* from turf grass in Israel. *Nematology* 5, 727-734. DOI: 10.1163/156854103322746904

OKA, Y., KARSSEN, G. & MOR, M. (2004). First report of the root knot nematode *Meloidogyne marylandi* on turfgrasses in Israel. *Plant Disease* 88, 309. DOI: 10.1094/PDIS.2004.88.3.309B

OKAMOTO, K. & YAEGASHI, T. (1981a). [Observation of six *Metoidogyne* species by scanning electron microscope. 1. *En face* views of second-stage larvae.] *Japanese Journal of Nematology* 10, 35-42.

OKAMOTO, K. & YAEGASHI, T. (1981b). [Observation on six *Meloidogyne* species by scanning electron microscopy. 2. *En face* views of males.] *Japanese Journal of Nematology* 10, 43-51.

OKAMOTO, K., YAEGASHI, T. & TOIDA, Y. (1983). [Morphological differences among some populations of *Meloidogyne mali* from apple and mulberry.] *Japanese Journal of Nematology* 12, 26-32.

OLIVEIRA, D.S., OLIVEIRA, R.D.L., FREITAS, L.G. & SILVA, R.V. (2005). Variability of *Meloidogyne exigua* on coffee in the Zona da Mata of Minas Gerais State, Brazil. *Journal of Nematology* 37, 323-327.

OLIVEIRA, S.A., OLIVEIRA, C.M.G., MALEITA, C.M.N., SILVA, M.F.A., ABRANTES, I.M.O. & WILCKEN, S.R.S. (2018). First report of *Meloidogyne graminis* on golf courses turfgrass in Brazil. *PLoS ONE* 13, e0192397. DOI: 10.1371/journal.pone.0192397

OLTHOF, T.H.A. (1986). Damage to an apple orchard cover crop of creeping red fescue (*Festuca rubra*) associated with *Meloidogyne microtyla. Plant Disease* 70, 436-438. DOI: 10.1094/PD-70-436

OLTHOF, T.H.A. & POTTER, J.H. (1972). Relationship between population densities of *Meloidogyne hapla* and crop losses in summer maturing vegetables in Ontario. *Phytopathology* 62, 981-986. DOI: 10.1094/Phyto-62-981

ONKENDI, E.M. & MOLELEKI, L.N. (2013). Detection of *Meloidogyne enterolobii* in potatoes in South Africa and phylogenetic analysis based on intergenic region and the mitochondrial DNA sequences. *European Journal of Plant Pathology* 136, 1-5. DOI: 10.1007/s10658-012-0142-y

ONKENDI, E.M., KARIUKI, G.M., MARAIS, M. & MOLELEKI, L.N. (2014). The threat of root-knot nematodes (*Meloidogyne* spp.) in Africa: a review. *Plant Pathology* 63, 727-737. DOI: 10.1111/ppa.12202

OPPERMAN, C.H., BIRD, D., WILLIAMSON, V.M., ROKHSAR, D.S., BURKE, M., COHN, J., CROMER, J., DIENER, S., GAJAN, J., GRAHAM, S. ET AL. (2008). Sequence and genetic map of *Meloidogyne hapla*: A compact nematode genome for plant parasitism. *Proceedings of the National Academy of Sciences of the United States of America* 105, 14802-14807. DOI: 10.1073/pnas.0805946105

ORION, D. & COHN, E. (1975). A resistant response of *Citrus* roots to the root-knot nematode *Meloidogyne javanica. Marcellia* 38, 327-328.

ORTON WILLIAMS, K.J. (1972). *Meloidogyne javanica. CIH Descriptions of plant-parasitic nematodes.* Set 1, No. 3. Farnham Royal, UK, Commonwealth Agricultural Bureaux.

ORTON WILLIAMS, K.J. (1973). *Meloidogyne incognita. CIH Descriptions of plant-parasitic nematodes.* Set 2, No. 18. Farnham Royal, UK, Commonwealth Agricultural Bureaux.

ORTON WILLIAMS, K.J. (1974). *Meloidogyne hapla. CIH Descriptions of plant-parasitic nematodes.* Set 3, No. 31. Farnham Royal, UK, Commonwealth Agricultural Bureaux.

ORTON WILLIAMS, K.J. (1975). *Meloidogyne arenaria. CIH Descriptions of plant-parasitic nematodes.* Set 5, No. 62. Farnham Royal, UK, Commonwealth Agricultural Bureaux.

ORUI, Y. (1998). Identification of Japanese species of the genus *Meloidogyne* (Nematoda: Meloidogynidae) by PCR-RFLP analysis. *Applied Entomology and Zoology* 33, 43-51. DOI: 10.1303/aez.33.43

OSMAN, H.A., DICKSON, D.W. & SMART JR, G.C. (1985). Morphological comparisons of host races 1 and 2 of *Meloidogyne arenaria* from Florida. *Journal of Nematology* 17, 279-285.

PADGHAM, J.L., ABAWI, G.S., DUXBURY, J.M. & MAZID, M.A. (2004a). Impact of wheat on *Meloidogyne graminicola* populations in the rice-wheat system of Bangladesh. *Nematropica* 34, 183-190.

PADGHAM, J.L., DUXBURY, J.M., MAZID, A.M., ABAWI, G.S. & HOSSAIN, M. (2004b). Yield loss caused by *Meloidogyne graminicola* on lowland rainfed rice in Bangladesh. *Journal of Nematology* 36, 42-48.

PAGAN, C., COYNE, D., CARNEIRO, R., KARIUKI, G., LUAMBANO, N., AFFOKPON, A. & WILLIAMSON, V.M. (2015). Mitochondrial haplotype-based identification of ethanol-preserved root-knot nematodes from Africa. *Phytopathology* 105, 350-357. DOI: 10.1094/PHYTO-08-14-0225-R

PAGE, S.L.J. (1983). *Biological studies of the African cotton root nematode* Meloidogyne acronea. Ph.D. Thesis. University of London, London, UK.

PAGE, S.L.J. (1985). *Meloidogyne acronea. CIH Descriptions of plant-parasitic nematodes*, Set 8, No. 114. Farnham Royal, UK, Commonwealth Agricultural Bureaux.

PAGE, S.L.J. & BRIDGE, J. (1994). The African cotton-root nematode, *Meloidogyne acronea* – its pathogenicity and intra-generic infectivity with *Gossypium. Fundamental and Applied Nematology* 17, 67-73.

PAIS, C.S. & ABRANTES, I.M. DE O. (1989). Esterase and malate dehydrogenase phenotypes in Portuguese populations of *Meloidogyne* species. *Journal of Nematology* 21, 342-346.

PALMISANO, A.M. & AMBROGIONI, L. (2000). *Meloidogyne ulmi* sp. n., a root-knot nematode from elm. *Nematologia Mediterranea* 28, 279-293.

PALOMARES-RIUS, J.E., VOVLAS, N., TROCCOLI, A., LIEBANAS, G., LANDA, B.B. & CASTILLO, P.A. (2007). New root-knot nematode parasitizing sea rocket from Spanish Mediterranean coastal dunes: *Meloidogyne dunensis* n. sp. (Nematoda: Meloidogynidae). *Journal of Nematology* 39, 190-202.

PALOMARES-RIUS, J.E., CANTALAPIEDRA-NAVARRETE, C., ARCHIDONA-YUSTE, A., BLOK, V.C. & CASTILLO, P. (2017a). Mitochondrial genome diversity in dagger and needle nematodes (Nematoda: Longidoridae). *Scientific Reports* 7, 41813. DOI: 10.1038/srep41813

PALOMARES-RIUS, J.E., ESCOBAR, C., CABRERA, J., VOVLAS, A. & CASTILLO, P. (2017b). Anatomical alterations in plant tissues induced by plant-parasitic nematodes. *Frontiers in Plant Science* 8, 1987. DOI: 10.3389/fpls.2017.01987

PAN, C., HU, X. & LIN, J. (1999). Temporal fluctuations in *Meloidogyne fujianensis* parasitizing *Citrus reticulata* in Nanjing, China. *Nematologia Mediterranea* 27, 327-330.

PAN, C.S. (1985). [Studies on plant-parasitic nematodes on economically important crops in Fujian. III. Description of *Meloidogyne fujianensis* n. sp. (Nematoda: Meloidogynidae) infesting citrus in Nanjing County.] *Acta Zoologica Sinica* 31, 263-268.

PAN, C.S. & LIN, J. (1998). [A new record of host plants of *Meloidogyne fujianensis* and their observation by scanning electron microscope.] *Acta Parasitologica et Medica Entomologica Sinica* 5, 125-126.

PAN, C.S., LIN, J. & WANG, S.G. (1988). [Studies on plant-parasitic nematodes on economically important crops in Fujian. IV. *Meloidogyne fujianensis*.] *Acta Zoologica Sinica* 34, 305-308.

PAN, C.S., LIN, J., WANG, S.G. & WU, S.Z. (1994). Further studies on species of root-knot nematodes and their host-plants. *Wui Science Journal* 11, 149-157.

PANAHI, P. & BAROOTI, S. (2015). First report of northern root-knot nematode, *Meloidogyne hapla*, parasitic on oaks, *Quercus brantii* and *Q. infectoria* in Iran. *Journal of Nematology* 47, 86.

PANG, W., HAFEZ, S.L. & SUNDARARAJ, P. (2009). Pathogenicity of *Meloidogyne hapla* on onion. *Nematropica* 39, 225-233.

PANKAJ, S.H.K. & PRASAD, J.S. (2010). The rice root-knot nematode *Meloidogyne graminicola*: an emerging nematode pest of rice-wheat cropping system. *Indian Journal of Nematology* 40, 1-11.

PATEL, D.J., PATEL, B.A., PATEL, S.K., PATEL, R.L. & PATEL, R.G. (1999). Root-knot nematode, *Meloidogyne indica* on kagzi lime in north Gujarat. *Indian Journal of Nematology* 29, 197.

PATEL, H.R., PATEL, R.G., PATEL, B.A., VYAS, R.V., PATEL, J.G. & PATEL, D.J. (2003). Biodiversity of *Meloidogyne indica* – a key pest of kagzi lime makes castor (*Ricinus communis* L.) vulnerable. *Indian Journal of Nematology* 33, 174-176.

PATNAIK, N.C. & PADHI, N.N. (1987). Damage by rice root-knot nematode. *International Rice Research Newsletter* 12, 27.

PEDROCHE, N.B., VILLANUEVA, L.M. & DE WAELE, D. (2013). Plant-parasitic nematodes associated with semi-temperate vegetables in the highlands of Benguet Province, Philippines. *Archives of Phytopathology and Plant Protection* 46, 278-294. DOI: 10.1080/03235408.2012.739928

PENG, H., LONG, H., HUANG, W., LIU, J., CUI, J., KONG, L., HU, X., GU, J. & PENG, D. (2017). Rapid, simple and direct detection of *Meloidogyne hapla* from infected root galls using loop-mediated isothermal amplification combined with FTA technology. *Scientific Reports* 7, 44853. DOI: 10.1038/srep44853

PERICHI, G., CROZZOLI, R. & ALCANO, M. (2006). First report of *Meloidogyne graminis* (Nematoda: Tylenchida) in Venezuela. *Fitopatologia Venezolana* 19, 17-18.

PERRY, R.N. & MOENS, M. (2011). Introduction to plant-parasitic nematodes; modes of parasitism. In: Jones, J., Gheysen, G. & Fenoll, C. (Eds). *Genomics and molecular genetics of plant-nematode interactions*. Dordrecht, The Netherlands, Springer, pp. 3-20. DOI: 10.1007/978-94-007-0434-3_1

PERRY, R.N. & MOENS, M. (Eds) (2013). *Plant nematology*, 2nd edition. Wallingford, UK CABI Publishing.

PERRY, R.N., MOENS, M. & STARR, J.L. (Eds) (2009). *Root-knot nematodes*. Wallingford, UK, CABI.

PERSON-DEDRYVER, F. (1986). Incidence du nématode à galle *Meloidogyne naasi* en cultures céréalières, méthodes de lutte. In: *Les rotations céréalières intensives. Dix années d'études concertées INRA-ONIG-ITCF, 1973-1983*. Paris, France, INRA, pp. 175-187.

PERSON-DEDRYVER, F. & FISCHER, J. (1987). Grasses as hosts of *Meloidogyne naasi* Franklin: I. Variation in host status of species and varieties grown in France. *Nematologica* 33, 61-71. DOI: 10.1163/187529287X00218

PESSIA, E., POPA, A., MOUSSET, S., REZVOY, C., DURET, L. & MARAIS, G.A. (2012). Evidence for widespread GC-biased gene conversion in eukaryotes. *Genome Biology and Evolution* 4, 675-682. DOI: 10.1093/gbe/evs052

PETERSEN, D.J., ZIJLSTRA, C., WISHART, J., BLOK, V. & VRAIN, T.C. (1997). Specific probes efficiently distinguish root-knot nematode species using signature sequences in the ribosomal intergenic spacer. *Fundamental and Applied Nematology* 20, 619-626.

PHAM, T.B. (1990). [Root knot nematodes of vegetables and potatoes in Dalat (the Tei Nguen Plateau, Vietnam) and a description of *Meloidogyne cynariensis* – an artichoke parasite.] *Zoologicheskii Zhurnal* 69, 128-131.

PHAN, N.T., ORJUELA, J., DANCHIN, E.G.J., KLOPP, C., PERFUS-BARBEOCH, L., KOZLOWSKI, D.K., KOUTSOVOULOS, G.D., LOPEZ-ROQUES, C., BOUCHEZ, O., ZAHM, M. ET AL. (2020). Genome structure and content of the rice root-knot nematode (*Meloidogyne graminicola*). *Ecology and Evolution* 10, 1106-11021. DOI: 10.1002/ece3.6680

PHANI, V., BISHNOI, S., SHARMA, A., DAVIES, K.G. & RAO, U. (2018). Characterization of *Meloidogyne indica* (Nematoda: Meloidogynidae) parasitizing neem in India, with a molecular phylogeny of the species. *Journal of Nematology* 50, 387-398.

PIEPENBURG, O., WILLIAMS, C.H., STEMPLE, D.L. & ARMES, N.A. (2006). DNA detection using recombination proteins. *PLoS Biology* 4, e204. DOI: 10.1371/journal.pbio.0040204

PIGANEAU, G., GARDNER, M. & EYRE-WALKER, A. (2004). A broad survey of recombination in animal mitochondria. *Molecular Biology and Evolution* 21, 2319-2325. DOI: 10.1093/molbev/msh244

PLANTARD, O., VALETTE, S. & GROSS, M.F. (2007). The root-knot nematode producing galls on *Spartina alterniflora* belongs to the genus *Meloidogyne*: rejection of *Hypsoperine* and *Spartonema* spp. *Journal of Nematology* 39, 127-132.

PLOEG, A.T. & PHILLIPS, M.S. (2001). Damage to melon (*Cucumis melo* L.) cv. Durango by *Meloidogyne incognita* in Southern California. *Nematology* 3, 151-157. DOI: 10.1163/156854101750236277

PLOWRIGHT, R. & BRIDGE, J. (1990). Effects of *Meloidogyne graminicola* (Nematoda) on the establishment; growth and yield of rice cv. IR36. *Nematologica* 36, 81-89. DOI: 10.1163/002925990X00059

POGHOSSIAN, E.E. (1961). [A root-knot nematode new for the USSR from Armenia.] *Izvestiya Akademii Nauk Armyanskoı˘ SSR, Biologicheskie i Sel'skokhozyaı˘stvennye Nauki* 14, 95-97.

POGHOSSIAN, E.E. (1971). [New species of root knot nematode *Hypsoperine megriensis* n. sp. (Nematoda, Heteroderidae) from Armenia SSR.] *Doklady Akademii Nauk Armjanskoi SSR*, 306-312.

POKHAREL, R.R. (2009). Damage of root-knot nematode (*Meloidogyne graminicola*) to rice in fields with different soil types. *Nematologia Mediterranea* 37, 203-217.

POKHAREL, R.R., ABAWI, G.S., DUXBURY, J.M., SMAT, C.D., WANG, X. & BRITO, J.A. (2010). Variability and the recognition of two races in *Meloidogyne graminicola*. *Australasian Plant Pathology* 39, 326-333. DOI: 10.1071/AP09100

POORNIMA, K., SURESH, P., KALAIARASAN, P., SUBRAMANIAN, S. & RAMARAJU, K. (2016). Root knot nematode, *Meloidogyne enterolobii* in guava (*Psidium guajava* L.) a new record from India. *Madras Agricultural Journal* 103. 359-365.

POTTER, J.W. & OLTHOF, T.H.A. (1977). Effects of population densities of *Meloidogyne hapla* on growth and yield of tomato. *Journal of Nematology* 9, 296-300.

POWERS, T.O. & HARRIS, T.S. (1993). A polymerase chain reaction for the identification of five major *Meloidogyne* species. *Journal of Nematology* 25, 1-6.

POWERS, T.O., MULLIN, P.G., HARRIS, T.S., SUTTON, L.A. & HIGGINS, R.S. (2005). Incorporating molecular identification of *Meloidogyne* spp. into a large-scale regional nematode survey. *Journal of Nematology* 37, 226-235.

POWERS, T.O., HARRIS, T.S., HIGGINS, R.S., MULLIN, P.G. & POWERS, K. (2018). Discovery and identification of *Meloidogyne* species using *COI* DNA barcoding. *Journal of Nematology* 50, 399-412. DOI: 10.21307/jofnem-2018-029

PRASAD, J.S., PANWAR, M.S. & RAO, Y.S. (1986). Screening of some rice cultivars against the root-knot nematode, *Meloidogyne graminicola*. *Indian Journal of Nematology* 16, 112-113.

PRIOR, T., TOZER, H., YALE, R., JONES, E.P., LAWSON, R., JUTSON, L., CORREIA, M., STUBBS, J., HOCKLAND, S. & KARSSEN, G. (2019). First report of *Meloidogyne mali* causing root galling to elm trees in the UK. *New Disease Reports* 39, 10. DOI: 10.5197/j.2044-0588.2019.039.010

PROT, J.-C. & MATIAS, D.M. (1995). Effects of water regime on the distribution of *Meloidogyne graminicola* and other root-parasitic nematodes in a rice field toposequence and pathogenicity of *M. graminicola* on rice cultivar UPL Ri5. *Nematologica* 41, 219-228. DOI: 10.1163/003925995X00189

PROT, J.-C., VILLAMMEVA, L.M. & GERGON, E.B. (1994). The potential of increased nitrogen supply to mitigate growth and yield reductions of upland rice cultivars UPL Ri-5 caused by *Meloidogyne graminicola*. *Fundamental and Applied Nematology* 17, 445-454.

QIU, J.J., WESTERDAHL, B.B., ANDERSON, C. & WILLIAMSON, V.M. (2006). Sensitive PCR detection of *Meloidogyne arenaria*, *M. incognita*, and *M. javanica* extracted from soil. *Journal of Nematology* 38, 434-441.

QUADER, M., RILEY, I.T. & WALKER, G.E. (2002). Damage threshold of *Meloidogyne incognita* for the establishment of grapevines. *International Journal of Nematology* 12, 125-130.

QUÉNÉHERVÉ, P., GODEFROID, M., MEGE, P. & MARIE-LUCE, S. (2011). Diversity of *Meloidogyne* spp. parasitizing plants in Martinique Island, French West Indies. *Nematropica* 41, 191-199.

RADEWALD, J.D., PYEATT, L., SHIBUYA, F. & HUMPHREY, W. (1970). *Meloidogyne naasi*, a parasite of turfgrass in southern California. *Plant Disease Reporter* 54, 940-942.

RAMÍREZ-SUÁREZ, A., ROSAS-HERNÁNDEZ, L. & ALCASIO-RANGE, S. (2014). First report of the root-knot nematode *Meloidogyne enterolobii*, parasitizing watermelon from Veracruz, Mexico. *Plant Disease* 98, 428. DOI: 10.1094/PDIS-06-13-0636-PDN

RAMÍREZ-SUÁREZ, A., ALCASIO, S., ROSAS-HERNÁNDEZ, L., LÓPEZ-BUENFIL, J.A. & BRITO, J.A. (2016). First report of *Meloidogyne enterolobii* infecting columnar cacti *Stenocereus queretaroensis* in Jalisco, Mexico. *Plant Disease* 100, 1506. DOI: 10.1094/PDIS-11-15-1272-PDN

RAMMAH, A. & HIRSCHMANN, H. (1988). *Meloidogyne mayaguensis* n. sp. (Meloidogynidae), a root-knot nematode from Puerto Rico. *Journal of Nematology* 20, 58-69.

RAMMAH, A. & HIRSCHMANN, H. (1990a). Morphological comparison of three host races of *Meloidogyne javanica*. *Journal of Nematology* 22, 56-68.

RAMMAH, A. & HIRSCHMANN, H. (1990b). *Meloidogyne morocciensis* n. sp. (Meloidogyninae), a root-knot nematode from Morocco. *Journal of Nematology* 22, 279-291.

RANDIG, O., BONGIOVANNI, M., CARNEIRO, R.M.D.G. & CASTAGNONE-SERENO, P. (2002a). Genetic diversity of root-knot nematodes from Brazil and development of SCAR markers specific for the coffee-damaging species. *Genome* 45, 862-870. DOI: 10.1139/g02-054

RANDIG, O., BONGIOVANNI, M., CARNEIRO, R.M.D.G., SARAH, J.-L. & CASTAGNONE-SERENO, P. (2002b). A species-specific satellite DNA family in the genome of the coffee root-knot nematode *Meloidogyne exigua*: application to molecular diagnostics of the parasite. *Molecular Plant Pathology* 3, 431-437. DOI: 10.1046/j.1364-3703.2002.00134.x

RANDIG, O., CARNEIRO, R.M.D.G. & CASTAGNONE-SERENO, P. (2004). Identificação das principais espécies de *Meloidogyne* parasitas do cafeeiro no Brasil com marcadores SCAR-café em multiplex PCR. *Nematologia Brasileira* 28, 1-10.

RAO, G.N. (1970). Tea pests in southern India and their control. *Pest Article & News Summaries* 16, 667-672.

RAO, Y.S., JENA, R.N. & PRASAD, K.S.K. (1977). Infectivity of two isolates of *Meloidogyne graminicola* in rice. *Indian Journal of Nematology* 7, 98-99.

RAU, G.J. & FASSULIOTIS, G. (1965). *Hypsoperine spartinae* sp. n., a gall-forming nematode on the roots of smooth cordgrass. *Proceedings of the Helminthological Society of Washington* 32, 159-162.

RAVICHANDRA, N.G. (2019). New report of root-knot nematode (*Meloidogyne enterolobii*) on guava from Karnataka, India. *EC Agriculture* 5.9, 504-506.

REIGHARD, G.L., HENDERSON, W.G., SCOTT, S.O. & SUBBOTIN, S.A. (2019). First report of the root-knot nematode, *Meloidogyne floridensis* infecting Guardian® peach rootstock in South Carolina, USA. *Journal of Nematology* 51, e2019-61. DOI: 10.21307/jofnem-2019-061

RENCO, M. & MURIN, J. (2013). Soil nematode assemblages in natural European peatlands of the Horna Orava protected landscape area, Slovakia. *Wetlands* 33, 459-470. DOI: 10.1007/s13157-013-0403-3

RICH, J.R., BRITO, J.A., KAUR, R. & FERRELL, J.A. (2008). Weed species as hosts of *Meloidogyne*: a review. *Nematropica* 39, 157-185.

RIFFLE, J. & KUNTZ, J. (1967). Pathogenicity and host range of *Meloidogyne ovalis*. *Phytopathology* 57, 104.

RIFFLE, J.W. (1963). *Meloidogyne ovalis* (Nematoda: Heteroderidae), a new species of root-knot nematode. *Proceedings of the Helminthological Society of Washington* 30, 287-292.

RIVOAL, R. & COOK, R. (1993). Nematode pest of cereals. In: Evans, K., Trudgill, D. & Webster, J.L. (Eds). *Plant parasitic nematodes in temperate agriculture*. Wallingford, UK, CAB International, pp. 259-303.

ROBERTS, P.A. (1995). Conceptual and practical aspects of variability in root-knot nematodes related to host plant resistance. *Annual Review of Phytopathology* 33, 199-221. DOI: 10.1146/annurev.py.33.090195.001215

ROBERTSON, L., DÍEZ-ROJO, M., LÓPEZ-PÉREZ, J.A., PIEDRA BUENA, A., ESCUER, M., LÓPEZ CEPERO, J., MARTÍNEZ, C. & BELLO, A. (2009). New host races of *Meloidogyne arenaria*, *M. incognita* and *M. javanica* from horticultural regions of Spain. *Plant Disease* 93, 180-184. DOI: 10.1094/PDIS-93-2-0180

Rodiuc, N., Vieira, P., Banora, M.Y. & Engler, J. de A. (2014). On the track of transfer cell formation by specialized plant-parasitic nematodes. *Frontiers in Plant Science*. May 5;5:160. DOI: 10.3389/fpls.2014.00160

Rodrigues, A., Abrantes, I.M.D., Melillo, M.T. & Bleve-Zacheo, T. (2000). Ultrastructural response of coffee roots to root-knot nematodes, *Meloidogyne exigua* and *M. megadora*. *Nematropica* 30, 201-210.

Rodrigues, H.C.S., Borges, C.T., Navroski, R., Soares, V.N., Gadotti, G.I. & Meneghello, G.E. (2017). Effect of chemical treatment on physiological quality of seed and control of *Meloidogyne javanica* in watermelon plants. *Australian Journal of Crop Science* 11, 18-24. DOI: 10.21475/ajcs.2017.11.01.pne165

Rodríguez-Kabana, R. & Robertson, D.G. (1987). Vertical distribution of *Meloidogyne arenaria* juvenile populations in a peanut field. *Nematropica* 17, 199-208.

Rodríguez-Kabana, R., Ivey, H. & Backman, P.A. (1987). Peanut-cotton rotations for the management of *Meloidogyne arenaria*. *Journal of Nematology* 19, 484-486.

Roese, A.D. & Oliveira, R.D.L. (2004). Capacidade reprodutiva de *Meloidogyne paranaensis* em especies de plantas daninhas. *Nematologia Brasileira* 28, 137-141.

Rohan, T.C., Aalders, L.T., Bell, N.L. & Shah, F.A. (2016). First report of *Melodogyne fallax* hosted by *Trifolium repens* (white clover): implications for pasture and crop rotations in New Zealand. *Australasian Plant Disease Notes* 11, 14. DOI: 10.1007/s13314-016-0201-x

Rohini, K., Ekanayake, H.M. & Di Vito, M. (1986). Life cycle and multiplication of *Meloidogyne incognita* on tomato and eggplant seedlings. *Tropical Agriculturist* 142, 59-68.

Roldi, M., Dias-Arieira, C.R., da Silva, S., Dorigo, O.F. & Machado, A.C.Z. (2017). Control of *Meloidogyne paranaensis* in coffee plants mediated by silicon. *Nematology* 19, 245-250. DOI: 10.1163/15685411-00003044

Roy, A.K. (1973). Reaction of some rice cultivars to the attack of *Meloidogyne graminicola*. *Indian Journal of Nematology* 3, 72-73.

Rubinoff, D. & Holland, B. (2005). Between two extremes: mitochondrial DNA is neither the panacea nor the nemesis of phylogenetic and taxonomic inference. *Systematic Biology* 54, 952-961. DOI: 10.1080/10635150500234674

Rusinque, L., Nóbrega, F., Cordeiro, L., Serra, C. & Inácio, M.L. (2021). First detection of *Meloidogyne luci* (Nematoda: Meloidogynidae) parasitizing potato in the Azores, Portugal. *Plants* 10, 99. DOI: 10.3390/plants10010099

RUTTER, W.B., SKANTAR, A.M., HANDOO, Z.A., MUELLER, J.D., AULTMAN, S.P. & AGUDELO, P. (2019). *Meloidogyne enterolobii* found infecting root-knot nematode resistant sweetpotato in South Carolina, United States. *Plant Disease* 103, 775. DOI: 10.1094/PDIS-08-18-1388-PDN

RYBARCZYK-MYDŁOWSKA, K., VAN MEGEN, H., VAN DEN ELSEN, S., MOOYMAN, P., KARSSEN, G., BAKKER, J. & HELDER, J. (2014). Both SSU rDNA and RNA polymerase II data recognise that root knot nematodes arose from migratory Pratylenchidae, but probably not from one of the economically high-impact lesion nematodes. *Nematology* 16, 125-136. DOI: 10.1163/15685411-00002750

RYSS, A.Y. (1988). *Parasitic root nematodes of the family Pratylenchidae (Tylenchida) of the world fauna*. Leningrad, USSR, Nauka.

SABIR, N. & GAUR, H.S. (2005). Comparison of host preferences of *Meloidogyne triticoryzae* and four Indian populations of *M. graminicola*. *International Journal of Nematology* 15, 230-237.

SAHOO, N.K. & GANGULY, S. (2000). Morphological characterisation of five Indian populations of root-knot nematode *Meloidogyne javanica* (Treub, 1885) Chitwood, 1949. *Indian Journal of Nematology* 30, 71-85.

SAHOO, N.K., GANGULY, S. & EAPEN, S.J. (2000). Description of *Meloidogyne piperi* sp. n. (Nematoda: Meloidogynidae) isolated from the roots of *Piper nigrum* in South India. *Indian Journal of Nematology* 30, 203-209.

SALALIA, R., WALIA, R.K., SOMVANSHI, V.S., KUMAR, P. & KUMAR, A. (2017). Morphological, morphometric, and molecular characterization of intraspecific variations within Indian populations of *Meloidogyne graminicola*. *Journal of Nematology* 49, 254-267.

SALAZAR, L., GÓMEZ, M., FLORES, L. & GÓMEZ-ALPÍZAR, L. (2013). First report of *Meloidogyne marylandi* infecting bermudagrass in Costa Rica. *Plant Disease* 97, 1005. DOI: 10.1094/PDIS-11-12-1079-PDN

SANCHEZ, W.L., CROW, W.T., HABTEWELD, A. & MENDES, M.L. (2018). First report of *Meloidogyne graminis* infecting limpograss (*Hemarthria altissima*). *Journal of Nematology* 50, 655.

SANCHO, C.L. & SALAZAR, L. (1985). Nematodos parasitos del arroz (*Oryza sativa* L.) en el sureste de Costa Rica. *Agronomia Costarricense* 9, 161-163.

SANCHO, C.L., SALAZAR, L. & LÓPEZ, R. (1987). Efecto de la densidad inicial del inóculo sobre la patogenicidad de *Meloidogyne salasi* en tres cultivares de arroz. *Agronomia Costarricense* 11, 233-238.

SANDGROUND, J. (1923). "*Oxyuris incognita*" or *Heterodera radicicola*? *Journal of Parasitology* 10, 92-94.

SANTIAGO, D.C., KRZYZANOWSKI, A.A. & HOMOCHIN, M. (2000). Behavior of *Ilex paraguariensis* st Hilaire, 1822 to *Meloidogyne incognita* and *M. paranaensis* and their influence on development of plantlets. *Brazilian*

Archives of Biology and Technology 43, 139-142. DOI: 10.1590/S1516-89132000000200001

SANTO, G.S. (1994). Biology and management of root-knot nematodes on potato in the Pacific Northwest. In: Zehner, G.W., Powelson, M.L., Jansson, R.K. & Raman, K.V. (Eds). *Advances in potato pest biology and management*. St Paul, MN, USA, APS Press, pp. 193-201.

SANTO, G.S. & O'BANNON, J.H. (1981). Pathogenicity of the Columbia root-knot nematode (*Meloidogyne chitwoodi*) on wheat, corn, oat and barley. *Journal of Nematology* 13, 548-550.

SANTO, G.S. & PINKERTON, J.N. (1985). A second host race of *Meloidogyne chitwoodi* discovered in Washington. *Plant Disease* 69, 631.

SANTOS, B.B. (1988). Nematóides do gênero *Meloidogyne* Goeldi e algumas plantas hospedeiras do Estado do Paraná. *Revista de Agricultura* 63, 37-43.

SANTOS, D., CORREIA, A., ABRANTES, I. & MALEITA, C. (2019). New host and records in Portugal for the root-knot nematode *Meloidogyne luci*. *Journal of Nematology* 51, 1-4. DOI: 10.21307/jofnem-2019-003

SANTOS, D., SILVA, P.M. DA, ABRANTES, I. & MALEITA, C. (2020). Tomato *Mi*-1.2 gene confers resistance to *Meloidogyne luci* and *M. ethiopica*. *European Journal of Plant Pathology* 156, 571-580. DOI: 10.1007/s10658-019-01907-8

SANTOS, J.M. (1997). *Estudos das principais espécies de* Meloidogyne *Goeldi que infectam o cafeeiro no Brasil com descrição de* Meloidogyne goeldii *sp. n.* Tese de Doutorado, Universidade Estadual Paulista, UNESP, Botucatu, SP, Brasil.

SANTOS, J.M., MATTOS, C., BARRÉ, L. & FERRAZ, S. (1992). *Meloidogyne exigua*, sério patógeno da seringueira nas plantações E. Michelin, em Rondonópolis, MT. *Resumos: XVI Congresso Brasileiro de Nematologia, Lavras, 24-28 fevereiro 1992*, p. 75.

SANTOS, M.C.V. DOS, ALMEIDA, M.T.M. & COSTA, S.R. (2020). First report of *Meloidogyne naasi* parasitizing turfgrass in Portugal. *Journal of Nematology* 52, 1-4. DOI: 10.21307/jofnem-2020-088

SANTOS, M.F.A. DOS, MATTOS, V. DE S., MONTEIRO, J.M.S., ALMEIDA, M.R.A., JORGE JR, A.S., CARES, J.E., CASTAGNONE-SERENO, P., COYNE, D. & CARNEIRO, R.M.D.G. (2018a). Diversity of *Meloidogyne* spp. from peri-urban areas of sub-Saharan Africa and their genetic similarity with populations from the Latin America. *Physiological and Molecular Plant Pathology* 105, 110-118. DOI: 10.1016/j.pmpp.2018.08.004

SANTOS, M.F.A., CORREA, V.R., PEIXOTO, J.R., MATTOS, V.S., SILVA, J.G.P., MOITA, A.W., SALGADO, S.M.L., CASTAGNONE-SERENO, P. & CARNEIRO, R.M.D.G. (2018b). Genetic variability of *Meloidogyne paranaensis* populations and their aggressiveness to susceptible coffee genotypes. *Plant Pathology* 67, 193-201. DOI: 10.1111/ppa.12718

SANTOS, M.F.A., MATTOS, V.S., GOMES, A.C.M.M., MONTEIRO, J.M.S., SOUZA, D.A., TORRES, C.A.R., SALGADO, S.M.L., CASTAGNONE SERENO, P. & CARNEIRO, R.M.D.G. (2020). Integrative taxonomy of *Meloidogyne paranaensis* populations belonging to different esterase phenotypes. *Nematology* 22, 453-468. DOI: 10.1163/15685411-00003316

SANTOS, M.S.N. DE A. (1968). *Meloidogyne ardenensis* n. sp. (Nematoda: Heteroderidae), a new British species of root-knot nematode. *Nematologica* 13(1967), 593-598. DOI: 10.1163/187529267X00418

SANTOS, M.S.N. DE A., ABRANTES, I.M. DE O., RODRIGUES, A.C.F. DE O., ESPIRITO SANTO, S.N. & JOAQUIM, P. (1992). Root-knot nematodes in coffee in the Democratic Republic of São Tomé e Príncipe. *Nematologica* 38, 434.

SAPKOTA, R., SKANTAR, A.M. & NICOLAISEN, M. (2016). A TaqMan real-time PCR assay for detection of *Meloidogyne hapla* in root galls and in soil. *Nematology* 18, 147-154. DOI: 10.1163/15685411-00002950

SASANELLI, N. (1994). Tables of nematode-pathogenicity. *Nematologia Mediterranea* 22, 153-157.

SASANELLI, N. & DI VITO, M. (1992). The effect of *Meloidogyne incognita* on growth on sunflower in pots. *Nematologia Mediterranea* 20, 9-12.

SASANELLI, N., DI VITO, M. & ZACCHEO, G. (1992a). Population densities of *Meloidogyne incognita* and growth of cabbage in pots. *Nematologia Mediterranea* 20, 21-23.

SASANELLI, N., VOVLAS, N. & D'ADDABBO, T. (1992b). Influence of *Meloidogyne javanica* on growth of sunflower. *Afro-Asian Journal of Nematology* 2, 84-88.

SASANELLI, N., D'ADDABBO, T. & PIERANGELI, D. (1996). The effect of *Meloidogyne incognita* on the growth of *Catalpa bignonioides*. *Nematologia Mediterranea* 24, 175-178.

SASANELLI, N., D'ADDABBO, T. & LEMOS, R.M. (2002). Influence of *Meloidogyne javanica* on growth of olive cuttings in pots. *Nematropica* 32, 59-63.

SASANELLI, N., D'ADDABBO, T. & LIŠKOVÁ, M. (2006). Influence of the root-knot nematode *Meloidogyne incognita* race 1 on growth of grapevine. *Helminthologia* 43, 168-170. DOI: 10.2478/s11687-006-0031-z

SASSER, J.N. (1954). Identification and host-parasite relationships of certain root-knot nematodes (*Meloidogyne* spp.). *University of Maryland, Agricultural Experimental Station Technical Bulletin* A-77.

SASSER, J.N. (1977). Worldwide dissemination and importance of the root-knot nematodes, *Meloidogyne* spp. *Journal of Nematology* 9, 26-29.

SASSER, J.N. & CARTER, C.C. (Eds) (1985). *An advanced treatise on* Meloidogyne. *Vol. I. Biology and control.* Raleigh, NC, USA, North Carolina State University Graphics.

SASSER, J.N. & FRECKMAN, D.W. (1987). A world perspective on nematology: the role of the Society. In: Veech, J.A. & Dickson, D.W. (Eds). *Vistas on nematology*. Hyattsville, MD, USA, Society of Nematology, pp. 7-14.

SASSER, J.N., EISENBACK, J.D., CARTER, C.C. & TRIANTAPHYLLOU, A.C. (1983). The international *Meloidogyne* project – its goals and accomplishments. *Annual Review of Phytopathology* 21, 271-288. DOI: 10.1146/annurev.py.21.090183.001415

SATO, K., KADOTA, Y., GAN, P., BINO, T., UEHARA, T., YAMAGUCHI, K., ICHIHASHI, Y., MAKI, N., IWAHORI, H., SUZUKI, T. ET AL. (2018). High-quality genome sequence of the root-knot nematode *Meloidogyne arenaria* genotype A2-O. *Genome Announcements* 6, e00519-18. DOI: 10.1128/genomeA.00519-18

SEGEREN, H.A. & SANCHIT, M.L. (1984). Observations on *Meloidogyne oryzae* Maas, Sanders & Dede, 1978 in irrigated rice in Surinam. *Surinaamse Landbow* 32, 51-59.

SEINHORST, J.W. (1965). The relation between nematode density and damage to plants. *Nematologica* 11, 137-154. DOI: 10.1163/187529265X00582

SEINHORST, J.W. (1979). Nematodes and growth of plants: formalization of the nematode-plant system. In: Lamberti, F. & Taylor, C.E. (Eds). *Root-knot nematodes (*Meloidogyne *species). Systematics, biology and control.* London, UK, Academic Press, pp. 231-256.

SEINHORST, J.W. (1995). The reduction of the growth and weight of plants by a second and later generations of nematodes. *Nematologica* 41, 592-602. DOI: 10.1163/003925995X00530

SEINHORST, J.W. (1998). The common relation between population density and plant weight in pot and microplot experiments with various nematode plant combinations. *Fundamental and Applied Nematology* 21, 459-468.

SEKORA, N.S., CROW, W.T. & MEKETE, T. (2012). First report of *Meloidogyne marylandi* infecting bermudagrass in Florida. *Plant Disease* 96, 1583-1584. DOI: 10.1094/PDIS-06-12-0544-PDN

SELISKAR, D.M. & HUETTEL, R.N. (1992). Nematode involvement in the dieout of *Ammophila breviligulata* (Poaceae) on the mid-Atlantic coastal dunes of the United States. *Journal of Coastal Research* 9, 97-103.

SEMBLAT, J.P., ROSSO, M.N., HUSSEY, R.S., ABAD, P. & CASTAGNONE-SERENO, P. (2001). Molecular cloning of a cDNA encoding an amphid-secreted putative avirulence protein from the root-knot nematode *Meloidogyne incognita*. *Molecular Plant-Microbe Interactions* 14, 72-79. DOI: 10.1094/MPMI.2001.14.1.72

SHAFFIEE, M.F. & JENKINS, W.R. (1963). Host-parasite relationships of *Capsicum frutescens* and *Pratylenchus penetrans*, *Meloidogyne incognita acrita*, and *M. hapla*. *Phytopathology* 53, 325-328.

SHAGALINA, L., IVANOVA, T. & KRALL, E. (1985). [Two new species of gall nematodes of the genus *Meloidogyne* (Nematoda: Meloidogynidae)-parasites of trees and shrubbery.] *Proceedings of the Academy of Sciences of the Estonian SSR, Biology* 34, 279-288.

SHAH, F.A., FALLOON, R.E. & BULMAN, S.R. (2010). Nightshade weeds (*Solanum* spp.) confirmed as hosts of the potato pathogens *Meloidogyne fallax* and *Spongospora subterranea* f. sp. *subterranea. Australasian Plant Pathology* 39, 492-498. DOI: 10.1071/AP10059

SHANER, G., STROMBERG, E.L. & GEORGE, H.L. (1992). Nomenclature and concepts of pathogenicity and virulence. *Annual Review of Phytopathology* 30, 47-66. DOI: 10.1146/annurev.py.30.090192.000403

SHARMA, R.D. (1981). Pathogenicity of *Meloidogyne javanica* to bean (*Phaseolus vulgaris* L.). *Sociedade Brasileira de Nematologia* 5, 137-144.

SHARMA, S.B., REDDY, M.V., SINGH, O., REGO, T.J. & SINGH, U. (1995). Tolerance in chickpea to *Meloidogyne javanica. Fundamental and Applied Nematology* 18, 197-203.

SHARMA POUDYAL, D.S., POKHAREL, R.R., SHRESTHA, S.M. & KHATRI-CHETRI, G.B. (2005). Effect of inoculum density of rice root knot nematode on growth of rice cv. Masuli and nematode development. *Australasian Plant Pathology* 34, 181-185. DOI: 10.1071/AP05011

SHARPE, R.H., HESSE, C.O., LOWNSBERRY, B.A., PERRY, V.G. & HANSEN, C.J. (1969). Breeding peaches for root-knot nematode resistance. *Journal of the American Society for Horticultural Science* 94, 209-212.

SHERIDAN, J.E. & GRBAVAC, N. (1979). Cereal root-knot nematode, *Meloidogyne naasi* Franklin, on barley in New Zealand. *Australasian Plant Pathology* 8, 53-54. DOI: 10.1071/APP9790053a

SHERMAN, W.B. & LYRENE, P.M. (1983). Improvement of peach rootstock resistant to root-knot nematodes. *Proceedings of the Florida State Horticultural Society* 96, 207-208.

SHERMAN, W.B., LYRENE, P.M. & HANSCHE, P.E. (1981). Breeding peach rootstocks resistant to root-knot nematode. *HortScience* 64, 523-524.

SHOKOOHI, E., PARASTAR, Z., PANAHI, H., ABBASPOUR, S., FOURIE, H. & MARAIS, M. (2016). First report of *Meloidogyne hispanica* in Iran. *Australasian Plant Disease Notes* 11, 16. DOI: 10.1007/s13314-016-0202-9

SHUJUN, L. & ZHANG, Y. (1991). [A new species of the *Meloidogyne* (*Meloidogyne actinidiae*) in Henan.] *Acta Agriculturae Universitatis Henanensis* 25, 251-254.

SIDDIQI, M.R. (1986). *Tylenchida parasites of plants and insects*, 1st edition. Farnham Royal, UK, Commonwealth Agricultural Bureaux.

SIDDIQI, M.R. (2000). *Tylenchida parasites of plants and insects*, 2nd edition. Wallingford, UK, CAB International.

SIDDIQI, M.R. & BOOTH, W. (1991). *Meloidogyne* (*Hypsoperine*) *mersa* sp. n. (Nematoda: Tylenchina) attacking *Sonneratia alba* trees in mangrove forest in Brunei Darussalam. *Afro-Asian Journal of Nematology* 1, 212-220.

SIDDIQUI, I. & TAYLOR, D. (1970). The biology of *Meloidogyne naasi*. *Nematologica* 16, 133-143. DOI: 10.1163/187529270X00568

SIKORA, R.A. (1988). Plant-parasitic nematodes of wheat and barley in temperate and temperate semi-arid regions – a comparative analysis. In: Saxena, M.C., Sikora, R.A. & Srivastava, J.P. (Eds). *Nematodes parasitic to cereals and legumes in temperate semi-arid regions*. Aleppo, Syria, ICARDA, pp. 46-48.

SILVA, J.G.P., FURLANETTO, C., ALMEIDA, M.R.A., ROCHA, D.B., MATTOS, V.S., CORREA, V.R. & GARNEIRO, R.M.D.G. (2014). Occurrence of *Meloidogyne* spp. in Cerrado vegetations and reaction of native plants to *Meloidogyne javanica*. *Journal of Phytopathology* 162, 449-455. DOI: 10. 1111/jph.12211

SILVA, M.C.L. (2014). *Identificação e Caracterização de espécies de* Meloidogyne *em áreas agrícolas e dispersão de* M. enterolobii *em pomares de goiabeira no estado do Ceará*. Tese de Doutorado, Universidade Federal do Ceará, Brasil.

SILVA, M.C.L. & SANTOS, C.D.G. (2012). Identificação de espécies de *Meloidogyne* ocorrendo naturalmente em culturas no estado do Ceará por meio de eletroforese em gel de poliacrilamida. *Tropical Plant Pathology* 37 (Supplement), 45° Congresso Brasileiro de Fitopatologia-Manaus, AM.

SILVA, W.R., MACHACA-CALSIN, C.P. & GOMES, C.B. (2020). First report of the root-knot nematode, *Meloidogyne morocciensis* infecting peach in Southern Brazil. *Journal of Nematology* 52, e2020-32. DOI: 10.21307/jofnem-2020-032

SIMARD, L., BÉLAIR, G., POWERS, T., TREMBLAY, N. & DIONNE, J. (2008). Incidence and population density of plant-parasitic nematodes on golf courses in Ontario and Quebec, Canada. *Journal of Nematology* 40, 241-251.

SINGH, K.P., KHANNA, A.S. & HEMA (2018). First report and morphological characterization of *Meloidogyne javanica* from Una district of Himachal Pradesh. *Journal of Entomology and Zoology Studies* 6, 2131-2136.

SINGH, S., KHURMA, U.R. & LOCKHART, P.J. (2010). Weed hosts of root-knot nematodes and their distribution in Fiji. *Weed Technology* 24, 607-612. DOI: 10.2307/40891303

SINGH, S.P. (1969). A new plant parasitic nematode *Meloidogyne lucknowica* n. sp. from the root galls of *Luffa cylindrica* (sponge gourd) in India. *Zoologischer Anzeiger* 182, 259-270.

SIPES, B., SCHMITT, D., XU, K. & SERRACIN, M. (2005). Esterase polymorphism in *Meloidogyne konaensis*. *Journal of Nematology* 37, 438-443.

SIPES, B.S. & ARAKAKI, A.S. (1997). Root-knot nematode management in dryland taro with tropical cover crops. *Journal of Nematology* 29, 721-724.

ŠIRCA, S., UREK, G. & KARSSEN, G. (2004). First report of the root-knot nematode *Meloidogyne ethiopica* on tomato in Slovenia. *Plant Disease* 88, 680. DOI: 10.1094/PDIS.2004.88.6.680C

SIVAPALAN, P. (1972). Nematode pests of tea. In: Webster, J.M. (Ed.). *Economic nematology*. New York, NY, USA, Academic Press, pp. 285-310.

SIVAPALAN, P. (1978). Investigations on root-knot nematodes in Sri Lanka under International *Meloidogyne* Project. *Kasetsart Journal* 12, 14-24.

SKANTAR, A.M. & CARTA, L.K. (2000). Amplification of Hsp90 homologs from plant-parasitic nematodes using degenerate primers and ramped annealing PCR. *BioTechniques* 29, 1182-1185. DOI: 10.2144/00296bm05

SKANTAR, A.M. & CARTA, L.K. (2004). Molecular characterization and phylogenetic evaluation of the *Hsp90* gene from selected nematodes. *Journal of Nematology* 36, 466-480.

SKANTAR, A.M., CARTA, L.K. & HANDOO, Z.A. (2008). Molecular and morphological characterization of an unusual *Meloidogyne arenaria* population from traveler's tree, *Ravenala madagascariensis*. *Journal of Nematology* 40, 179-189.

SLEDGE, E.B. & GOLDEN, A.M. (1964). *Hypsoperine graminis* (Nematoda: Heteroderidae), a new genus and species of plant-parasitic nematode. *Proceedings of the Helminthological Society of Washington* 31, 83-88.

SMITH, E.F. (1896). A bacterial disease of the tomato, eggplant and Irish potato (Bacillus solanacearum n. sp.). *United States Department of Agricultue, Division of Vegetable Physiology and Pathology Bulletin* 12, 1-26.

SMITH, E.F. & TOWNSEND, C.O. (1907). A plant tumor of bacterial origin. *Science* 25, 671-673. DOI: 10.1126/science.25.643.671

SMITH, I.M., MCNAMARA, D.G., SCOTT, P.R. & HOLDERNESS, M. (Eds) (1997). *Quarantine pests for Europe*, 2nd edition. Wallingford, UK, CAB International/EPPO.

SMITH, T., BRITO, J.A., HAN, H., KAUR, R., CETINTAS, R. & DICKSON, D.W. (2015). Identification of the peach root-knot nematode, *Meloidogyne floridensis*, using mtDNA PCR-RFLP. *Nematropica* 45, 138-143.

SOARES, M.R.C., MATTOS, V.S., LEITE, R.R., GOMES, A.C.M.M., GOMES, C.B., CASTAGNONE-SERENO, P., DIAS-ARIEIRA, C.R. & CARNEIRO, R.M.D.G. (2021). Integrative taxonomy of *Meloidogyne graminicola* populations with different esterase phenotypes parasitising rice in Brazil. *Nematology*, in press. DOI: 10.1163/15685411-bja10065

SOHRABI, E., MAAFI, Z.T., PANAHI, P. & BAROOTI, S. (2015). First report of northern root-knot nematode, *Meloidogyne hapla*, parasitic on oaks, *Quercus brantii* and *Q. infectoria* in Iran. *Journal of Nematology* 47, 86-86.

SOMVANSHI, V.S., TATHODE, M., SHUKLA, R.N. & RAO, U. (2018). Nematode genome announcement: a draft genome for rice root-knot nematode, *Meloidogyne graminicola*. *Journal of Nematology* 50, 111-116. DOI: 10.21307/jofnem-2018-018

SONG, Z.Q., CHENG, F.X., ZHANG, D.Y., LIU, Y., CHEN, X.W. & DAI, X. (2017a). First report of *Meloidogyne javanica* infecting hemp (*Cannabis sativa*) in China. *Plant Disease* 101, 842. DOI: 10.1094/PDIS-10-16-1537-PDN

SONG, Z.Q., ZHANG, D.Y., LIU, Y. & CHENG, F.X. (2017b). First report of *Meloidogyne graminicola* on rice (*Oryza sativa*) in Hunan Province, China. *Plant Disease* 101, 2153. DOI: 10.1094/PDIS-06-17-0844-PDN

SONG, Z.Q., YANG, X., ZHANG, X.W., LUAN, M.B., GUO, B., LIU, C.N., PAN, J.P. & MEI, S.Y. (2021). Rapid and visual detection of *Meloidogyne hapla* using recombinase polymerase amplification combined with a lateral flow dipstick (RPA-LFD) assay. *Plant Disease*. DOI: 10.1094/PDIS-06-20-1345-RE

SORIANO, I.R. & REVERSAT, G. (2003). Management of *Meloidogyne graminicola* and yield of upland rice in South-Luzon, Philippines. *Nematology* 5, 879-884. DOI: 10.1163/156854103773040781

SORIANO, I.R.S., PROT, J.-C. & MATIAS, D.M. (2000). Expression of tolerance for *Meloidogyne graminicola* in rice cultivars as affected by soil type and flooding. *Journal of Nematology* 32, 309-317.

SOUTHEY, J.F. (1993). Nematode pests of ornamental and bulb crops. In: Evans, K., Trudgill, D.L. & Webster, J.M. (Eds). *Plant parasitic nematodes in temperate agriculture*. Wallingford, UK, CAB International, pp. 463-500.

SOUZA, R.M., NOGUEIRA, M.S., LIMA, I.M., MELARATO, M. & DOLINSKI, C.M. (2006). Manejo do nematóide das galhas da goiabeira em São da Barra (RJ) e relato de novos hospedeiros. *Nematologia Brasileira* 30, 165-169.

SOUZA, R.M., VOLPATO, A.R. & VIANA, A.P. (2008). Epidemiology of *Meloidogyne exigua* in an upland coffee plantation in Brazil. *Nematologia Mediterranea* 36, 13-17.

SPAULL, V.W. (1977). *Meloidogyne propora* n. sp. (Nematoda: Meloidogynidae) from Aldabra Atoll, Western Indian Ocean, with a note on *M. javanica* (Treub). *Nematologica* 23, 177-186. DOI: 10.1163/187529277X00525

SPERANDIO, C.A. & MONTEIRO, A.R. (1991). Ocorrência de *Meloidogyne graminicola* em arroz irrigado no Rio Grande do Sul. *Nematologia Brasileira* 15, 24.

STANTON, J., HUGALL, A. & MORITZ, C. (1997). Nucleotide polymorphisms and an improved PCR-based mtDNA diagnostic for parthenogenetic root-knot nematodes (*Meloidogyne* spp.). *Fundamental and Applied Nematology* 20, 261-268.

STANTON, J.M. & O'DONNELL, W.E. (1998). Assessment of the North Carolina differential host test for identification of Australian populations of root-knot nematodes (*Meloidogyne* spp.). *Australasian Plant Pathology* 27, 104-111. DOI: 10.1071/AP98013

STARR, J.L., TOMASZEWSKI, E.K., MUNDO-OCAMPO, M. & BALDWIN, J.G. (1996). *Meloidogyne partityla* on pecan: isozyme phenotypes and other hosts. *Journal of Nematology* 28, 565-568.

STEFANELO, D.R., SANTOS, M.F.A. DOS, MATTOS, V.S., BRAGHINI, M.T., MENDONÇA, J.S.F., CARES, J.E. & CARNEIRO, R.M.D.G. (2019). *Meloidogyne izalcoensis* parasitizing coffee in Minas Gerais state: the first record in Brazil. *Tropical Plant Pathology* 44, 209-212. DOI: 10.1007/s40858-018-0251-z

STEPHAN, Z.A. (1983). The effect of different densities of *Meloidogyne ardenensis* and of three populations of *M. hapla* on the growth of tomato at four soil temperatures. *Nematologia Mediterranea* 11, 93-100.

STEPHAN, Z.A. & TRUDGILL, D.L. (1982). Population fluctuations, life cycle of root-knot nematode, *Meloidogyne ardenensis* in Cupar, Scotland, and the effect of temperature on its development. *Revue de Nématologie* 5, 281-284.

STONE, G.E. & SMITH, R.E. (1898). *Nematode worms*. Bulletin 55. MA, USA, Hatch Experiment Station of the Massachusetts Agricultural College.

STRAJNAR, P., ŠIRCA, S., GERIČ STARE, B. & UREK, G. (2009). Characterisation of the root-knot nematode, *Meloidogyne ethiopica* Whitehead, 1968, from Slovenia. *Russian Journal of Nematology* 17, 135-142.

STURHAN, D. (1976). Freilandvorkommen von *Meloidogyne*-Arten in der Bundesrepublik Deutschland. *Nachrichtenblatt des Deutschen Pflanzenschutzdienstes* 28, 113-117.

SUBBOTIN, S.A. (2019). Recombinase polymerase amplification assay for rapid detection of the root-knot nematode *Meloidogyne enterolobii*. *Nematology* 21, 243-251. DOI: 10.1163/15685411-00003210

SUBBOTIN, S.A. & BURBRIDGE, J. (2021). Sensitive, accurate and rapid detection of the northern root-knot nematode, *Meloidogyne hapla*, using Recombinase Polymerase Amplification assays. *Plants* 10, 336. DOI: 10.3390/plants10020336

SUBBOTIN, S.A. & KIM, D. (2021). Molecular characterisation and phylogenetic relationships of sedentary nematodes of the genus *Meloinema* Choi & Geraert, 1974 (Nematoda: Tylenchida). *Nematology* 23. DOI: 10.1163/15685411-bja10056

SUBBOTIN, S.A., STURHAN, D., CHIZHOV, V.N., VOVLAS, N. & BALDWIN, J.G. (2006). Phylogenetic analysis of Tylenchida Thorne, 1949 as inferred from D2 and D3 expansion fragments of the 28S rRNA gene sequences. *Nematology* 8, 455-474. DOI: 10.1163/156854106778493420

SUMITA, K., DAS, D. & CHOUDHURY, B.N. (2018). Morphological and morphometric variations among the population of root-knot nematode(s) of Assam. *International Journal of Current Microbiology and Applied Sciences* 7, 1701-1708. DOI: 10.20546/ijcmas.2018.707.201

SUN, L., ZHUO, K., LIN, B., WANG, H. & LIAO, J. (2014). The complete mitochondrial genome of *Meloidogyne graminicola* (Tylenchina): a unique gene arrangement and its phylogenetic implications. *PLoS ONE* 9, e98558. DOI: 10.1371/journal.pone.0098558

SURESH, P., POORNIMA, K., KALAIARASAN, P., SIVAKUMAR, M. & SUBRAMANIAN, S. (2017). Occurrence of barley root knot nematode, *Meloidogyne naasi* in orange jessamine (*Cestrum aurantiacum* L.) in Nilgiris, Tamil Nadu, India: A new record. *Journal of Entomology and Zoology Studies* 5, 629-634.

SUSIČ, N., ŠIRCA, S., UREK, G. & GERIČ STARE, B. (2020a). *Senecio vulgaris* L. recorded as a new host plant for the root-knot nematode *Meloidogyne luci*. *Acta Agriculturae Slovenica* 115/2, 495-497. DOI: 10.14720/aas.2020.115.2.1514

SUSIČ, N., KOUTSOVOULOS, G., RICCIO, C., DANCHIN, E.G., BLAXTER, M.L., LUNT, D.H., STRAJNAR, P., ŠIRCA, S., UREK, G. & GERIČ STARE, B. (2020b). Genome sequence of the root-knot nematode *Meloidogyne luci*. *Journal of Nematology* 52, e2020-25. DOI: 10.21307/jofnem-2020-025

SZITENBERG, A., SALAZAR-JARAMILLO, L., BLOK, V.C., LAETSCH, D.R., JOSEPH, S., WILLIAMSON, V.M., BLAXTER, M.L. & LUNT, D.H. (2017). Comparative genomics of apomictic root-knot nematodes: hybridization, ploidy, and dynamic genome change. *Genome Biology and Evolution* 9, 2844-2861. DOI: 10.1093/gbe/evx201

TANDINGAN DE LEY, I., DE LEY, P., VIERSTRAETE, A., KARSSEN, G., MOENS, M. & VANFLETEREN, J. (2002). Phylogenetic analyses of *Meloidogyne* small subunit rDNA. *Journal of Nematology* 34, 319-327.

TAO, Y., XU, C.L., YUAN, C.F., WANG, H.H., LIN, B., ZHUO, K. & LIAO, J.L. (2017). *Meloidogyne aberrans* sp. nov. (Nematoda: Meloidogynidae), a new root-knot nematode parasitizing kiwifruit in China. *PLoS ONE* 12, e0182627. DOI: 10.1371/journal.pone.0182627

TARTE, R. (1981). Informe sobre el progreso de la investigación para el proyecto Internacional de *Meloidogyne* en Panamá, 1976-1978. In: Tarte, R. (Ed.). *Memorias de la segunda conferencia regional de planeamiento del proyecto Internacional de* Meloidogyne. *Región I.* Panama City, Panama, International *Meloidogyne* Project, pp. 27-51.

TAYLOR, A.L. (1987). Identification and estimation of root-knot nematode species in mixed populations. *Bulletin 12*. Gainesville, FL USA, Florida Department of Agriculture and Consumer Services, Division of Plant Industry.

TAYLOR, A.L. & SASSER, J.N. (1978). *Biology, identification, and control of root-knot nematodes (*Meloidogyne *species).* Raleigh, NC, USA, Department of Plant Pathology, North Carolina State University and the United States Agency for International Development.

TAYLOR, A.L., DROPKIN, V.H. & MARTIN, G.C. (1955). Perineal patterns of root-knot nematodes. *Phytopathology* 45, 26-34.

TAYLOR, A.L., SASSER, J.N. & NELSON, L.A. (1982). *Relationship of climate and soil characteristics to geographical distribution of* Meloidogyne *species in agricultural soils.* Raleigh, NC, USA, Department of Plant Pathology, North Carolina State University and the United States Agency for International Development.

TAYLOR, C.E. (1990). Nematode interactions with other pathogens. *Annals of Applied Biology* 116, 405-416. DOI: 10.1111/j.1744-7348.1990.tb06622.x

TAYLOR, D.P., MALEK, R.B. & EDWARDS, D.I. (1971). The barley root-knot nematode; it's not a problem..., yet. *Crops & Soils* 23, 14-16.

TEILLET, A., DYBAL, K., KERRY, B.R., MILLER, A.J., CURTIS, R.H.C. & HEDDEN, P. (2013). Transcriptional changes of the root-knot nematode *Meloidogyne incognita* in response to *Arabidopsis thaliana* root signals. *PLoS ONE* 8, e61259. DOI: 10.1371/journal.pone.0061259

TENENTE, G.C.M.V., DE LEY, P., TANDINGAN DE LEY, I., KARSSEN, G. & VANFLETEREN, J.R. (2004). Sequence analysis of the D2/D3 region of the large subunit rDNA from different *Meloidogyne* isolates. *Nematropica* 34, 1-12.

TERENTEVA, T.G. (1965). [*Meloidogyne kirjanovae* n. sp. (Nematoda: Heteroderidae).] *Materialy Nauchnoi Konferentsii Vsesoyuznogo Obschestva Gel'mintologov* 4, 277-281.

THODEN, T.C., KORTHALS, G.W., VISSER, J. & VAN GASTEL-TOPPER, W. (2012). A field study on the host status of different crops for *Meloidogyne minor* and its damage potential on potatoes. *Nematology* 14, 277-284. DOI: 10.1163/156854111X594965

THOMAS, P.R. & BROWN, D.J.F. (1981). Some hosts of a *Meloidogyne ardenensis* population found in Scotland. *Plant Pathology* 30, 147-151. DOI: 10.1111/j.1365-3059.1981.tb01246.x

THOMAS, S.H., FUCHS, J.M. & HANDOO, Z.A. (2001). First report of *Meloidogyne partityla* on pecan in New Mexico. *Plant Disease* 85, 1030. DOI: 10.1094/PDIS.2001.85.9.1030B

THORNE, G. (1961). *Principles of nematology*. New York, NY, USA, McGraw-Hill Book Co. Inc.

TIAN, Z.L., BARSALOTE, E.M., LI, X.L., CAI, R.H. & ZHENG, J.W. (2017). First report of root-knot nematode, *Meloidogyne graminicola*, on rice in Zhejiang, Eastern China. *Plant Disease* 101, 2152. DOI: 10.1094/PDIS-06-17-0832-PDN

TIAN, Z.L., MARIA, M., BARSALOTE, E.M., CASTILLO, P. & ZHENG, J.W. (2018). Morphological and molecular characterization of the rice root-knot nematode, *Meloidogyne graminicola* Golden & Birchfield, 1965 occurring in Zhejiang, China. *Journal of Integrative Agriculture* 17, 2724-2733. DOI: 10.1016/S2095-3119(18)61971-9

TIGANO, M., SIQUEIRA, K. DE, CASTAGNONE-SERENO, P., MULET, K., QUEIROZ, P., SANTOS, M. DOS, TEIXEIRA, C., ALMEIDA, M., SILVA, J. & CARNEIRO, R.M.D.G. (2010). Genetic diversity of the root-knot nematode *Meloidogyne enterolobii* and development of a SCAR marker for this guava-damaging species. *Plant Pathology* 59, 1054-1061. DOI: 10.1111/j.1365-3059.2010.02350.x

TIGANO, M.S., CARNEIRO, R.M.D.G., JEYAPRAKASH, A., DICKSON, D.W. & ADAMS, B.J. (2005). Phylogeny of *Meloidogyne* spp. based on 18S rDNA and the intergenic region of mitochondrial DNA sequences. *Nematology* 7, 851-862. DOI: 10.1163/156854105776186325

TIWARI, S., PANDEY, S., SINGH CHAUHAN, P.S. & PANDEY, R. (2017). Bio-control agents in co-inoculation manages root knot nematode (*Meloidogyne incognita* (Kofoid & White) Chitwood) and enhances essential oil content in *Ocimum basilicum* L. *Industrial Crops and Products* 97, 292-301. DOI: 10.1016/j.indcrop.2016.12.030

TOIDA, Y. (1991). Mulberry damages caused by a root-knot nematode, *Meloidogyne mali* indigenous to Japan. *Japan Agricultural Research Quarterly* 24, 300-305.

TOIDA, Y. & YAEGASHI, T. (1984). Description of *Meloidogyne suginamiensis* n. sp. (Nematoda: Meloidogynidae) from Mulberry in Japan. *Japanese Journal of Nematology* 12, 49-57.

TOMALOVA, I., IACHIA, C., MULET, K. & CASTAGNONE-SERENO, P. (2012). The *map*-1 gene family in root-knot nematodes, *Meloidogyne* spp.: A set of taxonomically restricted genes specific to clonal species. *PLOS ONE* 7, e38656. DOI: 10.1371/journal.pone.0038656

TOPALOVIĆ, O., MOORE, J.F., JANSSEN, T., BERT, W. & KARSSEN, G. (2017). An early record of *Meloidogyne fallax* from Ireland. *ZooKeys* 643, 33-52. DOI: 10.3897/zookeys.643.11266

TOWNSHEND, J.L. & POTTER, J.W. (1978). Yield losses among forage legumes infected with *Meloidogyne hapla*. *Canadian Journal of Plant Science* 58, 939-943.

TOWNSHEND, J.L. & POTTER, J.W. (1986). *Meloidogyne microtyla*: pathogenicity to orchard cover grasses, survival in stored soil, and reproductivity after storage. *Plant Disease* 70, 438-440.

TOWNSHEND, J.L., POTTER, J.W. & DAVIDSON, T.R. (1984). Some monocotyledonous and dicotyledonous hosts of *Meloidogyne microtyla*. *Plant Disease* 68, 7-10.

TOYOTA, K., SHIRAKASHI, T., SATO, E., WADA, S. & MIN, Y.Y. (2008). Development of a real-time PCR method for the potato-cyst nematode *Globodera rostochiensis* and the root-knot nematode *Meloidogyne incognita. Soil Science and Plant Nutrition* 54, 72-76. DOI: 10.1111/j.1747-0765.2007.00212.x

TREUB, M. (1885). Onderzoekingen over Sereh-Ziek Suikkeriet gedaan in s'Lands Plantentium te Buitenzorg. *Mededeelingen uit's Lands Plantentium, Batavia* 2, 1-39.

TRIANTAPHYLLOU, A.C. (1966). Polyploidy and reproductive patterns in the root-knot nematode *Meloidogyne hapla. Journal of Morphology* 118, 403-413.

TRIANTAPHYLLOU, A.C. (1969). Gametogenesis and the chromosomes of two root-knot nematodes, *Meloidogyne graminicola* and *M. naasi. Journal of Nematology* 1, 62-71.

TRIANTAPHYLLOU, A. (1973). Gametogenesis and reproduction of *Meloidogyne graminis* and *M. ottersoni* (Nematoda: Heteroderidae). *Journal of Nematology* 5, 84-87.

TRIANTAPHYLLOU, A.C. (1979). Cytogenetics of root-knot nematodes. In: Lamberti, F. & Taylor, C.E. (Eds). *Root-knot nematodes (*Meloidogyne *species) systematics, biology and control.* New York, NY, USA, Academic Press, pp. 85-109.

TRIANTAPHYLLOU, A.C. (1981). Oogenesis and the chromosomes of the parthenogenetic root-knot nematode *Meloidogyne incognita. Journal of Nematology* 13, 95-104.

TRIANTAPHYLLOU, A.C. (1983). Cytogenetic aspects of nematode evolution. In: Stone, A.R., Platt, H.M. & Khalil, L.F. (Eds). *Concepts in nematode systematics.* London, UK, Academic Press, pp. 55-71.

TRIANTAPHYLLOU, A.C. (1984). Polyploidy in meiotic parthenogenetic populations of *Meloidogyne hapla* and a mechanism of conversion to diploidy. *Revue de Nématologie* 7, 65-72.

TRIANTAPHYLLOU, A.C. (1985a). Gametogenesis and the chromosomes of *Meloidogyne nataliei*: not typical of othe root-knot nematodes. *Journal of Nematology* 17, 1-5.

TRIANTAPHYLLOU, A.C. (1985b). Cytogenetics, cytotaxonomy and phylogeny of root-knot nematodes. In: Sasser, J.N. & Carter, C.C. (Eds). *An advanced treatise on* Meloidogyne. *Vol. I. Biology and control.* Raleigh, NC, USA, North Carolina State University Graphics, pp. 113-126.

TRIANTAPHYLLOU, A.C. (1987a). Cytogenetic status of *Meloidogyne (Hypsoperine) spartinae* in relation to other *Meloidogyne* species. *Journal of Nematology* 19, 1-7.

TRIANTAPHYLLOU, A.C. (1987b). Genetics of nematode parasitism of plants. In: Veech, A. & Dickson, D.W. (Eds). *Vistas on nematology*. Hyattsville, MD, USA, Society of Nematologists, pp. 354-371.

TRIANTAPHYLLOU, A.C. (1990). Cytogenetic status of *Meloidogyne kikuyensis* in relation to other root-knot nematode. *Revue de Nématologie* 13, 175-180.

TRIANTAPHYLLOU, A.C. & HIRSCHMANN, H. (1997). Evidence of direct polyploidization in the mitotic parthenogenetic *Meloidogyne microcephala*, through doubling of its somatic chromosome number: *Fundamental and Applied Nematology* 20, 385-391.

TRIANTAPHYLLOU, A.C. & SASSER, J.N. (1960). Variation in perineal patterns and host specificity of *Meloidogyne incognita*. *Phytopathology* 50, 724-735.

TRINH, Q.P., LE, T.M.L., NGUYEN, T.D., NGUYEN, H.T., LIÉBANAS, G. & NGUYEN, T.A.D. (2018). *Meloidogyne daklakensis* n. sp. (Nematoda: Meloidogynidae), a new root-knot nematode associated with Robusta coffee (*Coffea canephora* Pierre ex A. Froehner) in the Western Highlands, Vietnam. *Journal of Helminthology* 93, 242-254. DOI: 10.1017/S0022149X18000202

TRISCIUZZI, N., TROCCOLI, A., VOVLAS, N., CANTALAPIEDRA-NAVARRETE, C., PALOMARES-RIUS, J.E. & CASTILLO, P. (2014). Detection of the camellia root-knot nematode *Meloidogyne camelliae* Golden in Japanese camellia bonsai imported into Italy: integrative diagnosis, parasitic habits and molecular phylogeny. *European Journal of Plant Pathology* 138, 231-235. DOI: 10.1007/s10658-013-0337-x

TRUDGILL, D.L. & BLOK, V.C. (2001). Apomictic, polyphagous root-knot nematodes: exceptionally successful and damaging biotrophic root pathogens. *Annual Review of Phytopathology* 39, 53-77. DOI: 10.1146/annurev.phyto.39.1.53

TRUDGILL, D.L., BALA, G., BLOK, V.C., DAUDI, A., DAVIES, K.G., GOWEN, S.R., FARGETTE, M., MADULU, J.D., MATEILLE, T., MWAGENI, W. ET AL. (2000). The importance of tropical root-knot nematodes (*Meloidogyne* spp.) and factors affecting the utility of *Pasteuria penetrans* as a biocontrol agent. *Nematology* 2, 823-845. DOI: 10.1163/156854100750112789

TZORTZAKAKIS, E.A., ANASTASIADIS, A.I., SIMOGLOU, K.B., CANTALAPIEDRA-NAVARRETE, C., PALOMARES-RIUS, J.E. & CASTILLO, P. (2014). First report of the root-knot nematode, *Meloidogyne hispanica*, infecting sunflower in Greece. *Plant Disease* 98, 703. DOI: 10.1094/PDIS-08-13-0833-PDN

UPADHYAY, V., BHARDWAJ, N.R., NEELAM & SAJEESH, P.K. (2014). *Meloidogyne graminicola* (Golden & Birchfield): threat to rice production. *Research Journal of Agriculture and Forestry Sciences* 2, 31-36.

VAISH, S.S., PANDEY, S.K. & SINGH, K.P. (2012). First report of the root-knot disease of barley caused by *Meloidogyne graminicola* from India. *Current Nematology* 23, 77-79.

VAN DER BEEK, J.G. & KARSSEN, G. (1997). Interspecific hybridization of meiotic parthenogenetic *Meloidogyne chitwoodi* and *M. fallax*. *Phytopathology* 87, 1061-1066. DOI: 10.1094/PHYTO.1997.87.10.1061

VAN DER BEEK, J.G., LOS, J.A. & PIJNACKER, L.P. (1998). Cytology of parthenogenesis of five *Meloidogyne* species. *Fundamental and Applied Nematology* 21, 393-399.

VAN DER BEEK, J.G., MAAS, P.W.TH., JANSSEN, G.J.W., ZIJLSTRA, C. & VAN SILFHOUT, C.H. (1999). A pathotype system to describe intraspecific variation in pathogenicity of *Meloidogyne chitwoodi*. *Journal of Nematology* 31, 386-392.

VAN DER SOMMEN, A.T.C., DEN NIJS, L.J.M.F. & KARSSEN, G. (2005). The root-knot nematode *Meloidogyne fallax* on strawberry in the Netherlands. *Plant Disease* 89, 526. DOI: 10.1094/PD-89-0526A

VAN DER WURFF, A.W.G., JANSE, J., KOK, C.J. & ZOON, F.C. (2010). Biological control of root knot nematodes in organic vegetable and flower greenhouse cultivation – state of science: report of a study over the period 2005-2010. Bleiswijk, The Netherlands, Wageningen UR Greenhouse Horticulture.

VAN GUNDY, S.D. (1985). Ecology of *Meloidogyne* spp. – emphasis on environmental factors affecting survival and pathogenicity. In: Sasser, J.N. & Carter, C.C. (Eds). *An Advanced treatise on* Meloidogyne. *Vol. 1. Biology and control.* Raleigh, NC, USA, North Carolina State University Graphics, pp. 177-182.

VAN HALTEREN, P. (1972). Root-knot nematode infestation (*Meloidogyne* sp.) in irrigated rice at the SML Wageningen Rice Scheme. *De Surinaamse Landbouw* 20, 42-43.

VAN MEGEN, H., VAN DEN ELSEN, S., HOLTERMAN, M., KARSSEN, G., MOOYMAN, P., BONGERS, T., HOLOVACHOV, O., BAKKER, J. & HELDER, J. (2009). A phylogenetic tree of nematodes based on about 1200 full-length small subunit ribosomal DNA sequences. *Nematology* 11, 927-950. DOI: 10.1163/156854109X456862

VAN MEGGELEN, J., KARSSEN, G., JASSEN, G.J.W., VERKERK-BAKER, B. & JANSSEN, R. (1994). A new race of *Meloidogyne chitwoodi* Golden, O'Bannon, Santo & Finley, 1980? *Fundamental and Applied Nematology* 17, 93-96.

VANDENBOSSCHE, B., VIAENE, N., SUTTER, N.D., MAES, M., KARSSEN, G. & BERT, W. (2011). Diversity and incidence of plant-parasitic nematodes in Belgian turf grass. *Nematology* 13, 245-256. DOI: 10.1163/138855410X517084

Vela, M.D., Giné, A., López-Gómez, M., Sorribas, F.J., Ornat, C., Verdejo-Lucas, S. & Talavera, M. (2014). Thermal time requirements of root-knot nematodes on zucchini-squash and population dynamics with associated yield losses on spring and autumn cropping cycles. *European Journal of Plant Pathology* 140, 481-490. DOI: 10.1007/s10658-014-0482-x

Viaene, N.M. & Abawi, G.S. (1996). Damage threshold of *Meloidogyne hapla* to lettuce in organic soil. *Journal of Nematology* 28, 537-545.

Viaene, N.M. & Abawi, G.S. (1998). Management of *Meloidogyne hapla* on lettuce in organic soil with Sudan grass as a cover crop. *Plant Disease* 82, 945-952. DOI: 10.1094/PDIS.1998.82.8.945

Viaene, N.M., Simoens, P. & Abawi, G.S. (1997). SeinFit, a computer program for the estimation of the Seinhorst equation. *Journal of Nematology* 29, 474-477.

Viaene, N.M., Wiseborn, D.B. & Karssen, G. (2007). First report of the root-knot nematode *Meloidogyne minor* on turfgrass in Belgium. *Plant Disease* 91, 908. DOI: 10.1094/PDIS-91-7-0908B

Villain, L., Salgado, S.M.L. & Trinh, P.Q. (2018). Nematode parasites of coffee and cocoa. In: Sikora, R.A., Coyne, D., Hallmann, J. & Timper, P. (Eds). *Plant parasitic nematodes in subtropical and tropical agriculture*, 3rd edition. Wallingford, UK, CAB International, pp. 536-583.

Villain, L., Sarah, J.L., Hernández, A., Bertrand, B., Anthony, F., Lashermes, P., Charmetant, P., Anzueto, F., Figueroa, P. & Carneiro, R.M.D.G. (2013). Diversity of root-knot nematodes associated with coffee orchards in Central America. *Nematropica* 43, 194-206.

Vovlas, N. & Di Vito, M. (1991). Effect of root-knot nematodes *Meloidogyne incognita* and *M. javanica* on the growth of coffee (*Coffea arabica* L.) in pots. *Nematologia Mediterranea* 19, 253-258.

Vovlas, N. & Inserra, R.N. (1979). New host records of *Meloidogyne naasi* from Italy. *Plant Disease Reporter* 63, 644-646.

Vovlas, N. & Inserra, R.N. (1996). Distribution and parasitism of root-knot nematodes on *Citrus*. *Nematology Circular* No. 217. Gainesville, FL, USA, Florida Department of Agriculture & Consumer Services Division of Plant Industry.

Vovlas, N., Di Vito, M. & Grammatikaki, G. (1993). Growth response of *in vitro* produced banana plantlets to *Meloidogyne javanica* in pots. *Nematropica* 23, 203-208.

Vovlas, N., Simões, N.J.O., Sasanelli, N., Santos, M.C.V. dos & Abrantes, I.M. de O. (2004a). Host-parasite relationships in tobacco plants infected with a root-knot nematode (*Meloidogyne incognita*) population from the Azores. *Phytoparasitica* 32, 167-173. DOI: 10.1007/BF02979783

VOVLAS, N., LIÉBANAS, G. & CASTILLO, P. (2004b). SEM studies on the Mediterranean olive root-knot nematode, *Meloidogyne baetica*, and histopathology on two additional natural hosts. *Nematology* 6, 749-754. DOI: 10.1163/1568541042843540

VOVLAS, N., MIDSUF, D., LANDA, B.B. & CASTILLO, P. (2005). Pathogenicity of the root-knot nematode *Meloidogyne javanica* on potato. *Plant Pathology* 54, 657-664. DOI: 10.1111/j.1365-3059.2005.01244.x

VOVLAS, N., TROCCOLI, A., MINUTO, A., BRUZZONE, C., SASANELLI, N. & CASTILLO, P. (2008a). Pathogenicity and host-parasite relationships of *Meloidogyne arenaria* in sweet basil. *Plant Disease* 92, 1329-1335. DOI: 10.1094/PDIS-92-9-1329

VOVLAS, N., LUCARELLI, G., SASANELLI, N., TROCCOLI, A., PAPAJOVA, I., PALOMARES-RIUS, J.E. & CASTILLO, P. (2008b). Pathogenicity and host-parasite relationships of the root-knot nematode *Meloidogyne incognita* on celery. *Plant Pathology* 57, 981-987. DOI: 10.1111/j.1365-3059.2008.01843.x

VOVLAS, N., TROCCOLI, A., MINUTO, A., BRUZZONE, C. & CASTILLO, P. (2010). Parasitism of the root-knot nematode *Meloidogyne hapla* on peony in Northern Italy. *Nematologia Mediterranea* 38, 219-221.

VRAIN, T.C. (1982). Relationship between *Meloidogyne hapla* density and damage to carrots in organic soils. *Journal of Nematology* 14, 50-57.

VRAIN, T.C., WAKARCHUK, D.A., LEVESQUE, A.C. & HAMILTON, R.I. (1992). Intraspecific rDNA Restriction Fragment Length Polymorphism in the *Xiphinema americanum* group. *Fundamamental and Applied Nematology* 15, 563-573.

WALIULLAH, S., BELL, J., JAGDALE, G., STACKHOUSE, T., HAJIHASSANI, A., BRENNEMAN, T. & ALI, MD.E. (2020). Rapid detection of pecan root-knot nematode, *Meloidogyne partityla*, in laboratory and field conditions using loop-mediated isothermal amplification. *PLoS ONE* 15(6), e0228123. DOI: 10.1371/journal.pone.0228123

WALKER, N. (2014). First report of *Meloidogyne marylandi* infecting bermudagrass in Oklahoma. *Plant Disease* 98, 1286. DOI: 10.1094/PDIS-04-14-0399-PDN

WALLACE, H.R. (1983). Interactions between nematodes and other factors on plants. *Journal of Nematology* 15, 221-227.

WANG, G.F., XIAO, L.Y., LUO, H.G., PENG, D.L. & XIAO, Y.N. (2017). First report of *Meloidogyne graminicola* on rice in Hubei Province of China. *Plant Disease* 101, 1056. DOI: 10.1094/PDIS-12-16-1805-PDN

WANG, J.L., GU, J.F., GAO, F.F., ZHANG, H., HE, J., SHAO, F. & CHEN, X-F. (2013). [Identification of *Meloidogyne camelliae* intercepted in *Camelliae* sp. from Japan.] *Plant Quarantine* 27, 70-74.

WATANABE, T., MASUMURA, H., KIOKA, Y., NOGUCHI, K., MIN, Y.Y., MURAKAMI, R. & TOYOTA, K. (2013). Development of a direct quantitative detection method for *Meloidogyne incognita* and *M. hapla* in andosol and analysis of relationship between the initial population of *Meloidogyne* spp. and yield of eggplant in an andosol. *Nematological Research* 43, 21-29. DOI: 10.3725/jjn.43.21

WEI, H.Y., WANG, X., LI, H.M., SUN, W.R. & GU, J.F. (2016). Loop-mediated isothermal amplification assay for rapid diagnosis of *Meloidogyne mali*. *Journal of Plant Protection* 43, 260-266.

WEN, G.Y. & CHEN, T.A. (1976). Ultrastructure of the spicules of *Pratylenchus penetrans*. *Journal of Nematology* 8, 69-74.

WESEMAEL, W.M.L. & MOENS, M. (2008). Quality damage on carrots (*Daucus carota* L.) caused by the root-knot nematode *Meloidogyne chitwoodi*. *Nematology* 10, 261-270. DOI: 10.1163/156854108783476368

WESEMAEL, W.M.L., VIAENE, N. & MOENS, M. (2011). Root-knot nematodes (*Meloidogyne* spp.) in Europe. *Nematology* 13, 3-16. DOI: 10.1163/138855410X526831

WESEMAEL, W.M.L., TANING, L.M., VIAENE, N. & MOENS, M. (2014). Life cycle and damage of the root-knot nematode *Meloidogyne minor* on potato, *Solanum tuberosum*. *Nematology* 16, 185-192. DOI: 10.1163/15685411-00002756

WESTOLL, T.S. & STODDART, D.R. (1971). A discussion on the results of the Royal Society expedition to Aldabra 1967-68. *Philosophical Transactions of the Royal Society of London, B* 260, 1-654.

WESTPHAL, A., MAUNG, Z.T.Z., DOLL, D.A., YAGHMOUR, M.A., CHITAMBAR, J.J. & SUBBOTIN, S.A. (2019). First report of the peach root-knot nematode, *Meloidogyne floridensis* infecting almond on root-knot nematode resistant 'Hansen 536' and 'Bright's Hybrid 5' rootstocks in California, USA. *Journal of Nematology* 51, e2019-61. DOI: 10.21307/jofnem-2019-002

WHEELER, T.A. & STARR, J.L. (1987). Incidence and economic importance of plant-parasitic nematodes on peanut in Texas. *Peanut Science* 14, 94-96. DOI: 10.3146/i0095-3679-14-2-11

WHITEHEAD, A.G. (1960). The root-knot nematodes of East Africa. I. *Meloidogyne africana* n. sp., a parasite of arabica coffee (*Coffea arabica* L.). *Nematologica* 4(1959), 272-278. DOI: 10.1163/187529259X00471

WHITEHEAD, A.G. (1968). Taxonomy of *Meloidogyne* (Nematodea: Heteroderidae) with descriptions of four new species. *Transactions of the Zoological Society of London* 31, 263-401.

WHITEHEAD, A.G. (1969). The distribution of root-knot nematodes (*Meloidogyne* spp.) in tropical Africa. *Nematologica* 15, 315-333. DOI: 10.1163/187529269X00362

WHITEHEAD, A.G. & KARIUKI, L. (1960). Root-knot nematode surveys of cultivated areas in Africa. *East African Agricultural and Forestry Journal* 26, 87-90.

WILLERS, P. (1997). First record of *Meloidogyne mayaguensis* Rammah and Hirschmann, 1988: Heteroderidae on commercial crops in the Mpumalanga province, South Africa. *Inligtingsbulletin-Instituut vir Tropiese en Subtropiese Gewasse* 294, 19-20.

WILLIAMSON, V.M. (1991). Molecular techniques for nematode species identification. In: Nickle, W.R. (Ed.). *Manual of agricultural nematology*. New York, NY, USA, Marcel Dekker Inc., pp. 107-123.

WILLIAMSON, V.M., CASWELL-CHEN, E.P., WESTERDAHL, B.B., WU, F.F. & CARYL, G. (1997). A PCR assay to identify and distinguish single juveniles of *Meloidogyne hapla* and *M. chitwoodi*. *Journal of Nematology* 29, 9-15.

WINDHAM, G.L. & PEDERSON, G.A. (1992). Comparison of reproduction by *Meloidogyne graminicola* and *M. incognita* on *Trifolium* species. *Journal of Nematology* 24, 257-261.

WISHART, J., PHILLIPS, M.S. & BLOK, V.C. (2002). Ribosomal intergenic spacer: a polymerase chain reaction diagnostic for *Meloidogyne chitwoodi*, *M. fallax*, and *M. hapla*. *Phytopathology* 92, 884-892. DOI: 10.1094/PHYTO.2002.92.8.884

WOBALEM, Y. & VIAENE, N. (2005). Plant-parasitic nematodes and their antagonists in greenhouse-grown vegetables in Flanders. *57th International Symposium on Crop Protection, Ghent*, p. 75. [Abstr.]

WU, Y. (2011). [*Identification of a new species of root-knot nematode parasitizing citrus in Fujian province, China*.] Master's Thesis. Fuzhou, China, Fujian Agriculture and Forestry University.

XING, L. & WESTPHAL, A. (2012). Predicting damage of *Meloidogyne incognita* on watermelon. *Journal of Nematology* 44, 127-133.

XU, J.H., LIU, P.L., MENG, Q.P. & LONG, H. (2004). Characterisation of *Meloidogyne* species from China using isozyme phenotypes and amplified mitochondrial DNA restriction fragment length polymorphism. *European Journal of Plant Pathology* 110, 309-315. DOI: 10.1023/B:EJPP.0000019800.47389.31

YABUUCHI, E., KOSAKO, Y., YANO, I., HOTTA, H. & NISHIUCHI, Y. (1995). Transfer of two *Burkholderia* and an *Alcaligen*es species to *Ralstonia* gen. nov.: proposal of *Ralstonia pickettii* (Ralston, Palleroni and Douderoff 1973) comb. nov., *Ralstonia solanacearum* (Smith 1896) comb. nov. and *Ralstonia eutropha* (Davis 1969) comb. nov. *Microbiology and Immunology* 39, 897-904. DOI: 10.1111/j.1348-0421.1995.tb03275.x

YADAV, B.C., VELUTHAMBI, K. & SUBRAMANIAM, K. (2006). Host-generated double stranded RNA induces RNAi in plant parasitic nematodes and protects the host from infection. *Molecular and Biochemical Parasitology* 148, 219-222. DOI: 10.1016/j.molbiopara.2006.03.013

YAEGASHI, T. & OKAMOTO, K. (1981). Observation of six *Meloidogyne* species by scanning electron microscope. 2. *En face* views of males. *Japanese Journal of Nematology* 10, 43-51.

YANG, B.J. & EISENBACK, J.D. (1983). *Meloidogyne enterolobii* n. sp. (Meloidogynidae), a root-knot nematode parasitizing Pacara earpot tree in China. *Journal of Nematology* 15, 381-391.

YANG, B.J., HU, K.J. & XU, B.W. (1988a). [A new species of root-knot nematode *Meloidogyne lini* n. sp. parasitizing rice.] *Journal of Yunnan Agricultural University* 3, 11-17.

YANG, B.J., WANG, Q. & FENG, R. (1988b). [*Meloidogyne kongi* n. sp. (Nematoda: Meloidogynidae) a root-knot nematode parasitizing *Citrus* sp. in Guangxi China.] *Journal of Guangxi Agricultural College* 7, 1-9.

YANG, B.J., HU, K.J., CHEN, H. & ZHU, W.S. (1990). [A new species of root-knot nematode *Meloidogyne jianyangensis* n. sp. parasitizing mandarin orange.] *Acta Phytopathologica Sinica* 20, 259-264.

YANG, Y., HU, X., LIU, P., CHEN, L., PENG, H., WANG, Q. & ZHANG, Q. (2021). A new root-knot nematode, *Meloidogyne vitis* sp. nov. (Nematoda: Meloidogynidae), parasitizing grape in Yunnan. *PLoS ONE* 16, e0245201. DOI: 10.1371/journal.pone.0245201

YASSIN, A.M. & ZEIDAN, A.B. (1982). The root-knot nematodes in the Sudan. *Proceedings of the 3rd Research and Planning Conference on root-knot nematodes* Meloidogyne *spp, 13-17 September 1982, Coimbra, Portugal Region VII* (International *Meloidogyne* Project), pp. 131-135.

YE, W.M., KOENNING, S.R., ZHUO, K. & LIAO, J.L. (2013). First report of *Meloidogyne enterolobii* on cotton and soybean in North Carolina, United States. *Plant Disease* 97, 1262. DOI: 10.1094/PDIS-03-13-0228-PDN

YE, W.M., ZENG, Y. & KERNS, J. (2015). Molecular characterisation and diagnosis of root-knot nematodes (*Meloidogyne* spp.) from turfgrasses in North Carolina, USA. *PLoS ONE* 10, e0143556. DOI: 10.1371/journal.pone.0143556

YE, W.M., ROBBINS, R.T. & KIRKPATRICK, T. (2019). Molecular characterization of root-knot nematodes (*Meloidogyne* spp.) from Arkansas, USA. *Scientific Reports* 9, 15680. DOI: 10.1038/s41598-019-52118-4

YEATES, G.W. (2010). Phylum Nematoda: roundworms, eelworms. In: Gordon, D.P. (Ed.). *New Zealand inventory of biodiversity. Vol. 2. Kingdom Animalia. Chaetognatha, Ecdysozoa, Ichnofossils*. Christchurch, New Zealand, Canterbury University Press, pp. 480-493.

YIK, C.-P. & BIRCHFIELD, W. (1979). Host studies and reactions of rice cultivars to *Meloidogyne graminicola*. *Phytopathology* 69, 497-499. DOI: 10.1094/Phyto-69-497

YORK, P.A. (1980). Relationship between cereal root-knot nematode *Meloidogyne naasi* and growth and grain yield of spring barley. *Nematologica* 26, 220-229. DOI: 10.1163/187529280X00116

YOUNG, L.D. (1975). *A* Meloidogyne *sp. on American beachgrass in North Carolina*. M.S. Thesis. North Carolina State University, USA.

YOUNG, L.D. & LUCAS, L.T. (1977). Hosts of *Meloidogyne* sp. on American beachgrass. *Plant Disease Reporter* 61, 776-777.

YOUNG, M.J. & SHERMAN, W.B. (1977). Evaluation of peach root-stock for root-knot nematode resistance. *Proceedings of the Florida State Horticultural Society* 90, 241-242.

ZAHID, M.I., NOBBS, J., STANTON, J.M., GURR, G.M., HODDA, M., NIKANDROW, A. & FULKERSON, W.J. (2000). First record of *Meloidogyne trifoliophila* in Australia. *Australasion Plant Pathology* 29, 280. DOI: 10.1071/AP00052

ZAHID, M.I., NOBBS, J., GURR, G.M., HODDA, M., NIKANDROW, A., FULKERSON, W.J. & NICOL, H.I. (2001). Effect of the clover root-knot nematode (*Meloidogyne trifoliophila*) on growth of white clover. *Nematology* 3, 437-446. DOI: 10.1163/156854101753250764

ZASADA, I.A., RIGA, E., PINKERTON, J.N., WILSON, J.H. & SCHREINER, P.R. (2012). Plant-parasitic nematodes associated with grapevines, *Vitis vinifera*, in Washington and Idaho. *American Journal of Enology and Viticulture* 63, 522-528. DOI: 10.5344/ajev.2012.12062

ZHANG, F. & SCHMITT, D.P. (1994). Host status of 32 plants species to *Meloidogyne konaensis. Supplement to Journal of Nematology* 26(4S) 26, 744-748.

ZHANG, L. & GLEASON, C. (2019). Loop-mediated isothermal amplification for the diagnostic detection of *Meloidogyne chitwoodi* and *M. fallax. Plant Disease* 103, 12-18. DOI: 10.1094/PDIS-01-18-0093-RE

ZHANG, S. (1993). [*Meloidogyne mingnanica* n. sp. (Meloidogynidae) parasitizing *Citrus* in China.] *Journal of Fujian Agricultural University (Natural Sciences Edition)* 22, 69-76.

ZHANG, S. & WENG, Z. (1991). [Identification of root-knot nematode species in Fujian.] *Journal of Fujian Agricultural College* 20, 158-164.

ZHANG, S., GAO, R. & WENG, Z. (1990). [*Meloidogyne citri* n. sp. (Meloidogynidae), a new root-knot nematode parasitizing citrus in China.] *Journal of Fujian Agricultural College* 19, 305-311.

ZHANG, Y. (1983). [A root-knot nematode, *Meloidogyne sinensis* n. sp. from potatoes in China.] *San Don University Bulletin* 2, 89-95.

ZHANG, Y. & SU, C. (1986). [A new species of the genus *Meloidogyne* from Shandong Province, China (Tylenchida: Meloidogynidae).] *Journal of Shandong University* 21, 95-102.

ZHAO, H.H., LIU, W.Z., LIANG, C. & DUAN, Y.X. (2001). [*Meloidogyne graminicola*, a new record species from China.] *Acta Phytopathologica Sinica* 5, 184-189.
ZHAO, Y.-L., RUAN, W.-B., YU, L., ZHANG, J.-Y., FU, J.-M., SHAIN, E.B., HUANG, X.-T. & WANG, J.-G. (2010). Combining maxRatio analysis with real-time PCR and its potential application for the prediction of *Meloidogyne incognita* in field samples. *Journal of Nematology* 42, 166-172.
ZHAO, Z.Q., HO, W., GRIFFIN, R., SURREY, M., TAYLOR, R., AALDERS, L., BELL, N.L., XU, Y.M. & ALEXANDER, B.J.R. (2017). First record of the root knot nematode, *Meloidogyne minor* in New Zealand with description, sequencing information and key to known species of *Meloidogyne* in New Zealand. *Zootaxa* 4231, 203-218. DOI: 10.11646/zootaxa.4231.2.4
ZHENG, L., LIN, M. & ZHENG, M.H. (1990). [Occurrence and identification of a new disease of the citrus Donghai root-knot nematode *Meloidogyne donghaiensis* sp. nov. in coast sand soil of Fujian in China]. *Journal of Fujian Academy of Agricultural Sciences* 5, 56-63.
ZHONG, S., ZENG, H.-C. & JIN, Z.-Q. (2017). Influences of different tillage and residue management systems on soil nematode community composition and diversity in the tropics. *Soil Biology and Biochemistry* 107, 234-243. DOI: 10.1016/j.soilbio.2017.01.007
ZHOU, E. & STARR, J.L. (2003). A comparison of the damage functions, root galling, and reproduction of *Meloidogyne incognita* on resistant and susceptible cotton cultivars. *The Journal of Cotton Science* 7, 224-230.
ZHOU, Q.-J., CAI, Y., GU, J.-F., WANG, X. & CHEN, J. (2017). Rapid and sensitive detection of *Meloidogyne mali* by loop-mediated isothermal amplification combined with a lateral flow dipstick. *European Journal of Plant Pathology* 148, 755-769. DOI: 10.1007/s10658-016-1130-4
ZHOU, X., LIU, G.K., XIAO, S. & ZHANG, S.S. (2015). First report of *Meloidogyne graminicola* infecting banana in China. *Plant Disease* 99, 420. DOI: 10.1094/PDIS-08-14-0810-PDN
ZHUO, K., HU, M.X., WANG, H.H., TANG, Z.L., SHAO, X.Y. & LIAO, J.L. (2011). Identification of *Meloidogyne graminis* on golf greens. *Acta Prataculturae Sinica* 20, 253-256.
ZIJLSTRA, C. (1997). A fast PCR assay to identify *Meloidogyne hapla*, *M. chitwoodi* and *M. fallax*, and to sensitively differentiate them from each other and from *M. incognita* in mixtures. *Fundamental and Applied Nematology* 20, 505-511.
ZIJLSTRA, C. (2000). Identification of *Meloidogyne chitwoodi*, *M. fallax* and *M. hapla* based on SCAR-PCR: a powerful way of enabling reliable PCR-based techniques for *Meloidogyne* identification of populations or individuals that share common traits. *European Journal of Plant Pathology* 106, 283-290. DOI: 10.1023/A:1008765303364

ZIJLSTRA, C. & VAN HOOF, R.A. (2006). A multiplex real-time polymerase chain reaction (TaqMan) assay for the simultaneous detection of *Meloidogyne chitwoodi* and *M. fallax*. *Phytopathology* 96, 1255-1262. DOI: 10.1094/PHYTO-96-1255

ZIJLSTRA, C., LEVER, A.E.M., UENK, B.J. & VAN SILFHOUT, C.H. (1995). Differences between ITS regions of isolates of root-knot nematodes *Meloidogyne hapla* and *M. chitwoodi*. *Phytopathology* 85, 1231-1237. DOI: 10.1094/Phyto-85-1231

ZIJLSTRA, C., UENK, B.J. & VAN SILFHOUT, C.H. (1997). A reliable, precise method to differentiate species of root-knot nematodes in mixtures on the basis of ITS-RFLPs. *Fundamental and Applied Nematology* 20, 59-63.

ZIJLSTRA, C., DONKERS-VENNE, D.T.H.M. & FARGETTE, M. (2000). Identification of *Meloidogyne incognita, M. javanica* and *M. arenaria* using sequence characterised amplified region (SCAR) based PCR assays. *Nematology* 2, 847-853. DOI: 10.1163/156854100750112798

ZIJLSTRA, C., VAN HOOF, R. & DONKERS-VENNE, D. (2004). A PCR test to detect the cereal root-knot nematode *Meloidogyne naasi*. *European Journal of Plant Pathology* 110, 855-860. DOI: 10.1007/s10658-004-2492-6

Nematology Monographs & Perspectives, 2021, Vol. 14, 849-852

Index of Latin Nematode Names

Nematology Monographs & Perspectives, 2021, Vol. 14, 853-857

Index of Latin Plant Names

Printed in the United States
by Baker & Taylor Publisher Services